Ergebnisse der Mathematik und ihrer Grenzgebiete

Volume 46

3. Folge

A Series of Modern Surveys in Mathematics

Springer
Berlin
Heidelberg
New York
Hong Kong
London
Milan
Paris
Tokyo

Michel Talagrand

Spin Glasses: A Challenge for Mathematicians

Cavity and Mean Field Models

Springer

Michel Talagrand
Equipe d'Analyse Fonctionnelle
Institut de Mathématiques, UMR 7586 – CNRS
Université Pierre et Marie Curie – Paris 6
Boîte 186 – 4, Place Jussieu
75252 Paris Cedex 05
France
e-mail: mit@ccr.jussieu.fr

Cataloging-in-Publication Data applied for

Bibliographic information published by Die Deutsche Bibliothek

Die Deutsche Bibliothek lists this publication in the Deutsche Nationalbibliografie; detailed bibliographic data is available in the Internet at <http://dnb.ddb.de>.

Mathematics Subject Classification (2000): 82D30

ISSN 0071-1136
ISBN 3-540-00356-8 Springer-Verlag Berlin Heidelberg New York

Springer-Verlag Berlin Heidelberg New York
a member of BertelsmannSpringer Science+Business Media GmbH

Typeset by the author using a Springer TEX macro package.
Printed on acid-free paper 44/3142LK - 5 4 3 2 1 0

To the memory of my grand father
Clément Talagrand

Contents

0. Introduction

Consider a large finite collection $(X_k)_{k \le M}$ of random variables. What can we say about the largest of them? More generally, what can we say about the "few largest" of them? When the variables X_k are probabilistically independent, everything is rather easy. This is no longer the case when the variables are correlated. Even when the variables are identically distributed, the answer depends very much on their correlation structure. What are the correlation structures of interest? Most of the familiar correlation structures in Probability are low-dimensional, or even "one-dimensional". This is because they model random phenomena indexed by time, or, equivalently, by the real line, a one-dimensional object. In contrast with these familiar situations, the correlation structures considered here will be "high-dimensional" – in a sense that will soon become clear – and will create new and truly remarkable phenomena.

A natural idea to study the few largest elements of a given realization of the family $(X_k)_{k \le M}$ is to weigh these elements, giving large weights to the large elements. Considering a parameter β, weights proportional to $\exp \beta X_k$ seem to be particularly appropriate, that is, the weight of k is

$$\frac{\exp \beta X_k}{\sum_{i \le M} \exp \beta X_i} \,. \tag{0.1}$$

These weights define a probability measure on the index set. Under an appropriate normalization, one can expect that this probability measure will be essentially supported by the indices k for which X_k is approximately a certain value $x(\beta)$. This is because the number of variables X_k close to a given large value x should decrease as x increases, while the corresponding weights increase, so that an optimal compromise should be reached at a certain level. The number $x(\beta)$ will increase with β. Thus we have a kind of "scanner" that enables us to look at the values of the family $(X_k)_{k \le M}$ close to the (large) number $x(\beta)$, and this scanner is tuned with the parameter β. We must stress an essential point here. We are interested in a *typical* realization of the family (X_k). This can be very different (and much harder to understand) than what happens in average of all realizations.

In an apparently unrelated direction, let us consider a physical system that can be in a (finite) number of possible configurations. In each configuration, the system has a given energy. It is maintained by the outside world at a given temperature, and is subject to thermal fluctuations. If we observe the system after it has been left undisturbed for a long time, what is the probability to observe it in a given configuration?

The system we will mostly consider is $\Sigma_N = \{-1,1\}^N$, where N is a (large) integer. A *configuration* $\boldsymbol{\sigma} = \{\sigma_1, \ldots, \sigma_N\}$ is an element of Σ_N. It tells us the values of the N "spins" σ_i, each of which can take the values ± 1. When in the configuration $\boldsymbol{\sigma}$, the system has an energy $H_N(\boldsymbol{\sigma})$. Thus H_N is simply a real-valued function on Σ_N. It is called the Hamiltonian of the system. We consider a parameter β (that physically represents the inverse of the temperature). We weigh each configuration proportionally to its Boltzmann factor $\exp(-\beta H_N(\boldsymbol{\sigma}))$. This defines Gibbs' measure, a probability measure on Σ_N given by

$$G_N(\{\boldsymbol{\sigma}\}) = \frac{\exp(-\beta H_N(\boldsymbol{\sigma}))}{Z_N} \tag{0.2}$$

where the normalizing factor Z_N is given by

$$Z_N = Z_N(\beta) = \sum \exp(-\beta H_N(\boldsymbol{\sigma})) \, . \tag{0.3}$$

The summation is of course over $\boldsymbol{\sigma}$ in Σ_N. The factor Z_N is called the *partition function*. Statistical mechanics asserts that Gibbs' measure represents the probability of observing a configuration $\boldsymbol{\sigma}$ after the system has reached equilibrium with an infinite heat bath at temperature $1/\beta$. (Thus the expression "high temperature" will mean "β small" while the expression "low temperature" will mean "β large".)

The reason for the minus sign in Boltzmann's factor $\exp(-\beta H_N(\boldsymbol{\sigma}))$ is that the system favors low (and not high) energy configurations. It should be stressed that the (considerable...) depth of the innocuous looking definition (0.2) stems from the normalizing factor Z_N. This factor, the partition function, is the sum of many terms of widely different orders of magnitude, and it is unclear how to estimate it. The (few) large values become more important as β increases, and predominate over the more numerous smaller values. Thus the problem of understanding Gibbs' measure gets typically harder for large β (low temperature) than for small β (high temperature).

The energy levels $H_N(\boldsymbol{\sigma})$ are closely related to the "interactions" between the spins. When we try to model a situation of "disordered interactions" these energy levels will become random variables. Gibbs' measure will become random. There are two levels of randomness (a probabilist's paradise). The "disorder", that is the randomness of the energy levels $H_N(\boldsymbol{\sigma})$ is given with our sample system. It does not evolve with the thermal fluctuations. It is frozen, or "quenched" as the physicists say. The word "glass" of the expression "spin glasses" conveys (among many others) this idea of frozen disorder.

In modeling disordered interactions between the spins, the problem is thus to understand Gibbs' measure for a typical realization of the disorder. As we explained, this is closely related to the problem of understanding the large values among a typical realization of the family $(-H_N(\boldsymbol{\sigma}))$. This family is correlated. One reason for the choice of the index set Σ_N is that it is suitable to create extremely interesting correlation structures with simple formulas.

The mathematical objects studied in this book are of a rather fundamental nature. Yet they have been discovered by physicists. At the beginning of an already long story are "real" spin glasses, alloys with strange magnetic properties, which are of considerable interest, both experimentally and theoretically. It is believed that their remarkable properties arise from a kind of disorder among the interactions of magnetic impurities. To explain (at least qualitatively) the behavior of real spin glasses, theoretical physicists have invented a number of models. They fall into two broad categories: the "realistic" models, where the interacting atoms are located at the vertices of a lattice, and where the strength of the interaction between two atoms decreases when their distance increases; and the "mean field" models, where the geometric location of the atoms in space is forgotten, and where each atom interacts with all the others. The mean-field models are of special interest to mathematicians because they are very basic mathematical objects. Yet some physicists predict extremely intricate structures arising from these objects. (As for the "realistic" models, they appear to be intractable at the moment.) Moreover, they believe that their ideas are applicable in a wide range of situations. The breadth, and the ambition, of these physicists' work can in particular be admired in the book "Spin glass theory and beyond" by M. Mézard, G. Parisi, M.A. Virasoro, and the book "Field theory, Disorder and Simulation" by G. Parisi. The methods used by the physicists are probably best described here as a combination of heuristic arguments and numerical simulation. They are probably reliable, but they have no claim to rigor, and it is often not even clear how to give a precise mathematical formulation to some of the central predictions.

It is rather paradoxical for a mathematician like the author to see simple, basic mathematical objects being studied by the methods of theoretical physics. It is also very surprising, given the obvious importance of what the physicists have done, and the breadth of the paths they have opened, that mathematicians have not tried earlier to prove their conjectures.

This book will study through mathematical methods some of these fascinating mathematical objects discovered by the physicists. The reader has already been explained *all* the statistical mechanics (s)he needs to know. No knowledge whatsoever of physics is required or probably even useful to read it. The vast body of knowledge and techniques from statistical mechanics has little relevance because the overwhelming feature is the randomness of the interactions. (This randomness seems in particular to preclude the existence of "infinite-volume limiting objects".) Most of the book is devoted to finding ways to estimate apparently hopelessly complicated random quantities.

The main purpose of the book is to bring this wonderful new area of mathematics discovered by the physicists to the attention of the mathematical community, in terms that are easier for a mathematician to understand than the original work of the physicists. Despite considerable efforts, in particular from the author, the frontier of this new area is not very far from the basic principles. (In fact, it could be argued that this book contains very few good ideas.) The future will tell whether the methods needed to make further progress on the main issues need to be invented or already exist in other areas of mathematics. Meanwhile, we aim to make this work accessible to readers with diverse backgrounds. Great care has been taken to ensure that the most important material be accessible with *no* prerequisite, beside the most basic ideas of probability theory. For much of the book, the most advanced tool that is not proved in complete detail is Hölder's inequality. The book is long because it attempts to cover what is known about the main models, and studying different models seems presently the only way to convey this irresistible feeling that, at some deep level, there is a global picture waiting to be revealed. Reading the first two chapters should however be sufficient to get a very good idea of what spin glasses are about.

There is of course no way of telling when (and if) progress will be made on the really difficult questions. To provide an immediate incitement to the brilliant young (or even not so young) minds willing to seriously learn this topic the author has stated as research problems a number of interesting questions (the solution of which would likely deserve to be published) that he believes are within the reach of the already established methods, but that he purposely did not, and will not, try to solve. (On the other hand, there is ample warning about the potentially truly difficult problems.)

The material presented in this book covers a large part (but not all) of what the author knew about the topic at the time of writing. It is divided in three distinct parts. The longest of these consists of Chapters 1 to 5. In these chapters four important models are studied through unified methods. In Chapter 1, Derrida's Random Energy Model (REM), a "toy model" serves as an introduction. Chapter 2 studies in great detail the Sherrington-Kirkpatrick model (SK), the "original" spin glass. This model serves as an introduction to the basic ideas. In Chapters 3 and 4, we study the perception capacity model and in Chapter 5 the Hopfield model, both important in the theory of neural networks. The four models of the first four chapters are studied in order of increasing difficulty of their high-temperature phases. This allows a progressive introduction of more and more sophisticated versions of the same ideas. Extreme care has been taken to explain in great detail the first occurrence of each idea. In fact, the author feels that too much detail is better than not enough, so he is simply giving as much detail as he had the energy to do. Of course the details get fewer as the arguments get repeated and the author's energy gets depleted. Delicate topics (such as central limit theorems) are approached through increasingly more difficult versions. In a nutshell, it is advised to progress somewhat linearly through Chapters 1 to 5, and the reader attacking first Chapter 5 does it at her/his own risk.

The second part of the book consists of Chapter 6, that describes the behavior in a large portion of the low temperature phase of a cousin of the SK model, the p-spin interaction model. It can be read independently of Chapters 3, 4 and 5.

The last part of the book consists of Chapters 7 and 8. Chapter 7 studies a common generalization of the diluted SK model, and of the K-sat problem (a fundamental question of computer science), while Chapter 8 is motivated by the assignment problem. The high temperature phase of these models share some common features with the five previous cases, but require distinctly different methods. Chapters 7 and 8 can be read independently of Chapters 3 to 6, and of most of Chapter 2.

Finally, an appendix lists, and often proves, the basic facts of probability theory used throughout this work.

At some point I must apologize for the countless typos, inaccuracies, or downright mistakes that this volume is bound to contain. Arguing that towards the end of the writing I could read only with the help of a magnifying glass would be a paltry excuse. The real culprit is of course my dysfunctional brain, but there is little I can do to fix it. I am grateful to A. van Enter, V. Gayrard, R. Latala, M. Sortais, S. Tindel and specially to A. Hanen and D. Panchenko for laboring through the first versions and suggesting literally hundreds of corrections and improvements. Of course, all remaining mistakes are my sole responsibility.

Special thanks are due to Erwin Bolthausen, who introduced me to this area. This gave me so much happiness (and unhappiness too!) for what has already been a large part of my scientific life.

It is customary for authors, at the end of an introduction, to warmly thank their spouse for having granted them the peaceful time needed to complete their work. I find that these thanks are far too universal and overly enthusiastic to be believable. Yet, I must say that in the present case even what would sound for the reader as exaggerated thanks would not truly reflect the extraordinary privileges I have enjoyed. Be jealous, reader, for I yet have to hear the words I dread the most: "Now is not the time to work".

1. A Toy Model, the REM

1.1 High Temperature

Given a (random) Hamiltonian $(-H_N(\boldsymbol{\sigma}))$ on Σ_N, the partition function

$$Z_N = Z_N(\beta) = \sum_{\boldsymbol{\sigma}} \exp(-\beta H_N(\boldsymbol{\sigma})) \;, \tag{1.1}$$

where the summation is over all values $\boldsymbol{\sigma}$ in Σ_N, is of crucial importance. It is better studied on a logarithmic scale, through the quantity $N^{-1} \log Z_N(\beta)$.

Proposition 1.1.1. *The function $\beta \mapsto N^{-1} \log Z_N(\beta)$ is a (random) convex function.*

Proof. If f is any positive (measurable) function on a space provided with a positive measure μ, the map $\beta \mapsto \log \int f^\beta \, d\mu$ is convex, as follows from Hölder's inequality. Taking μ the counting measure on Σ_N and $f(\boldsymbol{\sigma}) = \exp(-H_N(\boldsymbol{\sigma}))$ proves the proposition. $\square$

As will be demonstrated in Section 2.12, this simple fact has amazing consequences.

The normalization will be chosen so that $N^{-1} \log Z_N(\beta)$ is a quantity of order 1. It will turn out that the fluctuations of this random quantity will be of lower order. The physicists express this very important fact by saying that the random quantity $N^{-1} \log Z_N(\beta)$ is *self-averaging.*

Thus, a good measure of the random quantity $N^{-1} \log Z_N(\beta)$ is the non-random quantity

$$p_N(\beta) = \frac{1}{N} \mathsf{E} \log Z_N(\beta) \;. \tag{1.2}$$

Here, and everywhere, E denotes expectation in the disorder, that is, in the randomness of the Hamiltonian. The quantity $p_N(\beta)$ is closely related to the "free energy", a quantity important in physics that is defined as $-\beta^{-1} p_N(\beta)$. The factor $-1/\beta$ is a nuisance in mathematics, and we will not use the free energy. Actually, the minus sign in Boltzmann's factor $\exp(-\beta H_N(\boldsymbol{\sigma}))$ is also a nuisance for mathematics. This nuisance greatly decreases if one gets the habit to think that the object of interest is $-H_N(\boldsymbol{\sigma})$, and that the minus sign cannot be separated from H_N.

As we just explained, $\mathsf{E}N^{-1}\log Z_N(\beta)$ is a correct way to measure the size of the random quantity $N^{-1}\log Z_N(\beta)$. However, $\mathsf{E}Z_N(\beta)$ is *not* in general the correct way to measure the size of $Z_N(\beta)$. This is because a large part of $\mathsf{E}Z_N(\beta)$ comes from the contribution of very rare events.

We are not interested in very rare events, but in typical events. In probabilistic terms, the median of $Z_N(\beta)$ is much smaller than its mean, even when measured on a logarithmic scale. On that scale, the median of $Z_N(\beta)$ and its mean are fairly well represented respectively by

$$p_N(\beta) = \frac{1}{N}\mathsf{E}\log Z_N(\beta) \quad \text{and} \quad \frac{1}{N}\log \mathsf{E}Z_N(\beta) .$$

These quantities can be related through Jensen's inequality. This inequality asserts that if φ is a convex function, for any random variable (r.v.) X, we have

$$\varphi(\mathsf{E}X) \le \mathsf{E}\varphi(X) . \tag{1.3}$$

Since the function log is concave, we get

$$p_N(\beta) \le \frac{1}{N}\log \mathsf{E}Z_N(\beta) . \tag{1.4}$$

In general, equality does not hold, even in the limit $N \to \infty$. Sadly, while it is easy to evaluate $\mathsf{E}Z_N(\beta)$, it is much less clear how to evaluate $p_N(\beta)$. The reader can discover in Section 2.15 the remarkable method the physicists have devised in order to bypass this difficulty.

Let us spell out the following consequence of Proposition 1.1.1.

Proposition 1.1.2. *The function $\beta \mapsto p_N(\beta)$ is convex.*

After these general considerations, we turn to the study of Derrida's Random Energy Model (REM). In this model, we assume

$$\begin{gathered}\text{The random variables } (H_N(\boldsymbol{\sigma})) \\ \text{are independent centered Gaussian with } \mathsf{E}H_N^2(\boldsymbol{\sigma}) = N/2 .\end{gathered} \tag{1.5}$$

All the Gaussian r.v. we consider are centered, so by "Gaussian r.v." we always mean "centered Gaussian r.v.".

Should, by extraordinary circumstances, a reader be ignorant of what are Gaussian r.v., this is explained in the Appendix, where all the properties of these variables that we will use can be found. The reason for the factor 1/2 in the normalization $\mathsf{E}H_N^2(\boldsymbol{\sigma}) = N/2$ is for easier comparison with the Sherrington-Kirkpatrick model of Chapter 2.

As explained in the introduction, a probabilist can view the introduction of Boltzmann's weights as a mean to ultimately understand the effect of the correlation among the many r.v. $H_N(\boldsymbol{\sigma})$. Since there is no correlation here, this is not a model of real interest. In fact, we do not need ideas from statistical mechanics to understand the family (1.5). Rather, the point of the exercise is to understand how we see this simple situation through Boltzmann's weights.

The search of upper bounds for $p_N(\beta)$ will turn out to be a central question. The following principle, while simple, is very useful.

Proposition 1.1.3. *Consider M centered Gaussian r.v. $(g_i)_{i\le M}$. We do* not *assume that they are independent. We assume that $\mathsf{E}\, g_i^2 \le \tau^2$ for each $i \le M$. Then the following occur:*

$$\mathsf{E}\log\left(\sum_{i\le M}\exp\beta\, g_i\right) \le \frac{\beta^2\tau^2}{2} + \log M \tag{1.6}$$

$$\mathsf{E}\max_{i\le M} g_i \le \tau\sqrt{2\log M} \tag{1.7}$$

$$\beta \ge \sqrt{2\log M}/\tau \Rightarrow \mathsf{E}\log\left(\sum_{i\le M}\exp\beta\, g_i\right) \le \beta\tau\sqrt{2\log M}\,. \tag{1.8}$$

Proof. The most important property of a Gaussian r.v. g is that

$$\mathsf{E}\exp\beta\, g = \exp\left(\frac{1}{2}\beta^2\,\mathsf{E}\, g^2\right)\,. \tag{1.9}$$

Using Jensen's inequality (1.3) and the concavity of the logarithm, we get

$$\begin{aligned}\mathsf{E}\log\left(\sum_{i\le M}\exp\beta\, g_i\right) &\le \log\left(\mathsf{E}\sum_{i\le M}\exp\beta\, g_i\right)\\ &\le \log\left(M\exp\left(\frac{1}{2}\beta^2\tau^2\right)\right)\,,\end{aligned} \tag{1.10}$$

and this proves (1.6).

Also,

$$\beta\max_{i\le M} g_i \le \log\left(\sum_{i\le M}\exp\beta\, g_i\right)\,,$$

so that, using (1.6),

$$\beta\,\mathsf{E}\max_{i\le M} g_i \le \mathsf{E}\log\left(\sum_{i\le M}\exp\beta\, g_i\right) \le \frac{\beta^2\tau^2}{2} + \log M\,.$$

Taking $\beta = \sqrt{2\log M}/\tau$ yields (1.7). Consider now the function

$$\varphi(\beta) = \mathsf{E}\log\left(\sum_{i\le M}\exp\beta\, g_i\right)\,.$$

Then

$$\varphi'(\beta) = \mathsf{E}\,\frac{\sum_{i\le M} g_i \exp\beta g_i}{\sum_{i\le M}\exp\beta g_i} \le \mathsf{E}\max_{i\le M} g_i\ .$$

If we combine with (1.6), this yields (1.8). Indeed, the derivative of the function $\beta \to \beta^2\tau^2/2 + \log M$ is equal to $\tau\sqrt{2\log M}$ for $\beta = \sqrt{2\log M}/\tau$, and the tangent line to its graph at this point goes through the origin. □

Proposition 1.1.4. *Assume that the r.v. $H_N(\boldsymbol{\sigma})$ are centered Gaussian (but not necessarily independent) and that*

$$\forall\boldsymbol{\sigma},\ \mathsf{E}H_N^2(\boldsymbol{\sigma}) \le \frac{N}{2}\ . \tag{1.11}$$

Then we have

$$p_N(\beta) \le \frac{\beta^2}{4} + \log 2 \tag{1.12}$$

and if $\beta \ge 2\sqrt{\log 2}$ we have

$$p_N(\beta) \le \beta\sqrt{\log 2}\ . \tag{1.13}$$

Proof. Use Proposition 1.1.3 with $M = 2^N$ and $\tau = \sqrt{N/2}$. □

When $\beta > 2\sqrt{\log 2}$, (1.13) gives a first example of a situation where

$$p_N(\beta) = \frac{1}{N}\mathsf{E}\log Z_N(\beta) \ne \frac{1}{N}\log\mathsf{E}\,Z_N(\beta) = \frac{\beta^2}{4} + \log 2\ .$$

In the present situation, it is both simple and instructive to explain the reason behind this. If g is a centered Gaussian r.v., such that $\tau^2 = \mathsf{E}\,g^2$ is large, the contribution to

$$\mathsf{E}\exp\beta g = \frac{1}{\sqrt{2\pi}\tau}\int\exp\left(\beta t - \frac{t^2}{2\tau^2}\right)\,\mathrm{d}t$$

comes essentially from the values of g that are about $\beta\tau^2$ (which is the value of t that maximizes $\beta t - t^2/2\tau^2$). Most of $\mathsf{E}Z_N(\beta)$ comes from the event that one of the r.v. $-H_N(\boldsymbol{\sigma})$ is about $N\beta/2$. This event occurs however very rarely; the standard estimate (A.37) on the tails of a Gaussian r.v. $\mathsf{P}(g \ge t) \le \exp(-t^2/2\,\mathsf{E}\,g^2)$ for $t > 0$ shows that the probability of this event is at most $2^N\exp(-\beta^2 N/4)$ and this is $\ll 1$ if $\beta > 2\sqrt{\log 2}$.

The bounds of Proposition 1.1.4 represent a kind of worst case, and are not sharp in general. Indeed, if $H_N(\boldsymbol{\sigma})$ does not depend on $\boldsymbol{\sigma}$, then $p_N(\beta) = \log 2$ for each β. But in the case of the REM, the bounds of Proposition 1.1.4 are asymptotically optimal.

Proposition 1.1.5. *In the case of the REM, we have*

$$\lim_{N\to\infty} p_N(\beta) = \frac{\beta^2}{4} + \log 2 \quad \text{if} \quad \beta \le 2\sqrt{\log 2} \tag{1.14}$$

$$\lim_{N\to\infty} p_N(\beta) = \beta\sqrt{\log 2} \quad \text{if} \quad \beta \ge 2\sqrt{\log 2}\,. \tag{1.15}$$

Proof. We consider a parameter s, to be chosen later, and

$$X = \mathrm{card}\{\boldsymbol{\sigma} \in \Sigma_N \; ; \; -H_N(\boldsymbol{\sigma}) \ge sN\}\,. \tag{1.16}$$

Thus

$$\mathsf{E}X = 2^N a\,, \tag{1.17}$$

where $a = a_N = \mathsf{P}(-H_N(\boldsymbol{\sigma}) \ge sN)$ is independent of $\boldsymbol{\sigma}$. Now

$$X^2 = \mathrm{card}\{(\boldsymbol{\sigma}^1, \boldsymbol{\sigma}^2) \in \Sigma_N^2 \; ; \; -H_N(\boldsymbol{\sigma}^1) \ge sN, -H_N(\boldsymbol{\sigma}^2) \ge sN\}\,,$$

and

$$\mathsf{E}X^2 = \sum \mathsf{P}(-H_N(\boldsymbol{\sigma}^1) \ge sN, -H_N(\boldsymbol{\sigma}^2) \ge sN)\,,$$

where the summation is over all values of $\boldsymbol{\sigma}^1, \boldsymbol{\sigma}^2$ in Σ_N. When $\boldsymbol{\sigma}^1 = \boldsymbol{\sigma}^2$, we have

$$\mathsf{P}(-H_N(\boldsymbol{\sigma}^1) \ge sN, -H_N(\boldsymbol{\sigma}^2) \ge sN) = \mathsf{P}(-H_N(\boldsymbol{\sigma}^1) \ge sN) = a\,.$$

When $\boldsymbol{\sigma}^1 \ne \boldsymbol{\sigma}^2$, by independence we have

$$\mathsf{P}(-H_N(\boldsymbol{\sigma}^1) \ge sN, -H_N(\boldsymbol{\sigma}^2) \ge sN) = \mathsf{P}(-H_N(\boldsymbol{\sigma}^1) \ge sN)^2 = a^2\,,$$

so that

$$\mathsf{E}X^2 = 2^N a + 2^N(2^N - 1)\,a^2$$

and

$$\mathsf{E}(X - \mathsf{E}X)^2 = \mathsf{E}X^2 - (\mathsf{E}X)^2 \le 2^N a\,.$$

By Markov's inequality

$$\mathsf{P}(Y \ge t) \le \frac{\mathsf{E}Y}{t} \tag{1.18}$$

(true for a non-negative random variable), if we define

$$A = \{X \le 2^{N-1} a\}\,,$$

we have

$$\mathsf{P}(A) \le \mathsf{P}((X - \mathsf{E}X)^2 \ge 2^{2N-2} a^2) \le \frac{4\mathsf{E}(X - \mathsf{E}X)^2}{2^{2N} a^2} \le \frac{4}{2^N a}.$$

On the complement A^c of A, we have

$$Z_N \geq X \exp \beta N s \geq 2^{N-1} a \exp \beta N s ,$$

so that

$$\frac{1}{N} \log Z_N \geq \frac{N-1}{N} \log 2 + \frac{1}{N} \log a + \beta s . \tag{1.19}$$

To get a lower bound for Z_N on A, we consider an arbitrary element $\boldsymbol{\sigma}_0$ of Σ_N and we write

$$Z_N \geq \exp(-\beta H_N(\boldsymbol{\sigma}_0))$$

so that

$$\begin{aligned} \frac{1}{N} \mathsf{E}(\mathbf{1}_A \log Z_N) &\geq -\frac{\beta}{N} \mathsf{E}(\mathbf{1}_A H_N(\boldsymbol{\sigma}_0)) \\ &\geq -\frac{\beta}{N} \mathsf{E}|H_N(\boldsymbol{\sigma}_0)| = -\sqrt{\frac{1}{\pi}} \frac{\beta}{\sqrt{N}} . \end{aligned} \tag{1.20}$$

In this inequality, the exact value $\sqrt{1/\pi}$ is not relevant. It only matters that it is a fixed number. It is clearer to rewrite (1.20) as

$$\frac{1}{N} \mathsf{E}(\mathbf{1}_A \log Z_N) \geq -\frac{L\beta}{\sqrt{N}} .$$

In this notation, and throughout the book, L denotes a universal constant, that is, a number. The value of L need not be the same at each occurrence. This convention might be disturbing at first, but it is very convenient when the actual exact value is irrelevant, as is often the case in our estimates.

Combining (1.19) and (1.20), we get

$$p_N(\beta) \geq \left(1 - \frac{4}{2^N a}\right) \left(\frac{N-1}{N} \log 2 + \frac{1}{N} \log a + \beta s\right) - \frac{L\beta}{\sqrt{N}} . \tag{1.21}$$

Now, if g is a centered Gaussian r.v. with $\mathsf{E}\, g^2 = \tau^2$, by (A.37) and (A.38) we have, for $t > \tau$,

$$\frac{\tau}{Lt} \exp\left(-\frac{t^2}{2\tau^2}\right) \leq \mathsf{P}(g \geq t) \leq \exp\left(-\frac{t^2}{2\tau^2}\right) ,$$

and using this for $t = sN$, $\tau^2 = N/2$, we get

$$\frac{1}{Ls\sqrt{N}} \exp(-Ns^2) \leq a_N = \mathsf{P}(-H_N(\boldsymbol{\sigma}) \geq sN) \leq \exp(-Ns^2) .$$

If we fix $s < \sqrt{\log 2}$, independently of N, we see that $2^N a_N \to \infty$, so that (1.21) gives

$$\liminf_{N \to \infty} p_N(\beta) \geq \log 2 + \beta s - s^2 .$$

Now we optimize over s. If $\beta < 2\sqrt{\log 2}$, we take $s = \beta/2$. If $\beta \geq 2\sqrt{\log 2}$, we let $s \to \sqrt{\log 2}$. □

Our objective was to prove Proposition 1.1.5 in the simplest possible way, so we have not tried to get a rate of convergence, as will be done later in more complicated situations. Of course, very precise results can be obtained [35].

What about the structure of Gibbs' measure? When $\beta < 2\sqrt{\log 2}$, it is simple to show that if one looks at a fixed number of spins (or, at least, at not too many of them), Gibbs' measure essentially cannot be distinguished from the uniform measure. That is, given an integer n, and $\eta_1, \ldots, \eta_n = \pm 1$, we have

$$G_N(\{\boldsymbol{\sigma}\,;\ \forall i \leq n,\ \sigma_i = \eta_i\}) \rightarrow 2^{-n}$$

as $N \rightarrow \infty$. Results along this line can be found in [153]. Since we will later prove more challenging results of the same nature, we do not reproduce the proofs.

1.2 Low Temperature

Our purpose is to study Gibbs' measure for the REM when $\beta > 2\sqrt{\log 2}$ and N is large. Rather than (somewhat) resembling the uniform measure on Σ_N, Gibbs' measure is, in a sense made precise below, "supported by a sequence of configurations". Moreover, the (random) Gibbs' weights of these configurations (once suitably reordered) have a remarkable limit distribution, a so-called Poisson-Dirichlet distribution. The properties of this distribution, some of which arise very naturally in connection with the REM, will play a fundamental role in Chapter 6. The present section illustrates clearly the difference between high and low temperature behavior, but its material will not really be needed before Chapter 6. Should a reader find that matters get too sophisticated, (s)he is invited to jump to Section 2.1.

The difference of behavior between the cases $\beta < 2\sqrt{\log 2}$ and $\beta > 2\sqrt{\log 2}$ ultimately has its roots in a very simple fact, which we explain first. Let us try to apply the Boltzmann weights $\exp(-\beta x_i)$ to a sequence of numbers $(x_i)_{i \leq 2^N}$. What happens when $x_i = \log i$? The answer is very different depending on whether $\beta < 1$ or $\beta > 1$. When $\beta < 1$, the normalizing factor

$$Z_N = \sum_{i \leq 2^N} \exp(-\beta x_i) = \sum_{1 \leq i \leq 2^N} i^{-\beta}$$

goes to infinity as N grows and the weight of each x_i for Gibbs' measure becomes exponentially small (that is, this weight is at most $\exp(-N/C(\beta))$). On the other hand, when $\beta > 1$, Z_N remains bounded with N, so that the Gibbs weight of a given x_i does not go to zero as $N \rightarrow \infty$.

What happens for the REM is that if we suitably reorder the 2^N values $(-H_N(\boldsymbol{\sigma}))$, (for a typical realization of the randomness), they strongly resemble the sequence

$$\left(c_N - \frac{\log i}{2\sqrt{\log 2}}\right)_{i\geq 1}$$

where the centering constant c_N does not influence the previous analysis. (The centering constant is $c_N = Na_N$ where a_N is defined in (1.26).)

Before we can state a convergence result, we need to construct the limiting object, the Poisson-Dirichlet distribution Λ_m, where m is a parameter, $0 < m < 1$. This construction requires some elementary facts about Poisson point processes. The reader who is not familiar with these will find in the Appendix all that is needed to know about these for the following discussion.

We consider the measure μ on $\mathbb{R}^+$ of density x^{-m-1} with respect to Lebesgue's measure. Since $m > 0$, for every $\varepsilon > 0$, we have

$$\mu\left([\varepsilon, \infty)\right) < \infty \tag{1.22}$$

and since $m < 1$ we have

$$\int_0^1 x\, \mathrm{d}\mu(x) = \int_0^1 x^{-m} \mathrm{d}x < \infty\,. \tag{1.23}$$

We consider a Poisson point process Π of intensity measure μ. (Go to the Appendix if that name makes you uncomfortable.) This is a random countable subset of $\mathbb{R}^+$, and by (1.22), $\Pi \cap [\varepsilon, \infty)$ is a.s. finite for every $\varepsilon > 0$, so that we can enumerate the points belonging to Π as a non-decreasing sequence $(u_\alpha)_{\alpha\geq 1}$. We have

$$\mathsf{E}\left(\sum u_\alpha \mathbf{1}_{\{u_\alpha \leq 1\}}\right) = \int_0^1 x\, \mathrm{d}\mu(x) < \infty$$

and since $\mathsf{E}\,\mathrm{card}\{\alpha; u_\alpha \geq 1\} = \mu([1,\infty))$ is finite, the sum $S = \sum_{\gamma\geq 1} u_\gamma$ is finite a.s., so that we can consider the sequence

$$v_\alpha = \frac{u_\alpha}{S} = \frac{u_\alpha}{\sum_{\gamma\geq 1} u_\gamma}\,. \tag{1.24}$$

This is a non-negative, non-increasing sequence that sums to 1.

Let us denote by $\mathcal{S}$ the set of non-negative, non-increasing sequences of sum at most 1. This set is provided with a natural topology, the weakest topology that makes the maps $\mathbf{x} \to x_\alpha$ continuous, where $\mathbf{x} = (x_\alpha)_{\alpha\geq 1}$. This topology can be defined by distances such as

$$\mathrm{d}(\mathbf{x}, \mathbf{y}) = \sum_{\alpha\geq 1} 2^{-\alpha} |x_\alpha - y_\alpha|$$

or

$$\mathrm{d}(\mathbf{x}, \mathbf{y}) = \left(\sum_{\alpha\geq 1} (x_\alpha - y_\alpha)^2\right)^{1/2}\,. \tag{1.25}$$

The reason we allow in $\mathcal{S}$ sequences of sum at most 1 rather than only sequences of sum 1 is to make $\mathcal{S}$ compact, but this is really a secondary detail and a matter of taste.

The law of the sequence $(v_\alpha)_{\alpha \geq 1}$ is a probability measure Λ_m on $\mathcal{S}$, that we will call the Poisson-Dirichlet distribution of parameter m. This distribution is part of a two parameters family of distributions with truly remarkable properties. The paper [120], where references to earlier work can be found, explores the multiples facets of this object. With the notations of [120], we have $\Lambda_m = PD(m, 0)$.

Going back to the REM we can reorder the Gibbs' weights $G_N(\{\boldsymbol{\sigma}\})$ of the configurations as a non-decreasing sequence $(w_\alpha)_{\alpha \leq 2^N}$. We define $w_\alpha = 0$ if $\alpha > 2^N$ and the sequence $(w_\alpha)_{\alpha \geq 1}$ then belongs to $\mathcal{S}$.

Theorem 1.2.1. *If $\beta > 2\sqrt{\log 2}$, as $N \to \infty$ the law of the sequence $(w_\alpha)_{\alpha \geq 1}$ converges to Λ_m, where $m = 2\sqrt{\log 2}/\beta$.*

What this means is that, given a continuous function f on $\mathcal{S}$, we have

$$\lim_{N \to \infty} \mathsf{E} f((w_\alpha)_{\alpha \geq 1}) = \mathsf{E} f((v_\alpha)_{\alpha \geq 1}) .$$

If we consider an integer n, and the function $f((x_\alpha)_{\alpha \geq 1}) = 1 - \sum_{\alpha \leq n} x_\alpha$, then we see that given $\varepsilon > 0$, we can find n such that $\mathsf{E} \sum_{\alpha \leq n} w_\alpha \geq 1 - \varepsilon$ for N large enough. Almost all the mass of Gibbs' measure is supported by the n configurations with the largest weights.

To compute Gibbs' measure for the REM, we can replace the numbers $H_N(\boldsymbol{\sigma})$ by $H'_N(\boldsymbol{\sigma}) = H_N(\boldsymbol{\sigma}) + N a_N$, where

$$N a_N^2 = \log\left(\frac{2^N}{\sqrt{N}}\right) . \tag{1.26}$$

Let us denote by $(h_\alpha)_{\alpha \leq 2^N}$ the numbers $-H'_N(\boldsymbol{\sigma})$ when enumerated in non-increasing order. The leading idea of Theorem 1.2.1 is that given any number b in $\mathbb{R}$, the trace of the set $(h_\alpha)_{\alpha \leq 2^N}$ on the half line $[b, \infty)$ asymptotically resembles a Poisson point process of intensity measure ν_b, where ν_b has a density

$$\frac{1}{\sqrt{\pi}} \exp(-2t\sqrt{\log 2})\, \mathbf{1}_{\{t \geq b\}} \tag{1.27}$$

with respect to Lebesgue's measure. This is expressed in the following lemma.

Lemma 1.2.2. *If $|\nu_b|$ denotes the total mass of ν_b, for every integer k we have*

$$\lim_{N \to \infty} \mathsf{P}(\operatorname{card}\{\alpha;\ h_\alpha \geq b\} = k) = \frac{|\nu_b|^k}{k!} e^{-|\nu_b|} . \tag{1.28}$$

Conditionally upon the fact that $\operatorname{card}\{\alpha;\ h_\alpha \geq b\} = k$, *the set $\{h_1, \ldots, h_k\}$ has the same distribution as the set $\{X_1, \ldots, X_k\}$, where $X_1, \ldots, X_k$ are independent with law $\nu_{b,N}$ of density proportional to*

$$\exp\left(-2\,t\,a_N - \frac{t^2}{N}\right) \mathbf{1}_{\{t\geq b\}}\,, \tag{1.29}$$

where a_N is defined as in (1.26).

Proof. Since

$$\mathsf{P}(-H_N(\boldsymbol{\sigma}) \geq x) = \frac{1}{\sqrt{\pi N}} \int_x^\infty \exp\left(-\frac{t^2}{N}\right) dt$$

we have

$$\begin{aligned}\mathsf{P}(-H'_N(\boldsymbol{\sigma}) \geq x) &= \mathsf{P}(-H_N(\boldsymbol{\sigma}) \geq N a_N + x) \\ &= \frac{1}{\sqrt{\pi N}} \int_{N a_N + x}^\infty \exp\left(-\frac{t^2}{N}\right) dt \\ &= \frac{1}{\sqrt{\pi N}} \int_x^\infty \exp\left(-\frac{(t+Na_N)^2}{N}\right) dt\,.\end{aligned}$$

Now,

$$\begin{aligned}\exp\left(-\frac{(t+Na_N)^2}{N}\right) &= \exp\left(-\frac{t^2}{N} - 2\,t\,a_N - N a_N^2\right) \\ &= 2^{-N}\sqrt{N}\exp\left(-\frac{t^2}{N} - 2\,t\,a_N\right)\end{aligned}$$

and thus

$$\mathsf{P}(-H'_N(\boldsymbol{\sigma}) \geq x) = \frac{2^{-N}}{\sqrt{\pi}} \int_x^\infty \exp\left(-2\,t\,a_N - \frac{t^2}{N}\right) dt\,. \tag{1.30}$$

If we denote by d_N the quantity (1.30) when $x = b$, we get

$$\mathsf{P}(\mathrm{card}\{\boldsymbol{\sigma}\,;\ -H'_N(\boldsymbol{\sigma}) \geq b\} = k) = \binom{2^N}{k} d_N^k (1-d_N)^{2^N - k}\,,$$

and (1.28) follows from the fact that $2^N d_N \to |\nu_b|$ through a straightforward estimate.

As for (1.29), it holds since conditionally upon the fact that $-H'_N(\boldsymbol{\sigma}) \geq b$, $-H'_N(\boldsymbol{\sigma})$ has law $\nu_{b,N}$ as is shown by (1.30). □

The next lemma brings forward the relation between Λ_m and the measure ν of density

$$\frac{1}{\sqrt{\pi}} \exp(-2t\sqrt{\log 2})$$

with respect to Lebesgue's measure.

Lemma 1.2.3. *Consider a Poisson point process of intensity measure ν, and $(c_\alpha)_{\alpha\geq 1}$ an enumeration in non-increasing order of the random set this process generates. Then the sequence*

$$v_\alpha = \frac{\exp \beta c_\alpha}{\sum_{\gamma\geq 1} \exp \beta c_\gamma}$$

has distribution Λ_m, where $m = 2\sqrt{2}/\beta$.

Proof. We observe that, given a number c, we have $v_\alpha = u_\alpha / \sum_{\gamma\geq 1} u_\gamma$, where $u_\alpha = c \exp \beta c_\alpha$. The sequence $(u_\alpha)_{\alpha\geq 1}$ is (a reordering of) a Poisson point process of intensity measure μ, where μ is the image of ν under the map $x \to c \exp \beta x$. Thus, by definition of the image measure, for a Borel set A, we have

$$\mu(A) = \frac{1}{\sqrt{\pi}} \int_B \exp(-2t\sqrt{\log 2})\, dt$$

where $B = \{t;\ c \exp \beta t \in A\}$. Making the change of variables $x = c \exp \beta t$, we get, using that $2\sqrt{\log 2}/\beta = m$

$$\mu(A) = c' \int_A x^{-m-1} dx$$

where $c' = c^m/\beta\sqrt{\pi}$, so that $c' = 1$ with an appropriate choice of c; the result then follows from the definition of Λ_m. □

Another important idea is that to study the distribution of the weights $(w_\alpha)_{\alpha\geq 1}$ with a given accuracy, it suffices to choose $-b$ large enough and to study those of the numbers h_α that are $\geq b$. To make the idea precise, we define

$$e_\alpha = \exp \beta h_\alpha\,;\ Z = \sum_{\alpha\leq 2^N} e_\alpha\,;\ Z_b = \sum_{h_\alpha \geq b} e_\alpha$$

and we set

$$w_\alpha^b = \begin{cases} 0 & \text{if } \ h_\alpha < b \\ e_\alpha/Z_b & \text{if } \ h_\alpha \geq b\,. \end{cases}$$

The dependence on N of these quantities is kept implicit.

Lemma 1.2.4. *We have*

$$\sum_{\alpha\geq 1} |w_\alpha - w_\alpha^b| = 2\,\frac{Z - Z_b}{Z}\,.$$

Proof. Let $A = \{\alpha;\ h_\alpha \geq b\}$. If $\alpha \in A$, then

$$w_\alpha - w_\alpha^b = e_\alpha \left(\frac{1}{Z} - \frac{1}{Z_b}\right) = e_\alpha \frac{Z_b - Z}{Z Z_b}$$

and thus

$$\sum_{\alpha \in A} |w_\alpha - w_\alpha^b| = \frac{Z - Z_b}{Z} \,.$$

If $\alpha \notin A$, then $|w_\alpha - w_\alpha^b| = w_\alpha$, and

$$\sum_{\alpha \notin A} w_\alpha = \frac{1}{Z} \sum_{\alpha \notin A} e_\alpha = \frac{Z - Z_b}{Z} \,.$$

□

Lemma 1.2.5. *Given $\varepsilon > 0$, we can find b such that*

$$\limsup_{N \to \infty} \mathsf{P}\left(\frac{Z - Z_b}{Z} \geq \varepsilon \right) \leq \varepsilon \,. \tag{1.31}$$

Proof. First, given $\varepsilon > 0$, we show that there exists $\eta > 0$ such that, for N large enough, we have

$$\mathsf{P}(Z \leq \eta) \leq \frac{\varepsilon}{2} \,. \tag{1.32}$$

To see this, we use that

$$Z \leq \exp \beta x \Rightarrow \operatorname{card}\{\boldsymbol{\sigma}\,;\ -H'_N(\boldsymbol{\sigma}) \geq x\} = \operatorname{card}\{\alpha\,;\ h_\alpha \geq x\} = 0$$

and we use (1.28) with $k = 0$ and x instead of b.

To complete the proof, we show that given $\eta, \varepsilon > 0$, we can find b such that, for N large enough

$$\mathsf{P}(Z - Z_b \geq \varepsilon\, \eta) \leq \frac{\varepsilon}{2} \tag{1.33}$$

and together with (1.32) this proves (1.31). To prove (1.33), we compute

$$\mathsf{E}(Z - Z_b) = 2^N \mathsf{E}(\exp(-\beta H'_N(\boldsymbol{\sigma}))\, \mathbf{1}_{\{-H'_N(\boldsymbol{\sigma}) \leq b\}})$$

and, using (A.44)

$$\mathsf{E}(Z - Z_b) \leq \beta \int_{-\infty}^{b} (\exp \beta\, x)\, 2^N \mathsf{P}(-H'_N(\boldsymbol{\sigma}) \geq x)\, \mathrm{d}x \,.$$

Now, using (1.30),

$$\begin{aligned} 2^N \mathsf{P}(\{-H'_N(\boldsymbol{\sigma}) \geq x\}) &\leq \frac{1}{\sqrt{\pi}} \int_x^\infty \exp(-2\,t\,a_N)\, \mathrm{d}t \\ &= \frac{1}{2\sqrt{\pi}\, a_N} \exp(-2\,x\,a_N) \end{aligned}$$

and thus

$$\mathsf{E}(Z - Z_b) \le \frac{\beta}{2\sqrt{\pi}\,a_N} \int_{-\infty}^{b} \exp x(\beta - 2\,a_N)\,\mathrm{d}x\,.$$

Since $\beta > 2\sqrt{\log 2} \ge 2\,a_N$, we can find b with $\mathsf{E}(Z - Z_b) \le \varepsilon^2\,\eta/2$, and this proves (1.33) by Markov's inequality. □

Proof of Theorem 1.2.1. Using the notation of Lemma 1.2.3, we define

$$S = \sum_{\alpha \ge 1} \exp \beta\, c_\alpha\,;\ S_b = \sum_{c_\alpha \ge b} \exp \beta\, c_\alpha$$

and we define $v_\alpha^b = 0$ if $c_\alpha < b$, $v_\alpha^b = (\exp \beta\, c_\alpha)/S_b$ if $c_\alpha \ge b$.

Considering a continuous function f on $\mathcal{S}$, we write

$$\begin{aligned} |\mathsf{E}f((w_\alpha)) - \mathsf{E}f((v_\alpha))| &\le |\mathsf{E}f((w_\alpha)) - \mathsf{E}f((w_\alpha^b))| \\ &+ |\mathsf{E}f((w_\alpha^b)) - \mathsf{E}f((v_\alpha^b))| \\ &+ |\mathsf{E}f((v_\alpha^b)) - \mathsf{E}f((v_\alpha))|\,. \end{aligned} \tag{1.34}$$

Since f is continuous and $\mathcal{S}$ is compact, given $\eta > 0$, we can find $\varepsilon > 0$ such that

$$\mathrm{d}(\mathbf{x}, \mathbf{y}) \le \sqrt{\varepsilon} \Rightarrow |f(\mathbf{x}) - f(\mathbf{y})| \le \eta$$

where d is the distance (1.25). In particular

$$\sum_{\alpha \ge 1} |x_\alpha - y_\alpha| \le \varepsilon \Rightarrow |f(\mathbf{x}) - f(\mathbf{y})| \le \eta\,.$$

Using Lemmas 1.2.4, 1.2.5 we see that we can find b (with $-b$ large enough) such that, for N large enough,

$$\mathsf{P}(|f((w_\alpha)) - f((w_\alpha^b))| \ge \eta) \le \eta$$

and thus, for N large enough,

$$|\mathsf{E}f((w_\alpha)) - \mathsf{E}f((w_\alpha^b))| \le \eta + 2\,\eta \max|f|\,.$$

We can repeat this argument for the last terms of (1.34), and thus all we have to do is to prove that, for any value of b,

$$\lim_{N \to \infty} |\mathsf{E}f(w_\alpha^b) - \mathsf{E}f(v_\alpha^b)| = 0\,. \tag{1.35}$$

The sequence (v_α^b) is constructed from the points c_α, $c_\alpha \ge b$, the way the sequence (w_α^b) is generated from the points h_α, $h_\alpha \ge b$. The probability that there are k points $c_\alpha \ge b$ is given by the right-hand side of (1.28). Conditionally upon this event, the set $\{c_1, \dots, c_k\}$ is distributed like the set $\{Y_1, \dots, Y_k\}$ where $Y_1, \dots, Y_k$ are independent r.v., with a density proportional to $\exp(-2t\sqrt{\log 2})$ on the half line $[b, \infty)$. Thus (1.35) follows from Lemma 1.2.2. □

We now turn to the exciting part of this section, some remarkable relations satisfied by the Poisson-Dirichlet distribution Λ_m, relations that will simply follow from Theorem 1.2.1.

We will always denote average with respect to Gibbs' measure G_N by $\langle\cdot\rangle$. Thus, for a function f on Σ_N,

$$\langle f\rangle = \frac{1}{Z_N(\beta)}\sum_{\boldsymbol{\sigma}} f(\boldsymbol{\sigma})\exp(-\beta H_N(\boldsymbol{\sigma}))\ . \tag{1.36}$$

It should be helpful to observe that for a subset A of Σ_N we have

$$G_N(A) = \langle \mathbf{1}_A\rangle\ .$$

We start with an innocuous observation.

Lemma 1.2.6. *If $\beta > 2\sqrt{\log 2}$ we have*

$$\lim_{N\to\infty}\mathsf{E}\left\langle\left|\frac{H_N(\boldsymbol{\sigma})}{N}+\sqrt{\log 2}\right|\right\rangle = 0\ . \tag{1.37}$$

Proof. Given $\varepsilon > 0$, we write

$$\mathsf{E}\left\langle\left|\frac{H_N(\boldsymbol{\sigma})}{N}+\sqrt{\log 2}\right|\right\rangle \le \varepsilon\sqrt{2}+\mathrm{I}+\mathrm{II}$$

where

$$\mathrm{I} = \mathsf{E}\left\langle\left|\frac{H_N(\boldsymbol{\sigma})}{N}+\sqrt{\log 2}\right|\mathbf{1}_{\{|H_N(\boldsymbol{\sigma})|\ge N(1+\varepsilon)\sqrt{\log 2}\}}\right\rangle$$

$$\begin{aligned}\mathrm{II} &= \mathsf{E}\left\langle\left|\frac{H_N(\boldsymbol{\sigma})}{N}+\sqrt{\log 2}\right|\mathbf{1}_{\{-N(1+\varepsilon)\sqrt{\log 2}\le -H_N(\boldsymbol{\sigma})\le N(1-\varepsilon)\sqrt{\log 2}\}}\right\rangle\\ &\le (2+\varepsilon)\sqrt{\log 2}\,\mathsf{E}\left\langle\mathbf{1}_{\{-H_N(\boldsymbol{\sigma})\le N(1-\varepsilon)\sqrt{\log 2}\}}\right\rangle\ .\end{aligned}$$

We observe that given $\varepsilon > 0$ and b, for N large enough, we have

$$\left\langle\mathbf{1}_{\{-H_N(\boldsymbol{\sigma})\le N(1-\varepsilon)\sqrt{\log 2}\}}\right\rangle \le \frac{Z-Z_b}{Z}$$

It then follows from (1.31) that $\mathrm{II}\to 0$ as $N\to\infty$. To prove that $\mathrm{I}\to 0$, we observe that if

$$U_N = \max_{\boldsymbol{\sigma}}\left|\frac{H_N(\boldsymbol{\sigma})}{N}\right|$$

then

$$\begin{aligned}\mathrm{I} &\le (\mathsf{E}(U_N+\sqrt{\log 2})\,\mathbf{1}_{\{U_N\ge(1+\varepsilon)\sqrt{\log 2}\}})\\ &\le \mathsf{E}((U_N+\sqrt{\log 2})^2)^{1/2}\mathsf{P}(U_N\ge(1+\varepsilon)\sqrt{\log 2})^{1/2})\ ,\end{aligned} \tag{1.38}$$

using the Cauchy-Schwarz inequality. Now, if $t > 0$,

$$\mathsf{P}(U_N \geq t) \leq \sum_{\boldsymbol{\sigma}} \mathsf{P}(|H_N(\boldsymbol{\sigma})| \geq tN) \leq 2^{N+1} \exp(-Nt^2)$$

and using (A.7.1) to control $\mathsf{E}\, U_N^2$, this implies from (1.38) that $\mathrm{I} \to 0$. $\square$

In particular (1.37) implies that

$$\lim_{N\to\infty} \mathsf{E}\left\langle -\frac{H_N(\boldsymbol{\sigma})}{N}\right\rangle = \sqrt{\log 2}\,. \tag{1.39}$$

It is remarkable that integration by parts of this trivial relation yields very nice information. Integration by parts will be one of our main tools, and we will learn to use it with increasing sophistication. The basic formula is that if F is a smooth function (of moderate growth), if g is a (centered) Gaussian r.v., then

$$\mathsf{E}\, g F(g) = \mathsf{E}\, g^2\, \mathsf{E}\, F'(g)\,. \tag{1.40}$$

(See (A.6.4) for a proof.) To apply this formula to $\mathsf{E}\langle -H_N(\boldsymbol{\sigma})\rangle$, we write the explicit value of the bracket $\langle\cdot\rangle$:

$$\mathsf{E}\langle -H_N(\boldsymbol{\sigma})\rangle = \sum_{\boldsymbol{\sigma}} \mathsf{E}\left(-H_N(\boldsymbol{\sigma})\frac{\exp(-\beta H_N(\boldsymbol{\sigma}))}{Z_N(\beta)}\right)\,. \tag{1.41}$$

What makes matters simple here is that when $\boldsymbol{\sigma}' \neq \boldsymbol{\sigma}$, $H_N(\boldsymbol{\sigma}')$ and $H_N(\boldsymbol{\sigma})$ are independent. Thus we can use (1.40) in (1.41) conditionally on $H_N(\boldsymbol{\sigma}')$ when $\boldsymbol{\sigma}' \neq \boldsymbol{\sigma}$. That is, we take $g = -H_N(\boldsymbol{\sigma})$, and for $\boldsymbol{\sigma}' \neq \boldsymbol{\sigma}$ we treat $H_N(\boldsymbol{\sigma}')$ as a constant. The appropriate function F is given by

$$F(x) = \frac{\exp \beta x}{\exp \beta x + Z'}$$

where $Z' = \sum_{\boldsymbol{\rho}\neq\boldsymbol{\sigma}} \exp(-\beta H_N(\boldsymbol{\rho}))$. Thus

$$F'(x) = \frac{\beta \exp \beta x}{\exp \beta x + Z'} - \frac{\beta \exp 2\beta x}{(\exp \beta x + Z')^2}$$

and (1.40) gives

$$\begin{aligned}&\mathsf{E}\left(-H_N(\boldsymbol{\sigma})\frac{\exp(-\beta H_N(\boldsymbol{\sigma}))}{Z_N(\beta)}\right)\\ &= \frac{\beta N}{2}\mathsf{E}\left(\frac{\exp(-\beta H_N(\boldsymbol{\sigma}))}{Z_N(\beta)} - \frac{\exp(-2\beta H_N(\boldsymbol{\sigma}))}{Z_N(\beta)^2}\right),\end{aligned}$$

so that by (1.41),

$$\mathsf{E}\left\langle -\frac{H_N(\boldsymbol{\sigma})}{N}\right\rangle = \frac{\beta}{2}\left(1 - \mathsf{E}\sum_{\boldsymbol{\sigma}} G_N^2(\{\boldsymbol{\sigma}\})\right)\,, \tag{1.42}$$

and, by (1.39),

$$\lim_{N\to\infty} \mathsf{E}\sum_{\sigma} G_N^2(\{\sigma\}) = 1 - \frac{2\sqrt{\log 2}}{\beta} .$$

If we combine this with Theorem 1.2.1, we obtain the following.

Proposition 1.2.7. *If the sequence* $(v_\alpha)_{\alpha\ge 1}$ *follows the Poisson-Dirichlet distribution* Λ_m, *it satisfies*

$$\mathsf{E}\sum_{\alpha\ge 1} v_\alpha^2 = 1 - m . \tag{1.43}$$

As pointed out, the distribution Λ_m is very well understood [120] and there are other methods (some of which will be used in Chapter 6) to prove (1.43). What is particularly pleasant, however, is that the relations that are the most important for us can be obtained through a generalization of the argument leading to Proposition 1.2.7. These relations concern the quantities $S^{(m)}(n_1, \ldots, n_k)$, where m is the parameter in Λ_m, k is an integer and $n_1, \ldots, n_k$ are integers ≥ 2. They are defined by

$$S^{(m)}(n_1, \ldots, n_k) = \mathsf{E}\prod_{s\le k}\left(\sum_{\alpha\ge 1} v_\alpha^{n_s}\right) \tag{1.44}$$

where the random sequence $(v_\alpha)_{\alpha\ge 1}$ follows the distribution Λ_m.

The importance of the numbers $S^{(m)}(n_1, \ldots, n_k)$ is that they completely determine the distribution Λ_m. To see this, consider a random sequence $(w_\alpha)_{\alpha\ge 1}$ valued in $\mathcal{S}$. Consider the random measure μ on $[0,1]$ that gives mass w_α^2 to the point w_α. Thus the moments of μ are given by

$$X_k = \int x^k \mathrm{d}\mu(x) = \sum_{\alpha\ge 1} w_\alpha^{k+2} .$$

The measure μ is determined by its moments X_n $(n \ge 0)$, so that the distribution of the sequence (w_α) is determined by the joint distribution of the variables X_n, which is in turn determined by the numbers

$$\mathsf{E}(X_{n_1} \cdots X_{n_k}) = \mathsf{E}\left(\prod_{s\le k}\sum_{\alpha\ge 1} w_\alpha^{n_s+2}\right) . \tag{1.45}$$

As a consequence, if we know that the numbers of (1.45) converge to the corresponding numbers $S^{(m)}(n_1+2, \ldots, n_k+2)$, we know that the law in $\mathcal{S}$ of the random sequence $(w_\alpha)_{\alpha\ge 1}$ converges to Λ_m.

Before we proceed further, we need to introduce one of the recurring ideas of this book, the idea of *replicas*. That is, given an integer n, we will consider the space Σ_N^n, endowed with the product measure $G_N^{\otimes n}$. Each factor Σ_N of

Σ_N^n is a copy, a "replica" of Σ_N. Each copy of G_N involves the *same* disorder, i.e. the same realization of the family $(H_N(\boldsymbol{\sigma}))$. What we call replicas is often called "real replicas" in physics, to distinguish them from replicas of the so-called "replica method" described in Section 2.15. If f is a real-valued function on Σ_N^n, we denote by $\langle f \rangle$ its integral with respect to $G_N^{\otimes n}$. This notation does not indicate the value of n. This does not create confusion, and is in fact very convenient. A function on Σ_N^n can always be seen as a function on $\Sigma_N^{n'}$ when $n' \geq n$.

The generic point of Σ_N^n will be denoted by $(\boldsymbol{\sigma}^1, \ldots, \boldsymbol{\sigma}^n)$.

Consider now a function f_N on Σ_N^n, and assume $|f_N| \leq 1$. Then, from (1.37), we see that

$$\left| \mathsf{E}\left\langle -\frac{H_N(\boldsymbol{\sigma}^1)}{N} f_N(\boldsymbol{\sigma}^1, \ldots, \boldsymbol{\sigma}^n) \right\rangle - \sqrt{\log 2}\, \mathsf{E}\langle f_N(\boldsymbol{\sigma}^1, \ldots, \boldsymbol{\sigma}^n) \rangle \right|$$
$$\leq \mathsf{E}\left\langle \left| -\frac{H_N(\boldsymbol{\sigma})}{N} - \sqrt{\log 2} \right| \right\rangle \to 0 \,. \tag{1.46}$$

It is a remarkable (and rather mysterious) fact that integration by parts of (1.46) yields relations much more powerful than (1.43). To proceed with the integration by parts, we write

$$\mathsf{E}\left\langle -\frac{H_N(\boldsymbol{\sigma}^1)}{N} f_N(\boldsymbol{\sigma}^1, \ldots, \boldsymbol{\sigma}^n) \right\rangle$$
$$= \sum f_N(\boldsymbol{\sigma}^1, \ldots, \boldsymbol{\sigma}^n)\, \mathsf{E}\left(-\frac{H_N(\boldsymbol{\sigma}^1)}{N} \frac{\exp\left(-\beta \sum_{\ell \leq n} H_N(\boldsymbol{\sigma}^\ell)\right)}{Z_N^n(\beta)} \right) . \tag{1.47}$$

This follows from the definition of $\langle \cdot \rangle$. The summation is of course over all possible values $\boldsymbol{\sigma}^1, \ldots, \boldsymbol{\sigma}^n$ in Σ_N. In the last term of (1.47), we use (1.40) at $H_N(\boldsymbol{\sigma}')$ given for $\boldsymbol{\sigma}' \neq \boldsymbol{\sigma}^1$, as we did in (1.41). We need of course to distinguish which of the configurations $\boldsymbol{\sigma}^\ell$, $1 \leq \ell \leq n$ are equal to $\boldsymbol{\sigma}^1$, and which are different. Assuming that f_N does *not* depend on the disorder, we find that (1.47) is equal to

$$\frac{\beta}{2} \sum f_N(\boldsymbol{\sigma}^1, \ldots, \boldsymbol{\sigma}^n)\, \mathsf{E}\left(\sum_{\ell \leq n} \mathbf{1}_{\{\boldsymbol{\sigma}^1 = \boldsymbol{\sigma}^\ell\}} \frac{\exp\left(-\beta \sum_{\ell \leq n} H_N(\boldsymbol{\sigma}^\ell)\right)}{Z_N^n(\beta)} \right.$$
$$\left. - \frac{n \exp\left(-\beta \sum_{\ell \leq n} H_N(\boldsymbol{\sigma}^\ell) - \beta H_N(\boldsymbol{\sigma}^1)\right)}{Z_N^{n+1}(\beta)} \right) , \tag{1.48}$$

where the summation is as in (1.47).

To express conveniently this quantity in terms of Gibbs' averages, we introduce one more replica $\boldsymbol{\sigma}^{n+1}$, a device that will be constantly used. We see that the quantity (1.48) is

$$\frac{\beta}{2}\Big(\sum_{\ell\le n} \mathsf{E}\langle f_N(\boldsymbol{\sigma}^1,\ldots,\boldsymbol{\sigma}^n)\,\mathbf{1}_{\{\boldsymbol{\sigma}^1=\boldsymbol{\sigma}^\ell\}}\rangle - n\,\mathsf{E}\langle f_N(\boldsymbol{\sigma}^1,\ldots,\boldsymbol{\sigma}^n)\,\mathbf{1}_{\{\boldsymbol{\sigma}^1=\boldsymbol{\sigma}^{n+1}\}}\rangle\Big)\,.$$

Typically, we will lighten notation and write this as

$$\frac{\beta}{2}\Big(\sum_{\ell\le n} \mathsf{E}\langle f_N\,\mathbf{1}_{\{\boldsymbol{\sigma}^1=\boldsymbol{\sigma}^\ell\}}\rangle - n\,\mathsf{E}\langle f_N\,\mathbf{1}_{\{\boldsymbol{\sigma}^1=\boldsymbol{\sigma}^{n+1}\}}\rangle\Big)\,.$$

Going back to (1.46), and setting $m = 2\sqrt{\log 2}/\beta$, we get

$$\lim_{N\to\infty}\left(\sum_{\ell\le n} \mathsf{E}\langle f_N\,\mathbf{1}_{\{\boldsymbol{\sigma}^1=\boldsymbol{\sigma}^\ell\}}\rangle - n\,\mathsf{E}\langle f_N\,\mathbf{1}_{\{\boldsymbol{\sigma}^1=\boldsymbol{\sigma}^{n+1}\}}\rangle - m\,\mathsf{E}\langle f_N\rangle\right) = 0\,,$$

and thus, since $\mathbf{1}_{\{\boldsymbol{\sigma}^1=\boldsymbol{\sigma}^\ell\}} = 1$ if $\ell = 1$, and assuming that the limits exist,

$$\begin{aligned}\lim_{N\to\infty} \mathsf{E}\langle f_N\,\mathbf{1}_{\{\boldsymbol{\sigma}^1=\boldsymbol{\sigma}^{n+1}\}}\rangle &= \frac{1-m}{n}\lim_{N\to\infty}\mathsf{E}\langle f_N\rangle\\ &\quad+\frac{1}{n}\sum_{2\le\ell\le n}\lim_{N\to\infty}\mathsf{E}\langle f_N\,\mathbf{1}_{\{\boldsymbol{\sigma}^1=\boldsymbol{\sigma}^\ell\}}\rangle\,. \end{aligned}\tag{1.49}$$

This is a special case of the fundamental Ghirlanda-Guerra relations that will be developed in Section 2.12 an will play a crucial role in Chapter 6.

Before we exploit these relations, we tidy up the notation. The idea of considering for each N a function f_N on a product Σ_N^n will occur constantly; it is convenient to remove the index N, a convention that is harmless since typically f will be"defined by the same formula for each N". We will simply say that "we consider a function f on Σ_N^n".

Consider now an integer k, and integers $n_1,\ldots,n_k$. Set $n = n_1+\cdots+n_k$. We decompose the set $\{1,\ldots,n\}$ into the disjoint union of sets $I_1,\ldots,I_k$, with card $I_s = n_s$ for $s\le k$.

We consider the function f_s on Σ_N^n defined by

$$f_s(\boldsymbol{\sigma}^1,\ldots,\boldsymbol{\sigma}^n) = 1 \quad \text{if} \quad \forall\,\ell,\ell'\in I_s\,,\ \boldsymbol{\sigma}^\ell = \boldsymbol{\sigma}^{\ell'}$$

and $f_s(\boldsymbol{\sigma}^1,\ldots,\boldsymbol{\sigma}^n) = 0$ otherwise. Thus

$$\langle f_s\rangle = \sum_{\boldsymbol{\sigma}} G_N(\{\boldsymbol{\sigma}\})^{n_s}\,.$$

We consider the function $f = \prod_{s\le n} f_s$. The functions f_s depend on disjoint sets of replicas, so that

$$\langle f\rangle = \prod_{s\le k}\langle f_s\rangle = \prod_{s\le k}\sum_{\boldsymbol{\sigma}} G_N(\{\boldsymbol{\sigma}\})^{n_s}\,.$$

Thus

$$S_N(n_1,\dots,n_k) := \mathsf{E}\langle f\rangle = \mathsf{E}\prod_{s\le k}\sum_{\boldsymbol\sigma} G_N(\{\boldsymbol\sigma\})^{n_s}\ . \tag{1.50}$$

We will interpret (1.49) in terms of these quantities. Assuming without loss of generality that $1 \in I_1$, we claim first that

$$\mathsf{E}\langle f\,\mathbf{1}_{\{\boldsymbol\sigma^1=\boldsymbol\sigma^{n+1}\}}\rangle = S_N(n_1+1,\dots,n_k)\ . \tag{1.51}$$

This is simply because

$$f\,\mathbf{1}_{\{\boldsymbol\sigma^1=\boldsymbol\sigma^{n+1}\}} = f_1' \prod_{2\le s\le k} f_s\ ,$$

where

$$f_1' = f_1\,\mathbf{1}_{\{\boldsymbol\sigma^1=\boldsymbol\sigma^{n+1}\}}\ ,$$

and that $f_1' = 1$ provided

$$\boldsymbol\sigma^{n+1} = \boldsymbol\sigma^1 = \boldsymbol\sigma^\ell\ ,\ \forall \ell\in I_1\ .$$

and is 0 otherwise.

Next, we observe that

$$f\,\mathbf{1}_{\{\boldsymbol\sigma^1=\boldsymbol\sigma^\ell\}} = f$$

if $\ell\in I_1$. Finally, if $\ell \in I_s$, $s\ge 2$, we observe that

$$f_1 f_s \mathbf{1}_{\{\boldsymbol\sigma^1=\boldsymbol\sigma^\ell\}} = 0$$

unless all $\boldsymbol\sigma^\ell$, $\ell\in I_1\cup I_s$ are equal. That is, $f\,\mathbf{1}_{\{\boldsymbol\sigma^1=\boldsymbol\sigma^\ell\}}$ is constructed like f, except that one has merged the sets I_1 and I_s, so that

$$\mathsf{E}\langle f\,\mathbf{1}_{\{\boldsymbol\sigma^1=\boldsymbol\sigma^\ell\}}\rangle = S_N(n_1+n_s,n_2,\dots,n_{s-1},n_{s+1},\dots,n_k)\ .$$

To lighten notation, we will write this last term as $S_N(n_1+n_s,\dots,n_k)$, leaving implicit the fact that the sth term n_s has been omitted; and similarly for $S^{(m)}(n_1+n_s,\dots,n_k)$.

Collecting the terms, the relation (1.49) implies

$$\begin{aligned}\lim_{N\to\infty} S_N(n_1+1,n_2,\dots,n_k) &= \frac{n_1-m}{n}\lim_{N\to\infty} S_N(n_1,\dots,n_k)\\ &\quad+\sum_{2\le s\le k}\frac{n_s}{n}\lim_{N\to\infty} S_N(n_1+n_s,\dots,n_k)\ .\end{aligned}$$

Using Theorem 1.2.1, we then have the following

Proposition 1.2.8. *For all positive integers $k, n_1,\dots,n_k$, we have*

$$\begin{aligned}S^{(m)}(n_1+1,\dots,n_k) &= \frac{n_1-m}{n}S^{(m)}(n_1,\dots,n_k)\\ &+\sum_{2\le s\le n}\frac{n_s}{n}S^{(m)}(n_1+n_s,n_2,\dots,n_{s-1},n_{s+1},,\dots,n_k)\ ,\end{aligned} \tag{1.52}$$

where $n = n_1+\cdots+n_k$.

It is good to observe that the terms on the right-hand side of (1.52) are less complicated than the term on the left-hand side, because $\sum_{s\leq k} n_s$ has been decreased. Thus (1.52), together with the facts that $S^{(m)}(1,\ldots,1) = 1$, that $S^{(m)}(2) = 1 - m$ by (1.43), and that $S^{(m)}(n_1,\ldots,n_k)$ is symmetric in its arguments completely determines all the numbers $S^{(m)}(n_1,\ldots,n_k)$, and hence the law of Λ_m.

1.3 Notes and Comments

One of the purposes of these notes and comments is to give more complete references than is done in the text. Among the large number of non-rigorous papers written by physicists, I am quoting only those that I did try to read. Typically, I find that these papers are very useful in suggesting results one would like to prove, but are much less helpful in suggesting ways to prove them. Among the papers that (attempt to) prove rigorous results I have performed no systematic search, and I am simply quoting those that appear to be the best known (be it for good or bad reasons!). I have read very few of them in any detail, simply because I did not think that they were likely to contain ideas that would be useful for my purpose. Therefore the discussion of references is going to be woefully incomplete.

The REM is introduced in [48]. I learned the exact inequality (1.12) (that is, without an error term) in [67]. It is impossible to recommend too highly the reading of Guerra's papers. Besides fundamental ideas that influenced this book in a major way they contain numerous innocuous-looking observations that are in fact very sharp and hard to discover.

One can find a different proof of Proposition 1.1.5 in [153]. I have tried not to repeat those notes, and some material there is not reproduced in the present book.

I simply sat down with a pen to figure out a proof of Theorem 1.2.1, feeling that it must have been proved at least 10 years ago. A. Bovier, who designed independently his own proof while writing his notes [19], interestingly pointed out to me that there does not seem to exist a previous complete proof of this result (although it was probably known to Ruelle [121]). The basic ideas of the proof (such as separating the large values) go back to P. Lévy and are typical of the study of normalized sums of triangular arrays of random variables. In that particular instance, there is certainly no claim that ideas from physics bring any new understanding to probability.

Starting with Lemma 1.2.6, the idea is to derive and apply the Ghirlanda-Guerra identities for the REM, taking advantage of the fact that it is easier in this case than in general to show that the Hamiltonian is "nearly constant".

I was pleased to have thought of (1.50), but then I learned that Kingman had developed similar ideas much earlier [86], [87]. Interestingly, the relations of Proposition 1.2.8 do not seem to have been observed previously. For generalization of these relations, see [119].

2. The Sherrington-Kirkpatrick Model

2.1 Introduction

Consider a large population of individuals (or atoms) that we label from 1 to N. We assume that each individual knows all the others. The feelings of the individual i towards the individual j are measured by a number g_{ij} that can be positive, or, unfortunately, negative. We assume symmetry, $g_{ij} = g_{ji}$, so in practice we will consider only the numbers $(g_{ij})_{i<j}$. We are trying to model a situation where these feelings are random. We are not trying to make realistic assumptions, but rather to find the simplest possible model; so we will assume that the numbers $(g_{ij})_{i<j}$ are independent r.v. Since we are aiming for simplicity, we will assume that these r.v. are standard Gaussian.

A very important feature of this model (called frustration, in physics) is that if $g_{ij} > 0$, $g_{jk} > 0$ (that is, i and j are friends, and j and k are friends), then i and k are just as likely to be enemies as they are to be friends. The interactions (g_{ij}) describe a very complex social situation.

Let us now choose the goal of separating the population in two classes, putting, as much as possible, friends together and enemies apart. It should be obvious that at best this can be done very imperfectly: some friends will be separated and some enemies will cohabit. To introduce a quantitative way to measure how well we have succeeded, it is convenient to assign to each individual i a number $\sigma_i \in \{-1, 1\}$, thereby defining two classes of individuals. Possibly the simplest measure of how well these two classes unite friends and separate enemies is the quantity

$$\sum_{i<j} g_{ij} \sigma_i \sigma_j \ . \tag{2.1}$$

Trying to make this large incites to take σ_i and σ_j of the same sign when $g_{ij} > 0$, and of opposite signs when $g_{ij} < 0$.

Despite the simplicity of the expression (2.1), the optimization problem of finding the maximum of this quantity (for a typical realization of the g_{ij}) over the configuration $\boldsymbol{\sigma} = (\sigma_1, \ldots, \sigma_N)$ appears to be of extreme difficulty, and little is rigorously known about it.

Such an optimization problem is called in physics a zero temperature problem, because at zero temperature a system is always found in its configuration of lowest energy. To a zero temperature problem is often associated a version of the problem "with a temperature", here the problem corresponding to the Hamiltonian

$$-H_N(\boldsymbol{\sigma}) = \frac{1}{\sqrt{N}} \sum_{i<j} g_{ij}\, \sigma_i \sigma_j \,. \tag{2.2}$$

The purpose of the normalization factor $N^{-1/2}$ will be apparent in (2.6) below. As in the case of the REM, the energies of configurations are centered Gaussian r.v., but if $\boldsymbol{\sigma}^1 \neq \boldsymbol{\sigma}^2$, the energies of $\boldsymbol{\sigma}^1$, $\boldsymbol{\sigma}^2$ are not independent. In fact

$$\begin{aligned} \mathsf{E}(H_N(\boldsymbol{\sigma}^1)\, H_N(\boldsymbol{\sigma}^2)) &= \frac{1}{N} \sum_{i<j} \sigma_i^1 \sigma_j^1\, \sigma_i^2 \sigma_j^2 \\ &= \frac{N}{2} \left(\frac{1}{N} \sum_{i \leq N} \sigma_i^1 \sigma_i^2 \right)^2 - \frac{1}{2} \,. \end{aligned} \tag{2.3}$$

Here we see the first occurrence of a quantity that plays an essential part in the sequel, namely

$$R_{1,2} = R_{1,2}(\boldsymbol{\sigma}^1, \boldsymbol{\sigma}^2) = \frac{1}{N} \sum_{i \leq N} \sigma_i^1 \sigma_i^2 \,. \tag{2.4}$$

This quantity is called the *overlap* of the configurations $\boldsymbol{\sigma}^1, \boldsymbol{\sigma}^2$, because the closer it is to 1, the closer they are to each other. Let us denote by $\mathrm{d}(\boldsymbol{\sigma}^1, \boldsymbol{\sigma}^2)$ the Hamming distance of $\boldsymbol{\sigma}^1, \boldsymbol{\sigma}^2$, that is the proportion of coordinates where $\boldsymbol{\sigma}^1, \boldsymbol{\sigma}^2$ differ. Then

$$R_{1,2} = 1 - 2\mathrm{d}(\boldsymbol{\sigma}^1, \boldsymbol{\sigma}^2) \,, \tag{2.5}$$

and this shows that $R_{1,2}$, and hence the correlation of the family $(H_N(\boldsymbol{\sigma}))$, is closely related to the structure of the space (Σ_N, d). This structure is very rich, and this explains why the simple expression (2.2) suffices to create a complex situation. Let us also note that (2.3) implies

$$\mathsf{E}\, H_N^2(\boldsymbol{\sigma}) = \frac{N-1}{2} \tag{2.6}$$

and this is motivates the normalization we have chosen for the REM.

In the heuristic considerations leading to (2.2), we have made the assumption that any two individuals interact. This is certainly not the case if the individuals are atoms in a macroscopic sample of matter. Thus (2.2) is not, by any means, a realistic model for disordered interaction among atoms. A more realistic model would locate the atoms at the vertices of a lattice (e.g. $\mathbb{Z}^2$) and would assume that the strength of the interaction between two atoms decreases as their distance increases. The problem is that these models, while more interesting from the point of view of physics, are also extremely difficult. Even if one makes the simplifying assumption that an atom interacts only with its nearest neighbors, they are so difficult that, at the time of this writing, no consensus has been reached among physicists regarding their probable behavior. It is to bypass this difficulty that Sherrington and Kirkpatrick introduced the Hamiltonian (2.2), where the geometric location of the atoms is forgotten and where they all interact with each other. Such a simplification is called a "mean field approximation", and the corresponding models are called "mean field models".

The Hamiltonian (2.2) presents a very special symmetry: Gibbs' measure is invariant by the transformation $\boldsymbol{\sigma} \to -\boldsymbol{\sigma}$. This special situation is somewhat misleading. In order not to get hypnotized by special features, we will consider the version of (2.2) "with an external field", i.e. where the Hamiltonian is

$$-H_N(\boldsymbol{\sigma}) = \frac{1}{\sqrt{N}} \sum_{i<j} g_{ij}\, \sigma_i \sigma_j + h \sum_{i \le N} \sigma_i \,. \tag{2.7}$$

The last term represents the action of an "external field", that is a magnetic field created by an apparatus outside the sample of matter we study. The external field favors the + spins over the − spins. With the Hamiltonian (2.7), the Boltzmann factors become

$$\exp\left(\frac{\beta}{\sqrt{N}} \sum_{i<j} g_{ij}\, \sigma_i \sigma_j + \beta h \sum_{i \le N} \sigma_i\right) \,. \tag{2.8}$$

The coefficient βh of $\sum_{i \le N} \sigma_i$ makes perfect sense to a physicist. However, when one looks at the mathematical structure of (2.8), one sees that the two terms $N^{-1/2} \sum_{i<j} g_{ij}\, \sigma_i \sigma_j$ and $\sum_{i \le N} \sigma_i$ look of different natures. Therefore, it would be much more convenient to have unrelated coefficients in front of these terms. For example, it is more cumbersome to take derivatives in β when using the factors (2.8) than when using the factors

$$\exp\left(\frac{\beta}{\sqrt{N}} \sum_{i<j} g_{ij}\, \sigma_i \sigma_j + h \sum_{i \le N} \sigma_i\right) \,.$$

Thus for the sake of mathematical clarity, it is better to abandon the physical point of view of having a "main parameter" β. Rather, we will think of the Hamiltonian as depending upon parameters $\beta, h, \ldots$ That is, we will write

$$-H_N(\boldsymbol{\sigma}) = -H_N(\boldsymbol{\sigma}, \beta, h) = \frac{\beta}{\sqrt{N}} \sum_{i<j} g_{ij}\, \sigma_i \sigma_j + h \sum_{i \le N} \sigma_i \tag{2.9}$$

for the Hamiltonian of the Sherrington-Kirkpatrick model. The Boltzmann factors become

$$\exp(-H_N(\boldsymbol{\sigma}, \beta, h))$$

or, with simpler notation, when β and h are implicit it becomes

$$\exp(-H_N(\boldsymbol{\sigma})) . \tag{2.10}$$

Once one has survived the initial surprise of not seeing the customary term β in (2.10), this notation works out appropriately. The formulas familiar to the physicists can be recovered by replacing our term h by βh.

2.2 The SK Model with no External Field at High Temperature

This is a very special model, which we will study through very special methods. These methods are simple and instructive, but they do not extend to more challenging situations. The model is however not trivial, and is a natural first step beyond the REM.

It follows from Proposition 1.1.4 (a one-line proof using Jensen's inequality (1.3)) that $p_N(\beta) \le \beta^2/4 + \log 2$, where p_N is defined in (1.2).

Theorem 2.2.1. [4] *If* $\beta < 1$, *we have*

$$\lim_{N\to\infty} p_N(\beta) = \frac{\beta^2}{4} + \log 2 . \tag{2.11}$$

We will see in Section 2.9 that (2.11) is not true for $\beta > 1$, in contrast with the case of the REM, where (2.11) holds for $\beta \le 2\sqrt{\log 2}$.

The meaning of (2.11) is that Jensen's inequality, when applied to $Z_N(\beta)$, is essentially an equality. The median and the mean of Z_N do not really differ, a most unexciting situation.

A truly special feature of the present model is that, if $\beta < 1$,

$$\mathsf{E}\, Z_N^2(\beta) \le K(\beta)\, (\mathsf{E}\, Z_N(\beta))^2 , \tag{2.12}$$

and we will prove this first. In (2.12), $K(\beta)$ denotes a number depending upon β only. We recall our convention to denote by L a universal constant, that need not be the same at each occurrence. In a similar spirit, we denote by K a number that does not depend upon N, but that might depend upon other parameters of the system, such as $\beta, h, \ldots$ When it clarifies matters, we indicate on which parameters K might depend. Thus for example $K(h)$ does not depend on β. The value of K need not be the same at each occurrence.

We start to apply our convention, justified in the previous section, to see β as a parameter on which H_N depends, that is

$$-H_{N,\beta}(\boldsymbol{\sigma}) = \frac{\beta}{\sqrt{N}} \sum_{i<j} g_{ij}\, \sigma_i\, \sigma_j ,$$

and consequently, rather than (2.3), we have

$$\mathsf{E}\, H_N(\boldsymbol{\sigma}^1)\, H_N(\boldsymbol{\sigma}^2) = \frac{\beta^2}{2}\,(N R_{1,2}^2 - 1) \tag{2.13}$$

and therefore

$$\mathsf{E}\, H_N^2(\boldsymbol{\sigma}) = \frac{\beta^2}{2}\,(N-1)\,. \tag{2.14}$$

Thus, (using (1.9))

$$\mathsf{E}\, Z_N(\beta) = 2^N \exp\frac{\beta^2}{4}\,(N-1)\,. \tag{2.15}$$

Lemma 2.2.2. *If $\gamma + \beta^2 < 1$, then*

$$\mathsf{E} \sum_{\boldsymbol{\sigma}^1,\boldsymbol{\sigma}^2} \exp\left(-H_N(\boldsymbol{\sigma}^1) - H_N(\boldsymbol{\sigma}^2) + \frac{\gamma N}{2}\, R_{1,2}^2\right)$$
$$\le \frac{1}{\sqrt{1-\beta^2-\gamma}}\,(\mathsf{E}\, Z_N(\beta))^2\,. \tag{2.16}$$

Proof. It follows from (2.13) that

$$\mathsf{E}((-H_N(\boldsymbol{\sigma}^1) - H_N(\boldsymbol{\sigma}^2))^2) = \beta^2(N-2) + \beta^2 N R_{1,2}^2\,,$$

so that, using (1.9) and (2.15), the left-hand side of (2.16) is at most

$$(\mathsf{E} Z_N(\beta))^2\, 2^{-2N} \sum_{\boldsymbol{\sigma}^1,\boldsymbol{\sigma}^2} \exp\frac{\delta N}{2} R_{1,2}^2\,,$$

where $\delta = \beta^2 + \gamma$. Given any value of $\boldsymbol{\sigma}^2$, we have

$$2^{-N}\sum_{\boldsymbol{\sigma}^1}\exp\frac{\delta N}{2}\left(\frac{1}{N}\sum_{i\le N}\sigma_i^1\sigma_i^2\right)^2 = 2^{-N}\sum_{\boldsymbol{\sigma}}\exp\frac{\delta N}{2}\left(\frac{1}{N}\sum_{i\le N}\sigma_i\right)^2$$

and the result follows from (A.5). □

If $\gamma = 0$, (2.16) implies that

$$\mathsf{E} Z_N^2(\beta) \le \frac{1}{\sqrt{1-\beta^2}}\,(\mathsf{E} Z_N(\beta))^2\,. \tag{2.17}$$

To take advantage of (2.17), we will use a simple, yet important argument, known as the second moment method, that goes back to the work of Paley and Zygmund on trigonometric series.

Proposition 2.2.3. *Consider a r.v. $X \ge 0$. Then*

$$\mathsf{P}\left(X \ge \frac{1}{2}\mathsf{E} X\right) \ge \frac{1}{4}\,\frac{(\mathsf{E} X)^2}{\mathsf{E} X^2}\,. \tag{2.18}$$

Proof. If $A = \{X \geq \mathsf{E}X/2\}$, then, since $X \leq \mathsf{E}X/2$ on the complement A^c of A we have

$$\mathsf{E}X = \mathsf{E}(X\mathbf{1}_A) + \mathsf{E}(X\mathbf{1}_{A^c}) \leq \mathsf{E}(X\mathbf{1}_A) + \frac{1}{2}\,\mathsf{E}X\ .$$

Thus, using the Cauchy-Schwarz inequality,

$$\frac{1}{2}\,\mathsf{E}X \leq \mathsf{E}(X\mathbf{1}_A) \leq (\mathsf{E}X^2)^{1/2}\,\mathsf{P}(A)^{1/2}\ .$$

□

From (2.17) and (2.18) we see that

$$\mathsf{P}\left(Z_N(\beta) \geq \frac{1}{2}\,\mathsf{E}Z_N(\beta)\right) \geq \frac{1}{4}\sqrt{1-\beta^2}\ , \tag{2.19}$$

so that, by (2.15),

$$\mathsf{P}\left(\frac{1}{N}\log Z_N(\beta) \geq \frac{N-1}{N}\left(\frac{\beta^2}{4} + \log 2\right)\right) \geq \frac{1}{4}\sqrt{1-\beta^2}\ . \tag{2.20}$$

Now we realize that (2.11) would follow if we knew that

$$\lim_{N\to\infty} \mathsf{E}\left(\frac{1}{N}\log Z_N(\beta) - p_N(\beta)\right)^2 = 0\ .$$

Actually, a fundamental property of Gaussian r.v. immediately implies much more. Here is a convenient formulation.

Theorem 2.2.4. *Consider a Lipschitz function F on $\mathbb{R}^M$, of Lipschitz constant A. That is, we assume that, given $\mathbf{x}, \mathbf{y}$ in $\mathbb{R}^M$, we have*

$$|F(\mathbf{x}) - F(\mathbf{y})| \leq A\,\mathrm{d}(\mathbf{x},\mathbf{y})\ , \tag{2.21}$$

where d denotes the Euclidean distance. If $g_1, \ldots, g_M$ are independent standard Gaussian r.v., and if $\mathbf{g} = (g_1, \ldots, g_M)$, then for each $t > 0$ we have

$$\mathsf{P}(|F(\mathbf{g}) - \mathsf{E}F(\mathbf{g})| \geq t) \leq 2\exp\left(-\frac{t^2}{4A^2}\right)\ . \tag{2.22}$$

The remarkable part of this statement is that (2.22) does not depend on M. Theorem 2.2.4 will be proved in Section 2.4. It is a typical occurrence of the "concentration of measure phenomenon".

Corollary 2.2.5. *For any β, any $t > 0$, we have*

$$\mathsf{P}\left(\left|\frac{1}{N}\log Z_N(\beta) - p_N(\beta)\right| \geq t\right) \leq 2\exp\left(-\frac{Nt^2}{2\beta^2}\right)\ . \tag{2.23}$$

This is true even for $\beta > 1$, and is one of the few facts known in that range.

Proof. We consider $M = N(N-1)/2$. For $\boldsymbol{\sigma}$ in Σ_N, we set

$$\mathbf{a}(\boldsymbol{\sigma}) = \left(\frac{\beta}{\sqrt{N}}\sigma_i\sigma_j\right)_{i<j},$$

that is seen as a point of $\mathbb{R}^M$, so that, if $\mathbf{g}$ is as in Theorem 2.2.4, we have

$$-H_N(\boldsymbol{\sigma}) = \mathbf{g}\cdot\mathbf{a}(\boldsymbol{\sigma}), \tag{2.24}$$

where the dot represents the dot product in $\mathbb{R}^M$. On $\mathbb{R}^M$, we consider the function

$$F(\mathbf{x}) = \frac{1}{N}\log\left(\sum_{\boldsymbol{\sigma}}\exp(\mathbf{a}(\boldsymbol{\sigma})\cdot\mathbf{x})\right). \tag{2.25}$$

Thus, the law of $F(\mathbf{g})$ coincides with the law of $N^{-1}\log Z_N$. To prove (2.23), it suffices to show that F satisfies (2.21) with $A = \beta/\sqrt{2N}$. To see this, we note that the Euclidean norm of $\mathbf{a}(\boldsymbol{\sigma})$ is not larger than $\beta\sqrt{N/2}$, so that, given $\mathbf{x}, \mathbf{y}$ in $\mathbb{R}^M$,

$$|\mathbf{a}(\boldsymbol{\sigma})\cdot\mathbf{x} - \mathbf{a}(\boldsymbol{\sigma})\cdot\mathbf{y}| \le \beta\sqrt{\frac{N}{2}}\,\mathrm{d}(\mathbf{x},\mathbf{y}),$$

and thus

$$\mathbf{a}(\boldsymbol{\sigma})\cdot\mathbf{x} \le \mathbf{a}(\boldsymbol{\sigma})\cdot\mathbf{y} + \beta\sqrt{\frac{N}{2}}\,\mathrm{d}(\mathbf{x},\mathbf{y}).$$

Hence

$$\sum_{\boldsymbol{\sigma}}\exp(\mathbf{a}(\boldsymbol{\sigma})\cdot\mathbf{x}) \le \exp\left(\beta\sqrt{\frac{N}{2}}\,\mathrm{d}(\mathbf{x},\mathbf{y})\right)\sum_{\boldsymbol{\sigma}}\exp(\mathbf{a}(\boldsymbol{\sigma})\cdot\mathbf{y}),$$

so that

$$F(\mathbf{x}) \le F(\mathbf{y}) + \frac{\beta}{\sqrt{2N}}\,\mathrm{d}(\mathbf{x},\mathbf{y}).$$

□

Remark 2.2.6. If rather than (2.24) we had

$$-H_N(\boldsymbol{\sigma}) = \mathbf{g}\cdot\mathbf{a}(\boldsymbol{\sigma}) + b(\boldsymbol{\sigma})$$

where $b(\boldsymbol{\sigma})$ is a number, there is nothing to change in the previous argument. There is nothing to change either if in the definition of Z_N we replace the summation $\sum_{\boldsymbol{\sigma}}$ by summation over a subset of Σ_N.

Proof of Theorem 2.2.1. By (2.23), if $t > 0$, we have

$$\mathsf{P}\left(\frac{1}{N}\log Z_N(\beta) \le p_N(\beta) + \frac{t\beta}{\sqrt{N}}\right) \ge 1 - 2\exp\left(-\frac{t^2}{2}\right) . \tag{2.26}$$

Thus, if

$$\exp\left(-\frac{t^2}{2}\right) < \frac{1}{8}\sqrt{1-\beta^2} ,$$

(2.26) and (2.20) imply that the events

$$\left\{\frac{1}{N}\log Z_N(\beta) \le p_N(\beta) + \frac{t\beta}{\sqrt{N}}\right\}$$

and

$$\left\{\frac{1}{N}\log Z_N(\beta) \ge \frac{N-1}{N}\left(\frac{\beta^2}{4} + \log 2\right)\right\}$$

intersect, so that

$$p_N(\beta) \ge \frac{N-1}{N}\left(\frac{\beta^2}{4} + \log 2\right) - \frac{t\beta}{\sqrt{N}}$$

and thus

$$p_N(\beta) \ge \frac{\beta^2}{4} + \log 2 - \frac{K(\beta)}{\sqrt{N}} . \tag{2.27}$$

□

One can rephrase (a consequence of) (2.23) by saying that the r.v. $N^{-1}\log Z_N(\beta)$ has fluctuations of order at most $N^{-1/2}$. In most of the cases we will consider later, this will be the true order of these fluctuations. Here, however, the situation is very special, and the fluctuations are of order N^{-1}. This is not an important part of the global picture, but the proofs are rather pretty. Upper and lower bounds will be treated separately.

Theorem 2.2.7. *Given $\beta < 1$ there is a constant $K(\beta)$ such that for each N, and each $t > 0$ we have*

$$\mathsf{P}\left(\log Z_N(\beta) \le N\left(\frac{\beta^2}{4} + \log 2\right) - t\right) \le K(\beta)\exp\left(-\frac{t^2}{K(\beta)}\right) . \tag{2.28}$$

We observe that by Jensen's inequality we have

$$Z_N \ge 2^N \exp\left(2^{-N}\sum_{\sigma}\sum_{i<j}\frac{\beta}{\sqrt{N}} g_{ij}\,\sigma_i\,\sigma_j\right) = 2^N$$

so that $\log Z_N \ge N\log 2 \ge 0$, and by (A.42) we have

$$
\begin{aligned}
\mathsf{E}\frac{1}{N}\log Z_N &= \int_0^\infty \mathsf{P}\left(\frac{1}{N}\log Z_N \ge u\right) du \\
&\ge \int_0^{\log 2+\beta^2/4} \mathsf{P}\left(\frac{1}{N}\log Z_N \ge \frac{\beta^2}{4} + \log 2 - t\right) dt \\
&\ge \log 2 + \frac{\beta^2}{4} - \int_0^\infty K(\beta)\exp\left(-\frac{N^2t^2}{K(\beta)}\right) dt \\
&\ge \log 2 + \frac{\beta^2}{4} - \frac{K(\beta)}{N} ,
\end{aligned}
$$

where we have used (2.28) on the second inequality. Since $p_N(\beta) \le \beta^2/4 + \log 2$, we get

$$
\left| p_N(\beta) - \left(\frac{\beta^2}{4} + \log 2\right)\right| \le \frac{K(\beta)}{N} , \tag{2.29}
$$

a much improved version of (2.11). Together with (2.29), (2.28) implies

$$
\mathsf{P}\left(\frac{1}{N}\log Z_N(\beta) \le p_N(\beta) - t\right) \le K(\beta)\exp\left(-\frac{N^2t^2}{K(\beta)}\right) ,
$$

which shows the gain over (2.23) of a factor N in the exponent, when one deals with deviations of $N^{-1}\log Z_N(\beta)$ below $p_N(\beta)$.

Before moving on to the proof of (2.28), let us ask what are the interesting ways to look at a sequence of random variables, (such as $N^{-1}\log Z_N(\beta)$). Three rather different ways come to mind. One can look for limit laws; one can look for inequalities true for all N, such as in (2.28); or one can be interested in "large deviations", such as in the following.

Research problem 2.2.8. (Level 2) Given $t > 0$, $\beta < 1$, prove the existence of the limit

$$
\lim_{N\to\infty} \frac{1}{N^2} \log \mathsf{P}\left(\frac{1}{N}\log Z_N(\beta) \le \frac{\beta^2}{4} + \log 2 - t\right) \tag{2.30}
$$

and compute it. More generally, given $0 \le \alpha < 1$, prove the existence of the limit

$$
\lim_{N\to\infty} \frac{1}{N^{2(1-\alpha)}} \log \mathsf{P}\left(\frac{1}{N}\log Z_N(\beta) \le \frac{\beta^2}{4} + \log 2 - \frac{t}{N^\alpha}\right) .
$$

To help the newcomer to the area, the research problems are ranked level 1 to level 3. The solution to level 1 problems should be suitable for publication, but the author feels that they can be successfully attacked by the methods explained in the book, or simple extensions of these methods. To put it another way, the author feels that he would be rather likely to solve them fast if he tried (which he won't). Level 2 problems are more likely to require ingredients substantially beyond what is found in the book. On the other hand these problems do not touch what seems to be the central issues of spin glass theory, and there is no particular reason to think that they are very hard. Simply, they have not been tried. Level 3 problems seem to touch essential issues, and there is currently no way of telling how difficult they might be. It goes without saying that this classification is based on the author's current understanding, and comes with no warranty whatsoever.

Coming back to Problem 2.2.1, all the results we will present deal with typical realizations of the disorder. Every single question of the type of (2.30), that asks "how rarely *exactly* does it happen (in the limit) that some typical feature of the problem deviates by a certain quantity from its typical value" is wide open.

We now start the preparations for the proof of Theorem 2.2.7. Since $R_{1,2}$ is a function of $\boldsymbol{\sigma}^1$ and $\boldsymbol{\sigma}^2$, we can consider the Gibbs average $\langle \exp \gamma N R_{1,2}^2/2 \rangle$, where $\langle \cdot \rangle$ means here average in an order 2 replica, i.e. average for $G_N^{\otimes 2}$ on Σ_N^2. By definition of the Gibbs' average, we have

$$\begin{aligned} Z_N^2(\beta) \left\langle \exp \frac{\gamma N}{2} R_{1,2}^2 \right\rangle \\ = \sum_{\boldsymbol{\sigma}^1, \boldsymbol{\sigma}^2} \exp \left(-H_N(\boldsymbol{\sigma}^1) - H_N(\boldsymbol{\sigma}^2) + \frac{\gamma N}{2} R_{1,2}^2 \right) . \end{aligned} \tag{2.31}$$

Thus (2.16) and (2.31) imply that

$$\mathsf{E} \left(Z_N^2(\beta) \left\langle \exp \frac{\gamma N}{2} R_{1,2}^2 \right\rangle \right) \le \frac{1}{\sqrt{1-\beta^2-\gamma}} (\mathsf{E} Z_N(\beta))^2 . \tag{2.32}$$

Lemma 2.2.9. *If $\beta < 1$, we have*

$$\mathsf{P} \left(Z_N \ge \frac{1}{2} \mathsf{E} Z_N, N \langle R_{1,2}^2 \rangle \le K \right) \ge \frac{1}{K} . \tag{2.33}$$

According to our previous convention that K denotes a number independent of N, this statement means "Given $\beta < 1$, there exists a number K depending only on β, such that...". We have also lightened the notation by not indicating the dependence in β of Z_N.

Proof. We set $\gamma = (1-\beta^2)/2$. Using Markov's inequality (1.18), we see from (2.32) that

$$\mathsf{P} \left(Z_N^2 \left\langle \exp \frac{\gamma N}{2} R_{1,2}^2 \right\rangle \le t (\mathsf{E} Z_N)^2 \right) \ge 1 - \frac{1}{t\sqrt{1-\beta^2-\gamma}} .$$

Combining with (2.19), and using the inequality $\mathsf{P}(A \cap B) \ge \mathsf{P}(A) + \mathsf{P}(B) - 1$, we see by taking t suitably large that

$$\mathsf{P} \left(Z_N^2 \left\langle \exp \frac{\gamma N}{2} R_{1,2}^2 \right\rangle \le K (\mathsf{E} Z_N)^2 ;\ Z_N \ge \frac{1}{2} \mathsf{E} Z_N \right) \ge \frac{1}{K} .$$

Now, since $e^x \geq x$, when $Z_N \geq \mathsf{E} Z_N/2$, we have

$$Z_N^2 \left\langle \exp \frac{\gamma N}{2} R_{1,2}^2 \right\rangle \leq K(\mathsf{E} Z_N)^2 \Rightarrow N\langle R_{1,2}^2 \rangle \leq K \; . \tag{2.34}$$

□

Our convention of the number K not being the same at each occurrence is very convenient when dealing with complicated estimates, but it might require some effort from the reader to get used to it. What (2.34) means is that

$$Z_N^2 \left\langle \exp \frac{\gamma N}{2} R_{1,2}^2 \right\rangle \leq K(\mathsf{E} Z_N)^2 \Rightarrow N\langle R_{1,2}^2 \rangle \leq \frac{8K}{\gamma} \; .$$

This convention of the "varying K" can get sometimes disturbing, such as if one writes

$$2K + 1 = K \; ,$$

which, while technically correct, probably requires either a training as a computer scientist, or, better, a biological mutation to be comfortable with. We will try to refrain from such excesses. At other times, this convention makes matters obscure, and one does need to distinguish between the various constants. We denote by $K_1, K_2, \ldots$ specific constants, the values of which remain fixed throughout a given section (but not from a section to the next).

The proof of Theorem 2.2.7 uses a simple (but unintuitive) algebraic identity. This (and similar identities) play an important role.

Lemma 2.2.10. *We have*

$$\langle R_{1,2}^2 \rangle = \frac{1}{N^2} \sum_{i,j} \langle \sigma_i \sigma_j \rangle^2 \; . \tag{2.35}$$

Because of the global symmetry of Gibbs' measure (that is invariant by the transformation $\boldsymbol{\sigma} \to -\boldsymbol{\sigma}$) we have $\langle \sigma_i \rangle = 0$ so that, if $i \neq j$

$$\langle \sigma_i \sigma_j \rangle = \langle \sigma_i \sigma_j \rangle - \langle \sigma_i \rangle \langle \sigma_j \rangle$$

is the correlation of the spins σ_i, σ_j, when seen as r.v. on the probability space (Σ_N, G_N). Thus (2.35) relates the Gibbs average of the overlaps with the "spatial average" of the correlations.

Proof. We have

$$N R_{1,2} = \sum_{i \leq N} \sigma_i^1 \sigma_i^2 \; ,$$

so that

$$N^2 R_{1,2}^2 = \sum_{i,j \leq N} \sigma_i^1 \sigma_i^2 \sigma_j^1 \sigma_j^2 = \sum_{i,j \leq N} \sigma_i^1 \sigma_j^1 \, \sigma_i^2 \sigma_j^2 \; .$$

Now

$$\langle \sigma_i^1 \sigma_j^1 \, \sigma_i^2 \sigma_j^2 \rangle = \langle \sigma_i^1 \sigma_j^1 \rangle \langle \sigma_i^2 \sigma_j^2 \rangle = \langle \sigma_i \sigma_j \rangle^2 \; . \tag{2.36}$$

This is because on $(\Sigma_N^2, G_N^{\otimes 2})$, the replicas 1 and 2 are probabilistically independent. The last brackets in (2.36) represent an average over Σ_N, not Σ_N^2.

□

A useful function of replicas was seen at work in (2.36). They transform products of Gibbs averages into a single Gibbs average, and it is much easier to take expectation of the latter. A typical example is the following generalization of (2.36):

$$\langle f_1(\boldsymbol{\sigma})\rangle \,\langle f_2(\boldsymbol{\sigma})\rangle = \langle f_1(\boldsymbol{\sigma}^1)\, f_2(\boldsymbol{\sigma}^2)\rangle\ , \tag{2.37}$$

which is a consequence of the fact that for two independent r.v. X_1, X_2, one has $\mathsf{E}X_1\,\mathsf{E}X_2 = \mathsf{E}X_1X_2$.

Our next lemma uses the notation of Theorem 2.2.4.

Lemma 2.2.11. *Consider a closed subset B of $\mathbb{R}^M$, and set*

$$\mathrm{d}(\mathbf{x}, B) = \inf\{\mathrm{d}(\mathbf{x}, \mathbf{y})\,;\ \mathbf{y} \in B\}\,,$$

the Euclidean distance from $\mathbf{x}$ to B. Then for $t > 0$, we have

$$\mathsf{P}\left(\mathrm{d}(\mathbf{g}, B) \ge t + 2\sqrt{\log \frac{2}{\mathsf{P}(\mathbf{g}\in B)}}\right) \le 2\exp\left(-\frac{t^2}{4}\right)\ . \tag{2.38}$$

Proof. The function $F(\mathbf{x}) = \mathrm{d}(\mathbf{x}, B)$ satisfies (2.21) with $A = 1$, so that for all $t > 0$ by (2.22) we have

$$\mathsf{P}(|\mathrm{d}(\mathbf{g}, B) - \mathsf{E}\mathrm{d}(\mathbf{g}, B)| \ge t) \le 2\exp\left(-\frac{t^2}{4}\right)\ . \tag{2.39}$$

If $t = \mathsf{E}\mathrm{d}(\mathbf{g}, B)$ we have

$$\mathsf{P}(\mathbf{g} \in B) \le \mathsf{P}(|\mathrm{d}(\mathbf{g}, B) - \mathsf{E}\mathrm{d}(\mathbf{g}, B)| \ge t)\ ,$$

and combining with (2.39) we get

$$t = \mathsf{E}\mathrm{d}(\mathbf{g}, B) \Rightarrow \mathsf{P}(\mathbf{g}\in B) \le 2\exp\left(-\frac{t^2}{4}\right)\ ,$$

so that

$$\mathsf{E}\mathrm{d}(\mathbf{g}, B) \le 2\sqrt{\log \frac{2}{\mathsf{P}(\mathbf{g}\in B)}}\ ,$$

and combining with (2.39) gives (2.38). □

Proof of Theorem 2.2.7 We set $M = N(N-1)/2$, and we consider the Gaussian r.v. $(g_{ij})_{i<j}$ used to define H_N as a point $\mathbf{g}$ in $\mathbb{R}^M$. Quite naturally, we view Z_N as a function $Z_N(\mathbf{g})$ of $\mathbf{g}$. The meaning of Lemma 2.2.9 is that

if K is large enough, the subset B of $\mathbb{R}^M$ given by

$$\mathbf{g} \in B \Leftrightarrow Z_N(\mathbf{g}) \geq \frac{1}{2}\,\mathsf{E}Z_N\,;\;\; N\langle R_{1,2}^2\rangle \leq K \tag{2.40}$$

satisfies

$$\mathsf{P}(\mathbf{g} \in B) \geq \frac{1}{K}\,. \tag{2.41}$$

It is of course understood that in (2.40), the Gibbs' measure used in the bracket $\langle\cdot\rangle$ corresponds to the realization $\mathbf{g}$ of the disorder. Next, we show that

$$\log Z_N(\mathbf{g}) \geq N\left(\frac{\beta^2}{4} + \log 2\right) - K(1 + \mathrm{d}(\mathbf{g}, B))\,. \tag{2.42}$$

Since

$$\log Z_N(\mathbf{g}') \geq \log\left(\frac{1}{2}\mathsf{E}Z_N\right) = (N-1)\left(\frac{\beta^2}{4} + \log 2\right)$$

for $\mathbf{g}' \in B$, to prove (2.42) it suffices to show that

$$\log Z_N(\mathbf{g}) \geq \log Z_N(\mathbf{g}') - K\mathrm{d}(\mathbf{g}, \mathbf{g}') \tag{2.43}$$

whenever $\mathbf{g}' \in B$. To see this, we observe that if we denote by $\langle\cdot\rangle'$ average with respect to the Gibbs' measure relative to $\mathbf{g}'$, then we have the algebraic identity

$$Z_N(\mathbf{g}) = Z_N(\mathbf{g}')\left\langle \exp\left(\frac{\beta}{\sqrt{N}}\sum_{i<j}(g_{ij} - g'_{ij})\,\sigma_i\,\sigma_j\right)\right\rangle'\,. \tag{2.44}$$

This is obvious by writing the definition of $\langle\cdot\rangle'$.

We use Jensen's inequality for the average $\langle\cdot\rangle'$ to see that

$$\left\langle \exp\frac{\beta}{\sqrt{N}}\sum_{i<j}(g_{ij} - g'_{ij})\,\sigma_i\,\sigma_j\right\rangle' \geq \exp\frac{\beta}{\sqrt{N}}\sum_{i<j}(g_{ij} - g'_{ij})\,\langle\sigma_i\,\sigma_j\rangle'\,. \tag{2.45}$$

Now, since $\mathrm{d}^2(\mathbf{g}, \mathbf{g}') = \sum_{i<j}(g_{ij} - g'_{ij})^2$, the Cauchy-Schwarz inequality and Lemma 2.2.10 imply

$$\sum_{i<j}(g_{ij} - g'_{ij})\,\langle\sigma_i\,\sigma_j\rangle' \geq -\mathrm{d}(\mathbf{g}, \mathbf{g}')\,(N^2\langle R_{1,2}^2\rangle')^{1/2} \geq -K\sqrt{N}\,\mathrm{d}(\mathbf{g}, \mathbf{g}')$$

using (2.40) and the fact that $\mathbf{g}' \in B$. Combining with (2.44) and (2.45), this proves (2.43), and hence (2.42). From (2.42) we see that

$$\begin{aligned}\log Z_N(\mathbf{g}) \leq N\left(\frac{\beta^2}{4} + \log 2\right) - t \Rightarrow \mathrm{d}(\mathbf{g}, \mathbf{B}) &\geq \frac{t - K_1}{K_1}\\ &\geq 2\sqrt{\log\frac{2}{\mathsf{P}(\mathbf{g} \in B)}} + \frac{t - K_2}{K_1}\,.\end{aligned}$$

Thus it follows from (2.38) that, if $t \geq K_2$,

$$\mathsf{P}\left(\log Z_N(\mathbf{g}) \leq N\left(\frac{\beta^2}{4} + \log 2\right) - t\right) \leq 2\exp\left(-\frac{(t-K_2)^2}{4K_1^2}\right) . \tag{2.46}$$

To conclude the proof we note that if K is large enough, then

$$t \geq K_2 \Rightarrow 2\exp\left(-\frac{(t-K_2)^2}{4K_1^2}\right) \leq K\exp\left(-\frac{t^2}{K}\right) ,$$

and

$$0 \leq t \leq K_2 \Rightarrow K\exp\left(-\frac{t^2}{K}\right) \geq 1 .$$

□

Theorem 2.2.12. *If $\beta < 1$ we have*

$$\mathsf{E}\left\langle \exp\frac{1-\beta^2}{8} NR_{1,2}^2 \right\rangle \leq K . \tag{2.47}$$

This inequality means that we have a very good control of the overlap $R_{1,2}$ when seen as a function of the two configurations $\boldsymbol{\sigma}^1, \boldsymbol{\sigma}^2$, weighted by their Gibbs' weights. It will become gradually clear why this type of information is of importance. It will require significant work to extend (2.47) to other situations, but for the time being we can enjoy a simple proof based on Theorem 2.2.7.

Proof. We use the Cauchy-Schwarz inequality for $\langle\cdot\rangle$ to get

$$\begin{aligned}\left\langle \exp\frac{1-\beta^2}{8} NR_{1,2}^2 \right\rangle &\leq \left\langle \exp\frac{1-\beta^2}{4} NR_{1,2}^2 \right\rangle^{1/2} \\ &= \frac{1}{Z_N}\left(Z_N^2\left\langle \exp\frac{1-\beta^2}{4} NR_{1,2}^2 \right\rangle\right)^{1/2} .\end{aligned}$$

We then use the Cauchy-Schwarz inequality for E and (2.16) with $\gamma = (1-\beta^2)/2$ to get

$$\begin{aligned}\mathsf{E}\left\langle \exp\frac{1-\beta^2}{8} NR_{1,2}^2 \right\rangle &\leq \left(\mathsf{E}\frac{1}{Z_N^2}\right)^{1/2}\left(\mathsf{E}\left(Z_N^2\left\langle \exp\frac{1-\beta^2}{4} NR_{1,2}^2 \right\rangle\right)\right)^{1/2} \\ &\leq \left(\mathsf{E}\frac{1}{Z_N^2}\right)^{1/2}(K(\mathsf{E}Z_N)^2)^{1/2} \\ &= K\left(\mathsf{E}\left(\frac{(\mathsf{E}Z_N)^2}{Z_N^2}\right)\right)^{1/2} .\end{aligned}$$

It follows from (2.28) that we have, for $t > 1$,

$$\mathsf{P}\left(\frac{\mathsf{E}Z_N}{Z_N} > t\right) \le K \exp\left(-\frac{(\log t)^2}{K}\right)$$

and this implies (using (A.42)) that

$$\mathsf{E}\left(\left(\frac{\mathsf{E}Z_N}{Z_N}\right)^2\right) < K .$$

□

We now turn to the study of the deviations of $\log Z_N$ *above* $N(\beta^2/4 + \log 2)$.

Theorem 2.2.13. *If $\beta < 1$ we have*

$$0 < t < \sqrt{N} \Rightarrow \mathsf{P}\left(\log Z_N \ge N\left(\frac{\beta^2}{4} + \log 2\right) + t\right) \le K \exp\left(-\frac{t^2}{K}\right) . \qquad (2.48)$$

Just as Theorem 2.2.7, this is a very special result. Although the proof is pretty and instructive, it is rather involved and it is probably better skipped at first reading. First of all, we will show that the restriction $t < \sqrt{N}$ is essentially necessary. This is simply because if $\boldsymbol{\sigma}_0$ is an arbitrary configuration, then

$$\log Z_N \ge -H_N(\boldsymbol{\sigma}_0) ,$$

so that, since $H_N(\boldsymbol{\sigma}_0)$ is Gaussian with $\mathsf{E}H_N^2(\boldsymbol{\sigma}_0) = \beta^2(N-1)/2$, we see from (A.38) that

$$\mathsf{P}\left(\log Z_N \ge N\left(\frac{\beta^2}{4} + \log 2\right) + N\right) \ge \exp(-KN)$$

so that (2.48) cannot hold for $t \ge K\sqrt{N}$.

The proof of (2.48) relies upon moment estimates, with a very interesting twist, and we explain this first. We start with the formula

$$Z_N^n = \sum_{\boldsymbol{\sigma}} e_n(\boldsymbol{\sigma}^1, \dots, \boldsymbol{\sigma}^n) , \qquad (2.49)$$

where

$$e_n(\boldsymbol{\sigma}^1, \dots, \boldsymbol{\sigma}^n) = \exp \frac{\beta}{\sqrt{N}} \sum_{i<j} g_{ij} \left(\sum_{\ell \le n} \sigma_i^\ell \sigma_j^\ell\right) ,$$

and where $\sum_{\boldsymbol{\sigma}}$ denotes, as throughout the entire argument, that a summation is over $\boldsymbol{\sigma}^1, \ldots, \boldsymbol{\sigma}^n$ in Σ_N. Now

$$\begin{aligned}\mathsf{E}\left(\sum_{i<j} g_{ij}\left(\sum_{\ell\le n}\sigma_i^\ell\sigma_j^\ell\right)\right)^2 &= \sum_{i<j}\left(\sum_{\ell\le n}\sigma_i^\ell\sigma_j^\ell\right)^2\\ &= \sum_{\ell,\ell'}\sum_{i<j}\sigma_i^\ell\sigma_i^{\ell'}\,\sigma_j^\ell\sigma_j^{\ell'}\\ &= \frac{1}{2}\sum_{\ell,\ell'}\left(\left(\sum_{i\le N}\sigma_i^\ell\sigma_i^{\ell'}\right)^2 - N\right)\\ &= \frac{1}{2}nN(N-n) + N^2\sum_{\ell<\ell'}R_{\ell,\ell'}^2\,,\end{aligned}$$

where

$$R_{\ell,\ell'} = \frac{1}{N}\sum_{i\le N}\sigma_i^\ell\sigma_i^{\ell'}\,. \tag{2.50}$$

Thus

$$\mathsf{E}\,e_n(\boldsymbol{\sigma}^1,\ldots,\boldsymbol{\sigma}^n) = \exp\left(\frac{\beta^2}{4}n(N-n) + \frac{\beta^2}{2}N\sum_{\ell<\ell'}R_{\ell,\ell'}^2\right)\,,$$

and

$$\mathsf{E}Z_N^n = \left(\exp\frac{\beta^2}{4}n(N-n)\right)\sum_{\boldsymbol{\sigma}}\exp\left(\frac{\beta^2}{2}N\sum_{\ell<\ell'}R_{\ell,\ell'}^2\right)\,. \tag{2.51}$$

If we restrict the summation $\sum_{\boldsymbol{\sigma}}$ to the case $\boldsymbol{\sigma}^1 = \cdots = \boldsymbol{\sigma}^n$, we see that

$$\mathsf{E}Z_N^n \ge 2^N\exp\frac{\beta^2}{4}n(N-n)\exp\frac{\beta^2}{4}Nn(n-1)$$

and thus

$$\mathsf{E}Z_N^n \ge (\mathsf{E}Z_N)^n\,2^{-N(n-1)}\exp\frac{\beta^2}{4}n(n-1)(N-1)\,,$$

so that if $\beta^2 n > 4\log 2$, it is no longer true that $\mathsf{E}Z_N^n \le K(\beta)(\mathsf{E}Z_N)^n$. There is a "moment explosion".

Our proof of Theorem 2.2.13 relies upon the fact that this moment explosion is created by a small set of configurations, and that if we remove it the moments of Z_N become manageable.

Given a number $x > 0$, we consider the set

$$B_n(x) = \left\{(\boldsymbol{\sigma}^1,\ldots,\boldsymbol{\sigma}^n)\in\Sigma_N^n\,;\ \sum_{\ell<\ell'}R_{\ell,\ell'}^2 \le x\right\}\,, \tag{2.52}$$

and its complement $B_n^c(x)$.

Proposition 2.2.14. *If $\beta < 1$, there is a constant $K(\beta)$ and a constant $c(\beta) > 0$ such that if we set $B_n = B_n(c(\beta))$, and if $n \le \sqrt{N}/K(\beta)$, we have*

$$\mathsf{E}\left(\sum_{B_n} e_n(\boldsymbol{\sigma}^1,\dots,\boldsymbol{\sigma}^n)\right) \le K(\beta)^{n^2}(\mathsf{E}Z_N)^n \tag{2.53}$$

$$\begin{aligned}\mathsf{P}\left(\sum_{B_n^c} e_n(\boldsymbol{\sigma}^1,\dots,\boldsymbol{\sigma}^n) \ge \exp\left(-\frac{N}{K(\beta)}\right)\max(Z_N^n,(\mathsf{E}Z_N)^n)\right)\\ \le \exp\left(-\frac{N}{K(\beta)}\right).\end{aligned} \tag{2.54}$$

The notation $\sum_B$ means of course that the summation is restricted to $(\boldsymbol{\sigma}^1,\dots,\boldsymbol{\sigma}^n)$ in B. The meaning of (2.53) is that if we restrict the summation to B_n, we can control these "partial moments of Z_N". The meaning of (2.54) is that the part of the summation that we have not taken into account in (2.53) gives typically a very small contribution (but this contribution is *not* small in average).

To provide motivation, we deduce Theorem 2.2.13 from Proposition 2.2.14. We set

$$\overline{Z}_N = \max(Z_N, \mathsf{E}Z_N)\,.$$

Proof of Theorem 2.2.13. Let

$$U = \sum_{B_n} e_n(\boldsymbol{\sigma}^1,\dots,\boldsymbol{\sigma}^n)\,;\ U^c = \sum_{B_n^c} e_n(\boldsymbol{\sigma}^1,\dots,\boldsymbol{\sigma}^n)\,.$$

We consider $t > 0$. Since $Z_N^n = U + U^c$, we have either $U \ge Z_N^n/2$ or else $U^c \ge Z_N^n/2$. When $Z_N \ge e^t\,\mathsf{E}Z_N$, we have $Z_N \ge \mathsf{E}Z_N$, so that $Z_N = \overline{Z}_N$, and hence we have either

$$U \ge \frac{e^{nt}}{2}(\mathsf{E}Z_N)^n \quad \text{or} \quad U^c \ge \frac{\overline{Z}_N^n}{2}\,.$$

Thus

$$\mathsf{P}(Z_N \ge e^t\,\mathsf{E}Z_N) \le \mathsf{P}\left(U \ge \frac{e^{nt}}{2}(\mathsf{E}Z_N)^n\right) + \mathsf{P}\left(U^c \ge \frac{1}{2}\overline{Z}_N^n\right)\,.$$

When $n \le \sqrt{N}/K(\beta)$, we use (2.53) and Markov inequality to get

$$\mathsf{P}\left(U \ge \frac{e^{nt}}{2}(\mathsf{E}Z_N)^n\right) \le 2\frac{K^{n^2}}{e^{nt}} = 2\left(\frac{K^n}{e^t}\right)^n\,.$$

Moreover, by (2.54),

$$\mathsf{P}\left(U^c \ge \frac{1}{2}\overline{Z}_N^n\right) \le \exp\left(-\frac{N}{K}\right)\,,$$

so that

$$\mathsf{P}(Z_N \geq e^t\, \mathsf{E}Z_N) \leq 2\left(\frac{K_1^n}{e^t}\right)^n + \exp\left(-\frac{N}{K}\right) . \tag{2.55}$$

Consider now the largest integer n such that $K_1^n \leq e^{t/2}$. Since we can assume $K_1 \geq e$, we have $n \leq t$, so that $n \leq \sqrt{N}/K(\beta)$ for $t < \sqrt{N}/K(\beta)$. Thus when $t \leq \sqrt{N}/K(\beta)$ we can use (2.55) for this value of n, and we get

$$\begin{aligned}\mathsf{P}(Z_N \geq e^t\, \mathsf{E}Z_N) &\leq 2\exp\left(-\frac{nt}{2}\right) + \exp\left(-\frac{N}{K}\right)\\ &\leq 2\exp\left(-\frac{t^2}{K} + \frac{t}{2}\right) + \exp\left(-\frac{N}{K}\right)\end{aligned}$$

because $n+1 \geq t/K$ by definition of n, and hence $n \geq t/K - 1$. Thus, for $t \leq \sqrt{N}/K(\beta)$ we have

$$\mathsf{P}(Z_N \geq e^t\, \mathsf{E}Z_N) \leq K \exp\left(-\frac{t^2}{K}\right)$$

and (2.48) follows since $\mathsf{E}Z_N \leq \exp N(\log 2 + \beta^2/4)$. □

Proof of (2.53). Using the formula before (2.51), it suffices to prove that

$$\sum_{B_n} \exp N\frac{\beta^2}{2}\sum_{\ell<\ell'} R_{\ell,\ell'}^2 \leq 2^{nN}\, K^{n^2} . \tag{2.56}$$

Let us denote by P_0 the uniform probability on Σ_N^n, and by E_0 the corresponding expectation. Then, recalling that $B_n = B_n(c(\beta))$ is given by (2.52), (2.56) simply means that

$$\mathsf{E}_0\left(\exp\left(N\frac{\beta^2}{2}\sum_{\ell<\ell'} R_{\ell,\ell'}^2\right)\mathbf{1}_{\{\sum R_{\ell,\ell'}^2 \leq c(\beta)\}}\right) \leq K^{n^2} , \tag{2.57}$$

and using the formula (A.44)

$$\mathsf{E}F(X)\,\mathbf{1}_{\{F(X)\leq F(c)\}} = \int_{-\infty}^{c} F'(t)\,\mathsf{P}(X \geq t)\,\mathrm{d}t \leq F(0) + \int_0^c F'(t)\,\mathsf{P}(X \geq t)\,\mathrm{d}t.$$

For $c = Nc(\beta)$, $F(x) = \exp \beta^2 x/2$ and $X = N\sum_{\ell<\ell'} R_{\ell,\ell'}^2$, we get that the left-hand side of (2.57) is at most

$$1 + \frac{\beta^2}{2}\int_0^{Nc(\beta)} f(t)\exp\frac{\beta^2 t}{2}\,\mathrm{d}t , \tag{2.58}$$

where

$$f(t) = \mathsf{P}_0\left(N\sum_{\ell<\ell'} R_{\ell,\ell'}^2 \geq t\right) .$$

We use Corollary A.4.6 with $1-2\varepsilon=\sqrt{\beta}$ and $u=\beta t$ to see that

$$f(t) \le K^{n^2} \exp\left(-\frac{\beta t}{2}\left(1-L\sqrt{\frac{t}{N}}\right)\right),$$

and thus the quantity (2.58) is not larger than

$$K^{n^2}\int_0^{Nc(\beta)} \exp\left(-\frac{t}{2}(\beta-\beta^2)+L\beta t\sqrt{\frac{t}{N}}\right)dt \le K^{n^2},$$

provided that we have chosen $c(\beta)$ small enough so that the exponent is at most $-t(\beta-\beta^2)/4$ for $t \le Nc(\beta)$. □

We now fix $c(\beta)$ small enough that (2.53) holds, and we prepare for the proof of (2.54).

Lemma 2.2.15. *We have*

$$\mathsf{E}\left(\sum_{D_n} e_n(\boldsymbol{\sigma}^1,\ldots,\boldsymbol{\sigma}^n)\right) \le (\mathsf{E}Z_N)^n K^{n^2}\exp\left(-\frac{N}{K}\right), \tag{2.59}$$

where

$$D_n = \left\{(\boldsymbol{\sigma}^1,\ldots,\boldsymbol{\sigma}^n)\,;\ \frac{1}{3}c(\beta) \le \sum_{\ell<\ell'} R_{\ell,\ell'}^2 \le c(\beta)\right\}.$$

Proof. This is a simple variation on the proof of (2.53). □

For a subset I of $\{1,\ldots,n\}$, we define

$$D_I = \left\{(\boldsymbol{\sigma}^1,\ldots,\boldsymbol{\sigma}^n)\,;\ \frac{c(\beta)}{3} \le \sum_{\ell,\ell'\in I,\ell<\ell'} R_{\ell,\ell'}^2 \le c(\beta)\right\}$$

when $\operatorname{card} I \ge 3$. When $\operatorname{card} I = 2$, $I=\{\ell,\ell'\}$, we define

$$D_I = \left\{(\boldsymbol{\sigma}^1,\ldots,\boldsymbol{\sigma}^n)\,;\ R_{\ell,\ell'}^2 \ge \frac{c(\beta)}{3}\right\}.$$

Lemma 2.2.16. *We have*

$$B_n^c \subset \bigcup D_I\,, \tag{2.60}$$

where the union is taken over all subsets I of $\{1,\ldots,n\}$ with at least two elements.

Proof. Given $(\boldsymbol{\sigma}^1, \ldots, \boldsymbol{\sigma}^n)$ in B_n^c, we consider the smallest integer $n' \geq 2$ such that there exists a subset I of $\{1, \ldots, n\}$ with $\operatorname{card} I = n'$ for which

$$\sum_{\ell,\ell' \in I, \ell < \ell'} R_{\ell,\ell'}^2 \geq \frac{c(\beta)}{3} .$$

The fact that $(\boldsymbol{\sigma}^1, \ldots, \boldsymbol{\sigma}^n)$ belongs to B_n^c is used to show that there exists at least one set I as above, namely $I = \{1, \ldots, n\}$.

If $n' = \operatorname{card} I = 2$, then $(\boldsymbol{\sigma}^1, \ldots, \boldsymbol{\sigma}^n) \in D_I$ and we are done. If $n' = \operatorname{card} I \geq 3$, the minimality of n' implies that given a subset I' of I with $\operatorname{card} I' = n' - 1$ we have

$$\sum_{\ell,\ell' \in I', \ell < \ell'} R_{\ell,\ell'}^2 \leq \frac{c(\beta)}{3} ,$$

and averaging over all the n' possible choices of I' we get

$$\frac{n'-2}{n'} \sum_{\ell,\ell' \in I, \ell < \ell'} R_{\ell,\ell'}^2 \leq \frac{c(\beta)}{3}$$

and since $(n'-2)/n' \geq 1/3$ we get $(\boldsymbol{\sigma}^1, \ldots, \boldsymbol{\sigma}^n) \in D_I$. □

Proof of (2.54). By Lemma 2.2.16 it suffices to prove that for each subset I of $\{1, \ldots, n\}$ with at least two elements, we have, since $n \leq \sqrt{N}/K \leq N/K$,

$$\mathsf{P}\left(\sum_{D_I} e_n(\boldsymbol{\sigma}^1, \ldots, \boldsymbol{\sigma}^n) \geq \exp\left(-\frac{N}{K}\right) \overline{Z}_N^n\right) \leq \exp\left(-\frac{N}{K}\right) . \tag{2.61}$$

Without loss of generality we assume $I = \{1, \ldots, n'\}$, so that

$$\sum_{D_I} e_n(\boldsymbol{\sigma}^1, \ldots, \boldsymbol{\sigma}^n) = Z_N^{n-n'} \sum_{D'} e_{n'}(\boldsymbol{\sigma}^1, \ldots, \boldsymbol{\sigma}^{n'}) ,$$

where

$$D' = \left\{(\boldsymbol{\sigma}^1, \ldots, \boldsymbol{\sigma}^{n'}) ; \ \frac{c(\beta)}{3} \leq \sum_{\ell < \ell' \leq n'} R_{\ell,\ell'}^2 \leq c(\beta)\right\}$$

if $n' \geq 3$ and

$$D' = \left\{(\boldsymbol{\sigma}^1, \boldsymbol{\sigma}^2) ; \ R_{1,2}^2 \geq \frac{c(\beta)}{3}\right\}$$

if $n' = 2$.

Since $\overline{Z}_N \geq Z_N$ and $\overline{Z}_N \geq \mathsf{E} Z_N$, the left-hand side of (2.61) is at most

$$\mathsf{P}\left(\sum_{D'} e_{n'}(\boldsymbol{\sigma}^1, \ldots, \boldsymbol{\sigma}^{n'}) \geq \exp\left(-\frac{N}{K}\right) (\mathsf{E} Z_N)^{n'}\right) .$$

To prove (2.61), we first assume $n' \geq 3$. We appeal to (2.59) with n' rather than n, to see that

$$\mathsf{P}\left(\sum_{D'} e_{n'}(\boldsymbol{\sigma}^1,\ldots,\boldsymbol{\sigma}^{n'}) \geq t(\mathsf{E}Z_N)^{n'}\right) \leq \frac{K^{n'^2}}{t} \exp\left(-\frac{N}{K}\right) . \tag{2.62}$$

It is here that the condition $n^2 \leq N/K(\beta)$ is really needed. Under this condition, the right-hand side of (2.62) is at most $\exp(-N/K)/t$, and a suitable choice of t yields (2.61).

When $n' = 2$, we use (2.16) with $\gamma = (1-\beta^2)/2$ to obtain

$$\exp \frac{N\gamma c(\beta)}{6} \mathsf{E}\left(\sum_{D'} e_2(\boldsymbol{\sigma}^1,\boldsymbol{\sigma}^2)\right) \leq K(\mathsf{E}Z_N)^2 ,$$

so that (2.61) follows again from Markov's inequality.

This concludes the proof of Theorem 2.2.13. □

We turn to the study of Gibbs' measure. First, we consider what happens if we consider only a few spins. Given a subset I of $\{1,\ldots,N\}$, we denote by G_I the image of G_N under the map $\boldsymbol{\sigma} \mapsto (\sigma_i)_{i\in I}$. We denote by μ_I the uniform probability measure on $\{-1,1\}^I$, and by $|G_I - \mu_I|$ the total variation distance between G_I and μ_I, given by

$$|G_I - \mu_I| = \sup_{|f|\leq 1} \left| \int f \mathrm{d}G_I - \int f \mathrm{d}\mu_I \right| = \sum_{\mathbf{x}} |G_I(\mathbf{x}) - \mu_I(\mathbf{x})| .$$

It is not difficult (see Section 2.4) to show that

$$\mathsf{E}|G_I - \mu_I| \leq \frac{K(n,\beta)}{\sqrt{N}} ,$$

if $I = \{1,\ldots,n\}$ and $\beta < 1$. One can, however, do much better.

Proposition 2.2.17. *If $\beta < 1$, we have*

$$\mathsf{E}|G_I - \mu_I| \leq K(\beta) \frac{\operatorname{card} I}{\sqrt{N}} .$$

We refer to [134], Theorem 1.8, to see how to deduce this from Theorem 2.2.12.

If, rather than studying how well μ_I approximates G_I for one given set I, we look for a uniform approximation on sets of a given size, there is apparently nothing known.

Research problem 2.2.18. (Level 1) Find (optimal, if you can) pairs of sequences $a_N \to \infty$, $b_N \to 0$ such that

$$\mathsf{P}\left(\sup_{\operatorname{card} I \leq a_N} |G_I - \mu_I| \geq b_N\right) \to 0 .$$

It seems certain that this is connected to the following

Research problem 2.2.19. (Level 1) For a typical realization of Gibbs' measure, study the family of numbers $(\langle\sigma_i \sigma_j\rangle_{i<j})$. More generally, study the family of numbers $(\langle\sigma_{i_1} \ldots \sigma_{i_p}\rangle)_{i_1<\cdots<i_p}$.

As we will see in the next section, we understand very well the overlaps. These are however "spatial averages". For example, we see from Theorem 2.2.12 that $\mathsf{E}\langle R_{1,2}^2\rangle \le K/N$, and using Lemma 2.2.10, we get $\mathsf{E}\langle\sigma_1 \sigma_2\rangle^2 \le K/N$. This, however, does not tell us what the order of $\max_{i<j}\langle\sigma_i \sigma_j\rangle^2$ is. A related question one could ask is, given $t > 0$, what is the typical proportion of couples $i < j$ for which $\langle\sigma_i \sigma_j\rangle \ge t/\sqrt{N}$ as $N \to \infty$.

In a related but different direction, we have the following

Research problem 2.2.20. (Level 2) If $I = \{1, \ldots, n\}$, find, for $t > 0$, satisfactory bounds for

$$\mathsf{P}(|G_I - \mu_I| \ge t/\sqrt{N}) .$$

Some (unsatisfactory) bounds are given in [134], where the problem is discussed. It is argued there that this problem is connected with the following

Conjecture 2.2.21. (Level 2) For $\beta \le 1$, the matrix $(\langle\sigma_i \sigma_j\rangle)_{i<j}$ has a bounded operator norm, that is

$$\sup_N \mathsf{E}\left(\sup_{\|\mathbf{x}\|\le 1} \sum_{i<j} x_i\, x_j\, \langle\sigma_i \sigma_j\rangle\right) < \infty , \tag{2.63}$$

where the supremum is taken over all sequences $\mathbf{x} = (x_i)_{i\le N}$ with $\sum_{i\le N} x_i^2 \le 1$.

The heuristic motivation is that the interaction term $\beta\, g_{ij}\, \sigma_i\sigma_j/\sqrt{N}$ between σ_i and σ_j in $-H_N$ should provide the main contribution to $\langle\sigma_i \sigma_j\rangle$, so that the matrix $\langle\sigma_i \sigma_j\rangle$ should resemble the matrix $(g_{ij}/\sqrt{N})$, which is known to be of bounded operator norm. The only apparent approach to (2.63) at this moment relies on the cavity method discussed in the next section, but considerable technical skill seems required to make it work. See (2.118) for a partial result.

Having briefly commented on the behavior of Gibbs' measure when focusing on "few spins", we now look at it globally. It is not surprising that, as soon as $\beta > 0$, this measure is extremely different from the uniform measure. Let us recall the definition of the Hamming distance on Σ_N

$$\mathrm{d}(\boldsymbol{\sigma}^1, \boldsymbol{\sigma}^2) = \frac{1}{N}\,\mathrm{card}\{i \le N\,;\; \sigma_i^1 \ne \sigma_i^2\} = \frac{1}{2}(1 - R_{1,2}) .$$

We refer the reader to [153] for the proof of the following.

Proposition 2.2.22. *If $\beta > 0$, with overwhelming probability we can find subsets A, B of Σ_N with*

$$G_N(A) \geq \frac{1}{3}, \ \mu_N(B) \geq \frac{1}{3}, \ \mathrm{d}(A,B) \geq \frac{1}{2}\min(1,\beta^2)\,,$$

where μ_N denotes the uniform measure on Σ_N.

In Proposition 2.2.22, as anywhere else, we say that an event occurs with overwhelming probability when the probability of failure is at most $K\exp(-N/K)$.

We end this section by formulating a conjecture which seems to touch the real issues. The real issue is the understanding, of the "energy landscape geometry "for a typical realization of the disorder, or, in other words, the structure of the set consisting of all the configurations for which $-H_N(\boldsymbol{\sigma})$ really contributes to Gibbs' measure. We do not really know what are the fruitful questions in that respect. The following formalizes the idea that the set of such configurations $\boldsymbol{\sigma}$ should be made up of well separated pieces.

Conjecture 2.2.23. (Level 3) If $\beta > 0$ we can find $\theta > 0$ such that with overwhelming probability we can find two subsets A, B of Σ_N such that

$$G_N(A) \geq \frac{1}{4}, \ G_N(B) \geq \frac{1}{4}, \ \mathrm{d}(A,B) \geq \theta\,.$$

One can also formulate stronger versions, e.g. asking to find $a(N)$ sets $(A_k)_{k\leq a(N)}$ with $G_N(A_k) \geq 1/(2\,a(N))$ and $\mathrm{d}(A_k, A_{k'}) \geq \theta$ for $k \neq k'$.

The meaning of this conjecture is better understood if one knows the following important result: nice probability measures μ on Σ_N (such as product measures) satisfy

$$\mu(A) \geq \frac{1}{4}, \ \mu(B) \geq \frac{1}{4} \Rightarrow \mathrm{d}(A,B) \leq \frac{L}{\sqrt{N}}\,.$$

The research problems mentioned in this section can be reformulated in the more difficult situations that we will consider later where they of course remain open.

2.3 The Cavity Method

In this section we introduce the main tool of this work, the cavity method, and we put it to work in the simple situation of Section 2.2. Originally, the cavity method is simply induction over N. To reduce to a system with one less spin, one removes a spin, creating a "cavity".

The basic idea is to regroup in the Hamiltonian the terms that depend on σ_N:

$$-H_N(\boldsymbol{\sigma}) = \frac{\beta}{\sqrt{N}} \sum_{i<j\leq N-1} g_{ij}\,\sigma_i\,\sigma_j + \sigma_N \left(\frac{\beta}{\sqrt{N}} \sum_{i\leq N-1} g_{i,N}\,\sigma_i\right) . \tag{2.64}$$

The first term on the right looks like the Hamiltonian of an $(N-1)$-spin system, but where β is replaced by β^- such that

$$\frac{\beta^-}{\sqrt{N-1}} = \frac{\beta}{\sqrt{N}}$$

i.e.

$$\beta^- = \beta\sqrt{1-\frac{1}{N}} . \tag{2.65}$$

We will write $\boldsymbol{\rho} = (\sigma_1, \ldots, \sigma_{N-1})$ for a configuration of Σ_{N-1}. For simplicity we write $g_i = g_{i,N}$, and

$$g(\boldsymbol{\rho}) = \frac{\beta}{\sqrt{N}} \sum_{i\leq N-1} g_i\,\sigma_i . \tag{2.66}$$

Thus (2.64) means that

$$-H_N(\boldsymbol{\sigma}) = -H_{N-1}(\boldsymbol{\rho}) + \sigma_N\, g(\boldsymbol{\rho}) , \tag{2.67}$$

where of course

$$-H_{N-1}(\boldsymbol{\rho}) = \frac{\beta}{\sqrt{N}} \sum_{1\leq i<j\leq N-1} g_{ij}\,\sigma_i\,\sigma_j . \tag{2.68}$$

We denote by $\langle\cdot\rangle_-$ an average with respect to the Gibbs measure on Σ_{N-1} with Hamiltonian H_{N-1}.

If f is a function on Σ_N, the following algebraic identity holds

$$\langle f\rangle = \frac{\langle \mathrm{Av}\, f \exp \sigma_N\, g(\boldsymbol{\rho})\rangle_-}{Z} \tag{2.69}$$

where

$$Z = \langle \mathrm{Av}\, \exp \sigma_N\, g(\boldsymbol{\rho})\rangle_- = \langle \mathrm{ch}\, g(\boldsymbol{\rho})\rangle_- .$$

(The reader should not confuse Z with the partition function Z_N.) In (2.69), Av means average over $\sigma_N = \pm 1$. Thus $\mathrm{Av}(f(\boldsymbol{\sigma}) \exp \sigma_N g(\boldsymbol{\rho}))$ is a function of $\boldsymbol{\rho}$ only, and in the numerator of (2.69) it is then averaged for $\langle\cdot\rangle_-$. The proof of (2.69) is immediate when one replaces $\langle\cdot\rangle$ and $\langle\cdot\rangle_-$ by their definition. Most often we will work with functions on Σ_N^n for $n \geq 2$. In this case (2.69) becomes

$$\langle f\rangle = \frac{\langle \mathrm{Av}\, f \exp \sum_{\ell\leq n} \sigma_N^\ell\, g(\boldsymbol{\rho}^\ell)\rangle_-}{Z^n} \tag{2.70}$$

where Av now means average over $\sigma_N^1 = \pm 1, \ldots, \sigma_N^n = \pm 1$. The function

$$\mathrm{Av}\, f \exp \sum_{\ell \le n} \sigma_N^\ell\, g(\boldsymbol{\rho}^\ell) = \mathrm{Av}(f(\boldsymbol{\sigma}^1, \ldots, \boldsymbol{\sigma}^n) \exp \sum_{\ell \le n} \sigma_N^\ell\, g(\boldsymbol{\rho}^\ell))$$

depends only upon $\boldsymbol{\rho}^1, \ldots, \boldsymbol{\rho}^n$ and $\langle \cdot \rangle_-$ means that this is integrated with respect to the nth-power of Gibbs' measure relative to H_{N-1}.

To lighten notation, throughout the entire book, we write

$$\varepsilon = \sigma_N \; ; \; \varepsilon_\ell = \sigma_N^\ell$$

so that (2.70) reads as

$$\mathsf{E}\langle f \rangle = \mathsf{E}\, \frac{\langle \mathrm{Av}\, f \exp \sum_{\ell \le n} \varepsilon_\ell\, g(\boldsymbol{\rho}^\ell) \rangle_-}{Z^n} \,. \tag{2.71}$$

The main feature of this fundamental formula is that the r.v. $g(\boldsymbol{\rho}^\ell)$ are probabilistically independent of the randomness (disorder) in $\langle \cdot \rangle_-$. Thus we will be able to first average over these r.v. $g(\boldsymbol{\rho}^\ell)$ conditionally upon $\langle \cdot \rangle_-$, that is, at given $\langle \cdot \rangle_-$. The problem, however, is to design a way to do this.

If we keep in mind the fact that (2.71) means that

$$\mathsf{E}\langle f \rangle = \mathsf{E}\, \frac{\left\langle \mathrm{Av}\, f \exp\left(\sum_{\ell \le n} \varepsilon_\ell\, u \sum_{i \le N-1} \frac{1}{\sqrt{N}}\, g_i\, \sigma_i^\ell \right) \right\rangle_-}{\left\langle \mathrm{Av}\, \exp\left(\varepsilon\, u \sum_{i \le N-1} \frac{1}{\sqrt{N}}\, g_i\, \sigma_i \right) \right\rangle_-^n}$$

for $u = \beta$, a natural idea is to attempt to take the derivative in u. We however prepare better the future by setting $u = \beta\sqrt{t}$ and taking the derivative in t. Let us write

$$\mathcal{E}_{n,t} = \mathcal{E}_{n,t}(\boldsymbol{\rho}^1, \ldots, \boldsymbol{\rho}^n) = \exp\left(\sqrt{t} \sum_{\ell \le n} \varepsilon_\ell\, g(\boldsymbol{\rho}^\ell) \right) ,$$

$\mathcal{E}_t = \mathcal{E}_{1,t}$, and

$$Z_t = \langle \mathrm{Av}\, \mathcal{E}_t \rangle_- = \langle \mathrm{ch}(\sqrt{t}\, g(\boldsymbol{\rho})) \rangle_- = \langle \mathrm{Av}\, \exp(\varepsilon \sqrt{t}\, g(\boldsymbol{\rho})) \rangle_- \,.$$

We define, for a function f on Σ_N^n,

$$\langle f \rangle_t = \frac{\langle \mathrm{Av}\, f\, \mathcal{E}_{n,t} \rangle_-}{Z_t^n} \tag{2.72}$$

$$\nu_t(f) = \mathsf{E}\langle f \rangle_t \,. \tag{2.73}$$

Thus

$$\nu_1(f) = \mathsf{E}\langle f \rangle \; ; \; \nu_0(f) = \mathsf{E}\langle \mathrm{Av}\, f \rangle_- \,. \tag{2.74}$$

To lighten notation we will write ν rather than ν_1, that is

$$\nu(f) = \mathsf{E}\langle f \rangle , \tag{2.75}$$

and from now on we write $\nu(f)$ instead of $\mathsf{E}\langle f \rangle$ whenever possible, even if we are not using the cavity method.

The underlying idea is that $\nu_0(f)$ should be simpler than $\nu(f)$ while, on the other hand, one should be able to relate these quantities, for example through the relation

$$\nu(f) - \nu_0(f) = \int_0^1 \frac{\mathrm{d}}{\mathrm{d}t} (\nu_t(f)) \, \mathrm{d}t \tag{2.76}$$

or higher-order versions.

To illustrate that $\nu_0(f) = \mathsf{E}\langle \mathrm{Av}\, f \rangle_-$ is easier to compute than $\nu(f)$, we observe e.g. that

$$\nu_0(\varepsilon_1 \varepsilon_2 R_{1,2}) = \frac{1}{N} \tag{2.77}$$

because $\mathrm{Av}(\varepsilon_1 \varepsilon_2 R_{1,2}) = 1/N$. More generally, the reason why $\nu_0(f)$ can be simpler to evaluate than $\nu(f)$ is that for ν_0, the last spin has been made independent of the first $N-1$ spins. If f^- is a function of $\boldsymbol{\rho}^1, \ldots, \boldsymbol{\rho}^n$ only and f' a function of $\varepsilon_1, \ldots, \varepsilon_n$ only, the function $f = f^- f'$ on Σ_N^n is such that

$$\nu_0(f) = \mathsf{E}\langle \mathrm{Av}\, f^- f' \rangle_- = \mathrm{Av}\, f' \mathsf{E}\langle f^- \rangle_- = \mathrm{Av}\, f' \, \nu_0(f^-) \, . \tag{2.78}$$

This is of particular interest if $\mathrm{Av}\, f' = 0$.

The idea of "making" the last spin independent of the first $N-1$ spins is very fruitful, and will occur in each of the following chapters. To lighten the vocabulary, a function f^- on Σ_N^n that depends only on $\boldsymbol{\rho}^1, \ldots, \boldsymbol{\rho}^n$ but not on $\varepsilon_1, \ldots, \varepsilon_n$ will simply be described as "a function on Σ_{N-1}^n". This is consistent with the natural identification of a function on Σ_{N-1}^n as a function on Σ_N^n.

To use (2.76), we need to compute $\mathrm{d}(\nu_t(f))/\mathrm{d}t$. To lighten notation, we will write

$$\nu_t'(f) = \frac{\mathrm{d}}{\mathrm{d}t} \nu_t(f) \, . \tag{2.79}$$

We set

$$R_{\ell,\ell'}^- = \frac{1}{N} \sum_{i \le N-1} \sigma_i^\ell \sigma_i^{\ell'} \, . \tag{2.80}$$

Proposition 2.3.1. *For a function f on Σ_N^n, and $t > 0$, we have*

$$\begin{aligned} \nu_t'(f) = \beta^2 & \sum_{1 \le \ell < \ell' \le n} \nu_t(f \varepsilon_\ell \varepsilon_{\ell'} R_{\ell,\ell'}^-) \\ & - \beta^2 n \sum_{\ell \le n} \nu_t(f \varepsilon_\ell \varepsilon_{n+1} R_{\ell,n+1}^-) \\ & + \beta^2 \frac{n(n+1)}{2} \nu_t(f \varepsilon_{n+1} \varepsilon_{n+2} R_{n+1,n+2}^-) \, , \end{aligned} \tag{2.81}$$

and

$$\nu_t'(f) = \beta^2 \sum_{1\le\ell<\ell'\le n} \nu_t(f\,\varepsilon_\ell\,\varepsilon_{\ell'}\,R_{\ell,\ell'})$$
$$- \beta^2 n \sum_{\ell\le n} \nu_t(f\,\varepsilon_\ell\,\varepsilon_{n+1}\,R_{\ell,n+1})$$
$$+ \beta^2 \frac{n(n+1)}{2} \nu_t(f\,\varepsilon_{n+1}\,\varepsilon_{n+2}\,R_{n+1,n+2})\,. \tag{2.82}$$

When talking about "a function on Σ_N^n" we mean that $f(\boldsymbol{\sigma}^1,\ldots,\boldsymbol{\sigma}^n)$ depends only on the values of $\boldsymbol{\sigma}^1,\ldots,\boldsymbol{\sigma}^n$ (e.g. $f = R_{1,2}$, or $f = R_{1,2}\,R_{1,3}$) but *not* upon the disorder, i.e. the variables g_{ij}. Later, we will need to consider functions on Σ_N^n depending on the disorder, and to avoid confusion, they will be called "random functions on Σ_N^n".

It is instructive to observe that there are as many (that is, n^2) terms of the form $\pm\beta^2\nu_t(\varepsilon_\ell\,\varepsilon_{\ell'}\,R_{\ell,\ell'})$ in (2.81) with a plus sign than terms with a minus sign. Thus, (2.82) follows from (2.81) since

$$\varepsilon_\ell\,\varepsilon_{\ell'}\,R_{\ell,\ell'} = \frac{1}{N} + \varepsilon_\ell\,\varepsilon_{\ell'}\,R^-_{\ell,\ell'}\,.$$

A function on Σ_N^n can also be seen as a function on $\Sigma_N^{n'}$ for $n' > n$. It is an instructive exercise to show that the two formally different expressions one gets for $\nu_t'(f)$ in this case coincide. This relies upon the fact that

$$\nu_t(f\,\varepsilon_\ell\,\varepsilon_{\ell_1}\,R_{\ell,\ell_1}) = \nu_t(f\,\varepsilon_\ell\,\varepsilon_{\ell_2}\,R_{\ell,\ell_2})$$

if $\ell_1,\ell_2 > n$, $\ell_1,\ell_2 \ne \ell$.

Proof. We observe that for $t > 0$ we have

$$\frac{\mathrm{d}}{\mathrm{d}t}\,\mathcal{E}_{n,t} = \frac{1}{2\sqrt{t}} \sum_{\ell\le n} \varepsilon_\ell\, g(\boldsymbol{\rho}^\ell)\,\mathcal{E}_{n,t}$$

and thus

$$\nu_t'(f) = \frac{1}{2\sqrt{t}}\,\mathsf{E}\Bigg(\frac{\left\langle \mathrm{Av}\, f \sum_{\ell\le n} \varepsilon_\ell\, g(\boldsymbol{\rho}^\ell)\,\mathcal{E}_{n,t}\right\rangle_-}{Z_t^n}$$
$$- n\frac{\langle \mathrm{Av}\, f\,\mathcal{E}_{n,t}\rangle_-\,\langle \mathrm{Av}\,\varepsilon\, g(\boldsymbol{\rho})\,\mathcal{E}_t\rangle_-}{Z_t^{n+1}}\Bigg)\,. \tag{2.83}$$

In this type of calculation, it simplifies matters to use replicas in order to write a product of brackets as a single bracket; here we write

$$\langle \mathrm{Av}\, f\,\mathcal{E}_{n,t}\rangle_-\,\langle \mathrm{Av}\,\varepsilon\, g(\boldsymbol{\rho})\,\mathcal{E}_t\rangle_- = \langle \mathrm{Av}\, f\,\varepsilon_{n+1}\, g(\boldsymbol{\rho}^{n+1})\,\mathcal{E}_{n+1,t}\rangle_-\,,$$

and (2.83) becomes

$$\nu_t'(f) = \frac{1}{2\sqrt{t}} \mathsf{E}\left(\frac{\left\langle \mathrm{Av}\, f \sum_{\ell \le n} \varepsilon_\ell\, g(\boldsymbol{\rho}^\ell)\, \mathcal{E}_{n,t} \right\rangle_-}{Z_t^n} \right.$$
$$\left. - n \frac{\langle \mathrm{Av}\, f\, \varepsilon_{n+1}\, g(\boldsymbol{\rho}^{n+1})\, \mathcal{E}_{n+1,t} \rangle_-}{Z_t^{n+1}} \right) . \tag{2.84}$$

To make this formula useful, we need to integrate by parts to get rid of the quantities $g(\boldsymbol{\rho}^\ell)$. The first time one sees this, one might feel for this computation about the same enthusiasm as for going to the torture chamber. In fact, this is not as bad as it looks. Moreover, this computation, that will occur in various forms, is the cornerstone of the cavity method.

We start the computation by writing

$$\frac{\left\langle \mathrm{Av}\, f \sum_{\ell \le n} \varepsilon_\ell\, g(\boldsymbol{\rho}^\ell)\, \mathcal{E}_{n,t} \right\rangle_-}{Z_t^n}$$
$$= \frac{\beta}{\sqrt{N}} \sum_{i \le N-1} g_i \frac{\left\langle \mathrm{Av}\, f \sum_{\ell \le n} \varepsilon_\ell\, \sigma_i^\ell\, \mathcal{E}_{n,t} \right\rangle_-}{Z_t^n} . \tag{2.85}$$

We will then apply the integration by parts formula (A.40), that is $\mathsf{E}\, g F(g) = \mathsf{E}\, g^2\, \mathsf{E}\, F'(g)$ for $g = g_i$, at all g_j, $j \ne i$ fixed, and conditionally upon the disorder of $\langle \cdot \rangle_-$. That is, we are thinking of the coefficient of g_i in (2.85) as a function of g_i alone. The dependence on g_i is through $\mathcal{E}_{n,t}$ and $\mathcal{E}_t$, and we have

$$\frac{\partial\, \mathcal{E}_{n,t}}{\partial\, g_i} = \frac{\beta\sqrt{t}}{\sqrt{N}} \left(\sum_{\ell' \le n} \varepsilon_{\ell'}\, \sigma_i^{\ell'} \right) \mathcal{E}_{n,t} .$$

Thus, the derivative of the coefficient of g_i in (2.85) with respect to g_i is

$$\frac{\beta\sqrt{t}}{\sqrt{N}} \left(\frac{\left\langle \mathrm{Av}\, f \sum_{\ell \le n} \varepsilon_\ell\, \sigma_i^\ell \sum_{\ell' \le n} \varepsilon_{\ell'}\, \sigma_i^{\ell'}\, \mathcal{E}_{n,t} \right\rangle_-}{Z_t^n} \right.$$
$$\left. - n \frac{\left\langle \mathrm{Av}\, f \sum_{\ell \le n} \varepsilon_\ell\, \sigma_i^\ell \mathcal{E}_{n,t} \right\rangle_- \langle \mathrm{Av}\, \varepsilon\, \sigma_i\, \mathcal{E}_t \rangle_-}{Z_t^{n+1}} \right) \tag{2.86}$$
$$= \frac{\beta\sqrt{t}}{\sqrt{N}} \left(\left\langle f \sum_{\ell \le n} \varepsilon_\ell\, \sigma_i^\ell \sum_{\ell' \le n} \varepsilon_{\ell'}\, \sigma_i^{\ell'} \right\rangle_t - n \left\langle f \sum_{\ell \le n} \varepsilon_\ell\, \sigma_i^\ell\, \varepsilon_{n+1}\, \sigma_i^{n+1} \right\rangle_t \right) ,$$

where we have used replicas to replace the product of the two brackets occurring in the second term of the left-hand side of (2.86) by a single bracket. Using the integration by parts (as in formula (A.40)), we thus get

$$\mathsf{E}\, g_i \frac{\left\langle \mathrm{Av}\, f \sum_{\ell \le n} \varepsilon_\ell\, \sigma_i^\ell\, \mathcal{E}_{n,t} \right\rangle_-}{Z_t^n}$$

$$= \frac{\beta\sqrt{t}}{\sqrt{N}} \left(\sum_{\ell,\ell' \le n} \nu_t(f\, \varepsilon_\ell\, \varepsilon_{\ell'}\, \sigma_i^\ell\, \sigma_i^{\ell'}) - n \sum_{\ell \le n} \nu_t(f\, \varepsilon_\ell\, \varepsilon_{n+1}\, \sigma_i^\ell\, \sigma_i^{n+1}) \right) . \quad (2.87)$$

Thus, if we recall (2.80), we get from (2.85) and (2.87), that

$$\mathsf{E} \frac{\left\langle \mathrm{Av}\, f \sum_{\ell \le n} \varepsilon_\ell\, g(\boldsymbol{\rho}^\ell)\, \mathcal{E}_{n,t} \right\rangle_-}{Z_t^n}$$

$$= \beta^2 \sqrt{t} \left(\sum_{\ell,\ell' \le n} \nu_t(f\, \varepsilon_\ell\, \varepsilon_{\ell'}\, R^-_{\ell,\ell'}) - n \sum_{\ell \le n} \nu_t(f\, \varepsilon_\ell\, \varepsilon_{n+1}\, R^-_{\ell,n+1}) \right) .$$

A similar computation yields

$$\mathsf{E} \frac{\langle \mathrm{Av}\, f\, \varepsilon_{n+1}\, g(\boldsymbol{\rho}^{n+1})\, \mathcal{E}_{n+1,t} \rangle_-}{Z_t^{n+1}}$$

$$= \beta^2 \sqrt{t} \left(\sum_{\ell \le n+1} \nu_t(f\, \varepsilon_\ell\, \varepsilon_{n+1}\, R^-_{\ell,n+1}) - (n+1)\, \nu_t(f\, \varepsilon_{n+1}\, \varepsilon_{n+2}\, R^-_{n+1,n+2}) \right) .$$

We combine these formulas with (2.84). We note that, since for $\ell = \ell'$, we have $\varepsilon_\ell\, \varepsilon_{\ell'}\, R^-_{\ell,\ell'} = (N-1)/N$, we have

$$\sum_{\ell,\ell' \le n} \nu_t(f\, \varepsilon_\ell\, \varepsilon_{\ell'}\, R^-_{\ell,\ell'}) = 2 \sum_{\ell < \ell'} \nu_t(f\, \varepsilon_\ell\, \varepsilon_{\ell'}\, R^-_{\ell,\ell'}) + n\, \frac{N-1}{N}\, \nu_t(f)$$

$$\sum_{\ell \le n+1} \nu_t(f\, \varepsilon_\ell\, \varepsilon_{n+1}\, R^-_{\ell,n+1}) = \sum_{\ell \le n} \nu_t(f\, \varepsilon_\ell\, \varepsilon_{n+1}\, R^-_{\ell,n+1}) + \frac{N-1}{N}\, \nu_t(f) ,$$

and we get (2.81). □

Even though we have proved Proposition 2.3.1 only for $t > 0$, it is also true for $t = 0$, if $\nu_0'(t)$ then means the right derivative. This is because the expression for $\nu_t'(f)$ given by (2.81) is continuous in t at $t = 0$.

It follows from Theorem 2.2.12 that for each k we have

$$\nu(R_{1,2}^{2k}) \le \frac{K(k,\beta)}{N^k} \quad (2.88)$$

and, consequently since $|R_{1,2} - R^-_{1,2}| \le 1/N$,

$$\nu((R^-_{1,2})^{2k}) \le \frac{K(k,\beta)}{N^k} . \tag{2.89}$$

We can then expect, by Hölder's inequality, that each factor $R^-_{\ell,\ell'}$ in (2.81) "contributes as $N^{-1/2}$", so that $\nu'_t(f)$ will have a tendency to be small. It is useful to note the following consequence of Hölder's inequality. If $\tau_1, \tau_2 > 0$, $1/\tau_1 + 1/\tau_2 = 1$, then

$$\nu_t(f_1 f_2) \le \nu_t(|f_1|^{\tau_1})^{1/\tau_1} \, \nu_t(|f_2|^{\tau_2})^{1/\tau_2} . \tag{2.90}$$

This is proved by applying Hölder's inequality to $\langle\cdot\rangle_t$ and then to E. We cannot directly use (2.89), because it holds for ν rather than for ν_t. This is why the following observation is so useful.

Proposition 2.3.2. *If f is a non-negative function on n replicas, we have*

$$\nu_t(f) \le \exp(2\beta^2 n^2(1-t))\, \nu(f) . \tag{2.91}$$

Proof. Since $|R_{\ell,\ell'}| \le 1$, we see from (2.82) that

$$\nu'_t(f) \ge -2\,\beta^2\, n^2\, \nu_t(f)$$

so that

$$\frac{\mathrm{d}}{\mathrm{d}t} \log \nu_t(f) \ge -2\,\beta^2\, n^2$$

and, integrating between t and 1,

$$\log \nu(f) - \log \nu_t(f) \ge -2\,\beta^2\, n^2(1-t) .$$

□

For $k \ge 1$, we denote by $\nu_t^{(k)}(f)$ the kth derivative of the function $t \mapsto \nu_t(f)$.

Proposition 2.3.3. *If $\beta < 1$, and f is a function on Σ_N^n, then*

$$|\nu_t^{(k)}(f)| \le \frac{K(\beta,n,k)}{N^{k/2}} \nu(f^2)^{1/2} . \tag{2.92}$$

Proof. Iteration of (2.82) shows that $\nu_t^{(k)}(f)$ is the sum of terms of the type

$$\pm\,\beta^{2k}\, \nu_t(f\, \varepsilon_{\ell_1}\, \varepsilon_{\ell_2}\, R_{\ell_1,\ell_2}\, \varepsilon_{\ell_3}\, \varepsilon_{\ell_4}\, R_{\ell_3,\ell_4} \cdots \varepsilon_{\ell_{2k-1}}\, \varepsilon_{\ell_{2k}}\, R_{\ell_{2k-1},\ell_{2k}}) ,$$

where $\ell_{2s-1} < \ell_{2s}$ for $1 \le s \le k$. We bound these by

$$\nu_t(|f|\, |R_{\ell_1,\ell_2}| \cdots |R_{\ell_{2k-1},\ell_{2k}}|) .$$

We then use (2.91), (2.90) and (2.88). □

Corollary 2.3.4. *If* $\beta < 1$ *we have*

$$|\nu(f) - \nu_0(f)| \le \frac{K}{\sqrt{N}} \nu(f^2)^{1/2} \tag{2.93}$$

$$|\nu(f) - \nu_0(f) - \nu_0'(f)| \le \frac{K}{N} \nu(f^2)^{1/2} . \tag{2.94}$$

Proof. To prove (2.93), combine (2.76) with (2.92) for $k = 1$; proceed similarly for (2.94). □

Here is a first illustration of the power of this methodology.

Proposition 2.3.5. *If* $\beta < 1$ *we have*

$$\left|\nu(R_{1,2}^2) - \frac{1}{N(1-\beta^2)}\right| \le \frac{K}{N^{3/2}} . \tag{2.95}$$

Proof. A basic feature of the previous construction is that the last coordinate plays a special role. It is then fruitful to "put as much weight as possible on this coordinate", by writing

$$\nu(R_{1,2}^2) = \frac{1}{N} \sum_{i \le N} \nu(\sigma_i^1 \sigma_i^2 R_{1,2}) = \nu(\varepsilon_1 \varepsilon_2 R_{1,2}) ,$$

using symmetry among $i = 1, \ldots, N$. (This argument will be referred as "using symmetry among sites".) Thus

$$\nu(R_{1,2}^2) = \frac{1}{N} + \nu(\varepsilon_1 \varepsilon_2 R_{1,2}^-) . \tag{2.96}$$

Next, we use (2.94) and (2.89) to write

$$|\nu(\varepsilon_1 \varepsilon_2 R_{1,2}^-) - \nu_0(\varepsilon_1 \varepsilon_2 R_{1,2}^-) - \nu_0'(\varepsilon_1 \varepsilon_2 R_{1,2}^-)| \le \frac{K}{N^{3/2}} . \tag{2.97}$$

We use (2.78) to see that $\nu_0(\varepsilon_1 \varepsilon_2 R_{1,2}^-) = 0$. We use (2.81) with $n = 2$ to see that

$$\begin{aligned}\nu_0'(\varepsilon_1 \varepsilon_2 R_{1,2}^-) &= \beta^2 \nu_0((R_{1,2}^-)^2) - 2\beta^2 \sum_{\ell \le 2} \nu_0(\varepsilon_1 \varepsilon_2 \varepsilon_\ell \varepsilon_3 R_{1,2}^- R_{\ell,3}^-) \\ &\quad + 3\beta^2 \nu_0(\varepsilon_1 \varepsilon_2 \varepsilon_3 \varepsilon_4 R_{1,2}^- R_{3,4}^-) \\ &= \beta^2 \nu_0((R_{1,2}^-)^2) ,\end{aligned}$$

applying (2.78) again. Now, using (2.93) and (2.89), we get

$$|\nu_0((R_{1,2}^-)^2) - \nu((R_{1,2}^-)^2)| \le \frac{K}{N^{3/2}} .$$

Finally, we write $R^-_{\ell,\ell'} = R_{\ell,\ell'} - N^{-1}\varepsilon_\ell\varepsilon_{\ell'}$, from which it should be obvious that

$$|\nu((R^-_{1,2})^2) - \nu(R^2_{1,2})| \le \frac{K}{N^{3/2}} .$$

Thus (2.97) implies that

$$|\nu(\varepsilon_1\varepsilon_2 R^-_{1,2}) - \beta^2\nu(R^2_{1,2})| \le \frac{K}{N^{3/2}} ,$$

and comparing with (2.96) this finishes the proof. □

Research problem 2.3.6. (Level 1) Compute higher order expansions for $\nu(R^2_{1,2})$.

For example, find $c(\beta)$ such that

$$\left|\nu(R^2_{1,2}) - \frac{1}{N(1-\beta^2)} - \frac{c(\beta)}{N^{3/2}}\right| \le \frac{K}{N^2} .$$

Of course, it is very unclear whether manageable formulas can be found for the general term of the expansion. This is related to Research problem 2.3.15 below.

At this stage one can say that the important ideas of this section have already been explained. The method of Proposition 2.3.5 can however be pushed much further, culminating in Theorem 2.3.9, which itself is the first of a series increasingly more difficult "central limit theorems". These theorems are not of fundamental importance in the big picture. Rather, they are used to illustrate in a meaningful case how precise our understanding can become if we really work at it. We have made a great effort to write these results in a way that none of them requires a big jump from the preceding one. Thus, as a first step towards the general statement, we will generalize Proposition 2.3.5 to higher moments.

We denote by $a(k)$ the kth moment of a standard Gaussian r.v. Thus, $a(0) = 1$, $a(1) = 0$, $a(2) = 1$, and integration by parts shows that $a(k) = (k-1)\,a(k-2)$ if $k \ge 2$.

Proposition 2.3.7. *If $k \ge 0$, $\beta < 1$, then*

$$\left|\nu(R^k_{1,2}) - \frac{a(k)}{(N(1-\beta^2))^{k/2}}\right| \le \frac{K}{N^{(k+1)/2}} . \tag{2.98}$$

Here, the number K depends on k. One may ask what is the best possible dependence on k; one may also ask for higher order expansions.

To prove a relation such as (2.98), we need not pay attention to any quantity W such that

$$|W| \le K\,N^{-(k+1)/2} . \tag{2.99}$$

Such a quantity is traditionally denoted by

$$O(N^{-(k+1)/2}) .$$

This notation is too cumbersome for massive use.

Definition 2.3.8. *We denote by $O(k+1)$ a quantity W as in (2.99).*

Thus, using (2.89) and (2.88)

$$\frac{1}{N^\ell} = O(2\ell)\,; \quad \nu((R^-_{\ell,\ell'})^k) = O(k)\,; \quad \nu(R^k_{\ell,\ell'}) = O(k) \tag{2.100}$$

and of course

$$O(k_1)\,O(k_2) = O(k_1+k_2) . \tag{2.101}$$

Proof of Proposition 2.3.7 We closely follow the scheme of Proposition 2.3.5. We use symmetry between sites to get

$$\begin{aligned}\nu(R^k_{1,2}) &= \nu(\varepsilon_1\,\varepsilon_2\,R^{k-1}_{1,2})\\ &= \nu(\varepsilon_1\,\varepsilon_2(R^-_{1,2})^{k-1}) + \frac{(k-1)}{N}\,\nu((R^-_{1,2})^{k-2})\\ &+ \sum_{\ell\geq 2}\frac{1}{N^\ell}\binom{k-1}{\ell}\,\nu((\varepsilon_1\,\varepsilon_2)^{\ell+1}\,(R^-_{1,2})^{k-\ell-1})\,,\end{aligned} \tag{2.102}$$

by writing $R_{1,2} = R^-_{1,2} + N^{-1}\,\varepsilon_1\,\varepsilon_2$ and expanding the power.

Using (2.100) and (2.101), we have

$$\frac{1}{N^\ell}\,\nu((\varepsilon_1\,\varepsilon_2)^{\ell+1}\,(R^-_{1,2})^{k-\ell-1}) = O(2\ell)\,O(k-\ell-1) = O(k+\ell-1) = O(k+1)$$

for $\ell \geq 2$. Writing $R^-_{1,2} = R_{1,2} - \varepsilon_1\varepsilon_2/N$ and expanding, we see in a similar manner that

$$\nu((R^-_{1,2})^{k-2}) = \nu(R^{k-2}_{1,2}) + O(k-1) \tag{2.103}$$

and thus

$$\frac{1}{N}\,\nu\,((R^-_{1,2})^{k-2}) = \frac{1}{N}\,\nu(R^{k-2}_{1,2}) + O(k+1)\,,$$

so that (2.102) gives

$$\nu(R^k_{1,2}) = \nu(\varepsilon_1\,\varepsilon_2(R^-_{1,2})^{k-1}) + \frac{(k-1)}{N}\,\nu(R^{k-2}_{1,2}) + O(k+1) . \tag{2.104}$$

We use (2.94) (and (2.89)) to write

$$\begin{aligned}\nu(\varepsilon_1\,\varepsilon_2(R^-_{1,2})^{k-1}) &= \nu_0(\varepsilon_1\,\varepsilon_2(R^-_{1,2})^{k-1}) + \nu_0'(\varepsilon_1\,\varepsilon_2(R^-_{1,2})^{k-1}) + O(k+1)\\ &= \beta^2\,\nu_0((R^-_{1,2})^k) + O(k+1)\\ &= \beta^2\,\nu(R^k_{1,2}) + O(k+1)\end{aligned}$$

by proceeding exactly as in the case $k = 1$ of Proposition 2.3.5, and using (2.103) for k rather than $k-2$ in the last line. We then get from (2.104) that

$$(1-\beta^2)\,\nu(R_{1,2}^k) = \frac{k-1}{N}\,\nu(R_{1,2}^{k-2}) + O(k+1)\,,$$

from which (2.98) follows by induction over k. □

The extra effort to get to a really general statement is now reasonable.

Theorem 2.3.9. *Consider an integer n, and for $1 \le \ell < \ell' \le n$ consider non-negative integers $k(\ell,\ell')$. Consider $k = \sum_{\ell<\ell'} k(\ell,\ell')$. Then*

$$\left|\nu\Big(\prod_{\ell<\ell'} R_{\ell,\ell'}^{k(\ell,\ell')}\Big) - \prod_{\ell<\ell'} \frac{a(k(\ell,\ell'))}{(N(1-\beta^2))^{k(\ell,\ell')/2}}\right| \le \frac{K}{N^{(k+1)/2}}\,. \tag{2.105}$$

Compared with Proposition 2.3.7, one gets the extra information that

$$\nu\Big(\prod_{\ell<\ell'} R_{\ell,\ell'}^{k(\ell,\ell')}\Big) = \prod_{\ell<\ell'} \nu(R_{\ell,\ell'}^{k(\ell,\ell')}) + O(k+1)\,, \tag{2.106}$$

a remarkable fact in itself.

It is instructive to interpret Theorem 2.3.9 as a central limit theorem. (One can also give formal statements, but they do not seem to bring anything more than (2.105), except severe headache.) Namely, for the typical disorder, a finite family of functions

$$\widehat{R}_{\ell,\ell'} = \sqrt{N(1-\beta^2)}\,R_{\ell,\ell'}$$

on $(\Sigma_N^n, G_N^{\otimes n})$ asymptotically looks like an independent family of standard centered Gaussian r.v. To see this, consider

$$U = \prod_{1\le\ell<\ell'\le n} \widehat{R}_{\ell,\ell'}^{k(\ell,\ell')}\,.$$

Then (2.105) implies that

$$\mathsf{E}(\langle U\rangle - \mathsf{E}\langle U\rangle)^2 = \mathsf{E}\langle U\rangle^2 - (\mathsf{E}\langle U\rangle)^2 \le \frac{K}{\sqrt{N}}\,.$$

This is obvious if one uses replicas to write $\langle U\rangle^2$ as a single bracket and then (2.106). Thus

$$\mathsf{E}|\langle U\rangle - \nu(U)| \le (\mathsf{E}(\langle U\rangle - \mathsf{E}\langle U\rangle)^2)^{1/2} \le \frac{K}{N^{1/4}}\,,$$

and, combining with (2.105),

$$\mathsf{E}\left|\langle U\rangle - \prod_{\ell<\ell'} a(k(\ell,\ell'))\right| \le \frac{K}{N^{1/4}} \,.$$

Thus, for the typical disorder, $\langle U\rangle$ is almost $\prod_{\ell<\ell'} a(k(\ell,\ell'))$.

Proof of Theorem 2.3.9. Without loss of generality, we assume $k(1,2) \ge 1$. For each integer $1 \le v \le k$ we consider integers $\ell(v)$ and $\ell'(v)$ such that

$$\prod_{\ell<\ell'} R_{\ell,\ell'}^{k(\ell,\ell')} = \prod_{v\le k} R_{\ell(v),\ell'(v)}$$

and, without loss of generality, we assume that

$$(\ell(v),\ell'(v)) = (1,2) \Leftrightarrow v \le k(1,2) \,.$$

We write

$$R(v) = R_{\ell(v),\ell'(v)} \,, \quad R^-(v) = R^-_{\ell(v),\ell'(v)} \,, \quad \varepsilon(v) = \varepsilon_{\ell(v)}\,\varepsilon_{\ell'(v)}$$

so that $R(v) = R^-(v) + \varepsilon(v)/N$. We use symmetry between sites to write

$$\nu\left(\prod_{\ell<\ell'} R_{\ell,\ell'}^{k(\ell,\ell')}\right) = \nu\left(\prod_{v\le k} R(v)\right) = \nu\left(\varepsilon(1)\prod_{2\le v\le k} R(v)\right) \,. \tag{2.107}$$

We expand the product

$$\prod_{2\le v\le k} R(v) = \prod_{2\le v\le k}\left(R^-(v) + \frac{\varepsilon(v)}{N}\right) \,. \tag{2.108}$$

In each of the $k-1$ factors, we can choose either the term $R^-(v)$ (henceforth called the big term) or the term $\varepsilon(v)/N$ (henceforth called the small term). These $k-1$ choices result in 2^{k-1} terms.

When we choose the small term in at least two factors the resulting contribution is $O(k+1)$. To see this we repeat the argument of Proposition 2.3.7 and we use Hölder's inequality. If we choose the small term in exactly ℓ factors, the resulting contribution is

$$O(2\ell)\,O(k-1-\ell) = O(k+\ell-1) = O(k+1)$$

for $\ell \ge 2$.

Thus, we need to consider only the contributions where we have chosen the small term in at most one factor, and this gives

$$\begin{aligned}\nu\left(\varepsilon(1)\prod_{2\le v\le k} R(v)\right) &= \nu\left(\varepsilon(1)\prod_{2\le v\le k} R^-(v)\right)\\ &+ \frac{1}{N}\sum_{2\le v\le k} \nu\left(\varepsilon(1)\,\varepsilon(v)\prod_{u} R^-(u)\right) + O(k+1)\end{aligned} \tag{2.109}$$

where the product is for $2 \le u \le k$, $u \ne v$. Since there are $k-2$ terms in the product $\prod_u R^-(u)$, (2.93) shows that

$$\nu\left(\varepsilon(1)\,\varepsilon(v)\prod_u R^-(u)\right) = \nu_0\left(\varepsilon(1)\,\varepsilon(v)\prod_u R^-(u)\right) + O(k-1)\,. \quad (2.110)$$

Since $\mathrm{Av}(\varepsilon_1\,\varepsilon_2\,\varepsilon_{\ell(v)}\,\varepsilon_{\ell'(v)})$ is zero unless $(\ell(v),\ell'(v)) = (1,2)$, we see from (2.78) that the term (2.110) is $O(k-1)$ unless $v \le k(1,2)$. We also note that in that case, if $k(1,2) \ge 2$, then

$$\nu\left(\varepsilon(1)\,\varepsilon(v)\prod_u R^-(u)\right) = \nu\left(\prod_{3\le v\le k} R^-(u)\right) \quad (2.111)$$

and (2.109) becomes (using (2.93) again)

$$\begin{aligned}\nu\left(\varepsilon(1)\prod_{2\le v\le k} R^-(v)\right) = \nu\left(\varepsilon(1)\prod_{2\le v\le k} R^-(v)\right) &+ \frac{k(1,2)-1}{N}\,\nu\left(\prod_{3\le v\le k} R^-(v)\right)\\ &+ O(k+1)\,. \end{aligned} \quad (2.112)$$

This remains true even in the case $k(1,2) = 1$, even though (2.111) need not be true then.

To help the reader understand further generalizations of Theorem 2.3.9, we point out that the key point in (2.112) is the seemingly trivial fact that if $\ell_1 \le \ell_2$, then $\mathrm{Av}(\varepsilon_1\,\varepsilon_2\,\varepsilon_{\ell_1}\,\varepsilon_{\ell_2}) = 0$ unless $\ell_1 = 1$, $\ell_2 = 2$.

We then proceed as in the case of Proposition 2.3.7 to get, using similar expansions as in (2.108),

$$\begin{aligned}\nu\left(\varepsilon(1)\prod_{2\le v\le k} R^-(v)\right) &= \beta^2\nu\left(R(1)\prod_{2\le v\le k} R(v)\right) + O(k+1)\\ &= \beta^2\nu\left(\prod_{1\le v\le k} R(v)\right) + O(k+1)\end{aligned}$$

and

$$\frac{1}{N}\,\nu\left(\prod_{3\le v\le k} R^-(v)\right) = \frac{1}{N}\,\nu\left(\prod_{3\le v\le k} R(v)\right) + O(k+1)\,.$$

Thus, from (2.107) and (2.112) we get

$$(1-\beta^2)\,\nu\left(\prod_{\ell,\ell'} R_{\ell,\ell'}^{k(\ell,\ell')}\right) = \frac{k(1,2)-1}{N}\,\nu\left(\prod_{3\le v\le k} R(v)\right) + O(k+1)\ . \quad (2.113)$$

We can now prove (2.105) by induction over k, observing that, when $k(1,2) \ge 2$, then

$$\prod_{3\le v\le k} R(v) = \prod_{\ell,\ell'} R_{\ell,\ell'}^{m(\ell,\ell')}$$

where $m(\ell,\ell') = k(\ell,\ell')$ if $(\ell,\ell') \neq (1,2)$ and $m(1,2) = k(1,2)-2$. And when $k(1,2) = 1$ (2.113) finishes the induction. □

Research problem 2.3.10. (Level 1) Given an integer n, prove a central limit theorem for the r.v. $(\langle\sigma_i\,\sigma_j\rangle)_{1\le i<j\le n}$.

Research problem 2.3.11. (Level 1) Generalize Theorem 2.3.9 by introducing the "multiple overlaps" such as

$$R_{1,2,3,\ldots,s} = \frac{1}{N}\sum_{i\le N}\sigma_i^1\ldots\sigma_i^s\ .$$

We now turn to a discussion of Conjecture 2.2.21, and the related question of going beyond the accuracy of Theorem 2.3.9. We consider the matrix $\mathcal{M} = (\langle\sigma_i\,\sigma_j\rangle)_{i,j}$. A standard way to control the operator norm $\|\mathcal{M}\|$ of $\mathcal{M}$ is through the traces of the powers of $\mathcal{M}$. Since $\mathcal{M}$ is symmetric, its norm is the absolute value of its largest eigenvalue. The sum of the kth powers of the eigenvalues is the trace of $\mathcal{M}^k$, so that, for k even

$$\|\mathcal{M}\| \le (\text{trace } \mathcal{M}^k)^{1/k}\ . \quad (2.114)$$

The trace of $\mathcal{M}^k$ has a nice expression.

Lemma 2.3.12. *We have, for all positive integers k,*

$$\text{Trace } \mathcal{M}^k = N^k\langle R_{1,2}\,R_{2,3}\ldots R_{k,1}\rangle\ . \quad (2.115)$$

Proof. It is immediate by induction over k that the entries of $\mathcal{M}^k$ are

$$N^{k-1}\langle\sigma_i^1\,R_{1,2}\,R_{2,3}\ldots R_{k-1,k}\,\sigma_j^k\rangle\ .$$

□

Corollary 2.3.13. *For all even integers k, we have*

$$\mathsf{E}\,\|\mathcal{M}\| \le (\mathsf{E}\,N^k \langle R_{1,2}\,R_{2,3}\dots R_{k,1}\rangle)^{1/k}\,. \tag{2.116}$$

Proof. Combine (2.114) and (2.115). □

We know from Theorem 2.3.9 that, if $k \ge 3$ and $\beta < 1$,

$$|\mathsf{E}\,N^k \langle R_{1,2}\,R_{2,3}\dots R_{k,1}\rangle| \le K(k,\beta)\,N^{k/2}\,.$$

In the rest of this section, we will improve this inequality into

$$|\mathsf{E}\,N^k \langle R_{1,2}\,R_{2,3}\dots R_{k,1}\rangle| \le K(k,\beta)\,N\,, \tag{2.117}$$

which implies, together with (2.116) that

$$\mathsf{E}\,\|\mathcal{M}\| \le K(k,\beta)\,N^{1/k}\,.$$

In a further effort, we will then analyze the value of $K(k,\beta)$ provided by our arguments, and we will reach the conclusion that

$$\mathsf{E}\,\|\mathcal{M}\| \le K(\beta)(\log N)^L\,. \tag{2.118}$$

This material is probably too sophisticated for the first time reader.

Let us consider a fixed family $\mathcal{I}$ of pairs (ℓ,ℓ'). Each pair might be repeated several times. We consider

$$T_{\mathcal{I}} = \prod_{(\ell,\ell')\in\mathcal{I}} NR_{\ell,\ell'}\,,$$

and we aim at computing the order of magnitude of $\nu(T_{\mathcal{I}})$ as $N \to \infty$. A first observation is that $\nu(T_{\mathcal{I}}) = 0$ unless the following holds

$$\begin{aligned}&\text{Each integer } \ell_0 \text{ appears an even number of times}\\ &\text{as an element of a pair } (\ell,\ell') \in \mathcal{I}\,.\end{aligned} \tag{2.119}$$

This is because if ℓ_0 appears an odd number of times, then the transformation $\boldsymbol{\sigma}^{\ell_0} \to -\boldsymbol{\sigma}^{\ell_0}$ changes the sign of $T_{\mathcal{I}}$, while leaving $\langle\cdot\rangle$ invariant, so that $\nu(T_{\mathcal{I}}) = 0$.

Thus we assume that (2.119) is satisfied. Let us say that $\mathcal{I}$ is a *cycle* if it consists of the pairs $(\ell_1,\ell_2),(\ell_2,\ell_3),\dots,(\ell_p,\ell_1)$, where the elements $\ell_1,\dots,\ell_p$ are all different. A cycle satisfies condition (2.119), and so does a union of cycles. Conversely if $\mathcal{I}$ satisfies condition (2.119), it is a union of cycles. To see this, it suffices to show that if $\mathcal{I}$ satisfies (2.119), it contains a cycle $\mathcal{I}_1$, because $\mathcal{I}\backslash\mathcal{I}_1$ will still satisfy (2.119). To construct a cycle in $\mathcal{I}$, we start with a pair (ℓ_1,ℓ_2) in $\mathcal{I}$ and we recursively choose integers ℓ_s such that $(\ell_{s-1},\ell_s) \in \mathcal{I}$ and $\ell_s \notin (\ell_1,\ell_2,\dots,\ell_{s-1})$, as long as we can. When the construction stops, we pick $r \in \{1,\dots,s-1\}$ such that $(\ell_s,\ell_r) \in \mathcal{I}$. The existence of r follows from the fact that ℓ_s must occur twice as an element of a pair of $\mathcal{I}$. Then $(\ell_r, \ell_{r+1}), \ldots, (\ell_{s-1}, \ell_s), (\ell_s, \ell_r)$ is the required cycle in $\mathcal{I}$.

In general, the cycle decomposition of a family $\mathcal{I}$ satisfying (2.119) is not unique. For example,

$$\mathcal{I} = \{(1,2),(1,2),(2,3),(2,3),(3,1),(3,1)\}$$

is the union of two cycles of length 3, and is also the union of three cycles of length 2. We denote by $C(\mathcal{I})$ the *maximum* number of cycles in which $\mathcal{I}$ can be decomposed.

Theorem 2.3.14. *We have*

$$|\nu(T_{\mathcal{I}})| \leq K(\beta, \mathcal{I})\, N^{C(\mathcal{I})} \,. \tag{2.120}$$

Research problem 2.3.15. (Level 2) Prove that the limit

$$\lim_{N \to \infty} N^{-C(\mathcal{I})}\, \nu(T_{\mathcal{I}})$$

exists, *and* find a manageable expression for it.

Proving existence of the limit should be a mere exercise; there is no guarantee however that the limit is a simple function of the "geometry" of $\mathcal{I}$.

In order to establish Theorem 2.3.14, we will prove a more general statement.

Proposition 2.3.16. *Given an integer s, given $\beta_0 < 1$, given a family $\mathcal{I}$, we have*

$$2\,C(\mathcal{I}) \leq \operatorname{card}\mathcal{I} - s \Rightarrow \forall \beta \leq \beta_0\,, |\nu(T_{\mathcal{I}})| \leq K(\beta_0, \mathcal{I}, s)\, N^{(\operatorname{card}\mathcal{I} - s)/2} \,. \tag{2.121}$$

The case $s < 0$ is allowed. This case does not make much sense since $2C(\mathcal{I}) \leq \operatorname{card}\mathcal{I}$, but will be useful in the proof.

The choice $s = \operatorname{card}\mathcal{I} - 2\,C(\mathcal{I})$ shows that this is more general than Theorem 2.3.14. The point of the formulation is to allow induction on s. The case $s \leq 0$ follows from (2.88). The charm of the proof is that it uses a touch of combinatorics, and that combinatorics seem needed to attack Conjecture 2.2.21.

Proof. We assume that (2.121) holds for each value $s' < s$, and we prove by induction over $\operatorname{card}\mathcal{I}$ that it holds for s. Thus, in the course of the proof, we can use (2.121) for a family $\mathcal{I}_0$, and a number s_0 if *either* $s_0 < s$ (in which case $\operatorname{card}\mathcal{I}_0$ can be whatever we want) *or* $s_0 = s$ and $\operatorname{card}\mathcal{I}_0 < \operatorname{card}\mathcal{I}$. We can also use (2.121) for s_0 and $\mathcal{I}_0$ as before, but for $N-1$ rather than N.

We fix $\beta_0 < 1$ and we denote throughout the proof a number K depending only on $s, \mathcal{I}, \beta_0$ and that may vary between occurrences. We assume that $\beta \leq \beta_0$.

Given a family $\mathcal{I}_1$ of pairs (ℓ,ℓ'), we set

$$T^-_{\mathcal{I}_1} = \prod_{(\ell,\ell')\in\mathcal{I}_1} (NR^-_{\ell,\ell'}) \ .$$

Keeping in mind that

$$NR_{\ell,\ell'} = \sum_{i\le N} \sigma_i^\ell \sigma_i^{\ell'} \ ; \ NR^-_{\ell,\ell'} = \sum_{i\le N-1} \sigma_i^\ell \sigma_i^{\ell'} \ ,$$

we see that $T^-_{\mathcal{I}_1}$ is obtained from $T_{\mathcal{I}_1}$ by replacing N by $N-1$. The principle of the proof is to establish the inequality

$$|\nu(T_{\mathcal{I}}) - \beta^2 \nu_0(T^-_{\mathcal{I}})| \le KN^{(\mathrm{card}\mathcal{I}-s)/2} \ . \tag{2.122}$$

The quantity $\nu_0(T^-_{\mathcal{I}})$ is of the same nature as $\nu(T_{\mathcal{I}})$, except that one has replaced N by $N-1$ and β by β^-, where β^- is given by (2.65). We observe that $\beta^- < \beta_0$. If we were following the spirit of the proof of Theorem 2.3.9, we would complement (2.122) by the inequality

$$|\nu_0(T^-_{\mathcal{I}}) - \nu(T_{\mathcal{I}})| \le KN^{(\mathrm{card}\mathcal{I}-s)/2} \ . \tag{2.123}$$

This inequality, combined with (2.122) would give

$$(1-\beta^2)\,|\nu(T_{\mathcal{I}})| \le KN^{(\mathrm{card}\mathcal{I}-s)/2}$$

concluding the proof. The inequality (2.123) is probably true, but since the proof would require work, we will use only (2.122). We consider the quantity

$$A(N_0,\beta_0,\mathcal{I},s) = \sup\{N^{-(\mathrm{card}\mathcal{I}-s)/2}|\nu(T_{\mathcal{I}})| \ ; \ \beta \le \beta_0 \ , \ N \le N_0\} \ ,$$

and we observe (since $\beta^- \le \beta$) that (2.122) implies

$$A(N_0,\beta_0,\mathcal{I},s) \le \beta_0^2 \, A(N_0,\beta_0,\mathcal{I},s) + K \ ,$$

so that $A(N_0,\beta_0,\mathcal{I},s) \le K$, which is the required inequality. So all we have to do is to prove (2.122).

This follows the spirit of the proof of Theorem 2.3.9, except that we have to use higher order expansions.

Without loss of generality we assume $(1,2) \in \mathcal{I}$, and we set $\mathcal{I}' = \mathcal{I}\backslash\{(1,2)\}$. (If the pair $(1,2)$ occurs several times in $\mathcal{I}$, we remove it just once.) As always, we use symmetry between sites to write

$$\nu(T_{\mathcal{I}}) = \nu(f) \ , \tag{2.124}$$

where

$$f = N\,\varepsilon_1\,\varepsilon_2\,T_{\mathcal{I}'} \ . \tag{2.125}$$

We write

$$\left|\nu(f) - \sum_{k=0}^{s} \frac{1}{k!}\, \nu_0^{(k)}(f)\right| \le \sup_{0\le t\le 1} |\nu_t^{(s+1)}(f)| \,. \tag{2.126}$$

Since f contains $\operatorname{card}\mathcal{I}$ factors N and $\operatorname{card}\mathcal{I}-1$ factors $R_{\ell,\ell'}$, it follows from (2.88) that

$$\nu(f^2)^{1/2} \le K N^{(\operatorname{card}\mathcal{I}+1)/2} \,,$$

and from Proposition 2.3.3 we see that

$$|\nu_t^{(s+1)}(f)| \le K N^{(\operatorname{card}\mathcal{I}-s)/2} \,. \tag{2.127}$$

We now examine the term $\nu_0^{(k)}(f)$. It is a sum of terms

$$\pm\, \beta^{2k} \nu_0(\varepsilon_1\, \varepsilon_2\, \varepsilon_{\mathcal{I}_1}\, T_{\mathcal{I}'}\, T^-_{\mathcal{I}_1}\, N^{1-k}) \,, \tag{2.128}$$

where $\mathcal{I}_1$ is a family of k pairs (ℓ,ℓ'), and where

$$\varepsilon_{\mathcal{I}_1} = \prod_{(\ell,\ell')\in\mathcal{I}_1} \varepsilon_\ell\, \varepsilon_{\ell'} \,.$$

(It might help to observe that $N^{-\operatorname{card}\mathcal{I}_1}\, T^-_{\mathcal{I}_1} = \prod_{(\ell,\ell')\in\mathcal{I}_1} R^-_{\ell,\ell'}$.)

We will compute (2.128) using (2.78), but we must first bring out the dependence of $T_{\mathcal{I}'}$ in the variables ε_ℓ. For this we write each factor $NR_{\ell,\ell'}$ of $T_{\mathcal{I}'}$ as $\varepsilon_\ell\, \varepsilon_{\ell'} + NR^-_{\ell,\ell'}$, and we expand the resulting product. We get

$$T_{\mathcal{I}'} = \sum_{\mathcal{I}_2} \varepsilon_{\mathcal{I}'\setminus\mathcal{I}_2}\, T^-_{\mathcal{I}_2} \,,$$

where the sum is over all subsets $\mathcal{I}_2$ of $\mathcal{I}'$. Thus, we are now concerned with terms

$$\begin{aligned} &\pm\, \beta^{2k}\, N^{1-k}\, \nu_0(\varepsilon_1\, \varepsilon_2\, \varepsilon_{\mathcal{I}_1}\, \varepsilon_{\mathcal{I}'\setminus\mathcal{I}_2}\, T^-_{\mathcal{I}_2}\, T^-_{\mathcal{I}_1}) \\ &= \pm\, \beta^{2k}\, N^{1-k}\, \nu_0(\varepsilon_{\mathcal{I}_1}\, \varepsilon_{\mathcal{I}\setminus\mathcal{I}_2}\, T^-_{\mathcal{I}_1\cup\mathcal{I}_2}) \,, \end{aligned} \tag{2.129}$$

because $\varepsilon_1\, \varepsilon_2\, \varepsilon_{\mathcal{I}'\setminus\mathcal{I}_2} = \varepsilon_{\mathcal{I}\setminus\mathcal{I}_2}$ and $T^-_{\mathcal{I}_1}\, T^-_{\mathcal{I}_2} = T^-_{\mathcal{I}_1\cup\mathcal{I}_2}$. Now (2.78) shows that the quantity (2.129) is zero unless $\varepsilon_{\mathcal{I}_1}\, \varepsilon_{\mathcal{I}\setminus\mathcal{I}_2}$ is identically 1. Since $(1,2)\in\mathcal{I}\setminus\mathcal{I}_2$, we must have

$$k + \operatorname{card}(\mathcal{I}\setminus\mathcal{I}_2) \ge 2 \,. \tag{2.130}$$

We first examine the case where there is equality. A first possibility is that $\operatorname{card}(\mathcal{I}\setminus\mathcal{I}_2) = 1 = k$. Then, for $\varepsilon_{\mathcal{I}_1}\, \varepsilon_{\mathcal{I}\setminus\mathcal{I}_2}$ to be identically 1, we must have $\mathcal{I}_1 = \mathcal{I}\setminus\mathcal{I}_2 = \{(1,2)\}$ so that $\mathcal{I}_1\cup\mathcal{I}_2 = \mathcal{I}$, and the term is

$$\beta^2\, \nu_0(T^-_{\mathcal{I}}) \,.$$

A second possibility is that $k = 0$ and $\operatorname{card}(\mathcal{I}\setminus\mathcal{I}_2) = 2$. For $\varepsilon_{\mathcal{I}\setminus\mathcal{I}_2}$ to be identically 1, we must have $\mathcal{I}_2 = \mathcal{I}\setminus\{(1,2),(1,2)\}$. Since $\mathcal{I}$ is the union of

$\mathcal{I}_2$ and a cycle, we have $C(\mathcal{I}) \geq C(\mathcal{I}_2) + 1$, i.e. $C(\mathcal{I}_2) \leq C(\mathcal{I}) - 1$. Also $\operatorname{card}\mathcal{I}_2 = \operatorname{card}\mathcal{I} - 2$, so that the condition $2C(\mathcal{I}) \leq \operatorname{card}\mathcal{I} - s$ implies

$$2C(\mathcal{I}_2) \leq \operatorname{card}\mathcal{I}_2 - s .$$

Since $\operatorname{card}\mathcal{I}_2 < \operatorname{card}\mathcal{I}$, we can use (2.121) for the same value of s, for β^- rather than β and $N-1$ rather than N to see that the absolute value of the corresponding term in (2.129) is bounded by

$$|N\nu_0(T^-_{\mathcal{I}_2})| \leq KN^{1+(\operatorname{card}\mathcal{I}_2 - s)/2} = KN^{(\operatorname{card}\mathcal{I} - s)/2} .$$

Now we examine the case where

$$k + \operatorname{card}(\mathcal{I}\backslash\mathcal{I}_2) \geq 3 , \tag{2.131}$$

and we want to study

$$N^{1-k}\,\nu_0(T^-_{\mathcal{I}_1\cup\mathcal{I}_2}) . \tag{2.132}$$

We only have to consider the case where the family $\mathcal{I}_1 \cup \mathcal{I}_2$ satisfies (2.119), for otherwise, as already pointed out, the quantity (2.132) is zero. In that case, $\mathcal{I}_1 \cup \mathcal{I}_2$ is a union of cycles. We claim that

$$C(\mathcal{I}_1 \cup \mathcal{I}_2) \leq C(\mathcal{I}) + k - 1 . \tag{2.133}$$

To see this, we consider a decomposition of $\mathcal{I}_1 \cup \mathcal{I}_2$ into cycles. Let n_1 (resp. n_2) be the number of cycles that contain at least one pair of $\mathcal{I}_1$ (resp. that consist entirely of pairs of $\mathcal{I}_2$). Then $n_1 \leq \operatorname{card}\mathcal{I}_1 = k$. Also, $C(\mathcal{I}) \geq n_2 + 1$. This is because the set obtained by removing from $\mathcal{I}$ all the elements of $\mathcal{I}_2$ making up the n_2 cycles consisting only of elements of $\mathcal{I}_2$ is not empty, since it contains $(1,2)$. Thus

$$n_1 + n_2 \leq C(\mathcal{I}) + k - 1 ,$$

and this proves (2.133). It follows from (2.133) that

$$\begin{aligned} 2C(\mathcal{I}_1 \cup \mathcal{I}_2) &\leq 2C(\mathcal{I}) + 2k - 2 \\ &\leq \operatorname{card}\mathcal{I} + 2k - s - 2 \\ &= \operatorname{card}(\mathcal{I}_1 \cup \mathcal{I}_2) - s' , \end{aligned} \tag{2.134}$$

where

$$\begin{aligned} s' &= s + 2 + \operatorname{card}(\mathcal{I}_1 \cup \mathcal{I}_2) - \operatorname{card}\mathcal{I} - 2k \\ &= s + 2 - \operatorname{card}(\mathcal{I}\backslash\mathcal{I}_2) - k . \end{aligned} \tag{2.135}$$

Thus, under (2.131) we have $s' \leq s - 1$, and by (2.134) we can use (2.121) to obtain

$$N^{1-k}|\nu_0(T^-_{\mathcal{I}_1\cup\mathcal{I}_2})| \leq KN^{1-k+(\operatorname{card}(\mathcal{I}_1\cup\mathcal{I}_2)-s')/2} .$$

Now, (2.135) shows that

$$1 - k + \frac{1}{2}(\mathrm{card}(\mathcal{I}_1 \cup \mathcal{I}_2) - s') = \frac{1}{2}(\mathrm{card}\,\mathcal{I} - s) , \tag{2.136}$$

and this finishes the proof of (2.122) and of Proposition 2.3.16. □

For the truly energetic reader, we now explain how to carry out an explicit value for the constant $K(\beta_0, \mathcal{I}, s)$ in (2.121).

Proposition 2.3.17. *Given $\beta_0 < 1$, an integer $s \geq 0$, a family $\mathcal{I}$, with $m = \mathrm{card}\,\mathcal{I}$, we have*

$$\begin{aligned} &2C(\mathcal{I}) \leq m - s \Rightarrow \forall\, \beta \leq \beta_0 ,\\ &|\nu(T_{\mathcal{I}})| \leq K(\beta_0)^{m+7s}(m+s)^{m+7s} N^{(m-s)/2} . \end{aligned} \tag{2.137}$$

Strange quantities such as $m + 7s$ are no magic. They merely correspond to a convenient choice.

In the particular case where $C(\mathcal{I}) = 1$, using this for $s = m - 2$ improves (2.117) into

$$|\mathsf{E}\, N^k \langle R_{1,2}\, R_{2,3}, \ldots, R_{k,1} \rangle| \leq (K(\beta)k)^{8k} N$$

and by (2.116) this gives

$$\mathsf{E}\, \|\mathcal{M}\| \leq (K(\beta)k)^8 N^{1/k} .$$

Taking k about $\log N$ leads to (2.118).

The proof of (2.137) consists simply in repeating the proof of (2.121), while attempting to estimate more precisely the various contributions.

First, the inequality $\exp Nx^2 \geq N^k x^{2k}/k!$ and Theorem 2.2.12 show that the inequality (2.88) can be improved into

$$\nu(R_{1,2}^{2k}) \leq K(\beta_0)^k \frac{k!}{N^k} \leq K(\beta_0)^k \frac{k^k}{N^k} , \tag{2.138}$$

and (since $(a+b)^m \leq 2^{m-1}(a^m + b^m)$ for $a, b \geq 0$) that (2.89) can be improved into

$$\nu((R_{1,2}^-)^{2k}) \leq K(\beta_0)^k \frac{k^k}{N^k} . \tag{2.139}$$

In particular we have the inequality

$$|\nu(T_{\mathcal{I}})| \leq N^m \prod_{(\ell,\ell') \in \mathcal{I}} (\nu(R_{\ell,\ell'}^{2m}))^{1/2m} \leq K(\beta_0)^m\, m^{m/2}\, N^{m/2} . \tag{2.140}$$

Let us observe that (2.140) also holds for ν_t, as is seen by applying (2.91) with $n = 2$ to each factor $\nu(R_{\ell,\ell'}^{2m})$. It follows from (2.140) that Proposition 2.3.17 holds for $s = 0$, and we can assume $s \geq 1$. Next, when f is a function on n replicas, we can describe $\nu_t'(f)$ as the sum of $2n^2$ terms

$$\pm \beta^2 \nu_t(f \varepsilon_\ell \varepsilon_{\ell'} R^-_{\ell,\ell'}) , \tag{2.141}$$

if we count the terms in (2.81) "with their order of multiplicity"; e.g. $-\beta^2 \nu_t(f \varepsilon_\ell \varepsilon_{n+1} R^-_{\ell,n+1})$ occurs n times if $\ell \le n$. The terms (2.141) involve the function $f \varepsilon_\ell \varepsilon_{\ell'} R^-_{\ell,\ell'}$, that might depend on $n+2$ replicas. By iteration, $\nu_t^{(k)}(f)$ is the sum of at most

$$2^k (n(n+2)\cdots(n+2k-2))^2 \le 2^k (n+2k)^{2k} \tag{2.142}$$

terms

$$\pm \beta^{2k} \nu_t(f \varepsilon_{\mathcal{I}_1} T^-_{\mathcal{I}_1} N^{-k}) ,$$

where we use the notation of (2.128) for $\mathcal{I}_1$. We apply this to $f = N \varepsilon_1 \varepsilon_2 T_{\mathcal{I}'}$, $k = s+1$. Obviously we have $n \le 2m$, so the total number of terms is at most

$$2^{s+1}(2(m+s+1))^{2(s+1)} \le (8(m+s))^{2(s+1)} .$$

Every term contains $k - 1 = s$ factors N^{-1}, and since $\operatorname{card} \mathcal{I}_1 = m-1$, it contains $m-1+s+1 = m+s$ factors $N\, R^-_{\ell,\ell'}$, so that, using (2.140) for ν_t, it is at most

$$K(\beta_0)^{m+s}(m+s)^{(m+s)/2} N^{(m-s)/2}.$$

We have improved (2.127) into

$$|\nu_t^{(s+1)}(f)| \le K_0(\beta_0)^{m+s}(m+s)^{m+7s} N^{(m-s)/2} \tag{2.143}$$

since, s being at least 1, the following holds true

$$\frac{m+s}{2} + 2(s+1) \le m + 7s .$$

Now we examine the contribution of the terms (2.129). We denote by $K_1(\beta_0)$ the constant of (2.137). When there is equality in (2.130), we have seen that, besides the term $\beta^2 \nu_0(T^-_{\mathcal{I}})$, the only possible contribution is $N \nu_0^-(T^-_{\mathcal{I}_2})$ where $\mathcal{I}_2 = \mathcal{I} \backslash \{(1,2),(1,2)\}$, and that $2C(\mathcal{I}_2) \le \operatorname{card} \mathcal{I}_2 - s$. Using the induction hypothesis (2.137) with $m-2$ rather than m and the same value of s we get

$$N|\nu_0(T^-_{\mathcal{I}_2})| \le K(\beta_0)^{m+7s-2}(m+s)^{m+7s} N^{(m-s)/2} . \tag{2.144}$$

Now we examine the contribution of the terms (2.129) (or, equivalently, $\pm \beta^{2k} N^{1-k} \nu_0(T^-_{\mathcal{I}_1 \cup \mathcal{I}_2})$) when

$$r = k + \operatorname{card}(\mathcal{I} \backslash \mathcal{I}_2) \ge 3 . \tag{2.145}$$

We use the induction hypothesis (2.137) with

$$m' = \operatorname{card}(\mathcal{I}_1 \cup \mathcal{I}_2) \le k + m - 1 \tag{2.146}$$

and (see (2.135))

$$s' = \max(0, s_1)\ ; \quad s_1 = s + 2 - r\ .$$

Thus $s' < s$, and, if $s_1 \geq 0$,

$$m' + s' = m + s + 1 + k - r \leq m + s$$

since $r \geq k + 1$, while, if $s_1 \leq 0$

$$m' + s' = m' \leq k + m - 1 \leq m + s\ .$$

Since $r \geq 3$, we have $6r \geq 2k + 14$, so that, if $s_1 > 0$,

$$m' + 7s' \leq m + k - 1 + 7s - 7r + 14 \leq m + 7s - (k + r + 1)\ .$$

If $s_1 \leq 0$, we observe that

$$m' + r \leq \operatorname{card}(\mathcal{I}_1) + \operatorname{card}(\mathcal{I}_2) + k + \operatorname{card}(\mathcal{I}\backslash\mathcal{I}_2) \leq 2k + m$$

and thus

$$m' + 7s' = m' \leq 2k + m - r \leq m + 7s - (k + r + 1)$$

since $s \geq k$ and $s \geq 1$. Proceeding as in (2.136), it follows from (2.137) that

$$N^{1-k}|\nu_0(T^-_{\mathcal{I}_1\cup\mathcal{I}_2})| \leq K_1(\beta_0)^{m+7s-(k+r+1)}(m+s)^{m+7s-(k+r+1)}N^{(m-s)/2}\ .$$

Now we have to count how many such terms occur. Counting terms with their multiplicity, by (2.142), and since $n \leq 2m$, there are at most (since $k \leq r$)

$$2^k(2(m+k))^{2k} \leq 8^r(m+k)^{2k}$$

choices for $\mathcal{I}_1$.

Since $\mathcal{I}_2$ is determined by $\mathcal{I}'\backslash\mathcal{I}_2$, and since $\operatorname{card}\mathcal{I}'\backslash\mathcal{I}_2 = r - k - 1$, there are at most m^{r-k-1} choices for $\mathcal{I}_2$.

Thus there are at most $8^r(m+k)^{r+k-1}$ choices for the pair $(\mathcal{I}_1, \mathcal{I}_2)$. Since we can assume $K_1(\beta_0) \geq 8$, the total contribution of the terms (2.132) for given values of k, r is at most

$$K_1(\beta_0)^{m+7s-1}(m+s)^{m+7s-2}N^{(m-s)/2}$$

and since $k \leq s$, $r \leq k + m \leq s + m$, the total contribution over all values of k, $r \leq s + m$ is at most

$$K_1(\beta_0)^{m+7s-1}(m+s)^{m+7s}N^{(m-s)/2}\ .$$

Collecting these estimates, we have found that

$$\nu(T_{\mathcal{I}}) \leq \beta^2\nu_0(T^-_{\mathcal{I}}) + (K_0(\beta_0)^{m+7s-1} + 2K_1(\beta_0)^{m+7s-1})(m+s)^{m+7s}N^{(m-s)/2}\ ,$$

and if we choose $K_1(\beta_0) = 3K_0(\beta_0)/(1-\beta_0^2)$, we can then complete the proof as in Proposition 2.3.16. □

2.4 The Replica-Symmetric Solution and the Smart Path Method

In this section, we go back to the Hamiltonian (2.9),

$$-H_N(\boldsymbol{\sigma}) = \frac{\beta}{\sqrt{N}} \sum_{i<j} g_{ij}\, \sigma_i \sigma_j + h \sum_{i \le N} \sigma_i \,.$$

Quite naturally, we set

$$Z_N = Z_N(\beta, h) = \sum_{\boldsymbol{\sigma}} \exp(-H_N(\boldsymbol{\sigma}))$$

$$p_N(\beta, h) = \frac{1}{N} \mathsf{E} \log Z_N(\beta, h) \,.$$

To make the point that the case $h \neq 0$ brings in new challenges, we state the following, where z is a standard Gaussian r.v.

Theorem 2.4.1. *There exists β_0 such that for all $\beta \le \beta_0$ and for all $h > 0$, we have*

$$\lim_{N \to \infty} p_N(\beta, h) = \frac{\beta^2}{4}(1-q)^2 + \log 2 + \mathsf{E} \log \mathrm{ch}(\beta z \sqrt{q} + h) \,, \tag{2.147}$$

where q is the unique solution of the equation

$$q = \mathsf{E}\, \mathrm{th}^2(\beta z \sqrt{q} + h) \,. \tag{2.148}$$

For $h > 0$, the equation (2.148) has a unique solution whatever the value the value of $\beta \ge 0$. The situation is different for $h = 0$. If $\beta \le 1$, the equation (2.148) has 0 as unique solution, while if $\beta > 1$ it has two solutions, one of them being zero. The case $h = 0$ is not interesting because probably (2.147) holds only for $\beta \le 1$. The case $h = 0$ and $\beta < 1$ has already been studied in detail, so we always assume $h > 0$, and we denote by q the unique solution of (2.148).

The formula (2.147) is called by physicists the "replica-symmetric solution" (for the SK model) and Section 2.15 contains a sketch of how they derive it. We will not follow the physicist's usage of calling (2.147) a "solution" even when it is known to be wrong. We will call (2.147) *the replica symmetric formula* (for the SK model). The remarkable character of this formula should be self apparent. This formula is not magic, but reflects an underlying structure, much of which can be expressed in the following statement.

Theorem 2.4.2. *If h, β_0, q and $\beta \le \beta_0$ are as in Theorem 2.4.1, then we have*

$$\mathsf{E}\langle (R_{1,2} - q)^2 \rangle \le \frac{L}{N} \,. \tag{2.149}$$

A similar structure will appear in the models of Chapters 3 and 5.

It is very simple to see that

$$\lim_{N\to\infty}\frac{1}{N}\log \mathsf{E}Z_N=\frac{\beta^2}{4}+\log 2+\log\operatorname{ch} h$$

and that, unless $h = 0$, this is *not* the value (2.147). Thus, if $h \neq 0$ and $\beta \leq \beta_0$,

$$\lim_{N\to\infty}\frac{1}{N}\mathsf{E}\log Z_N \neq \lim_{N\to\infty}\frac{1}{N}\log \mathsf{E}Z_N\ ,$$

an intrinsically more interesting situation than $h = 0$.

The methods of Section 2.2 do not work anymore, and we will use the cavity method. If we define

$$-H_{N-1}(\boldsymbol{\rho})=\frac{\beta}{\sqrt{N}}\sum_{i<j\leq N-1} g_{ij}\,\sigma_i\,\sigma_j+h\sum_{i\leq N-1}\sigma_i \tag{2.150}$$

and $g(\boldsymbol{\rho})$ as in (2.66), then, for a function f on Σ_N^n, (2.70) becomes

$$\langle f\rangle=\frac{\Big\langle \mathrm{Av} f\exp\sum_{\ell\leq n}\sigma_N^\ell(g(\boldsymbol{\rho}^\ell)+h)\Big\rangle}{Z^n}\ , \tag{2.151}$$

where

$$Z=\langle \operatorname{ch}(g(\boldsymbol{\rho})+h)\rangle_-\ . \tag{2.152}$$

As usual, we write ε_ℓ for σ_N^ℓ, so that (2.151) reads

$$\langle f\rangle=\frac{\Big\langle \mathrm{Av} f\exp\sum_{\ell\leq n}\varepsilon_\ell(g(\boldsymbol{\rho}^\ell)+h)\Big\rangle_-}{Z^n}\ . \tag{2.153}$$

In Section 2.3 we took the derivative of the left-hand side in the quantity β occurring in $g(\boldsymbol{\rho})$. This worked beautifully. The reason was that this derivation created factors $R_{\ell,\ell'}$, which were small. This will no longer be the case. As (2.149) indicates, it would be much more desirable to get factors $R_{\ell,\ell'} - q$. The method we will use relies on the same (very old...) idea as in Section 2.3. We will find a path between the situation we want to study and a simpler situation and we will control derivatives along this path. The key to the effectiveness of such a method is the appropriate choice of the path. Usually this choice (in contrast with the situation of Section 2.3) is not the obvious one, and can be found only after the structure of the problem has been well understood. To insist upon the fact that the choice of the path is the real issue, we will call this method the *smart path method*.

Before explaining what to do with (2.153), we will give two illustrations of the method.

Proposition 2.4.3. (Slepian's lemma) *Consider a function F on $\mathbb{R}^M$, that has order 2 derivatives of moderate growth. Assume that*

$$\forall i \neq j \, , \; \frac{\partial^2 F}{\partial x_i \partial x_j} \geq 0 \, . \tag{2.154}$$

Consider two centered jointly Gaussian families

$$\mathbf{u} = (u_i)_{i \leq M} \, , \; \mathbf{v} = (v_i)_{i \leq M} \, ,$$

and assume that

$$\forall i \leq M \, , \; \mathsf{E} u_i^2 = \mathsf{E} v_i^2 \, ; \; \forall i \neq j \, , \; \mathsf{E} u_i u_j \geq \mathsf{E} v_i v_j \, . \tag{2.155}$$

Then

$$\mathsf{E} F(\mathbf{u}) \geq \mathsf{E} F(\mathbf{v}) \, .$$

Proof. We replace the family $\mathbf{v}$ by a family with the same distribution that is independent of the family $\mathbf{u}$. Thus, without loss of generality, we can assume that the families $\mathbf{u}$ and $\mathbf{v}$ are independent. The appropriate path between the families $\mathbf{u}$ and $\mathbf{v}$ is given by $\mathbf{u}(t) = (u_i(t))_{i \leq M}$, where

$$u_i(t) = \sqrt{t} \, u_i + \sqrt{1-t} \, v_i \, , \tag{2.156}$$

so that $\mathbf{u} = \mathbf{u}(1)$ and $\mathbf{v} = \mathbf{u}(0)$. We consider the function

$$\varphi(t) = \mathsf{E} F(\mathbf{u}(t))$$

so that, for $0 < t < 1$,

$$\varphi'(t) = \mathsf{E} \sum_{i \leq M} \frac{\mathrm{d}}{\mathrm{d}t} u_i(t) \frac{\partial F}{\partial x_i} (\mathbf{u}(t)) \, .$$

Now

$$\frac{\mathrm{d}}{\mathrm{d}t} u_i(t) = \frac{1}{2\sqrt{t}} u_i - \frac{1}{2\sqrt{1-t}} v_i$$

satisfies

$$\mathsf{E} \left(\left(\frac{\mathrm{d}}{\mathrm{d}t} u_i(t) \right) u_j(t) \right) = \frac{1}{2} (\mathsf{E} \, u_i u_j - \mathsf{E} \, v_i v_j)$$

and use of integration by parts as in (A.41) yields

$$\varphi'(t) = \frac{1}{2} \sum_{i,j} (\mathsf{E} \, u_i u_j - \mathsf{E} \, v_i v_j) \, \mathsf{E} \frac{\partial^2 F}{\partial x_i \partial x_j} (\mathbf{u}(t)) \geq 0 \, . \tag{2.157}$$

□

As a second application, we prove Theorem 2.2.4.

Proof of Theorem 2.2.4. It should be obvious (using regularization by convoluting with a smooth function) that we can assume that F is twice differentiable (a condition that is satisfied in all the cases where we use this result). We consider a parameter s, and, given s, we consider the function G on $\mathbb{R}^{2M}$ given by

$$G((y_i)_{i\le 2M}) = \exp s(F((y_i)_{i\le M}) - F((y_{i+M})_{i\le M})) \,.$$

We consider independent standard Gaussian r.v. $(u_i)_{i\le 2M}$. We consider a family $(v_i)_{i\le 2M}$ such that the r.v $(v_i)_{i\le M}$ are independent standard Gaussian, independent of the sequence $(u_i)_{i\le 2M}$, and such that $v_i = v_{i-M}$ if $i \ge M+1$. Thus $v_i = v_{i+M}$ if $i \le M$. We note that

$$\mathsf{E}\, u_i u_j - \mathsf{E}\, v_i v_j = 0$$

unless $j = i + M$ or $i = j + M$ in which case

$$\mathsf{E}\, u_i u_j - \mathsf{E}\, v_i v_j = -1 \,.$$

We consider $\mathbf{u}(t)$ as in (2.156), and $\varphi(t) = \mathsf{E}\, G(\mathbf{u}(t))$. Using the first part of (2.157) for G rather than F, we get

$$\varphi'(t) = -\mathsf{E} \sum_{i\le M} \frac{\partial^2 G}{\partial y_i\, \partial y_{i+M}}(u(\mathbf{t})) \,. \tag{2.158}$$

We compute

$$\frac{\partial^2 G}{\partial y_i\, \partial y_{i+M}}(\mathbf{y}) = -s^2 \frac{\partial F}{\partial x_i}((y_i)_{i\le M}) \frac{\partial F}{\partial x_i}((y_{i+M})_{i\le M})\, G(\mathbf{y}) \,. \tag{2.159}$$

Now, (2.21) implies

$$\forall \mathbf{x} \in \mathbb{R}^M \,, \; \sum_{i\le M} \left(\frac{\partial F}{\partial x_i}(\mathbf{x})\right)^2 \le A^2$$

and (2.158), (2.159) and the Cauchy-Schwarz inequality show that

$$\varphi'(t) \le s^2 A^2 \varphi(t) \,. \tag{2.160}$$

The choice of the family $(v_i)_{i\le 2M}$ ensures that $\varphi(0) = 1$, so that (2.160) implies that $\varphi(1) \le \exp s^2 A^2$, i.e.

$$\mathsf{E} \exp s(F(u_1, \ldots, u_M) - F(u_{M+1}, \ldots, u_{2M})) \le \exp s^2 A^2 \,.$$

We use Jensen's inequality while taking expectation in $u_{M+1}, \ldots, u_{2M}$, so that, going back to the notation $\mathbf{g}$ of Theorem 2.2.4 we have

$$\mathsf{E} \exp s(F(\mathbf{g}) - \mathsf{E} F(\mathbf{g})) \le \exp s^2 A^2 \,.$$

Using Markov's inequality we get that for $t > 0$

$$\mathsf{P}(F(\mathbf{g}) - \mathsf{E}F(\mathbf{g}) \geq t) \leq \exp(s^2 A^2 - st)$$

and

$$\mathsf{P}(F(\mathbf{g}) - \mathsf{E}F(\mathbf{g}) \geq t) \leq \exp\left(-\frac{t^2}{4A^2}\right)$$

by taking $s = t/2A^2$. Applying the same inequality to $-F$ completes the proof. □

We now go back to (2.153). We set

$$g_t(\boldsymbol{\rho}) = \sqrt{t}\, g(\boldsymbol{\rho}) + \beta\sqrt{1-t}\, z\sqrt{q} \tag{2.161}$$

where z is a standard Gaussian r.v., independent of all the previously considered r.v., and where $q \geq 0$ is a number to be determined later.

It should be apparent that the idea is that we try to replace $g(\boldsymbol{\rho})$ by $\beta\, z\sqrt{q}$; but why this is the right thing to do will become clear only gradually.

We consider the probability measure $\langle\cdot\rangle_t$ on Σ_N given by

$$\langle f\rangle_t = \frac{\langle \mathrm{Av} f \exp \varepsilon(g_t(\boldsymbol{\rho}) + h)\rangle_-}{Z_t}\,, \tag{2.162}$$

where

$$Z_t = \langle \mathrm{Av} \exp \varepsilon(g_t(\boldsymbol{\rho}) + h)\rangle_- = \langle \mathrm{ch}(g_t(\boldsymbol{\rho}) + h)\rangle_-\,. \tag{2.163}$$

Another way to express these formulas is by saying that $\langle\cdot\rangle_t$ is Gibbs' measure relative to the Hamiltonian

$$-H_{N,t}(\boldsymbol{\sigma}) = -H_{N-1}(\boldsymbol{\rho}) + \sigma_N(g_t(\boldsymbol{\rho}) + h)\,.$$

We denote also by $\langle\cdot\rangle_t$ the powers of this measure on Σ_N^n, so that, for a function f on Σ_N^n, we have

$$\langle f\rangle_t = \frac{\langle \mathrm{Av} f \mathcal{E}_{n,t}\rangle_-}{Z_t^n}\,, \tag{2.164}$$

where

$$\mathcal{E}_{n,t} = \exp\sum_{\ell\leq n} \varepsilon_\ell(g_t(\boldsymbol{\rho}^\ell) + h)\,. \tag{2.165}$$

As in Section 2.2, we write

$$\nu_t(f) = \mathsf{E}\,\langle f\rangle_t\,,\ \nu_t'(f) = \frac{\mathrm{d}}{\mathrm{d}t}\nu_t(f)\,.$$

Thus $\nu_1(f) = \nu(f) = \mathsf{E}\,\langle f\rangle$.

The reason why ν_0 is simpler than ν_1 is that it decouples the last coordinate from the other ones. We set

$$Y = \beta\, z\sqrt{q} + h\,. \tag{2.166}$$

Lemma 2.4.4. *For any function f^- on Σ_{N-1}^n and any subset I of $\{1,\ldots,n\}$ we have*

$$\nu_0\left(f^-\prod_{i\in I}\varepsilon_i\right) = \mathsf{E}(\operatorname{th}Y)^{\operatorname{card}I}\nu_0(f^-) = \nu_0\left(\prod_{i\in I}\varepsilon_i\right)\nu_0(f^-)\,. \tag{2.167}$$

Proof. Since $g_0(\boldsymbol{\rho}) + h = Y$, from (2.164) we have

$$\langle f\rangle_0 = \frac{\left\langle \operatorname{Av} f^-\prod_{i\in I}\varepsilon_i\exp\left(\sum_{\ell\le n}\varepsilon_\ell Y\right)\right\rangle_-}{Z_0^n}$$

and from (2.163) we have $Z_0 = \operatorname{ch}Y$. Now, since f^- does not depend on $\varepsilon_1,\ldots,\varepsilon_n$, we have

$$\operatorname{Av} f^-\prod_{i\in I}\varepsilon_i\exp\left(\sum_{\ell\le n}\varepsilon_\ell Y\right) = f^-(\operatorname{sh}Y)^{\operatorname{card}I}(\operatorname{ch}Y)^{n-\operatorname{card}I}\,,$$

so that

$$\left\langle f^-\prod_{i\in I}\varepsilon_i\right\rangle_0 = \langle f^-\rangle_-\,(\operatorname{th}Y)^{\operatorname{card}I}\,. \tag{2.168}$$

Since Y is independent of the randomness occurring in $\langle\cdot\rangle_-$, we get

$$\nu_0\left(f^-\prod_{i\in I}\varepsilon_i\right) = \mathsf{E}(\langle f^-\rangle_-)\,\mathsf{E}(\operatorname{th}Y)^{\operatorname{card}I}\,.$$

Now, using (2.168) when $I=\emptyset$, we get $\langle f^-\rangle_- = \langle f^-\rangle_0$. □

We now calculate ν_t'.

Proposition 2.4.5. *For $0<t<1$ and for all functions f on Σ_N^n we have*

$$\begin{aligned}\nu_t'(f) = \beta^2 &\sum_{1\le\ell<\ell'\le n}\nu_t(f\,\varepsilon_\ell\,\varepsilon_{\ell'}(R^-_{\ell,\ell'}-q))\\ &-\beta^2 n\sum_{\ell\le n}\nu_t(f\,\varepsilon_\ell\,\varepsilon_{n+1}(R^-_{\ell,n+1}-q))\\ &+\beta^2\frac{n(n+1)}{2}\nu_t(f\,\varepsilon_{n+1}\,\varepsilon_{n+2}(R^-_{n+1,n+2}-q))\end{aligned} \tag{2.169}$$

and

$$\begin{aligned}\nu_t'(f) = \beta^2 &\sum_{1\le\ell<\ell'\le n}\nu_t(f\,\varepsilon_\ell\,\varepsilon_{\ell'}(R_{\ell,\ell'}-q))\\ &-\beta^2 n\sum_{\ell\le n}\nu_t(f\,\varepsilon_\ell\,\varepsilon_{n+1}(R_{\ell,n+1}-q))\\ &+\beta^2\frac{n(n+1)}{2}\nu_t(f\,\varepsilon_{n+1}\,\varepsilon_{n+2}(R_{n+1,n+2}-q))\,.\end{aligned} \tag{2.170}$$

This should be compared with Proposition 2.3.1, which corresponds to the special case $q = h = 0$.

Proof. It is a variation of that of Proposition 2.3.1, which should be mastered first. We have

$$\frac{\mathrm{d}}{\mathrm{d}t}\mathcal{E}_{n,t} = \left(\frac{1}{2\sqrt{t}}\sum_{\ell\le n}\varepsilon_\ell\, g(\boldsymbol{\rho}^\ell) - \frac{\beta\, z\sqrt{q}}{2\sqrt{1-t}}\sum_{\ell\le n}\varepsilon_\ell\right)\mathcal{E}_{n,t}\ ,$$

so that rather than (2.84) we have

$$\nu_t'(f) = \mathrm{I} - \mathrm{II}\ , \tag{2.171}$$

where I is the right-hand side of (2.84) and where

$$\mathrm{II} = \frac{\beta\sqrt{q}}{2\sqrt{1-t}}\,\mathsf{E}\left(z\frac{\left\langle \mathrm{Av} f\sum_{\ell\le n}\varepsilon_\ell\,\mathcal{E}_{n,t}\right\rangle_-}{Z_t^n} - nz\frac{\langle \mathrm{Av} f\varepsilon_{n+1}\,\mathcal{E}_{n+1,t}\rangle_-}{Z_t^{n+1}}\right)\ .$$

The term I is integrated by parts as in Proposition 2.3.1. The integration by parts of II is similar, and yet much easier. Using that

$$\frac{\partial\mathcal{E}_{n,t}}{\partial z} = \beta\sqrt{q}\,\sqrt{1-t}\sum_{\ell\le n}\varepsilon_\ell\,\mathcal{E}_{n,t}\ ,$$

we find

$$\begin{aligned}\mathrm{II} = \frac{\beta^2 q}{2}\Bigg(&\nu_t\bigg(f\Big(\sum_{\ell\le n}\varepsilon_\ell\Big)^2\bigg) - n\,\nu_t\Big(f\,\varepsilon_{n+1}\sum_{\ell\le n}\varepsilon_\ell\Big)\\ &- n\,\nu_t\bigg(f\,\varepsilon_{n+1}\Big(\sum_{\ell\le n+1}\varepsilon_\ell\Big)\bigg) + n(n+1)\,\nu_t(f\,\varepsilon_{n+1}\,\varepsilon_{n+2})\Bigg)\\ = \beta^2 q\Bigg(&\sum_{\ell<\ell'}\nu_t(f\,\varepsilon_\ell\,\varepsilon_{\ell'}) - n\sum_{\ell\le n}\nu_t(f\,\varepsilon_\ell\,\varepsilon_{n+1}) + \frac{n(n+1)}{2}\,\nu_t(f\,\varepsilon_{n+1}\,\varepsilon_{n+2})\Bigg)\ .\end{aligned}$$

□

It seems worthwhile to mention that there is a general principle at work in Proposition 2.4.5, a general principle that will be used on and on. Suppose we have a Hamiltonian of the type

$$-H_N(\boldsymbol{\sigma}) = -H_{N-1}(\boldsymbol{\rho}) + \sigma_N(g(\boldsymbol{\rho}) + h)$$

where the r.v. $g(\boldsymbol{\rho})$ are Gaussian and independent of the disorder occurring in H_{N-1}. We are trying to replace $g(\boldsymbol{\rho})$ by a variable $Y(\boldsymbol{\rho})$, where the variables

$(Y(\boldsymbol{\rho}))$ are jointly Gaussian, and independent of the variables $H_{N-1}(\boldsymbol{\rho})$ and $g(\boldsymbol{\rho})$. We then consider the Hamiltonian

$$-H_{N-1}(\boldsymbol{\rho}) + \sigma_N(\sqrt{t}\, g(\boldsymbol{\rho}) + \sqrt{1-t}\; Y(\boldsymbol{\rho}) + h) \;,$$

and the corresponding Gibbs measure $\langle\cdot\rangle_t$ on Σ_N.

We write $\nu_t(f) = \mathsf{E}\,\langle f\rangle_t$. Let us define

$$\Delta_{\ell,\ell'} = \Delta(\boldsymbol{\rho}^\ell, \boldsymbol{\rho}^{\ell'}) = \mathsf{E}\, g(\boldsymbol{\rho}^\ell)\, g(\boldsymbol{\rho}^{\ell'}) - \mathsf{E}\, Y(\boldsymbol{\rho}^\ell)\; Y(\boldsymbol{\rho}^{\ell'}) \,.$$

Then, if f is a function on Σ_N^n, we have

$$\begin{aligned}\frac{\mathrm{d}}{\mathrm{d}t}\, \nu_t(f) = \frac{1}{2}\Bigg(\sum_{\ell\le n} \nu_t(\Delta_{\ell,\ell}\, f) - n\, \nu_t(\Delta_{n+1,n+1} f)\Bigg) + \sum_{1\le\ell<\ell'\le n} \nu_t(\Delta_{\ell,\ell'}\, \varepsilon_\ell\, \varepsilon_{\ell'}\, f) \\ - n \sum_{\ell\le n} \nu_t(\Delta_{\ell,n+1}\, \varepsilon_\ell\, \varepsilon_{n+1}\, f) \\ + \frac{n(n+1)}{2}\, \nu_t(\Delta_{n+1,n+2}\, \varepsilon_{n+1}\, \varepsilon_{n+2}\, f) \,. \qquad (2.172)\end{aligned}$$

The proof of (2.172) is similar to the proof of (2.170), but the integration by parts, that uses (A.41), is more delicate. To apply (A.41), one must think of $\langle f\rangle_t$ as a function of the Gaussian variables $g(\boldsymbol{\rho})$, $Y(\boldsymbol{\rho})$ for $\boldsymbol{\rho}$ in Σ_{N-1}, and this is cumbersome. We have presented the proof above (and will present other direct proofs of special uses of (2.172)) in order not to scare the reader with the proof of (2.172). But, in the long run, the reader might realize that it is a good investment to prove (2.172).

Proposition 2.4.6. *If f is a non-negative function on Σ_N^n, we have*

$$\nu_t(f) \le \exp(4\, n^2 \beta^2)\, \nu(f) \,. \qquad (2.173)$$

Proof. Identical to that of Proposition 2.3.2, using now that $|R_{\ell,\ell'} - q| \le 2$ since we assume $|q| \le 1$. □

Proposition 2.4.7. *Consider a function f on Σ_N^n, and τ_1, τ_2 with $1/\tau_1 + 1/\tau_2 = 1$. Then we have*

$$|\nu(f) - \nu_0(f)| \le 2\, n^2 \beta^2 \exp(4\, n^2 \beta^2)\, \nu(|f|^{\tau_1})^{1/\tau_1}\, \nu(|R_{1,2} - q|^{\tau_2})^{1/\tau_2} \,. \qquad (2.174)$$

Proof. We have

$$|\nu(f) - \nu_0(f)| = \left|\int_0^1 \nu_t'(f)\, \mathrm{d}t\right| \le \sup_{0<t<1} |\nu_t'(f)| \,.$$

Now, by Hölder's inequality for ν_t, (as in (2.90)), we have

$$|\nu_t(f\, \varepsilon_\ell\, \varepsilon_{\ell'} (R_{\ell,\ell'} - q))| \le \nu_t(|f||R_{\ell,\ell'} - q|) \le \nu_t(|f|^{\tau_1})^{1/\tau_1}\, \nu_t(|R_{\ell,\ell'} - q|^{\tau_2})^{1/\tau_2}$$

and thus by (2.170),

$$|\nu_t'(f)| \le 2n^2\beta^2\, \nu_t(|f|^{\tau_1})^{1/\tau_1}\, \nu_t(|R_{1,2}-q|^{\tau_2})^{1/\tau_2}\,.$$

We then use (2.173) for $|f|^{\tau_1}$ and $|R_{1,2}-q|^{\tau_2}$. □

Proposition 2.4.8. (Latala-Guerra) *The function $\Psi(x) = \mathsf{E}\,\mathrm{th}^2(z\sqrt{x}+h)/x$ is decreasing on $\mathbb{R}^+$. Consequently for $h>0$ there is a unique solution to (2.148).*

Proof. We have, writing $Y = z\sqrt{x}+h$,

$$x^2\,\Psi'(x) = \mathsf{E}\left(z\sqrt{x}\,\frac{\mathrm{th}\,Y}{\mathrm{ch}^2 Y} - \mathrm{th}^2 Y\right) = \mathsf{E}\left(Y\,\frac{\mathrm{th}\,Y}{\mathrm{ch}^2 Y} - \mathrm{th}^2 Y\right) - h\,\mathsf{E}\,\frac{\mathrm{th}\,Y}{\mathrm{ch}^2 Y}\,.$$

Now, for $x \ge 0$, we have $x \le \mathrm{sh}\,x$, so that $2x \le \mathrm{sh}\,2x$ and $x \le \mathrm{sh}\,x\,\mathrm{ch}\,x$, so $x\,\mathrm{sh}\,x \le \mathrm{ch}\,x\,\mathrm{sh}^2 x$ for all x, and dividing by $\mathrm{ch}^3 x$ we get $x\,\mathrm{th}\,x/\mathrm{ch}^2 x \le \mathrm{th}^2 x$. Thus all we have to prove is that $\mathsf{E}\,\mathrm{th}\,Y/\mathrm{ch}^2 Y \ge 0$. Now

$$\begin{aligned}\mathsf{E}\,\frac{\mathrm{th}\,Y}{\mathrm{ch}^2 Y} &= \frac{1}{\sqrt{2\pi}}\int \frac{\mathrm{th}\,Y}{\mathrm{ch}^2 Y}\, e^{-z^2/2}\,\mathrm{d}z \\ &= \frac{1}{\sqrt{2\pi x}}\int \frac{\mathrm{th}\,Y}{\mathrm{ch}^2 Y}\exp\left(-\frac{Y^2}{2x} + \frac{hY}{x} - \frac{h^2}{2x}\right)\mathrm{d}Y\,.\end{aligned}$$

Making the change of variable $Y' = -Y$, and combining with the above we get

$$\mathsf{E}\,\frac{\mathrm{th}\,Y}{\mathrm{ch}^2 Y} = \frac{1}{\sqrt{2\pi x}}\int \frac{\mathrm{th}\,Y\,\mathrm{sh}(h\,Y/x)}{\mathrm{ch}^2 Y}\exp\left(-\frac{Y^2}{2x} - \frac{h^2}{2x}\right)\mathrm{d}Y \ge 0\,.$$

□

We now take for q the root of (2.148), and to illustrate why this is a good choice we prove Theorem 2.4.2.

Proof of Theorem 2.4.2. We use symmetry among sites to write

$$\nu((R_{1,2}-q)^2) = \nu(f) \tag{2.175}$$

where

$$f = (\varepsilon_1\,\varepsilon_2 - q)(R_{1,2}-q) = \frac{1}{N}(\varepsilon_1\,\varepsilon_2 - q)^2 + (\varepsilon_1\,\varepsilon_2 - q)(R^-_{1,2}-q)\,.$$

The key point is that Lemma 2.4.4 implies

$$\nu_0((\varepsilon_1\,\varepsilon_2 - q)(R^-_{1,2}-q)) = (\mathsf{E}\,\mathrm{th}^2 Y - q)\,\nu_0(R^-_{1,2}-q) = 0$$

because (2.148) means that $q = \mathsf{E}\,\mathrm{th}^2 Y$. Thus, using Lemma 2.4.4 again

$$\begin{aligned}\nu_0(f) &= \frac{1}{N}\,\nu_0((\varepsilon_1\,\varepsilon_2 - q)^2) \\ &= \frac{1}{N}\,\nu_0(1 - 2\,\varepsilon_1\,\varepsilon_2\,q + q^2) \\ &= \frac{1}{N}\,(1 - 2q\,\mathsf{E}\,\mathrm{th}^2\,Y + q^2) = \frac{1}{N}\,(1 - q^2)\,. \end{aligned} \tag{2.176}$$

We now use (2.174) with $\tau_1 = \tau_2 = 2$ and $n = 2$. Since $|\varepsilon_1\,\varepsilon_2 - q| \le 2$, we get

$$|\nu(f) - \nu_0(f)| \le 16\,\beta^2 \exp 16\beta^2\,\nu((R_{1,2} - q)^2)$$

and using (2.175), (2.176)

$$\nu((R_{1,2} - q)^2) \le \frac{1}{N} + 16\,\beta^2 \exp 16\beta^2\,\nu((R_{1,2} - q)^2)\,.$$

Thus, if β_0 is chosen so that

$$16\,\beta_0^2 \exp 16\beta_0^2 \le \frac{1}{2}$$

we get

$$\nu((R_{1,2} - q)^2) \le \frac{2}{N}\,.$$

□

Even when $h = 0$, it is very unclear how to prove (2.149) in all the range $\beta < 1$ using the previous method. This is related to an apparently intrinsic difficulty, that will be discussed in Section 2.9.

While condition (2.149) is easy to write, it is not so clear what it really means. It might not even be so clear what were the ideas behind the somewhat magical proofs, and this is what we will try to explain now.

From Jensen's inequality, we get

$$(\langle R_{1,2}\rangle - q)^2 \le \langle (R_{1,2} - q)^2\rangle$$

so that (2.149) implies

$$\mathsf{E}((\langle R_{1,2}\rangle - q)^2) \le \frac{L}{N}\,. \tag{2.177}$$

Now

$$\langle R_{1,2}\rangle = \frac{1}{N}\sum_{i\le N}\langle \sigma_i^1\,\sigma_i^2\rangle = \frac{1}{N}\sum_{i\le N}\langle \sigma_i\rangle^2 \tag{2.178}$$

is a very natural quantity. The vector

$$\mathbf{b} = (\langle \sigma_i\rangle)_{i\le N} \tag{2.179}$$

is the barycentre of G_N, when we think of G_N as a probability measure on $\mathbb{R}^N$. If $\|\mathbf{b}\|$ denotes the Euclidean norm of $\mathbf{b}$, then (2.178) means that

$$\langle R_{1,2}\rangle = \frac{\|\mathbf{b}\|^2}{N} , \tag{2.180}$$

and it is natural that this quantity should play a role. We can view (2.177) as a kind of law of large numbers: averaging takes place among the quantities $\langle\sigma_i\rangle^2$.

Here is another consequence of (2.149).

Proposition 2.4.9. *If $\beta < \beta_0$ we have*

$$\mathsf{E}((\langle\sigma_1\,\sigma_2\rangle - \langle\sigma_1\rangle\,\langle\sigma_2\rangle)^2) \le \frac{L}{N} . \tag{2.181}$$

The quantity $\langle\sigma_1\,\sigma_2\rangle - \langle\sigma_1\rangle\,\langle\sigma_2\rangle$ is clearly of importance: it is the correlation between the spins σ_1, σ_2 when seen as functions on (Σ_N, G_N). (The physicists call truncated correlation what a probabilist calls correlation. We will follow the probabilistic terminology.)

Proof. From (2.149) we have

$$\mathsf{E}\langle(R_{1,3} - R_{1,4} - R_{2,3} + R_{2,4})^2\rangle \le \frac{L}{N}$$

or, equivalently, that

$$\mathsf{E}\left\langle\left(\frac{(\boldsymbol{\sigma}^1 - \boldsymbol{\sigma}^2)\cdot(\boldsymbol{\sigma}^3 - \boldsymbol{\sigma}^4)}{N}\right)^2\right\rangle \le \frac{L}{N} . \tag{2.182}$$

Letting $a_i = (\sigma_i^1 - \sigma_i^2)(\sigma_i^3 - \sigma_i^4)/N$, we observe that, using symmetry between sites,

$$\begin{aligned}\mathsf{E}\left\langle\left(\frac{(\boldsymbol{\sigma}^1 - \boldsymbol{\sigma}^2)\cdot(\boldsymbol{\sigma}^3 - \boldsymbol{\sigma}^4)}{N}\right)^2\right\rangle &= \mathsf{E}\left\langle\left(\sum_{i\le N} a_i\right)^2\right\rangle \\ &= N\mathsf{E}\langle a_N^2\rangle + N(N-1)\,\mathsf{E}\langle a_1 a_2\rangle .\end{aligned} \tag{2.183}$$

Now

$$\begin{aligned}\langle a_1\,a_2\rangle &= \frac{1}{N^2}\langle(\sigma_1^1 - \sigma_1^2)(\sigma_2^1 - \sigma_2^2)(\sigma_1^3 - \sigma_1^4)(\sigma_2^3 - \sigma_2^4)\rangle \\ &= \frac{1}{N^2}\langle(\sigma_1^1 - \sigma_1^2)(\sigma_2^1 - \sigma_2^2)\rangle^2 ,\end{aligned}$$

and

$$\begin{aligned}\langle(\sigma_1^1 - \sigma_1^2)(\sigma_2^1 - \sigma_2^2)\rangle &= \langle\sigma_1^1\,\sigma_2^1\rangle - \langle\sigma_1^1\,\sigma_2^2\rangle - \langle\sigma_1^2\,\sigma_2^1\rangle + \langle\sigma_1^2\,\sigma_2^2\rangle \\ &= 2(\langle\sigma_1\,\sigma_2\rangle - \langle\sigma_1\rangle\langle\sigma_2\rangle) ,\end{aligned}$$

so that (2.183) implies

$$\begin{aligned}
&4\left(1-\frac{1}{N}\right)\mathsf{E}(\langle\sigma_1\sigma_2\rangle-\langle\sigma_1\rangle\langle\sigma_2\rangle)^2\\
&=-\frac{1}{N}\mathsf{E}\langle((\varepsilon_1-\varepsilon_2)(\varepsilon_3-\varepsilon_4))^2\rangle+\mathsf{E}\left\langle\left(\frac{(\boldsymbol{\sigma}^1-\boldsymbol{\sigma}^2)\cdot(\boldsymbol{\sigma}^3-\boldsymbol{\sigma}^4)}{N}\right)^2\right\rangle\\
&\le\mathsf{E}\left\langle\left(\frac{(\boldsymbol{\sigma}^1-\boldsymbol{\sigma}^2)\cdot(\boldsymbol{\sigma}^3-\boldsymbol{\sigma}^4)}{N}\right)^2\right\rangle . \qquad (2.184)
\end{aligned}$$

Thus (2.181) follows from (2.182). □

In fact, quite a lot more than (2.181) is true. If we look only at a fixed number of spins, Gibbs' measure is asymptotically a product measure.

Theorem 2.4.10. *Consider an integer n, and the projection $G_{N,n}$ of Gibbs' measure on $\{-1,1\}^n$, that is, the law of $(\sigma_1,\dots,\sigma_n)$ under G_N. Consider the random measure μ_n on $\{-1,1\}^n$, determined by the fact that it is a product measure and that, for $i\le n$, $\int x_i\,\mathrm{d}\mu_n(\mathbf{x})=\langle\sigma_i\rangle$, where $\mathbf{x}$ denotes the generic point of $\{-1,1\}^n$. Then, if $\beta\le\beta_0$,*

$$\mathsf{E}\,|G_{N,n}-\mu_n|^2\le\frac{K(n)}{N}\,, \qquad (2.185)$$

where $|G_{N,n}-\mu_n|$ denotes the total variation distance between $G_{N,n}$ and μ_n.

The proof will use only (2.149), so that Theorem 2.4.10 will hold in other models where (2.149) holds.

As mentioned in Section 2.2, there are all kinds of research problems related to Theorem 2.4.10.

Proof. We write

$$\begin{aligned}
|G_{N,n}-\mu_n|^2&=\left(\sum_{\mathbf{x}}|G_{N,n}(\{\mathbf{x}\})-\mu_n(\{\mathbf{x}\})|\right)^2\\
&\le 2^n\sum_{\mathbf{x}}(G_{N,n}(\{\mathbf{x}\})-\mu_n(\{\mathbf{x}\}))^2\,, \qquad (2.186)
\end{aligned}$$

where the summation is over $\mathbf{x}$ in $\{-1,1\}^n$. The equality is by definition of the total variation distance, and the inequality follows from the Cauchy-Schwarz inequality. If $\mathbf{y}=(\sigma_1,\dots,\sigma_n)$ and $\mathbf{x}=(\eta_1,\dots,\eta_n)$, we have

$$\mathbf{1}_{\{\mathbf{x}\}}(\mathbf{y})=2^{-n}\prod_{i\le n}(1+\sigma_i\eta_i)=2^{-n}\sum_I\sigma_I\eta_I\,,$$

where the summation is over $I\subset\{1,\dots,n\}$, and where $\sigma_I=\prod_{i\in I}\sigma_i$, $\eta_I=\prod_{i\in I}\eta_i$. Thus

$$G_{N,n}(\{\mathbf{x}\}) = \int \mathbf{1}_{\{\mathbf{x}\}}(\mathbf{y})\, \mathrm{d}G_{N,n}(\mathbf{y}) = 2^{-n} \sum_I \eta_I \int \sigma_I \, \mathrm{d}G_{N,n}$$
$$= 2^{-n} \sum_I \eta_I \langle \sigma_I \rangle \,, \tag{2.187}$$

by definition of $G_{N,n}$. Since μ_n is a product measure,

$$\int \sigma_I \, \mathrm{d}\mu_n = \prod_{i \in I} \int \sigma_i \, \mathrm{d}\mu_n = \prod_{i \in I} \langle \sigma_i \rangle \,,$$

and, as in (2.187)

$$\mu_n(\{\mathbf{x}\}) = 2^{-n} \sum_I \eta_I \prod_{i \in I} \langle \sigma_i \rangle \,.$$

Thus, to prove (2.185), it suffices to prove that for each set I we have

$$\mathsf{E}\left(\left(\langle \sigma_I \rangle - \prod_{i \in I} \langle \sigma_i \rangle\right)^2\right) \le \frac{K}{N} \,. \tag{2.188}$$

We can assume that $I = \{1, \ldots, p\}$, and we prove (2.188) by induction over p. The case $p = 2$ is the content of Proposition 2.4.9. For the induction step from $p-1$ to p, it suffices to prove that, since $|\langle \sigma_p \rangle| \le 1$,

$$\mathsf{E}((\langle \sigma_I \rangle - \langle \sigma_1 \ldots \sigma_{p-1} \rangle \langle \sigma_p \rangle)^2) \le \frac{K}{N} \,,$$

or, equivalently, setting $\dot{\sigma}_i = \sigma_i - \langle \sigma_i \rangle$, that

$$\mathsf{E}(\langle \sigma_1 \ldots \sigma_{p-1} \, \dot{\sigma}_p \rangle^2) \le \frac{K}{N} \,. \tag{2.189}$$

The proof of this uses an elaborate version of the idea of the proof of Proposition 2.4.9. Considering $M = N(N-1)\ldots(N-p)$, symmetry between sites shows that

$$M\mathsf{E}(\langle \sigma_1 \ldots \sigma_{p-1} \, \dot{\sigma}_p \rangle^2) \le \sum_{i_1, \ldots, i_p} \mathsf{E}(\langle \sigma_{i_1} \ldots \sigma_{i_{p-1}} \, \dot{\sigma}_{i_p} \rangle^2) \tag{2.190}$$

where the summation is over all values of $i_1, \ldots, i_p \le N$.

Using replicas,

$$N^{-p} \sum_{i_1, \ldots, i_p} \mathsf{E}(\langle \sigma_{i_1} \ldots \sigma_{i_{p-1}} \, \dot{\sigma}_{i_p} \rangle^2)$$
$$= N^{-p} \sum_{i_1, \ldots, i_p} \mathsf{E}\langle \sigma^1_{i_1} \sigma^2_{i_1} \ldots \sigma^1_{i_{p-1}} \sigma^2_{i_{p-1}} \dot{\sigma}^1_{i_p} \dot{\sigma}^2_{i_p} \rangle$$
$$= \mathsf{E}\left\langle R_{1,2}^{p-1} \frac{\dot{\boldsymbol{\sigma}}^1 \cdot \dot{\boldsymbol{\sigma}}^2}{N} \right\rangle$$
$$= \mathsf{E}\left\langle (R_{1,2}^{p-1} - q^{p-1}) \frac{\dot{\boldsymbol{\sigma}}^1 \cdot \dot{\boldsymbol{\sigma}}^2}{N} \right\rangle \,, \tag{2.191}$$

because $\langle \dot{\boldsymbol{\sigma}}^1 \cdot \dot{\boldsymbol{\sigma}}^2 \rangle = 0$. Using the inequality $|x^{p-1} - y^{p-1}| \le (p-1)|x-y|$ for $|x|, |y| \le 1$, and the Cauchy-Schwarz inequality we get

$$\begin{aligned} &\mathsf{E}\left\langle (R_{1,2}^{p-1} - q^{p-1}) \frac{\dot{\boldsymbol{\sigma}}^1 \cdot \dot{\boldsymbol{\sigma}}^2}{N} \right\rangle \\ &\le K(p)\,(\mathsf{E}\langle (R_{1,2} - q)^2 \rangle)^{1/2}\, \mathsf{E}\left(\left\langle \left(\frac{\dot{\boldsymbol{\sigma}}^1 \cdot \dot{\boldsymbol{\sigma}}^2}{N} \right)^2 \right\rangle \right)^{1/2} . \end{aligned} \tag{2.192}$$

Now, using Jensen's inequality for $\langle \cdot \rangle$, we see that

$$\begin{aligned} &\left\langle \left(\frac{(\boldsymbol{\sigma}^1 - \langle \boldsymbol{\sigma}^1 \rangle) \cdot (\boldsymbol{\sigma}^2 - \langle \boldsymbol{\sigma}^2 \rangle)}{N} \right)^2 \right\rangle \\ &\le \left\langle \left(\frac{(\boldsymbol{\sigma}^1 - \boldsymbol{\sigma}^3) \cdot (\boldsymbol{\sigma}^2 - \boldsymbol{\sigma}^4)}{N} \right)^2 \right\rangle \le L\langle (R_{1,2} - q)^2 \rangle \end{aligned}$$

as in the proof of (2.182). Thus the right-hand side of (2.192) is at most $K(p)\,\mathsf{E}\langle (R_{1,2} - q)^2 \rangle$. Combining (2.190) with (2.192) and (2.149) proves (2.189).

□

We can certainly not expect that, when we look at all the spins, G_N will be close to a product measure in a global sense. Still, we can hope that there is enough independence among the maps $\boldsymbol{\sigma} \mapsto \sigma_i$ to have central limit theorems. The variance of σ_i is $\langle \sigma_i^2 \rangle - \langle \sigma_i \rangle^2 = 1 - \langle \sigma_i \rangle^2$. Thus, we might hope that for the typical disorder and numbers $(s_i)_{i \le N}$ the law of

$$\boldsymbol{\sigma} \mapsto \frac{1}{\sqrt{N}} \sum_{i \le N} s_i \,\sigma_i \tag{2.193}$$

under G_N would be approximately (non centered) Gaussian of mean

$$\frac{1}{\sqrt{N}} \sum_{i \le N} s_i \langle \sigma_i \rangle = \frac{1}{\sqrt{N}} \mathbf{s} \cdot \mathbf{b}$$

and of variance

$$\frac{1}{N} \sum_{i \le N} s_i^2 (1 - \langle \sigma_i \rangle^2) ,$$

the sum of the variances of the maps $\sigma_i \mapsto N^{-1/2}\, s_i\, \sigma_i$. When the quantities s_i are random, $\mathsf{E}\, s_i^2 = 1$, uncorrelated to the randomness of $\langle \cdot \rangle$, we can expect averaging to take place,

$$\frac{1}{N} \sum_{i \le N} s_i^2 (1 - \langle \sigma_i \rangle^2) \simeq 1 - \frac{1}{N} \sum_{i \le N} \langle \sigma_i \rangle^2 \simeq 1 - q .$$

It turns out that the above indeed happens. If the s_i are independent standard Gaussian, independent of the randomness in $\langle\cdot\rangle$, then for the typical value of $\langle\cdot\rangle$ and of the (s_i), the map

$$\boldsymbol{\sigma} \mapsto \frac{1}{\sqrt{N}} \mathbf{s} \cdot \boldsymbol{\sigma} = \frac{1}{\sqrt{N}} \sum_{i \le N} s_i \sigma_i$$

has a law under G_N which is nearly Gaussian, of mean $N^{-1/2}\,\mathbf{s}\cdot\mathbf{b}$ and of variance $1-q$. Much of what we do up to Chapter 5 included is motivated by this idea. The physicists express this property by saying that "the local fields are Gaussian". The name arises from the fact that the term $\beta N^{-1/2}\sum_{i\le N} s_i \sigma_i$ can be seen as the action on the last spin of a magnetic field due to the first N spins in an $(N+1)$-spin model (see (2.70)).

To put things in a different way, given a real-valued function u one should expect that

$$\left\langle u\left(\frac{1}{\sqrt{N}} \mathbf{s} \cdot \boldsymbol{\sigma}\right)\right\rangle \simeq \mathsf{E}_\xi\, u\left(\frac{1}{\sqrt{N}} \mathbf{s} \cdot \mathbf{b} + \xi\sqrt{1-q}\right)$$

where ξ is standard Gaussian independent of all the other r.v. considered and where E_ξ denotes integration in ξ only. Since ξ is independent of the randomness in $\langle\cdot\rangle$, one should also expect that, if f is a function on Σ_N, then

$$\left\langle f\, u\left(\frac{1}{\sqrt{N}} \mathbf{s} \cdot \boldsymbol{\sigma}\right)\right\rangle \simeq \langle f\rangle\, \mathsf{E}_\xi\, u\left(\frac{1}{\sqrt{N}} \mathbf{s} \cdot \mathbf{b} + \xi\sqrt{1-q}\right) . \tag{2.194}$$

We will show soon that this follows very simply from (2.149), but first, let us motivate the construction of ν_t. Used for $\langle\cdot\rangle_-$ rather than for $\langle\cdot\rangle$, and for $u(x) = \exp(\beta x + h)$, (2.194) shows that

$$\langle f \exp\varepsilon(g(\boldsymbol{\rho}) + h)\rangle_- \simeq \langle f\rangle_- \exp\varepsilon(g(\mathbf{c}) + h) \exp\frac{\beta^2}{2}(1-q),$$

where

$$g(\mathbf{c}) = \frac{\beta}{\sqrt{N}} \sum_{i \le N-1} g_i \langle\sigma_i\rangle_- \tag{2.195}$$

and thus

$$\langle \mathrm{Av} f \exp\varepsilon(g(\boldsymbol{\rho}) + h)\rangle_- \simeq \exp\frac{\beta^2}{2}(1-q)\langle \mathrm{Av} f \exp\varepsilon(g(\mathbf{c}) + h)\rangle_- .$$

When $f = 1$,

$$\langle \mathrm{ch}(g(\boldsymbol{\rho}) + h)\rangle_- \simeq \exp\frac{\beta^2}{2}(1-q)\,\mathrm{ch}(g(\mathbf{c}) + h)$$

so that

$$\langle f\rangle = \frac{\langle \mathrm{Av} f \exp\varepsilon(g(\boldsymbol{\rho}) + h)\rangle_-}{\langle \mathrm{ch}(g(\boldsymbol{\rho}) + h)\rangle_-} \simeq \frac{\langle \mathrm{Av} f \exp\varepsilon(g(\mathbf{c}) + h)\rangle_-}{\mathrm{ch}(g(\mathbf{c}) + h)} . \tag{2.196}$$

Given $\langle\cdot\rangle$ (or, more formally, conditionally on $\langle\cdot\rangle$), $g(\mathbf{c})$ is a centered Gaussian r.v. of variance

$$\frac{\beta^2}{N}\sum_{i\le N-1}\langle\sigma_i\rangle_-^2$$

and, hopefully, this is about $\beta^2 q$ (see (2.177) and (2.179)) so that we can replace $g(\mathbf{c})$ by $\beta\sqrt{q}\,z$, where z is standard Gaussian independent of all the other r.v. That is, (2.196) reads

$$\mathsf{E}\langle f\rangle \simeq \mathsf{E}\frac{\langle \mathrm{Av} f \exp\varepsilon(\beta z\sqrt{q}+h)\rangle_-}{\mathrm{ch}(\beta z\sqrt{q}+h)} = \nu_0(f)\ .$$

It is this formula that motivates the definition of ν_0. The definition of ν_t then simply uses the natural path from ν_0 to ν.

It is remarkable that there is a very simple proof of (2.194). It is based on the observation that if g_1, g_2 are jointly (centered) Gaussian r.v. with $\mathsf{E}g_1^2 = \mathsf{E}g_2^2 = 1$, then $\mathsf{E}\,u(g_1)\,u(g_2)$ depends only on $\mathsf{E}\,g_1\,g_2$ (because the joint law of (g_1, g_2) depends only on $\mathsf{E}\,g_1g_2$). Thus for a certain function θ,

$$\mathsf{E}\,u(g_1)\,u(g_2) = \theta(\mathsf{E}\,g_1\,g_2)\ ,$$

and the function θ is well behaved under very mild conditions. We write

$$X_1 = \left\langle fu\left(\frac{1}{\sqrt{N}}\mathbf{s}\cdot\boldsymbol{\sigma}\right)\right\rangle\ ;\ X_2 = \langle f\rangle\,\mathsf{E}_\xi\,u\left(\frac{1}{\sqrt{N}}\mathbf{s}\cdot\mathbf{b}+\xi\sqrt{1-q}\right)\ .$$

We compute $\mathsf{E}_s(X_1 - X_2)^2$, where E_s denotes expectation in s only. Using replicas,

$$X_1^2 = \left\langle f(\boldsymbol{\sigma}^1)\,f(\boldsymbol{\sigma}^2)\,u\left(\frac{1}{\sqrt{N}}\mathbf{s}\cdot\boldsymbol{\sigma}^1\right)u\left(\frac{1}{\sqrt{N}}\mathbf{s}\cdot\boldsymbol{\sigma}^2\right)\right\rangle$$

so that

$$\mathsf{E}_s X_1^2 = \langle f(\boldsymbol{\sigma}^1)\,f(\boldsymbol{\sigma}^2)\,\theta(R_{1,2})\rangle\ .$$

If we denote by ξ_1, ξ_2 two independent standard Gaussian r.v., then

$$\begin{aligned}&\mathsf{E}_s\left(\mathsf{E}_\xi\,u\left(\frac{1}{\sqrt{N}}\mathbf{s}\cdot\mathbf{b}+\xi\sqrt{1-q}\right)\right)^2\\ &= \mathsf{E}_{s,\xi_1,\xi_2}\,u\left(\frac{1}{\sqrt{N}}\mathbf{s}\cdot\mathbf{b}+\xi_1\sqrt{1-q}\right)\,u\left(\frac{1}{\sqrt{N}}\mathbf{s}\cdot\mathbf{b}+\xi_2\sqrt{1-q}\right)\\ &\simeq \theta(\|\mathbf{b}\|^2/N) = \theta(\langle R_{1,2}\rangle)\end{aligned}$$

because if we define $g_j = N^{-1/2}\mathbf{s}\cdot\mathbf{b}+\xi_j\sqrt{1-q}$ for $j=1,2$, then the couple (g_1, g_2) is jointly Gaussian, $\mathsf{E}g_1^2 \simeq 1$, $\mathsf{E}g_2^2 \simeq 1$ and $\mathsf{E}\,g_1\,g_2 = \|\mathbf{b}\|^2/N$. Thus

$$\mathsf{E}_s X_2^2 \simeq \langle f(\boldsymbol{\sigma}^1)\,f(\boldsymbol{\sigma}^2)\rangle\,\theta(\langle R_{1,2}\rangle)\ .$$

In a similar manner,

$$\mathsf{E}_s X_1 X_2 \simeq \langle f(\boldsymbol{\sigma}^1) f(\boldsymbol{\sigma}^2) \theta(\boldsymbol{\sigma}^1 \cdot \mathbf{b}/N) \rangle .$$

It then follows from (2.149) (if f is bounded and θ smooth) that $\mathsf{E}(X_1 - X_2)^2$ is small.

One can ask all kinds of questions related to the previous circle of ideas.

Research problem 2.4.11. (Level 1^+) For a given typical realization of G_N, describe the sequences $(t_i)_{i \le N}$ such that the law under G_N of $\boldsymbol{\sigma} \mapsto \sum_{i \le N} t_i \sigma_i$ is approximately Gaussian.

It would be nice if all the sequences such that $\max_{i \le N} |t_i| / (\sum_{i \le N} t_i^2)^{1/2}$ is small would work.

After this long digression, we return to the study of Gibbs' measure. When we look at a fixed number of spins, Theorem 2.4.10 shows that we would understand G_N if we understood the random sequence $(\langle \sigma_i \rangle)_{i \le N}$. Certainly this whole sequence is a complicated object: yet, if we look at a fixed number of terms, the situation is again as simple as possible.

Theorem 2.4.12. *Given $\beta \le \beta_0$, and an integer n, we can find independent standard Gaussian r.v. $(z_i)_{i \le n}$ such that*

$$\mathsf{E} \sum_{i \le n} (\langle \sigma_i \rangle - \mathrm{th}(\beta z_i \sqrt{q} + h))^2 \le \frac{K(n,h)}{N} . \tag{2.197}$$

In particular, the sequence $(\langle \sigma_1 \rangle, \dots, \langle \sigma_n \rangle)$ is asymptotically independent, identically distributed, and the limit law of $\langle \sigma_i \rangle$ is the law of $\mathrm{th}(\beta z \sqrt{q} + h)$; in addition (2.197) provides a precise rate for this.

A crucial ingredient is the following, where we recall that $g(\mathbf{c})$ is defined in (2.195).

Lemma 2.4.13. *We have, for $\beta \le \beta_0$,*

$$\mathsf{E}((\langle \sigma_N \rangle - \mathrm{th}(g(\mathbf{c}) + h))^2) \le \frac{L}{N} \tag{2.198}$$

$$\mathsf{E}((\langle \sigma_1 \rangle - \langle \sigma_1 \rangle_-)^2) \le \frac{L}{N} . \tag{2.199}$$

It is possible to prove this by adapting our sketch of proof of (2.194); but we prefer to give a proof that we can adapt to other models, and this will be better done at the end of Section 2.5.

Lemma 2.4.14. *We can find a standard Gaussian r.v. z, that depends only on the r.v. $(g_{ij})_{i<j \le N}$, but is probabilistically independent of the r.v. $(g_{ij})_{i<j \le N-1}$, and such that*

$$\mathsf{E}((\langle \sigma_N \rangle - \mathrm{th}(\beta z \sqrt{q} + h))^2) \le \frac{K(h)}{N} .$$

In order to prove this, we need a simple fact.

Lemma 2.4.15. *Denote by $q_- = q_-(N)$ the solution of (2.148) when β is replaced by $\beta_- = \beta\sqrt{1-N^{-1}}$. Then, if $\beta \le \beta_0$, we have*

$$|q - q_-| \le \frac{L}{N} \,. \tag{2.200}$$

Proof. If one views (2.148) as

$$q(\beta) = F(\beta, q(\beta))$$

then

$$q'(\beta) = \frac{\frac{\partial F}{\partial \beta}(\beta, q(\beta))}{1 - \frac{\partial F}{\partial q}(\beta, q(\beta))}$$

and a straightforward calculation shows that $|q'(\beta)| \le L\beta$ for $\beta \le \beta_0$. □

Proof of Lemma 2.4.14. We can and do assume $h \ne 0$, since the result is obvious if $h = 0$. Then the solution q of (2.148) is bounded from below uniformly on $\beta \le \beta_0$.

We define

$$z = \frac{1}{\|\mathbf{c}\|} \sum_{i \le N-1} g_i \langle \sigma_i \rangle_- \,,$$

where $\|\mathbf{c}\|^2 = \sum_{i \le N-1} \langle \sigma_i \rangle_-^2$. Thus z depends only upon the r.v. $(g_{ij})_{i<j\le N}$ (recall that $g_i = g_{i,N}$ for $i \le N-1$). Conditionally upon the numbers $\langle \sigma_i \rangle_-$, the law of z is standard Gaussian. (That is, when z is seen as a function of $g_1, \dots, g_{N-1}$ only.) As strange as it may seem the first time one thinks about this, this does imply that z is probabilistically independent of the r.v. $(g_{ij})_{i<j\le N-1}$.

If we use (2.198) and the inequality $|\mathrm{th}\, x - \mathrm{th}\, y| \le |x - y|$, we see that all that is required to prove is that

$$\mathsf{E}\left(z\sqrt{q} - \frac{1}{\sqrt{N}} \sum_{i \le N-1} g_i \langle \sigma_i \rangle_- \right)^2 \le \frac{K(h)}{N} \,. \tag{2.201}$$

Taking first the expectation in $(g_i)_{i \le N-1}$, the left-hand side of (2.201) is

$$\mathsf{E}\left(\sum_{i \le N-1} \langle \sigma_i \rangle_-^2 \left(\frac{\sqrt{q}}{\|\mathbf{c}\|} - \frac{1}{\sqrt{N}} \right)^2 \right) = \mathsf{E}\left(\sqrt{q} - \frac{\|\mathbf{c}\|}{\sqrt{N}} \right)^2 .$$

Now,

$$\left(\sqrt{q} - \frac{\|\mathbf{c}\|}{\sqrt{N}} \right)^2 = \frac{\left(q - \|\mathbf{c}\|^2/N\right)^2}{\left(\sqrt{q} + \|\mathbf{c}\|/\sqrt{N}\right)^2} \le \frac{1}{q}\left(q - \frac{\|\mathbf{c}\|^2}{N} \right)^2 ,$$

so that it suffices to prove that

$$\mathsf{E}\left(\left(q - \frac{\|\mathbf{c}\|^2}{N}\right)^2\right) = \mathsf{E}((q - \langle R^-_{1,2}\rangle_-)^2) \le \frac{L}{N}, \tag{2.202}$$

which follows from (2.200) and (2.177). □

Remark. This argument does not yield, as one would wish, a number K independent of h in (2.197), and it is unclear if this could be achieved.

Proof of Theorem 2.4.12. The proof goes by induction over n. As a part of the induction hypothesis, we assume that the r.v. $(z_i)_{i\le n}$ are functions of the variables $(g_{ij})_{i<j\le N}$. For $n = 1$, this follows from Lemma 2.4.14. For the induction step from n to $n+1$, we apply the induction hypothesis to the $(N-1)$-spin system with Hamiltonian H_{N-1}, where H_{N-1} is given by (2.150). This amounts to replacing β by $\beta_- \le \beta$. We then get from (2.197) that

$$\sum_{i\le n} \mathsf{E}(\langle\sigma_i\rangle_- - \mathrm{th}(\beta_-\, z_i\sqrt{q_-} + h))^2 \le \frac{K(n,h)}{N}, \tag{2.203}$$

where the variables $(z_i)_{i\le n}$ depend only on $(g_{ij})_{i<j\le N-1}$. We observe that

$$(\sqrt{q_-} - \sqrt{q})^2 = \frac{(q_- - q)^2}{(\sqrt{q_-} + \sqrt{q})^2} \le \frac{(q_- - q)^2}{q_- + q} \le |q_- - q| \le \frac{L}{N}$$

by (2.200), so that

$$(\beta_-\sqrt{q_-} - \beta\sqrt{q})^2 \le \frac{L}{N}$$

and

$$\mathsf{E}((\mathrm{th}(\beta_-\, z_i\sqrt{q_-} + h) - \mathrm{th}(\beta\, z_i\sqrt{q} + h))^2 \le \frac{L}{N}.$$

Combining with (2.199) and (2.203) we get

$$\sum_{i\le n} \mathsf{E}(\langle\sigma_i\rangle - \mathrm{th}(\beta\, z_i\sqrt{q} + h))^2 \le \frac{K(n,h)}{N}.$$

To finish the proof, we appeal to Lemma 2.4.14. The variable $z = z_{n+1}$ there is probabilistically independent of the r.v. $(z_i)_{i\le n}$ because these are functions of the r.v. $(g_{ij})_{i<j\le N-1}$. □

We now turn to the proof of Theorem 2.4.1. Given $q \ge 0$, β, h we consider

$$F(\beta,h,q) = \frac{\beta^2}{4}(1-q)^2 + \log 2 + \mathsf{E}\log\mathrm{ch}(\beta\, z\sqrt{q} + h). \tag{2.204}$$

We define

$$\mathrm{SK}(\beta,h) = F(\beta,h,q) \tag{2.205}$$

where q satisfies (2.148).

Lemma 2.4.16. *We have*

$$\frac{\partial F}{\partial q}(\beta, h, q) = 0 \Leftrightarrow q \text{ satisfies (2.148)} \tag{2.206}$$

$$\frac{\partial\, \mathrm{SK}}{\partial \beta}(\beta, h) = \frac{\beta}{2}(1 - q^2)\,. \tag{2.207}$$

Of course, (2.207) takes out some of the mystery of (2.205). We use the short-hand notation

$$Y = \beta z \sqrt{q} + h\,. \tag{2.208}$$

Proof. We have

$$\frac{\partial}{\partial q} F(\beta, h, q) = -\frac{\beta^2}{2}(1-q) + \mathsf{E}\frac{\beta}{2\sqrt{q}} z \operatorname{th} Y = -\frac{\beta^2}{2}(1-q) + \frac{\beta^2}{2}\mathsf{E}\frac{1}{\operatorname{ch}^2 Y}$$

by integration by parts. Since $\operatorname{th}^2 x = 1 - \operatorname{ch}^{-2} x$, this proves (2.206).

We now write $q = q(\beta, h)$ the solution of (2.148), so that

$$\mathrm{SK}(\beta, h) = F(\beta, h, q(\beta, h))$$

and

$$\begin{aligned}\frac{\partial\, \mathrm{SK}}{\partial \beta}(\beta, h) &= \frac{\partial F}{\partial \beta}(\beta, h, q(\beta, h)) + \frac{\partial q}{\partial \beta}(\beta, h)\frac{\partial F}{\partial q}(\beta, h, q(\beta, h)) \\ &= \frac{\partial F}{\partial \beta}(\beta, h, q(\beta, h))\end{aligned}$$

by (2.206). Going back to the notation q rather than $q(\beta, h)$, we thus have

$$\frac{\partial\, \mathrm{SK}}{\partial \beta}(\beta, h) = \frac{\beta}{2}(1-q)^2 + \mathsf{E} z\sqrt{q} \operatorname{th} Y = \frac{\beta}{2}(1-q)^2 + \beta q \mathsf{E}\frac{1}{\operatorname{ch}^2 Y}$$

using integration by parts. Now

$$\mathsf{E}\frac{1}{\operatorname{ch}^2 Y} = 1 - \mathsf{E}\operatorname{th}^2 Y = 1 - q$$

by (2.148), so that

$$\frac{\partial\, \mathrm{SK}}{\partial \beta}(\beta, h) = \beta\left(\frac{1}{2} - q + \frac{q^2}{2} + q - q^2\right) = \frac{\beta}{2}(1 - q^2)\,. \tag{2.209}$$

□

The following is fundamental.

Theorem 2.4.17. *We have*

$$\frac{\partial p_N}{\partial \beta}(\beta, h) = \frac{\beta}{2}(1 - \mathsf{E}\langle R_{1,2}^2\rangle)\,. \tag{2.210}$$

Proof. Let us denote by

$$B(\boldsymbol{\sigma}) = B_N(\beta, h, \boldsymbol{\sigma}) = \exp(-H_N(\boldsymbol{\sigma}))$$

the Boltzmann factor, and let us observe that

$$\frac{\partial B(\boldsymbol{\sigma})}{\partial \beta} = \frac{1}{\sqrt{N}} \sum_{i<j} g_{ij}\, \sigma_i\, \sigma_j\, B(\boldsymbol{\sigma})$$

and that (considering $B(\boldsymbol{\sigma})$ as a function of the variables g_{ij})

$$\frac{\partial B(\boldsymbol{\sigma})}{\partial\, g_{ij}} = \frac{\beta}{\sqrt{N}} \sigma_i\, \sigma_j\, B(\boldsymbol{\sigma})\,. \tag{2.211}$$

Now

$$p_N(\beta, h) = \frac{1}{N} \mathsf{E} \log Z_N = \frac{1}{N} \mathsf{E} \log \sum_{\boldsymbol{\sigma}} B(\boldsymbol{\sigma})\,,$$

and thus

$$\frac{\partial\, p_N}{\partial \beta}(\beta, h) = \frac{1}{N^{3/2}} \sum_{i<j} \sum_{\boldsymbol{\sigma}} \sigma_i\, \sigma_j\, \mathsf{E}\left(g_{ij} \frac{B(\sigma)}{Z_N}\right)\,. \tag{2.212}$$

We will then integrate by parts in g_{ij}, at all the other r.v. $g_{i'j'}$ fixed. Since, by (2.211)

$$\frac{\partial}{\partial g_{ij}}\left(\frac{B(\boldsymbol{\sigma})}{Z_N}\right) = \frac{\beta}{\sqrt{N}}\left(\frac{\sigma_i\, \sigma_j\, B(\boldsymbol{\sigma})}{Z_N} - \frac{B(\boldsymbol{\sigma}) \sum_{\boldsymbol{\sigma}'} \sigma'_i\, \sigma'_j\, B(\boldsymbol{\sigma}')}{Z_N^2}\right)$$

we get from (2.212) that

$$\begin{aligned}
\frac{\partial\, p_N}{\partial \beta}(\beta, h) &= \frac{\beta}{N^2} \sum_{i<j} \mathsf{E} \sum_{\boldsymbol{\sigma}} \left(\frac{B(\boldsymbol{\sigma})}{Z_N} - \frac{\sum_{\boldsymbol{\sigma}'} \sigma_i\, \sigma_j\, \sigma'_i\, \sigma'_j B(\boldsymbol{\sigma}) B(\boldsymbol{\sigma}')}{Z_N^2}\right) \\
&= \frac{\beta}{N^2} \sum_{i<j} \mathsf{E}(1 - \langle \sigma_i\, \sigma_j\, \sigma'_i\, \sigma'_j \rangle)\,.
\end{aligned}$$

Since for $i = j$ the term in the above summation is zero, we get

$$\begin{aligned}
\frac{\partial\, p_N}{\partial \beta}(\beta, h) &= \frac{\beta}{2}\left(\frac{1}{N^2} \sum_{i,j \le N} \mathsf{E}(1 - \langle \sigma_i\, \sigma_j\, \sigma'_i\, \sigma'_j \rangle\right) \\
&= \frac{\beta}{2}\left(1 - \mathsf{E}\left\langle \frac{1}{N^2} \sum_{i,j \le N} \sigma_i\, \sigma'_i\, \sigma_j\, \sigma'_j \right\rangle\right) \\
&= \frac{\beta}{2}\left(1 - \mathsf{E}\left\langle \left(\frac{\boldsymbol{\sigma} \cdot \boldsymbol{\sigma}'}{N}\right)^2 \right\rangle\right)\,.
\end{aligned}$$

□

It should probably be mentioned that (2.210) is a special case of a general principle. If we consider a Hamiltonian

$$-H_N(\boldsymbol{\sigma}) = \beta A_N(\boldsymbol{\sigma}) + A'_N(\boldsymbol{\sigma}) ,$$

where the family $(A_N(\boldsymbol{\sigma}))$ is jointly Gaussian and probabilistically independent of the variables $A'_N(\boldsymbol{\sigma})$, then

$$\frac{1}{N}\frac{\partial}{\partial\beta}\mathsf{E}\log Z_N = \frac{\beta}{N}\left(\mathsf{E}\langle \mathsf{E}\, A_N^2(\boldsymbol{\sigma})\rangle - \mathsf{E}\langle \mathsf{E}(A_N(\boldsymbol{\sigma}^1)A_N(\boldsymbol{\sigma}^2))\rangle\right) .$$

This can be shown using integration by parts (A.41). The direct proof of (2.210) was presented because it requires less abstraction.

Theorem 2.4.18. *For $\beta \le \beta_0$ we have*

$$|p_N(\beta,h) - \mathrm{SK}(\beta,h)| \le \frac{L}{N} . \tag{2.213}$$

Proof. Combining (2.207) and (2.210), it is enough to show that (2.213) is true for $\beta = 0$ and that

$$|\mathsf{E}\langle R_{1,2}^2\rangle - q^2| \le \frac{L}{N} . \tag{2.214}$$

Now

$$\mathsf{E}\langle R_{1,2}^2\rangle - q^2 = \mathsf{E}\langle (R_{1,2}-q)^2\rangle + 2q(\mathsf{E}\langle R_{1,2}\rangle - q)$$

so that (2.214) follows from (2.149) and the inequality

$$|\mathsf{E}\langle R_{1,2} - q\rangle| \le \frac{L}{N} \tag{2.215}$$

that will be proved in Lemma 2.5.4.

As for the case $\beta = 0$, it is obvious that

$$p_N(0,h) = \log(2\operatorname{ch} h) = \mathrm{SK}(0,h) .$$

□

Our proof of Theorem 2.4.1 does not tell us how we could have discovered the expression of $\mathrm{SK}(\beta,h)$, even if we had discovered (2.210) and we knew (2.149). The proof of the following result shows how to discover the expression for $\mathrm{SK}(\beta,h)$. We fix h once and for all.

Theorem 2.4.19. *Assume that $p_\infty(x) = \lim_{N\to\infty} p_N(x,h)$ exists for x in a neighborhood of β and that $p'_\infty(\beta)$ exists. Assume that*

$$\lim_{N\to\infty} \nu((R_{1,2}-q)^2) = 0 . \tag{2.216}$$

Then we have

$$\lim_{N\to\infty} p_N(\beta,h) = \mathrm{SK}(\beta,h) .$$

Proof. The starting point is the algebraic identity

$$Z_{N+1}(\beta_+, h) = Z_N(\beta, h)\, 2 \left\langle \mathrm{ch}\left(\frac{\beta}{\sqrt{N}} \sum_{i \le N} g_{i,N+1}\, \sigma_i + h\right)\right\rangle \tag{2.217}$$

where

$$\beta_+ = \beta\sqrt{1 + \frac{1}{N}} \simeq \beta\left(1 + \frac{1}{2N}\right) . \tag{2.218}$$

From (2.194) we see that

$$\left\langle \mathrm{ch}\left(\frac{\beta}{\sqrt{N}} \sum_{i \le N} g_{i,N+1}\, \sigma_i + h\right)\right\rangle$$
$$= \exp\frac{\beta^2}{2}(1-q)\,\mathrm{ch}\left(\frac{\beta}{\sqrt{N}} \sum_{i \le N} g_{i,N+1} \langle \sigma_i\rangle + h\right) + \mathcal{R} .$$

Here and below, $\mathcal{R}$ denotes a quantity such that $\lim_{N\to\infty} \mathsf{E}\,|\mathcal{R}| = 0$.

Thus (since $N^{-1}\sum_{i\le N} \langle\sigma_i\rangle^2 = \langle R_{1,2}\rangle = q + \mathcal{R}$)

$$\mathsf{E}\log\left\langle \mathrm{ch}\left(\frac{\beta}{\sqrt{N}} \sum_{i \le N} g_{i,N+1}\, \sigma_i + h\right)\right\rangle$$
$$= \frac{\beta^2}{2}(1-q) + \mathsf{E}\log\mathrm{ch}(\beta\, z\sqrt{q} + h) + \mathcal{R} . \tag{2.219}$$

Hence (2.217) gives

$$\begin{aligned}(N+1)\, p_{N+1}(\beta_+, h) = N p_N(\beta, h) &+ \frac{\beta^2}{2}(1-q) \\ &+ \mathsf{E}\log 2\,\mathrm{ch}(\beta\, z\sqrt{q} + h) + \mathcal{R} .\end{aligned} \tag{2.220}$$

Now, by the mean value theorem, we have

$$p_{N+1}(\beta_+, h) = p_{N+1}(\beta, h) + (\beta_+ - \beta)\,\frac{\partial\, p_N}{\partial \beta}(\beta', h) \tag{2.221}$$

where $\beta < \beta' < \beta_+$. We claim that

$$\lim_{N\to\infty} \frac{\partial p_N}{\partial \beta}(\beta', h) = \lim_{N\to\infty} \frac{\partial p_N}{\partial \beta}(\beta, h) = \frac{\beta}{2}(1-q^2) . \tag{2.222}$$

The second equality follows from (2.210) and (2.216). To prove the first equality, we show if $\beta_N \to \beta$, then $\lim_{N\to\infty} \partial p_N/\partial\beta(\beta_N, h) = p'_\infty(\beta)$. Indeed, consider $\beta_2 > \beta_1 > \beta$. Then, by convexity, and if N is large enough that $\beta_N < \beta_1$, we have

$$\frac{\partial p_N}{\partial \beta}(\beta_N, h) \le \frac{p_N(\beta_2, h) - p_N(\beta_1, h)}{\beta_2 - \beta_1} ,$$

so that

$$\limsup_{N\to\infty} \frac{\partial p_N}{\partial \beta}(\beta_N, h) \le \frac{p_\infty(\beta_2) - p_\infty(\beta_1)}{\beta_2 - \beta_1} . \tag{2.223}$$

Letting $\beta_1 \to \beta$, and then $\beta_2 \to \beta$, we see that the left-hand side of (2.223) is at most $p'_\infty(\beta)$. The same argument changing β in $-\beta$ completes the proof of (2.222).

It follows from (2.155) that

$$p_{N+1}(\beta_+, h) = p_{N+1}(\beta, h) + \frac{\beta^2}{4N}(1 - q^2) + \mathcal{R} ,$$

using (2.218) and (2.222). Substitution in (2.220) gives

$$(N+1)\, p_{N+1}(\beta, h) + \frac{\beta^2}{4}(1 - q^2) = N p_N(\beta, h) + \frac{\beta^2}{2}(1 - q) + \log 2 \\ + \mathsf{E} \log \operatorname{ch}(\beta z \sqrt{q} + h) + \mathcal{R} .$$

The result follows by summation. □

Remark. The reader finding the proof of (2.219) too informal should prove this relation using the smart path method.

Theorem 2.4.20. *If $\beta \le \beta_0$ we have*

$$\mathsf{E}\left(\langle\sigma_N\rangle - \operatorname{th}\left(\frac{\beta}{\sqrt{N}} \sum_{i\le N-1} g_i \langle\sigma_i\rangle + h - \beta^2(1-q)\langle\sigma_N\rangle\right)\right)^2 \le \frac{L}{N} . \tag{2.224}$$

By symmetry among sites, and writing $g_{ij} = g_{ji}$ if $i > j$, we get that for each i we have

$$\langle\sigma_i\rangle \approx \operatorname{th}\left(\frac{\beta}{\sqrt{N}} \sum_{j\ne i} g_{ij}\langle\sigma_j\rangle + h - \beta^2(1-q)\langle\sigma_i\rangle\right) . \tag{2.225}$$

These are the so called TAP equations, that (hopefully) determine the components $\langle\sigma_i\rangle$ of $\mathbf{b}$. The physicists have no qualms writing exact equalities in (2.225). It is however not so clear that (2.225) is true simultaneously for each i.

Research problem 2.4.21. (Level 1) Prove that

$$\mathsf{E} \max_{i\le N} \left| \langle\sigma_i\rangle - \operatorname{th}\left(\frac{\beta}{\sqrt{N}} \sum_{j\ne i} g_{ij}\langle\sigma_j\rangle + h - \beta^2(1-q)\langle\sigma_i\rangle\right)\right| \tag{2.226}$$

becomes small for N large. Find a competent rate of convergence.

It is probably a long but straightforward task to prove that if in (2.224) the second moment is replaced by a fourth moment, one obtains a bound L/N^2; that would show that the quantity (2.226) is at most $LN^{-1/4}$; but it might be quite a lot harder to reach a rate such as $\sqrt{\log N}/\sqrt{N}$. (But, conceivably, the rate in (2.224) is not optimal.)

Using (2.198), to prove Theorem 2.4.20, it suffices to prove the following.

Proposition 2.4.22. *If $\beta \le \beta_0$ we have*

$$\mathsf{E}\left(\frac{1}{\sqrt{N}}\sum_{i\le N-1} g_i(\langle\sigma_i\rangle - \langle\sigma_i\rangle_-) - \beta(1-q)\langle\sigma_N\rangle\right)^2 \le \frac{L}{N}\,. \tag{2.227}$$

Proof. This proof is rather straightforward, and also very tedious. For this reason, we will give only part of it.

Consider the quantity $R_1 = N^{-1}\sum_{i\le N-1}\sigma_i^1\langle\sigma_i\rangle_-$. Before we start the main computation, we need as a preliminary result that

$$\mathsf{E}\,\langle(R_1-q)^2\rangle \le \frac{L}{N} \tag{2.228}$$

$$|\mathsf{E}\,\langle R_1 - q\rangle| \le \frac{L}{N}\,. \tag{2.229}$$

To see this, we observe that, even though R_1 depends on the disorder through $\langle\sigma_i\rangle_-$, it does not depend on the r.v. $(g_i)_{i\le N}$. This is all that is required for (2.169) to hold, and as in the proof of Proposition 2.3.2, we see that

$$\nu((R_1-q)^2) \le L\,\nu_0((R_1-q)^2) = L\,\mathsf{E}\,\langle(R_1-q)^2\rangle_- \le L\,\mathsf{E}\,\langle(R_{1,2}^- - q)^2\rangle_-\,,$$

By Jensen's inequality the fact that the last term is $\le L/N$ follows from (2.149) (used on the $(N-1)$-spin system) and (2.200). This proves (2.228). To prove (2.229), we first see that $|\nu_0(R_1-q)| \le L/N$, using (2.200) and (2.215) for the $(N-1)$-spin system. Then we show that $|\nu_0'(R_1-q)| \le L/N$ using (2.149) (again for the $(N-1)$-spin system) and finally we show that $|\nu_t''(R_1-1)| \le L/N$.

To establish (2.227) it suffices to prove that

$$\mathsf{E}\sum_{i\le N-1}\frac{g_i}{\sqrt{N}}\left(\langle\sigma_i\rangle - \langle\sigma_i\rangle_-\right)\langle\sigma_N\rangle = \beta(1-q)\,\mathsf{E}\,\langle\sigma_N\rangle^2 + O(2) \tag{2.230}$$

$$\mathsf{E}\sum_{i\le N-1}\frac{g_i^2}{N}\left(\langle\sigma_i\rangle - \langle\sigma_i\rangle_-\right)^2 = O(2) \tag{2.231}$$

$$\begin{aligned}\mathsf{E}\sum_{i\ne j}\frac{g_i\,g_j}{N^2}&\left(\langle\sigma_i\rangle - \langle\sigma_i\rangle_-\right)\left(\langle\sigma_j\rangle - \langle\sigma_j\rangle_-\right)\\ &= \beta^2(1-q)^2\,\mathsf{E}\,\langle\sigma_N\rangle^2 + O(2)\,.\end{aligned} \tag{2.232}$$

Here, as usual, $O(2)$ denotes a quantity bounded by L/N. The proof of these relations relies on integration by parts, and to perform this we observe that (abusing notation)

$$\frac{\partial}{\partial g_i}\langle f\rangle = \frac{\beta}{\sqrt{N}}\left(\langle \sigma_N\,\sigma_i\, f\rangle - \langle \sigma_N\,\sigma_i\rangle\langle f\rangle\right). \tag{2.233}$$

Thus, the left-hand side of (2.230) is I + II, where

$$\mathrm{I} = \frac{\beta}{N}\,\mathsf{E}\sum_{i\le N-1}\left(\langle\sigma_N\rangle - \langle\sigma_N\,\sigma_i\rangle\langle\sigma_i\rangle\right)\langle\sigma_N\rangle \tag{2.234}$$

$$\mathrm{II} = \frac{\beta}{N}\,\mathsf{E}\sum_{i\le N-1}\left(\langle\sigma_i\rangle - \langle\sigma_i\rangle_-\right)\left(\langle\sigma_i\rangle - \langle\sigma_N\rangle\,\langle\sigma_i - \sigma_N\rangle\right).$$

Using replicas, we have

$$\begin{aligned}\mathrm{I} &= \frac{\beta}{N}\,\mathsf{E}\sum_{i\le N-1}\left(\langle\sigma_N^1\,\sigma_N^2\rangle - \langle\sigma_N^1\,\sigma_N^2\,\sigma_i^2\,\sigma_i^3\rangle\right)\\ &= \beta\,\mathsf{E}\left(\langle\sigma_N^1\,\sigma_N^2\rangle - \langle\sigma_N^1\,\sigma_N^2\,R_{2,3}\rangle\right) + O(2)\\ &= \beta\left(\nu(R_{1,2}) - \nu(R_{1,2}\,R_{2,3})\right) + O(2)\end{aligned}$$

by symmetry between sites. Now

$$\begin{aligned}\nu(R_{1,2}\,R_{2,3}) &= q^2 + 2\,q\,\nu(R_{1,2}-q) + \nu((R_{1,2}-q)(R_{2,3}-q))\\ &= q^2 + O(2)\end{aligned} \tag{2.235}$$

using (2.149) and (2.215), and thus $\mathrm{I} = \beta(q-q^2)+O(2)$. In a similar manner we have

$$\mathrm{II} = \beta\,\nu(R_{1,2} - R_{1,2}\,\sigma_N^1\,\sigma_N^3 - R_1 + R_1\,\sigma_N^1\,\sigma_N^3) + O(2). \tag{2.236}$$

By symmetry among sites, we have

$$\nu\left(\left(R_1 + \frac{1}{N}\,\sigma_N^1\langle\sigma_N\rangle_-\right)\sigma_N^1\,\sigma_N^3\right) = \nu\left(\left(R_1 + \frac{1}{N}\,\sigma_N^1\langle\sigma_N\rangle_-\right)R_{1,3}\right)$$

and thus $\nu(R_1\,\sigma_N^1\,\sigma_N^3) = \nu(R_1\,R_{1,3}) + O(2)$. Hence, by (2.236) we have

$$\mathrm{II} = \beta\,\nu(R_{1,2} - R_{1,2}\,R_{1,3} - R_1 + R_1\,R_{1,3}) + O(2)$$

and proceeding as in (2.235), using (2.228) and (2.229) we see that $\mathrm{II} = O(2)$. Since

$$\mathsf{E}\,\langle\sigma_N\rangle^2 = \mathsf{E}\,\langle R_{1,2}\rangle = q + O(2)$$

this proves (2.230).

The proof of (2.231) is straightforward, using (2.199), the fact that

$$\mathsf{E}(g^2F(g)) = \mathsf{E}g^2\,\mathsf{E}(F(g) + F''(g))$$

and that each integration by parts brings an extra factor $1/\sqrt{N}$.

The proof of (2.232) requires two integrations by parts. It is *really* tedious, as there are more than 20 terms, but it is entirely similar to that of (2.230). □

2.5 Exponential Inequalities

The goal of this section is to control the higher moments of $R_{1,2} - q$.

Theorem 2.5.1. *There exists $\beta_0 > 0$ such that if $\beta < \beta_0$, we have*

$$\forall k \geq 0\,, \quad \nu((R_{1,2} - q)^{2k}) = \mathsf{E}\langle (R_{1,2} - q)^{2k} \rangle \leq \left(\frac{Lk}{N}\right)^k . \tag{2.237}$$

Of course here q is as in (2.148). The equality $\exp x^2 = \sum_{k \geq 1} x^{2k}/k!$ and the fact that $(k/L)^k \leq k! \leq k^k$ show that (2.237) is equivalent to a statement of the type

$$\nu\left(\exp \frac{N}{L}(R_{1,2} - q)^2\right) \leq L\,, \tag{2.238}$$

hence the title of this section. We have proved (2.237) for $k = 1$ in Section 2.4. We will proceed by induction over k. The induction hypothesis will be that

$$\forall \ell\,,\ 1 \leq \ell \leq k\,,\ \nu((R_{1,2} - q)^{2\ell}) \leq \left(\frac{L_0\,\ell}{N}\right)^\ell . \tag{2.239}$$

According to our convention, L_0 denotes a fixed number. We will show that, if L_0 has been chosen large enough, knowing (2.239) we can prove that

$$\nu((R_{1,2} - q)^{2k+2}) \leq \frac{L_0^{k+1}(k+1)^{k+1}}{N^{k+1}}\,, \tag{2.240}$$

and this will complete the proof of Theorem 2.5.1. The argument is an elaborate version of the argument of Theorem 2.4.2.

A first observation is that we can assume $k \leq N - 1$. Indeed, since we can assume $L_0 \geq 4$, for $\ell \geq N$ we have

$$\nu((R_{1,2} - q)^{2\ell}) \leq 4^\ell \leq L_0^\ell \left(\frac{\ell}{N}\right)^\ell .$$

Lemma 2.5.2. *We assume (2.239) and $k \leq N - 1$. Then, if $L_0 \geq 4$ we have*

$$\forall j \leq 2k\,,\ \nu(|R_{1,2} - q|^j) \leq \left(\frac{L_0(j+1)}{N}\right)^{j/2} \tag{2.241}$$

$$\nu((R_{1,2}^- - q)^{2k}) \leq 3\left(\frac{L_0(k+1)}{N}\right)^k . \tag{2.242}$$

Proof. To prove (2.241), if j is even we use (2.239) for $\ell = j/2$. If j is odd, say $j = 2\ell - 1$, we use Hölder's inequality to write

$$\nu(|R_{1,2}-q|^j) \le (\nu((R_{1,2}-q)^{2\ell}))^{j/2\ell} \le \left(\left(\frac{L_0\,\ell}{N}\right)^{\ell}\right)^{j/2\ell}$$
$$= \left(\frac{L_0\,\ell}{N}\right)^{j/2} = \left(\frac{L_0(j+1)}{2N}\right)^{j/2} .$$

To prove (2.242), we expand the expression

$$(R^-_{1,2}-q)^{2k} = \left(R_{1,2}-q-\frac{\varepsilon_1\varepsilon_2}{N}\right)^{2k}$$

to get, using (2.241),

$$\nu((R^-_{1,2}-q)^{2k}) \le \sum_{0\le j\le 2k}\binom{2k}{j}\frac{1}{N^{2k-j}}\left(\frac{L_0(j+1)}{N}\right)^{j/2}$$
$$\le \sum_{0\le j\le 2k}\binom{2k}{j}\frac{1}{N^{2k-j}}\left(\frac{L_0(k+1)}{N}\right)^{j/2}$$
$$= \left(\frac{1}{N}+\sqrt{\frac{L_0(k+1)}{N}}\right)^{2k}$$
$$= \left(\frac{L_0(k+1)}{N}\right)^{k}\left(1+\frac{1}{\sqrt{NL_0(k+1)}}\right)^{2k} .$$

Since $k \le N$, we have $(1+1/\sqrt{NL_0(k+1)})^{2k} \le \exp(2\sqrt{k/NL_0}) \le e \le 3$ if we assume $L_0 \ge 4$. □

Proof of Theorem 2.5.1. We have to prove (2.240). Using symmetry between sites, we write

$$\nu((R_{1,2}-q)^{2k+2}) = \nu(f) \tag{2.243}$$

where

$$f = (\varepsilon_1\varepsilon_2-q)(R_{1,2}-q)^{2k+1} . \tag{2.244}$$

We use (2.174) with $n=2$, $\tau_1=(2k+2)/(2k+1)$ and $\tau_2=2k+2$ to get

$$\nu(f) \le \nu_0(f) + 16\beta^2\exp 16\beta^2\nu((R_{1,2}-q)^{2k+2}) ,$$

so that, if $\beta\le\beta_0$ and

$$32\beta_0^2\exp 16\beta_0^2 \le 1 , \tag{2.245}$$

we have $\nu((R_{1,2}-q)^{2k+2}) \le \nu_0(f) + \nu((R_{1,2}-q)^{2k+2})/2$, and hence

$$\nu((R_{1,2}-q)^{2k+2}) \le 2\,\nu_0(f) = 2\,\nu_0((\varepsilon_1\varepsilon_2-q)(R_{1,2}-q)^{2k+1}) \tag{2.246}$$
$$= 2\,\nu_0((\varepsilon_1\varepsilon_2-q)((R_{1,2}-q)^{2k+1}-(R^-_{1,2}-q)^{2k+1}))$$

since $\nu_0((\varepsilon_1\varepsilon_2 - q)(R^-_{1,2} - q)^{2k+1}) = 0$. Using the inequality $|x^{2k+1} - y^{2k+1}| \le (2k+1)|x-y|(x^{2k} + y^{2k})$, and assuming $\exp(16\,\beta_0^2) \le 2$, we get

$$\begin{aligned}\nu((R_{1,2} - q)^{2k+2}) &\le 4\,\frac{(2k+1)}{N}\,\nu_0((R_{1,2} - q)^{2k} + (R^-_{1,2} - q)^{2k}) \\ &\le 16\,\frac{k+1}{N}\,\nu((R_{1,2} - q)^{2k} + (R^-_{1,2} - q)^{2k})\end{aligned}$$

using (2.173) and the fact that $2k+1 \le 2(k+1)$ in the last inequality. Using (2.239) for $\ell = k$ and (2.242), we see that

$$\nu((R_{1,2} - q)^{2k+2}) \le 64\,\frac{k+1}{N}\left(\frac{L_0(k+1)}{N}\right)^k = \left(\frac{L_0(k+1)}{N}\right)^{k+1}$$

if $L_0 = 64$. This concludes the proof of (2.240). □

Proposition 2.5.3. *If $\beta \le \beta_0$, we have, for a function f on Σ_N^n,*

$$|\nu(f) - \nu_0(f)| \le \frac{K(n)}{\sqrt{N}}\,\nu(f^2)^{1/2} \tag{2.247}$$

$$|\nu(f) - \nu_0(f) - \nu_0'(f)| \le \frac{K(n)}{N}\,\nu(f^2)^{1/2}\,. \tag{2.248}$$

Proof. The proof is almost identical to that of Corollary 2.3.4, now that we know (2.237). □

Proposition 2.5.3 will be our main tool to prove central limit theorems in the next section. As a warm up, we prove (2.215).

Lemma 2.5.4. *We have*

$$|\nu(R_{1,2} - q)| \le L/N\,.$$

Proof. We use symmetry among the sites to get

$$\nu(R_{1,2} - q) = \nu(f)$$

for $f = \varepsilon_1\varepsilon_2 - q$. Thus, by (2.248),

$$|\nu(R_{1,2} - q) - \nu_0(f) - \nu_0'(f)| \le \frac{L}{N}\,. \tag{2.249}$$

By (2.167) and since $q = \mathsf{E}\,\mathrm{th}^2 Y$ we have $\nu_0(f) = 0$. By (2.169) we have

$$\begin{aligned}\nu_0'(f) &= \beta^2\nu_0((\varepsilon_1\varepsilon_2 - q)\,\varepsilon_1\varepsilon_2(R^-_{1,2} - q)) \\ &\quad - 4\,\beta^2\nu_0((\varepsilon_1\varepsilon_2 - q)\,\varepsilon_1\varepsilon_3(R^-_{1,3} - q)) \\ &\quad + 3\,\beta^2\nu_0((\varepsilon_1\varepsilon_2 - q)\,\varepsilon_3\varepsilon_4(R^-_{1,4} - q))\,.\end{aligned}$$

If we set $\widehat{q} = \mathsf{E}\,\mathrm{th}^4 Y$, we get from (2.167) that

$$\nu_0'(f) = \beta^2(1 - 4\,q + 3\,\widehat{q})\,\nu_0(R_{1,2}^- - q)\,.$$

Using (2.247) we have

$$|\nu_0(R_{1,2}^- - q) - \nu(R_{1,2} - q)| \le \frac{L}{N}\,,$$

so that (2.249) implies that

$$(1 - \beta^2(1 - 4\,q + 3\,\widehat{q}))\,|\nu(R_{1,2} - q)| \le \frac{L}{N}\,.$$

Also, since $\widehat{q} \le q$, we have $\beta^2(1 - 4\,q + 3\,\widehat{q}) \le 1/4$ for $\beta \le 1/2$. □

Let us point out for further use that the previous argument only relies on (2.149), i.e. the case $k = 1$ of (2.237). (This is because when $|f| \le 2$, (2.247) and (2.248) require only (2.149).) The quantity $1 - \beta^2(1 - 4\,q + 3\,\widehat{q})$ will be important later.

In the rest of this section, we prove Lemma 2.4.13. More generally, given a function f on Σ_{N-1}^n, and a number $p \le n$, we would like to bound

$$A = \mathsf{E}\left(\left\langle f^- \prod_{\ell \le p} \varepsilon_\ell \right\rangle - \langle f^- \rangle_-\,\mathrm{th}^p(g(\mathbf{c}) + h)\right)^2\,.$$

Consider the variables $g_t(\boldsymbol{\rho})$ as given in (2.161), and consider $g_t(\mathbf{c})$ that is defined similarly. We consider the function

$$\varphi(t) = \mathsf{E}((U(t) - V(t))^2)$$

where

$$U(t) = \left\langle f^- \prod_{\ell \le p} \varepsilon_\ell \right\rangle_t$$

$$V(t) = \frac{\left\langle \mathrm{Av} f^- \prod_{\ell \le p} \varepsilon_\ell \exp \sum_{\ell \le n} \varepsilon_\ell (g_t(\mathbf{c}) + h) \right\rangle_-}{\langle \mathrm{ch}(g_t(\mathbf{c}) + h)\rangle_-^n} = \langle f^- \rangle_-\,\mathrm{th}^p(g_t(\mathbf{c}) + h)\,.$$

Thus $V(t)$ is defined as $U(t)$, except that all the occurrences of $g_t(\boldsymbol{\rho}^\ell)$ are replaced by $g_t(\mathbf{c})$. Since $g_0(\mathbf{c}) = g_0(\boldsymbol{\rho}) = \beta\, z\sqrt{q}$, we have $U(0) = V(0)$, and $\varphi(0) = 0$. We write $\varphi(t) = \sum_{j \le 3} \varphi_j(t)$, where

$$\varphi_1(t) = \mathsf{E}(U(t)^2)\,;\ \ \varphi_2(t) = -2\,\mathsf{E}(U(t)\,V(t))\,;\ \ \varphi_3(t) = \mathsf{E}(V(t)^2)\,.$$

Using replicas we see that $\varphi_1(t) = \nu_t(f^\sim)$, where

$$f^\sim(\boldsymbol{\sigma}^1, \ldots, \boldsymbol{\sigma}^{2n}) = \prod_{\ell \le p} \varepsilon_\ell \, \varepsilon_{n+\ell} \, f^-(\boldsymbol{\sigma}^1, \ldots, \boldsymbol{\sigma}^n) \, f^-(\boldsymbol{\sigma}^{n+1}, \ldots, \boldsymbol{\sigma}^{2n}) \, .$$

We can use (2.248), so that if $|f^-| \le 1$ (and thus $|f^\sim| \le 1$) we have

$$|\varphi_1(1) - \varphi_1(0) - \varphi_1'(0)| \le \frac{L}{N} \, . \tag{2.250}$$

We now show that if $p = n = 1$ and $f^- = 1$ (resp. $f^- = \sigma_1$) we have $|\varphi_1'(0)| \le L/N$. In this case we have $f^\sim = \varepsilon_1 \varepsilon_2$ (resp. $f^\sim = \sigma_1^1 \sigma_1^2 \, \varepsilon_1 \varepsilon_2$) and we compute $\varphi_1'(0) = \nu_0'(f^\sim)$ using (2.169) and (2.167). When $f^- = 1$, $\nu_0'(f^\sim)$ is sum of terms $\pm \beta^2 \mathsf{E} \, \mathrm{th}^a \, Y \, \nu_0(R^-_{\ell,\ell'} - q)$, ($a = 0, 2$ or 4) so that by (2.247) and Lemma 2.5.4, we see that $|\varphi_1'(0)| \le L/N$. When $f^- = \sigma_1^1 \sigma_1^2$ the term $\nu_0(R^-_{\ell,\ell'} - q)$ is replaced by $\nu_0(\sigma_1^1 \sigma_1^2 (R^-_{\ell,\ell'} - q))$. Now

$$|\nu_0(\sigma_1^1 \sigma_1^2 (R^-_{\ell,\ell'} - q)) - \nu_0(\sigma_1^1 \sigma_1^2 (R_{\ell,\ell'} - q))| \le \frac{L}{N}$$

and using (2.247)

$$|\nu_0(\sigma_1^1 \sigma_1^2 (R^-_{\ell,\ell'} - q)) - \nu(\sigma_1^1 \sigma_2^2 (R_{\ell,\ell'} - q))| \le \frac{L}{N} \, .$$

Using symmetry between sites,

$$\nu(\sigma_1^1 \sigma_1^2 (R_{\ell,\ell'} - q)) = \nu(R_{1,2}(R_{\ell,\ell'} - q)) = \nu((R_{1,2} - q)(R_{\ell,\ell'} - q)) + q \, \nu(R_{\ell,\ell'} - q) \, .$$

Thus, using (2.239) and Lemma 2.5.4 we see that $|\varphi_1'(0)| \le L/N$.

From (2.250) we see that (when $p = n = 1$, and $f^- = 1$ or $f^- = \sigma_1$) we have

$$|\varphi_1(1) - \varphi_1(0)| \le \frac{L}{N} \, . \tag{2.251}$$

Thus, since $A = \varphi(1)$, and $\varphi(0) = 0$, we will have proved Lemma 2.4.13 if we can prove that (2.251) also holds for φ_2 and φ_3. For a function f on Σ_N^n, let us define

$$\langle f \rangle_t' = \frac{\left\langle \mathrm{Av} f \exp \sum_{\ell \le n} \varepsilon_\ell (g_t(\mathbf{c}) + h) \right\rangle_-}{(\mathrm{ch}(g_t(\mathbf{c}) + h))^n} \, , \tag{2.252}$$

so that, using replicas

$$\varphi_3(t) = \mathsf{E} \, \langle f^\sim \rangle_t' \, .$$

We have to learn how to compute the derivatives of this quantity, and this is done by repeating the computation leading to (2.169). The result of this computation is obvious, because the difference between (2.252) and the definition of $\langle f \rangle_t$ is that all occurrences of quantities $g(\boldsymbol{\rho}^\ell)$ are replaced by $g(\mathbf{c})$. Thus formula (2.169) remains valid, provided one replaces $\nu_t(\cdot)$ by $\mathsf{E}\langle \cdot \rangle_t'$ and

$R_{\ell,\ell'}$ by $\mathbf{c}\cdot\mathbf{c}/N = \|\mathbf{c}\|^2/N$. To prove that (2.250) holds for φ_3 rather than for φ_1, it suffices to prove that $\psi(t) := \mathsf{E}\langle(\|\mathbf{c}\|^2/N - q)^2\rangle'_t \le L/N$. It follows from (2.202) that this is true for $t = 0$; and it is straightforward to see that $|\psi'(t)| \le L\psi(t)$.

To prove that $|\varphi'_3(0)| \le L/N$, we recall that if f^- is a function on Σ^n_{N-1}, then

$$\left\langle f^-\left(\frac{\|\mathbf{c}\|^2}{N} - q\right)\right\rangle_- = \langle f^-(R^-_{n+1,n+2} - q)\rangle_- \,,$$

and thus

$$\mathsf{E}\left\langle f^-\left(\frac{\|\mathbf{c}\|^2}{N} - q\right)\right\rangle = \nu_0(f^-(R^-_{n+1,n+2} - q))$$

and this shows that the same analysis as for φ_1 carries over.

The case of φ_2 represents a "mixed" situation between the cases of φ_1 and φ_3. We write $\varphi_2(t) = \mathsf{E}\langle f^\sim\rangle''_t$, where $\langle\cdot\rangle''_t$ is defined in a manner similar to ν_t, but with some of the terms $g(\boldsymbol{\rho}^\ell)$ replaced by $g(\mathbf{c})$. (Roughly speaking, those for $n+1 \le \ell \le 2n$.) In (2.190) some of the terms $R^-_{\ell,\ell'}$ have to be replaced by $N^{-1}\sum_{i\le N-1}\sigma_i^\ell\langle\sigma_i\rangle_-$ or by $\|\mathbf{c}\|^2/N$. The analysis, however, carries over similarly. □

2.6 Second Moment Computations and the Almeida-Thouless Line

In this section, we compute the average of the square of the correlation between two spins, a result of fundamental importance. We then turn to more challenging results, such as the computation of $\mathsf{E}\langle(R_{1,2} - q)^2\rangle$. The results of this section will be used crucially in Sections 2.8 and 2.9.

Throughout this section we set

$$\widehat{q} = \mathsf{E}\,\mathrm{th}^4 Y \tag{2.253}$$

where, as usual, $Y = \beta z\sqrt{q} + h$.

Proposition 2.6.1. *If $\beta \le \beta_0$ we have*

$$\left|\mathsf{E}\left\langle\left(\frac{(\boldsymbol{\sigma}^1 - \boldsymbol{\sigma}^2)\cdot(\boldsymbol{\sigma}^3 - \boldsymbol{\sigma}^4)}{N}\right)^2\right\rangle - \frac{4(1 - 2q + \widehat{q})}{N(1 - \beta^2(1 - 2q + \widehat{q}))}\right| \le \frac{L}{N^{3/2}}\,. \tag{2.254}$$

One can note that

$$1 - 2q + \widehat{q} = \mathsf{E}\frac{1}{\mathrm{ch}^4 Y}\,.$$

This nice expression seems accidental, and, in the long range it seems better to keep q and $\widehat{q}$ as parameters.

Since a square is positive, (2.254) can hold only when

$$\beta^2(1-2q+\widehat{q}) = \beta^2\,\mathsf{E}\frac{1}{\operatorname{ch}^4 Y} < 1\,. \tag{2.255}$$

It is believed that the domain of parameters (2.255) is exactly the domain where the replica-symmetric formula (and, more generally the other results of Sections 2.4 to 2.6) is valid. This question will be the object of Sections 2.8 and 2.9. The boundary of the domain (2.255) is called the A-T line.

Corollary 2.6.2. *If $\beta \le \beta_0$, we have*

$$\left|\mathsf{E}((\langle\sigma_1\sigma_2\rangle - \langle\sigma_1\rangle\langle\sigma_2\rangle)^2) - \frac{\beta^2(1-2q+\widehat{q})^2}{N(1-\beta^2(1-2q+\widehat{q}))}\right| \le \frac{L}{N^{3/2}}\,. \tag{2.256}$$

Proof. We observe that

$$(\varepsilon_1-\varepsilon_2)^2(\varepsilon_3-\varepsilon_4)^2 = 4(1-\varepsilon_1\varepsilon_2)(1-\varepsilon_3\varepsilon_4)\,,$$

so that by (2.174) and (2.167) we have

$$|\nu((\varepsilon_1-\varepsilon_2)^2(\varepsilon_3-\varepsilon_4)^2) - 4(1-2q+\widehat{q})| \le \frac{L}{\sqrt{N}}\,. \tag{2.257}$$

The result then follows from (2.184) and (2.254). □

Proposition 2.6.3. *Consider a function f^- on Σ^n_{N-1}. Then*

$$\begin{aligned}&\nu_0'((\varepsilon_1-\varepsilon_2)(\varepsilon_3-\varepsilon_4)\,f^-)\\ &= \beta^2(1-2q+\widehat{q})\,\nu_0(f^-(R^-_{1,3}-R^-_{1,4}-R^-_{2,3}+R^-_{2,4}))\end{aligned} \tag{2.258}$$

and

$$\begin{aligned}&|\nu((\varepsilon_1-\varepsilon_2)(\varepsilon_3-\varepsilon_4)\,f^-) - \beta^2(1-2q+\widehat{q})\,\nu(f^-(R_{1,3}-R_{1,4}-R_{2,3}+R_{2,4}))|\\ &\quad \le \frac{K(n)}{N}\,\nu((f^-)^4)^{1/4}\,.\end{aligned} \tag{2.259}$$

With minimal extra work, one could get $\nu((f^-)^2)^{1/2}$ in the error term; but this makes no difference for our purposes.

Proof. We use (2.167) to see that

$$\nu_0((\varepsilon_1-\varepsilon_2)(\varepsilon_3-\varepsilon_4)\,\varepsilon_\ell\,\varepsilon_{\ell'}\,f^-(R^-_{\ell,\ell'}-q)) = a(\ell,\ell')\,\nu_0(f^-(R^-_{\ell,\ell'}-q))$$

where

$$a(\ell,\ell') = \nu_0((\varepsilon_1-\varepsilon_2)(\varepsilon_3-\varepsilon_4)\,\varepsilon_\ell\,\varepsilon_{\ell'})\,.$$

Thus when $\ell < \ell'$, $a(\ell,\ell') = 0$ unless $\ell \in \{1,2\}$, $\ell' \in \{3,4\}$, and

$$\begin{aligned} a(1,3) = a(2,4) = -a(1,4) = -a(2,3) &= \nu_0((1-\varepsilon_1\varepsilon_2)(1-\varepsilon_3\varepsilon_4)) \\ &= 1 - 2\nu_0(\varepsilon_1\varepsilon_2) + \nu_0(\varepsilon_1\varepsilon_2\varepsilon_3\varepsilon_4) \\ &= 1 - 2q + \widehat{q} \,. \end{aligned}$$

We then use (2.169) to get (2.258). Using (2.248) and (2.258) we get

$$\begin{aligned} &\big|\nu((\varepsilon_1-\varepsilon_2)(\varepsilon_3-\varepsilon_4)\, f^-) \\ &\qquad -\beta^2(1-2q+\widehat{q})\,\nu_0(f^-(R^-_{1,3}-R^-_{1,4}-R^-_{2,3}+R^-_{2,4}))\big| \\ &\le \frac{K(n)}{N}\,\nu((f^-)^2)^{1/2} \,. \end{aligned}$$

Also,

$$\begin{aligned} &|\nu_0(f^-(R^-_{1,3}-R^-_{1,4}-R^-_{2,3}+R^-_{2,4})) - \nu_0(f^-(R_{1,3}-R_{1,4}-R_{2,3}+R_{2,4}))| \\ &\le \frac{K(n)}{N}\,\nu_0(|f^-|) = \frac{K(n)}{N}\,\nu(|f^-|) \,. \end{aligned}$$

Using that

$$R_{1,3}-R_{1,4}-R_{2,3}+R_{2,4} = (R_{1,3}-q)-(R_{1,4}-q)-(R_{2,3}-q)+(R_{2,4}-q) \,,$$

(2.247) and the Cauchy-Schwarz inequality we get

$$\begin{aligned} &|\nu_0(f^-(R_{1,3}-R_{1,4}-R_{2,3}+R_{2,4}) - \nu(f^-(R_{1,3}-R_{1,4}-R_{2,3}+R_{2,4}))| \\ &\le \frac{K(n)}{\sqrt{N}}\,\nu((f^-)^4)^{1/4}\,\nu((R_{1,2}-q)^4)^{1/4} \le \frac{K(n)}{N}\,\nu((f^-)^4)^{1/4} \,. \end{aligned}$$

Combining these proves (2.259). □

Proof of Proposition 2.6.1. This proof should be totally predictable to the reader. We observe that

$$\frac{(\boldsymbol{\sigma}^1-\boldsymbol{\sigma}^2)\cdot(\boldsymbol{\sigma}^3-\boldsymbol{\sigma}^4)}{N} = R_{1,3}-R_{1,4}-R_{2,3}+R_{2,4} \tag{2.260}$$

so that, using symmetry between sites

$$\nu\left(\left(\frac{(\boldsymbol{\sigma}^1-\boldsymbol{\sigma}^2)\cdot(\boldsymbol{\sigma}^3-\boldsymbol{\sigma}^4)}{N}\right)^2\right) = \nu(f) \,, \tag{2.261}$$

where

$$f = (\varepsilon_1-\varepsilon_2)(\varepsilon_3-\varepsilon_4)(R_{1,3}-R_{1,4}-R_{2,3}+R_{2,4}) \,. \tag{2.262}$$

Thus

$$\nu(f) = \frac{1}{N}\nu((\varepsilon_1 - \varepsilon_2)^2(\varepsilon_3 - \varepsilon_4)^2) + \nu((\varepsilon_1 - \varepsilon_2)(\varepsilon_3 - \varepsilon_4)\, f^-) \qquad (2.263)$$

where

$$f^- = R^-_{1,3} - R^-_{1,4} - R^-_{2,3} + R^-_{2,4} \,.$$

We observe that $\nu((f^-)^4)^{1/4} \le LN^{-1/2}$, and that

$$|\nu(f^-(R_{1,3} - R_{1,4} - R_{2,3} + R_{2,4})) - \nu((R_{1,3} - R_{2,3} - R_{1,4} + R_{2,4})^2)| \le \frac{L}{N^{3/2}} \,.$$

Combining with (2.259), (2.257) and (2.263), we get

$$\left|(1 - \beta^2(1 - 2\,q + \widehat{q}))\,\nu\left(\left(\frac{(\boldsymbol{\sigma}^1 - \boldsymbol{\sigma}^2)\cdot(\boldsymbol{\sigma}^3 - \boldsymbol{\sigma}^4)}{N}\right)^2\right) - \frac{4}{N}(1 - 2\,q + \widehat{q})\right|$$
$$\le \frac{L}{N^{3/2}} \,. \qquad (2.264)$$

□

The purpose of the present section is not only to make second moment computations, but also to perform them in a manner that will pave the way for the higher moment computations of the next section. The interesting phenomenon is that when $h \neq 0$ the quantities $R_{\ell,\ell'} - q$ are not the proper generalization of the quantities $R_{\ell,\ell'}$ when $h = 0$. It is much more fruitful to consider the quantities

$$T_{\ell,\ell'} = \frac{(\boldsymbol{\sigma}^\ell - \mathbf{b})\cdot(\boldsymbol{\sigma}^{\ell'} - \mathbf{b})}{N}\;;\; T_\ell = \frac{(\boldsymbol{\sigma}^\ell - \mathbf{b})\cdot\mathbf{b}}{N}\;;\; T = \frac{\mathbf{b}\cdot\mathbf{b}}{N} - q\,, \qquad (2.265)$$

where, as usual, $\mathbf{b} = \langle\boldsymbol{\sigma}\rangle = (\langle\sigma_i\rangle)_{i\le N}$. This is because these quantities turn out to be "independent" in the sense of (2.106). On the other hand, they allow to reconstruct the quantities $R_{\ell,\ell'}$ by the formula

$$R_{\ell,\ell'} - q = T_{\ell,\ell'} + T_\ell + T_{\ell'} + T \,. \qquad (2.266)$$

Proposition 2.6.4. *If $\beta \le \beta_0$, we have*

$$|\nu(T^2_{1,2}) - A^2| \le \frac{L}{N^{3/2}} \qquad (2.267)$$

where

$$A^2 = \frac{1 - 2\,q + \widehat{q}}{N(1 - \beta^2(1 - 2\,q + \widehat{q}))} \,. \qquad (2.268)$$

Proof. One must keep in mind that $T_{1,2}$ is NOT a function on Σ_N^n, because it depends on the disorder through $\mathbf{b}$. But we have

$$\nu(T_{1,2}^2) = \nu\left(\frac{(\boldsymbol{\sigma}^1 - \boldsymbol{\sigma}^3)\cdot(\boldsymbol{\sigma}^2 - \boldsymbol{\sigma}^4)}{N}\,\frac{(\boldsymbol{\sigma}^1 - \boldsymbol{\sigma}^5)\cdot(\boldsymbol{\sigma}^2 - \boldsymbol{\sigma}^6)}{N}\right),$$

or to match better with the notation of Proposition 2.6.3

$$\nu(T_{1,2}^2) = \nu\left(\frac{(\boldsymbol{\sigma}^1 - \boldsymbol{\sigma}^2)\cdot(\boldsymbol{\sigma}^3 - \boldsymbol{\sigma}^4)}{N}\,\frac{(\boldsymbol{\sigma}^1 - \boldsymbol{\sigma}^5)\cdot(\boldsymbol{\sigma}^3 - \boldsymbol{\sigma}^6)}{N}\right) = \nu(f)\,,$$

using symmetry among sites, where

$$\begin{aligned} f &= (\varepsilon_1 - \varepsilon_2)(\varepsilon_3 - \varepsilon_4)(R_{1,3} - R_{1,6} - R_{5,3} + R_{5,6}) \\ &= \frac{1}{N}(\varepsilon_1 - \varepsilon_2)(\varepsilon_3 - \varepsilon_4)(\varepsilon_1 - \varepsilon_5)(\varepsilon_3 - \varepsilon_6) + (\varepsilon_1 - \varepsilon_2)(\varepsilon_3 - \varepsilon_4)\,f^-\,, \end{aligned}$$

for $f^- = R_{1,3}^- - R_{1,6}^- - R_{5,3}^- + R_{5,6}^-$. We observe that

$$\begin{aligned} &\langle(\varepsilon_1 - \varepsilon_2)(\varepsilon_3 - \varepsilon_4)(\varepsilon_1 - \varepsilon_5)(\varepsilon_3 - \varepsilon_6)\rangle_0 \\ &= \langle(\varepsilon_1 - \varepsilon_2)(\varepsilon_1 - \varepsilon_5)\rangle_0^2 = \langle 1 - \varepsilon_1\varepsilon_5 - \varepsilon_1\varepsilon_2 + \varepsilon_2\varepsilon_5\rangle_0^2 \\ &= (1 - \operatorname{th}^2 Y)^2 = 1 - 2\operatorname{th}^2 Y + \operatorname{th}^4 Y \end{aligned} \tag{2.269}$$

so that

$$\nu_0((\varepsilon_1 - \varepsilon_2)(\varepsilon_3 - \varepsilon_4)(\varepsilon_1 - \varepsilon_5)(\varepsilon_3 - \varepsilon_6)) = 1 - 2q + \widehat{q}\,.$$

One then proceeds exactly as in Proposition 2.6.1 to prove (2.267). □

Proposition 2.6.5. *If $\ell < \ell'$ and $(\ell, \ell') \neq (1,2)$, we have*

$$|\nu(T_{1,2}\,T_{\ell,\ell'})| \leq \frac{L}{N^{3/2}}\,. \tag{2.270}$$

For any ℓ we have

$$|\nu(T_{1,2}\,T_{\ell})| \leq \frac{L}{N^{3/2}} \tag{2.271}$$

$$|\nu(T_{1,2}\,T)| \leq \frac{L}{N^{3/2}}\,. \tag{2.272}$$

Proof. In fact we have much better, since all the left-hand sides are 0. To see this, e.g. in the case of (2.271), if $\ell \neq 1$ we simply integrate in the first replica, using linearity. □

This easy success is accidental, as this method would not work e.g. to study $\nu(T_{1,2}T_{2,3}T_{1,3})$. We pointed out in Theorem 2.3.9 that the nice behavior of the family $R_{\ell,\ell'}$ arose from the fact that $\nu_0(\varepsilon_1\varepsilon_2\,\varepsilon_\ell\varepsilon_{\ell'}) = 0$ if $(\ell,\ell') \neq (1,2)$. The following lemma explains, in a similar manner, the joint behavior of the family $T_{\ell,\ell'}$, T_ℓ, T.

Lemma 2.6.6. *Consider* $\ell < \ell'$, *and 4 distinct indices* $j(1), j(2), j(3), j(4)$, *none of which is one of the indices* $1, 2, \ell, \ell'$. *Then, we have*

$$(\ell,\ell') \neq (1,2) \Rightarrow \nu_0((\varepsilon_1-\varepsilon_{j(1)})(\varepsilon_2-\varepsilon_{j(2)})(\varepsilon_\ell-\varepsilon_{j(3)})(\varepsilon_{\ell'}-\varepsilon_{j(4)})) = 0 \quad (2.273)$$

$$\nu_0((\varepsilon_1 - \varepsilon_{j(1)})(\varepsilon_2 - \varepsilon_{j(2)})(\varepsilon_\ell - \varepsilon_{j(3)})\,\varepsilon_{j(4)}) = 0 \quad (2.274)$$

$$\nu_0((\varepsilon_1 - \varepsilon_{j(1)})(\varepsilon_2 - \varepsilon_{j(2)})(\varepsilon_{j(3)}\,\varepsilon_{j(4)} - q)) = 0 \quad (2.275)$$

$$\nu_0((\varepsilon_1 - \varepsilon_{j(1)})(\varepsilon_2 - \varepsilon_{j(2)})\,\varepsilon_{j(3)}\,\varepsilon_{j(4)}) = 0 \quad (2.276)$$

$$\nu_0((\varepsilon_1 - \varepsilon_{j(1)})\,\varepsilon_{j(2)}\,(\varepsilon_{j(3)}\,\varepsilon_{j(4)} - q)) = 0\,. \quad (2.277)$$

Proof. The proofs are very similar, so we prove only (2.273). Assuming e.g. that $\ell \neq 1$, we see that

$$\begin{aligned}&\langle(\varepsilon_1 - \varepsilon_{j(1)})(\varepsilon_2 - \varepsilon_{j(2)})(\varepsilon_\ell - \varepsilon_{j(3)})(\varepsilon_{\ell'} - \varepsilon_{j(4)})\rangle_0 \\ &= \langle\varepsilon_1 - \varepsilon_{j(1)}\rangle_0\,\langle(\varepsilon_2 - \varepsilon_{j(2)})(\varepsilon_\ell - \varepsilon_{j(3)})(\varepsilon_{\ell'} - \varepsilon_{j(4)})\rangle_0 = 0\,.\end{aligned}$$

□

Second proof of Proposition 2.6.5. Even though the result is obvious as already observed, it seems useful for pedagogical reasons to repeat in this simple setting the basic argument. We again follow the scheme of proof of Proposition 2.6.1. We consider the indices $j(1),\ldots,j(4)$ as in Lemma 2.6.6. Then

$$\begin{aligned}\nu(T_{1,2}\,T_{\ell,\ell'}) &= \nu\left(\frac{(\boldsymbol{\sigma}^1 - \boldsymbol{\sigma}^{j(1)})\cdot(\boldsymbol{\sigma}^2 - \boldsymbol{\sigma}^{j(2)})}{N}\,\frac{(\boldsymbol{\sigma}^\ell - \boldsymbol{\sigma}^{j(3)})\cdot(\boldsymbol{\sigma}^{\ell'} - \boldsymbol{\sigma}^{j(4)})}{N}\right) \\ &= \nu(f)\end{aligned}$$

for

$$\begin{aligned}f &= (\varepsilon_1 - \varepsilon_{j(1)})(\varepsilon_2 - \varepsilon_{j(2)})(R_{\ell,\ell'} - R_{\ell,j(4)} - R_{\ell',j(3)} + R_{j(3),j(4)}) \\ &= \frac{1}{N}(\varepsilon_1 - \varepsilon_{j(1)})(\varepsilon_2 - \varepsilon_{j(2)})(\varepsilon_\ell - \varepsilon_{j(3)})(\varepsilon_{\ell'} - \varepsilon_{j(4)}) \\ &\quad + (\varepsilon_1 - \varepsilon_{j(1)})(\varepsilon_2 - \varepsilon_{j(2)})\,f^-\,,\end{aligned}$$

where $f^- = R^-_{\ell,\ell'} - R^-_{\ell,j(4)} - R^-_{\ell',j(3)} + R^-_{j(3),j(4)}$. We use (2.248), (2.273) and (2.259) (with indices 1, $j(1)$, 2, $j(2)$ rather than 1, 2, 3, 4) to obtain

$$|\nu(T_{1,2}\,T_{\ell,\ell'}) - \beta^2(1-2\,q+\widehat{q})\,\nu((R_{1,2}-R_{1,j(2)}-R_{j(1),2}+R_{j(1),j(2)})\,f^-)| \leq \frac{L}{N^{3/2}}$$

so that

$$(1 - \beta^2(1 - 2\,q + \widehat{q}))\,|\nu(T_{1,2}\,T_{\ell,\ell'})| \leq \frac{L}{N^{3/2}}\,.$$

One could prove (2.271) and (2.272) in an entirely similar manner. □

Proposition 2.6.7. *If $\beta < \beta_0$, we have*

$$|\nu(T_1^2) - B^2| \leq \frac{L}{N^{3/2}} \tag{2.278}$$

where

$$B^2 = \frac{1}{N} \frac{q - \widehat{q}}{(1 - \beta^2(1 - 2\,q + \widehat{q}))(1 - \beta^2(1 - 4\,q + 3\,\widehat{q}))} \,. \tag{2.279}$$

Moreover

$$|\nu(T_1\,T_2)| \leq \frac{L}{N^{3/2}}\,;\ |\nu(T_1\,T)| \leq \frac{L}{N^{3/2}}\,. \tag{2.280}$$

This, of course, looks a bit scary. The complication is purely algebraic, but the reader worried by (2.279) should go directly to Proposition 2.6.11. The actual value of B will never be relevant. The occurrence of the quantity $\beta^2(1-4\,q + 3\,\widehat{q})$ is explained by the following, which resembles Proposition 2.6.3.

Proposition 2.6.8. *Consider a function f^- on Σ_{N-1}^n. Then*

$$\begin{aligned}\nu_0'((\varepsilon_1 - \varepsilon_2)\,\varepsilon_3\,f^-) &= \beta^2(1-q)\,\nu_0(f^-(R_{1,3}^- - R_{2,3}^-)) \\ &\quad + \beta^2(q - \widehat{q})\Bigg[\sum_{4\leq\ell\leq n} \nu_0(f^-(R_{1,\ell}^- - R_{2,\ell}^-)) \\ &\quad - n\,\nu_0(f^-(R_{1,n+1}^- - R_{2,n+1}^-))\Bigg]\,. \end{aligned} \tag{2.281}$$

Moreover, if f^- does not depend on $\boldsymbol{\rho}^3$, we have

$$\begin{aligned}&\Bigg|\nu((\varepsilon_1 - \varepsilon_2)\,\varepsilon_3\,f^-) - \beta^2(1 - 4\,q + 3\,\widehat{q})\,\nu(f^-(R_{1,3} - R_{2,3})) \\ &\qquad - \beta^2(q-\widehat{q})\sum_{4\leq\ell\leq n} \nu(f^-(R_{1,\ell} - R_{2,\ell} - R_{1,n+1} + R_{2,n+1}))\Bigg| \\ &\qquad \leq \frac{K(n)}{N}\,\nu((f^-)^4)^{1/4}\,. \end{aligned} \tag{2.282}$$

Proof. Just as for Proposition 2.6.3, this follows from (2.169). For $\ell < \ell'$, let us consider

$$a(\ell,\ell') = \beta^2\,\nu_0((\varepsilon_1 - \varepsilon_2)\,\varepsilon_3\,\varepsilon_\ell\,\varepsilon_{\ell'})\,.$$

Then $a(\ell,\ell') = 0$ if $\ell = 1$, $\ell' = 2$, or if $\ell \geq 3$. Moreover

$$a(1,3) = -a(2,3) = \beta^2(1-q)$$

$$a(1,\ell') = -a(2,\ell') = \beta^2(q - \widehat{q}) \quad \text{if} \quad \ell' \geq 4\,.$$

To get (2.281) we apply (2.169) to $(\varepsilon_1 - \varepsilon_2)\,\varepsilon_3\, f^-$, and we regroup the terms for $\ell = 1$ with the terms for $\ell = 2$.

When f^- does not depend on $\boldsymbol{\rho}_3$, we have

$$\nu_0(f^-(R^-_{1,n+1} - R^-_{2,n+1})) = \nu_0(f^-(R^-_{1,3} - R^-_{2,3}))$$

and the right-hand side of (2.281) becomes

$$\beta^2(1 - 4\,q + 3\,\widehat{q})\,\nu_0(f^-(R^-_{1,3} - R^-_{2,3})) + \sum_{4\le\ell\le n} \beta^2(q - \widehat{q})\nu_0(f^-(R^-_{1,\ell} - R^-_{2,\ell} - R^-_{1,n+1} + R^-_{2,n+1}))\,.$$

The rest of the proof is as in Proposition 2.6.3. □

Proof of Proposition 2.6.7. To prove (2.279), we write

$$\begin{aligned}\nu(T_1^2) &= \nu\left(\frac{(\boldsymbol{\sigma}^1 - \boldsymbol{\sigma}^2)\cdot\boldsymbol{\sigma}^3}{N}\,\frac{(\boldsymbol{\sigma}^1 - \boldsymbol{\sigma}^4)\cdot\boldsymbol{\sigma}^5}{N}\right)\\ &= \nu((\varepsilon_1 - \varepsilon_2)\,\varepsilon_3\,(R_{1,5} - R_{4,5}))\\ &= \frac{1}{N}\,\nu((\varepsilon_1 - \varepsilon_2)\,\varepsilon_3\,(\varepsilon_1 - \varepsilon_4)\,\varepsilon_5) + \nu((\varepsilon_1 - \varepsilon_2)\,\varepsilon_3(R^-_{1,5} - R^-_{4,5}))\,.\end{aligned}$$

We have

$$\begin{aligned}\nu_0((\varepsilon_1 - \varepsilon_2)\,\varepsilon_3\,(\varepsilon_1 - \varepsilon_4)\,\varepsilon_5) &= \mathsf{E}\langle(1 - \varepsilon_1\varepsilon_4 - \varepsilon_1\varepsilon_2 + \varepsilon_2\varepsilon_4)\,\varepsilon_3\varepsilon_5\rangle_0\\ &= \mathsf{E}(\mathrm{th}^2\,Y - \mathrm{th}^4\,Y) = q - \widehat{q}\,. \end{aligned}\tag{2.283}$$

Thus, by (2.247), (2.248) and (2.282), used for $n = 5$, $f^- = R^-_{1,5} - R^-_{4,5}$, we get

$$\begin{aligned}\nu(T_1^2) = \frac{1}{N}\,(q - \widehat{q}) &+ \beta^2(1 - 4\,q + 3\,\widehat{q})\,\nu((R_{1,3} - R_{2,3})(R^-_{1,5} - R^-_{4,5}))\\ &+ \beta^2(q - \widehat{q}) \sum_{4\le\ell\le5} \nu((R^-_{1,5} - R^-_{4,5})(R_{1,\ell} - R_{2,\ell} - R_{1,6} + R_{2,6}))\\ &+ O(3)\,,\end{aligned}\tag{2.284}$$

where, according to our conventions, $O(3)$ denotes a quantity U such that $|U| \le KN^{-3/2}$ (in fact here $|U| \le LN^{-3/2}$). From (2.284) we get

$$\begin{aligned}(1 - \beta^2(1 - 4\,q + 3\,\widehat{q}))\,\nu(T_1^2) = \frac{1}{N}\,(q - \widehat{q}) + \beta^2(q - \widehat{q}) \sum_{\ell=4,5} \nu((R_{1,5} - R_{4,5})\\ (R_{1,\ell} - R_{2,\ell} - R_{1,6} + R_{2,6})) + O(3)\,.\end{aligned}\tag{2.285}$$

To compute this last term, we use (2.266) to write

$$R_{1,5} - R_{4,5} = T_{1,5} - T_{4,5} + T_1 - T_4$$

$$R_{1,\ell} - R_{2,\ell} - R_{1,6} + R_{2,6} = T_{1,\ell} - T_{2,\ell} - T_{1,6} + T_{2,6}$$

and by Proposition 2.6.5, we have

$$\sum_{\ell=4,5} \nu((T_{1,5} - T_{4,5} + T_1 + T_4)(T_{1,\ell} - T_{2,\ell} - T_{1,6} + T_{2,6}))$$
$$= \nu(T_{1,5}^2) + O(3) = A^2 + O(3) \tag{2.286}$$

using Proposition 2.6.4. Thus (2.285) yields

$$(1 - \beta^2(1 - 2q + \widehat{q}))\,\nu(T_1^2) = \frac{1}{N}(q - \widehat{q}) + \beta^2(q - \widehat{q})\,A^2 + O(3) \tag{2.287}$$

and this is (2.278).

Both quantities (2.280) are 0, as in the case of Proposition 2.6.5. To repeat yet another time the basic scheme the case $\nu(T_1 T_2)$, we write

$$\nu(T_1 T_2) = \nu(T_1 T_4) = \nu\left(\frac{(\boldsymbol{\sigma}^1 - \boldsymbol{\sigma}^2)\cdot\boldsymbol{\sigma}^3}{N}\,\frac{(\boldsymbol{\sigma}^4 - \boldsymbol{\sigma}^5)\cdot\boldsymbol{\sigma}^6}{N}\right). \tag{2.288}$$

If we now use (2.276), rather than (2.284) we get

$$\nu(T_1 T_2) = \beta^2(1 - 4q + 3\widehat{q})\,\nu((R_{1,2} - R_{2,3})(R^-_{4,6} - R^-_{5,6}))$$
$$+ \beta^2(q - \widehat{q}) \sum_{4\le\ell\le6} \nu((R^-_{4,6} - R^-_{5,6})(R_{1,\ell} - R_{2,\ell} - R_{1,7} + R_{2,7}))$$
$$+ O(3)\,, \tag{2.289}$$

and rather than (2.285) we get

$$(1 - \beta^2(1 - 4q + 3\widehat{q}))\,\nu(T_1 T_2) = \beta^2(q - \widehat{q}) \sum_{4\le\ell\le6} \nu((R_{4,6} - R_{5,6})$$
$$(R_{1,\ell} - R_{2,\ell} - R_{1,7} + R_{2,7})) + O(3)\,. \tag{2.290}$$

Now

$$R_{4,6} - R_{5,6} = T_{4,6} - T_{5,6} + T_4 - T_5$$
$$R_{1,\ell} - R_{2,\ell} - R_{1,7} + R_{2,7} = T_{1,\ell} - T_{2,\ell} - T_{1,7} + T_{2,7}\,,$$

so that Proposition 2.6.5 shows that

$$(1 - \beta^2(1 - 4q + 3\widehat{q}))\,\nu(T_1 T_2) = O(3)\,.$$

The case of $\nu(T_1 T)$ could be treated following exactly the same pattern. □

We are ready for the last step.

Proposition 2.6.9. *If $\beta \le \beta_0$, we have*

$$|\nu(T^2) - C^2| \le \frac{L}{N^{3/2}}\,, \tag{2.291}$$

where C^2 is given by

$$\begin{aligned}&(1-\beta^2(1-4\,q+3\,\widehat{q}))\,C^2\\&=\frac{\widehat{q}-q^2}{N}+\beta^2(\widehat{q}-q^2)\,A^2+2\beta^2(2\,q+q^2-3\,\widehat{q})\,B^2\,.\end{aligned}\tag{2.292}$$

Corollary 2.6.10. *We have*

$$|\mathsf{E}\langle(R_{1,2}-q)^2\rangle-(A^2+2B^2+C^2)|\le\frac{L}{N^{3/2}}\,.$$

Proof. Use (2.266), Propositions 2.6.4, 2.6.5, 2.6.7 and 2.6.9. □

Proof of Proposition 2.6.9. We write, following the usual pattern,

$$\begin{aligned}\nu(T^2)&=\nu((R_{1,2}-q)(R_{3,4}-q))\\&=\nu((\varepsilon_1\varepsilon_2-q)(R_{3,4}-q))\\&=\frac{1}{N}\,\nu((\varepsilon_1\varepsilon_2-q)\varepsilon_3\varepsilon_4)+\nu((\varepsilon_1\varepsilon_2-q)(R^-_{3,4}-q))\,.\end{aligned}\tag{2.293}$$

We observe that

$$\nu_0((\varepsilon_1\varepsilon_2-q)\varepsilon_3\varepsilon_4)=\mathsf{E}(\mathrm{th}^4\,Y-q\,\mathrm{th}^2\,Y)=\widehat{q}-q^2\,.\tag{2.294}$$

To study $\nu((\varepsilon_1\varepsilon_2-q)(R^-_{3,4}-q))$ we use (2.248), for $f=(\varepsilon_1\varepsilon_2-q)(R^-_{3,4}-q)$; we use (2.169) and (2.167) to compute $\nu_0'(f)$. This gives

$$\begin{aligned}\nu((\varepsilon_1\varepsilon_2-q)(R^-_{3,4}-q))=&\sum_{1\le\ell<\ell'\le4}b(\ell,\ell')\,\nu((R_{\ell,\ell'}-q)(R_{3,4}-q))\\&-4\sum_{\ell\le4}b(\ell,5)\,\nu((R_{\ell,5}-q)(R_{3,4}-q))\\&+10\,b(5,6)\,\nu((R_{5,6}-q)(R_{3,4}-q))\\&+O(3)\,,\end{aligned}\tag{2.295}$$

where $b(\ell,\ell')=\beta^2\,\nu_0((\varepsilon_1\varepsilon_2-q)\,\varepsilon_\ell\,\varepsilon_{\ell'})$. Thus $b(1,2)=\beta^2(1-q^2)$, $b(\ell,\ell')=\beta^2\,q(1-q)$ if $\ell\le2$, $\ell'\ge3$ and $b(\ell,\ell')=\beta^2(\widehat{q}-q^2)$ if $\ell\ge3$.

We now observe that, using (2.266), Propositions 2.6.4, 2.6.5, 2.6.7, we know how to compute $\nu((R_{\ell,\ell'}-q)(R_{\ell_1,\ell_2}-q))$ as a function of $\nu(T^2)$:

$$\nu((R_{\ell,\ell'}-q)^2)=A^2+2\,B^2+\nu(T^2)+O(3)$$

$$\nu((R_{\ell,\ell'}-q)(R_{\ell_1,\ell_2}-q))=B^2+\nu(T^2)+O(3)$$

if $\mathrm{card}(\{\ell,\ell'\}\cap\{\ell_1,\ell_2\})=1$, and if $\{\ell,\ell'\}\cap\{\ell_1,\ell_2\}=\emptyset$, then

$$\nu((R_{\ell,\ell'}-q)(R_{\ell_1,\ell_2}-q))=\nu(T^2)\,.$$

We substitute these expressions in the right-hand side of (2.295) and we collect the terms. The coefficient of $\nu(T^2)$ is

$$\begin{aligned}\sum_{1\le\ell<\ell'\le 4} b(\ell,\ell') - 4\sum_{\ell\le 4} b(\ell,5) + 10\,b(5,6) &= \beta^2(1-q+4q(1-q)+(\widehat{q}-q^2)\\ &\quad - 4(2q(1-q)+2(\widehat{q}-q^2)) + 10(\widehat{q}-q^2))\\ &= \beta^2(1-4q+3\widehat{q})\;.\end{aligned}$$

The coefficient of A^2 is $\beta^2(\widehat{q}-q^2)$, and the coefficient of B^2 is

$$\begin{aligned}&\sum_{\ell=1,2,\ell'=3,4} b(\ell,\ell') + 2b(3,4) - 4(b(3,5)+b(4,5))\\ &= \beta^2(4(q-q^2)+2(\widehat{q}-q^2)-8(\widehat{q}-q^2)) = 2\beta^2(2q+q^2-3\widehat{q})\;.\end{aligned}$$

From (2.293), (2.294) and (2.295) we get

$$\begin{aligned}\nu(T^2) &= \frac{1}{N}(\widehat{q}-q^2) + \beta^2(1-4q+3\widehat{q})\,\nu(T^2) + \beta^2(\widehat{q}-q^2)\,A^2\\ &\quad + 2\beta^2(2q+q^2-3\widehat{q})\,B^2 + O(3)\end{aligned} \tag{2.296}$$

and this implies the result. □

The techniques we have developed in this section will be of importance in Sections 2.8 and 2.9. We will need to use them for values of the parameters where we do not (yet) know that $\nu((R_{1,2}-q)^2)$ is small.

Proposition 2.6.11. *Consider* any *value of β and of h, and q the unique solution of (2.148). Then we have*

$$\begin{aligned}&\left|(1-\beta^2(1-2q+\widehat{q}))\,\nu\left(\left(\frac{(\boldsymbol{\sigma}^1-\boldsymbol{\sigma}^2)\cdot(\boldsymbol{\sigma}^3-\boldsymbol{\sigma}^4)}{N}\right)^2\right) - \frac{4}{N}(1-2q+\widehat{q})\right|\\ &\le K(\beta,h)\left(\frac{1}{N^{3/2}} + \nu(|R_{1,2}-q|^3)\right)\;.\end{aligned} \tag{2.297}$$

Proof. The proof of this extension of (2.264) is obtained by examining how the error terms arose. The error terms in (2.248) are bounded by $\sup_t |\nu_t''(f)|$, i.e. by a sum of terms

$$K\,\nu(|f|\,|R_{\ell,\ell'}-q|\,|R_{\ell_1,\ell_2}-q|) \le K\,\nu(|R_{1,2}-q|^3) \tag{2.298}$$

in the case where f is given by (2.262), and using Hölder's inequality. The other terms, as in (2.257) are of the type

$$\frac{1}{N}\nu(|R_{1,2}-q|) \le \frac{1}{N^{3/2}} + \nu(|R_{1,2}-q|^3) \tag{2.299}$$

using the inequality $ab \le a^{3/2} + b^3$ for $a, b \ge 0$. □

In the same manner, the error terms of the other results of this section are all of the type (2.298), (2.299) or K/N^2. Thus we have the following.

Proposition 2.6.12. *Consider any values of* β, h. *Assume that*

$$|1-\beta^2(1-2\,q+\widehat{q})|>0\,;\;\;|1-\beta^2(1-4\,q+3\,\widehat{q})|>0\,. \tag{2.300}$$

Then there is a constant $K(\beta,h)$ *such that*

$$\begin{aligned}&|\nu((R_{1,2}-q)^2)-(A^2+2\,B^2+C^2)|\\&\leq K(\beta,h)\left(\frac{1}{N^{3/2}}+\nu(|R_{1,2}-q|^3)\right)\,.\end{aligned} \tag{2.301}$$

Moreover, $K(\beta,h)$ *remains bounded as* β *remains bounded and the quantities (2.300) remain bounded below.*

What matters in this statement is not the actual value of $A^2+2\,B^2+C^2$, but the fact that this quantity decays as $1/N$. The reader is referred to Proposition 2.8.1 for applications.

2.7 Central Limit Theorems

In this section, we perform for higher moments the work that was done in the previous sections for moments of order 2. We will reach the conclusion that for the typical disorder, any given family of variables

$$\{N^{1/2}(R_{\ell,\ell'}-\langle R_{1,2}\rangle)\}_{1\leq\ell<\ell'\leq n}$$

is asymptotically Gaussian on the probability space (Σ_N^n, G_N^n), with an explicit covariance structure that is independent of the disorder. Moreover, the random variable $N^{1/2}(\langle R_{1,2}\rangle-q)$ is asymptotically Gaussian. The results of this section will be used only in Sections 3.6 and 5.10.

Somehow it is true that all the ideas required for the present section already occur in Theorem 2.3.9 or in Section 2.6, and that the rest is a simple matter of writing things down. The basic idea, as in Section 2.6, is "to handle first all expressions containing a term $T_{\ell,\ell'}$, then all expressions containing T_ℓ, and finally to take care of T ". We recall the definitions (2.265), (2.268), and also that $a(k)=\mathsf{E}\,g^k$ where g is standard Gaussian.

Theorem 2.7.1. *Assume* $\beta\leq\beta_0$. *Consider an integer* n, *and for* $1\leq\ell<\ell'\leq n$ *consider integers* $k(\ell,\ell')\geq 0$. *Set* $k=\sum_{\ell<\ell'}k(\ell,\ell')$. *Then we have*

$$\left|\nu\left(\prod_{\ell<\ell'}T_{\ell,\ell'}^{k(\ell,\ell')}\right)-\prod_{\ell<\ell'}a(k(\ell,\ell'))\,A^k\right|\leq\frac{K(k)}{N^{(k+1)/2}}\,. \tag{2.302}$$

Proof. It will follow closely the proof of Theorem 2.3.9. The central fact that $\nu_0(\varepsilon_1\varepsilon_2\,\varepsilon_\ell\varepsilon_{\ell'})=0$ if $(\ell,\ell')\neq(1,2)$ will be replaced by (2.273). We consider for any integer $v \le k$ two integers $\ell(v)$, $\ell'(v)$ such that

$$\prod_{\ell<\ell'} T_{\ell,\ell'}^{k(\ell,\ell')} = \prod_{1\le v\le k} T_{\ell(v),\ell'(v)}$$

and

$$(\ell(v),\ell'(v)) = (1,2) \Leftrightarrow v \le k(1,2)\,. \tag{2.303}$$

We assume without loss of generality that $k(1,2) \ge 1$. We consider, for $1 \le v \le k$, integers $j(v)$ and $j'(v)$, all distinct and greater than n. Thus

$$\begin{aligned}\nu\left(\prod_{\ell<\ell'} T_{\ell,\ell'}^{k(\ell,\ell')}\right) &= \nu\left(\prod_{1\le v\le k} T_{\ell(v),\ell'(v)}\right)\\ &= \nu\left(\prod_{1\le v\le k} U(v)\right)\end{aligned} \tag{2.304}$$

where

$$U(v) = \frac{(\boldsymbol{\sigma}^{\ell(v)} - \boldsymbol{\sigma}^{j(v)})\cdot(\boldsymbol{\sigma}^{\ell'(v)} - \boldsymbol{\sigma}^{j'(v)})}{N}\,.$$

Consider

$$\varepsilon(v) = (\varepsilon_{\ell(v)} - \varepsilon_{j(v)})(\varepsilon_{\ell'(v)} - \varepsilon_{j'(v)})\,,$$

so that, by symmetry among sites,

$$\nu\left(\prod_{\ell<\ell'} T_{\ell,\ell'}^{k(\ell,\ell')}\right) = \nu\left(\varepsilon(1)\prod_{2\le v\le k} U(v)\right)\,. \tag{2.305}$$

We write $U(v) = U^-(v) + \varepsilon(v)/N$ (this defines $U^-(v)$), and we expand the product

$$\prod_{2\le v\le k}\left(U^-(v) + \frac{\varepsilon(v)}{N}\right)\,. \tag{2.306}$$

This creates 2^{k-1} terms, since in each factor we can choose either the term $U^-(v)$, or the term $\varepsilon(v)/N$. We recall the notation $O(k)$ for a quantity V such that $|V| \le KN^{-k/2}$. Just as in proof of Theorem 2.3.9, we see that each term of the expansion of (2.306) where we have chosen $\varepsilon(v)/N$ in at least two of the factors is $O(k+1)$. So we need only to consider the other terms. That is, we have

$$\nu\left(\varepsilon(1)\prod_{2\le v\le k} U(v)\right) = \nu\left(\varepsilon(1)\prod_{2\le v\le k} U^-(v)\right) + \mathrm{I} + O(k+1) \tag{2.307}$$

$$\mathrm{I} = \frac{1}{N} \sum_{2 \le u \le k} \nu \left(\varepsilon(1)\, \varepsilon(u) \prod_{v \ne u} U^-(v) \right) ,$$

where $\prod_{v \ne u}$ means that the product ranges over $2 \le v \le k$, $v \ne u$. We use (2.247) and (2.167) to see that

$$\frac{1}{N} \nu \left(\varepsilon(1)\, \varepsilon(u) \prod_{v \ne u} U^-(v) \right) = \frac{1}{N} \nu_0(\varepsilon(1)\, \varepsilon(u))\, \nu_0 \left(\prod_{v \ne u} U^-(v) \right) + O(k+1) ,$$

because there are $k-2$ factors $U^-(v)$. It follows from (2.273) and (2.303) that $\nu_0(\varepsilon(1)\,\varepsilon(u)) = 0$ unless $u \le k(1,2)$; if $u \le k(1,2)$ then, by (2.269)

$$\nu_0(\varepsilon(1)\,\varepsilon(u)) = 1 - 2\,q + \widehat{q} \,.$$

Using again (2.303) and (2.247) it follows that if $k(1,2) \ge 2$,

$$\mathrm{I} = \frac{k(1,2)-1}{N} (1 - 2\,q + \widehat{q})\, \nu \left(\prod_{3 \le v \le k} U^-(v) \right) + O(k+1) , \tag{2.308}$$

and $\mathrm{I} = 0$ if $k(1,2) = 1$. Next, we use (2.248) and (2.259) (with the indices 1, $j(1)$, 2, $j'(1)$ rather than 1, 2, 3, 4) to see that

$$\begin{aligned} &\nu_0' \left(\varepsilon(1) \prod_{2 \le v \le k} U^-(v) \right) \\ &= \beta^2 (1 - 2\,q + \widehat{q})\, \nu \left((R_{1,2} - R_{1,j'(1)} - R_{j(1),2} + R_{j(1),j'(1)}) \prod_{2 \le v \le k} U^-(v) \right) \\ &+ 0(k+1) \\ &= \beta^2 (1 - 2\,q + \widehat{q})\, \nu \left(U(1) \prod_{2 \le v \le k} U^-(v) \right) + O(k+1) \,. \end{aligned} \tag{2.309}$$

We proceed as in Proposition 2.3.7 to show that in (2.308) and (2.309) we can replace all terms $U^-(v)$ by $U(v)$ with error $0(k+1)$. Combining (2.305) to (2.309) we get

$$\begin{aligned} &((1 - \beta^2 (1 - 2\,q + \widehat{q}))\, \nu \left(\prod_{\ell < \ell'} T_{\ell,\ell'}^{k(\ell,\ell')} \right) \\ &= \frac{k(1,2)-1}{N} (1 - 2\,q + \widehat{q})\, \nu \left(\prod_{3 \le v \le n} U(v) \right) + O(k+1) \,. \end{aligned} \tag{2.310}$$

Now

$$\nu\left(\prod_{3\le v\le n} U(v)\right) = \nu\left(\prod_{\ell<\ell'} T_{\ell,\ell'}^{n(\ell,\ell')}\right) ,$$

where $n(\ell,\ell') = k(\ell,\ell')$ if $(\ell,\ell') \neq (1,2)$, and $n(1,2) = k(1,2)-2$. (If $k(1,2) = 1$, the term does not occur.) Then (2.302) follows by induction over k from (2.310). □

We recall the notation B of (2.279).

Theorem 2.7.2. *Consider an integer n. For $1 \le \ell < \ell' \le n$, consider an integer $k(\ell,\ell')$; set $k_1 = \sum_{1\le\ell<\ell'\le n} k(\ell,\ell')$. For $1 \le \ell \le n$, consider integers $k(\ell)$, and set $k_2 = \sum_{\ell\le n} k(\ell)$. Then, if $k = k_1 + k_2$, we have*

$$\left|\nu\left(\prod_{1\le\ell<\ell'\le n} T_{\ell,\ell'}^{k(\ell,\ell')} \prod_{\ell\le n} T_\ell^{k(\ell)}\right) - \prod_{1\le\ell<\ell'\le n} a(k(\ell,\ell')) \prod_{\ell\le n} a(k(\ell))\, A^{k_1} B^{k_2}\right|$$
$$\le \frac{K(k)}{N^{(k+1)/2}} . \tag{2.311}$$

Proof. The proof goes by induction over k. At first, we consider the case $k_1 > 0$, and assume without loss of generality that $k(1,2) \ge 1$. Using (2.274) to control the contribution of the terms T_ℓ, this stage is nearly identical to the proof of Theorem 2.7.1, so we do not repeat it, and we prove the theorem for $k_1 = 0$, assuming that we already know it for $k_1 > 0$. For each integer $1 \le v \le k = k_2$, we consider integers $\ell(v) \le n$ such that

$$\prod_{1\le v\le k} T_{\ell(v)} = \prod_{1\le\ell\le n} T_\ell^{k(\ell)} \tag{2.312}$$

$$\ell(v) = 1 \Leftrightarrow v \le k(1) . \tag{2.313}$$

We assume without loss of generality that $k(1) \ge 1$. We consider for $1 \le v \le k$ integers $j(v)$ and $j'(v)$, all distinct and all greater than n. Thus

$$\nu\left(\prod_{\ell\le n} T_\ell^{k(\ell)}\right) = \nu\left(\prod_{1\le v\le k} T_{\ell(v)}\right) = \nu\left(\prod_{1\le v\le k} U(v)\right) , \tag{2.314}$$

where

$$U(v) = \frac{(\boldsymbol{\sigma}^{\ell(v)} - \boldsymbol{\sigma}^{j(v)}) \cdot \boldsymbol{\sigma}^{j'(v)}}{N} .$$

Consider

$$\varepsilon(v) = (\varepsilon_{\ell(v)} - \varepsilon_{j(v)})\, \varepsilon_{j'(v)} ,$$

so that, by symmetry among sites,

$$\nu\left(\prod_{\ell\le n} T_\ell^{k(\ell)}\right) = \nu\left(\varepsilon(1) \prod_{2\le v\le k} U(v)\right) . \tag{2.315}$$

We write $U(v) = U^-(v) + \varepsilon(v)/N$ (this defines $U^-(v)$), we expand the product

$$\prod_{2 \le v \le k} \left(U^-(v) + \frac{\varepsilon(v)}{N} \right) \tag{2.316}$$

and obtain that

$$\nu\left(\varepsilon(1) \prod_{2 \le v \le k} U(v)\right) = \nu\left(\varepsilon(1) \prod_{2 \le v \le k} U^-(v)\right) + \mathrm{I} + O(k+1) , \tag{2.317}$$

where

$$\mathrm{I} = \frac{1}{N} \sum_{2 \le u \le k} \nu\left(\varepsilon(1)\, \varepsilon(u) \prod_{v \ne u} U^-(v)\right) .$$

We use (2.313) and (2.276) to see that $\nu_0(\varepsilon(1)\, \varepsilon(u)) = 0$ if $u > k(1)$, and (2.283) to see that $\nu_0(\varepsilon(1)\, \varepsilon(u)) = q - \widehat{q}$ if $u \le k(1)$. Thus, as in Theorem 2.7.1, we get, if $k(1) \ge 2$,

$$\begin{aligned}
\mathrm{I} &= \frac{k(1)-1}{N} (q - \widehat{q})\, \nu\left(\prod_{3 \le v \le k} U^-(v) \right) + O(k+1) \\
&= \frac{k(1)-1}{N} (q - \widehat{q})\, \nu\left(\prod_{3 \le v \le k} U(v) \right) + O(k+1) \\
&= \frac{k(1)-1}{N} (q - \widehat{q})\, \nu\left(\prod_{3 \le v \le k} T_{\ell(v)} \right) + O(k+1) ,
\end{aligned} \tag{2.318}$$

proceeding as in Proposition 2.3.7.

We set $f^- = \prod_{2 \le v \le k} U^-(v)$, and we use (2.282) (with indices 1, $j(1)$, $j'(1)$ rather than 1, 2, 3) to see that, if n' is an integer larger than all the indices $j(v)$, $j'(v)$, we have

$$\begin{aligned}
\nu(\varepsilon(1)\, f^-) &= \beta^2 (1 - 4\,q + 3\,\widehat{q})\, \nu(f^-(R_{1,j'(1)} - R_{j(1),j'(1)})) \\
&\quad + \beta^2 (q - \widehat{q}) \sum_{\ell} \nu(f^-(R_{1,\ell} - R_{j(1),\ell} - R_{1,n'+1} + R_{j(1),n'+1})) \\
&\quad + 0(k+1) ,
\end{aligned} \tag{2.319}$$

and where the summation is over $2 \le \ell \le n'$, $\ell \ne j'(1)$. Setting $f = \prod_{2 \le v \le k} U(v)$, we see as in Proposition 2.3.7 that (2.319) implies

$$\begin{aligned}
\nu(\varepsilon(1)\, f^-) &= \beta^2 (1 - 4\,q + 3\widehat{q})\, \nu(f\, U(1)) \\
&\quad + \beta^2 (q - \widehat{q}) \sum \nu(f(R_{1,\ell} - R_{j(1),\ell} - R_{1,n'+1} + R_{j(1),n'+1})) \\
&\quad + 0(k+1) ,
\end{aligned} \tag{2.320}$$

where the summation is again over $2 \leq \ell \leq n'$, $\ell \neq j'(1)$. We now come to the new part of the argument, namely the computation of

$$\nu\left(\prod_{2\leq v\leq k} U(v)(R_{1,\ell} - R_{j(1),\ell} - R_{1,n'+1} + R_{j(1),n'+1})\right)$$
$$= \nu\left(\prod_{2\leq v\leq k} U(v)(T_{1,\ell} - T_{j(1),\ell} - T_{1,n'+1} + T_{j(1),n'+1})\right) . \quad (2.321)$$

In this expression, we can replace $U(v)$ by $T(v)$ whenever the sets

$$\{j(v), j'(v)\} \quad \text{and} \quad \{1, \ell, j(1), n'+1\}$$

are disjoint, or equivalently, whenever $\ell \notin \{j(v), j'(v)\}$.

Suppose first that ℓ is not one of the indices $j(v)$, $j'(v)$, $2 \leq v \leq k$. Then the quantity (2.321) is

$$\nu\left(\prod_{2\leq v\leq k} T_{\ell(v)}(T_{1,\ell} - T_{j(1),\ell} - T_{1,n'+1} + T_{j(1),n'+1})\right) .$$

There are k factors, one of which is of the type $T_{\ell,\ell'}$; we are thus within the range of our induction hypothesis ($k_1 = 1$), and we know that this term is $O(k+1)$ (since $a(1) = \mathsf{E}\, g = 0$).

Suppose now that $\ell \in \{j(w), j'(w)\}$ for a (necessarily unique) index $2 \leq w \leq k$. Consider $f' = \prod T_{\ell(v)}$, where in the product v ranges over $2 \leq v \leq k$, $v \neq w$. Then the term (2.321) is

$$\nu(f'\, U(w)\, (T_{1,\ell} - T_{j(1),\ell} - T_{1,n'+1} + T_{j(1),n'+1})) . \quad (2.322)$$

In order to use again the induction hypothesis, we write

$$U(w) = R_{\ell(w),j'(w)} - R_{j(w),j'(w)} = T_{\ell(w),j'(w)} - T_{j(w),j'(w)} + T_{\ell(w)} - T_{j(w)} .$$

By the induction hypothesis (case $k_1 \geq 1$), we see that all terms

$$\nu(f'\, T_{\ell_1}\, T_{\ell_1,\ell_2}) \quad \text{or} \quad \nu(f'\, T_{\ell_1,\ell_2}\, T_{\ell_3,\ell_4})$$

where $(\ell_1, \ell_2) \neq (\ell_3, \ell_4)$ are $O(k+1)$. When can one of the sets

$$\{\ell(w), j'(w)\} \quad \text{or} \quad \{j(w), j'(w)\}$$

be equal to one of the sets

$$\{1, \ell\}\, ,\ \{j(1), \ell\}\, ,\ \{1, n'+1\}\, ,\ \{j(1), n'+1\}\ ?$$

Since $\ell(w) \leq n$ and since all the indices $j(\ell)$, $j'(\ell)$ are distinct, greater than n and smaller than $n'+1$, the only possibility is that $\ell(w) = 1$, $\ell = j'(w)$.

Thus $w \le k(1)$, and the value of (2.322) is then

$$\nu(f' T^2_{1,j'(w)}) + O(k+1) . \tag{2.323}$$

When $k(1) = 1$, there is no integer $w \ge 2$, $w \le k(1)$, and the quantity (2.321) is $O(k+1)$. By (2.318), I is also $O(k+1)$, and by (2.315), (2.317), $\nu\left(\prod_{\ell \le n} T_\ell^{k(\ell)}\right) = O(k+1)$, and this finishes the induction step in that case.

When $k(1) \ge 2$, all the quantities (2.323) are equal to

$$\nu\Big(\prod_{3 \le v \le n} T_{\ell(v)} T_{1,n+1}\Big) .$$

This is because $f' = \prod_{3 \le v \le n} T_{\ell(v)}$, and because $\nu\Big(\prod_{3 \le v \le n} T_{\ell(v)}\, T_{1,p}\Big)$ does not depend on the index $p \ge n+1$. Since there are $k(1) - 1$ possible values of w, the quantity (2.320) is

$$\begin{aligned}
&\beta^2(1 - 4\,q + 3\,\widehat{q})\,\nu\left(\prod_{v \le k} U(v)\right) \\
&+ (k(1) - 1)\,\beta^2(q - \widehat{q})\,\nu\left(\prod_{3 \le v \le k} T_{\ell(v)}\, T^2_{1,n+1}\right) \\
&+ O(k+1) .
\end{aligned}$$

Now $\nu\left(\prod_{v \le k} U(v)\right) = \nu\left(\prod_{v \le k} T_\ell\right)$, so that combining (2.315), (2.317) and (2.318) we have obtained the relation

$$\begin{aligned}
(1 - \beta^2(1 - 4\,q + 3\,\widehat{q}))\,\nu\left(\prod_{\ell \le n} T_\ell^{k(\ell)}\right) &= (k(1) - 1)(q - \widehat{q})\left(\frac{1}{N}\,\nu\left(\prod_{\ell \le n} T_\ell^{n(\ell)}\right)\right. \\
&\quad \left. + \beta^2\,\nu\left(\prod_{\ell \le n} T_\ell^{n(\ell)}\, T^2_{1,n+1}\right)\right) \\
&\quad + O(k+1) ,
\end{aligned} \tag{2.324}$$

where $n(\ell) = k(\ell)$ if $\ell \ne 1$ and $n(1) = k(1) - 2$. We apply the induction hypothesis to the two terms on the right-hand side. We then use the relation $a(k) = (k-1)\,a(k-2)$ and

$$(1 - \beta^2(1 - 4\,q + 3\,\widehat{q}))\,B^2 = (q - \widehat{q})\left(\frac{1}{N} + \beta^2\,A^2\right)$$

to complete the induction step. □

As a consequence of Theorem 2.7.1, we see (with the notation of the theorem) that the quantity

$$V = N^{k/2} \left\langle \prod_{\ell,\ell' \le n} T_{\ell,\ell'}^{k(\ell,\ell')} \prod_{\ell \le n} T_\ell^{k(\ell)} \right\rangle$$

is essentially independent of the disorder. Indeed (as in the case of Theorem 2.3.9, using replicas), we see that

$$\mathsf{E}\,(V - \mathsf{E}\,V)^2 \le \frac{K}{\sqrt{N}} \,. \tag{2.325}$$

(This, of course, raises the question as to whether V itself satisfies a central limit theorem...) Writing for simplicity $\overline{q} = \|\mathbf{b}\|^2/N$ (so that $T = \overline{q} - q$) and observing that

$$R_{\ell,\ell'} - \overline{q} = T_{\ell,\ell'} + T_\ell + T_{\ell'} \,, \tag{2.326}$$

we conclude that if the product $\prod_{\ell,\ell'}(R_{\ell,\ell'} - \overline{q})$ contains k factors, the quantity

$$V = N^{k/2} \left\langle \prod_{\ell,\ell'} (R_{\ell,\ell'} - \overline{q}) \right\rangle$$

satisfies (2.325). It can be checked that the asymptotic value of V is $\mathsf{E}\Big(\prod_{\ell,\ell'} g_{\ell,\ell'}\Big)$ where the jointly Gaussian family $g_{\ell,\ell'}$ satisfies $\mathsf{E}\,g_{\ell,\ell'}^2 = N(A^2 + 2B^2)$, $\mathsf{E}\,g_{\ell,\ell'}\,g_{\ell_1,\ell_2} = NB^2$ if $\mathrm{card}(\{\ell,\ell'\} \cap \{\ell_1,\ell_2\}) = 1$, and $\mathsf{E}\,g_{\ell,\ell'}\,g_{\ell_1,\ell_2} = 0$ if $\{\ell,\ell'\} \cap \{\ell_1,\ell_2\} = \emptyset$.

We now describe the joint behavior of the family $T_{\ell,\ell'}$, T_ℓ, T. We recall the notation C of (2.292).

Theorem 2.7.3. *Consider an integer n. For $1 \le \ell < \ell' \le n$, consider integers $k(\ell,\ell')$. For $1 \le \ell \le n$, set $k_1 = \sum_{1 \le \ell < \ell' \le n} k(\ell,\ell')$ and $k_2 = \sum_{1 \le \ell \le n} k(\ell)$. Consider an integer k_3, and $k = k_1 + k_2 + k_3$. Then*

$$\Bigg| \nu\Bigg(\prod_{1 \le \ell < \ell' \le n} T_{\ell,\ell'}^{k(\ell,\ell')} \prod_{\ell \le n} T_\ell^{k(\ell)}\, T^{k_3} \Bigg) - \prod_{1 \le \ell < \ell' \le n} a(k(\ell,\ell'))$$
$$\prod_{\ell \le n} a(k(\ell))\, a(k_3)\, A^{k_1} B^{k_2} C^{k_3} \Bigg| \le \frac{K(k)}{N^{(k+1)/2}} \,. \tag{2.327}$$

Corollary 2.7.4. *We have*

$$|\mathsf{E}(T^k) - a(k)\, C^k| \le \frac{K(k)}{N^{(k+1)/2}} \,.$$

Proof of Theorem 2.7.3. The proof goes by induction over k, starting with $k = 0$, and has three steps. In the first step, we assume $k_1 > 0$. Using (2.274) and (2.275), this proof is almost identical to the proof of Theorem 2.7.1. In the second step, we assume $k_1 = 0$, $k_2 > 0$. Using (2.277), the proof is nearly identical to the proof of Theorem 2.7.2 (except that one defines $U(w) = R_{j(w),j'(w)} - q$ for $k_2 < w \le k$). Thus we consider the case $k_1 = k_2 = 0$, $k_3 = k$. We write

$$\nu(T^k) = \nu\left(\prod_{1\le v\le k}(R_{2v-1,2v} - q)\right) = \nu\left((\varepsilon_1\varepsilon_2 - q)\prod_{2\le v\le k}(R_{2v-1,2v} - q)\right),$$

and we proceed as before to obtain the relation

$$\nu(T^k) = \nu((\varepsilon_1\varepsilon_2 - q)\, f^-) + \frac{k-1}{N}(\widehat{q} - q^2)\,\nu(T^{k-2}) + O(k+1)\,, \tag{2.328}$$

where

$$f^- = \prod_{2\le v\le k}(R^-_{2v-1,2v} - q)\,.$$

(If $k = 1$, the last term does not appear.) To calculate $\nu((\varepsilon_1\varepsilon_2 - q)\, f^-)$, we use (2.248). We apply (2.167) to each term, (2.247) to replace ν_0 by ν, and we proceed as in Proposition 2.3.7 to replace $R^-_{\ell,\ell'}$ by $R_{\ell,\ell'}$. We get, setting $n = 2k$,

$$\begin{aligned}\nu((\varepsilon_1\varepsilon_2 - q)\, f^-) = &\sum_{\ell<\ell'\le n} b(\ell,\ell')\,\nu((R_{\ell,\ell'} - q)\, f)\\ &- n\sum_{\ell\le n} b(\ell, n+1)\,\nu((R_{\ell,n+1} - q)\, f)\\ &+ \frac{n(n+1)}{2}\, b(n+1, n+2)\,\nu((R_{n+1,n+2} - q)\, f)\\ &+ 0(k+1)\end{aligned} \tag{2.329}$$

where $f = \prod_{2\le v\le k}(R_{2v-1,2v} - q)$ and

$$b(\ell,\ell') = \beta^2\,\nu_0(\varepsilon_\ell\varepsilon_{\ell'}(\varepsilon_1\varepsilon_2 - q))\,.$$

Thus

$$\begin{aligned} b(1,2) &= \beta^2(1 - q^2)\,,\\ b(\ell,\ell') &= \beta^2(q - q^2) \quad \text{if} \quad \ell \le 2\,,\ \ell' \ge 3\\ b(\ell,\ell') &= \beta^2(\widehat{q} - q^2) \quad \text{if} \quad \ell \ge 3\,.\end{aligned}$$

In order to compute $\nu((R_{\ell,\ell'} - q)\, f)$ we can replace each term $R_{2v-1,2v} - q$ by T whenever $\{2v-1, 2v\} \cap \{\ell,\ell'\} = \emptyset$. Thus

$$\nu((R_{n+1,n+2} - q)\, f) = \nu(T^{k-1}(R_{n+1,n+2} - q)) = \nu(T^k)\,.$$

If w is the unique integer $\le k$ such that $\ell \in \{2w-1, 2w\}$, then for $w = 1$ we have $\nu((R_{\ell,n+1} - q)\, f) = \nu(T^k)$, whereas for $w \ge 2$ we have

$$\nu((R_{\ell,n+1} - q)\, f) = \nu((R_{\ell,n+1} - q)(R_{2w-1,2w} - q)\, T^{k-2})\,.$$

To compute this term, we use (2.266) to write

$$R_{\ell,n+1} - q = T_{\ell,n+1} + T_\ell + T_{n+1} + T$$

$$R_{2w-1,2w} - q = T_{2w-1,2w} + T_{2w-1} + T_{2w} + T \ ,$$

and we expand the product of these quantities. Since $a(1) = 0$, the induction hypothesis shows that

$$\nu((R_{\ell,n+1} - q)\, f) = \nu(T^k) + \nu(T_\ell^2\, T^{k-2}) + 0(k+1) \ .$$

To compute $\nu((R_{\ell,\ell'} - q)\, f)$, we first consider the case where for some $w \le n$, $\ell = 2w - 1$ and $\ell' = 2w$. In that case

$$\nu((R_{\ell,\ell'} - q)\, f) = \nu((R_{\ell,\ell'} - q)^2\, T^{k-2}) \ .$$

Using again (2.266), the induction hypothesis, and the fact that $a(1) = 0$, we get

$$\nu((R_{\ell,\ell'} - q)^2\, T^{k-2}) = \nu(T^k) + \nu(T_{\ell,\ell'}^2\, T^{k-2}) + \nu(T_\ell^2\, T^{k-2}) \\ + \nu(T_{\ell'}^2\, T^{k-2}) + O(k+1) \ .$$

When $\ell \in \{2w-1, 2w\}$, $\ell' \in \{2w'-1, 2w'\}$ for some $2 \le w < w'$, we have

$$\nu((R_{\ell,\ell'} - q)\, f) = \nu((R_{\ell,\ell'} - q)(R_{2w-1,2w} - q)(R_{2w'-1,2w'} - q)\, T^{k-3}) \ ,$$

and proceeding as before we get

$$\nu((R_{\ell,\ell'} - q)\, f) = \nu(T^k) + \nu(T_\ell^2\, T^{k-2}) + \nu(T_{\ell'}^2\, T^{k-2}) + O(k+1) \ .$$

If $w = 1$, we find instead $\nu(T^k) + \nu(T_{\ell'}^2\, T^{k-2}) + O(k+1)$.

It remains to regroup these terms as in the right-hand side of (2.329). The coefficient of $\nu(T^2)$ is

$$\sum_{\ell<\ell'\le n} b(\ell,\ell') - n \sum_{\ell\le n} b(\ell, n+1) + \frac{n(n+1)}{2}\, b(n+1, n+2)$$
$$= \beta^2\Big(1 - q^2 + 2(n-2)(q - q^2) + \frac{(n-2)(n-1)}{2}\,(\widehat{q} - q^2)$$
$$- 2n(q - q^2) - (n-2)\, n(\widehat{q} - q^2) + \frac{n(n+1)}{2}\,(\widehat{q} - q^2)\Big)$$
$$= \beta^2((1 - q^2) - 4(q - q^2) + 3(\widehat{q} - q^2)) = \beta^2(1 - 4\,q + 3\,\widehat{q}) \ .$$

We observe that $\nu(T_{\ell,\ell'}^2\, T^{k-2}) = \nu(T_{1,2}^2\, T^{k-2})$. The coefficient of $\nu(T_{1,2}^2\, T^{k-2})$ is

$$\sum_{2\le w\le k} b(2w-1, 2w) = (k-1)\, \beta^2(\widehat{q} - q^2) \ .$$

The coefficient of $\nu(T_1^2 T^{k-2})$ is

$$\beta^2(2(n-2)(q-q^2)+2\,\frac{(n-3)(n-2)}{2}\,(\widehat{q}-q^2)-(n-2)\,n(\widehat{q}-q^2))$$
$$=2\,\beta^2(k-1)(2\,q+q^2-3\,\widehat{q})\;,$$

since $n-2=2(k-1)$.

Combining these relations just as before, we get

$$(1-\beta^2(1-4\,q+3\,\widehat{q}))\,\nu(T^k)=(k-1)\Bigg(\frac{\widehat{q}-q^2}{N}\,\nu(T^{k-2})$$
$$+\,\beta^2(\widehat{q}-q^2)\,\nu(T_{1,2}^2\,T^{k-2})+2\,\beta^2(2\,q+q^2-3\,\widehat{q})\,\nu(T_1^2\,T^{k-2})\Bigg)\;.$$

One then finishes the proof using the induction hypothesis and (2.292). □

Theorem 2.7.3 is not the end of the story. It would be of interest to describe the joint behavior of the quantities

$$\frac{1}{N}\sum_{i\le N}\sigma_i^\ell-\mathsf{E}\,\mathrm{th}\,Y$$

(recentered magnetization) and how they interact with the overlaps.

Research problem 2.7.5. (Level 1) Generalize Theorem 2.7.3 to obtain a description of the joint behavior of the "multiple overlaps"

$$R_{\ell_1,\ldots,\ell_p}=\frac{1}{N}\sum_{i\le N}\sigma_i^{\ell_1}\ldots\sigma_i^{\ell_p}$$

for $p\ge 1$.

Of course the "level 1" rating does not mean that it is easy but, rather, that it should not require major new ideas.

2.8 Replica-Symmetry Breaking

The cavity method, as we used it, can hardly give any result unless one assumes $\beta\le\beta_0$ where β_0 is rather small. One obvious reason is that we were very crude. Proposition 2.4.6 uses the bound $|R_{1,2}-q|\le 2$! If we had more information on the behavior of $R_{1,2}$ before developing the cavity method, we could eventually do better. This will be a central idea later on. In the present section we will turn things around and argue that, when certain results of previous sections fail, $R_{1,2}$ must behave badly.

Proposition 2.8.1. *For values of β, h such that*

$$\beta^2 \mathsf{E} \frac{1}{\mathrm{ch}^4 Y} = \beta^2(1 - 2q + \widehat{q}) > 1 , \tag{2.330}$$

there exists a number $\delta > 0$ such that for N large enough, we have

$$\nu(|R_{1,2} - q|^3) \geq \delta \, \nu((R_{1,2} - q)^2) + N^{-3/2} . \tag{2.331}$$

Here of course q is defined as in (2.148).

Conjecture 2.8.2. (Level 2) We suspect in fact the stronger result:

$$\nu((R_{1,2} - q)^2) \geq \delta .$$

Lemma 2.8.3. *For each value of β, h, we have*

$$\beta^2(1 - 4q + 3\widehat{q}) < 1 . \tag{2.332}$$

Proof. Consider the function

$$\Phi(x) = \mathsf{E}\,\mathrm{th}^2(\beta z \sqrt{x} + h) . \tag{2.333}$$

Then Proposition 2.4.8 shows that $\Phi(x)/x$ is decreasing, so that

$$x \Phi'(x) - \Phi(x) < 0 ,$$

and since $q = \Phi(q)$, we have $\Phi'(q) < 1$. Now, using integration by parts, and writing as usual $Y = \beta z \sqrt{x} + h$,

$$\begin{aligned}
\Phi'(q) &= \beta \mathsf{E} \left(\frac{z}{\sqrt{q}} \frac{\mathrm{th}\, Y}{\mathrm{ch}^2 Y} \right) = \beta^2 \mathsf{E} \left(\frac{1}{\mathrm{ch}^4 Y} - 2 \frac{\mathrm{sh}^2 Y}{\mathrm{ch}^4 Y} \right) \\
&= \beta^2 \mathsf{E} \left(\frac{3}{\mathrm{ch}^4 Y} - \frac{2}{\mathrm{ch}^2 Y} \right) = \beta^2(3(1 - 2q + \widehat{q}) - 2(1 - q)) \\
&= \beta^2(1 - 4q + 3\widehat{q}) .
\end{aligned}$$

□

Proof of Proposition 2.8.1. Consider $\delta > 0$ such that

$$\nu(|R_{1,2} - q|^3) \leq \delta \, \nu((R_{1,2} - q)^2) + \frac{1}{N^{3/2}} . \tag{2.334}$$

By Lemma 2.8.3, condition (2.300) holds. We use (2.301) to see that

$$\nu((R_{1,2} - q)^2) \leq A^2 + 2B^2 + C^2 + \frac{K(\beta, h)}{N^{3/2}} + \delta K(\beta, h) \, \nu((R_{1,2} - q)^2) . \tag{2.335}$$

If $\delta K(\beta, h) \le 1/2$, we see from (2.335) that

$$\nu((R_{1,2} - q)^2) \le 2(A^2 + 2B^2 + C^2) + \frac{2K(\beta, h)}{N^{3/2}} . \tag{2.336}$$

Using (2.334) again, and since we can assume $\delta \le 2$,

$$\nu(|R_{1,2} - q|^3) \le 2\delta(A^2 + 2B^2 + C^2) + \frac{K(\beta, h)}{N^{3/2}} .$$

Now we use (2.297). Since we assume $1 - \beta^2(1 - 2q + \widehat{q}) \le 0$, since a square is non-negative, and since $(1 - 2q + \widehat{q}) = \mathsf{E}\,\mathrm{ch}^{-4} Y \ge 0$, the left-hand side of (2.297) is at least $4(1 - 2q + \widehat{q})/N$. This and (2.336) imply

$$\frac{4}{N}(1 - 2q + \widehat{q}) \le K(\beta, h)\left(\frac{1}{N^{3/2}} + \delta(A^2 + 2B^2 + C^2)\right) . \tag{2.337}$$

For N large this implies

$$\delta N(A^2 + 2B^2 + C^2) \ge \frac{1 - 2q + \widehat{q}}{K(\beta, h)} .$$

Since $N(A^2 + 2B^2 + C^2)$ does not depend on N, this shows that (2.334) fails for δ small enough and N large enough. □

We might think that possibly the bad behavior (2.331) comes from the fact that the solution q of (2.148) is simply not the correct value to consider. But this is not the case. The following Proposition expresses that if $R_{1,2}$ is nearly constant, then it is nearly equal to q.

Proposition 2.8.4. *Given parameters β, h, and $\delta > 0$, there exists $\xi > 0$ such that*

$$\forall x ,\ \nu((R_{1,2} - x)^2) < \xi \Rightarrow |x - q| \le \delta . \tag{2.338}$$

Proof. Let us assume

$$\nu((R_{1,2} - x)^2) < \xi . \tag{2.339}$$

Then by Jensen's inequality we have

$$(\nu(R_{1,2}) - x)^2 \le \xi . \tag{2.340}$$

By symmetry among sites, we have $\nu(R_{1,2}) = \nu(\varepsilon_1\varepsilon_2)$. Let us now use the smart path method, but with q replaced by x. It follows from (2.174) that

$$|\nu(\varepsilon_1\varepsilon_2) - \Phi(x)| \le K\,\nu((R_{1,2} - x)^2)^{1/2} ,$$

where Φ is given by (2.333). Using (2.339) and (2.340) we get $|x - \Phi(x)| \le K\sqrt{\xi}$. Thus for ξ small, x gets close to q. The result follows. □

In the rest of this section, we will formulate a consequence of Propositions 2.8.1 and 2.8.4 that has a nice physical interpretation, and that also connects naturally to probabilistic ideas.

To control $R_{1,2}$ as a function on Σ_N^2, it is natural to consider its Laplace transform $\langle \exp \lambda N R_{1,2} \rangle$, where λ is a parameter. It is natural to rescale, and to consider $N^{-1} \log \langle \exp \lambda N R_{1,2} \rangle$. One can then hope, as in Corollary 2.2.5, that this random quantity will have small fluctuations and that what really matters is the number

$$\psi_N(\lambda) = \frac{1}{N} \mathsf{E} \log \langle \exp \lambda N R_{1,2} \rangle \,. \tag{2.341}$$

(This of course also depends on β and h.) It is important to note that this function is convex (as in Proposition 1.1.1). Writing the definition of the bracket in (2.341), we see that

$$\psi_N(\lambda) = r_N(\beta, h, \lambda) - 2 p_N(\beta, h) \tag{2.342}$$

where

$$r_N(\beta, h, \lambda) = \frac{1}{N} \mathsf{E} \log Z_N(\beta, h, \lambda) \tag{2.343}$$

$$Z_N(\beta, h, \lambda) = \sum_{\boldsymbol{\sigma}^1, \boldsymbol{\sigma}^2} \exp(-H_N(\boldsymbol{\sigma}^1, \boldsymbol{\sigma}^2)) \tag{2.344}$$

$$-H_N(\boldsymbol{\sigma}^1, \boldsymbol{\sigma}^2) = -H_N(\boldsymbol{\sigma}^1) - H_N(\boldsymbol{\sigma}^2) + \lambda N R_{1,2} \,. \tag{2.345}$$

The quantity (2.345) has a simple interpretation; it is the Hamiltonian of a single system made of two copies of (Σ_N, G_N) that *interact* (or are coupled) through the term $\lambda N R_{1,2}$. Thus $\boldsymbol{\sigma}^1$ and $\boldsymbol{\sigma}^2$ are not independent r.v. for the Gibbs measure relative to $-H_N(\boldsymbol{\sigma}^1, \boldsymbol{\sigma}^2)$. Integration with respect to this Gibbs measure is denoted by $\langle \cdot \rangle_\lambda$. From (2.343) we get

$$\psi_N'(\lambda) = \frac{\partial r_N}{\partial \lambda}(\beta, h, \lambda) = \mathsf{E} \langle R_{1,2} \rangle_\lambda \,. \tag{2.346}$$

The meaning of the next theorem is that, if we knew that $\psi(\lambda) = \lim_{N \to \infty} \psi_N(\lambda)$ exists, then, under (2.330), this convex function would not be differentiable at $\lambda = 0$, or equivalently, its (left) derivative would not be continuous at $\lambda = 0$. Since we cannot (yet) prove the existence of the limit, the natural statement is a little more complicated.

Theorem 2.8.5. *Assume that the parameters β and h satisfy (2.330). Then*

$$\lim_{t \to 0} \liminf_{N \to \infty} (\psi_N'(t) - \psi_N'(-t)) > 0 \,. \tag{2.347}$$

Equivalently, there exists $\delta > 0$ such that, if $t > 0$, for N large enough we have

$$\psi_N'(t) \geq \delta + \psi_N'(-t) \,.$$

In (2.347), $\lim_{t\to 0}$ exists because $\psi'_N(t) - \psi'_N(-t)$ is non-decreasing.

The meaning of Theorem 2.8.5 is that a change in the value of λ, however small, is sufficient to significantly change the average value of $\langle R_{1,2}\rangle_\lambda$. It is a kind of instability, sometimes called "replica symmetry breaking".

To prove Theorem 2.8.5, we proceed by contradiction. We will assume

$$\lim_{t\to 0}\liminf_{N\to\infty}\,(\psi'_N(t) - \psi'_N(-t)) = 0\,. \tag{2.348}$$

Lemma 2.8.6. *Assume that, for a certain value of N and certain numbers q_N and η we have*

$$|\lambda| \le t \Rightarrow \left|\frac{1}{N}\,\mathsf{E}\log\langle\exp\lambda N(R_{1,2} - q_N)\rangle\right| \le \eta t\,. \tag{2.349}$$

Then we have

$$\mathsf{E}\,G_N^{\otimes 2}(\{|R_{1,2} - q_N| \ge 3\eta\}) \le L\exp\left(-\frac{N}{K}\right) \tag{2.350}$$

where K depends on β, η and t only.

Proof. If we copy the proof of Corollary 2.2.5, (see Remark 2.2.6) we get that, for $x > 0$

$$P\left(\left|\frac{1}{N}\log\langle\exp\lambda N(R_{1,2} - q_N)\rangle - \frac{1}{N}\,\mathsf{E}\log\langle\exp\lambda N(R_{1,2} - q_N)\rangle\right| \ge x\right)$$
$$\le 2\exp\left(-\frac{Nx^2}{L\beta^2}\right)\,. \tag{2.351}$$

Taking $x = \eta t$ and combining with (2.349), we see that

$$P\left(\frac{1}{N}\log\langle\exp(\pm tN(R_{1,2} - q_N))\rangle \ge 2\,\eta t\right) \le L\exp\left(-\frac{N}{K}\right)\,,$$

and thus

$$P(\langle\exp(\pm tN(R_{1,2} - q_N))\rangle \ge \exp 2N\eta t) \le L\exp\left(-\frac{N}{K}\right)\,. \tag{2.352}$$

Now

$$G_N^{\otimes 2}(\{R_{1,2} - q_N \ge 3\eta\}) \le \exp(-3Nt\eta)\langle\exp tN(R_{1,2} - q_N)\rangle$$
$$G_N^{\otimes 2}(\{R_{1,2} - q_N \le -3\eta\}) \le \exp(-3Nt\eta)\langle\exp(-tN(R_{1,2} - q_N))\rangle$$

and combining with (2.352) we get

$$P(G_N^{\otimes 2}(\{|R_{1,2} - q_N| \ge 3\eta\}) \ge \exp(-Nt\eta)) \le L\exp\left(-\frac{N}{K}\right)\,.$$

The result follows. □

Proposition 2.8.7. *Under (2.348), given $\delta > 0$ we can find $K = K(\delta) > 0$ and N arbitrarily large such that*

$$\mathsf{E}\, G_N^{\otimes 2}(\{|R_{1,2} - \psi_N'(0)| \geq \delta\}) \leq L \exp\left(-\frac{N}{K}\right) . \tag{2.353}$$

Even though we do not know how K depends on δ, this gives us an exponential control of $R_{1,2}$.

Proof. Given $\eta > 0$, by (2.348) we can find $t > 0$ such that $\liminf_N (\psi_N'(t) - \psi_N'(-t)) < \eta$. Thus we can find N arbitrarily large such that

$$\psi_N'(t) - \psi_N'(-t) \leq \eta . \tag{2.354}$$

In particular, if we set $q_N = \psi_N'(0)$, we have (since ψ_N' is non-decreasing)

$$|\lambda| \leq t \Rightarrow |\psi_N'(\lambda) - q_N| \leq \eta ,$$

and, since by definition we have $\psi_N(0) = 0$, we get

$$|\lambda| \leq t \Rightarrow |\psi_N(\lambda) - \lambda q_N| \leq \eta t .$$

This means that (2.349) holds, and hence also (2.350), so that the proof is finished by taking $\eta = \delta/3$. □

Lemma 2.8.8. *Under (2.348), given $\delta > 0$ we can find $K = K(\delta)$ and N arbitrarily large such that*

$$\mathsf{E}\, G_N^{\otimes 2}(\{|R_{1,2} - q| \geq \delta\}) \leq L \exp\left(-\frac{N}{K}\right) . \tag{2.355}$$

Proof. We consider $\xi \leq \delta$ satisfying (2.338). Using (2.353) for $\xi/2$ rather than δ, we see that for N large enough we have $\nu((R_{1,2} - q_N)^2) \leq \xi$, so that by (2.338) we have $|q_N - q| \leq \delta$, and using (2.353), we get

$$\mathsf{E}\, G_N^{\otimes 2}(\{|R_{1,2} - q| \geq 2\delta\}) \leq L \exp\left(-\frac{N}{K}\right) .$$

□

Proof of Theorem 2.8.5. It follows from (2.355) that for any $\delta > 0$ we have

$$\nu(|R_{1,2} - q|^3) \leq \delta\, \nu((R_{1,2} - q)^2) + L \exp\left(-\frac{N}{K}\right)$$

and for large N this contradicts Proposition 2.8.1. Thus (2.348) does not hold. □

2.9 Guerra's Replica Symmetric Bound

In Sections 2.4 to 2.7 we have gained a very precise control of the SK model when $\beta \leq \beta_0$, where β_0 is a small enough number. In Section 2.8, we have shown that this picture does not extend to the region $\beta^2(1 - 2\,q + \widehat{q}) > 1$. But does it extend to the entire region $\beta^2(1 - 2\,q + \widehat{q}) < 1$? And if this is the case, how can we expect to prove it?

Having observed that the cavity method does not seem to yield anything when β is large unless one starts with some extra information, it becomes natural to design the following line of attack. Let us fix $h > 0$ once and for all, and assume that for a certain value of β_1 we understand the system very well (say as we did in the previous sections for $\beta \leq \beta_0$). Then we will select $\beta_2 > \beta_1$, such that $\beta_2 - \beta_1$ is small enough. We somehow find a way of transferring some of our knowledge of the system at inverse temperature β_1 to the system at inverse temperature β_2, and we combine this knowledge with the cavity method to gain a complete understanding of the system at inverse temperature β_2. We then iterate this procedure to reach the A-T line $\beta^2(1 - 2\,q + \widehat{q}) = 1$.

There is unfortunately an intrinsic reason for which this new approach does not work. It is the possibility of "discontinuous transitions". In such a transition, as β increases, $R_{1,2}$ suddenly starts taking values very different from q. In that case, it need not be true that $\nu(|R_{1,2}-q|^3)$ is much smaller than $\nu((R_{1,2}-q)^2)$. They can be of the same order, and the cavity method yields no information. One might think at first that a "discontinuous transition" is an unbelievable pathology. Unfortunately this does not seem to be a pathology, but rather a common phenomenon. This phenomenon can be witnessed in a transparent way in the model studied in Chapter 6.

The major challenge in controlling the entire high-temperature region of the SK model is to prove that, as β increases, a discontinuous transition does not occur before one reaches the A-T line (this is also a major challenge for the models of Chapters 3 and 5). A natural method is to find a way of proving that $R_{1,2}$ "cannot take values far from q ". Related to this is the question of finding upper bounds for Z_N, or "parts" of Z_N (where the summation is restricted to a subset of Σ_N). Here is one of the very few results known about this potentially fundamental question. It uses the notation of (2.205).

Theorem 2.9.1. *For each value of β, h, we have*

$$p_N(\beta, h) \leq \mathrm{SK}(\beta, h)\,. \tag{2.356}$$

As a consequence, we have in particular $p_N(\beta, 0) < \log 2 + \beta^2/4$ if $\beta > 1$, since it is not hard to show that $\mathrm{SK}(\beta, 0) < \log 2 + \beta^2/4$ in that case.

In words, (2.356) means that the value proposed by the replica-symmetric formula is an upper bound of the true value of p_N.

Research problem 2.9.2. (Level3!) Prove that this is also true for the models of Chapters 3 and 5.

The difficulty is that, beside a crystal-clear idea, the proof of (2.356) also uses very special features of the SK model, and that it is unclear what to do in other cases.

The proof of (2.356) will be less magic if we introduce the Hamiltonian

$$-H_N(\boldsymbol{\sigma}) = \frac{\beta}{\sqrt{N}} \sum_{i<j} g_{ij}\, \sigma_i \sigma_j + \gamma \sum_{i \le N} g_i\, \sigma_i + h \sum_{i \le N} \sigma_i \,, \tag{2.357}$$

where g_i is a family of independent standard Gaussian r.v. independent of all other families previously considered. Compared to the standard Hamiltonian of the SK model we have added a "random external field" $\sum_{i \le N} g_i \sigma_i$. We will consider only the case $\gamma \ge 0$. We write

$$p_N(\beta, \gamma, h) = \frac{1}{N} \mathsf{E} \log Z_N \,, \tag{2.358}$$

where $Z_N = Z_N(\beta, \gamma, h)$ is the partition function of the Hamiltonian (2.357), i.e.

$$Z_N = \sum_{\boldsymbol{\sigma}} \exp(-H_N(\boldsymbol{\sigma})) \,.$$

We define

$$\mathrm{SK}(\beta, \gamma, h) = \frac{\beta^2}{4}(1 - q_0)^2 + \log 2 + \mathsf{E} \log \mathrm{ch}(\beta z \sqrt{q_0} + \gamma g + h) \,, \tag{2.359}$$

where z, g are independent standard Gaussian and where q_0 is the unique root of

$$q_0 = \mathsf{E}\, \mathrm{th}^2(\beta z \sqrt{q_0} + \gamma g + h) \,. \tag{2.360}$$

This solution is unique if either $\gamma \ne 0$ or $h \ne 0$, because (2.360) can be written as

$$q_0 = \mathsf{E}\, \mathrm{th}^2(z \sqrt{\beta^2 q_0 + \gamma^2} + h)$$

(since $z\sqrt{\beta^2 q_0 + \gamma^2}$ has the same law as $\beta z \sqrt{q_0} + \gamma g$), and by Proposition 2.4.8, the function $q \mapsto \mathsf{E}\, \mathrm{th}^2(z\sqrt{\beta^2 q + \gamma^2} + h)/q$ is decreasing.

Theorem 2.9.3. (Guerra's bound) *For any values of* β, γ, h *we have*

$$p_N(\beta, \gamma, h) \le \mathrm{SK}(\beta, \gamma, h) \,. \tag{2.361}$$

Taking $\gamma = 0$, this implies (2.356).

Proof. This proof is a magnificent use of the smart path method. The idea is to transfer the dangerous first term on the right of (2.357) to the much less dangerous second term. An important observation is that $\beta z\sqrt{q_0}+\gamma g$ is centered Gaussian, with

$$\mathsf{E}\left((\beta z\sqrt{q_0}+\gamma g)^2\right)=\beta^2 q_0+\gamma^2$$

so that in (2.359) and (2.360) we can replace $\beta z\sqrt{q_0}+\gamma g$ by $z\sqrt{\beta^2 q_0+\gamma^2}$. Thus, we will avoid a lot of trouble if we ensure that $\beta^2 q_0+\gamma^2$ is constant along the smart path. We fix β, γ, and the corresponding value of q_0. For $0\le t\le 1$, we consider

$$\beta(t)=\beta\sqrt{t}\,,\ \gamma(t)=\sqrt{\gamma^2+\beta^2 q_0(1-t)}\,, \tag{2.362}$$

so that

$$\beta^2(t)\,q_0+\gamma^2(t)=\beta^2 q_0+\gamma^2\,. \tag{2.363}$$

It follows from (2.360) that

$$q_0=\mathsf{E}\,\mathrm{th}^2(\beta(t)\,z\sqrt{q_0}+\gamma(t)\,g+h)\,,$$

so that the value q_0 does not vary along the smart path.

For $t=0$, we have $\beta(0)=0$, so that

$$\mathrm{SK}(\beta(0),\gamma(0),h)=\log 2+\mathsf{E}\log\mathrm{ch}(\gamma(0)\,g+h)$$

and, by (2.357),

$$Z_N=\prod_{i\le N}\left(\exp(\gamma(0)\,g_i+h)+\exp(-(\gamma(0)\,g_i+h)\right)=2^N\prod_{i\le N}\mathrm{ch}(\gamma(0)\,g_i+h)\,,$$

so that

$$p_N(0,\gamma(0),h)=\log 2+\mathsf{E}\log\mathrm{ch}(\gamma(0)\,g+h)=\mathrm{SK}(0,\gamma(0),h)\,. \tag{2.364}$$

Consider the functions $\eta(t)=\mathrm{SK}(\beta(t),\gamma(t),h)$ and $\xi(t)=p_N(\beta(t),\gamma(t),h)$. To prove (2.361), it suffices to prove that $\xi'(t)\le\eta'(t)$. The beauty of the argument is that this will be done without knowing anything on the system. To compute η', we note that by (2.363) we have

$$\eta'(t)=\frac{\beta^2}{4}(1-q_0)^2\,.$$

To compute ξ', we note that

$$\frac{\partial p_N}{\partial\beta}=\frac{1}{N}\mathsf{E}\left\langle\frac{1}{\sqrt{N}}\sum_{i<j}g_{ij}\,\sigma_i\,\sigma_j\right\rangle=\frac{\beta}{2}(1-\nu(R_{1,2}^2)) \tag{2.365}$$

$$\frac{\partial p_N}{\partial\gamma}=\frac{1}{N}\mathsf{E}\left\langle\sum_{i\le N}g_i\,\sigma_i\right\rangle=\gamma\,(1-\nu(R_{1,2}))\,, \tag{2.366}$$

where of course $\langle\cdot\rangle$ is Gibbs' measure relative to the Hamiltonian (2.357). In (2.365) we use (2.210), and in (2.365) a similar (but simpler) integration by parts. Since

$$\frac{\mathrm{d}\beta(t)}{dt} = \frac{\beta^2}{2\,\beta(t)}\,, \quad \frac{\mathrm{d}\gamma(t)}{dt} = -\frac{\beta^2 q_0}{2\,\gamma(t)}\,,$$

we get

$$\xi'(t) = \frac{\beta^2}{4}\left(1 - \nu(R_{1,2}^2)\right) - \frac{\beta^2 q_0}{2}\left(1 - \nu(R_{1,2})\right)\,,$$

so that (O miracle!)

$$\eta'(t) - \xi'(t) = \frac{\beta^2}{4}\,\nu((R_{1,2} - q_0)^2) \geq 0\,.$$

□

We now go back to the problem of proving that the values of $R_{1,2}$ far from q do not matter. We consider a standard normal r.v. z and $Y = z\sqrt{y} + h$. We consider the functions

$$W(x,y) = 2\log 2 + \mathsf{E}\log(\operatorname{ch} x\operatorname{ch}^2 Y + \operatorname{sh} x\operatorname{sh}^2 Y) \tag{2.367}$$

$$F(u,y) = \inf_x (W(x,y) - xu)\,. \tag{2.368}$$

Given a number u, $|u| \leq 1$, we consider

$$Z_N(\beta,\gamma,u) = \sum_{R_{1,2}=u} \exp(-H_N(\boldsymbol{\sigma}^1) - H_N(\boldsymbol{\sigma}^2)) \tag{2.369}$$

$$r_N(\beta,\gamma,u) = \frac{1}{N}\,\mathsf{E}\log Z_N(\beta,\gamma,u)\,, \tag{2.370}$$

where $H_N(\boldsymbol{\sigma})$ is given by (2.357) and where the summation in (2.369) is over all pairs $(\boldsymbol{\sigma}^1, \boldsymbol{\sigma}^2) \in \Sigma_N^2$ such that $R_{1,2} = u$ (it is always assumed that u is such that such pairs exist). Our main technical result is as follows.

Proposition 2.9.4. *For all values of u, all values of $\beta, \gamma, q \geq 0$, we have*

$$r_N(\beta,\gamma,u) \leq \frac{\beta^2}{2}\left((1-q)^2 + (u-q)^2\right) + F(u, \gamma^2 + \beta^2 q)\,. \tag{2.371}$$

Since we have

$$r_N(\beta,\gamma,u) \leq 2p_N(\beta,\gamma,h) \leq 2\mathrm{SK}(\beta,\gamma,h)$$

it would be desirable to obtain a bound for $r_N(\beta,\gamma,u)$ by a quantity that is obviously less that or equal to $2\mathrm{SK}(\beta,\gamma,h)$. This is not the case of the bound of Proposition 2.9.4 and is an obvious shortcoming of this result. This shortcoming will be repaired later, with much more work, in Theorem 2.11.14.

The first step of the proof is the case $\beta = 0$, that explains the occurrence of the function F.

Lemma 2.9.5. *We have* $r_N(0,\gamma,u) \le F(u,\gamma^2)$.

Proof. Given x we have

$$r_N(0,\gamma,u) = \frac{1}{N}\,\mathsf{E}\log(e^{Nux} Z_N(0,\gamma,u)) - xu\ ,$$

so that it suffices to prove that

$$\frac{1}{N}\,\mathsf{E}\log e^{Nux} Z_N(0,\gamma,u) \le W(x,\gamma^2)\ . \tag{2.372}$$

We have

$$\begin{aligned} e^{Nux} Z_N(0,\gamma,u) &= \sum_{R_{1,2}=u} \exp(Nx\,R_{1,2} - H_N(\boldsymbol{\sigma}^1) - H_N(\boldsymbol{\sigma}^2)) \\ &\le \sum \exp(Nx\,R_{1,2} - H_N(\boldsymbol{\sigma}^1) - H_N(\boldsymbol{\sigma}^2))\ , \end{aligned}$$

where now the summation is over all values of $\boldsymbol{\sigma}^1, \boldsymbol{\sigma}^2$ in Σ_N. Since

$$Nx\,R_{1,2} - H_N(\boldsymbol{\sigma}^1) - H_N(\boldsymbol{\sigma}^2) = \sum_{i\le N} (x\,\sigma_i^1\,\sigma_i^2 + \sigma_i^1(\gamma\,g_i + h) + \sigma_i^2(\gamma\,g_i + h))\,,$$

we have

$$\begin{aligned} e^{Nux} Z_N(0,\gamma,u) &\le \prod_{i\le N} \sum_{\varepsilon_1,\varepsilon_2=\pm 1} \exp(x\,\varepsilon_1\,\varepsilon_2 + \varepsilon_1(\gamma\,g_i + h) + \varepsilon_2(\gamma\,g_i + h)) \\ &= \prod_{i\le N} (4(\operatorname{ch} x \operatorname{ch}^2(\gamma\,g_i + h) + \operatorname{sh} x \operatorname{sh}^2(\gamma\,g_i + h)))\ . \end{aligned}$$

Taking logarithm and expectation proves (2.372). □

We will think of $Z_N(\beta,\gamma,u)$ as the partition function corresponding to the Hamiltonian $\exp(-H_N(\boldsymbol{\sigma}^1) - H_N(\boldsymbol{\sigma}^2))$, but when the space of configurations is the set $\Sigma_{N,u}$ of pairs $(\boldsymbol{\sigma}^1, \boldsymbol{\sigma}^2)$ in Σ_N^2 for which $R_{1,2} = u$. Since this does not seem to create any possibility of confusion, we will still denote by $\langle\cdot\rangle$ average for the corresponding Gibbs' measure $G_{N,u}$.

We will need a replica of the system $(\boldsymbol{\sigma}^1, \boldsymbol{\sigma}^2)$; we will denote it by $(\boldsymbol{\tau}^1, \boldsymbol{\tau}^2)$; we will also denote by $\langle\cdot\rangle$ average on $\Sigma_{N,u}^2$ with respect to $G_{N,u}^{\otimes 2}$. We will write

$$R(\boldsymbol{\sigma}^1, \boldsymbol{\tau}^1) = \frac{1}{N}\sum_{i\le N} \sigma_i^1\,\tau_i^1 \tag{2.373}$$

and similar definitions for $R(\boldsymbol{\sigma}^1, \boldsymbol{\tau}^2)$, etc. Let us note that by symmetry, we have

$$\nu(R(\boldsymbol{\sigma}^1, \boldsymbol{\tau}^1)^2) = \nu(R(\boldsymbol{\sigma}^1, \boldsymbol{\tau}^2)^2)\ , \tag{2.374}$$

where of course $\nu(\cdot) = \mathsf{E}\langle\cdot\rangle$.

Lemma 2.9.6. *We have*

$$\frac{\partial r_N}{\partial \beta}(\beta,\gamma,u) = \beta(1+u^2-2\nu(R(\boldsymbol{\sigma}^1,\boldsymbol{\tau}^1)^2)) \tag{2.375}$$

$$\frac{\partial r_N}{\partial \gamma}(\beta,\gamma,u) = 2\gamma(1+u-2\nu(R(\boldsymbol{\sigma}^1,\boldsymbol{\tau}^1))) \tag{2.376}$$

Proof. There is no new idea there. To prove (2.375), we write

$$\begin{aligned}\frac{\partial r_N}{\partial \beta}(\beta,\gamma,u) &= \frac{1}{N}\nu\left(\frac{1}{\sqrt{N}}\sum_{i<j} g_{ij}(\sigma_i^1\sigma_j^1+\sigma_i^2\sigma_j^2)\right)\\ &= \frac{2}{N^{3/2}}\nu\left(\sum_{i<j} g_{ij}\,\sigma_i^1\sigma_j^1\right),\end{aligned}$$

by symmetry. Integrating by parts as in the proof of Theorem 2.4.17, this is

$$\frac{2\beta}{N^2}\left(\nu\left(\sum_{i<j}\sigma_i^1\sigma_j^1(\sigma_i^1\sigma_j^1+\sigma_i^2\sigma_j^2)\right)-\nu\left(\sum_{i<j}\sigma_i^1\sigma_j^1(\tau_i^1\tau_j^1+\tau_i^2\tau_j^2)\right)\right)$$

$$=\beta\nu\left(\frac{1}{N^2}\sum_{i,j}\sigma_i^1\sigma_j^1(\sigma_i^1\sigma_j^1+\sigma_i^2\sigma_j^2)\right)-\beta\nu\left(\frac{1}{N^2}\sum_{i,j}\sigma_i^1\sigma_j^1(\tau_i^1\tau_j^1+\tau_i^2\tau_j^2)\right)$$

because the terms for $j=i$ cancel out. Then (2.375) follows using (2.374), since $R_{1,2}=u$ on $\Sigma_{N,u}$. The proof of (2.376) is similar. □

For $0\le t\le 1$, we define

$$\beta(t)=\beta\sqrt{t}\,;\ \gamma(t)=\sqrt{\gamma^2+\beta^2 q(1-t)}\,;\ \xi(t)=r_N(\beta(t),\gamma(t),u)\,.$$

Lemma 2.9.5 shows that to prove Proposition 2.9.4, it suffices to prove the following.

Lemma 2.9.7. *We have*

$$\xi'(t)\le\frac{\beta^2}{2}((1-q)^2+(u-q)^2)\,.$$

Proof. We proceed as in the proof of Theorem 2.9.3, using now Lemma 2.9.6 to get

$$\begin{aligned}\xi'(t) &= \frac{\beta^2}{2}(1+u^2-2\nu(R(\boldsymbol{\sigma}^1,\boldsymbol{\tau}^1)^2)-2q-2qu+4q\nu(R(\boldsymbol{\sigma}^1,\boldsymbol{\tau}^1)))\\ &= \frac{\beta^2}{2}(1-2q+u^2-2qu+2q^2)-\beta^2\nu((R(\boldsymbol{\sigma}^1,\boldsymbol{\tau}^1)-q)^2)\,.\end{aligned}$$

□

We define the region $\mathcal{R}'_h$ by $(\beta,\gamma) \in \mathcal{R}'_h$ if

$$\forall\, u \neq q_0, \quad \inf_q \left(\frac{\beta^2}{2}((1-q)^2 + (u-q)^2) + F(u, \gamma^2 + \beta^2 q) \right) < 2\,\mathrm{SK}(\beta,\gamma,h) \tag{2.377}$$

where $\mathrm{SK}(\beta,\gamma,h)$ is given by (2.359) and q_0 by (2.360). We define the region $\mathcal{R}_h$ by

$$(\beta,\gamma) \in \mathcal{R}_h \quad \text{if} \quad \forall\, v\,,\, 0 \le v \le \beta \quad \Longrightarrow \quad (v,\gamma) \in \mathcal{R}'_h\,. \tag{2.378}$$

As we will show later, we can control the SK model on this region.

Now, is it true that the region $\mathcal{R}_h$ coincides with the region

$$\beta^2\, \mathsf{E}\, \frac{1}{\mathrm{ch}^4(z\sqrt{\gamma^2+\beta^2 q_0}+h)} < 1\ ? \tag{2.379}$$

At first sight it is not obvious at all that there is any relationship whatsoever between (2.379) and (2.377). The next two pages are devoted to prove that under (2.379), then (2.377) is true when $|u - q_0|$ is small enough. The author perfomed this calculation before he invented the better bounds arising from Theorem 2.11.14, for which the corresponding result is much easier to see. So, this computation is now essentially useless, but we present it because the way condition (2.379) occurs looks nearly miraculous.

Lemma 2.9.8. *We have*

$$\frac{\partial W}{\partial x}(x,y) = \mathsf{E}\left(\frac{\mathrm{sh}\, x\, \mathrm{ch}^2 Y + \mathrm{ch}\, x\, \mathrm{sh}^2 Y}{\mathrm{ch}\, x\, \mathrm{ch}^2 Y + \mathrm{sh}\, x\, \mathrm{sh}^2 Y} \right) \tag{2.380}$$

$$\frac{\partial W}{\partial y}(x,y) = \mathsf{E}\left(\frac{e^x(\mathrm{sh}^2 Y + \mathrm{ch}^2 Y)}{\mathrm{ch}\, x\, \mathrm{ch}^2 Y + \mathrm{sh}\, x\, \mathrm{sh}^2 Y} - \frac{2\,e^{2x}\,\mathrm{sh}^2 Y\, \mathrm{ch}^2 Y}{(\mathrm{ch}\, x\, \mathrm{ch}^2 Y + \mathrm{sh}\, x\, \mathrm{sh}^2 Y)^2} \right). \tag{2.381}$$

Proof. The proof of (2.380) is straightforward. To prove (2.381) we write

$$\frac{\partial W}{\partial y}(x,y) = \mathsf{E}\left(\frac{z}{2\sqrt{y}}\, \frac{2\,e^x\, \mathrm{sh}\, Y\, \mathrm{ch}\, Y}{\mathrm{ch}\, x\, \mathrm{ch}^2 Y + \mathrm{sh}\, x\, \mathrm{sh}^2 Y} \right)$$

and we integrate by parts. □

To lighten notation we set

$$W_x = \frac{\partial}{\partial x} W(0, \gamma^2 + \beta^2 q_0)\,;\ W_{xx} = \frac{\partial^2 W}{\partial x^2}(0, \gamma^2 + \beta^2 q_0)$$

and we define W_y, W_{xy}, W_{yy} similarly. We set

$$\widehat{q}_0 = \mathsf{E}\,\mathrm{th}^4(z\sqrt{\gamma^2+\beta^2 q_0}+h)\,.$$

The following is a straightforward computation from (2.380) and (2.381).

Lemma 2.9.9. *We have*

$$W_x = q_0\,;\; W_y = 1 - q_0\,;\; W_{xx} = 1 - \widehat{q}_0\,;\; W_{xy} = -W_{yy} = 1 - 4\,q_0 + 3\,\widehat{q}_0\ .$$

Let us observe that

$$W_{xx} - W_{xy} = 3\,(q_0 - \widehat{q}_0) > 0 \tag{2.382}$$

and that (2.379) is equivalent to

$$\beta^2(W_{xx} + W_{xy}) < 2\ . \tag{2.383}$$

Consider now $u = q_0 + t$. Given λ, μ, we have

$$\inf_q \left(\frac{\beta^2}{2}((1-q)^2 + (q_0 + t - q)^2) + F(q_0 + t, \gamma^2 + \beta^2 q)\right) \le f(t)$$

where

$$f(t) = \frac{\beta^2}{2}((1-q_0-\lambda t)^2 + t^2(\lambda-1)^2) + W(\mu\,t, \gamma^2 + \beta^2 q_0 + \beta^2\lambda t) - \mu\,t(q_0+t)\ ,$$

as is seen by taking $q = q_0 + \lambda t$, and since $F(u,y) \le W(x,y) - ux$ for $x = \mu\,t$. Now

$$f(0) = \frac{\beta^2}{2}(1-q_0)^2 + W(0, \gamma^2 + \beta^2 q_0) = 2\,\mathrm{SK}(\beta,\gamma,h)\ ,$$

$$f'(0) = -\beta^2\lambda(1-q_0) + \mu\,W_x + \beta^2\lambda\,W_y - \mu\,q_0 = 0$$

since $W_x = q_0$ and $W_y = 1 - q_0$. Also, we have

$$f''(0) = \beta^2(\lambda^2 + (\lambda-1)^2) + \mu^2\,W_{xx} + 2\,\beta^2\mu\,\lambda\,W_{xy} + \beta^4\lambda^2 W_{yy} - 2\mu\ .$$

We will show that under (2.379) we can find μ, λ such that $f''(0) < 0$. Choosing $\mu = (1 - \beta^2\lambda\,W_{xy})/W_{xx}$ we get

$$f''(0) = \beta^2(\lambda^2 + (\lambda-1)^2) + \beta^4\lambda^2 W_{yy} - \frac{(1-\beta^2\lambda\,W_{xy})^2}{W_{xx}} = \frac{1}{W_{xx}}(a\,\lambda^2 + 2\,b\,\lambda + c)$$

where

$$a = 2\,\beta^2 W_{xx} + \beta^4(W_{xx}\,W_{yy} - W_{xy}^2)$$

$$b = -\beta^2(W_{xx} - W_{xy})\,;\; c = \beta^2 W_{xx} - 1\ .$$

All we have to show is that under (2.379) (or, equivalently, under (2.383)) we have

$$\inf_\lambda\,(a\,\lambda^2 + 2\,b\,\lambda + c) < 0\ .$$

This is obvious if $a \le 0$, since $b \ne 0$ by (2.382), so we can assume $a > 0$, in which case it suffices to show that $b^2 - ac > 0$. Now, replacing W_{yy} by $-W_{xy}$ in the expression of a, and using algebra, we get

$$b^2 - ac = W_{xx}(2 - \beta^2(W_{xx} + W_{xy}))(1 - \beta^2 W_{xy})\ .$$

We have already noticed that $1 - \beta^2 W_{xy} = 1 - \beta^2(1 - 4\,q_0 + 3\,\widehat{q}_0) > 0$, so that $b^2 - ac > 0$ exactly if (2.383) occurs. □

It is of course harder to analyze (2.377) when $u - q_0$ is not small. Numerical studies [156] indicate that the region $\mathcal{R}_h$ does not coincide with the region (2.379).

Theorem 2.9.10. *Assume that $(\beta, \gamma) \in \mathcal{R}_h$. Then we have*

$$p_N(\beta,\gamma,h) \ge \mathrm{SK}(\beta,\gamma,h) - \frac{K}{N}$$

and Theorem 2.7.3 is true for the value (β,γ,h) of the parameters.

Proof. The proof considers as self-evident the fact that all the results that were proved in the previous sections for $\gamma = 0$ also hold for $\gamma \ne 0$. The reader feeling uncomfortable with this statement should simply assume that $\gamma = 0$. In the present proof $K, K_0, \ldots$ will denote numbers depending only on β, γ, h. For $v \le \beta$, let us denote by $q(v)$ the unique solution of the equation

$$q(v) = \mathsf{E}\,\mathrm{th}^2\, Y$$

where $Y = z\sqrt{\gamma^2 + v^2 q(v)} + h$. It follows from the discussion preceding the present theorem that

$$\forall v \le \beta\,,\ v^2\,\mathsf{E}\frac{1}{\mathrm{ch}^4\, Y} < 1\,.$$

It then follows from Proposition 2.6.12 (as in the proof of (2.336)) that there exists a number $\delta > 0$ such that, for $v \le \beta$,

$$\nu(|R_{1,2} - q(v)|^3) \le \delta\nu((R_{1,2} - q(v))^2) + \frac{1}{N} \Rightarrow \nu((R_{1,2} - q(v))^2) \le \frac{K_0}{N}\,, \tag{2.384}$$

where ν is for the values v, γ and h of the parameters.

Let us write $\mathrm{SK}(v) = \mathrm{SK}(v,\gamma,h)$ and $p_N(v) = p_N(v,\gamma,h)$. Since $(v,\gamma) \in \mathcal{R}'_h$ for $v \le \beta$, it follows from (2.377) and compactness that we can find a number $\varepsilon > 0$ such that

$$\forall v \le \beta\,,\ |u - q(v)| \ge \delta \Rightarrow U(u,v) \le 2\,\mathrm{SK}(v) - 7\varepsilon\,,$$

where

$$U(u,v) = \inf_q \left(\frac{v^2}{2}((1-q)^2 + (q-u)^2) + F(u, \gamma^2 + v^2 q)\right)\,.$$

It thus follows from Proposition 2.9.4 that

$$\frac{1}{N}\mathsf{E}\log Z_N(v,\gamma,u) \le 2\,\mathrm{SK}(v) - 7\varepsilon\,.$$

By (an obvious extension of) Corollary 2.2.5, it follows that with overwhelming probability we have

$$\frac{1}{N}\log Z_N(v,\gamma,u) \le 2\,\mathrm{SK}(v) - 6\varepsilon$$

and thus

$$Z_N(v,\gamma,u) \le \exp N(2\,\mathrm{SK}(v) - 6\varepsilon)\ .$$

Thus, since there are at most $2N+1 \le 3N$ values of u to consider, with overwhelming probability we have

$$\sum_{|R_{1,2}-q(v)|\ge\delta} \exp(-H_N(\boldsymbol{\sigma}^1) - H_N(\boldsymbol{\sigma}^2)) \le 3N\exp N(2\,\mathrm{SK}(v) - 6\varepsilon)\ .$$

We also have, with overwhelming probability, that

$$Z_N(v) = \sum_{\boldsymbol{\sigma}} \exp(-H_N(\boldsymbol{\sigma})) \ge \exp N(p_N(v) - \varepsilon)$$

and thus, with overwhelming probability, we have

$$G_N^{\otimes 2}(|R_{1,2} - q(v)| \ge \delta) \le 3N\exp 2N(\mathrm{SK}(v) - p_N(v) - 2\varepsilon)\ .$$

It follows that if $\limsup_{N\to\infty}(\mathrm{SK}(v) - p_N(v)) \le \varepsilon$, then for N large enough the left inequality of (2.384) occurs.

We observe that $p'_N(v) = v(1-\mathsf{E}\langle R_{1,2}^2\rangle)/2 \ge 0$, and that (easily) $\mathrm{SK}'(v) \le v/2$. (Let us note that in order to make the argument work, we simply need uniform bounds on both these derivatives, a very weak information.) Thus, if $\limsup_{N\to\infty}(\mathrm{SK}(v) - p_N(v)) = 0$ uniformly on the interval $[0, v_0]$, then $\limsup_{N\to\infty}(\mathrm{SK}(v) - p_N(v)) \le \varepsilon$ uniformly on the interval $[0, v_1]$, where $v_1 = \min(\beta, v_0+\varepsilon/\beta)$. Thus, by (2.384), for N large enough we have $\nu((R_{1,2}-q(v))^2) \le K_0/N$ for all $v \le v_1$. The proof of Lemma 2.5.4 (see (2.248)) then shows that $|\nu(R_{1,2} - q(v))| \le K_1/N$, and the proof of (2.213) shows that $\mathrm{SK}(v) - p_N(v) \le K/N$ for $v \le v_1$. It should be obvious that these facts together imply that for N large enough (and hence for all N, with a different constant K) we have $\mathrm{SK}(v) - p_N(v) \le K/N$ and $\nu((R_{1,2} - q(v))^2) \le K/N$ for all $v \le \beta$.

To prove the assertion concerning Theorem 2.7.3 we need to generalize Proposition 2.6.12 to higher moments. This is better left to the reader. □

Research problem 2.9.11. (Level 2) If $(\beta,\gamma) \in \mathcal{R}_h$, does there exists K such that

$$\nu\left(\exp\frac{N}{K}(R_{1,2} - q_0)^2\right) \le K\ ?$$

2.10 Guerra-Toninelli Interpolation and the Thermodynamical Limit

Given two Hamiltonians $H_1(\boldsymbol{\sigma})$ and $H_2(\boldsymbol{\sigma})$ such that each of the families $(H_1(\boldsymbol{\sigma}))$ and $(H_2(\boldsymbol{\sigma}))$ is a jointly Gaussian family of r.v., it is natural to interpolate between them using the path

$$H_t(\boldsymbol{\sigma}) = \sqrt{t}\, H_1(\boldsymbol{\sigma}) + \sqrt{1-t}\, H_2(\boldsymbol{\sigma}) , \tag{2.385}$$

and to study the function

$$\varphi(t) = \frac{1}{N} \mathsf{E} \log Z_t = \frac{1}{N} \mathsf{E} \log \sum_{\boldsymbol{\sigma}} \exp(-H_t(\boldsymbol{\sigma})) , \tag{2.386}$$

the derivative of which can be computed through integration by parts. (To be more accurate, one should say that this idea is natural after one reads the beautiful paper [75].) Of course one wishes to assume, without loss of generality, that H_1 and H_2 are probabilistically independent. An equivalent way to look at the proof of Theorem 2.9.3 is to think that we interpolate between the Hamiltonian $H_1(\boldsymbol{\sigma})$ given by (2.9) and the Hamiltonian

$$-H_2(\sigma) = a \sum_{i \le N} g'_i \sigma_i + h \sum_{i \le N} \sigma_i ,$$

where the r.v. g'_i are independent standard Gaussian, independent of the r.v. occurring in the definition of H_1, and where $a^2 = \gamma^2 + \beta^2 q^2$. The reader having understood the computations in the proof of Theorem 2.9.3 will realize that he can easily prove that

$$\varphi'(t) = \frac{\beta^2}{4}(1-q)^2 - \frac{\beta^2}{4} \langle (R_{1,2} - q)^2 \rangle_t , \tag{2.387}$$

where $\langle \cdot \rangle_t$ refers of course to the Gibbs' measure with Hamiltonian (2.385). Of course, (2.387) implies that $\varphi(1) \le \beta^2(1-q)^2/4 + \varphi(0)$, which is exactly the statement of Theorem 2.9.3.

Theorem 2.10.1. *[75] For all values of β, h, the sequence $(p_N(\beta, h))_{N \ge 1}$ is superadditive, that is, for integers N_1, N_2,*

$$p_{N_1+N_2}(\beta, h) \ge \frac{N_1}{N_1 + N_2} p_{N_1}(\beta, h) + \frac{N_2}{N_1 + N_2} p_{N_2}(\beta, h) . \tag{2.388}$$

Consequently, $\lim_{N \to \infty} p_N(\beta, h)$ exists.

Comment. The same proof holds for $p_N(\beta, \gamma, h)$, but, for simplicity of notation, we consider only the case $\gamma = 0$.

Proof. We set $N = N_1 + N_2$. Consider the Hamiltonian

$$-H_2(\boldsymbol{\sigma}) = \frac{\beta}{\sqrt{N_1}} \sum_{1 \le i < j \le N_1} g'_{ij}\, \sigma_i \sigma_j + \frac{\beta}{\sqrt{N_2}} \sum_{N_1 < i < j \le N} g'_{ij}\, \sigma_i \sigma_j + h \sum_{i \le N} \sigma_i \,,$$

where the r.v. g'_{ij} are independent and independent of the g_{ij}. We observe that

$$\sum_{\boldsymbol{\sigma}} \exp(-H_2(\boldsymbol{\sigma})) = Z_1 Z_2 \,,$$

where

$$Z_1 = \sum_{\sigma_1,\ldots,\sigma_{N_1} = \pm 1} \exp\left(-\frac{\beta}{\sqrt{N_1}} \sum_{1 \le i < j \le N_1} g'_{ij}\, \sigma_i \sigma_j + h \sum_{i \le N_1} \sigma_i \right)$$

$$Z_2 = \sum_{\sigma_{N_1+1},\ldots,\sigma_N = \pm 1} \exp\left(-\frac{\beta}{\sqrt{N_2}} \sum_{N_1 < i < j \le N} g'_{ij}\, \sigma_i \sigma_j + h \sum_{N_1 < i \le N} \sigma_i \right) .$$

Thus, if φ is given by (2.386), we have

$$\varphi(0) = \frac{1}{N} \mathsf{E} \log Z_1 + \frac{1}{N} \mathsf{E} \log Z_2 = \frac{N_1}{N} p_{N_1}(\beta, h) + \frac{N_2}{N} p_{N_2}(\beta, h) \,.$$

Also, $\varphi'(t) = \mathrm{I} + \mathrm{II}$, where

$$\mathrm{I} = \frac{1}{2N\sqrt{t}} \mathsf{E} \langle -H_1(\boldsymbol{\sigma}) \rangle_t \,; \ \mathrm{II} = -\frac{1}{2N\sqrt{1-t}} \mathsf{E} \langle -H_2(\boldsymbol{\sigma}) \rangle_t \,.$$

As before we get

$$\mathrm{I} = \frac{\beta^2}{4} (1 - \mathsf{E} \langle R_{1,2}^2 \rangle_t) \,. \tag{2.389}$$

We set $R' = N_1^{-1} \sum_{i \le N_1} \sigma_i^1 \sigma_i^2$ and $R'' = N_2^{-1} \sum_{N_1 < i \le N} \sigma_i^1 \sigma_i^2$, and we observe that $N R_{1,2} = N_1 R' + N_2 R''$ $(= \sum_{i \le N} \sigma_i^1 \sigma_i^2)$. Inspection of the proof of (2.389) shows that

$$\mathrm{II} = -\frac{N_1}{N} \frac{\beta^2}{4} (1 - \mathsf{E} \langle R'^2 \rangle_t) - \frac{N_2}{N} \frac{\beta^2}{4} (1 - \mathsf{E} \langle R''^2 \rangle_t)$$

and thus

$$\varphi'(t) = \frac{\beta^2}{4} \mathsf{E} \left\langle -R_{1,2}^2 + \frac{N_1}{N} R'^2 + \frac{N_2}{N} R''^2 \right\rangle_t \ge 0$$

because

$$R_{1,2}^2 = \left(\frac{N_1}{N} R' + \frac{N_2}{N} R'' \right)^2 \le \frac{N_1}{N} R'^2 + \frac{N_2}{N} R''^2 \,,$$

since the function $x \mapsto x^2$ is convex. Thus $\varphi(1) \ge \varphi(0)$, and this proves (2.388). The last claim is standard.

Research problem 2.10.2. (Level 3) Find the exact rate of convergence of $p_N(\beta,h)$ to its limit.

Corollary 2.10.3. *For all values of β, h, the limit*

$$\frac{1}{N}\lim_{N\to\infty}\mathsf{E}\max_{\boldsymbol{\sigma}}\left(\frac{\beta}{\sqrt{N}}\sum_{i<j} g_{ij}\sigma_i\sigma_j + h\sum_{i\le N}\sigma_i\right)$$

exists.

Proof. We observe that if $U(\boldsymbol{\sigma})$ is a random function on Σ_N, then for $t>0$ we have

$$\left|\frac{1}{N}\mathsf{E}\max_{\boldsymbol{\sigma}} U(\boldsymbol{\sigma}) - \frac{1}{Nt}\mathsf{E}\log\sum_{\boldsymbol{\sigma}}\exp tU(\boldsymbol{\sigma})\right| \le \frac{\log 2}{t}\,. \tag{2.390}$$

This follows from the fact that

$$\exp\left(t\max_{\boldsymbol{\sigma}} U(\boldsymbol{\sigma})\right) \le \sum_{\boldsymbol{\sigma}}\exp(tU(\boldsymbol{\sigma})) \le 2^N\exp\left(t\max_{\boldsymbol{\sigma}} U(\boldsymbol{\sigma})\right)$$

by taking logarithm, expectation and dividing by Nt. Using this for $U(\boldsymbol{\sigma}) = -H_N(\boldsymbol{\sigma})$, and since $\lim_{N\to\infty} p_N(t\beta, th)$ exists for each t by Theorem 2.10.1, the result follows. □

Corollary 2.10.4. *Recall the Hamiltonian (2.357), and consider h as fixed. Then for each values of (β,γ) outside an exceptional set that is negligible for the two dimensional Lebesgue measure, the limits $\lim_{N\to\infty}\nu(R_{1,2}^2)$ and $\lim_{N\to\infty}\nu(R_{1,2})$ exist.*

Proof. The limit $p_\infty(\beta,\gamma,h) = \lim_{N\to\infty} p_N(\beta,\gamma,h)$ exists from Theorem 2.10.1 and is a convex function of β,γ. Given γ, the function $\beta\mapsto p_\infty(\beta,\gamma)$ is differentiable outside a countable set. In the proof of Theorem 2.4.19 it is shown that $\lim_{N\to\infty}\partial p_N/\partial\beta$ exists at each point where $p_\infty(\cdot,\gamma)$ is differentiable (a simple fact about convex functions). Since $\partial p_N/\partial\beta = \beta(1-\nu(R_{1,2}^2))/2$, the existence of $\lim_{N\to\infty}\nu(R_{1,2}^2)$ follows. The same argument works for $\nu(R_{1,2})$, differentiating now with respect to γ.

Remark. We do not know how to prove the existence of $\lim_{N\to\infty}\nu(R_{1,2})$ using only the Hamiltonian (2.9).

Even though the property of the Hamiltonian used in Theorem 2.10.1 is somewhat less special than the property used in Section 2.9, it is certainly not clear how this approach can be generalized. A first direction to look at is the case of two (or more) coupled copies of the SK model as considered in Section 2.8. The difficulty in this case is that some terms have the wrong sign, but it can be solved using the method of Section 2.12 of restricting to couples of configurations with a given overlap, see [76]. (This allows one to simplify the statement of Theorem 2.8.5, although it does not help much in proving it.) In Section 6.6 we will develop a tool that will enable us to apply the method of the present section to new cases. In Chapter 6 will also occur an interesting situation where knowing that the limit exists greatly helps in proving that it has the desired value.

Research problem 2.10.5. (Level 2) Extend Theorem 2.10.1 to the case of the Hamiltonians of Chapter 3.

2.11 Guerra's Broken Replica-Symmetry Bound and the Parisi Formula

In this section we prove a (deep) extension of Theorem 2.9.3. The proof relies on an extremely beautiful construction of F. Guerra [72]. As Theorem 2.9.3, it involves rather special properties of the SK model. As an application we will be able to compute the value of $\lim_{N\to\infty} p(\beta,\gamma,h)$ for all values of the parameters, an expression first discovered by G. Parisi.

We consider h and γ fixed once and for all.

Consider an integer k, and consider $q_0 = 0 \le q_1 \le q_2 \le \cdots \le q_k \le q_{k+1} = 1$. Consider $0 = m_0 \le m_1 \le \cdots \le m_k = 1$. By decreasing induction over $0 \le \ell \le k+1$, we define the functions $\psi_\ell(y)$ as follows. We set $\psi_{k+1}(y) = \mathrm{ch} y$. For $0 \le \ell \le k$ we then set

$$\psi_\ell(y) = (\mathsf{E}\, \psi_{\ell+1}^{m_\ell}(y + \beta\sqrt{q_{\ell+1} - q_\ell}\, z))^{1/m_\ell} , \tag{2.391}$$

where z is standard normal. When $m_\ell = 0$, this has to be understood as

$$\psi_\ell(y) = \exp \mathsf{E} \log \psi_{\ell+1}(y + \beta\sqrt{q_{\ell+1} - q_\ell}\, z)) . \tag{2.392}$$

Theorem 2.11.1. *(Guerra's broken replica-symmetry bound.) For each N, all choices of $k, q_0, \ldots, q_{k+1}, m_0, \ldots, m_k$ as above we have*

$$p_N(\beta,\gamma,h) \le \log 2 + \mathsf{E}\log\psi_0(\gamma g + h) - \frac{\beta^2}{4} \sum_{0\le\ell\le k} m_\ell(q_{\ell+1}^2 - q_\ell^2) \tag{2.393}$$

where g is standard normal.

To see that this is indeed a generalization of Theorem 2.9.3, we consider the case $k = 1$, $m_1 = 1$ and $q_1 = q$. Then

$$\psi_1(y) = \exp\frac{\beta^2(1-q)}{2}\,\mathrm{ch} y$$

$$\log\psi_0(y) = \frac{\beta^2(1-q)}{2} + \mathsf{E}\log\mathrm{ch}(y + \beta z\sqrt{q})$$

so that (2.393) implies

$$p_N(\beta,\gamma,h) \le \log 2 + \frac{\beta^2}{2}(1-q) + \mathsf{E}\log\operatorname{ch}(\beta z\sqrt{q}+\gamma g+h) - \frac{\beta^2}{4}(1-q^2) \quad (2.394)$$

and this is (2.361).

The right-hand side of (2.393) can be seen as a functional $\mathcal{P}(w,\beta,\gamma,h)$ of the function w such that $w(q) = m_\ell$ for $q_\ell \le q < q_{\ell+1}$. The physicists think of this function w as a parameter.

As an immediate consequence of Theorem 2.11.1, we have the following.

Theorem 2.11.2. *We have*

$$p_N(\beta,\gamma,h) \le \inf_w \mathcal{P}(w,\beta,\gamma,h) . \quad (2.395)$$

The proof of Theorem 2.11.1 relies on the following construction. We consider independent standard normal r.v. $(z_{i,\ell})$, $i \le N$, $0 \le \ell \le k$, independent of the r.v. (g_{ij}), and for $0 \le t \le 1$ we consider the Hamiltonian

$$\begin{aligned} -H_{N,t} &= \frac{\beta\sqrt{t}}{\sqrt{N}}\sum_{i<j} g_{ij}\,\sigma_i\,\sigma_j + \beta\sqrt{1-t}\sum_{i\le N,\,0\le p\le k} z_{p,i}\,\sigma_i\,\sqrt{q_{p+1}-q_p} \\ &\quad + \gamma\sum_{i\le N} g_i\,\sigma_i + h\sum_{i\le N}\sigma_i . \end{aligned} \quad (2.396)$$

For $0 \le \ell \le k$, we denote by E_ℓ expectation in the r.v. $(z_{i,p})$, for $i \le N$, $\ell \le p \le k$. We define

$$Z_{t,k+1} = \sum_{\boldsymbol{\sigma}} \exp(-H_{N,t}(\boldsymbol{\sigma}))$$

and by decreasing induction, for $\ell \le k$, we define

$$Z_{t,\ell} = (\mathsf{E}_\ell\, Z_{t,\ell+1}^{m_\ell})^{1/m_\ell} . \quad (2.397)$$

When $m_\ell = 0$, this is understood as

$$Z_{t,\ell} = \exp \mathsf{E}_\ell \log Z_{t,\ell+1} .$$

We define

$$\varphi(t) = \frac{1}{N}\,\mathsf{E}\log Z_{t,0} .$$

Lemma 2.11.3. *We have*

$$\varphi(1) = p_N(\beta,\gamma,h) \quad (2.398)$$

$$\varphi(0) = \mathsf{E}\,\psi_0(\gamma g + h) + \log 2 . \quad (2.399)$$

Proof. The proof of (2.398) is obvious since for $t = 1$, $Z_{t,k+1}$ does not depend on the r.v. $z_{i,\ell}$. To prove (2.399), we observe that

$$Z_{0,k+1} = 2^N \prod_{i \le N} \operatorname{ch}\left(\beta \sum_{p \le k} z_{p,i} \sqrt{q_{p+1} - q_p} + \gamma g_i + h\right)$$
$$= 2^N \prod_{i \le N} \psi_{k+1}\left(\beta \sum_{p \le k} z_{p,i} \sqrt{q_{p+1} - q_p} + \gamma g_i + h\right) .$$

Using independence, by decreasing induction on $\ell \le k$ we get that

$$Z_{0,\ell} = 2^N \prod_{i \le N} \psi_\ell \left(\beta \sum_{p \le \ell - 1} z_{p,i} \sqrt{q_{p+1} - q_p} + \gamma g_i + h\right)$$

and thus $Z_{0,0} = 2^N \prod_{i \le N} \psi_0(\gamma g_i + h)$, from which (2.399) follows. □

We now compute $\varphi'(t)$. This computation is broken down into in a series of simple lemmas, but that it works at all looks like a miracle. We denote by $\langle \cdot \rangle_t$ an average with respect to Gibbs' measure with Hamiltonian (2.396). We set, for $0 \le \ell \le k$

$$W_\ell = \frac{Z_{t,\ell+1}^{m_\ell}}{\mathsf{E}_\ell\, Z_{t,\ell+1}^{m_\ell}} . \tag{2.400}$$

We observe that $\mathsf{E}_\ell W_\ell = 1$, and thus since $\mathsf{E}(W_0 \cdots W_\ell) = \mathsf{E}(W_0 \cdots W_{\ell-1}$ $\mathsf{E}_\ell W_\ell)$ by induction we get

$$\mathsf{E}\,(W_0 \cdots W_\ell) = 1 . \tag{2.401}$$

Lemma 2.11.4. *We have*

$$\frac{1}{Z_{t,\ell}} \frac{\partial Z_{t,\ell}}{\partial t} = \mathsf{E}_\ell \left(W_\ell \frac{1}{Z_{t,\ell+1}} \frac{\partial Z_{t,\ell+1}}{\partial t}\right) . \tag{2.402}$$

Proof. If $m_\ell \ne 0$, we have

$$\log Z_{t,\ell} = \frac{1}{m_\ell} \log \mathsf{E}_\ell\, Z_{t,\ell+1}^{m_\ell}$$

so that

$$\frac{1}{Z_{t,\ell}} \frac{\partial Z_{t,\ell}}{\partial t} = \frac{1}{\mathsf{E}_\ell\, Z_{t,\ell+1}^{m_\ell}} \mathsf{E}_\ell \left(\frac{Z_{t,\ell+1}^{m_\ell}}{Z_{t,\ell+1}} \frac{\partial Z_{t,\ell+1}}{\partial t}\right)$$

and the result because $\mathsf{E}_\ell\, Z_{t,\ell+1}^{m_\ell}$ does not depend on the r.v., $z_{i,\ell'}$, $\ell' \ge \ell$. If $m_\ell = 0$, then (2.395) follows immediately by differentiating the relation $\log Z_{t,\ell} = \mathsf{E}_\ell \log Z_{t,\ell+1}$, since then $W_\ell = 1$. □

Corollary 2.11.5. *We have*

$$N\varphi'(t) = \mathsf{E}\left(W_0 \cdots W_k \frac{1}{Z_{t,k+1}} \frac{\partial Z_{t,k+1}}{\partial t}\right) . \tag{2.403}$$

Proof. We observe that W_ℓ does not depend on $z_{i,\ell'}$ if $\ell' > \ell$. Thus for any r.v. X we have

$$W_\ell \,\mathsf{E}_{\ell'}(X) = \mathsf{E}_{\ell'}(W_\ell X) . \tag{2.404}$$

Since $\mathsf{E}\,\mathsf{E}_\ell = \mathsf{E}$, we see using (2.402) that by (increasing) induction over ℓ we have

$$N\varphi'(t) = \mathsf{E}\frac{1}{Z_{t,0}}\frac{\partial Z_{t,0}}{\partial t} = \mathsf{E}\left(W_0 \cdots W_\ell \frac{1}{Z_{t,\ell+1}} \frac{\partial Z_{t,\ell+1}}{\partial t}\right) .$$

□

Lemma 2.11.6. *We have $N\varphi'(t) = \mathrm{I} + \mathrm{II}$, where*

$$\mathrm{I} = \frac{\beta}{2\sqrt{tN}} \sum_{i<j} \mathsf{E}\left(W_0 \cdots W_k \, g_{ij} \, \langle \sigma_i \, \sigma_j \rangle_t\right) \tag{2.405}$$

$$\mathrm{II} = -\frac{\beta}{2\sqrt{1-t}} \sum_{p \le k} \sqrt{q_{p+1} - q_p} \sum_{i \le N} \mathsf{E}\left(W_0 \cdots W_k \, z_{i,p} \, \langle \sigma_i \rangle_t\right) . \tag{2.406}$$

Proof. Combine (2.403) with the straightforward formula

$$\begin{aligned} \frac{1}{Z_{t,k+1}} \frac{\partial Z_{t,k+1}}{\partial t} &= \frac{\beta}{2\sqrt{Nt}} \sum_{i<j} g_{ij} \, \langle \sigma_i \, \sigma_j \rangle_t \\ &\quad - \frac{\beta}{2\sqrt{1-t}} \sum_{p \le k} \sqrt{q_{p+1} - q_p} \sum_{i \le N} z_{i,p} \langle \sigma_i \rangle_t . \end{aligned}$$

□

We now integrate by parts, so that

$$\mathsf{E}\left(W_0 \cdots W_k \, g_{ij} \, \langle \sigma_i \, \sigma_j \rangle_t\right) = \sum_{0 \le \ell \le k+1} \mathrm{III}(i,j,\ell) \tag{2.407}$$

where (slightly abusing notation)

$$\mathrm{III}(i,j,k+1) = \mathsf{E}\left(W_0 \cdots W_k \frac{\partial}{\partial g_{ij}} \langle \sigma_i \, \sigma_j \rangle_t\right) \tag{2.408}$$

and for $\ell \le k$

$$\mathrm{III}(i,k,\ell) = \mathsf{E}\left(W_0 \cdots W_{\ell-1} \frac{\partial W_\ell}{\partial g_{ij}} W_{\ell+1} \cdots W_k \, \langle \sigma_i \, \sigma_j \rangle_t\right) . \tag{2.409}$$

Similarly, we have

$$\mathsf{E}\,(W_0 \cdots W_k\, z_{i,p}\, \langle \sigma_i \rangle_t) = \sum_{0 \le \ell \le k+1} \mathrm{IV}(i,p,\ell) \tag{2.410}$$

where

$$\mathrm{IV}(i,p,k+1) = \mathsf{E}\left(W_0 \cdots W_k \, \frac{\partial}{\partial z_{i,p}} \langle \sigma_i \rangle_t \right)$$

and, for $\ell \le k$

$$\mathrm{IV}(i,p,\ell) = \mathsf{E}\left(W_0 \cdots W_{\ell-1} \, \frac{\partial W_\ell}{\partial z_{i,p}}\, W_{\ell+1} \cdots W_k \, \langle \sigma_i \rangle_t \right) .$$

We have the familiar-looking formulas

$$\frac{\partial}{\partial g_{ij}} \langle \sigma_i\, \sigma_j \rangle_t = \frac{\beta \sqrt{t}}{\sqrt{N}} (1 - \langle \sigma_i\, \sigma_j \rangle_t^2) \tag{2.411}$$

$$\frac{\partial}{\partial z_{i,p}} \langle \sigma_i \rangle_t = \beta \sqrt{1-t}\, \sqrt{q_{p+1} - q_p}\, (1 - \langle \sigma_i \rangle_t^2) . \tag{2.412}$$

Lemma 2.11.7. *We have*

$$\frac{1}{Z_{t,\ell}} \frac{\partial Z_{t,\ell}}{\partial g_{ij}} = \mathsf{E}_\ell \left(W_\ell \, \frac{1}{Z_{t,\ell+1}} \, \frac{\partial Z_{t,\ell+1}}{\partial g_{ij}} \right) \tag{2.413}$$

$$\frac{1}{Z_{t,\ell}} \frac{\partial Z_{t,\ell}}{\partial g_{ij}} = \mathsf{E}_\ell \left(W_\ell \cdots W_k \, \frac{1}{Z_{t,k+1}} \, \frac{\partial Z_{t,k+1}}{\partial g_{ij}} \right) . \tag{2.414}$$

Proof. The proof of (2.413) is similar to the proof of (2.402) and (2.414) follows by iteration as in the proof of (2.403). □

Given t, for a function f on Σ_N, we define $\gamma_\ell(f) = \mathsf{E}_\ell(W_\ell \cdots W_k \, \langle f \rangle_t)$ if $\ell \le k$ and $\gamma_{k+1}(f) = \langle f \rangle_t$. We observe that

$$\mathsf{E}_\ell(W_\ell\, \gamma_{\ell+1}(f)) = \gamma_\ell(f) . \tag{2.415}$$

Indeed, by (2.404) we have $W_\ell\, \gamma_{\ell+1}(f) = \mathsf{E}_{\ell+1}(W_\ell \cdots W_k \, \langle f \rangle_t)$, and (2.415) follows since $\mathsf{E}_\ell\, \mathsf{E}_{\ell+1} = \mathsf{E}_\ell$.

Since

$$\frac{1}{Z_{t,k+1}} \, \frac{\partial Z_{t,k+1}}{\partial g_{ij}} = \beta \sqrt{\frac{t}{N}} \, \langle \sigma_i\, \sigma_j \rangle_t ,$$

we can rewrite (2.414) as

$$\frac{1}{Z_{t,\ell}} \frac{\partial Z_{t,\ell}}{\partial g_{ij}} = \beta \sqrt{\frac{t}{N}} \, \gamma_\ell(\sigma_i\, \sigma_j) . \tag{2.416}$$

Lemma 2.11.8. *For $\ell \le k$ we have*

$$\frac{\partial W_\ell}{\partial g_{ij}} = m_\ell \beta \sqrt{\frac{t}{N}} W_\ell(\gamma_{\ell+1}(\sigma_i \sigma_j) - \gamma_\ell(\sigma_i \sigma_j)) . \tag{2.417}$$

Proof. From (2.400) we have

$$\begin{aligned}\frac{\partial W_\ell}{\partial g_{ij}} &= m_\ell Z_{t,\ell+1}^{m_\ell-1} \frac{\partial Z_{t,\ell+1}}{\partial g_{ij}} \frac{1}{\mathsf{E}_\ell Z_{t,\ell+1}^{m_\ell}} \\ &- m_\ell Z_{t,\ell+1}^{m_\ell} \mathsf{E}_\ell \left(Z_{t,\ell+1}^{m_\ell-1} \frac{\partial Z_{t,\ell+1}}{\partial g_{ij}} \right) \frac{1}{(\mathsf{E}_\ell Z_{t,\ell+1}^{m_\ell})^2} \\ &= m_\ell W_\ell \left(\frac{1}{Z_{t,\ell+1}} \frac{\partial Z_{t,\ell+1}}{\partial g_{ij}} - \mathsf{E}_\ell \left(W_\ell \frac{1}{Z_{t,\ell+1}} \frac{\partial Z_{t,\ell+1}}{\partial g_{ij}} \right) \right) .\end{aligned}$$

Using (2.416), this is

$$m_\ell \beta \sqrt{\frac{t}{N}} W_\ell(\gamma_{\ell+1}(\sigma_i \sigma_j) - \mathsf{E}_\ell(W_\ell \gamma_{\ell+1}(\sigma_i \sigma_j)))$$

and the conclusion follows by (2.415). □

Lemma 2.11.9. *We have*

$$\frac{\partial W_\ell}{\partial z_{i,p}} = 0 \quad \text{if} \quad p > \ell . \tag{2.418}$$

$$\frac{\partial W_\ell}{\partial z_{i,p}} = \beta\sqrt{1-t}\sqrt{q_{p+1} - q_p}\, m_\ell W_\ell(\gamma_{\ell+1}(\sigma_i) - \gamma_\ell(\sigma_i)) \text{ if } p < \ell . \tag{2.419}$$

$$\frac{\partial W_p}{\partial z_{i,p}} = \beta\sqrt{1-t}\sqrt{q_{p+1} - q_p}\, m_p W_p \gamma_{p+1}(\sigma_i) . \tag{2.420}$$

Proof. Very similar to that of Lemma 2.11.8. □

Lemma 2.11.10. *If f_1, f_2 are functions on Σ_N, we have*

$$\mathsf{E}\,(W_0 \cdots W_k \gamma_\ell(f_1) \langle f_2 \rangle_t) = \mathsf{E}\,(W_0 \cdots W_{\ell-1} \gamma_\ell(f_1) \gamma_\ell(f_2)) . \tag{2.421}$$

For $\ell = 0$, the last term is $\mathsf{E}\,(\gamma_0(f_1)\gamma_0(f_2))$.

Proof. We have

$$\begin{aligned}&\mathsf{E}\,(W_0 \cdots W_k \gamma_\ell(f_1) \langle f_2 \rangle_t) \\ &= \mathsf{E}\,\mathsf{E}_\ell(W_0 \cdots W_{\ell-1} \gamma_\ell(f_1)(W_\ell \cdots W_k \langle f_2 \rangle_t)) \\ &= \mathsf{E}\,(W_0 \cdots W_{\ell-1} \gamma_\ell(f_1) \mathsf{E}_\ell(W_\ell \cdots W_k \langle f_2 \rangle_t))\end{aligned}$$

since $W_0, \ldots, W_{\ell-1}$ and $\gamma_\ell(f_1)$ do not depend on the r.v. $z_{i,p}$ for $p \ge \ell$. □

Lemma 2.11.11. *We have*

$$\varphi'(t) = -\frac{\beta^2}{4} \sum_{0\le \ell\le k} m_\ell (q_{\ell+1}^2 - q_\ell^2) \tag{2.422}$$
$$- \sum_{1\le\ell\le k} \frac{\beta^2}{4}(m_\ell - m_{\ell-1})\,\mathsf{E}\,(W_0\cdots W_{\ell-1}\,\gamma_\ell^{\otimes 2}((R_{1,2}-q_\ell)^2))\,.$$

Here, γ_ℓ is considered as a functional (= probability) on Σ_N, and $\gamma_\ell^{\otimes 2}$ is the product of two replicas. Combining (2.422) with Lemma 2.11.3 finishes the proof of Theorem 2.11.1.

Proof. We simply need to carefully collect all the terms. Using Lemma 2.11.6, (2.407) and (2.410) we have $N\varphi'(t) = \mathrm{I} + \mathrm{II}$

$$\mathrm{I} = \frac{\beta}{2\sqrt{tN}} \sum_{i<j} \sum_{\ell\le k+1} \mathrm{III}(i,j,l)$$

$$\mathrm{II} = -\frac{\beta}{2\sqrt{1-t}} \sum_{i\le N} \sum_{p\le k} \sum_{\ell\le k+1} \mathrm{IV}(i,p,\ell)\,.$$

By (2.411) and (2.408) we have

$$\frac{\beta}{2\sqrt{tN}}\,\mathrm{III}(i,j,k+1) = \frac{\beta^2}{2N}\,(1-\mathsf{E}\,(W_0\cdots W_k\,\gamma_{k+1}^2(\sigma_i\,\sigma_j)))$$

using (2.401). For $\ell \le k$, using (2.417) and (2.421) we get

$$\frac{\beta}{2\sqrt{tN}}\mathrm{III}(i,j,\ell) = \frac{m_\ell\,\beta^2}{2N}\,(\mathsf{E}\,(W_0\cdots W_\ell\,\gamma_{\ell+1}^2(\sigma_i\,\sigma_j))$$
$$- \mathsf{E}\,(W_0\cdots W_{\ell-1}\,\gamma_\ell^2(\sigma_i\,\sigma_j))).$$

Since $\gamma_{\ell+1}^2(\sigma_i\,\sigma_j) = \gamma_{\ell+1}^{\otimes 2}(\sigma_i^1\,\sigma_j^1\,\sigma_i^2\,\sigma_j^2)$, we get by summation over i,j

$$\frac{\mathrm{I}}{N} = \frac{\beta^2}{4}\Big[(1-\mathsf{E}\,(W_0\cdots W_k\,\gamma_{k+1}^{\otimes 2}(R_{1,2}^2))$$
$$+ \sum_{0\le\ell\le k} m_\ell(\mathsf{E}\,(W_0\cdots W_\ell\,\gamma_{\ell+1}^{\otimes 2}(R_{1,2}^2)) - \mathsf{E}\,(W_0\cdots W_{\ell-1}\,\gamma_\ell^{\otimes 2}(R_{1,2}^2))\Big]$$
$$= \frac{\beta^2}{4}\left(1 - \sum_{1\le\ell\le k}(m_\ell - m_{\ell-1})\,\mathsf{E}\,(W_0\cdots W_{\ell-1}\,\gamma_\ell^{\otimes 2}(R_{1,2}^2))\right) \tag{2.423}$$

using that $m_0 = 0$ and $m_k = 1$.

Writing for simplicity $A_\ell = \mathsf{E}\,(W_0\cdots W_{\ell-1}\,\gamma_\ell^2(\sigma_i))$, we see in a similar fashion that

$$\frac{\beta}{2\sqrt{1-t}}\,\sqrt{q_{p+1}-q_p}\;\mathrm{IV}(i,p,k+1) = \frac{\beta^2}{2}\,(q_{p+1}-q_p)(1-A_{k+1})\,.$$

For $p < \ell \le k$, we have

$$\frac{\beta}{2\sqrt{1-t}}\sqrt{q_{p+1}-q_p}\;\mathrm{IV}(i,p,\ell) = \frac{\beta^2}{2}(q_{p+1}-q_p)\,m_\ell(A_{\ell+1}-A_\ell)\,.$$

For $\ell < p$, we have $\mathrm{IV}(i,p,\ell) = 0$, and finally

$$\frac{\beta}{2\sqrt{1-t}}\sqrt{q_{\ell+1}-q_\ell}\;\mathrm{IV}(i,\ell,\ell) = \frac{\beta^2}{2}(q_{\ell+1}-q_\ell)\,m_\ell\,A_{\ell+1}\,.$$

Summation of these for $0 \le p \le k$, $0 \le \ell \le k+1$ gives (since $q_0 = 0$ and $q_{k+1} = 1$)

$$\begin{aligned}&\frac{\beta^2}{2}\left(1 - A_{k+1} + \sum_{0\le\ell\le k} q_{\ell+1}\,m_\ell\,A_{\ell+1} - \sum_{0\le\ell\le k} q_\ell\,m_\ell\,A_\ell\right)\\ &= \frac{\beta^2}{2}\left(1 - \sum_{1\le\ell\le k}(m_\ell - m_{\ell-1})\,q_\ell\,A_\ell\right)\end{aligned}$$

and thus by summation over i,

$$\frac{\mathrm{II}}{N} = -\frac{\beta^2}{2}\left(1 - \sum_{1\le\ell\le k}(m_\ell - m_{\ell-1})\,q_\ell\,\mathsf{E}\left(W_0\cdots W_{\ell-1}\,\gamma_\ell^{\otimes 2}(R_{1,2})\right)\right)\,.$$

Using the relation

$$\begin{aligned}\mathsf{E}\left(W_0\cdots W_{\ell-1}\,\gamma_\ell^{\otimes 2}(R_{1,2}^2)\right) &= \mathsf{E}\left(W_0\cdots W_{\ell-1}\,\gamma_\ell^{\otimes 2}((R_{1,2}-q_\ell)^2\right)\\ &\quad + 2\,q_\ell\,\mathsf{E}\left(W_0\cdots W_{\ell-1}\,\gamma_\ell^{\otimes 2}(R_{1,2})\right) - q_\ell^2\end{aligned}$$

in (2.423) we get the result. □

We recall the notation $\mathcal{P}(w,\beta,\gamma,h)$ of (2.395). We denote by q the (unique) root of the equation $q = \mathsf{E}\,\mathrm{th}^2(\beta z\sqrt{q} + \gamma g + h)$. We denote by $\mathcal{L}_h$ the "low temperature region"

$$\mathcal{L}_h = \{(\beta,\gamma)\,;\ \beta^2\,\mathsf{E}\,\mathrm{ch}^{-4}(\beta z\sqrt{q} + \gamma g + h) > 1\}\,. \tag{2.424}$$

Theorem 2.11.12. *[157] If $(\beta,\gamma) \in \mathcal{L}_h$, we have*

$$\mathrm{SK}(\beta,\gamma,h) > \inf \mathcal{P}(w,\beta,\gamma,h)\,.$$

In particular Theorem 2.11.1 shows that

$$\lim_{N\to\infty} p_N(\beta,\gamma,h) < \mathrm{SK}(\beta,\gamma,h) \quad \text{for} \quad (\beta,\gamma) \quad \text{in} \quad \mathcal{L}_h\,.$$

Proof. It suffices to show that we can find w with $\mathcal{P}(w,\beta,\gamma,h) < \mathrm{SK}(\beta,\gamma,h)$. We will take $k=2$, $q_1=q$, $q_2=q'>q$, $m_1=m$ and $m_2=1$. (Thus, the function w takes the values $0,m,1$.) We then find

$$\psi_2(y) = \exp\frac{\beta^2}{2}(1-q')\operatorname{ch} y$$

$$\psi_1(y) = \exp\frac{\beta^2}{2}(1-q')\left(\mathsf{E}\operatorname{ch}^m(y+\beta z'\sqrt{q'-q})\right)^{1/m}$$

$$\log\psi_0(y) = \frac{\beta^2}{2}(1-q') + \frac{1}{m}\mathsf{E}\log\mathsf{E}'\operatorname{ch}^m(y+\beta z\sqrt{q}+\beta z'\sqrt{q'-q})\,.$$

Here, z,z',g are independent standard Gaussian r.v. and E' denotes expectation in z' only. Writing

$$Y' = \beta\, z\sqrt{q} + \beta\, z'\sqrt{q'-q} + \gamma g + h\,,$$

we thus have

$$\mathcal{P}(w,\beta,\gamma,h) \le \log 2 + \frac{\beta^2}{4}(1-q')^2 - \frac{\beta^2}{4}m(q'^2-q^2) + \frac{1}{m}\mathsf{E}\log\mathsf{E}'\operatorname{ch}^m Y'\,. \tag{2.425}$$

Fixing β,γ,h, let us consider the right-hand side of (2.425) as a function $U(m,q')$. We note that $U(1,q)=\mathrm{SK}(\beta,\gamma,h)$. To show that we can find m such that $U(m,q')<\mathrm{SK}(\beta,\gamma,h)$ we compute the partial derivative in m at $m=1$, which we call $V(q')$. We have

$$V(q') = -\frac{\beta^2}{4}(q'^2-q^2) - \mathsf{E}\log\mathsf{E}'\operatorname{ch}Y' + \mathsf{E}\frac{\mathsf{E}'\log(\operatorname{ch}Y')\operatorname{ch}Y'}{\mathsf{E}'\operatorname{ch}Y'}\,.$$

When $q'=q$, the expectation E' disappears, so that $V(q)=0$. To prove that we can find q' with $V(q')>0$, we will show that $V'(q)=0$ and $V''(q)>0$. Since

$$\mathsf{E}'\operatorname{ch}Y' = \exp\frac{\beta^2}{2}(q'-q)\operatorname{ch}Y$$

where $Y=\beta\, z\sqrt{q}+\gamma g+h$ does not depend on q', we have

$$\begin{aligned} V(q') = &-\frac{\beta^2}{4}(q'^2-q^2) - \frac{\beta^2}{2}(q'-q) - \mathsf{E}\log\operatorname{ch}Y \\ &+ \exp\left(-\frac{\beta^2}{2}(q'-q)\right)\mathsf{E}\frac{\mathsf{E}'\log(\operatorname{ch}Y')\operatorname{ch}Y'}{\operatorname{ch}Y}\,. \end{aligned}$$

Using the formula $\mathrm{d}\mathsf{E}' f(z'\sqrt{x})/\mathrm{d}x = \mathsf{E}' f''(z'\sqrt{x})/2$, we have

$$\frac{\mathrm{d}}{\mathrm{d}q'}\mathsf{E}'\log(\operatorname{ch}Y')\operatorname{ch}Y' = \frac{\beta^2}{2}\mathsf{E}'\left(\operatorname{ch}Y'(1+\log\operatorname{ch}Y') + \frac{\operatorname{sh}^2 Y'}{\operatorname{ch}Y'}\right)$$

and thus

$$\begin{aligned}V'(\,q'\,) &= -\frac{\beta^2}{2}q' - \frac{\beta^2}{2} - \frac{\beta^2}{2}\exp\left(-\frac{\beta^2}{2}(q'-q)\right)\mathsf{E}\,\frac{\mathsf{E}'\log(\operatorname{ch}Y')\operatorname{ch}Y'}{\operatorname{ch}Y}\\ &\quad+\frac{\beta^2}{2}\exp\left(-\frac{\beta^2}{2}(q'-q)\right)\mathsf{E}\,\frac{\mathsf{E}'(\operatorname{ch}Y')\,(1+\log\operatorname{ch}Y')+\operatorname{sh}^2Y'\operatorname{ch}^{-1}Y')}{\operatorname{ch}Y}\\ &= -\frac{\beta^2}{2}q' - \frac{\beta^2}{2} + \frac{\beta^2}{2}\exp\left(-\frac{\beta^2}{2}(q'-q)\right)\mathsf{E}\,\frac{\mathsf{E}'(\operatorname{ch}Y'+\operatorname{sh}^2Y'\operatorname{ch}^{-1}Y')}{\operatorname{ch}Y}\,.\end{aligned}$$

In particular, since for $q'=q$ we have $Y'=Y$, we have

$$V'(q) = \frac{\beta^2}{2}(-q-1+\mathsf{E}\,(1+\operatorname{th}^2Y)) = 0$$

because $q=\mathsf{E}\operatorname{th}^2Y$. Proceeding in a similar manner, we get

$$V''(q') = -\frac{\beta^2}{2} + \frac{\beta^4}{4}\exp\left(-\frac{\beta^2}{2}(q'-q)\right)\mathsf{E}\,\frac{A}{\operatorname{ch}Y}$$

where $A=\mathsf{E}'(-f(Y')+f''(Y'))$, for $f(x)=\operatorname{ch}x+\operatorname{sh}^2x\operatorname{ch}^{-1}x=2\operatorname{ch}x-\operatorname{ch}^{-1}x$, so that $-f(x)+f''(x)=2\operatorname{ch}^{-3}x$. In particular

$$V''(q) = \frac{\beta^2}{2}\left(\beta^2\,\mathsf{E}\,\frac{1}{\operatorname{ch}^4Y}-1\right) > 0$$

since $(\beta,\gamma)\in\mathcal{L}_h$. □

Corollary 2.11.13. *There exists a subset* $\mathcal{N}_h$ *of* $(\mathbb{R}^+)^2$ *of Lebesgue measure zero such that for each value* (β,γ) *in* $\mathcal{L}_h\backslash\mathcal{N}_h$ *we have*

$$\lim_{N\to\infty}\nu((R_{1,2}-\nu(R_{1,2}))^2) > 0 \tag{2.426}$$

and hence $\lim_{N\to\infty}\nu((R_{1,2}-q)^2)>0$.

Proof. We define $\mathcal{N}_h$ as in the proof of Corollary 2.10.4, i.e. $\mathcal{N}_h^c$ is the set where $\partial p_\infty/\partial\gamma$ and $\partial p_\infty/\partial\beta$ exist, where $p_\infty(\beta,\gamma)=\lim_{N\to\infty}p_N(\beta,\gamma,h)$. Consider $(\beta,\gamma)\notin\mathcal{N}_h$, so that, as shown in the proof of Corollary 2.10.4, the limits $\lim_{N\to\infty}\nu(R_{1,2}^2)$ and $\lim_{N\to\infty}\nu(R_{1,2})$ exist. Let us assume that $\lim_{N\to\infty}\nu((R_{1,2}-\nu(R_{1,2}))^2)=0$. Then, by Proposition 2.8.4 we have $\lim_{N\to\infty}\nu((R_{1,2}-q)^2)=0$. By (an obvious extension of) Theorem 2.4.19, we have

$$\lim_{N\to\infty}p_N(\beta,\gamma,h)=\operatorname{SK}(\beta,\gamma,h)$$

so that $(\beta,\gamma)\notin\mathcal{L}_h$ by Theorem 2.11.12. The last assertion follows from the fact that

$$\inf_x\nu((R_{1,2}-x)^2)=\nu((R_{1,2}-\nu(R_{1,2}))^2)\,.$$

□

Interestingly, we do not see how to prove that there exists any value at all of β such that (2.426) occurs when $\gamma = 0$. (On the other hand, Proposition 2.8.1 is valid at every point of $\mathcal{L}_h$.)

Now that we are better armed, we can hope to improve on Proposition 2.9.4 and to control the entire high-temperature region of the SK model. Consider an integer k, and consider $0 = q_0 \le q_1 \le \cdots \le q_k \le q_{k+1} = 1$. Consider $0 = m_0 \le m_1 \le \cdots \le m_k = 1$, and consider $c_0, c_1, \cdots, c_k$ with $-1 \le c_\ell \le 1$ for $0 \le \ell \le k$. Consider $y_1, y_2, x \in \mathbb{R}$. By decreasing induction over $0 \le \ell \le k+1$, we define the function $\psi_\ell(y_1, y_2, x)$ as follows. First, we set

$$\psi_{k+1}(y_1, y_2, x) = \operatorname{ch} x \operatorname{ch} y_1 \operatorname{ch} y_2 + \operatorname{sh} x \operatorname{sh} y_1 \operatorname{sh} y_2 .$$

Then we set

$$\psi_\ell(y_1, y_2, x) = (\mathsf{E}\, \psi_{\ell+1}^{m_\ell}(y_1 + z_1, y_2 + z_2, x))^{1/m_\ell} ,$$

where (z_1, z_2) is a couple of jointly Gaussian centered r.v. such that $\mathsf{E}\, z_1^2 = \mathsf{E}\, z_2^2 = \beta^2(q_{\ell+1} - q_\ell)$ and $\mathsf{E}\, z_1 z_2 = \beta^2(q_{\ell+1} - q_\ell)\, c_\ell$. When $m_\ell = 0$, this of course has to be interpreted as in (2.392). We recall the notation $r_N(\beta, \gamma, u)$ of (2.370).

Theorem 2.11.14. *For the choice of parameters as above, we have*

$$\begin{aligned} r_N(\beta,\gamma,u) \le\ & 2\log 2 + \mathsf{E}\log \psi_0(\gamma g + h, \gamma g + h, x) - ux \\ & - \frac{\beta^2}{2} \sum_{1\le \ell\le k} m_\ell (q_{\ell+1}^2 - q_\ell^2) - \frac{\beta^2}{2} \sum_{1\le \ell\le k} m_\ell (d_{\ell+1}^2 - d_\ell^2) \\ & + \frac{\beta^2}{2}(u - d_{k+1})^2 \end{aligned} \tag{2.427}$$

where $d_\ell = \sum_{0\le p<\ell} c_p (q_{p+1} - q_p)$.

Corollary 2.11.15. *If $U(\beta, \gamma, u)$ denotes the infimum of the right-hand side over the choices of x, k, $(q_\ell)_{\ell\le k+1}$, $(m_\ell)_{\ell\le k+1}$ and $(c_\ell)_{\ell\le k+1}$ as above we have*

$$r_N(\beta, \gamma, u) \le U(\beta, \gamma, u) .$$

We consider two independent standard Gaussian r.v. z and z'. We denote by E' expectation in z' only. For $u \ge q$ we write $Y = \gamma g + h + \beta z\sqrt{q}$ and $Y' = Y + \beta z' \sqrt{u - q}$, where of course q is the solution of $q = \mathsf{E}\operatorname{th}^2 Y$. We consider the function

$$V(u) = V(u, \beta, \gamma, h) = -\frac{\beta^2}{4}(u^2 - q^2) - \mathsf{E}\log \mathsf{E}' \operatorname{ch} Y' + \mathsf{E}\frac{\mathsf{E}' \log(\operatorname{ch} Y')\operatorname{ch} Y'}{\operatorname{ch} Y'} \tag{2.428}$$

We define the region $\mathcal{H}_h$ as follows

$$\mathcal{H}_h = \{(\beta, \gamma)\,;\ \forall u,\ u > q \Rightarrow V(u, \beta, \gamma, h) < 0\} .$$

Theorem 2.11.16. *The conclusions of Theorem 2.9.10 are true in the region* $\mathcal{H}_h$.

This result describes the *entire* region where $\lim_{N\to\infty} p_N(\beta,\gamma,h) = \mathrm{SK}(\beta,\gamma,h)$, or more accurately, the interior of this region. Indeed, if for some $u > q$ we have $V(u) > 0$, the proof of Theorem 2.11.12 shows that $\mathrm{SK}(\beta,\gamma,h) > \inf \mathcal{P}(w,\beta,\gamma,h) \geq p_N(\beta,\gamma,h)$ by Theorem 2.11.1. It is shown in the proof of Theorem 2.11.12 that $\mathcal{H}_h \subset \{\beta^2 \mathsf{E}\,\mathrm{ch}^{-4} Y < 1\}$ and one can hope that the reverse inclusion holds. In fact, this proof shows that if $\beta^2 \mathsf{E}\,\mathrm{ch}^{-4} Y < 1$, then $V(q) = 0$ and $V'(q) < 0$, so that one can hope that $V(u) < 0$ for $u > q$, but it does not seem obvious how to prove this. Whether this is true or not, the correct high-temperature region has been found, and the natural definition of the low-temperature region should be

$$\{(\beta,\gamma)\,;\ \exists u\,,\ u > q \Rightarrow V(u,\beta,\gamma,h) > 0\}\,. \tag{2.429}$$

When we proved Theorem 2.9.10 we moved from $\beta = 0$ keeping γ fixed. Alternatively, we could have moved along the path $t \mapsto (\beta(t),\gamma(t))$ given by (2.362). (This path is somewhat more natural, but it does require to develop the cavity method also in the case $\gamma \neq 0$.) The advantage of using this path is that it is rather straightforward to see that if $(\beta,\gamma) \in \mathcal{H}_h$ then for all $t \leq 1$ we have $(\beta(t),\gamma(t)) \in \mathcal{H}_h$. This approach shows that to prove Theorem 2.11.16 it suffices to prove that for $(\beta,\gamma) \in \mathcal{H}_h$ and $u \neq q$ we have $U(\beta,\gamma,u) < 2\mathrm{SK}(\beta,\gamma,h)$. The following takes care of the values of u with $-q \leq u < q$.

Lemma 2.11.17. *If* $\beta^2 \mathsf{E}\mathrm{ch}^{-4} Y < 1$*, then for* $-q \leq u < q$ *we have* $U(\beta,\gamma,u) < 2\mathrm{SK}(\beta,\gamma,h)$.

Proof. Assume first $0 \leq u < q$. We choose $k = 2, q_1 = u, q_2 = q, m_1 = 0, m_2 = 1$ and $c_0 = 1, c_1 = c_2 = 0$, so that $d_1 = d_2 = d_3 = u$, and (2.427) yields

$$U(\beta,\gamma,u) \leq W(x) := 2\log 2 + \mathsf{E}\log \psi_0(g\gamma + h, g\gamma + h, x) - ux - \frac{\beta^2}{2}(1-q^2)\,.$$

(Of course, W depends also on β,γ,h,u, but this dependence is kept implicit.) It is straightforward to see that if z, z_1, z_2 are independent standard normal r.v., we have

$$\mathsf{E}\log \psi_0(g\gamma + h, g\gamma + h, x) = \beta^2(1-q) + \mathsf{E}\log(\mathrm{ch}x\mathrm{ch}Y_1\mathrm{ch}Y_2 + \mathrm{sh}x\mathrm{sh}Y_1\mathrm{sh}Y_2)\,,$$

where $Y_j = \gamma g + h + \beta z\sqrt{u} + \beta z_j\sqrt{q-u}$.

Since $\beta z\sqrt{u} + \beta z_j\sqrt{q-u}$ has the same distribution as $\beta z\sqrt{q}$, we have $W(0) = 2\mathrm{SK}(\beta,\gamma,h)$ and it suffices to prove that

$$W'(0) = \mathsf{E}\mathrm{th}Y_1\mathrm{th}Y_2 - u \neq 0\,.$$

We fix β, γ and h and we consider the right hand side as a function $f(u)$. Writing

$$Y_j' := \frac{\mathrm{d}}{\mathrm{d}u} Y_j = \frac{\beta z}{2\sqrt{u}} - \frac{\beta z_j}{2\sqrt{q-u}},$$

using integration by parts and the fact that $\mathsf{E} Y_j' Y_j = 0$ and $\mathsf{E} Y_1' Y_2 = \mathsf{E} Y_2' Y_1 = \beta^2/2$, we get

$$\begin{aligned} f'(u) &= \mathsf{E}\left(Y_1' \frac{\mathrm{th} Y_2}{\mathrm{ch}^2 Y_1} + Y_2' \frac{\mathrm{th} Y_1}{\mathrm{ch}^2 Y_2}\right) - 1 \\ &= \beta^2 \mathsf{E} \frac{1}{\mathrm{ch}^2 Y_1} \frac{1}{\mathrm{ch}^2 Y_2} - 1 \\ &\le \beta^2 \mathsf{E} \frac{1}{\mathrm{ch} Y_1^4} - 1 = \beta^2 \mathsf{E} \frac{1}{\mathrm{ch} Y^4} - 1 < 0 , \end{aligned}$$

using the Cauchy-Schwarz inequality and the fact that Y_1 is distributed like Y in the last line. Since $f(q) = 0$, we have $f(u) > 0$ when $u < q$.

Consider now the case where $-q \le u \le 0$. We choose $k = 2, q_1 = -u, q_2 = q, m_1 = 0, m_2 = 1$ and $c_0 = -1, c_1 = c_2 = 0$, so that $d_1 = d_2 = d_3 = u$, and we proceed very much as above. □

The following lemma proposes a choice of parameters in (2.427) when $u > q$.

Lemma 2.11.18. *For* $0 \le m \le 1$ *we have*

$$\begin{aligned} U(\beta,\gamma,u) \le W(m) &:= 2\log 2 + \beta^2(1-u) - \frac{\beta^2}{2}(1 - u^2 + 2m(u^2 - q^2)) \\ &\quad + \frac{1}{m} \mathsf{E} \log \mathsf{E}' \mathrm{ch}^{2m} Y' , \end{aligned} \tag{2.430}$$

where $Y' = \gamma g + h + \beta z\sqrt{q} + \beta z' \sqrt{u-q}$.

Proof. We choose $k = 2, x = 0, q_1 = q, q_2 = u, m_1 = m, m_2 = 1, c_0 = c_1 = 1, c_2 = 0$, so that $d_1 = q, d_2 = d_3 = u$, and we observe that

$$\mathsf{E} \log \psi_0(g\gamma + h, g\gamma + h, 0) = \beta^2(1-u) + \frac{1}{m} \mathsf{E} \log \mathsf{E}' \mathrm{ch}^{2m} Y' .$$

□

It is straightforward (but rather beautiful) that $W(1/2) = 2\mathrm{SK}(\beta,\gamma,h)$, so that to prove that $U(\beta,\gamma,u) < 2\mathrm{SK}(\beta,\gamma,h)$, it suffices to observe that

$$W'(\frac{1}{2}) = 4V(u) \ne 0 . \tag{2.431}$$

To complete the proof of Theorem 2.11.16, it suffices to handle the case where $u < -q$. In that case we chose $k = 2, x = 0, q_1 = q, q_2 = -u, m_1 = 1/2, m_2 = 1, c_0 = c_1 = -1, c_2 = 0$, so that $d_1 = -q, d_2 = d_3 = u$, and we get

$$U(\beta,\gamma,u) \le 2\log 2 + \beta^2(1 - |u|) - \frac{\beta^2}{2}(1 - q^2) + 2\mathsf{E} \log \mathsf{E}' \sqrt{\mathrm{ch} Y_1 \mathrm{ch} Y_2} ,$$

where for $j = 1, 2$, $Y_j = \gamma g + h + (-1)^j(\beta z\sqrt{q} + \beta z'\sqrt{|u| - q})$. We then obtain that $U(\beta, \gamma, u) < 2\mathrm{SK}(\beta, \gamma, h)$ by the Cauchy-Schwarz inequality, that gives here $\mathsf{E}'\sqrt{\mathrm{ch}Y_1 \mathrm{ch}Y_2} < (\mathsf{E}'\mathrm{ch}Y_1 \mathsf{E}'\mathrm{ch}Y_2)^{1/2} = \mathsf{E}'\mathrm{ch}Y_1$ because we can assume that $\gamma \neq 0$ or $h \neq 0$.

It is possible to develop the ideas of this proof to obtain the following.

Theorem 2.11.19. *(The Parisi Formula) We have*

$$\lim_{N\to\infty} p_N(\beta, \gamma, h) = \inf_w \mathcal{P}(w, \beta, \gamma, h) \,. \tag{2.432}$$

This theorem was proved less than a month before the book was to be printed, and we decided not to include the proof in order not to delay publication. It is a special case of a more general result [154]. One should probably look at this theorem not as the culmination of our efforts to understand the SK model, but rather as the first result of a new chapter, yet to be written, that will study in detail its low-temperature phase.

It seems certain that Theorem 2.11.19 allows one to improve on certain results proved elsewhere in this chapter, but we had no time to investigate this. For example, it should be possible to prove (2.426) in the entire region (2.429), by showing that $\partial p_\infty/\partial\gamma$ and $\partial p_\infty/\partial\beta$ exist at every point, and showing that $(\lim_{N\to\infty} \nu(R_{1,2}))^2 \neq \lim_{N\to\infty} \nu(R_{1,2}^2)$.

Proof of Theorem 2.11.14. This proof requires long calculations, similar to those of Theorem 2.11.1, so we will only sketch the argument. For $i \leq N$, $p \leq k$, we consider i.i.d. pairs $(z_{i,p}^1, z_{i,p}^2)$ of jointly Gaussian r.v., with $\mathsf{E}\,(z_{i,p}^1)^2 = \mathsf{E}\,(z_{i,p}^2)^2 = \beta^2(q_{p+1} - q_p)$ and $\mathsf{E}\, z_{i,p}^1 z_{i,p}^2 = c_p\,\beta^2(q_{p+1} - q_p)$. These pairs are all independent as i, p vary. We consider the Hamiltonian

$$\begin{aligned} -H_{N,t}(\boldsymbol{\sigma}^1, \boldsymbol{\sigma}^2) &= \sqrt{t}\,(-H_N(\boldsymbol{\sigma}^1) + -H_N(\boldsymbol{\sigma}^2)) \\ &\quad + \sum_{j=1,2} \left(\sqrt{1-t} \sum_{p\leq k} \sum_{i\leq N} z_{i,p}^j\, \sigma_i^j + \gamma \sum_{i\leq N} g_i\, \sigma_i^j + h \sum_{i\leq N} \sigma_i^j \right) \end{aligned} \tag{2.433}$$

on Σ_N^2, where $-H_N(\boldsymbol{\sigma}) = \sum_{i<j} \beta\, g_{ij}\, \sigma_i\, \sigma_j/\sqrt{N}$. We set

$$Z_{t,k+1} = \sum_{R_{1,2}=u} \exp(-H_{N,t}(\boldsymbol{\sigma}^1, \boldsymbol{\sigma}^2)) \,,$$

and we define $\varphi(t)$ as in the proof of Theorem 2.11.1. The case of $\varphi(0)$ is handled as before, and the theorem follows from the identity

$$\begin{aligned} \varphi'(t) &= \frac{\beta^2}{2}(u - d_{k+1})^2 - \frac{\beta^2}{2} \sum_{0\leq \ell\leq k} m_\ell(q_{\ell+1}^2 - q_\ell^2) - \frac{\beta^2}{2} \sum_{0\leq \ell\leq k} m_\ell(d_{\ell+1}^2 - d_\ell^2) \\ &\quad - \frac{\beta^2}{2} \sum_{1\leq \ell\leq k} (m_\ell - m_{\ell-1})\, \mathsf{E}\,(W_0 \cdots W_{\ell-1}\, \gamma_\ell^{\otimes 2}((R(\boldsymbol{\sigma}^1, \boldsymbol{\tau}^1) - q_\ell)^2) \\ &\quad - \frac{\beta^2}{2} \sum_{1\leq \ell\leq k} (m_\ell - m_{\ell-1})\, d_\ell\, \mathsf{E}\,(W_0 \cdots W_{\ell-1}\, \gamma_\ell^{\otimes 2}((R(\boldsymbol{\sigma}^1, \boldsymbol{\tau}^2) - d_\ell)^2) \,. \end{aligned}$$

Here, as in the proof of Proposition 2.9.4, $(\boldsymbol{\tau}^1, \boldsymbol{\tau}^2)$ denotes a copy of the system $(\boldsymbol{\sigma}^1, \boldsymbol{\sigma}^2)$. It is important to note that there is no longer symmetry between $\boldsymbol{\sigma}^1$ and $\boldsymbol{\sigma}^2$, because the corresponding randomness are different (since $z^1_{i,p} \neq z^2_{i,p}$ when $c_p < 1$) so that there is no reason why one should have $\langle R(\boldsymbol{\sigma}^1, \boldsymbol{\tau}^1)\rangle_t = \langle R(\boldsymbol{\sigma}^1, \boldsymbol{\tau}^2)\rangle_t$. □

2.12 The Ghirlanda-Guerra Identities

In Section 1.2 we have seen that in the REM, for $\beta > 2\sqrt{\log 2}$, only the configurations $\boldsymbol{\sigma}$ for which $-H_N(\boldsymbol{\sigma}) \simeq N\sqrt{\log 2}$ matter, a fact which results in the remarkable identities (1.52). The theme of this section is the following: it is a rather general fact that "the Hamiltonian is constant", and this implies remarkable identities.

It clarifies matters to give a general statement. We consider a Hamiltonian depending on a parameter x in the following manner:

$$H(\boldsymbol{\sigma}) = H_x(\boldsymbol{\sigma}) = H_1(\boldsymbol{\sigma}) + xH_2(\boldsymbol{\sigma}) \,, \tag{2.434}$$

where $\boldsymbol{\sigma} \in \Sigma_N$. Here H, H_1 and H_2 depend on N as usual, but the dependence is kept implicit. Of course we assume that H_1 and H_2 do not depend on x, though they might depend on other parameters. We assume that for $j = 1, 2$

$$-H_j(\boldsymbol{\sigma}) = \mathbf{g} \cdot \mathbf{a}_j(\boldsymbol{\sigma}) + b_j(\boldsymbol{\sigma}) \,, \tag{2.435}$$

where $b_j(\boldsymbol{\sigma})$ is a number, $\mathbf{a}_j(\boldsymbol{\sigma}) \in \mathbb{R}^M$, and $\mathbf{g}$ is a standard Gaussian vector valued in $\mathbb{R}^M$, (that is, the components of $\mathbf{g}$ are independent standard Gaussian). We control the size of H_j by

$$\forall \boldsymbol{\sigma} \,,\ \|\mathbf{a}_j(\boldsymbol{\sigma})\| \le \sqrt{N}\, A_j \ ;\ |b_j(\boldsymbol{\sigma})| \le N\, B_j \,, \tag{2.436}$$

where A_j, B_j are certain constants.

Theorem 2.12.1. *For each number a we have*

$$\int_{-a}^{a} \mathsf{E}\left\langle \left| \frac{H_2(\boldsymbol{\sigma})}{N} - \mathsf{E}\left\langle \frac{H_2(\boldsymbol{\sigma})}{N} \right\rangle \right| \right\rangle \mathrm{d}x \le \frac{K}{N^{1/4}} \tag{2.437}$$

where K stays bounded with a, A_1, A_2, B_1 and B_2 and does not depend on N.

It is understood that the bracket $\langle\cdot\rangle$ is computed for Gibbs' measure relative to the Hamiltonian $H_x(\boldsymbol{\sigma})$ given in (2.434).

The meaning of (2.437) is that $H_2(\boldsymbol{\sigma})/N$ is typically constant when $\boldsymbol{\sigma}$ is weighted for Gibbs' measure. The obvious drawback of (2.437) is that it allows exceptional values of x where the integrand might be large, and gives no control of these. Moreover, it seems likely that the left-hand side should decay as $1/\sqrt{N}$, not as $N^{-1/4}$.

We consider

$$Z_N(x) = \sum_{\boldsymbol{\sigma}} \exp(-H_x(\boldsymbol{\sigma}))$$

$$\theta(x) = \frac{1}{N} \log Z_N(x) \; ; \; \psi(x) = \mathsf{E}\,\theta(x) \, . \tag{2.438}$$

Lemma 2.12.2. *We have*

$$\psi''(x) = N\mathsf{E}\left\langle \left(\frac{H_2(\boldsymbol{\sigma})}{N} - \left\langle \frac{H_2(\boldsymbol{\sigma})}{N} \right\rangle \right)^2 \right\rangle .$$

Of course, it is understood here that Gibbs' measure is calculated for the Hamiltonian (2.434).

Proof. This is a straightforward calculation. We have

$$\psi'(x) = \frac{1}{N} \mathsf{E}\langle -H_2(\boldsymbol{\sigma})\rangle \tag{2.439}$$

$$\begin{aligned} \psi''(x) &= \frac{1}{N} \mathsf{E}(\langle H_2^2(\boldsymbol{\sigma})\rangle - \langle H_2(\boldsymbol{\sigma})\rangle^2) \\ &= N\,\mathsf{E}\left\langle \left(\frac{H_2(\boldsymbol{\sigma})}{N} - \left\langle \frac{H_2(\boldsymbol{\sigma})}{N} \right\rangle \right)^2 \right\rangle . \end{aligned} \tag{2.440}$$

□

Lemma 2.12.3. *We have*

$$|\psi'(x)| \le B_2 + 3A_2 \, . \tag{2.441}$$

Proof. From (2.439) we see that

$$|\psi'(x)| \le \frac{1}{N} \max_{\boldsymbol{\sigma}} |b_2(\boldsymbol{\sigma})| + \frac{1}{N} \mathsf{E}\,|\max_{\boldsymbol{\sigma}} \mathbf{g} \cdot \mathbf{a}_2(\boldsymbol{\sigma})| \, .$$

Now $|x| = x + 2\max(0, -x)$, so that

$$\begin{aligned} |\max_{\boldsymbol{\sigma}} \mathbf{g} \cdot \mathbf{a}_2(\boldsymbol{\sigma})| &= \max \mathbf{g} \cdot \mathbf{a}_2(\boldsymbol{\sigma}) + 2\max(0, -\max_{\boldsymbol{\sigma}} \mathbf{g} \cdot \mathbf{a}_2(\boldsymbol{\sigma})) \\ &\le \max \mathbf{g} \cdot \mathbf{a}_2(\boldsymbol{\sigma}) + 2\max(0, -\mathbf{g} \cdot \mathbf{a}_2(\boldsymbol{\sigma}_0)) \end{aligned}$$

where $\boldsymbol{\sigma}_0$ is a given arbitrary element of Σ_N. Thus, using (1.7),

$$\mathsf{E}\,|\max_{\boldsymbol{\sigma}} \mathbf{g} \cdot \mathbf{a}_2(\boldsymbol{\sigma})| \le 2NA_2 + \sqrt{\frac{2}{\pi}}\sqrt{N}\,A_2 \, .$$

□

Lemma 2.12.4. *Given* $0 < y \le 1$ *we have*

$$\begin{aligned}&\mathsf{E}(\theta'(x) - \psi'(x))^2 \\ &\le L\left(\left(\frac{\psi(x+y) + \psi(x-y) - 2\psi(x)}{y}\right)^2 + \frac{A_1^2 + (1+x^2)\,A_2^2}{Ny^2}\right). \end{aligned} \tag{2.442}$$

$$\begin{aligned}&\mathsf{E}|\theta'(x) - \psi'(x)| \\ &\le L\left(\psi'(x+y) - \psi'(x-y) + \frac{A_1 + (1+|x|)\,A_2}{y\sqrt{N}}\right). \end{aligned} \tag{2.443}$$

Proof. We use the convexity of the random function θ to write

$$\frac{\theta(x) - \theta(x-y)}{y} \le \theta'(x) \le \frac{\theta(x+y) - \theta(x)}{y}\,, \tag{2.444}$$

so that, setting

$$U = \frac{1}{y}(|\theta(x+y) - \psi(x+y)| + |\theta(x-y) - \psi(x-y)| + |\theta(x) - \psi(x)|)$$

we have

$$\frac{\psi(x) - \psi(x-y)}{y} - \psi'(x) - U \le \theta'(x) - \psi'(x) \le \frac{\psi(x+y) - \psi(x)}{y} - \psi'(x) + U\,.$$

Using now (2.444) for ψ rather than θ, we get

$$|\theta'(x) - \psi'(x)| \le \frac{\psi(x+y) + \psi(x-y) - 2\psi(x)}{y} + U\,. \tag{2.445}$$

An obvious adaptation of Corollary 2.2.5 (using $A = (A_1 + |z|\,A_2)\sqrt{N}$ rather than $A = \beta\sqrt{N/2}$ in the proof) shows that for any real z we have

$$\mathsf{E}(\theta(z) - \psi(z))^2 \le \frac{L}{N}\,(A_1^2 + z^2\,A_2^2)\,.$$

Thus

$$\mathsf{E}\,U^2 \le \frac{L}{Ny^2}\,(A_1^2 + (x^2+1)\,A_2^2)\,.$$

This proves (2.442). To prove (2.443), we use convexity again to see that

$$\psi(x+y) + \psi(x-y) - 2\psi(x) \le y(\psi'(x+y) - \psi'(x-y))\,.$$

The assertion now easily follows from the Cauchy-Schwarz inequality and (2.445). □

Proof of Theorem 2.12.1. We keep in mind that

$$\theta'(x) = \left\langle -\frac{H_2(\boldsymbol{\sigma})}{N} \right\rangle \; ; \; \psi'(x) = \mathsf{E}\left\langle -\frac{H_2(\boldsymbol{\sigma})}{N} \right\rangle .$$

We note that

$$\begin{aligned} &\int_{-a}^{a} (\psi'(x+y) - \psi'(x-y))\, \mathrm{d}x \\ &= \int_{a-y}^{a+y} \psi'(x)\, \mathrm{d}x - \int_{-a-y}^{-a+y} \psi'(x)\, \mathrm{d}x \le 4\, y(B_2 + 3A_2) \end{aligned}$$

using Lemma 2.12.3. Thus, by (2.443),

$$\mathsf{E} \int_{-a}^{a} |\theta'(x) - \psi'(x)|\, \mathrm{d}x \le L \left(\frac{a\, A_1 + (a + a^2)\, A_2}{y\sqrt{N}} + y(B_2 + A_2) \right) . \quad (2.446)$$

Using Lemmas 2.12.2 and 2.12.3, we see that

$$\begin{aligned} &\int_{-a}^{a} \mathsf{E} \left\langle \left(\frac{H_2(\boldsymbol{\sigma})}{N} + \theta'(x) \right)^2 \right\rangle \mathrm{d}x \\ &= \int_{-a}^{a} \mathsf{E} \left\langle \left(\frac{H_2(\boldsymbol{\sigma})}{N} - \left\langle \frac{H_2(\boldsymbol{\sigma})}{N} \right\rangle \right)^2 \right\rangle \mathrm{d}x \\ &= \frac{1}{N} \int_{-a}^{a} \psi''(x)\, \mathrm{d}x = \frac{1}{N} (\psi'(a) - \psi'(-a)) \le \frac{2B_2 + 6A_2}{N} . \quad (2.447) \end{aligned}$$

Thus the Cauchy-Schwarz inequality implies that

$$\int_{-a}^{a} \mathsf{E} \left\langle \left| \frac{H_2(\boldsymbol{\sigma})}{N} + \theta'(x) \right| \right\rangle \mathrm{d}x \le L \left(\frac{a(B_2 + A_2)}{N} \right)^{1/2} . \quad (2.448)$$

To conclude, we combine this with (2.446), taking $y = N^{-1/4}$. □

Consider a function f on Σ_N^n, and assume $|f| \le 1$. Then

$$\begin{aligned} &\left| \nu\left(\frac{H_2(\boldsymbol{\sigma}^1)}{N} f \right) - \nu\left(\frac{H_2(\boldsymbol{\sigma}^1)}{N} \right) \nu(f) \right| \\ &\le \nu\left(\left| \frac{H_2(\boldsymbol{\sigma}^1)}{N} - \nu\left(\frac{H_2(\boldsymbol{\sigma}^1)}{N} \right) \right| \right) \quad (2.449) \end{aligned}$$

will have a tendency to be small for most values of x, by Theorem 2.12.1. The miracle of the Ghirlanda-Guerra identities is that integration by parts of (2.449) yields powerful consequences. To illustrate this we consider the simplest case, where

$$-H_x(\boldsymbol{\sigma}) = \frac{\beta}{\sqrt{N}} \sum_{i<j} g_{ij}\, \sigma_i\, \sigma_j + h \sum_{i \le N} \sigma_i + x \sum_{i \le N} g_i\, \sigma_i \quad (2.450)$$

where $(g_i)_{i\le N}$ are independent standard Gaussian r.v. This Hamiltonian was already considered in (2.357). Thus (2.450) is of the type (2.434) with $H_2(\boldsymbol{\sigma}) = \sum_{i\le N} g_i \sigma_i$. Integration by parts as already performed several times yields in the present case

$$\nu\left(-\frac{H_2(\boldsymbol{\sigma})}{N}\right) = x(1-\nu(R_{1,2}))$$

$$\nu\left(-\frac{H_2(\boldsymbol{\sigma}^1)}{N} f\right) = x\left(\sum_{1\le \ell\le n} \nu(R_{1,\ell}\, f) - n\nu(R_{1,n+1}\, f)\right) .$$

Since $R_{1,1} = 1$, the left-hand side of (2.449) is

$$x\left|\nu(R_{1,2})\,\nu(f) + \sum_{2\le \ell\le n} \nu(R_{1,\ell}\, f) - n\nu(R_{1,n+1}\, f)\right| .$$

Thus Theorem 2.12.1 implies that

$$\nu(R_{1,n+1}\, f) = \frac{1}{n}\,\nu(R_{1,2})\,\nu(f) + \frac{1}{n}\sum_{2\le \ell\le n} \nu(R_{1,\ell}\, f) + \delta \,, \tag{2.451}$$

where δ is a quantity depending on f, N, x, β, h, and satisfying, for each number $a > 0$,

$$\lim_{N\to\infty} \int_{-a}^{a} |\delta(x)|\,\mathrm{d}x = 0 \,, \tag{2.452}$$

where it is understood that β, h and a remain fixed. In fact, there is even a rate of convergence in $N^{-1/4}$, but, since this is unlikely to be optimal, we will not be interested in rates of convergence. Throughout the rest of this section, δ will denote a quantity satisfying (2.452), that may not be the same at each occurrence. The relations (2.451) are called the Ghirlanda-Guerra identities.

When $n = 2$, $f = R_{1,2}$, (2.451) becomes

$$\nu(R_{1,3}\, R_{1,2}) = \frac{1}{2}\,\nu(R_{1,2})^2 + \frac{1}{2}\,\nu(R_{1,2}^2) + \delta \,.$$

When $n = 3$, $f = R_{2,3}$, since $\nu(R_{1,3}\, R_{2,3}) = \nu(R_{1,2}\, R_{2,3})$, (2.451) becomes

$$\nu(R_{1,4}\, R_{2,3}) = \frac{1}{3}\,\nu(R_{1,2})^2 + \frac{2}{3}\,\nu(R_{1,2}\, R_{2,3}) + \delta$$

and combining these yields

$$\nu(R_{1,4}\, R_{2,3}) = \frac{2}{3}\,\nu(R_{1,2})^2 + \frac{1}{3}\,\nu(R_{1,2}^2) + \delta$$

which we can write as

$$\nu((R_{1,2} - \nu(R_{1,2}))^2) = 3\,\mathsf{E}((\langle R_{1,2}\rangle - \mathsf{E}\langle R_{1,2}\rangle)^2) + \delta \,. \tag{2.453}$$

This has the following consequence, where we recall the notation $\mathcal{L}_h$ of (2.424).

Theorem 2.12.5. *Consider an open subset D of $(\mathbb{R}^+)^2$. Assume that for the Hamiltonian (2.450) we have*

$$\liminf_{N\to\infty} \int\int_D \mathsf{E}(\langle R_{1,2}\rangle - \mathsf{E}\langle R_{1,2}\rangle)^2 \,\mathrm{d}x\,\mathrm{d}\beta = 0 \,. \tag{2.454}$$

Then we have $D \cap \mathcal{L}_h = \emptyset$.

Thus, however large N, there are values of β, x for which the quantity $\langle R_{1,2}\rangle$ $(= \|\mathbf{b}\|^2/N)$ genuinely fluctuates with the randomness. (It is not self-averaging.) This seems to have been deemed surprising at first, even though it is quite apparent that such a phenomenon also occurs in the REM.

Proof. By (2.453), (2.454), we have

$$\liminf_{N\to\infty} \int\int_D \nu((R_{1,2} - \nu(R_{1,2}))^2)\,\mathrm{d}x\,\mathrm{d}\beta = 0 \,.$$

Together with (2.426) this proves that $D \cap \mathcal{L}_h = \emptyset$. □

Research problem 2.12.6. (Level2) For the Hamiltonian (2.7), prove that given h, if β_0 is large, we have

$$\liminf_{N\to\infty} \int_0^{\beta_0} \mathsf{E}((\langle R_{1,2}\rangle - \mathsf{E}\langle R_{1,2}\rangle)^2)\,\mathrm{d}\beta > 0 \,.$$

An easier question is to prove that

$$\liminf_{N\to\infty} \int_0^{\beta_0} \mathsf{E}((\langle R_{1,2}^2\rangle - \mathsf{E}\langle R_{1,2}^2\rangle)^2)\,\mathrm{d}\beta > 0 \,. \tag{2.455}$$

In the case $h = 0$, (2.455) is proved in [134].

2.13 The SK Model with d-Components Spins

A model where spins take only the values ± 1 could be an oversimplification. It is more physical to consider spins as vectors in $\mathbb{R}^3$ or $\mathbb{R}^d$. This is what we will do in this section. The corresponding model is of obvious interest. It has been investigated in less detail than the standard SK model, so many questions remain unanswered. On the other hand, this is somewhat specialized material, and it is unrelated to the rest of the book.

In the SK model with d-components spins, the individual spin σ_i is a vector $(\sigma_{i,1}, \ldots, \sigma_{i,d})$ of $\mathbb{R}^d$. We will denote by $(\cdot,\cdot)$ the dot product in $\mathbb{R}^d$, so that

$$(\sigma_i, \sigma_j) = \sum_{u\le d} \sigma_{i,u}\,\sigma_{j,u} \,.$$

The Hamiltonian is given by

$$-H_N = \frac{\beta}{\sqrt{N}} \sum_{i<j} g_{ij}(\sigma_i, \sigma_j) \tag{2.456}$$

where, of course, $(g_{ij})_{i<j}$ are independent standard normal. We can rewrite (2.456) as

$$-H_N = \frac{\beta}{\sqrt{N}} \sum_{u \le d} \sum_{i<j} g_{ij}\, \sigma_{i,u}\, \sigma_{j,u} \,, \tag{2.457}$$

a formula that is reminiscent of a Hamiltonian depending on d configurations $\boldsymbol{\sigma}_u = (\sigma_{1,u}, \ldots, \sigma_{N,u})$, each of which has the same energy as in the SK model. (But now, $\sigma_{i,\ell}$ varies in $\mathbb{R}$ rather than in $\{-1,1\}$.) In order to compare with the SK model, and to accommodate the case $\sigma_{i,u} = \pm 1$, we will assume that

$$\forall i\,, \quad \sum_{u \le d} \sigma_{i,u}^2 \le d \,, \tag{2.458}$$

or, in words, that σ_i belongs to the Euclidean ball B_d centered at 0, of radius $\sqrt{d}$. We consider a probability measure μ on B_d. We will define Gibbs' measure as the probability measure on B_d^N of density proportional to $\exp(-H_N)$ with respect to $\mu^{\otimes N}$. This is a generalization of the standard SK model, as is shown by the case $d = 1$, μ supported by $\{-1,1\}$, and having a density proportional to $\exp \varepsilon\, h$ with respect to the uniform measure on $\{-1,1\}$. Also of special interest is the case where $d = 2$, μ is supported by $\{-1,1\}^2$, and has a density proportional to

$$\exp(\varepsilon_1\, h + \varepsilon_2\, h + \lambda\, \varepsilon_1\, \varepsilon_2)$$

with respect to the uniform measure on $\{-1,1\}^2$. This is the case of "two coupled copies of the SK model" considered in Sections 2.8.

Let us define

$$Z_N = Z_N(\beta, \mu) = \int \exp(-H_N(\boldsymbol{\sigma}_1, \ldots, \boldsymbol{\sigma}_d))\, \mathrm{d}\mu(\boldsymbol{\sigma}_1) \ldots \mathrm{d}\mu(\boldsymbol{\sigma}_d) \,, \tag{2.459}$$

where

$$\boldsymbol{\sigma}_u = (\sigma_{i,u}, \ldots, \sigma_{N,u}) \,, \tag{2.460}$$

and where $-H_N(\boldsymbol{\sigma}_1, \ldots, \boldsymbol{\sigma}_d)$ is the Hamiltonian (2.456). The index in $\boldsymbol{\sigma}_u$ is a subscript, to insist on the fact that $\boldsymbol{\sigma}_1, \ldots, \boldsymbol{\sigma}_d$ are not replicas, but interacting configurations. (This notation is not consistent with the notation of Section 2.9, but is appropriate here.) Let us write

$$p_N(\beta, \mu) = \frac{1}{N} \mathsf{E} \log Z_N(\beta, \mu) \,. \tag{2.461}$$

One of our objectives is the computation of $\lim_{N \to \infty} p_N(\beta, \mu)$. It will be achieved when β is small enough. This computation has applications to the theory of "large deviations". For example, in the case of "two coupled copies of the SK model", it follows from (2.342) that computing $\lim_{N\to\infty} r_N(\beta, h, \lambda)$ amounts to computing

$$\lim_{N\to\infty} \frac{1}{N} \mathsf{E} \log\langle \exp \lambda N R_{1,2}\rangle \ . \tag{2.462}$$

"Concentration of measure" (as in Theorem 2.2.4) shows that $N^{-1} \log\langle \exp \lambda N R_{1,2}\rangle$ fluctuates little with the disorder. Thus computing (2.462) amounts to compute the value of $\langle \exp \lambda N R_{1,2}\rangle$ for the typical disorder. Since we can do this for every λ this is very much the same as computing $N^{-1} \log G_N^{\otimes 2}(\{R_{1,2} \geq q+a\})$ and $N^{-1} \log G_N^{\otimes 2}(\{R_{1,2} \leq q-a\})$ for $a > 0$. In summary, the result of (2.462) can be transfered in a result about the "large deviations of $R_{1,2}$" for the typical disorder. (On the other hand, we remind the reader of Research Problem 2.2.8.)

We will be able to compute $\lim_{N\to\infty} p_N(\beta, \mu)$ under the condition $L\beta d \leq 1$. Despite what one might think at first, this is a competent result, and there is little hope if $\beta d > 1$. This is because if μ gives mass 1/2 to two points $(\pm\sqrt{d}, 0, \ldots, 0)$, the corresponding model is "a clone" of the usual SK model at temperature βd.

The SK model with d-components spins offers new features compared with the standard SK model. One of these is that if μ is "spread out" then one can understand the system up to values of β much larger than $1/d$. For example, if μ is uniform on $\{-1, 1\}^d$, the model simply consists in d replicas of the SK model with $h = 0$, and we understand it for $\beta \leq 1$, independently of the value of d. Comparable results will be proved when μ is the uniform measure on the boundary of B_d.

The fundamental property of the SK model is (2.149), i.e. $\nu((R_{1,2}-q)^2) \leq L/N$ (for $\beta \leq \beta_0$). The situation is more complicated here. Consider $1 \leq u$, $v \leq d$, and set

$$R^{u,v} = \frac{1}{N} \sum_{i\leq N} \sigma_{i,u}\, \sigma_{i,v} \ . \tag{2.463}$$

This is a function of a single configuration $(\boldsymbol{\sigma}_1, \ldots, \boldsymbol{\sigma}_d)$. Consider now two configurations $(\boldsymbol{\sigma}_1^1, \ldots, \boldsymbol{\sigma}_d^1)$ and $(\boldsymbol{\sigma}_1^2, \ldots, \boldsymbol{\sigma}_d^2)$. Consider the function of these two configurations

$$R_{1,2}^{u,v} = \frac{1}{N} \sum_{i\leq N} \sigma_{i,u}^1\, \sigma_{i,v}^2 \ . \tag{2.464}$$

In the present context, similar to (2.149), we have the following.

Theorem 2.13.1. *If $L\beta d \leq 1$, we can find numbers $(q_{u,v})$, $(\rho_{u,v})$ such that*

$$\sum_{u,v\leq d} \nu((R^{u,v} - \rho_{u,v})^2) \leq \frac{K(d)}{N} \ , \tag{2.465}$$

$$\sum_{u,v\le d} \nu((R_{1,2}^{u,v} - q_{u,v})^2) \le \frac{K(d)}{N} \,. \tag{2.466}$$

Here $K(d)$ depends on d only; $\nu(\cdot) = \mathsf{E}\langle\cdot\rangle$, $\langle\cdot\rangle$ is an average for Gibbs' measure, over one configuration in (2.465), and over two configurations in (2.466).

In order to give a proper description of these numbers $(q_{u,v})_{1\le u,v\le d}$, $(\rho_{u,v})_{1\le u,v\le d}$, let us consider, for each symmetric positive definite matrix $(q_{u,v})_{1\le u,v\le d}$, a centered jointly Gaussian family $(Y_u)_{u\le d}$ with covariance

$$\mathsf{E}\, Y_u\, Y_v = \beta^2\, q_{u,v} \,. \tag{2.467}$$

For each family of real numbers $(\rho_{u,v})_{1\le u,v\le d}$ and for $x = (x_1,\dots,x_d)$ in $\mathbb{R}^d$, let us set further

$$\mathcal{E} = \mathcal{E}(x) = \exp\left(\sum_{u\le d} x_u\, Y_u + \frac{\beta^2}{2}\sum_{u,v\le d} x_u\, x_v(\rho_{u,v} - q_{u,v})\right) \,. \tag{2.468}$$

Theorem 2.13.2. *Assuming that $\beta L d \le 1$, and setting $Z = \int \mathcal{E}(x)\,\mathrm{d}\mu(x)$, we have*

$$q_{u,v} = \mathsf{E}\left(\frac{1}{Z^2}\int x_u\,\mathcal{E}(x)\,\mathrm{d}\mu(x)\int x_v\,\mathcal{E}(x)\,\mathrm{d}\mu(x)\right) \,, \tag{2.469}$$

$$\rho_{u,v} = \mathsf{E}\left(\frac{1}{Z}\int x_u\, x_v\,\mathcal{E}(x)\,\mathrm{d}\mu(x)\right) \,. \tag{2.470}$$

Theorem 2.13.3. *If $\beta L d \le 1$ we have*

$$\lim_{N\to\infty} p_N(\beta,\mu) = -\frac{\beta^2}{4}\sum_{u,v\le d}(\rho_{u,v}^2 - q_{u,v}^2) + \mathsf{E}\log\int \mathcal{E}(x)\,\mathrm{d}\mu(x) \tag{2.471}$$

where $\mathcal{E}(x)$ is given by (2.468).

We will later explain why the solutions to (2.469) and (2.470) exist and are unique for $L\beta d \le 1$, but let us accept this for the time being and turn to the fun part, the search for the "smart path".

We consider the quantities

$$\boldsymbol{\rho}_u = (\sigma_{1,u},\dots,\sigma_{N-1,u}) \,.$$

(One will distinguish the configuration $\boldsymbol{\rho}_u$ and the numbers $\rho_{u,v}$.)

$$g(\boldsymbol{\rho}_u) = \frac{\beta}{\sqrt{N}}\sum_{i\le N-1} g_i\,\sigma_{i,u}$$

$$g_t(\boldsymbol{\rho}_u) = \sqrt{t}\, g(\boldsymbol{\rho}_u) + \sqrt{1-t}\, Y_u \,.$$

We consider the Hamiltonian

$$-H_{N,t}(\boldsymbol{\sigma}_1,\ldots,\boldsymbol{\sigma}_d) = \frac{\beta}{\sqrt{N}} \sum_{u\le d} \sum_{i<j\le N-1} g_{i,j}\, \sigma_{i,u}\, \sigma_{j,u} + \sum_{u\le d} \sigma_{N,u} g_t(\boldsymbol{\rho}_u)$$
$$+ \frac{\beta^2}{2}(1-t) \sum_{u,v\le d} \sigma_{N,u}\, \sigma_{N,v}(\rho_{u,v} - q_{u,v}) \,. \qquad (2.472)$$

The last term is the new feature compared to the standard case.

The configuration space is now

$$S_N = \{(\sigma_1,\ldots,\sigma_N) \in B_d^N\} \,.$$

For a function f on S_N^n we write

$$\nu_t(f) = \mathsf{E}\langle f\rangle_t \,,$$

where $\langle\cdot\rangle_t$ denotes integration with respect to (the nth power of) the Gibbs' measure relative to the Hamiltonian (2.472). A function f on S_N^n depends on configurations $(\boldsymbol{\sigma}_u^1)_{u\le d}, (\boldsymbol{\sigma}_u^2)_{u\le d}, \ldots, (\boldsymbol{\sigma}_u^n)_{u\le d}$. We write

$$R_{\ell,\ell'}^{-,u,v} = \frac{1}{N} \sum_{i\le N-1} \sigma_{i,u}^{\ell}\, \sigma_{i,v}^{\ell'}$$

for $\ell,\ell' \le n$ and $u,v \le d$. As usual, we write $\nu_t'(f) = \frac{\mathrm{d}}{\mathrm{d}t}\nu_t(f)$. We write

$$q_{u,v}(\ell,\ell') = q_{u,v} \quad \text{if} \quad \ell \ne \ell'$$

$$q_{u,v}(\ell,\ell) = \rho_{u,v} \,.$$

Proposition 2.13.4. *We have*

$$\nu_t'(f) = \frac{\beta^2}{2} \sum_{\ell,\ell'\le n, u,v\le d} \nu_t(f\, \varepsilon_u^{\ell}\, \varepsilon_v^{\ell'}\, (R_{\ell,\ell'}^{-,u,v} - q_{u,v}(\ell,\ell')))$$
$$- n\,\beta^2 \sum_{\ell\le n+1, u,v\le d} \nu_t(f\, \varepsilon_u^{\ell}\, \varepsilon_v^{n+1} (R_{\ell,n+1}^{-,u,v} - q_{u,v}(\ell,n+1))) \qquad (2.473)$$
$$+ \frac{n(n+1)}{2}\beta^2 \sum_{u,v\le d} \nu_t(f\, \varepsilon_u^{n+1}\, \varepsilon_v^{n+2}(R_{n+1,n+2}^{-,u,v} - q_{u,v}(n+1,n+2))) \,.$$

Here we have set $\varepsilon_u^{\ell} = \sigma_{N,u}^{\ell}$.

Probably the first thing to do is to explain why this coincides with (2.169) in the case of the standard SK model. In such a case we have $d = 1$, $\varepsilon_1^{\ell} = \varepsilon_{\ell}$, $R_{\ell,\ell'}^{-,1,1} = R_{\ell,\ell'}^{-}$, $q_{1,1} = q$, $\rho_{1,1} = 1$ (because $x^2 = 1$ if $x \in \{-1,1\}$ cf. (2.470)). Let us also point out that $R_{\ell,\ell}^{-} = (N-1)/N$, so that the contribution of the case $\ell' = \ell$ in the first sum of the right-hand side of (2.473) cancels out with the contribution of the case $\ell = n+1$ in the second sum. We finally observe that $\sum_{\ell \neq \ell'} = 2\sum_{\ell < \ell'}$.

Proof. The new feature is the last term of the Hamiltonian (2.472), that creates the first two terms of the following (straightforward) relation

$$
\begin{aligned}
\nu_t'(f) = & -\frac{\beta^2}{2} \sum_{\ell \le n} \sum_{u,v \le d} \nu_t(f\, \varepsilon_u^\ell\, \varepsilon_v^\ell (\rho_{u,v} - q_{u,v})) \\
& + \frac{n\,\beta^2}{2} \sum_{u,v \le d} \mathsf{E}(\langle f \rangle_t \, \langle \varepsilon_u \, \varepsilon_v (\rho_{u,v} - q_{u,v}) \rangle_t) \\
& + \frac{1}{2} \sum_{\ell \le n} \sum_{u \le d} \nu_t \left(f\, \varepsilon_u^\ell \left(\frac{1}{\sqrt{t}} g(\rho_u^\ell) - \frac{1}{\sqrt{1-t}} Y_u \right) \right) \\
& - \frac{n}{2} \sum_{u \le d} \mathsf{E} \left(\langle f \rangle_t \left\langle \varepsilon_u \left(\frac{1}{\sqrt{t}} g(\rho_u) - \frac{1}{\sqrt{1-t}} Y_u \right) \right\rangle_t \right) . \qquad (2.474)
\end{aligned}
$$

One then has to perform an integration by parts in the last two terms of (2.474). This is done (after using replicas in the last term) exactly as in the case of (2.169). This integration by parts yields an expression similar to (2.473), except that one has $q_{u,v}$ rather than $q_{u,v}(\ell, \ell')$ everywhere. Combining this with the first two terms on the right of (2.474) yields (2.473). □

The proof of Theorem 2.13.1 will follow the proof of (2.149), but keeping the correct dependence in d requires a minimum of caution.

Corollary 2.13.5. *If $n = 2$, we have*

$$
\begin{aligned}
|\nu_t'(f)| \le L\beta^2 d\, \nu_t(f^2)^{1/2} \Bigg(& \nu_t \left(\sum_{u,v \le d} (R_{1,1}^{-,u,v} - \rho_{u,v})^2 \right)^{1/2} \\
& + \nu_t \left(\sum_{u,v \le d} (R_{1,2}^{-,u,v} - q_{u,v})^2 \right)^{1/2} \Bigg) . \qquad (2.475)
\end{aligned}
$$

$$
|\nu_t'(f)| \le L\beta^2 d^2 \nu_t(|f|) . \qquad (2.476)
$$

Proof. First, we observe that since we are assuming

$$
\sum_{u \le d} \sigma_{i,u}^2 \le d \qquad (2.477)
$$

for each i, we have

$$
\sum_{u \le d} (\varepsilon_u^\ell)^2 \le d . \qquad (2.478)
$$

Thus, by the Cauchy-Schwarz inequality

$$\begin{aligned}
&\sum_{u,v\le d} \varepsilon_u^\ell \varepsilon_v^{\ell'} (R_{\ell,\ell'}^{-,u,v} - q_{u,v}(\ell,\ell')) \\
&\le \left(\sum_{u,v\le d} (\varepsilon_u^\ell \varepsilon_v^{\ell'})^2\right)^{1/2} \left(\sum_{u,v\le d} (R_{\ell,\ell'}^{-,u,v} - q_{u,v}(\ell,\ell'))^2\right)^{1/2} \\
&\le d \left(\sum_{u,v\le d} (R_{\ell,\ell'}^{-,u,v} - q_{u,v}(\ell,\ell'))^2\right)^{1/2} .
\end{aligned}$$

Using the Cauchy-Schwarz inequality for ν_t in (2.473) yields (2.475).

In order to prove (2.476), it suffices to show that, for each ℓ, ℓ'

$$\left|\sum_{u,v\le d} \varepsilon_u^\ell \varepsilon_v^{\ell'} (R_{\ell,\ell'}^{-,u,v} - q_{u,v}(\ell,\ell'))\right| \le Ld^2 .$$

Using (2.478) and the Cauchy-Schwarz inequality, it suffices to prove that

$$\sum_{u,v\le d} (R_{\ell,\ell'}^{-,u,v})^2 \le d^2 \; ; \; \sum_{u,v\le d} q_{u,v}(\ell,\ell')^2 \le d^2 . \tag{2.479}$$

To prove this we observe that

$$\begin{aligned}
\sum_{u,v\le d} (R_{\ell,\ell'}^{-,u,v})^2 &= \sum_{u,v\le d} \left(\frac{1}{N} \sum_{i\le N-1} \sigma_{i,u}^\ell \sigma_{i,v}^{\ell'}\right)^2 \\
&\le \frac{1}{N} \sum_{u,v\le d} \sum_{i\le N-1} (\sigma_{i,u}^\ell \sigma_{i,v}^{\ell'})^2 \le d^2
\end{aligned}$$

by (2.458). Next, we observe by (2.469) that

$$q_{u,v}^2 \le \mathsf{E}\left(\frac{1}{Z^4}\left(\int x_u \,\mathcal{E}(x)\,\mathrm{d}\mu(x)\right)^2 \left(\int x_v \,\mathcal{E}(x)\,\mathrm{d}\mu(x)\right)^2\right) .$$

Using the Cauchy-Schwarz inequality,

$$\left(\int x_u \,\mathcal{E}(x)\,\mathrm{d}\mu(x)\right)^2 \le Z \int x_u^2 \,\mathcal{E}(x)\,\mathrm{d}\mu$$

so that

$$\sum_{u,v} q_{u,v}^2 \le \mathsf{E}\left(\frac{1}{Z^2} \int \sum_u x_u^2 \,\mathcal{E}(x)\,\mathrm{d}\mu(x) \int \sum_v x_v^2 \,\mathcal{E}(x)\,\mathrm{d}\mu(x)\right) \le d^2$$

since $\sum_{u\le d} x_u^2 \le d$ for x in the support of μ. The inequality $\sum_{u,v} \rho_{u,v}^2 \le d^2$ is similar. □

Proof of Theorem 2.13.1. In this proof we assume the existence of numbers $q_{u,v}$, $\rho_{u,v}$ satisfying (2.469) and (2.470). Using symmetry between sites, we have

$$A := \nu\left(\sum_{u,v\le d}(R_{1,2}^{u,v} - q_{u,v})^2\right) = \nu(f)\ , \tag{2.480}$$

where

$$f = \sum_{u,v\le d}(\varepsilon_u^1\varepsilon_v^2 - q_{u,v})(R_{1,2}^{u,v} - q_{u,v})\ .$$

Let us observe, using the Cauchy-Schwarz inequality and (2.478), (2.479) that

$$f^2 \le Ld^2\sum_{u,v\le d}(R_{1,2}^{u,v} - q_{u,v})^2\ . \tag{2.481}$$

Next, as in the case of the ordinary SK model, (2.469) implies that for a function f^- on S_{N-1}^n, we have

$$\nu_0((\varepsilon_u^1\varepsilon_v^2 - q_{u,v})\,f^-) = 0$$

and thus, as in the case of the ordinary SK model,

$$|\nu_0(f)| \le \frac{K(d)}{N}\ .$$

If $\beta d \le 1$, (2.476) implies $\nu_t(f) \le L\nu_1(f)$ when $f \ge 0$. Combining with (2.475), and the usual relation

$$\nu(f) \le \nu_0(f) + \sup_{0<t<1}|\nu_t'(f)|\ ,$$

we get that

$$\nu(f) \le \frac{K(d)}{N} + L\beta^2 d\,\nu(f^2)^{1/2}\left(\nu\left(\sum_{u,v\le d}(R_{1,2}^{-,u,v} - \rho_{u,v})^2\right)^{1/2}\right.$$
$$\left. + \nu\left(\sum_{u,v\le d}(R_{1,2}^{-,u,v} - q_{u,v})^2\right)^{1/2}\right)\ .$$

Using (2.480), (2.481) and the fact that a replacement of $R_{1,2}^{-,u,v}$ by $R^{u,v}$ creates an error term of at most $K(d)/N$, we get the relation

$$A \le \frac{K(d)}{N} + L\beta^2 d^2 A^{1/2}(B^{1/2} + A^{1/2})\ , \tag{2.482}$$

where A is defined in (2.480) and

$$B = \nu\left(\sum_{u,v\le d}(R^{u,v} - \rho_{u,v})^2\right)\ .$$

The same argument yields the relation

$$B \le \frac{K(d)}{N} + L\beta^2 d^2 B^{1/2}(B^{1/2} + A^{1/2})$$

and, if $L\beta^2 d^2 \le 1$, together with (2.482) this implies $A + B \le K(d)/N$. □

To prove Theorems 2.13.1 and 2.13.2, it remains to prove the existence of solutions to the equations (2.469) and (2.470). It seems to be a general fact that "the proof of the existence of solutions to the replica-symmetric equations is implicitly part of the proof of the validity of the replica-symmetric solution". What we mean here is that an argument proving the existence of a solution to (2.469) and (2.470) can be extracted from the smart path method as used in the proof of Theorem 2.13.1. The same phenomenon will occur in Chapters 3 and 5.

Consider a positive definite symmetric matrix $Q = (q_{u,v})_{u,v\le d}$, and a symmetric matrix $Q' = (\rho_{u,v})_{u,v\le d}$. Consider a centered jointly Gaussian family $(Y_u)_{u\le d}$ such that

$$\mathsf{E}\, Y_u Y_v = \beta^2 q_{u,v} \,.$$

Consider the matrices $T(Q,Q')$ and $T'(Q,Q')$ given by the left-hand sides of (2.469) and (2.470) respectively. The proof of the existence of a solution to (2.469) and (2.470) consists in showing that if we provide the set of pairs of matrices (Q,Q') as above with Euclidean distance (when seen as a subset of $(\mathbb{R}^{d^2})^2$), the map $(Q,Q') \mapsto (T(Q,Q'), T(Q,Q'))$ is a contraction provided $L\beta d \le 1$. (Thus it admits a unique fixed point.) To see this, considering another pair $(\widehat{Q},\widehat{Q}')$ of matrices, we move from the pair (Q,Q') to the pair $(\widehat{Q},\widehat{Q}')$ using the path $t \mapsto (Q(t), Q'(t))$, where

$$Q(t) = (t\, q_{u,v} + (1-t)\,\widehat{q}_{u,v})_{u,v\le d} \tag{2.483}$$

$$Q'(t) = (t\, \rho_{u,v} + (1-t)\,\widehat{\rho}_{u,v})_{u,v\le d} \,.$$

This is very closely related to the smart path used in the proof of Theorem 2.13.1, if one observes that (with obvious notation) the Gaussian process $Y_u(t)$ associated to $Q(t)$ is given by

$$Y_u(t) = \sqrt{t}\, Y_u + \sqrt{1-t}\, \widehat{Y}_u$$

where Y_u, $\widehat{Y}_u$ are assumed to be independent. This is simply because

$$\mathsf{E}\, Y_u(t)\, Y_v(t) = t\, \mathsf{E}\, Y_u Y_v + (1-t)\, \mathsf{E}\, \widehat{Y}_u \widehat{Y}_v \,.$$

All we have to do is to compute the derivative of the map $t \mapsto (Q(t), Q'(t))$ and to exhibit a convenient upper bound for the modulus of this derivative, depending on the distance between the pairs (Q,Q') and $(\widehat{Q},\widehat{Q}')$, i.e. on

$$\left(\sum_{u,v} (q_{u,v} - \widehat{q}_{u,v})^2 + (\rho_{u,v} - \widehat{\rho}_{u,v})^2\right)^{1/2} .$$

The estimates required are very similar to those of Corollary 2.13.5 and the details are better left to the reader.

Proof of Theorem 2.13.3. It follows the proof of Theorem 2.4.1. We start with the relation

$$\frac{\partial p_N}{\partial \beta}(\beta,\mu) = \frac{1}{N^{3/2}} \sum_{i<j} \sum_{u \le d} \mathsf{E}\left(g_{ij} \langle \sigma_{i,u}\, \sigma_{j,u} \rangle\right)$$
$$= \frac{\beta}{N^2} \sum_{i<j} \sum_{u,v \le d} \mathsf{E}\left(\langle \sigma_{i,u}\, \sigma_{j,u}\, \sigma_{i,v}\, \sigma_{j,v} \rangle - \langle \sigma^1_{i,u}\, \sigma^1_{j,u}\, \sigma^2_{i,v}\, \sigma^2_{j,v} \rangle\right)$$

using integration by parts. Thus

$$\left| \frac{\partial p_N}{\partial \beta}(\beta,\mu) - \frac{\beta}{2} \sum_{u,v} \mathsf{E}\langle (R^{u,v})^2 - (R^{u,v}_{1,2})^2 \rangle \right|$$
$$\le \frac{\beta^2}{N^2} \sum_{u,v} \left| \sum_{i \le N} \mathsf{E}\langle (\sigma_{i,u}\, \sigma_{i,v})^2 - (\sigma^1_{i,u}\, \sigma^2_{i,v})^2 \rangle \right| \le \frac{K(d)}{N}$$

and thus, by Theorem 2.13.1,

$$\left| \frac{\partial p_N}{\partial \beta}(\beta,\mu) - \frac{\beta}{2} \sum_{u,v} (\rho^2_{u,v} - q^2_{u,v}) \right| \le \frac{K(d)}{\sqrt{N}} \,.$$

(With some extra work one can probably reach the better rate $K(d)/N$.) Thus, all we have to check is that the derivative of

$$-\frac{\beta^2}{4} \sum_{u,v \le d} (\rho^2_{u,v} - q^2_{u,v}) + \mathsf{E} \log \int \mathcal{E}(x)\, \mathrm{d}\mu(x) \tag{2.484}$$

with respect to β is $\beta \sum_{u,v} (\rho^2_{u,v} - q^2_{u,v})/2$. This is a straightforward computation, once one observes that the relations (2.469) and (2.470) mean that the partial derivatives of the quantity (2.471) with respect to $q_{u,v}$, $\rho_{u,v}$ are zero. □

Let us now focus on the case where μ is the uniform measure μ_d on the boundary of B_d, that is, on the sphere of center 0 and radius $\sqrt{d}$ of $\mathbb{R}^d$.

Theorem 2.13.6. *If $\beta < 1/L$, then*

$$\lim_{N \to \infty} p_N(\beta, \mu_d) = \frac{\beta^2 d}{4} \,. \tag{2.485}$$

The difficult part of the theorem is to reach values of β independent of d. The striking feature is that given β, for d large, $N^{-1} \log \mathsf{E}\, Z_N(\beta, \mu_d)$ is of order $\beta^2 d^2$ (not $\beta^2 d$). This dramatic difference between $\mathsf{E}\log Z_N$ and $\log \mathsf{E}\, Z_N$ comes from a few configurations. This is closely related to what happens in Theorem 2.2.13. Thus, it is probably not realistic to attempt reading the following unless one has understood Theorem 2.2.13. As Theorem 2.2.13, Theorem 2.13.6 is a very special (and pretty) result, begging for a sweeping generalization.

Research problem 2.13.7. (Level2?) Can one compute the supremum β_d of the values of β for which (2.485) holds? Or, at least, what is $\lim_{d\to\infty} \beta_d$?

In Theorem 2.13.6 we need lower and upper bounds. Let us start with the easy part, namely the lower bounds.

Lemma 2.13.8. *Consider numbers $(r_i)_{i\le N}$. If $\sum_{i\le N} r_i^4 \le N/2$, then*

$$\frac{1}{N}\mathsf{E}\log 2^{-N}\sum \exp\left(\frac{1}{\sqrt{N}}\sum_{i<j} g_{ij}\,\eta_i\,\eta_j\,r_i\,r_j\right) \ge \frac{1}{4N}\sum_{i\le N} r_i^2 - \frac{L}{\sqrt{N}}\,, \tag{2.486}$$

where the summation is over $\eta_i = \pm 1$.

Proof. This is proved with a straightforward adaptation of the proof of (2.27). The key point is that provided $\sum_{i\le N} r_i^4 \le N/2$, it follows from (A.5) that

$$2^{-2N}\sum \exp\frac{1}{2N}\left(\sum_{i\le N}\eta_i^1\,\eta_i^2\,r_i^2\right)^2 \le 2\,,$$

where the summation is over $\eta_i^1, \eta_i^2 = \pm 1$, $i \le N$. □

To lighten notation, we write μ rather than μ_d.

Proposition 2.13.9. *If $\beta < 1/L$ we have*

$$\liminf_{N\to\infty} p_N(\beta,\mu) \ge \frac{d\beta^2}{4}\,.$$

Proof. Given $\sigma_1,\ldots,\sigma_N$ in S_d, we write

$$e(\sigma_1,\ldots,\sigma_N) = 2^{-Nd}\sum\exp\left(\frac{\beta}{\sqrt{N}}\sum_{i<j} g_{ij}\sum_{u\le d}\sigma_{i,u}\,\sigma_{j,u}\,\eta_i^u\,\eta_j^u\right) \tag{2.487}$$

where the summation is over $\eta_i^u = \pm 1$ (so there are 2^{Nd} terms in the summation). The symmetries of S_d imply that

$$Z_N = \int e(\sigma_1,\ldots,\sigma_N)\,\mathrm{d}\mu(\sigma_1)\ldots\mathrm{d}\mu(\sigma_N)\,.$$

Thus, by Jensen's inequality,

$$\mathsf{E}\log Z_N \geq \int \mathsf{E}\log e(\sigma_1,\ldots,\sigma_N)\,\mathrm{d}\mu(\sigma_1)\ldots\mathrm{d}\mu(\sigma_N)\,. \tag{2.488}$$

Now, (2.487) shows that

$$e(\sigma_1,\ldots,\sigma_N) = \prod_{u\leq d} e_u(\sigma_1,\ldots,\sigma_N)$$

where

$$e_u(\sigma_1,\ldots,\sigma_N) = 2^{-N}\sum \exp\left(\frac{\beta}{\sqrt{N}}\sum_{i<j} g_{ij}\sigma_{i,u}\,\sigma_{j,u}\,\eta_i\,\eta_j\right)$$

and where the summation is now over $\eta_i = \pm 1$, $i \leq N$. Thus (2.488) and symmetry show that

$$\mathsf{E}\log Z_N \geq d\int \mathsf{E}\log e_1(\sigma_1,\ldots,\sigma_N)\,\mathrm{d}\mu(\sigma_1)\ldots\mathrm{d}\mu(\sigma_N)\,. \tag{2.489}$$

Jensen's inequality (used for the summation over η_i) implies that

$$e_1(\sigma_1,\ldots,\sigma_N) \geq 1\,.$$

Lemma 2.13.8 shows that

$$\frac{\beta^2}{N}\sum_{i\leq N}\sigma_{i,1}^4 \leq \frac{1}{2} \Rightarrow \frac{1}{N}\mathsf{E}\log e_1(\sigma_1,\ldots,\sigma_N) \geq \frac{\beta}{4N}\left(\sum_{i\leq N}\sigma_{i,1}^2\right) - \frac{K(d)}{\sqrt{N}}\,.$$

Thus (2.489) implies

$$\frac{1}{N}\mathsf{E}\log Z_N \geq d\int_A \frac{\beta}{4N}\left(\sum_{i\leq N}\sigma_{i,1}^2\right)\mathrm{d}\mu(\sigma_1)\ldots\mathrm{d}\mu(\sigma_N) - \frac{K(d)}{\sqrt{N}}\,, \tag{2.490}$$

where

$$A = \left\{(\sigma_1,\ldots,\sigma_N)\in S_d^N\,;\ \frac{\beta^2}{N}\sum_{i\leq N}\sigma_{i,1}^4 \leq \frac{1}{2}\right\}\,.$$

Now $\int \sigma_1^8\,\mathrm{d}\mu(\sigma) \leq L$ (a very weak special case of Proposition 2.13.10 that we will prove next) so that if

$$f = \frac{1}{N}\sum_{i\leq N}\sigma_{i,1}^4$$

we get

$$\int f^2\,\mathrm{d}\mu^{\otimes N} - \left(\int f\,\mathrm{d}\mu^{\otimes N}\right)^2 \leq \frac{L}{N}$$

and thus, by Markov's inequality, if $L\beta \le 1$ we get

$$\mu^{\otimes N}(A^c) \le L/N \; . \tag{2.491}$$

We observe that

$$\int \sum_{u\le d} \sigma_u^2 \, \mathrm{d}\mu(\sigma) = d = d \int \sigma_1^2 \, \mathrm{d}\mu(\sigma) \; ,$$

so that (2.490) implies

$$\begin{aligned} \frac{1}{N} \mathsf{E} \log Z_N &\ge \frac{d\beta^2}{4} - \int_{A^c} \frac{\beta^2}{4N} \left(\sum_{i\le N} \sigma_{i,1}^2 \right) \mathrm{d}\mu(\sigma_1)\ldots\mathrm{d}\mu(\sigma_N) - \frac{K(d)}{\sqrt{N}} \\ &\ge \frac{d\beta^2}{4} - \frac{K(d)}{\sqrt{N}} \end{aligned} \tag{2.492}$$

using (2.491), the Cauchy-Schwarz inequality, and the fact that

$$\int \frac{1}{N^2} \left(\sum_{i\le N} \sigma_{i,1}^2 \right)^2 \mathrm{d}\mu(\sigma_1)\ldots\mathrm{d}\mu(\sigma_N) \le L$$

since $\int \sigma_1^4 \, \mathrm{d}\mu(\sigma) \le L$. □

Now we describe the special properties of $\mu = \mu_d$ on which Theorem 2.13.6 relies.

Proposition 2.13.10. a) *If the numbers $(a_u)_{u\le d}$ satisfy $|a_u| \le 1/4$, then*

$$\int \exp \left(\sum_{u\le d} a_u (x_u^2 - 1) \right) \mathrm{d}\mu(x) \le \exp L \sum_{u\le d} a_u^2 \; . \tag{2.493}$$

b) *If the numbers $(a_{u,v})_{u,v\le d}$ satisfy $a_{u,v} = a_{v,u}$, $a_{u,u} = 0$, $\sum_{u,v\le d} a_{u,v}^2 \le 1/16$ then*

$$\int \exp \left(\sum_{u,v\le d} a_{u,v} \, x_u \, x_v \right) \mathrm{d}\mu(x) \le \exp L \sum_{u,v\le d} a_{u,v}^2 \; . \tag{2.494}$$

Here of course $x = (x_u)_{u\le d}$.

Proof. We will proceed by comparison with Gaussian r.v. Consider independent standard Gaussian r.v. $(h_u)_{u\le d}$ and a r.v. x that is uniform over S_d and independent of $h_1, \ldots, h_d$. Then the $\mathbb{R}^d$-valued random vectors

$$(h_1,\ldots,h_d)\,;\; x\left(\frac{1}{d}\sum_{u\le d} h_u^2\right)^{1/2}, \tag{2.495}$$

have the same distribution. This is because the density of the distribution of $(h_1,\ldots,h_d)$ depends only on the distance to the origin, so that it suffices to observe that the distribution of the lengths of the vectors in (2.495) coincide.

Next, we observe that to prove (2.448) we can assume that $\sum_{u\le d} a_u = 0$. This is because if we set $a'_u = a_u - d^{-1}\sum_{v\le d} a_v$, we have $\sum_{u\le d} a_u = 0$, $\sum_{u\le d} a_u'^2 \le \sum_{u\le d} a_u^2$, and, using that $\sum_{u\le d} x_u^2 = d$ on S_d, we also have $\sum_{u\le d} a_u(x_u^2-1) = \sum_{u\le d} a'_u(x_u^2-1)$. Assuming that $\sum_{u\le d} a_u = 0$, we have $\int \sum_{u\le d} a_u(x_u^2-1)\,\mathrm{d}\mu(x) = 0$, and Jensen's inequality shows that

$$\int \exp\left(\sum_{u\le d} a_u(x_u^2-1)\right)\mathrm{d}\mu(x) \ge 1\,.$$

It then follows from Hölder's inequality that

$$\int \exp\left(r\sum_{u\le d} a_u\, x_u^2\right)\mathrm{d}\mu(x)$$

is an increasing function of r. Thus

$$\begin{aligned}
&\mathbf{1}_{\{\sum_{u\le d} h_u^2 \ge d\}} \int \exp\left(\sum_{u\le d} a_u(x_u^2-1)\right)\mathrm{d}\mu(x)\\
&\le \int \exp\left(\left(\frac{1}{d}\sum_{v\le d} h_v^2\right)\sum_{u\le d} a_u\, x_u^2 - \sum_{u\le d} a_u\right)\mathrm{d}\mu(x)\,.
\end{aligned} \tag{2.496}$$

As explained in (2.495), the vector

$$\left(x_u^2\left(\frac{1}{d}\sum_{v\le d} h_v^2\right)\right)_{u\le d},$$

has the same distribution as $(h_u^2)_{u\le d}$ when x varies uniformly over B_d. Thus, performing expectation in (2.496), we obtain

$$\begin{aligned}
&\mathsf{P}\left(\sum_{u\le d} h_u^2 \ge d\right)\int \exp\left(\sum_{u\le d} a_u(x_u^2-1)\right)\mathrm{d}\mu(x)\\
&\le \mathsf{E}\exp\left(\sum_{u\le d} a_u(h_u^2-1)\right) = \prod_{u\le d}\frac{1}{\sqrt{1-2a_u}}\exp(-a_u)\\
&\le \exp L\sum_{u\le d} a_u^2
\end{aligned} \tag{2.497}$$

because $(1-2t)^{-1/2}\exp(-t) \le \exp L t^2$ for $t \le 1/4$. Now (using the one dimensional central limit theorem), $\mathsf{P}\left(\sum_{u\le d} h_u^2 \ge d\right)$ is bounded below independently of d, and (2.497) gives

$$\int \exp\left(\sum_{u\le d} a_u(x_u^2-1)\right) \mathrm{d}\mu(x) \le L \exp L \sum_{u\le d} a_u^2 \,.$$

To finish the proof of (2.493) we appeal to a general fact (proved implicitly during the proof of Theorem A.2.1): if a r.v. X satisfies $\mathsf{E}X = 0$, $\mathsf{E}\exp tX \le L\exp t^2A^2$ for $|t| \le 1$, then for $|t| \le 1$, $\mathsf{E}\exp tX \le \exp Lt^2A^2$ (and we recall that $\int x_u^2 \,\mathrm{d}\mu_d(x) = 1$).

To prove (2.494) we use the fact that a symmetric matrix can be diagonalized in an orthogonal basis, and that the corresponding diagonal matrix $(b_{u,v})$ has the same trace and the same Hilbert-Schmidt norm, i.e.

$$\sum_{u,v\le d} b_{u,v}^2 = \sum_{u\le d} b_{u,u}^2 = \sum_{u,v\le d} a_{u,v}^2 \le \frac{1}{16}$$

$$\sum_{u\le d} b_{u,u} = \sum_{u\le d} a_{u,u} = 0 \,.$$

Writing $b_u = b_{u,u}$, we then see that $|b_u| \le 1/4$ for each u, and that

$$\sum_{u,v\le d} b_{u,v}\, x_u\, x_v = \sum_{u\le d} b_u\, x_u^2 = \sum_{u\le d} b_u(x_u^2-1) \,.$$

Thus (2.494) follows from (2.493). □

Corollary 2.13.11. a) *Consider numbers* $(a_u)_{u\le d}$ *with* $|a_u| \le 1/4$. *Then*

$$\int \exp\left(N\sum_{u\le d} a_u(R^{u,u}-1)\right) \mathrm{d}\mu(\sigma_1)\ldots\mathrm{d}\mu(\sigma_N) \le \exp LN \sum_{u\le d} a_u^2 \,. \quad (2.498)$$

b) *Consider numbers* $(a_{u,v})_{u\le v}$ *with*

$$a_{u,u} = 0\,,\; a_{u,v} = a_{v,u}\,,\; \sum_{u,v\le d} a_{u,v}^2 \le 1/16 \,. \quad (2.499)$$

Then

$$\int \exp\left(N\sum_{u,v\le d} a_{u,v}\, R^{u,v}\right) \mathrm{d}\mu(\sigma_1)\ldots\mathrm{d}\mu(\sigma_N) \le \exp LN \sum_{u,v\le d} a_{u,v}^2 \,. \quad (2.500)$$

Proof. Follows from Proposition 2.13.10 and the fact that

$$NR^{u,v} = \sum_{i \le N} \sigma_{i,u}\, \sigma_{i,v} \ .$$

□

Lemma 2.13.12. *Consider the set*

$$C = \left\{ (\sigma_1, \ldots, \sigma_N) \in B_d^N \ ; \ \sum_{u \ne v} (R^{u,v})^2 \le 1 \right\} . \tag{2.501}$$

Then

$$\int_C \exp\left(\frac{N}{L} \sum_{u \ne v} (R^{u,v})^2 \right) \mathrm{d}\mu(\sigma_1) \ldots \mathrm{d}\mu(\sigma_N) \le K(d) \ . \tag{2.502}$$

Proof We apply Proposition A.3.1 to the subspace $\mathcal{M}$ of $\mathbb{R}^{d^2}$ consisting of sequences $(x_{u,v})_{u,v \le d}$ with $x_{u,v} = x_{v,u}$ and $x_{u,u} = 0$, with $\varepsilon = 1/4$ and B such that $4B$ is the Euclidean unit ball of $\mathcal{M}$. We see that there exists a finite set A of families $(a_{u,v})$ satisfying (2.499) such that, given numbers $(x_{u,v})_{u,v}$ in $\mathcal{M}$, we have

$$\sup_A \sum_{u \ne v} a_{u,v}\, x_{u,v} \ge \frac{1}{8} \left(\sum_{u \ne v} x_{u,v}^2 \right)^{1/2} .$$

Thus if $t > 0$, $0 \le w \le 1$, we have

$$\begin{aligned} \mu^{\otimes N}\left(\left\{ \sum_{u \ne v} (R^{u,v})^2 \ge t \right\} \right) &= \mu^{\otimes N}\left(\left\{ \frac{1}{8} \left(\sum_{u \ne v} (R^{u,v})^2 \right)^{1/2} \ge \frac{\sqrt{t}}{8} \right\} \right) \\ &\le \sum_A \mu^{\otimes N}\left(\left\{ Nw \sum_{u \ne v} a_{u,v}\, R^{u,v} \ge \frac{Nw}{8} \sqrt{t} \right\} \right) \\ &\le \sum_A \exp\left(NLw^2 - \frac{Nw}{8} \sqrt{t} \right) , \end{aligned}$$

using (2.500) for $w\, a_{u,v}$ rather than $a_{u,v}$. Optimization over w shows that, for $t \le 1$,

$$\mu^{\otimes N}\left(\left\{ \sum_{u \ne v} R_{u,v}^2 \ge t \right\} \right) \le \operatorname{card} A \exp\left(-\frac{Nt}{L} \right) ,$$

and, using (A.43), this proves (2.502). □

Lemma 2.13.13. *We have*

$$\int \exp\left(\frac{N}{L}\sum_{u\le d}\min(|R^{u,u}-1|,(R^{u,u}-1)^2)\right)\mathrm{d}\mu(\sigma_1)\dots\mathrm{d}\mu(\sigma_N)\le K(d)\,. \tag{2.503}$$

Proof. If we restrict the integral to the set $\{\sum_u (R^{u,u}-1)^2\le 1\}$, the proof is identical to that of Lemma 2.13.12, using (2.498) rather than (2.500), so we consider only the integral on $C_1=\{\sum_u (R^{u,u}-1)^2\ge 1\}$. Let us denote by L_0 the constant L of (2.498). Then given numbers $(x_u)_{u\le d}$, we have

$$\sup\left\{\sum_{u\le d} a_u\,x_u - L_0\sum_{u\le d} a_u^2\right\}\ge\frac{1}{L}\sum_{u\le d}\min(|x_u|,x_u^2)$$

where the supremum is taken over sequences (a_u) with $|a_u|\le 1/4$. (This follows from the case $d=1$.) Thus, there is a finite set A of such sequences such that

$$\sum_{u\le d} x_u^2\ge 1\Rightarrow \sup_A\left\{\sum_{u\le d} a_u\,x_u - L_0\sum_{u\le d} a_u^2\right\}\ge\frac{1}{L}\sum_{u\le d}\min(|x_u|,x_u^2)\,.$$

This becomes obvious through a compactness argument, since only the sequences $(x_u)_{u\le d}$, $|x_u|\le 1$, $u\le d$ need to be considered (but it is wrong if one does not request $\sum x_u^2$ to stay away from zero). Thus

$$\begin{aligned}&\int_{C_1}\exp\frac{N}{L}\sum_{u\le d}\min(|R^{u,u}-1|,(R^{u,u}-1)^2)\,\mathrm{d}\mu(\sigma_1)\dots\mathrm{d}\mu(\sigma_N)\\ &\le\sum_A\int_{C_1}\exp\left(\sum_{u\le d}a_u(R^{u,u}-1)-L_0\sum_{u\le d}a_u^2\right)\mathrm{d}\mu(\sigma_1)\dots\mathrm{d}\mu(\sigma_N)\\ &\le\operatorname{card}A\end{aligned}$$

by (2.498). □

It is easy to show (as proved in Lemma A.4.1) that there exists an event Ω such that

$$\mathsf{P}(\Omega)\ge 1-\exp\left(-\frac{N}{L}\right) \tag{2.504}$$

and

$$\forall\,(y_i)_{i\le N}\,,\ \left|\sum_{i<j} g_{ij}\,y_i\,y_j\right|\le L\sqrt{N}\sum_{i\le N}y_i^2\,. \tag{2.505}$$

We will prove the following.

Proposition 2.13.14. *If $L\beta \le 1$, we have*

$$\mathsf{E}\,(\mathbf{1}_\Omega\, Z_N(\beta,\mu)) \le K(d) \exp \frac{N\beta^2 d}{4}\,. \tag{2.506}$$

Taking (2.504) into account, this implies that for $0 < t \le 1$, we have

$$\mathsf{P}\left(\frac{1}{N}\log Z_N \ge \frac{\beta^2 d}{4} + t\right) \le K(d) \exp\left(-\frac{Nt}{L}\right)\,. \tag{2.507}$$

Moreover,

$$\begin{aligned}\mathsf{E}\, Z_N &= \int \exp\left(\frac{\beta^2}{2N}\sum_{i<j}\sum_{u,v}\sigma_{i,u}\,\sigma_{i,v}\,\sigma_{j,u}\,\sigma_{j,v}\right) \mathrm{d}\mu(\sigma_1)\ldots\mathrm{d}\mu(\sigma_N)\\ &\le \int \exp\left(\frac{N\beta^2}{4}\sum_{u,v}(R^{u,v})^2\right)\mathrm{d}\mu(\sigma_1)\ldots\mathrm{d}\mu(\sigma_N) \le \exp\frac{N\beta^2 d^2}{4}\end{aligned}$$

(see (2.479)), thus for $t > 0$

$$\mathsf{P}\left(\frac{1}{N}\log Z_N \ge \frac{\beta^2 d^2}{4} + t\right) \le \exp(-Nt)$$

so that combining with (2.507) and using (A.43) this completes the proof of Theorem 2.13.6.

Proof of Proposition 2.13.14. Recalling the set C of (2.501) we show in a first step that if

$$Z_{N,C} = \int_C \exp(-H_N(\sigma_1,\ldots,\sigma_N))\,\mathrm{d}\mu(\sigma_1)\ldots\mathrm{d}\mu(\sigma_N)$$

then

$$\begin{aligned}\mathsf{E}(\mathbf{1}_\Omega\, Z_{N,C}) &= \int_C \mathsf{E}\left(\mathbf{1}_\Omega \exp\left(\frac{\beta}{\sqrt{N}}\sum_{i<j}\sum_{u\le d} g_{ij}\,\sigma_{i,u}\,\sigma_{j,u}\right)\right)\mathrm{d}\mu(\sigma_1)\ldots\mathrm{d}\mu(\sigma_N)\\ &\le K(d)\exp\frac{N\beta^2 d}{4}\,.\end{aligned} \tag{2.508}$$

Given a configuration $(\sigma_1,\ldots,\sigma_N)$ in C, let

$$U = \{u \le d\,;\ R^{u,u} \ge 2\}\,.$$

By (2.505), if Ω occurs, we have

$$\frac{1}{\sqrt{N}}\sum_{i<j} g_{ij}\,\sigma_{i,u}\,\sigma_{j,u} \le L\sum_{i\le N}\sigma_{i,u}^2 = LNR^{u,u}$$

and thus

$$\begin{aligned}
&\mathsf{E}\left(\mathbf{1}_\Omega \exp\left(\frac{\beta}{\sqrt{N}}\sum_{i<j} g_{ij}\sum_{u\le d}\sigma_{i,u}\,\sigma_{j,u}\right)\right)\\
&\le \exp\left(L\beta N\sum_{u\in U} R^{u,u}\right)\mathsf{E}\exp\left(\frac{\beta}{\sqrt{N}}\sum_{i<j} g_{ij}\sum_{u\notin U}\sigma_{i,u}\,\sigma_{j,u}\right)\\
&= \exp\left(L\beta N\sum_{u\in U} R^{u,u}+\frac{\beta^2}{2N}\sum_{i<j}\sum_{u,v\notin U}\sigma_{i,u}\,\sigma_{j,u}\,\sigma_{i,v}\,\sigma_{j,v}\right)\\
&\le \exp\left(L\beta N\sum_{u\in U} R^{u,u}+\frac{\beta^2 N}{4}\sum_{u,v\notin U}(R^{u,v})^2\right). \qquad (2.509)
\end{aligned}$$

We now observe that

$$\sum_{u\le d} R^{u,u} = \frac{1}{N}\sum_{i\le N}\sum_{u\le d}\sigma_{i,u}^2 = d\,.$$

Since $R^{u,u}\ge 2\ge 1$ for $u\in U$, we have

$$\sum_{u\notin U} R^{u,u}\le d-\operatorname{card} U$$

and

$$\begin{aligned}
\sum_{u\notin U}(R^{u,u})^2 &= \sum_{u\notin U}(R^{u,u}-1)^2+2\sum_{u\notin U}R^{u,u}-\operatorname{card} U^c\\
&\le \sum_{u\notin U}(R^{u,u}-1)^2+2(d-\operatorname{card} U)-\operatorname{card} U^c\\
&\le \sum_{u\notin U}(R^{u,u}-1)^2+d
\end{aligned}$$

since $\operatorname{card} U+\operatorname{card} U^c = d$. Also, for u in U we have $R^{u,u}\ge 2$, so that $R^{u,u}\le 2\,|R^{u,u}-1|$.

We have shown that the quantity (2.509) is bounded by

$$W\exp\left(LN\beta\sum_{u\in U}|R^{u,u}-1|+\frac{\beta^2 N}{4}\left(\sum_{u\notin U}(R^{u,u}-1)^2+\sum_{u\ne v}(R^{u,v})^2\right)\right),$$

where $W=\exp((N\beta^2 d)/4)$. Now, we have

$$\min(|R^{u,u}-1|,(R^{u,u}-1)^2)=\begin{cases}|R^{u,u}-1| & \text{if } u\in U\\ (R^{u,u}-1)^2 & \text{if } u\notin U\,.\end{cases}$$

Thus the quantity (2.509) is bounded by

$$W \exp\left(L\beta N\left(\sum_u \min(|R^{u,u}-1|,(R^{u,u}-1)^2)\right)+\frac{\beta^2 N}{4}\left(\sum_{u\neq v}(R^{u,v})^2\right)\right),$$

and (2.508) follows from (2.502), (2.503) and the Cauchy-Schwarz inequality.

Consider an orthogonal basis $W=(e_1,\ldots,e_d)$ of $\mathbb{R}^d$. Define

$$R_W^{u,v}=\frac{1}{N}\sum_{i\le N}(\sigma_i,e_u)(\sigma_i,e_v)\,,$$

$$C_W=\left\{(\sigma_1,\ldots,\sigma_N)\,;\;\sum_{u\neq v}(R_W^{u,v})^2\le 1\right\}\,. \tag{2.510}$$

Then (2.508) holds if we replace C by C_W.

To finish the argument, we show that there exists a finite set $\mathcal{W}$ of orthogonal bases of $\mathbb{R}^d$, with a cardinality independent of N, such that any configuration is contained in a set C_W, W in $\mathcal{W}$. (Thus we will have $Z_N\le\sum_{\mathcal{W}}Z_{N,C_W}$, and this will finish the proof.) To see this we observe that

$$R_W^{u,v}=Q(e_u,e_v)$$

where

$$Q(x,y)=\frac{1}{N}\sum_{i\le N}(\sigma_i,x)(\sigma_i,y)\,.$$

Thus it suffices to take $\mathcal{W}$ rich enough so that given any symmetric quadratic form Q on $\mathbb{R}^d$ for which $Q(x,y)\le d\,\|x\|\,\|y\|$ there exists W in $\mathcal{W}$ with $\sum_{u\neq v}Q(e_u,e_v)\le 1$ ("Q is nearly diagonalized by W"). □

It is possible to go quite beyond Theorem 2.13.6.

Theorem 2.13.15. *There exists a constant L with the following property. Consider a probability measure μ on S_d, and assume that μ has a density f with respect to μ_d. Then, if $|f-1|\le 1/L$, the replica-symmetric solution corresponding to μ holds for $\beta\le 1/L$.*

What we mean here is that

$$\lim_{N\to\infty}\sum_{u,v\le d}\nu((R^{u,v}-\rho_{u,v})^2)=0$$

$$\lim_{N\to\infty}\sum_{u,v\le d}\nu((R_{1,2}^{u,v}-\rho_{u,v})^2)=0$$

and that (2.471) holds.

We could not prove this result using the methods of this section so we refer the interested reader to the original paper [145].

2.14 A Research Problem: The Transition at $\beta = 1$

We understand well the SK model without external field for $\beta < 1$. On the other hand, the case $\beta > 1$ is so difficult that we do not even know where to start. A natural idea is thus to consider a temperature $\beta = \beta_N$ depending on N, and to find the slowest rate of growth where we do not understand what happens.

Theorem 2.14.1. *Assume that*

$$\lim_{N\to\infty} N^{1/3}(1-\beta_N^2) = \infty\,. \tag{2.511}$$

Then, given integers $k(\ell,\ell')$ for $1 \le \ell < \ell' \le n$, we have

$$\lim_{N\to\infty} \nu\left(\prod_{\ell<\ell'}((N(1-\beta_N^2))^{1/2}\,R_{\ell,\ell'})^{k(\ell,\ell')}\right) = \prod_{\ell<\ell'} a(k(\ell,\ell')) \tag{2.512}$$

where $a(k) = \mathsf{E}\,g^k$, g standard Gaussian. If

$$\lim_{N\to\infty} N^{1/3}(1-\beta_N^2) = c > 0 \tag{2.513}$$

then for each $k > 0$

$$\sup_N \nu((N(1-\beta_N^2)\,R_{1,2}^2)^k) < \infty\,, \tag{2.514}$$

but (2.512) fails, even if one replaces the normalizing factor $N(1-\beta_N^2)$ by $(\mathsf{E}\,\langle R_{1,2}^2\rangle)^{-1}$.

In other words, the orderly picture of Theorem 2.3.9 holds under (2.511) but not under (2.513). We do not understand precisely what happens under (2.513). The variable $(N(1-\beta_N^2))^{1/2}R_{1,2}$ is "sub Gaussian" by (2.514), but we do not know how to compute its moments.

Research problem 2.14.2. (Level3) Understand precisely what happens under (2.513).

Of course condition (2.513) is not particularly important. It occurs naturally if one looks at Theorem 2.3.9, but the case $\beta = 1$ is probably more important.

Research problem 2.14.3. (Level3) Understand the SK model for $\beta = 1$.

Apparently, the only precise result known there is the following.

Proposition 2.14.4. *If $\beta \le 1$, then*

$$\sqrt{N}\ \nu(R_{1,2}^2) \le L\ . \tag{2.515}$$

It is however not known whether for each k (or even when $k = 2$) we have

$$\sup_N N^{k/2} \nu(R_{1,2}^{2k}) \le K(k)\ .$$

In fact, it is not known either that $\sqrt{N}$ is the correct normalization factor in (2.515). Maybe it can be replaced by (say) $N^{2/3}$.

The proof of Theorem 2.14.1 relies on the cavity method. It is however harder to control the higher order terms when $\beta_N \to 1$; condition (2.513) corresponds exactly to the case where "these higher order terms are of the same size as the main terms". Another exciting feature will be that the cavity method yields really mysterious relations. We will start with the following.

Proposition 2.14.5. *If $\beta \le 1$, we have*

$$x \ge L\left(\frac{\log N}{N}\right)^{3/8} \Rightarrow \mathsf{E}\, G_N^{\otimes 2}(R_{1,2}^2 \ge x)$$
$$\le LN\exp\left(-\frac{Nx^2}{L}\right) + \exp\left(-\frac{N^{3/2}}{L\sqrt{\log N}}\, x^4\right) \tag{2.516}$$

Proof. The proof consists in rewriting the proof of Theorem 2.2.13, so the reader must have mastered this first. In a first step we prove that

$$\mathsf{E} Z_N^2 \left\langle \exp\frac{N}{12}\, R_{1,2}^4 \right\rangle \le (2N+1)(\mathsf{E} Z_N^2)\ . \tag{2.517}$$

The basic observation is that the function $\mathcal{I}(t)$ of (A.7) satisfies $\mathcal{I}(0) = \mathcal{I}'(0) = 0$, $\mathcal{I}'(t) = (1-t^2)^{-1} \ge 1 + t^2$, so that $\mathcal{I}(t) \ge t^2/2 + t^4/12$. Thus, mimicking the proof of (2.16), we get

$$\mathsf{E} Z_N^2 \left\langle \exp\frac{N}{12}\, R_{1,2}^4 \right\rangle \le \sum_{\boldsymbol{\sigma}^1,\boldsymbol{\sigma}^2} \exp\left(\frac{\beta^2}{2}(N-2) + \frac{N}{2}\, R_{1,2}^2 + \frac{N}{12}\, R_{1,2}^4\right)$$
$$\le (\mathsf{E} Z_N)^2\, 2^{-2N} \sum_{\boldsymbol{\sigma}^1,\boldsymbol{\sigma}^2} \exp N\mathcal{I}(R_{1,2})\ . \tag{2.518}$$

Now, using (A.9),

$$\sum_{\boldsymbol{\sigma}^1,\boldsymbol{\sigma}^2} \exp N\mathcal{I}(R_{1,2}) = \sum_k \exp\left(N\mathcal{I}\left(\frac{k}{N}\right)\right) \operatorname{card}\left\{(\boldsymbol{\sigma}^1,\boldsymbol{\sigma}^2); R_{1,2} = \frac{k}{N}\right\}$$
$$= 2^N \sum_k \exp\left(N\mathcal{I}\left(\frac{k}{N}\right)\right) \operatorname{card}\left\{\boldsymbol{\sigma}; \sum_{i\le N} \sigma_i = k\right\}$$
$$\le (2N+1)2^{2N} \tag{2.519}$$

and together with (2.518) this proves (2.517). Using (2.18), we get

$$\mathsf{P}\left(Z_N \geq \frac{1}{2}\,\mathsf{E}Z_N\right) \geq \frac{1}{4(2N+1)}$$

and thus using (2.517) we have $\mathsf{P}(B) \geq 1/LN$, where

$$B = \left\{Z_N \geq \frac{1}{2}\,\mathsf{E}Z_N; Z_N^2\left\langle\exp\frac{N}{12}\,R_{1,2}^4\right\rangle \leq LN^2(\mathsf{E}Z_N)^2\right\}\,.$$

On B, we have

$$\exp\frac{N}{12}\,\langle R_{1,2}^4\rangle \leq \left\langle\exp\frac{N}{12}\,R_{1,2}^4\right\rangle \leq LN^2$$

and thus

$$\langle R_{1,2}^2\rangle \leq \langle R_{1,2}^4\rangle^{1/2} \leq L\sqrt{\frac{\log N}{N}}\,,$$

so that, rather than (2.43) we get

$$\log Z_N(\mathbf{g}) \geq \log\mathsf{E}Z_N - L(1 + d(\mathbf{g}, B)(N\log N)^{1/4})\,. \tag{2.520}$$

Since $\mathsf{P}(B) \geq 1/LN$, by (2.38) we have, for $t > 0$

$$\mathsf{P}(d(\mathbf{g}, B) \geq t + L\sqrt{\log N}) \leq 2\exp\left(-\frac{t^2}{4}\right)$$

and combining with (2.520) we get that, if $u \geq LN^{1/4}(\log N)^{3/4}$, we have

$$\mathsf{P}(Z_N \leq e^{-u}\,\mathsf{E}Z_N) \leq \exp\left(-\frac{u^2}{L\sqrt{N}\log N}\right)\,. \tag{2.521}$$

We have

$$G_N^{\otimes 2}(\{R_{1,2}^2 \geq x\}) \leq \exp\left(-\frac{Nx^2}{12}\right)\left\langle\exp\frac{N}{12}\,R_{1,2}^4\right\rangle$$

so that

$$\begin{aligned}\mathsf{E}\,G_N^{\otimes 2}(\{R_{1,2}^2 \geq x\}) &\leq \exp\left(-\frac{Nx^2}{24}\right)\\ &\quad + \mathsf{P}\left(\left\langle\exp\frac{N}{12}\,R_{1,2}^4\right\rangle \geq \exp\left(\frac{Nx^2}{24}\right)\right)\,.\end{aligned} \tag{2.522}$$

Writing

$$\left\langle\exp\frac{N}{12}\,R_{1,2}^4\right\rangle = \frac{(\mathsf{E}Z_N)^2}{Z_N^2}\,\frac{Z_N^2}{(\mathsf{E}Z_N)^2}\left\langle\exp\frac{N}{12}\,R_{1,2}^4\right\rangle\,,$$

we obtain

$$\mathsf{P}\left(\left\langle \exp \frac{N}{12} R_{1,2}^4 \right\rangle \ge e^{2u}\right) \le \mathsf{P}\left(\frac{(\mathsf{E}Z_N)^2}{Z_N^2} \ge e^u\right) + \mathsf{P}\left(\frac{Z_N^2}{(\mathsf{E}Z_N)^2}\left\langle \exp \frac{N}{12} R_{1,2}^4 \right\rangle \ge e^u\right) .$$

Using (2.517) and (2.521), if $u \ge LN^{1/4}(\log N)^{3/4}$, we see that this is at most

$$\exp\left(-\frac{u^2}{L\sqrt{N}\log N}\right) + LNe^{-u} .$$

Using this for $u = Nx^2/48$, and combining with (2.522) we obtain (2.516). □

Proposition 2.14.6. *There exists a number L_0 with the following property. Given any integer k_0, we can find a number $K(k_0)$ such that*

$$\beta \le 1 - K(k_0)\left(\frac{\log N}{N}\right)^{3/8} \Rightarrow \forall\, k \le k_0\,,\ \nu(R_{1,2}^{2k}) \le \left(\frac{L_0 k}{N(1-\beta^2)}\right)^k . \tag{2.523}$$

The following argument contains an essentially new feature: the signs of certain terms will be taken into account, because we do not know how to properly control their absolute values.

Lemma 2.14.7. *We have*

$$\nu((R_{1,2}^-)^k\, R_{2,3}^-\, R_{1,3}^-) \le 0 . \tag{2.524}$$

Proof. We expand $(R_{1,2}^-)^k$ as a sum of terms $\sigma_{i_1}^1 \ldots \sigma_{i_k}^1\, \sigma_{i_1}^2 \ldots \sigma_{i_k}^2$, each of which is of the type $f(\boldsymbol{\sigma}^1)\, f(\boldsymbol{\sigma}^2)$. Now

$$\langle f(\boldsymbol{\sigma}^1)\, R_{1,3}^-\, f(\boldsymbol{\sigma}^2)\, R_{2,3}^-\rangle = \int \left(\int f(\boldsymbol{\sigma}^1)\, R^-(\boldsymbol{\sigma}^1, \boldsymbol{\sigma}^2)\, dG_N(\boldsymbol{\sigma}^1)\right)^2 dG_N(\boldsymbol{\sigma}^2) \ge 0 .$$

□

Here is another simple fact

Lemma 2.14.8. *We have*

$$\nu((R_{1,2}^-)^k) \le \nu(R_{1,2}^k) . \tag{2.525}$$

Proof. Expanding the power, we have

$$\nu((R^-_{1,2})^k) = N^{-k}\sum \nu(\sigma^1_{i_1}\,\sigma^2_{i_1}\cdots\sigma^1_{i_k}\,\sigma^2_{i_k}) = N^{-k}\sum \mathsf{E}\langle\sigma_{i_1}\cdots\sigma_{i_k}\rangle^2\,,$$

where the sum is over all choices $i_1,\ldots,i_k \le N-1$. And $\nu(R^k_{1,2})$ is given by the same expression, with the sum over all choices $i_1,\ldots,i_k \le N$. □

Proof of Proposition 2.14.6. As induction hypothesis we assume that

$$\forall \ell \le k\,,\ \nu(R^{2\ell}_{1,2}) \le \left(\frac{L_0\,\ell}{N(1-\beta^2)}\right)^{\ell}\,. \tag{2.526}$$

Using symmetry between sites, we have

$$\nu(R^{2k+2}_{1,2}) = \nu(\varepsilon_1\,\varepsilon_2\,R^{2k+1}_{1,2})\,. \tag{2.527}$$

Consider then $f = \varepsilon_1\,\varepsilon_2\,(R^-_{1,2})^{2k+1}$. Using the inequality $|x^{2k+1}-y^{2k+1}| \le 2(k+1)|x-y|(x^{2k}+y^{2k})$, we see that, if $f = \varepsilon_1\,\varepsilon_2\,(R^-_{1,2})^{2k+1}$, then

$$\begin{aligned}|\nu(\varepsilon_1\,\varepsilon_2\,R^{2k+1}_{1,2}) - \nu(f)| &\le \frac{2(k+1)}{N}\,\nu((R_{1,2})^{2k} + (R^-_{1,2})^{2k})\\ &\le \frac{4(k+1)}{N}\left(\frac{L_0 k}{N(1-\beta^2)}\right)^{k}\end{aligned} \tag{2.528}$$

using (2.526) and (2.525). Now

$$\nu(f) \le \nu_0(f) + \nu_0'(f) + \frac{1}{2}\,\nu_0''(f) + \sup_{0\le t\le 1}\nu_t^{(3)}(f)\,. \tag{2.529}$$

We recall that

$$\begin{aligned}\nu_t'(f) = \beta^2\big(\nu_t((R^-_{1,2})^{2k+2}) &- 4\nu_t(\varepsilon_1\,\varepsilon_3\,R^-_{2,3}(R^-_{1,2})^{2k+1})\\ &+ 3\nu_t(\varepsilon_1\,\varepsilon_2\,\varepsilon_3\,\varepsilon_4\,R^-_{1,4}(R^-_{1,2})^{2k+1})\big)\end{aligned} \tag{2.530}$$

(so that $\nu_0'(f) = \beta^2\nu_0((R^-_{1,2})^{2k+2})$).

Taking the derivative again shows that

$$\nu_0''(f) = -4\,\beta^4\,\nu_0(R^-_{1,3}\,R^-_{2,3}\,(R^-_{1,2})^{2k+1}) \le 0$$

by Lemma 2.14.7. Thus,

$$\nu(f) \le \beta^2\,\nu_0((R^-_{1,2})^{2k+2}) + \sup_{0<t\le 1}\nu_t^{(3)}(f)\,. \tag{2.531}$$

Now, by (2.82), $|\nu_t^{(3)}(f)|$ is bounded by a sum of terms

$$\nu_t((R^-_{1,2})^{2k+1}\,|R_{\ell_1,\ell_2}|\,|R_{\ell_3,\ell_4}|\,|R_{\ell_5,\ell_6}|)\,.$$

Also, since $|R_{\ell,\ell'}| \le 1$, by (2.82) we have $\nu_t(f') \le L\nu(f')$ if f' is a non-negative function on Σ_N^8. Thus $|\nu_t^{(3)}(f)|$ is bounded by a sum of terms

$$L\,\nu((R_{1,2}^-)^{2k+1}\,|R_{\ell_1,\ell_2}|\,|\,|R_{\ell_3,\ell_4}|\,|R_{\ell_5,\ell_6}|)\ . \tag{2.532}$$

Consider x as in (2.516) and let us write

$$C(x) = L\,N\exp\left(-\frac{Nx^2}{L}\right) + \exp\left(-\frac{N^{3/2}}{L\sqrt{\log N}}\,x^4\right)\ .$$

Using (2.516) on the last two factors of (2.532), using Hölder's inequality and (2.525), we get

$$|\nu_t^{(3)}(f)| \le L\,x\,\nu(R_{1,2}^{2k+2}) + L\,C(x)\ . \tag{2.533}$$

We now write

$$|\nu((R_{1,2}^-)^{2k+2}) - \nu_0((R_{1,2}^-)^{2k+2}) - \nu_0'((R_{1,2}^-)^{2k+2}))| \le \sup_{0<t<1} \nu_t''((R_{1,2}^-)^{2k+2})\ .$$

Since $\nu_0'((R_{1,2}^-)^{2k+2}) = 0$, we get as previously

$$\nu_0((R_{1,2}^-)^{2k+2}) \le \nu(R_{1,2}^{2k+2}) + L\,x\,\nu(R_{1,2}^{2k+2}) + L\,C(x)\ .$$

Combining with (2.529) and (2.533), we get

$$\nu(f) \le (\beta^2 + L\,x)\,\nu(R_{1,2}^{2k+2}) + L\,C(x)\ ,$$

and combining with (2.527) and (2.528) we get

$$(1-\beta^2 - L\,x)\,\nu(R_{1,2}^{2k+2}) \le \frac{4(k+1)}{N}\left(\frac{L_0 k}{N(1-\beta^2)}\right)^k + L\,C(x)\ . \tag{2.534}$$

It is a good idea to take $x = (1-\beta^2)/2L$. In order to make this possible, we should have $\beta^2 \le 1 - L((\log N)/N)^{3/8}$; moreover, if $\beta^2 \le 1 - K(k)((\log N)/N)^{3/8}$ for a $K(k)$ depending on k only, we have $L\,C(x) \le N^{-k+1}$, and (2.534) becomes

$$(1-\beta^2)\,\nu(R_{1,2}^{2k+2}) \le \frac{4(k+1)}{N}\left(\frac{L_0 k}{N(1-\beta)^2}\right)^k\ ,$$

which completes the induction. □

Proposition 2.14.6 is relevant to Theorem 2.14.1 because $3/8 > 1/3$. In fact, once we know (2.523) the proof of (2.512) is a straightforward adaptation of the proof of Theorem 2.3.9, if one interprets the quantities $O(k)$ as being quantities such that

$$|O(k)| \le \frac{K(k)}{(N(1-\beta_N^2))^{k/2}}\ .$$

For example, following the usual method one gets

$$\nu(R_{1,2}^2) = \frac{1}{N} + \beta^2 \nu(R_{1,2}^2) + O(3) , \tag{2.535}$$

and (2.511) is exactly what is required to ensure that $\lim_{N\to\infty} NO(3) = 0$, so that the term $O(3)$ is indeed an error term in (2.535), since it is of smaller order than $1/N$. The point of (2.513) is that this then stops to be true and that matters become much more complicated.

We now investigate the situation (2.513). Let $b_N = \nu(R_{1,2}^2)$, so that by Proposition 2.14.6 we have $b_N = O(N^{-2/3})$. (Here and until the end of this section, the notations O, o have their standard meaning.) Thus, roughly speaking, "each factor $R_{\ell,\ell'}$ counts as $N^{-1/3}$". Moreover, if we repeat the proof of the case $k = 1$ of Proposition 2.14.6 (using (2.520)) we get the relation

$$\nu(R_{1,2}^2) = \frac{1}{N} + \beta^2 \nu(R_{1,2}^2) - 2\beta^4 \nu(R_{1,2} R_{2,3} R_{1,3}) + O(N^{-4/3}) . \tag{2.536}$$

To prove that (2.512) does not hold, let us assume for contradiction that it does. Then

$$\nu(R_{1,2} R_{2,3} R_{1,3}) = o(b_N^{3/2}) = o(N^{-1}) \tag{2.537}$$

and (2.536) shows that

$$b_N = \frac{1}{c N^{2/3}} + o(N^{-2/3}) . \tag{2.538}$$

Now we try to compute $\nu(R_{1,2} R_{2,3} R_{1,3})$. As usual

$$\nu(R_{1,2} R_{2,3} R_{1,3}) = \nu(f)$$

where $f = \varepsilon_1 \varepsilon_2 R_{1,3} R_{2,3}$, and, as in (2.529),

$$\nu(f) = \nu_0(f) + \nu_0'(f) + \frac{1}{2}\nu_0''(f) + O(N^{-5/3}) . \tag{2.539}$$

It is easily seen that $\nu_0(f) = 1/N^2$. Also,

$$\begin{aligned}\nu_t'(f) = \beta^2 \big(&\nu_t(R_{1,2} R_{1,3} R_{2,3}) + 2\nu_t(\varepsilon_1\varepsilon_3 R_{1,3} R_{2,3}^2) \\ &- 6\nu_t(\varepsilon_1\varepsilon_4 R_{1,3} R_{2,3} R_{2,4}) - 3\nu_t(\varepsilon_1\varepsilon_2\varepsilon_3\varepsilon_4 R_{1,3} R_{2,3} R_{3,4}) \\ &+ 6\nu_t(\varepsilon_1\varepsilon_2\varepsilon_4\varepsilon_5 R_{1,3} R_{2,3} R_{4,5})\big) ,\end{aligned}$$

and, replacing $R_{\ell,\ell'}$ by $R_{\ell,\ell'}^- + \varepsilon_\ell \varepsilon_{\ell'}/N$, one sees that

$$\nu_0'(f) = \beta^2 \nu_0(R_{1,2} R_{1,3} R_{2,3}) + O(N^{-5/3}) .$$

Setting $f' = R_{1,2} R_{1,3} R_{2,3}$, we have

$$\nu(f') = \nu_0(f') + \nu_0'(f') + O(N^{-5/3}) .$$

Computation of $\nu_0'(f')$ (using again that $R_{\ell,\ell'} = R^-_{\ell,\ell'} + \varepsilon_\ell \varepsilon_{\ell'}/N$) shows that $\nu_0'(f') = O(N^{-5/3})$ and thus

$$\nu_0'(f) = \beta^2 \nu(R_{1,2} R_{1,3} R_{2,3}) + O(N^{-5/3}) .$$

We observe that

$$\nu_0''(f) = \beta^2 (2 \nu_0(R_{1,3}^2 R_{2,3}^2) - 6 \nu_0(R_{1,3} R_{1,4} R_{2,3} R_{2,4})) + O(N^{-5/3})$$

and, as previously,

$$\nu_0''(f) = 2\beta^2 (\nu(R_{1,3}^2 R_{2,3}^2) - 3 \nu(R_{1,3} R_{1,4} R_{2,3} R_{2,4})) + O(N^{-5/3}) .$$

Since $1 - \beta^2 = c N^{-1/3}$, by (2.537) and (2.539) we get

$$\nu(R_{1,3}^2 R_{2,3}^2) - 3 \nu(R_{1,3} R_{1,4} R_{2,3} R_{2,4}) = o(N^{-4/3}) . \tag{2.540}$$

Now, under (2.512), we have

$$\nu(R_{1,3} R_{1,4} R_{2,3} R_{2,4}) = o(b_N^{-4}) = o(N^{-4/3}) ,$$

so that (2.540) implies $\nu(R_{1,3}^2 R_{2,3}^2) = o(N^{-4/3})$. But then we cannot have simultaneously (2.538) and $\lim_{N\to\infty} b_N^{-2} \nu(R_{1,3}^2 R_{1,4}^2) = 1$. □

Let us now turn to the proof of Proposition 2.14.4

Lemma 2.14.9. *We have*

$$N(\nu(R_{1,2}^4) - 4\nu(R_{1,2}^2 R_{1,3}^2) + 3\nu(R_{1,2}^2 R_{3,4}^2)) = \frac{\partial}{\partial \beta} \nu(R_{1,2}^2) \le \frac{1}{\beta} . \tag{2.541}$$

Proof. We have

$$p_N'(\beta) = \frac{\beta}{2}(1 - \nu(R_{1,2}^2)) ,$$

and, since p_N is a convex function of β, we have

$$0 \le \frac{\mathrm{d}^2}{\mathrm{d}\beta^2} p_N(\beta) = \frac{1}{2}(1 - \nu(R_{1,2}^2)) - \frac{\beta}{2} \frac{\mathrm{d}}{\mathrm{d}\beta} \nu(R_{1,2}^2)$$

and thus

$$\frac{\mathrm{d}}{\mathrm{d}\beta} \nu(R_{1,2}^2) \le \frac{1}{\beta}(1 - \nu(R_{1,2}^2)) \le \frac{1}{\beta} .$$

The left-hand equality in (2.541) follows through computation using integration by parts. □

Lemma 2.14.10. *Let us write*

$$R_{1,2,3,4} = \frac{1}{N} \sum_{i \le N} \sigma_i^1 \sigma_i^2 \sigma_i^3 \sigma_i^4 .$$

Then if $\beta \le 1$, we have

$$\mathsf{E} G_N^{\otimes 2} \left(R_{1,2,3,4}^2 \ge L \left(\frac{\log N}{N} \right)^{3/8} \right) \le \frac{1}{N} . \tag{2.542}$$

Proof. The proof of (2.516) extends to the case of $R_{1,2,3,4}$ with essentially no modification. □

Proof of Proposition 2.14.4. We have $\nu(R_{1,2}^2) = \nu(f)$, where $f = \varepsilon_1 \varepsilon_2 R_{1,2}$, and, using (2.82) in the second line,

$$\nu(f) = \nu_0(f) + \int_0^1 \nu_t'(f)\,\mathrm{d}t\,,$$

$$\nu_t'(f) = \beta^2(\nu_t(R_{1,2}^2) - 4\,\nu_t(\varepsilon_2 \varepsilon_3 R_{1,2} R_{1,3}) + 3\,\nu_t(\varepsilon_1 \varepsilon_2 \varepsilon_3 \varepsilon_4 R_{1,2} R_{3,4}))\,.$$

Now,

$$\begin{aligned}\nu_t(\varepsilon_2 \varepsilon_3 R_{1,2} R_{1,3}) &= \mathsf{E}\,\langle \varepsilon_1 \varepsilon_2 R_{1,2} \varepsilon_1 \varepsilon_3 R_{1,3} \rangle_t \\ &= \mathsf{E} \int \left(\int \varepsilon_1 \varepsilon_2 R_{1,2}\, \mathrm{d}G_t(\boldsymbol{\sigma}^2) \right)^2 \mathrm{d}G_t(\boldsymbol{\sigma}^1) \\ &\geq \mathsf{E} \left(\int \int \varepsilon_1 \varepsilon_2 R_{1,2}\, \mathrm{d}G_t(\boldsymbol{\sigma}^1)\, \mathrm{d}G_t(\boldsymbol{\sigma}^2) \right)^2 \\ &= \nu_t(\varepsilon_1 \varepsilon_2 \varepsilon_3 \varepsilon_4 R_{1,2} R_{3,4}) \geq 0\end{aligned}$$

and thus $\nu_t'(f) \leq \beta^2\,\nu_t(R_{1,2}^2)$, so that

$$\nu(R_{1,2}^2) \leq \frac{1}{N} + \beta^2 \int_0^1 \varphi'(t)\,\mathrm{d}t\,, \tag{2.543}$$

where $\varphi(t) = \nu_t(R_{1,2}^2)$. Using integration by parts,

$$\begin{aligned}\nu(R_{1,2}^2) &\leq \frac{1}{N} + \beta^2 \left(\varphi(1) - \int_0^1 t\,\varphi'(t)\,\mathrm{d}t \right) \\ &\leq \frac{1}{N} + \beta^2 \left(\varphi(1) - \frac{1}{2}\varphi'(1) + \int_0^1 \frac{t^2}{2} \varphi''(t)\,\mathrm{d}t \right) \qquad (2.544) \\ &\leq \frac{1}{N} + \beta^2 \left(\varphi(1) - \frac{1}{2}\varphi'(1) + \frac{1}{6}\varphi''(1) - \int_0^1 \frac{t^3}{6} \varphi'''(t)\,\mathrm{d}t \right)\,.\end{aligned}$$

Now, according to (2.516), each factor $R_{\ell,\ell'}$ "counts as $((\log N)/N)^{3/16}$" so that

$$|\varphi'''(t)| \leq L \left(\frac{\log N}{N} \right)^{9/16} \nu(R_{1,2}^2) + \frac{L}{N}\,.$$

Also,

$$\varphi'(t) = \beta^2(\nu_t(\varepsilon_1 \varepsilon_2 R_{1,2}^3) - 4\,\nu_t(\varepsilon_1 \varepsilon_3 R_{1,2}^2 R_{1,3}) + 3\,\nu_t(\varepsilon_3 \varepsilon_4 R_{1,2}^2 R_{3,4}))$$

so that, using symmetry between sites,

$$\varphi'(1) = \frac{\beta^2}{N} (\nu(R_{1,2}^4) - 4\,\nu(R_{1,2}^2 R_{1,3}^2) + 3\,\nu(R_{1,2}^2 R_{3,4}^2))\,. \tag{2.545}$$

Moreover, using again symmetry between sites, we see that

$$\varphi''(1) = \beta^4(\nu(R_{1,2}^4) - 4\,\nu(R_{1,2}^2\,R_{1,3}^2) + 3\,\nu(R_{1,2}^2\,R_{3,4}^2)) + \mathrm{I}\,, \tag{2.546}$$

where I is a sum of terms of the type $\beta^4\,\nu(f')$, f' being a product of 5 terms of the type $R_{\ell,\ell'}$, or of 4 terms of the type $R_{\ell,\ell'}$ and 1 of the type $R_{\ell_1,\ell_2,\ell_3,\ell_4}$. Thus, using (2.542) and (2.516) we get

$$|\mathrm{I}| \le L\left(\frac{\log N}{N}\right)^{9/16} \nu(R_{1,2}^2)$$

and from (2.544) (2.545) and (2.546) we see that

$$\begin{aligned}\nu(R_{1,2}^2) \le \frac{1}{N} &+ \left(\beta^2 + L\left(\frac{\log N}{N}\right)^{9/16}\right)\nu(R_{1,2}^2)\\ &+ \left(\frac{\beta^4}{6} - \frac{\beta^2}{2N}\right)(\nu(R_{1,2}^4) - 4\,\nu(R_{1,2}^2\,R_{1,3}^2) + 3\,\nu(R_{1,2}^2\,R_{3,4}^2))\,.\end{aligned}$$

We can assume $\beta^4/6 - \beta^2/2N \ge 0$, so that (2.541) now yields

$$\nu(R_{1,2}^2) \le \frac{L}{N} + \left(\beta^2 + L\left(\frac{\log N}{N}\right)^{9/16}\right)\nu(R_{1,2}^2)\,.$$

It follows that if $\beta^2 \le 1 - 1/\sqrt{N}$, then $\nu(R_{1,2}^2) \le L/\sqrt{N}$. Finally (2.541) shows that this is still true for $\beta = 1$. □

2.15 The Physicist's Replica Method

At the root of this method is the following perfectly rigorous result.

Theorem 2.15.1. *Consider an integer $n \ge 1$. Then*

$$\begin{aligned}\lim_{N\to\infty} \frac{1}{N}\log \mathsf{E}\, Z_N^n(\beta,h) &= n\log 2 \\ &+ \max_q\left(\frac{n\beta^2}{4}(1-q)^2 - \frac{n^2\beta^2}{4}q^2 + \log \mathsf{E}\,\mathrm{ch}^n(\beta z\sqrt{q} + h)\right)\end{aligned} \tag{2.547}$$

where z is standard normal.

To prove this, proceeding as in (2.51), we have

$$\begin{aligned}\mathsf{E}\, Z_N^n &= \left(\exp\frac{\beta^2}{4}n(N-n)\right)\\ &\quad\sum_{\boldsymbol{\sigma}} \exp\left(\frac{\beta^2}{2N}\sum_{1\le\ell<\ell'\le n}(\boldsymbol{\sigma}^\ell\cdot\boldsymbol{\sigma}^{\ell'})^2 + h\sum_{\ell\le n, i\le N}\sigma_i^\ell\right)\,,\end{aligned} \tag{2.548}$$

where $\sum_{\boldsymbol{\sigma}}$ means that the summation is over $\boldsymbol{\sigma}^1, \ldots, \boldsymbol{\sigma}^n$ in Σ_N. Consider now i.i.d. Gaussian r.v. $(g_{\ell,\ell'})$ with $\mathsf{E}\, g_{\ell,\ell'}^2 = 1/N$. It follows from (A.36) that

$$\begin{aligned}
&\sum_{\boldsymbol{\sigma}} \exp\left(\frac{\beta^2}{2N} \sum_{1\le\ell<\ell'\le n} (\boldsymbol{\sigma}^\ell \cdot \boldsymbol{\sigma}^{\ell'})^2 + h \sum_{\ell\le n, i\le N} \sigma_i^\ell\right) \\
&= \mathsf{E} \sum_{\boldsymbol{\sigma}} \exp\left(\beta \sum_{1\le\ell<\ell'\le n} g_{\ell,\ell'}\, \boldsymbol{\sigma}^\ell \cdot \boldsymbol{\sigma}^{\ell'} + h \sum_{\ell\le n, i\le N} \sigma_i^\ell\right) \\
&= \mathsf{E}\left(\sum \exp\left(\beta \sum_{1\le\ell<\ell'\le n} g_{\ell,\ell'}\, \varepsilon_\ell\, \varepsilon_{\ell'} + h \sum_{\ell\le n} \varepsilon_\ell\right)\right)^N \\
&= \mathsf{E} \exp N A(\mathbf{g}) \ ,
\end{aligned}$$

where

$$A(\mathbf{g}) = \log \sum \exp\left(\beta \sum_{1\le\ell<\ell'\le n} g_{\ell,\ell'}\, \varepsilon_\ell\, \varepsilon_{\ell'} + h \sum_{\ell\le n} \varepsilon_\ell\right) .$$

Here and above, the summation is over $\varepsilon_\ell = \pm 1$, $\ell \le n$. Now,

$$\mathsf{E} \exp N A(g) = \left(\frac{N}{2\pi}\right)^{n(n-1)/2} \int \exp N \left(A(\mathbf{g}) - \frac{1}{2} \sum_{1\le\ell<\ell'\le n} g_{\ell,\ell'}^2\right) d\mathbf{g}$$

where the integral is taken with respect to the Lebesgue measure on $\mathbb{R}^{n(n-1)/2}$. Since $|A(\mathbf{g})| \le K(\|\mathbf{g}\| + 1)$, it is elementary to see that

$$\lim_{N\to\infty} \frac{1}{N} \log \mathsf{E} \exp N A(\mathbf{g}) = \max_{\mathbf{g}} \left(A(\mathbf{g}) - \frac{1}{2} \sum_{\ell<\ell'} g_{\ell,\ell'}^2\right)$$

and from (2.548) we get

$$\lim_{N\to\infty} \frac{1}{N} \log \mathsf{E}\, Z_N^n = \frac{\beta^2 n}{4} + \max_{\mathbf{g}} \left(A(\mathbf{g}) - \frac{1}{2} \sum_{\ell<\ell'} g_{\ell,\ell'}^2\right) . \tag{2.549}$$

Lemma 2.15.2. *The maximum of $(A(\mathbf{g}) - \sum_{\ell<\ell'} g_{\ell,\ell'}^2/2)$ is obtained at a point $\mathbf{g}$ where the coordinates $g_{\ell,\ell'}$ are all equal.*

Proof. It suffices to prove the corresponding statement for

$$B(\mathbf{g}) = \sum \exp\left(\beta \sum_{\ell<\ell'} g_{\ell,\ell'}\, \varepsilon_\ell\, \varepsilon_{\ell'} + h \sum_{\ell\le n} \varepsilon_\ell - \frac{1}{2} \sum_{\ell<\ell'} g_{\ell,\ell'}^2\right) .$$

The basic observation is that the 2×2 matrix $(\exp \beta\, g_{1,2}\, \varepsilon_1\, \varepsilon_2)$ for $\varepsilon_1, \varepsilon_2 = \pm 1$ is positive semi-definite, so that

$$\sum_{\varepsilon_1,\varepsilon_2=\pm 1} U(\varepsilon_1)\, V(\varepsilon_2) \exp \beta\, g_{1,2}\, \varepsilon_1\, \varepsilon_2$$
$$\leq \left(\sum_{\varepsilon_1,\varepsilon_2=\pm 1} U(\varepsilon_1)\, U(\varepsilon_2) \exp \beta\, g_{1,2}\, \varepsilon_1\, \varepsilon_2 \right)^{1/2}$$
$$\left(\sum_{\varepsilon_1,\varepsilon_2=\pm 1} V(\varepsilon_1)\, V(\varepsilon_2) \exp \beta\, g_{1,2}\, \varepsilon_1\, \varepsilon_2 \right)^{1/2} .$$

Fixing $\varepsilon_3, \ldots, \varepsilon_n$, we use this for

$$U(\varepsilon) = \exp \left(\varepsilon \left(\beta \sum_{3\leq \ell\leq n} g_{1,\ell}\, \varepsilon_\ell + h \right) - \frac{1}{2} \sum_{3\leq \ell\leq n} g_{1,\ell}^2 \right) ,$$

$$V(\varepsilon) = \exp \left(\varepsilon \left(\beta \sum_{3\leq \ell\leq n} g_{2,\ell}\, \varepsilon_\ell + h \right) - \frac{1}{2} \sum_{3\leq \ell\leq n} g_{2,\ell}^2 \right) .$$

We multiply the resulting inequality by

$$\exp \left(\beta \sum_{3\leq \ell<\ell'\leq n} g_{\ell,\ell'}\, \varepsilon_\ell\, \varepsilon_{\ell'} + h \sum_{3\leq \ell\leq n} \varepsilon_\ell - \frac{1}{2} \sum_{3\leq \ell<\ell'\leq n} g_{\ell,\ell'}^2 - \frac{1}{2} g_{1,2}^2 \right) .$$

Summing over $\varepsilon_3, \ldots, \varepsilon_n$, and, using the Cauchy-Schwarz inequality in the right-hand side we then get

$$B(\mathbf{g}) \leq B(\mathbf{g}')^{1/2}\, B(\mathbf{g}'')^{1/2}$$

where $\mathbf{g}'$ (resp. $\mathbf{g}''$) is obtained from $\mathbf{g}$ by replacing $g_{3,\ell}$ by $g_{2,\ell}$ (resp. $g_{2,\ell}$ by $g_{3,\ell}$). Thus, given $\mathbf{g}$, we can find $\mathbf{g}'$ such that $B(\mathbf{g}) \leq B(\mathbf{g}')$ and $g_{1,\ell} = g_{2,\ell}$ for $\ell \geq 3$. Iteration of this procedure yields the result. □

Thus, in (2.549) we can restrict the max to the case where $g_{\ell,\ell'} = \beta q$ for each $1 \leq \ell < \ell' \leq n$. We have

$$\sum \exp \left(\beta^2 q \sum_{\ell<\ell'} \varepsilon_\ell\, \varepsilon_{\ell'} + h \sum_{\ell\leq n} \varepsilon_\ell \right)$$
$$= \sum \exp \left(\frac{\beta^2 q}{2} \left(\sum_{\ell\leq n} \varepsilon_\ell \right)^2 - \frac{n\beta^2 q}{2} + h \sum_{\ell\leq n} \varepsilon_\ell \right)$$

$$= \sum \mathsf{E}\exp\left((\beta z\sqrt{q}+h)\sum_{\ell\le n}\varepsilon_\ell - \frac{n\beta^2 q}{2}\right)$$

$$= \exp\left(-\frac{n\beta^2 q}{2}\right)\,\mathsf{E}(2\operatorname{ch}(\beta z\sqrt{q}+h))^n\ ,$$

where z is standard Gaussian and where the summations are over $\varepsilon_1,\ldots,\varepsilon_n = \pm 1$. Thus

$$\max_{\mathbf{g}}\left(A(\mathbf{g}) - \frac{1}{2}\sum_{\ell<\ell'} g^2_{\ell,\ell'}\right)$$

$$= \max_q\left(-\frac{n\beta^2 q}{2} + n\log 2 + \log\mathsf{E}\operatorname{ch}^2(\beta z\sqrt{q}+h) - \frac{n(n-1)}{4}\beta^2 q^2\right)\ .$$

Combining this with (2.549) proves (2.547). □

It is a simple computation to see that the maximum in (2.547) is obtained for a value q_n such that

$$q_n = \frac{\mathsf{E}(\operatorname{ch}^n Y\operatorname{th}^2 Y)}{\mathsf{E}\operatorname{ch}^n Y} \tag{2.550}$$

where $Y = \beta z\sqrt{q_n} + h$.

Let us also observe that

$$\lim_{t\to 0_+}\frac{1}{t}\log\mathsf{E}\,Z_N^t = \mathsf{E}\log Z_N\ , \tag{2.551}$$

as follows from the fact that $Z_N^t \simeq 1 + t\log Z_N$ for small t.

Now we take a deep breath. We pretend that Theorem 2.15.1 is true not only for n integer, but for any number $n > 0$. We rewrite (2.551) as

$$\frac{1}{N}\mathsf{E}\log Z_N = \lim_{n\to 0}\frac{1}{Nn}\log\mathsf{E}\,Z_N^n\ .$$

Let us moreover pretend that we can exchange the limits $N\to\infty$ and $n\to 0$. Then presumably $q = \lim_{n\to 0} q_n$ exists, and (2.547) yields,

$$\lim_{N\to\infty}\frac{1}{N}\mathsf{E}\log Z_N = \frac{\beta^2}{4}(1-q)^2 + \log 2 + \mathsf{E}\log\operatorname{ch} Y$$

where $Y = \beta z\sqrt{q}+h$ and (2.550) becomes $q = \mathsf{E}\operatorname{th}^2 Y$.

When trying to justify this procedure one is tempted to think about analytic continuation. However the information contained in Theorem 2.15.1 about the large values of n seems to be completely irrelevant to the problem at hand. To get convinced of this, one can consider the case where $h = 0$ and $\beta < 1$; then $\lim_{n\to\infty} q_n = 1$, and this is hard to relate to the fact that $q = 0$.

Research problem 2.15.3. (Level 1) Prove that there exists $\beta_0 > 0$ such that if $\beta \le \beta_0$, then (2.547) holds whenever $n \in \mathbb{R}$, $n > 0$, n not necessarily integer. (When $n < 1$, it might be needed to replace the $\max_q$ by $\min_q$.)

It is also of (limited) interest to investigate whether (2.547) is true for $n < 0$. The function

$$\varphi_N(t) = \frac{1}{N} \log \mathsf{E}\, Z_N^t$$

is a convex function of t (by Hölder's inequality). Thus, if one proves the existence of $\varphi(t) = \lim_{N\to\infty} \varphi_N(t)$ in an interval containing zero, and if φ is differentiable at zero, then one obtains that

$$\varphi'(0) = \lim_{N\to\infty} \varphi'_N(0)$$

which would prove that the exchange of limits $N \to \infty$ and $n \to 0$ above is indeed justified.

It is probably not very difficult to carry out the previous "a posteriori" justification of the replica method, but this has a limited interest.

At the present time, it is difficult to see in the physicist's replica method more than a way to guess the correct formula. Moreover, the search for arguments making this method legitimate "a priori" does not look to this author as a promising area of investigation.

2.16 Notes and Comments

Section 2.1 draws heavily on the introduction of the book of Mézard, Parisi, Virasoro [99] (the one single part of this book I feel I really understand).

Theorem 2.2.1 is proved by Aizenman, Lebowitz and Ruelle in [4]. This paper relies on a "cluster expansion technique", which is a common tool in physics. Their method does not seem to extend to the case $h \ne 0$. Among other results, this paper contains the fundamental relation (2.210), and proves central limit theorems for Z_N and $\log Z_N$. It also contains interesting information about $\mathsf{E} \max_{\boldsymbol{\sigma}}(-H_N(\boldsymbol{\sigma}))$. About the same time Fröhlich and Zegarlinski [57] prove (as a consequence of a more general approach that is also based on a cluster expansion technique) that the spin correlations vanish if $\beta \le \beta_0$, even if $h \ne 0$. In fact they prove that

$$\mathsf{E}((\langle \sigma_1 \sigma_2 \rangle - \langle \sigma_1 \rangle \langle \sigma_2 \rangle)^2) \le \frac{L}{N} \,. \tag{2.552}$$

A later paper by Comets and Neveu [46] provides more elegant proofs of several of the main results of [4] using stochastic calculus. They put forward the crucial relation (2.12). Their method unfortunately does not appear to extend beyond the case $h = 0$. They prove a central limit theorem for the overlap $R_{1,2}$; most probably their method would yield Theorem 2.3.9. The point of my more complicated approach is that it prepares the ground for the case $h \ne 0$.

Theorem 2.2.4 is a special occurrence of the general phenomenon of concentration of measure. This phenomenon was first discovered by P. Lévy, and its importance was brought to light largely through the efforts of V.D. Milman [100]. It is arguably one of the truly great ideas of probability theory. More references, and applications to probability theory can be found in [132] and [133]. In the fundamental case of Gaussian measure, the optimal result is already obtained in [83], and Theorem 2.2.4 is a weak consequence of this result. Interestingly, it took almost 20 years after the paper [83] before results similar to Corollary 2.2.5 were obtained in the theory of disordered systems, by Pastur and Shcherbina [113], using martingale difference sequences. A very nice exposition of most of what is known about concentration of measure can be found in the book of M. Ledoux [90]

It was not immediately understood that, while the case $\beta < 1$, $h = 0$ of the SK model is not very difficult, the case $h \neq 0$ is an entirely different matter. The first rigorous attempt at justifying the mysterious formula (2.144) is apparently that of Pastur and Shcherbina [113]. They prove that this formula holds in the domain where

$$\lim_{N\to\infty} \mathrm{Var}\,(\langle R_{1,2}\rangle) = \lim_{N\to\infty} \mathsf{E}((\langle R_{1,2}\rangle - \mathsf{E}\,\langle R_{1,2}\rangle)^2) = 0\ , \tag{2.553}$$

but they do *not* prove that (2.553) is true for small β. Their proof required them to add a strange perturbation term to the Hamiltonian. The result was later clarified by Shcherbina [122], who used the Hamiltonian (2.452). Using arguments somewhat similar to those of the Ghirlanda-Guerra identities, she proved that (2.553) is equivalent (over a certain domain) to

$$\lim_{N\to\infty} \mathsf{E}\langle (R_{1,2} - \langle R_{1,2}\rangle)^2\rangle = 0\ . \tag{2.554}$$

She did not prove (2.554). She was apparently unaware that (2.554) is proved in [57] for small β. Since the paper [122] was not published, I was not aware of it and rediscovered its results in Section 4 of [134] with essentially the same proof. I also give a very simple proof of (2.552) for small β. Discovering this simple proof was an absolute disaster, because I wasted considerable energy trying to use the same principle in other situations, which invariably led to difficult proofs of suboptimal results. I will not describe in detail the contents of [134] or my other papers because this now does not seem so interesting anymore. I hope that the proofs presented here are much cleaner than those of these previous papers.

In a later paper Shcherbina [123] proved (2.146) in a remarkably large region containing in particular all values $\beta < 1$. The ideas of this paper are not really transparent to me. A later version [124] is more accessible, but I became aware of its existence too late to have the energy to analyze it. It would be very interesting to decide if this approach succeeds because of a special trick, or if it contains the germ of a powerful method. One should however point out that her use of relations similar to the Ghirlanda-Guerra identities seems to preclude obtaining the correct rates of convergence.

An expansion somewhat similar to (2.168) was proved [144], using a more complicated method that does not seem to extend to the model of Chapter 3. This paper proves weaker versions of many of the results of Section 2.4 to Section 2.8. The existence of the limits of quantities such as $N^{k/2}\mathsf{E}\langle A\rangle$, where A is the product of k terms of the type $R_{\ell,\ell'}$ is proved by a recursion method very similar to the one used here, but the limit is not computed explicitly.

I do not know who first used the "smart path method". The proof of Proposition 2.4.3 is due to J.P. Kahane [84] and that of Theorem 2.2.4 is due to G. Pisier [118]. I had known these papers since they appeared, but it took a very, very long time to realize that it was the route to take here. It was first used in this context in [140], and then systematically in [153]. Interestingly, Guerra and Toninelli arrived independently at the very similar idea of interpolating between Hamiltonians as in Section 2.10.

The reader might wonder about the purpose of (2.169), since we always use (2.168) instead. The idea is that, using symmetry between sites, we can get a nice expression for $\nu_1'(f)$. This idea will be used in Chapter 6. We do not use it here, because, besides controlling the quantities $R_{1,2}$, it requires controlling $R_{1,2,3,4} = N^{-1}\sum_{i\le N}\sigma_i^1\sigma_i^2\sigma_i^3\sigma_i^4$. To give a specific example, if $f = R_{1,2} - q$, we get from (2.169) that

$$\begin{aligned}
\nu_1'((\varepsilon_1\varepsilon_2 - q)f) &= \beta^2\nu((1-\varepsilon_1\varepsilon_2 q)(R_{1,2}-q)f)\\
&\quad - 4\beta^2\nu((\varepsilon_2\varepsilon_3 - q\varepsilon_1\varepsilon_3)(R_{1,3}-q)f)\\
&\quad + 3\beta^2\nu((\varepsilon_1\varepsilon_2\varepsilon_3\varepsilon_4 - q\varepsilon_3\varepsilon_4)(R_{3,4}-q)f)\\
&= \beta^2\nu((1-R_{1,2}q)(R_{1,2}-q)f)\\
&\quad - 4\beta^2\nu((R_{2,3} - qR_{1,3})(R_{1,3}-q)f)\\
&\quad + 3\beta^2\nu((R_{1,2,3,4} - qR_{3,4})(R_{3,4}-q)f)\,.
\end{aligned}$$

If we know that $\nu(|R_{1,2}-q|^3)^{1/3} \le L/\sqrt{N}$ and $\nu(|R_{1,2,3,4}-\widehat{q}|^3)^{1/3} \le L/\sqrt{N}$, we get

$$\begin{aligned}
\nu_1'((\varepsilon_1\varepsilon_2 - q)f) &= \beta^2(1-q^2)\nu((R_{1,2}-q)^2)\\
&\quad - 4\beta^2(q-q^2)\nu((R_{1,2}-q)(R_{2,3}-q))\\
&\quad + 3\beta^2(\widehat{q}-q^2)\nu((R_{1,2}-q)(R_{3,4}-q)) + O(3)
\end{aligned}$$

a relation that we may combine with

$$\nu((R_{1,2}-q)^2) = \nu((\varepsilon_1\varepsilon_2-q)f) = \nu_0((\varepsilon_1\varepsilon_2-q)f) + \nu_1'((\varepsilon_1\varepsilon_2-q)f) + O(3)\,.$$

In this way we have fewer error terms to control in the course of proving the central limit theorems presented here. The drawback is that one must prove first that $\nu((R_{1,2,3,4}-\widehat{q})^{2n}) \le K/N^n$ (which is not very difficult).

In the first version of this work I asked as a (level 1) research problem the question of proving at high enough temperature a central limit theorem for $N^{-1}\log Z_N$. I had saved this question for years (see [153]) as it seemed a perfect entry level problem. Unfortunately, two months after this chapter was widely circulated (in a version that already contained the central limit theorems of Section 2.6), the paper [74] came out, offering a solution to this question (as well as a proof of results very similar to those of Section 2.6). Still, the reader who truly wants to master the techniques of this book (e.g. for the purpose of proving a similar CLT for other models) should complete the details of the (rather pretty) calculation we outline now. Setting $U = N^{-1}\log Z_N - \mathrm{SK}(\beta, h)$, we will show that for a certain number $A(\beta, h)$ we have for each integer ℓ

$$\mathsf{E}\, U^{2\ell-1} = O(2\ell)\,;\ \mathsf{E}\, U^{2\ell} = (2\ell-1)(2\ell-3)\cdots A^{\ell}(\beta,h) + O(2\ell+1)\,. \tag{2.555}$$

We fix k (and h) and we set $f_k(\beta) = \mathsf{E}\, U^k$. Differentiation in β and integration by parts give

$$\begin{aligned} f_k'(\beta) &= \frac{k\beta}{2}\,\mathsf{E}((q^2 - \langle R_{1,2}^2\rangle)\,U^{k-1}) \\ &\quad + \frac{k(k-1)\beta}{2N}\,\mathsf{E}(\langle R_{1,2}^2\rangle\,U^{k-2}) + O(k+1)\,. \end{aligned} \tag{2.556}$$

It follows from Corollary 2.2.5 that $\mathsf{E}\,|U|^{\ell} = O(\ell)$, so that, combining with (2.239), we get that

$$\mathsf{E}\,(\langle (R_{1,2}-q)^2\rangle\,U^{k-1}) = O(k+1)\,;\ N^{-1}\mathsf{E}\,(\langle |R_{1,2}-q|\rangle\,U^{k-2}) = O(k+1)\,. \tag{2.557}$$

Substituting $R_{1,2}^2 = (R_{1,2}-q)^2 + 2(R_{1,2}-q)q + q^2$ in (2.556) then yields

$$f_k'(\beta) = -k\beta q\,\mathsf{E}(\langle R_{1,2}-q\rangle\,U^{k-1}) + \frac{k(k-1)\beta q^2}{2N}\,\mathsf{E}(U^{k-2}) + O(k+1)\,. \tag{2.558}$$

The computation of $\mathsf{E}(\langle R_{1,2}-q\rangle\,U^{k-1})$ is really nice. We observe, with the notation of Section 2.4 that $Z_N = 2Z_{N-1}\langle \mathrm{ch}(g(\boldsymbol{\sigma})+h)\rangle_-$, and we define

$$U_t = \frac{1}{N}\log(2Z_{N-1}) + \frac{1}{N}\log\langle \mathrm{ch}(g_t(\boldsymbol{\sigma})+h)\rangle_- - \mathrm{SK}(\beta,h)\,,$$

$$\varphi(t) = \mathsf{E}(\langle \varepsilon_1\,\varepsilon_2 - q\rangle_t\,U_t^{k-1})\,.$$

Using symmetry between sites we observe that

$$\varphi(1) = \mathsf{E}(\langle R_{1,2}-q\rangle\,U^{k-1})\,. \tag{2.559}$$

Taking the derivative of $\varphi(t)$ and integrating by parts brings out a factor $R_{\ell,\ell'} - q$ in each term, so that, by (2.557)

$$\varphi(1) = \varphi(0) + \varphi'(0) + O(k+1)\,. \tag{2.560}$$

There is a slight difficulty due to the fact that U_0 depends on Y. The following computation shows how to get rid of this dependence. We have

$$
\begin{aligned}
\varphi(0) &= \mathsf{E}(\langle \varepsilon_1 \varepsilon_2 - q\rangle_0 U_0^{k-1}) \\
&= \mathsf{E}\left((\mathrm{th}^2 Y - q) \left(\frac{1}{N}\log(2Z_{N-1}) + \frac{1}{N}\log\mathrm{ch}\, Y - \mathrm{SK}(\beta,h)\right)^{k-1}\right) \\
&= \frac{k-1}{N}\mathsf{E}\left((\mathrm{th}^2 Y - q)\log\mathrm{ch}\, Y \left(\frac{1}{N}\log(2Z_{N-1}) - \mathrm{SK}(\beta,h)\right)^{k-2}\right) \\
&\quad + O(k+1) \\
&= \frac{k-1}{N}\mathsf{E}\left((\mathrm{th}^2 Y - q)\log\mathrm{ch}\, Y\right)\, \mathsf{E}\left(\frac{1}{N}\log(2Z_{N-1}) - \mathrm{SK}(\beta,h)\right)^{k-2} \\
&\quad + O(k+1) \\
&= \frac{k-1}{N}\mathsf{E}((\mathrm{th}^2 Y - q)\log\mathrm{ch}\, Y)\,\mathsf{E}\, U^{k-2} + O(k+1)\,. \qquad (2.561)
\end{aligned}
$$

The dependence of U_t on $g_t(\boldsymbol{\sigma})$ is of order $1/N$ and it gives only lower contribution in the computation of $\varphi'(0)$, and we have

$$
\begin{aligned}
\varphi'(0) = \beta^2\, &\mathsf{E}(\langle (\varepsilon_1\varepsilon_2 - q)\,\varepsilon_1\varepsilon_2(R_{1,2} - q)\rangle_0 U_0^{k-1}) \\
&- 4\beta^2\,\mathsf{E}(\langle (\varepsilon_1\varepsilon_2 - q)\,\varepsilon_1\varepsilon_3(R_{1,3} - q)\rangle_0 U_0^{k-1}) \\
&+ 3\beta^2\,\mathsf{E}(\langle (\varepsilon_1\varepsilon_2 - q)\,\varepsilon_3\varepsilon_4(R_{3,4} - q)\rangle_0 U_0^{k-1}) + O(k+1)\,. \qquad (2.562)
\end{aligned}
$$

Proceeding as in (2.561) we get from (2.562) that

$$
\varphi'(0) = \beta^2(1 - 4q + 3\hat{q})\,\mathsf{E}(\langle R_{1,2} - q\rangle_0 U_0^{k-1}) + O(k+1) \qquad (2.563)
$$

and thus

$$
\varphi'(0) = \beta^2(1 - 4q + 3\hat{q})\,\varphi(1) + O(k+1) \qquad (2.564)
$$

and combining with (2.558), (2.559) and (2.560) this gives

$$
f_k'(\beta) = \frac{k(k-1)}{2}\, a(\beta)\, f_{k-2}(\beta)
$$

where

$$
Na(\beta) = \beta q^2 - \frac{2\beta q}{1 - \beta^2(1 - 4q + 3\hat{q})}\,\mathsf{E}((\mathrm{th}^2 Y - q)\log\mathrm{ch}\, Y)\,. \qquad (2.565)
$$

Setting $A(\beta,h) = \int_0^\beta a(x)\,\mathrm{d}x$, (2.555) then follows by induction. The one result of [74] that this approach does not give is the expression

$$
NA(\beta,h) = \mathsf{E}((\log\mathrm{ch}\, Y)^2) - (\mathsf{E}\log\mathrm{ch}\, Y)^2 - \frac{q^2\beta^2}{2}\,. \qquad (2.566)
$$

It can of course be checked that the derivative of the right-hand side of (2.566) is the right-hand side of (2.565), but this could have been hard to guess.

I am grateful to M. Mézard for having explained to me the idea of coupling two copies of the SK model, and the discontinuity this should produce beyond the A-T line. This led to Theorem 2.8.5 and later proved a basic idea in Section 2.8.

I do not know who first got the idea of using that the derivative of $N^{-1}\log Z_N$ has small fluctuations. I think I learned about it in [113]. There are several papers using this idea [114], [134], [135], [68], and the Ghirlanda-Guerra identities are the current final form of it (although in some cases it is possible to get more out of the same idea [152]). The power of these identities is not yet fully explained.

Guerra's bound of Theorem 2.9.3 is proved in [71] where Proposition 2.4.7 can also be found. (This lemma was also proved independently by R. Latala.) An earlier version of Theorem 2.9.10 can be found in [149]. The result there seems weaker than Theorem 2.9.10, and the proof is more complicated. In some sense however, this proof is very beautiful, and I was almost sad to discover the more efficient approach presented here. The material of Section 2.10 comes from [75], Theorem 2.11.1 from [72], and Theorem 2.11.12 from [157].

The material of Section 2.13 suggests the following (loosely formulated) question. Consider a probability measure μ on $B_d = \{\sigma; \sum_{u \le d} \sigma_u^2 \le d\}$ and assume that the barycenter of μ is 0. Is there a simple characteristic of μ that allows us to predict (approximately) the value c_μ of β for which (2.485) ceases to be true? One would like to think that if μ is "very spread" then $c_\mu \approx 1$, while if μ is very concentrated, $c_\mu \approx 1/d$. Does this make sense?

Another remarkable paper of Guerra [67] proves that at $\beta = 1$, one has $\nu(R_{1,2}^2) \le L/N^{1/4}$, a result he obtained very early. This motivated Section 2.14. Section 2.15 is essentially copied from [77].

The present work should make self-apparent the amount of energy already spent in trying to reach a mathematical understanding of mean field models related to spin glasses. Probably the most important idea as of today is the marvelous work of F. Guerra presented in Theorem 2.11.1. This approach in some sense breaks down the study of the low-temperature phase in a series of high-temperature problems, the key to Theorem 2.11.19. I feel confident now that a detailed understanding of the low-temperature behavior of the SK model is just a few years of hard work away, something that could certainly not be said when the present project started. The beauty and the breadth of the speculation about this behavior can be sampled in [99] and [109]. For a more mathematical paper related to the so called Parisi solution, one can also see [12], but this paper says nothing about the SK model itself.

It unfortunate that the methods of Sections 2.11 rely on very specific properties of the SK model. (See [154] for a wider class of models to which one can extend these methods and in particular Theorem 2.11.19.) This model, however fascinating and canonical, is a specific object, and as such its importance can be questioned. I feel that the appeal of the "theory" of spin glasses does not lie in any particular model, but rather in the apparent generality of the phenomenon it predicts. About this, we understand very little, which explains the title of the book.

3. The Capacity of the Perceptron: The Ising Case

3.1 Introduction

The name of this chapter comes from the theory of neural networks. What these are is not relevant to our purpose, which is to study the underlying mathematics. (An accessible introduction to neural networks is provided in [79].) Roughly speaking, the basic problem is as follows. What "proportion" of Σ_N is left when one intersects this set with many random half-spaces? A natural definition for a random half-space is a set $\{\mathbf{x} \in \mathbb{R}^N; \mathbf{x} \cdot \mathbf{v} \geq 0\}$ where the random vector $\mathbf{v}$ is uniform over the unit sphere of $\mathbb{R}^N$. More conveniently, and equivalently, one can consider the set $\{\mathbf{x} \in \mathbb{R}^N; \mathbf{x} \cdot \mathbf{g} \geq 0\}$, where $\mathbf{g}$ is a standard Gaussian vector, i.e. $\mathbf{g} = (g_i)_{i \leq N}$, where g_i are independent standard Gaussian. (Another natural definition will be considered in Section 3.8.) Consider now M such Gaussian vectors $\mathbf{g}_k$, all independent, the half-spaces

$$U_k = \{\mathbf{x};\ \mathbf{x} \cdot \mathbf{g}_k \geq 0\}$$

and

$$\Sigma_N \cap \bigcap_{k \leq M} U_k\ . \tag{3.1}$$

A given point of Σ_N has exactly a 50% chance to belong to U_k, so that

$$\mathsf{E}\,\mathrm{card}\left(\Sigma_N \cap \bigcap_{k \leq M} U_k\right) = 2^{N-M}\ . \tag{3.2}$$

The case of interest is when N becomes large and M is a proportion of N, $M/N \to \alpha > 0$. A consequence of (3.2) is that if $\alpha > 1$ the set (3.1) is typically empty when N is large . What is interesting is *not* however the expected value of the cardinality of the set (3.1), but rather the typical value of this cardinality, that is likely to be smaller.

Research problem 3.1.1. (Level 3!) Prove that there exists a number α^* and a function $\varphi : [0, \alpha^*) \to \mathbb{R}$ with the following properties

If $\alpha > \alpha^*$ then as $N \to \infty$ and $M/N \to \alpha$ the probability that the set

$$\text{(3.1) is not empty is at most } \exp(-N/K(\alpha))\,. \tag{3.3}$$

If $\alpha < \alpha^*$, $N \to \infty, M/N \to \alpha$, then

$$\frac{1}{N}\log\operatorname{card}\left(\Sigma_N \cap \bigcap_{k\le M} U_k\right) \to \varphi(\alpha) \tag{3.4}$$

in probability. Compute α^* and φ.

This problem is a typical example of a situation where one expects "regularity" as $N \to \infty$, but where it is unclear how to even start doing anything relevant. One of the main results of this chapter will be the proof of (3.4) when α is small, and the computation of $\varphi(\alpha)$ in that case. (We expect that the case of larger α is much more difficult.) As a corollary, we will prove that there exists a number $\alpha_0 < 1$ such that if $M = \lfloor \alpha N \rfloor$, $\alpha > \alpha_0$, then the set (3.1) is typically empty for N large.

One way to approach the (very difficult) problem mentioned above is to introduce a version "with a temperature". Rather than looking for configurations contained in each set U_k, we count for each configuration how many sets U_k contain it, and we weigh the configurations according to this number. In other words, we consider the Hamiltonian

$$-H_{N,M}(\boldsymbol{\sigma}) = \beta\operatorname{card}\{k \le M; \boldsymbol{\sigma} \in U_k\} = \beta\sum_{k\le M} \mathbf{1}_{\{\boldsymbol{\sigma}\in U_k\}}\,. \tag{3.5}$$

Since it changes nothing to add a constant to the Hamiltonian, rather than (3.5) we could also use the Hamiltonian

$$-H_{N,M}(\boldsymbol{\sigma}) = -\beta\sum_{k\le M} \mathbf{1}_{\{\boldsymbol{\sigma}\in U_k^c\}}\,. \tag{3.6}$$

One then hopes that, for large β, Gibbs' measure will bring some information on the set (3.1). (In principle, it could be much more rare that a configuration belongs to *all* the sets U_k than that it belongs to, say, 99% of them; but one hopes that such a pathology will not occur.)

More generally, we will consider Hamiltonians of the type

$$-H_{N,M}(\boldsymbol{\sigma}) = \sum_{k\le M} u\left(\frac{1}{\sqrt{N}}\sum_{i\le N} g_{i,k}\,\sigma_i\right)\,, \tag{3.7}$$

where u is a function, and where $(g_{i,k})$ are independent standard normal. The role of the factor $N^{-1/2}$ is to make the quantity $N^{-1/2}\sum_{i\le N} g_{i,k}\,\sigma_i$ typically of order 1. There is no parameter β in the right-hand side of (3.7), since this parameter can be thought of as being included in the function u.

We will assume that u is piecewise continuous. (Borel measurable suffices.) In order to attack the proof of (3.4), even for small α, it will be required to consider the Hamiltonian (3.5), or better, (3.6), for very large β. A natural condition on u will be that u is bounded above, but not necessarily below. This however creates serious difficulties. Thus, in a first stage (up to Section 3.9) we will assume that u is bounded,

$$\forall x\,,\ |u(x)| \le D\,. \tag{3.8}$$

This hypothesis is convenient, although one could do with less. Probably a condition

$$\forall x\,,\ |u(x)| \le D(1+|x|^\gamma)$$

where $\gamma < 2$ would yield the same results. But the case $u(x) = \beta\, x^2$, that is the object of Chapter 5, is fundamentally different.

The purpose of the chapter up to section 3.7 is to reach a very good understanding of the Gibbs' measure governed by (3.7) provided, for some large enough number L, we have

$$L\alpha\, e^{LD} \le 1\,. \tag{3.9}$$

This corresponds to the condition $\beta \le \beta_0$ of Chapter 2. In this chapter, when we write "if $L\alpha\, e^{LD} \le 1$" this of course means "there exists a number L such that, if $L\alpha\, e^{LD} \le 1\ldots$".

In Sections 3.9 to 3.11, we will then undertake the further effort of dealing with the case where u is only assumed to be bounded above, in order to prove (3.4) for small α.

Let us also point out that Section 3.8 introduces a simple technique that is required reading for Chapter 5

3.2 The Replica-Symmetric Equations and the Smart Path

We consider two independent standard Gaussian r.v. z, ξ. Given $x \in \mathbb{R}$, $y > 0$ we consider

$$\Phi(x,y) = \frac{\mathsf{E}\,\xi \exp u(x+\xi y)}{y\,\mathsf{E}\exp u(x+\xi y)}\,. \tag{3.10}$$

When u is differentiable, integration by parts of the numerator yields

$$\Phi(x,y) = \frac{\mathsf{E}\,u'(x+\xi y)\exp u(x+\xi y)}{\mathsf{E}\exp u(x+\xi y)}\,, \tag{3.11}$$

a slightly less mysterious formula.

We consider the system of equations

$$q = \mathsf{E}\,\mathrm{th}^2(z\sqrt{r}) \tag{3.12}$$

$$r = \alpha \mathsf{E} \Phi^2(z\sqrt{q}, \sqrt{1-q}) \,. \tag{3.13}$$

The mystery behind these equations will only gradually disappear. Our first goal is very modest.

Proposition 3.2.1. *Under (3.9) the equations (3.12) and (3.13) have a solution.*

Proof. If we substitute (3.12) into (3.13) we obtain an equation of the type $r = \alpha f(r)$. We will find a number r_0 such that

$$|r| \le r_0 \Rightarrow \alpha f(r) \le r_0 \,, \; \alpha f'(r) \le \frac{1}{2} \tag{3.14}$$

and this will prove the existence of a solution to the equation $r = \alpha f(r)$. We take for r_0 the largest number such that

$$|r| \le r_0 \Rightarrow q = \mathsf{E}\,\mathrm{th}^2(z\sqrt{r}) \le \frac{1}{2} \,. \tag{3.15}$$

Since, denoting by E_ξ integration in ξ only, we have

$$\Phi(z\sqrt{q}, \sqrt{1-q}) = \frac{\mathsf{E}_\xi\, \xi \exp u(z\sqrt{q} + \xi\sqrt{1-q})}{\sqrt{1-q}\, \mathsf{E}_\xi \exp u(z\sqrt{q} + \xi\sqrt{1-q})} \,, \tag{3.16}$$

we have

$$|\Phi(z\sqrt{q}, \sqrt{1-q})| \le \frac{\mathsf{E}\,|\xi|\, e^{2D}}{\sqrt{1-q}} \le \sqrt{\frac{2}{\pi}} \frac{e^{2D}}{\sqrt{1-q}} \,,$$

and, since $1 - q \ge 1/2$,

$$\alpha \mathsf{E} \Phi^2(z\sqrt{q}, \sqrt{1-q}) \le \frac{4\,\alpha}{\pi} e^{4D} \,,$$

so that if $4\,\alpha\, e^{4D} \le r_0\pi$, we have the first part of (3.14). For the second part, we recall the formula

$$\frac{\mathrm{d}}{\mathrm{d}x} \mathsf{E}\, w(z\sqrt{x}) = \frac{1}{2} \mathsf{E}\, w''(z\sqrt{x}) \tag{3.17}$$

that shows in particular that $\left|\frac{\mathrm{d}}{\mathrm{d}r} \mathsf{E}\,\mathrm{th}^2(z\sqrt{r})\right| \le L$. Since by (3.15) we know that $1 - q \ge 1/2$, all we have to do is to show that over the region $q > 0$, $y > 1/\sqrt{2}$, the partial derivatives of order 1 of the function

$$q, y \to \mathsf{E} \Phi^2(z\sqrt{q}, y)$$

exist and are uniformly bounded by $L\, e^{LD}$. Using (3.17), it suffices to show that the partial derivatives of order at most 2 of the functions

$$(x, y) \to \mathsf{E}\, \xi \exp u(x + \xi y) \,; \; (x, y) \to \mathsf{E} \exp u(x + \xi y) \tag{3.18}$$

exist and are bounded by $L\exp LD$ over the domain $y \geq 1/\sqrt{2}$. But this is obvious if we write these functions as integrals and make a change of variable, e.g.

$$\begin{aligned} \mathsf{E}\exp u(x+\xi y) &= \frac{1}{\sqrt{2\pi}}\int \exp\left(-\frac{t^2}{2}+u(x+ty)\right)\mathrm{d}t \\ &= \frac{1}{\sqrt{2\pi}\,y}\int \exp\left(-\frac{1}{2}\left(\frac{v-x}{y}\right)^2+u(v)\right)\mathrm{d}v\,. \quad (3.19)\end{aligned}$$

□

A few more lines of work would suffice to prove that the solution to the equations (3.12) and (3.13) is actually unique. This however will not be needed.

We now turn to the definition of the smart path. Given $0 \leq t \leq 1$ we set

$$S_{k,t}(\boldsymbol{\sigma}) = \frac{1}{\sqrt{N}}\sum_{i\leq N-1} g_{i,k}\,\sigma_i + \sqrt{\frac{t}{N}}\,g_{N,k}\,\sigma_N \tag{3.20}$$

and we consider the Hamiltonian

$$-H_{N,M,t}(\boldsymbol{\sigma}) = \sum_{k\leq M} u(S_{k,t}(\boldsymbol{\sigma})) + \sigma_N\sqrt{1-t}\,Y \tag{3.21}$$

where $Y = z\sqrt{r}$, r is as in (3.12), (3.13) and z is a standard Gaussian r.v. independent of all the other r.v. involved.

To make this definition less mysterious one can observe that if one assumes that u has enough differentiability, then

$$\sum_{k\leq M} u(S_{k,1}(\boldsymbol{\sigma})) = \sum_{k\leq M} u(S_{k,0}(\boldsymbol{\sigma})) + \sigma_N\sum_{k\leq M}\frac{g_{N,k}}{\sqrt{N}}\,u'(S_{k,0}(\boldsymbol{\sigma})) + \cdots$$

so that (3.21) is reminiscent of what we did in Chapter 2.

To simplify notation, we write g_k for $g_{N,k}$ and when we consider replicas we write $S^\ell_{k,t}$ for $S_{k,t}(\boldsymbol{\sigma}^\ell)$. We write $\langle\cdot\rangle_t$ Gibbs' measure with Hamiltonian (3.21), and we denote the powers of this measure by the same notation. If f is a function on Σ_N^n, we write

$$\nu_t(f) = \mathsf{E}\langle f\rangle_t\,;\ \nu_t'(f) = \frac{\mathrm{d}}{\mathrm{d}t}\,\nu_t(f)\,. \tag{3.22}$$

We also write $\nu(f) = \nu_1(f) = \mathsf{E}\langle f\rangle$, where $\langle\cdot\rangle$ refers to Gibbs' measure with Hamiltonian (3.7). As in Chapter 2, we write $\varepsilon_\ell = \sigma_N^\ell$.

Lemma 3.2.2. *Given a function f^- on Σ_{N-1}^n, and a subset I of $\{1,\ldots,n\}$, we have*

$$\nu_0\left(f^-\prod_{\ell\in I}\varepsilon_\ell\right) = \mathsf{E}((\mathrm{th}\,Y)^{\mathrm{card}\,I})\,\nu_0(f^-) = \nu_0\left(\prod_{\ell\in I}\varepsilon_\ell\right)\nu_0(f^-)\,.$$

The proof is identical to that of Lemma 2.4.4. The Hamiltonian $H_{N,M,0}$ decouples the last spin from the first $N-1$ spins.

We now turn to the computation of $\nu_t'(f)$. One of the nice features of the present model is that, even though one assumes only minimal regularity for u, everything behaves as if u were very smooth. It is however simpler to write intelligible formulas when we assume u differentiable, which we do from this point on until further notice. Let us point out that assuming u differentiable is a very mild hypothesis. This is because our results will hold for every u and every N, so that as N increases, u can depend on N. A strong hypothesis would be a control of the derivative of u uniformly over N; we must not do this if we want to be able to approximate the case (3.5).

Proposition 3.2.3. *Assume that u is twice differentiable. Write $\alpha = M/N$. Then, for a function f on Σ_N^n, we have*

$$\nu_t'(f) = \mathrm{I} + \mathrm{II} + \mathrm{III} \tag{3.23}$$

$$\mathrm{I} = \frac{\alpha}{2}\sum_{\ell\le n} \nu_t((u'' + u'^2)(S_{M,t}^\ell)\, f) - \frac{\alpha}{2}\, n\, \nu_t((u'' + u'^2)(S_{M,t}^{n+1})\, f) \tag{3.24}$$

$$\begin{aligned}\mathrm{II} = {}& \alpha \sum_{1\le\ell<\ell'\le n} \nu_t(\varepsilon_\ell\, \varepsilon_{\ell'}\, u'(S_{M,t}^\ell)\, u'(S_{M,t}^{\ell'})\, f) \\ & - \alpha\, n \sum_{\ell\le n} \nu_t(\varepsilon_\ell\, \varepsilon_{n+1}\, u'(S_{M,t}^\ell)\, u'(S_{M,t}^{n+1})\, f) \\ & + \alpha\, \frac{n(n+1)}{2}\, \nu_t(\varepsilon_{n+1}\, \varepsilon_{n+2}\, u'(S_{M,t}^{n+1})\, u'(S_{M,t}^{n+2})\, f)\,.\end{aligned} \tag{3.25}$$

$$\begin{aligned}\mathrm{III} = -r\Bigg(& \sum_{1\le\ell<\ell'\le n} \nu_t(\varepsilon_\ell\, \varepsilon_{\ell'}\, f) - n\sum_{\ell\le n} \nu_t(\varepsilon_\ell\, \varepsilon_{n+1}\, f) \\ & + \frac{n(n+1)}{2}\, \nu_t(\varepsilon_{n+1}\, \varepsilon_{n+2}\, f)\Bigg)\,.\end{aligned} \tag{3.26}$$

The reader should take heart. As in the case of Proposition 2.4.5, this formula will turn out to be much easier to use than one might think at first.

Proof. If one considers the case of a Hamiltonian $H_t(\boldsymbol{\sigma})$ depending on a parameter t, and one denotes by $\langle\cdot\rangle_t$ the corresponding Gibbs' measure, one has the general fact that

$$\frac{\mathrm{d}}{\mathrm{d}t}\langle f\rangle_t = -\sum_{\ell\le n}\left\langle \frac{\mathrm{d}}{\mathrm{d}t} H_t(\boldsymbol{\sigma}^\ell)\, f\right\rangle_t + n\left\langle \frac{\mathrm{d}}{\mathrm{d}t} H_t(\boldsymbol{\sigma}^{n+1})\, f\right\rangle_t\,. \tag{3.27}$$

This is immediate from the expression of $\langle f\rangle_t$ as a function of $H_{N,M,t}$, that is

$$\langle f\rangle_t = \frac{\sum_{\boldsymbol{\sigma}^1,\ldots,\boldsymbol{\sigma}^n} f(\boldsymbol{\sigma}^1,\ldots,\boldsymbol{\sigma}^n)\exp\left(-\sum_{\ell\le n} H_t(\boldsymbol{\sigma}^\ell)\right)}{\left(\sum_{\boldsymbol{\sigma}}\exp(-H_t(\boldsymbol{\sigma}))\right)^n}\,.$$

In the present case we obtain, writing g_k for $g_{N,k}$

$$\begin{aligned}\nu_t'(f) &= \frac{1}{2}\sum_{\ell\le n}\mathsf{E}\left\langle \varepsilon_\ell\left(\sum_{k\le M}\frac{g_k}{\sqrt{t}\sqrt{N}}\,u'(S_{k,t}^\ell) - \frac{Y}{\sqrt{1-t}}\right) f\right\rangle_t \\ &\quad - \frac{n}{2}\mathsf{E}\left\langle \varepsilon_{n+1}\left(\sum_{k\le M}\frac{g_k}{\sqrt{t}\sqrt{N}}\,u'(S_{k,t}^{n+1}) - \frac{Y}{\sqrt{1-t}}\right) f\right\rangle_t\,. \qquad (3.28)\end{aligned}$$

We observe the symmetry between the values of $k\le M$. Since $M/\sqrt{N} = \alpha\sqrt{N}$, we get

$$\nu_t'(f) = \mathrm{IV} + \mathrm{V} \qquad (3.29)$$

$$\mathrm{IV} = \frac{\alpha}{2}\sqrt{\frac{N}{t}}\left(\sum_{\ell\le n}\mathsf{E}\,g_M\langle \varepsilon_\ell\, u'(S_{M,t}^\ell)\, f\rangle_t - n\,\mathsf{E}\,g_M\langle\varepsilon_{n+1}\, u'(S_{M,t}^{n+1})\, f\rangle_t\right)\,.$$

$$\mathrm{V} = -\frac{1}{2}\frac{1}{\sqrt{1-t}}\left(\sum_{\ell\le n}\mathsf{E}\,Y\langle\varepsilon_\ell\, f\rangle_t - n\,\mathsf{E}\,Y\langle\varepsilon_{n+1}\, f\rangle_t\right)\,.$$

We now integrate by parts. If we think of $\langle\varepsilon_\ell\, u'(S_{M,t}^\ell)\, f\rangle_t$ as a function of g_M we see (by writing the cumbersome explicit formula for $\langle\varepsilon_\ell\, u'(S_{M,t}^\ell)\, f\rangle_t$) that

$$\begin{aligned}\frac{\partial}{\partial g_M}\langle\varepsilon_\ell\, u'(S_{M,t}^\ell)\, f\rangle_t &= \sqrt{\frac{t}{N}}\Bigg(\langle u''(S_{M,t}^\ell)\, f\rangle_t \\ &+ \sum_{\ell'\le n}\langle\varepsilon_\ell\,\varepsilon_{\ell'}\, u'(S_{M,t}^\ell)\, u'(S_{M,t}^{\ell'})\, f\rangle_t - n\langle\varepsilon_\ell\,\varepsilon_{n+1}\, u'(S_{M,t}^\ell)\, u'(S_{M,t}^{n+1})\, f\rangle_t\Bigg)\end{aligned}$$

and, similarly,

$$\begin{aligned}\frac{\partial}{\partial g_M}\langle\varepsilon_{n+1}\, u'(S_{M,t}^{n+1})\, f\rangle_t = \sqrt{\frac{t}{N}}\Bigg(&\langle u''(S_{M,t}^{n+1})\, f\rangle_t \\ &+ \sum_{\ell'\le n+1}\langle\varepsilon_{\ell'}\,\varepsilon_{n+1}\, u'(S_{M,t}^{\ell'})\, u'(S_{M,t}^{n+1})\, f\rangle_t \\ &- (n+1)\langle\varepsilon_{n+1}\,\varepsilon_{n+2}\, u'(S_{M,t}^{n+1})\, u'(S_{M,t}^{n+2})\, f\rangle_t\Bigg)\,.\end{aligned}$$

Since g_M is independent of all the other r.v. involved, we can use the previous formulas and the ubiquitous relation $\mathsf{E}(g_M\, F(g_M)) = \mathsf{E}(F'(g_M))$ to compute

IV. Regrouping the terms we get IV = I + II. (The term I is singled out because it plays a special role later.) In a similar manner, and using now that $\mathsf{E}\,Y^2 = r$, we get V = III. □

It is not clear at all how to use Proposition 3.2.3, in particular because u' and u'' can be huge. Finding a way to do this is the purpose of the next section.

3.3 The Main Estimate

It seems a good idea, if one wants to make sense of formulas like (3.23), to make explicit the dependence of $\langle\cdot\rangle_t$ on $S_{M,t}$. We consider the Hamiltonian

$$-H_{N,M-1,t} = \sum_{k\le M-1} u(S_{k,t}(\boldsymbol{\sigma})) + \sigma_N\sqrt{1-t}\,Y \tag{3.30}$$

and we denote averaging with respect to the corresponding Gibbs' measure by $\langle\cdot\rangle_{t,\sim}$. When $t=1$, this is denoted by $\langle\cdot\rangle_\sim$. We have identities such as

$$\langle f\,u''(S^\ell_{M,t})\rangle_t = \frac{\left\langle f\,u''(S^\ell_{M,t})\exp\left(\sum_{\ell'\le n} u(S^{\ell'}_{M,t})\right)\right\rangle_{t,\sim}}{\left\langle \exp\left(\sum_{\ell\le n} u(S^{\ell}_{M,t})\right)\right\rangle_{t,\sim}}\,. \tag{3.31}$$

Now $\langle\cdot\rangle_{t,\sim}$ does not depend on $S_{M,t}$.

In other words, the variables $g_{i,M}$ are independent of the randomness in $\langle\cdot\rangle_{t,\sim}$. Thus we can hope to appeal to (2.194) to compute the numerator and the denominator of (3.31). There is a serious obstacle to pass however. Even if we would control the size of u'', the second moment argument of Section 2.4 that proves (2.194) would not be accurate enough to give a useful result. (Anyway, we do not control the size of u''.) To pass this obstacle we will use again the smart path method, but we must first learn to be much more sophisticated about integration by parts.

Definition 3.3.1. *We say that a centered jointly Gaussian family $h_1,\ldots,h_n$ of r.v. is* widely spread *if for each $\ell\le n$ we can find numbers $b_{k,\ell}$ such that we can write*

$$h_\ell = h'_\ell + \sum_{k\ne\ell} b_{k,\ell}\,h_k\,, \tag{3.32}$$

where h'_ℓ is independent of the r.v. h_k for $k\ne\ell$, and where

$$\mathsf{E}\,h'^2_\ell \ge \frac{1}{16}\,. \tag{3.33}$$

It can be shown that there is always a unique decomposition as in (3.32) (see the proof of Lemma 3.3.3), so the meaningful part of this definition is condition (3.33).

Lemma 3.3.2. *Consider a centered jointly Gaussian family of r.v.* $(h_\ell)_{\ell\geq 1}$*, with* $\mathsf{E}\, h_\ell^2 \leq 1$ *for each* ℓ*. Consider two integers* $n \leq n'$ *and assume that the sequence* $h_1, \dots, h_n$ *is widely spread. Then, given a bounded infinitely differentiable function* F *on* $\mathbb{R}^n$*, for all integers* $r_1, \dots, r_{n'}$*,* $s_1, \dots, s_n$ *we have*

$$\left| \mathsf{E} \left(h_1^{r_1} \dots h_{n'}^{r_{n'}} \frac{\partial^s F}{\partial x_1^{s_1} \dots \partial x_n^{s_n}} (h_1, \dots, h_n) \right) \right| \leq K \sup |F| \,, \tag{3.34}$$

where $s = s_1 + \dots + s_n$ *and where* K *depends only on* $s_1, \dots, s_n$*,* $r_1, \dots, r_{n'}$*.*

It is interesting to observe that if g is standard Gaussian, it is not true that for a function $F(x_1, x_2)$ of two variables one has

$$\left| \mathsf{E}\, \frac{\partial F}{\partial x_1} (g, g) \right| \leq K \sup |F| \,.$$

To see this, one can take $F(x_1, x_2) = f(x_1 - x_2)$ where $f'(0)$ is large.

Proof. We proceed by induction on s. The result is obvious for $s = 0$. For the induction step from $s-1$ to s, we can assume without loss of generality that $s_1 > 1$. Since the sequence $h_1, \dots, h_n$ is widely spread, we can write

$$h_1 = h + \sum_{2 \leq k \leq n} b_k\, h_k \tag{3.35}$$

where h is independent of $h_2, \dots, h_n$ and $\mathsf{E}\, h^2 \geq 1/16$.

For $\ell \leq n'$, we write

$$h_\ell = c_\ell\, h + h'_\ell \tag{3.36}$$

where h'_ℓ is independent of h. (Thus $c_\ell = 0$ and $h_\ell = h'_\ell$ for $2 \leq \ell \leq n$.) This is done by taking $c_\ell = \mathsf{E}\,(h\, h_\ell)/\mathsf{E}\, h^2$ (so that $|c_\ell| \leq 4$) since then if we set $h'_\ell = h_\ell - c_\ell\, h$ we have $\mathsf{E}\, h\, h'_\ell = 0$. Using the decomposition (3.36), the fact that $|c_\ell| \leq 4$, $\mathsf{E}\,((h'_\ell)^2) \leq 1$, and expanding the product $\prod_{\ell \leq n'} (c_\ell\, h + h'_\ell)^{r_\ell}$ we see that all we have to do is to bound quantities of the type

$$\mathsf{E} \left(h^r \prod_{\ell \leq n'} (h'_\ell)^{r'_\ell} \frac{\partial^s F}{\partial x_1^{s_1} \dots \partial x_n^{s_n}} (h_1, \dots, h_n) \right) \,, \tag{3.37}$$

where the variables h'_ℓ are independent of h.

Given a number a and a function f, integration by parts shows that

$$\mathsf{E}\,(h(h+a)^r f(h+a)) = \mathsf{E}\, h^2\, \mathsf{E}\,(r(h+a)^{r-1} f(h+a)) + \mathsf{E}\, h^2\, \mathsf{E}\,((h+a)^r f'(h+a))$$

so that

$$\begin{aligned} \mathsf{E}\,((h+a)^r f'(h+a)) &= -r\, \mathsf{E}\,((h+a)^{r-1} f(h+a)) \\ &\quad + \frac{1}{\mathsf{E}\, h^2}\, \mathsf{E}\,(h(h+a)^r f(h+a)) \,. \end{aligned} \tag{3.38}$$

We use this for $a = \sum_{2 \le k \le n} b_k h_k$,

$$f(x) = \frac{\partial^{s-1} F}{\partial x_1^{s_1 - 1} \partial x_2^{s_2} \ldots \partial x_n^{s_n}} (x, h_2, \ldots, h_n)$$

at $h_2, \ldots, h_n, h'_1, \ldots, h'_{n'}$ fixed. This is possible because h is independent of these variables. Since the family $h, h_1, \ldots, h_n, h'_1, \ldots, h'_{n'}$ is jointly Gaussian the required bound follows from the induction hypothesis. □

We consider a fixed number $0 < t < 1$, a new independent sequence $(g_i)_{i \le N}$ of standard Gaussian r.v. and for $\boldsymbol{\sigma}$ in Σ_N we set

$$g(\boldsymbol{\sigma}) = \frac{1}{\sqrt{N}} \sum_{i \le N-1} g_i \, \sigma_i + \sqrt{\frac{t}{N}} \, g_N \, \sigma_N \,. \tag{3.39}$$

If we write $g_i = g_{i,M}$, we see that (3.39) is simply a new name for $S_{M,t}(\boldsymbol{\sigma})$.

Lemma 3.3.3. *Consider a Gibbs' measure G on Σ_N such that the Hamiltonian H of G satisfies*

$$-H(\boldsymbol{\sigma}) = -H_0(\boldsymbol{\sigma}) + \sigma_N \, y \tag{3.40}$$

where y is a number and where

$$\forall \boldsymbol{\sigma} \in \Sigma_N, \quad |H_0(\boldsymbol{\sigma})| \le \frac{N}{40} \,. \tag{3.41}$$

(Here $H_0(\boldsymbol{\sigma})$ is allowed to depend on σ_N.) Then we have

$$G^{\otimes n}(\{(\boldsymbol{\sigma}^1, \ldots, \boldsymbol{\sigma}^n) \in \Sigma_N^n \,;\ \text{the sequence } g(\boldsymbol{\sigma}^1), \ldots, g(\boldsymbol{\sigma}^n) \text{ is not widely spread}\}) \le L^n \exp\left(-\frac{N}{20}\right) \,. \tag{3.42}$$

Proof. For $\mathbf{x}$ in $\mathbb{R}^N$, we set

$$g(\mathbf{x}) = \frac{1}{\sqrt{N}} \sum_{i \le N-1} x_i \, g_i + \sqrt{\frac{t}{N}} \, x_N \, g_N \,.$$

We consider the dot product $(\cdot \mid \cdot)$ on $\mathbb{R}^N$ given by

$$(\mathbf{x} \mid \mathbf{y}) = \mathsf{E}\, g(\mathbf{x})\, g(\mathbf{y}) = \frac{1}{N} \sum_{i \le N-1} x_i \, y_i + \frac{t}{N} \, x_N \, y_N \,. \tag{3.43}$$

Let us fix $\boldsymbol{\sigma}^2, \ldots, \boldsymbol{\sigma}^n$, and denote by W the linear subspace of $\mathbb{R}^N$ they span, so $\dim W \le n - 1$. Consider Q, the orthogonal projection on W for the dot product $(\cdot \mid \cdot)$. That is, if $\mathbf{x} \in \mathbb{R}^N$, then $Q(\mathbf{x}) \in W$ and

$$(\mathbf{x} - Q(\mathbf{x}) \mid \mathbf{y}) = 0$$

whenever $\mathbf{y} \in W$. Thus if we write

$$g(\mathbf{x}) = h + g(Q(\mathbf{x}))$$

where $h = g(\mathbf{x}) - g(Q(\mathbf{x}))$, then $g(Q(\mathbf{x}))$ is a linear combination of the variables $g(\boldsymbol{\sigma}^\ell)$, $\ell = 2, \ldots, n$, and h is independent of these.

We will prove that

$$G\left(\left\{\boldsymbol{\sigma}; \mathsf{E}\left((g(\boldsymbol{\sigma}) - g(Q(\boldsymbol{\sigma})))^2\right) \le \frac{1}{16}\right\}\right) \le L^n \exp\left(-\frac{N}{20}\right) . \tag{3.44}$$

Fubini's theorem then implies that the measure, for $G^{\otimes n}$, of the set of n-tuples $(\boldsymbol{\sigma}^1, \ldots, \boldsymbol{\sigma}^n)$ for which the sequence $h_\ell = g(\boldsymbol{\sigma}^\ell)$ fails (3.32) for $\ell = 1$ is at most $L^n \exp(-N/20)$. Repeating this for each coordinate proves (3.42).

To prove (3.44) we consider

$$\|\mathbf{x}\| = (\mathbf{x} \mid \mathbf{x})^{1/2} = (\mathsf{E}\, g^2(\mathbf{x}))^{1/2} ,$$

so that the left-hand side of (3.44) is

$$G\left(\left\{\boldsymbol{\sigma}; \|\boldsymbol{\sigma} - Q(\boldsymbol{\sigma})\| \le \frac{1}{4}\right\}\right) . \tag{3.45}$$

Since $\|\boldsymbol{\sigma}\| \le 1$, we have $\|Q(\boldsymbol{\sigma})\| \le 1$. Proposition A.3.1 implies that there is a subset A of W with $\operatorname{card} A \le L^n$ such that

$$\forall \mathbf{y} \in W\,,\ \|\mathbf{y}\| \le 1 \Rightarrow \exists \mathbf{a} \in A\,,\ \|\mathbf{y} - \mathbf{a}\| \le \frac{1}{4} .$$

Thus, if $\|\boldsymbol{\sigma} - Q(\boldsymbol{\sigma})\| \le 1/4$, we have $\|\boldsymbol{\sigma} - \mathbf{a}\| \le 1/2$ for some $\mathbf{a}$ in A and the quantity (3.45) is bounded by

$$L^n \max_{\mathbf{a} \in A} G\left(\left\{\boldsymbol{\sigma}\,;\ \|\boldsymbol{\sigma} - \mathbf{a}\|^2 \le \frac{1}{4}\right\}\right) . \tag{3.46}$$

We have

$$\|\boldsymbol{\sigma} - \mathbf{a}\|^2 \ge \frac{1}{N} \sum_{i \le N-1} (\sigma_i - a_i)^2 .$$

Assuming for clarity that $a_i \le 0$ for each i, we have

$$\sum_{i \le N-1} (\sigma_i - a_i)^2 \ge \operatorname{card}\{i \le N-1; \sigma_i = 1\} = \frac{N-1}{2} + \frac{1}{2} \sum_{i \le N-1} \sigma_i$$

because

$$\sum_{i \le N-1} \sigma_i = \operatorname{card}\{i \le N-1; \sigma_i = 1\} - \operatorname{card}\{i \le N-1; \sigma_i = -1\} .$$

Thus

$$\begin{aligned}\mathrm{card}\left(\left\{\boldsymbol{\sigma}\,;\ \|\boldsymbol{\sigma}-\mathbf{a}\|^2\le\frac{1}{4}\right\}\right)&\le \mathrm{card}\left(\left\{\boldsymbol{\sigma}\,;\ \sum_{i\le N-1}\sigma_i\le-\frac{N-2}{2}\right\}\right)\\&\le 2^N\exp\left(-\frac{(N-2)^2}{8N}\right)\end{aligned}\tag{3.47}$$

from the subgaussian inequality (A.3). Now we have

$$\exp\left(-\frac{(N-2)^2}{8N}\right)\le L\exp\left(-\frac{N}{10}\right)\,. \tag{3.48}$$

Hence, using (3.46) to (3.48), to finish the proof of (3.44) it suffices to show that if A is a subset of Σ_N that does not depend on the last coordinate, we have

$$G(A)\le\exp\left(\frac{N}{20}\right)2^{-N}\,\mathrm{card}\,A\,. \tag{3.49}$$

By (3.40) and (3.41), we have

$$\sum_{\sigma\in A}\exp(-H(\boldsymbol{\sigma}))\le\exp\frac{N}{40}\sum_{\sigma\in A}\exp\sigma_N\,y=\exp\frac{N}{40}\,\mathrm{ch}\,y\,\mathrm{card}\,A \tag{3.50}$$

because A does not depend on the last coordinate. Also,

$$\sum_{\sigma}\exp(-H(\boldsymbol{\sigma}))\ge 2^N\exp\left(-\frac{N}{40}\right)\mathrm{ch}\,y\,. \tag{3.51}$$

Comparing (3.50) and (3.51) proves (3.49). □

We consider now independent standard Gaussian r.v. $z,\xi_1,\xi_2,\ldots$, that are independent of the r.v. $(g_{i,k})$, and we set

$$\theta_\ell=z\sqrt{q}+\xi_\ell\sqrt{1-q}\,. \tag{3.52}$$

In a subsequent use of the smart path method, we will need the variables

$$g_{v,\ell}=g_{v,\ell}(\boldsymbol{\sigma}^\ell)=\sqrt{v}\,g(\boldsymbol{\sigma}^\ell)+\sqrt{1-v}\,\theta_\ell\,, \tag{3.53}$$

where $0\le v\le 1$.

Lemma 3.3.4. *Assume $q\le 1/2$. Consider configurations $\boldsymbol{\sigma}^1,\ldots,\boldsymbol{\sigma}^n$. Assume that the sequence $g(\boldsymbol{\sigma}^1),\ldots,g(\boldsymbol{\sigma}^n)$ is widely spread. Then the sequence $g_{v,1}(\boldsymbol{\sigma}^1),\ldots,g_{v,n}(\boldsymbol{\sigma}^n)$ is widely spread.*

Proof. We want to prove that the sequence $(g_{v,\ell}(\boldsymbol{\sigma}^\ell))$ satisfies (3.32). To simplify notations, we assume $\ell = 1$. Since the sequence $(g(\boldsymbol{\sigma}^\ell))$ satisfies (3.32) we can write

$$g(\boldsymbol{\sigma}^1) = h_0 + \sum_{k\geq 2} b_k\, g(\boldsymbol{\sigma}^k) \tag{3.54}$$

where h_0 is independent of $g(\boldsymbol{\sigma}^k)$, $k \geq 2$ and $\mathsf{E}\, h_0^2 \geq 1/16$. We set $h = h_0/(\mathsf{E}\, h_0^2)^{1/2}$, so that h is independent of $g(\boldsymbol{\sigma}^k)$, $k \geq 2$, and, by (3.54),

$$\mathsf{E}\,(g(\boldsymbol{\sigma}^1)\, h) = (\mathsf{E}\, h_0^2)^{1/2} \geq \frac{1}{4}\,.$$

We consider $h_1 = \sqrt{v}\, h + \sqrt{1-v}\, \xi_1$, so that

$$\begin{aligned}\mathsf{E}\,(g_{v,1}(\boldsymbol{\sigma}^1)\, h_1) &= v\, \mathsf{E}\,(g(\boldsymbol{\sigma}^1)\, h) + (1-v)\, \mathsf{E}\,(\theta_\ell\, \xi_\ell) \\ &\geq \frac{v}{4} + (1-v)\sqrt{1-q} \geq \frac{1}{4}\,.\end{aligned}$$

Now $\mathsf{E}\, h_1^2 = 1$ and h_1 is independent of the r.v. $g_{v,k}(\boldsymbol{\sigma}^k)$ for $k \geq 2$. We can write

$$g_{v,1}(\boldsymbol{\sigma}^1) = h_2 + \sum_{k\geq 2} b'_k\, g_{v,k}(\boldsymbol{\sigma}^k)$$

where h_2 is independent of $g_{v,k}(\boldsymbol{\sigma}^k)$ for $k \geq 2$. Then

$$\mathsf{E}\, h_2\, h_1 = \mathsf{E}\,(g_{v,1}(\boldsymbol{\sigma}^1)\, h_1) \geq \frac{1}{4}$$

and thus

$$\frac{1}{16} \leq \mathsf{E}\, h_1^2\, \mathsf{E}\, h_2^2 \leq \mathsf{E}\, h_2^2\,.$$

□

Corollary 3.3.5. *Consider a Gibbs' measure G as in Lemma 3.3.3. Then*

$$G^{\otimes n}(\{(\boldsymbol{\sigma}^1, \ldots, \boldsymbol{\sigma}^n) \text{ the sequence } g_{v,1}, \ldots, g_{v,n} \text{ is not widely spread}\})$$
$$\leq L^n \exp\left(-\frac{N}{20}\right)\,.$$

We will use this when G corresponds to the Hamiltonian (3.30), so that to ensure (3.41) we assume

$$40\,\alpha\, D \leq 1\,. \tag{3.55}$$

Then conditions (3.40) and (3.41) hold for *any* realization of the randomness in (3.30).

Now that we have the required tools to integrate by parts, we turn to the computation of the expected value of quantities such as (3.31). This expected value will be first taken conditionally on $\langle\cdot\rangle_{t,\sim}$.

We start with a boundedness principle. We recall the notation (3.53), and we denote by E_ξ expectation in $\xi_1, \xi_2, \ldots$ only.

Lemma 3.3.6. *Consider an integer n, integers $s_1, \ldots, s_n$, a number B. Consider a function W on $\mathbb{R}^n$, a r.v. Z. We assume*

$$Z \geq e^{-nD} \; ; \; |W| \leq e^{nD} \, . \tag{3.56}$$

Consider the function

$$U = \frac{\partial^s W}{\partial x_1^{s_1} \ldots \partial x_n^{s_n}} \tag{3.57}$$

for $s = s_1 + \cdots + s_n$, and assume

$$|U| \leq B \, . \tag{3.58}$$

Consider a non-random Gibbs' measure G on Σ_N with a Hamiltonian that satisfies (3.40), (3.41) and denote by $\langle \cdot \rangle$ an average with respect to G or its powers. Then, for a function f on Σ_N^n that satisfies $|f| \leq 1$, we have

$$\left| \mathsf{E} \frac{1}{Z} \langle f \, \mathsf{E}_\xi U(g_{v,1}, \ldots, g_{v,n}) \rangle \right| \leq K \left(e^{2nD} \langle |f| \rangle + B \exp \left(-\frac{N}{40} \right) \right) \, , \tag{3.59}$$

where K depends only on $n, s_1, \ldots, s_n$.

Comment. This lemma will be used for the measure $G = \langle \cdot \rangle_{t,\sim}$, conditionally on this measure. This is possible because the r.v. $g_{v,\ell}$ are probabilistically independent of the randomness of $\langle \cdot \rangle_{t,\sim}$, and this explains why we can assume that G is non-random in the lemma.

Proof. We write $T = U(g_{v,1}, \ldots, g_{v,n})$. We have

$$\mathsf{E} \left(\frac{\langle f \, \mathsf{E}_\xi T \rangle}{Z} \right) \leq \mathsf{E} \left(\left(\frac{\langle f \, \mathsf{E}_\xi T \rangle}{Z} \right)^2 \right)^{1/2} \leq e^{nD} \mathsf{E} \left(\langle f \, \mathsf{E}_\xi T \rangle^2 \right)^{1/2} \tag{3.60}$$

using (3.56). Consider the function f' on Σ_N^{2n} given by

$$f' = f(\boldsymbol{\sigma}^{n+1}, \ldots, \boldsymbol{\sigma}^{2n})$$

and consider

$$T' = U(g_{v,n+1}, \ldots, g_{v,2n}) \, .$$

A basic observation is that

$$\langle f \, \mathsf{E}_\xi T \rangle^2 = \langle f \, f' \, \mathsf{E}_\xi T T' \rangle \, . \tag{3.61}$$

To see this, we first observe that

$$\mathsf{E}_\xi T T' = \mathsf{E}_\xi T \, \mathsf{E}_\xi T' \, .$$

This is because $\xi_1, \xi_2, \ldots$ are independent, and that (at $g(\boldsymbol{\sigma})$, z fixed) T depends only on $\xi_1, \ldots, \xi_n$, while T' depends only on $\xi_{n+1}, \ldots, \xi_{2n}$. Next, we observe that $f' \mathsf{E}_\xi T'$ is obtained from $f \mathsf{E}_\xi T$ by replacing $\boldsymbol{\sigma}^1, \ldots, \boldsymbol{\sigma}^n$ by $\boldsymbol{\sigma}^{n+1}, \ldots, \boldsymbol{\sigma}^{2n}$ respectively, so that (3.61) is the usual linearization of squares of brackets with replicas. From (3.61) we get

$$\mathsf{E}\left(\langle f \mathsf{E}_\xi T\rangle^2\right) = \mathsf{E}\,\langle f f' \mathsf{E}_\xi T T'\rangle = \langle f f' \mathsf{E} T T'\rangle \tag{3.62}$$

because $\langle \cdot \rangle$ is non-random.

Consider the set

$$C = \{(\boldsymbol{\sigma}^1, \ldots, \boldsymbol{\sigma}^{2n});\ \text{the sequence } g_{v,1}, \ldots, g_{v,2n} \text{ is } \textit{not} \text{ widely spread}\}\ .$$

Then Corollary 3.3.5 shows that

$$G^{\otimes 2n}(C) \le L^n \exp\left(-\frac{N}{20}\right)\ ,$$

and thus by (3.58) we have

$$\langle f f' \mathsf{E}(T T')\, \mathbf{1}_C\rangle \le B^2 L^n \exp\left(-\frac{N}{20}\right)\ . \tag{3.63}$$

Let

$$W'(x_1, \ldots, x_{2n}) = W(x_{n+1}, \ldots, x_{2n})$$
$$U'(x_1, \ldots, x_{2n}) = U(x_{n+1}, \ldots, x_{2n})\ ,$$

so that, seeing U, U' as functions on $\mathbb{R}^{2n}$, we have

$$T T' = U U'(g_{v,1}, \ldots, g_{v,2n})\ ,$$

while

$$U U' = \frac{\partial^{2s}}{\partial x_1^{s_1} \ldots \partial x_n^{s_n}\, \partial x_{n+1}^{s_1} \ldots \partial x_{2n}^{s_n}} (W W')\ .$$

Thus, we can use Lemma 3.3.2 (with $2n$ rather than n, and $F = W W'$) to see that

$$|\mathbf{1}_{C^c}\, \mathsf{E} T T'| \le K e^{2nD}\ ,$$

where K depends only on $n, s_1, \ldots, s_n$. Hence

$$\langle f f'\, \mathbf{1}_{C^c}\, \mathsf{E} T T'\rangle \le K e^{2nD} \langle |f|\,|f'|\rangle = K e^{2nD} \langle |f|\rangle^2\ .$$

Combining with (3.62) and (3.63), we get

$$\mathsf{E}(\langle f \mathsf{E}_\xi T\rangle^2) \le K e^{2nD} \langle |f|\rangle^2 + B^2 L^n \exp\left(-\frac{N}{20}\right)\ . \tag{3.64}$$

Combining this with (3.60) and the inequality $\sqrt{a+b} \le \sqrt{a} + \sqrt{b}$ finishes the proof. □

At this stage the reader might be convinced that the terms obtained in Proposition 3.2.3 will not be huge. This is however not enough. We need to compute them, or, at least, to approximate them. We recall the notation (3.39), (3.52) and (3.53).

Theorem 3.3.7. *Consider an integer m and integers $s_1, \ldots, s_m$ with $s = s_1 + \cdots + s_m \leq 2$. Consider $\tau_1, \tau_2 > 1$, with $1/\tau_1 + 1/\tau_2 = 1$. Consider a non-random Gibbs' measure G with a Hamiltonian that satisfies (3.40), (3.41), and denote by $\langle \cdot \rangle$ an average with respect to G. Consider the functions*

$$V(x_1, \ldots, x_m) = \exp \sum_{\ell \leq m} u(x_\ell) \tag{3.65}$$

$$U = \frac{\partial^s V}{\partial x_1^{s_1} \ldots \partial x_m^{s_m}} \,. \tag{3.66}$$

Assume that for a certain number D',

$$\forall \ell \leq 4 \,, \; |u^{(\ell)}| \leq D' \,. \tag{3.67}$$

Then, if f is a function on Σ_N^m with $|f| \leq 1$ we have

$$\mathsf{E} \frac{\langle f\, U(g(\boldsymbol{\sigma}^1), \ldots, g(\boldsymbol{\sigma}^m))\rangle}{\langle V(g(\boldsymbol{\sigma}^1), \ldots, g(\boldsymbol{\sigma}^m))\rangle} = \langle f \rangle \, \mathsf{E} \frac{U(\theta_1, \ldots, \theta_m)}{(\mathsf{E}_\xi \exp u(\theta_1))^m} + \mathcal{R} \tag{3.68}$$

where

$$|\mathcal{R}| \leq K \, e^{2(m+2)D} \Bigg(\langle |f|^{\tau_1} \rangle^{1/\tau_1} \left(\langle |R_{1,2} - q|^{\tau_2} \rangle^{1/\tau_2} + \frac{1}{N} \right) + D'^{\,4} \exp\left(-\frac{N}{40} \right) \Bigg) \,, \tag{3.69}$$

and where K depends only on m.

Comment. As Lemma 3.3.6, this will be used for $G = \langle \cdot \rangle_{t,\sim}$, conditionally on the randomness in G.

We observe the relation

$$\langle V(g(\boldsymbol{\sigma}^1), \ldots, g(\boldsymbol{\sigma}^m))\rangle = \langle \exp u(g(\boldsymbol{\sigma}))\rangle^m \,. \tag{3.70}$$

We will use the smart path method. Consider the function

$$\varphi(v) = \mathsf{E} \left(\frac{1}{Z_v^m} \langle f\, U(g_{v,1}, \ldots, g_{v,m})\rangle \right) \tag{3.71}$$

where

$$Z_v = \mathsf{E}_\xi \langle \exp u(g_{v,1})\rangle \,. \tag{3.72}$$

Since $g_{1,\ell} = g(\boldsymbol{\sigma}^\ell)$, $g_{0,\ell} = \theta_\ell$, (3.70) shows that the left-hand side of (3.68) is $\varphi(1)$ and the first term on the right-hand side is $\varphi(0)$.

We will use the bound

$$|\varphi(1) - \varphi(0)| \leq \sup_{0<v<1} |\varphi'(v)| . \tag{3.73}$$

Our first task is to compute $\varphi'(v)$. This is a mechanical computation. The outcome looks bad at first sight, but once we get used to it, it will turn out to be quite manageable. We observe that

$$\mathsf{E}\, g^2(\boldsymbol{\sigma}) = \frac{N-1}{N} + \frac{t}{N} = 1 + \frac{t-1}{N} . \tag{3.74}$$

For $\ell \neq \ell'$, we write

$$\Delta_{\ell,\ell'} = \mathsf{E}\, g(\boldsymbol{\sigma}^\ell)\, g(\boldsymbol{\sigma}^{\ell'}) - q = R_{\ell,\ell'} - q + \frac{t-1}{N}\, \sigma_N^\ell \sigma_N^{\ell'} . \tag{3.75}$$

We define

$$V_1(x) = u'(x) \exp u(x)\, ;\ V_2(x) = (u'^2(x) + u''(x)) \exp u(x) . \tag{3.76}$$

Proposition 3.3.8. *We have*

$$\varphi'(v) = \mathrm{I} + \mathrm{II} + \mathrm{III} + \mathrm{IV} \tag{3.77}$$

$$\begin{aligned} \mathrm{I} = \frac{t-1}{2N} \Bigg(& \sum_{\ell \leq m} \mathsf{E}\, \frac{1}{Z_v^m} \left\langle f \frac{\partial^2 U}{\partial x_\ell^2} (g_{v,1}, \ldots, g_{v,m}) \right\rangle \\ & - m\, \mathsf{E}\, \frac{1}{Z_v^{m+1}} \langle f\, U(g_{v,1}, \ldots, g_{v,m})\, V_2(g_{v,m+1}) \rangle \Bigg) \end{aligned} \tag{3.78}$$

$$\mathrm{II} = \sum_{\ell < \ell' \leq m} \mathsf{E}\, \frac{1}{Z_v^m} \left\langle f\, \Delta_{\ell,\ell'} \frac{\partial^2 U}{\partial x_\ell \partial x_{\ell'}} (g_{v,1}, \ldots, g_{v,m}) \right\rangle \tag{3.79}$$

$$\mathrm{III} = -m \sum_{\ell \leq m} \mathsf{E}\, \frac{1}{Z_v^{m+1}} \left\langle f\, \Delta_{\ell,m+1} \frac{\partial U}{\partial x_\ell} (g_{v,1}, \ldots, g_{v,m})\, V_1(g_{v,m+1}) \right\rangle \tag{3.80}$$

$$\begin{aligned} \mathrm{IV} = \frac{m(m+1)}{2} \mathsf{E} \Bigg(& \frac{1}{Z_v^{m+2}} \\ & \times \langle f\, \Delta_{m+1,m+2}\, U(g_{v,1}, \ldots, g_{v,m})\, V_1(g_{v,m+1})\, V_1(g_{v,m+2}) \rangle \Bigg) . \end{aligned} \tag{3.81}$$

The terms II to IV are reminiscent of Proposition 3.2.3 and the factor $1/N$ in the term I will make it unimportant. The special form of U will allow significant simplifications of this expression, but it seems better to perform first a formal computation.

Proof. We write

$$g'_{v,\ell} = \frac{\mathrm{d}}{\mathrm{d}v} g_{v,\ell} = \frac{1}{2\sqrt{v}} g(\boldsymbol{\sigma}^\ell) - \frac{1}{2\sqrt{1-v}} \theta_\ell , \tag{3.82}$$

so that

$$\varphi'(v) = \mathrm{V} + \mathrm{VI} \tag{3.83}$$

$$\mathrm{V} = \mathsf{E}\left(\frac{1}{Z_v^m} \sum_{\ell \le m} \left\langle g'_{v,\ell} f \frac{\partial U}{\partial x_\ell}(g_{v,1}, \ldots, g_{v,m})\right\rangle\right) \tag{3.84}$$

$$\mathrm{VI} = -m\,\mathsf{E}\left(\frac{1}{Z_v^{m+1}} \frac{d\,Z_v}{dv} \langle f\, U(g_{v,1}, \ldots, g_{v,m})\rangle\right) .$$

At first one would think that the expectation E_ξ in Z_v is going to create complications and to prevent us from using replicas as we did e.g. in Proposition 3.2.3. This is not the case; the expectation E_ξ behaves very well with respect to replicas (as was used already in the proof of (3.62)). We have

$$\frac{\mathrm{d}\,Z_v}{\mathrm{d}v} = \mathsf{E}_\xi \langle g'_{v,1}\, V_1(g_{v,1})\rangle = \mathsf{E}_\xi \langle g'_{v,m+1}\, V_1(g_{v,m+1})\rangle .$$

Since ξ_{m+1} is independent of $\xi_1, \ldots, \xi_m$, we get that

$$\begin{aligned}\mathrm{VI} &= -m\,\mathsf{E}\left(\frac{1}{Z_v^{m+1}} \mathsf{E}_\xi(\langle g'_{v,m+1}\, V_1(g_{v,m+1})\rangle \langle f\, U(g_{v,1}, \ldots, g_{v,m})\rangle)\right) \\ &= -m\,\mathsf{E}\left(\frac{1}{Z_v^{m+1}} \langle g'_{v,m+1}\, V_1(g_{v,m+1})\rangle \langle f\, U(g_{v,1}, \ldots, g_{v,m})\rangle\right)\end{aligned}$$

because Z_v does not depend on ξ and thus

$$\mathrm{VI} = -m\,\mathsf{E}\left(\frac{1}{Z_v^{m+1}} \langle f\, g'_{v,m+1}\, U(g_{v,1}, \ldots, g_{v,m})\, V_1(g_{v,m+1})\rangle\right) . \tag{3.85}$$

We then recall the value of $g'_{v,\ell}$

$$g'_{v,\ell} = \frac{1}{2\sqrt{v}}\left(\frac{1}{\sqrt{N}} \sum_{i \le N-1} g_i \sigma_i^\ell + \sqrt{\frac{t}{N}}\, g_N \sigma_N^\ell\right) - \frac{1}{2\sqrt{1-v}} \theta_\ell$$

and we can integrate by parts in (3.84) and (3.85) as in Proposition 2.4.5, using formulas such as

$$\frac{\partial Z_v}{\partial g_i} = \sqrt{\frac{v}{N}}\, \mathsf{E}_\xi \langle \sigma_i^1\, V_1(g_{v,1})\rangle$$

if $i \le N-1$ and proceeding as in the proof of (3.85) to obtain the formula as stated. □

Proof of Theorem 3.3.7. We use (3.77), and to bound $\varphi'(v)$ we apply Lemma 3.3.6 to each term from (3.78) to (3.81). We detail this use in the case of the term (3.80). The others are similar. We use Lemma 3.3.6 with $n = m+1$, $Z = Z_v^{m+1}$, $W(x_1, \ldots, x_{m+1}) = \exp \sum_{\ell \le m+1} u(x_\ell)$, with base measure $\langle \cdot \rangle_{t,\sim}$. We observe that

$$\frac{\partial U}{\partial x_\ell}(x_1, \ldots x_m)\, V_1(x_{m+1})$$
$$= \frac{\partial^{s+1} W}{\partial x_1^{s_1} \ldots \partial x_\ell^{s_\ell+1} \ldots \partial x_m^{s_m} \partial x_{m+1}}(x_1, \ldots, x_{m+1})\,. \tag{3.86}$$

Let us define $w_\ell(x)$ by

$$w_\ell(x) \exp u(x) = \frac{\mathrm{d}^\ell}{\mathrm{d}x^\ell} \exp u(x)\,,$$

so that $w_1(x) = u'(x)$, $w_2(x) = u'^2(x) + u''(x)$, etc. and, under (3.67) we have $|w_k| \le LB^k$ for $k \le 4$. The left-hand side of (3.86) is

$$w_{s_1}(x_1) \ldots w_{s_\ell+1}(x_\ell) \ldots w_{s_m}(x_m)\, w_1(x_{m+1})\, W(x_1, \ldots, x_{m+1}) \tag{3.87}$$

and since $\sum_{\ell \le m} s_\ell \le 2$, this is bounded by $LD'^4 \exp(m+1)D$.

In this manner we bound the term III of (3.80) by

$$K\, e^{2(m+1)D} \left(\sum_{\ell \le m} \mathsf{E}\langle |f|\, |\Delta_{\ell,m+1}| \rangle + D'^4 \exp\left(-\frac{N}{40}\right)\right)\,.$$

Use of the inequality

$$|\Delta_{\ell,\ell'}| \le |R_{\ell,\ell'} - q| + \frac{1}{N}$$

and of Hölder's inequality finishes the proof. □

3.4 The Replica-Symmetric Solution

In this section we will prove results corresponding to Theorems 2.4.1 and 2.4.2 under a very mild control of the derivatives of u. We will later on give a separate argument to show that this mild condition is not required. The number L^* will be determined in (3.187).

Theorem 3.4.1. *There exists a constant L with the following property. Assume $|u| \le D$, $L\alpha\, e^{LD} \le 1$, where $\alpha = M/N$. Assume that*

$$\forall \ell \le 4\,,\ |u^{(\ell)}| \le L^* e^{L^* D} \exp\left(\frac{N}{250}\right)\,. \tag{3.88}$$

Then, if q is a solution of (3.12) and (3.13), we have

$$\mathsf{E}\langle (R_{1,2} - q)^2 \rangle \le \frac{L}{N}\,. \tag{3.89}$$

This implies that (if u satisfies the above conditions) the solution of the equations (3.12) and (3.13) is unique.

We write

$$p_{N,M}(u) = \frac{1}{N}\,\mathsf{E}\log Z_{N,M}$$

where $Z_{N,M}$ is the partition function of the Hamiltonian (3.7),

$$Z_{N,M} = \sum_{\boldsymbol{\sigma}} \exp(-H_{N,M}(\boldsymbol{\sigma}))\;.$$

Theorem 3.4.2. *Under the conditions of Theorem 3.4.1, we have*

$$\Big| p_{N,M}(u) + \frac{1}{2} r(1-q) - \mathsf{E}\log(2\operatorname{ch}(z\sqrt{r})) - \alpha\,\mathsf{E}\log \mathsf{E}_\xi \exp u(z\sqrt{q} + \xi\sqrt{1-q}) \Big| \le \frac{L}{N}\;. \tag{3.90}$$

We start preparations for the proof of Theorem 3.4.1. We recall the measure $\langle\cdot\rangle_{t,\sim}$ corresponding to the Hamiltonian of (3.30).

Lemma 3.4.3. *If f is a function on m replicas, $f \ge 0$, we have*

$$\langle f\rangle_{t,\sim} \le e^{2mD}\langle f\rangle_t\;. \tag{3.91}$$

Proof. We have

$$\langle f\rangle_t = \frac{\left\langle f \exp\sum_{\ell\le m} u(S^\ell_{M,t})\right\rangle_{t,\sim}}{\left\langle \exp\sum_{\ell\le m} u(S^\ell_{M,t})\right\rangle_{t,\sim}} \ge e^{-2mD}\langle f\rangle_{t,\sim}$$

because $|u| \le D$. □

We recall the smart path ν_t of Section 3.2.

Lemma 3.4.4. *Consider a function f on Σ_N^2, $0 \le f \le 1$. Then under (3.88), if $L\alpha \exp LD \le 1$, we have*

$$\nu_t(f) \le 2\,\nu(f) + 2\exp\left(-\frac{N}{L}\right)\;. \tag{3.92}$$

Proof. We compute $\nu'_t(f)$ through Proposition 3.2.3. The r.v. $S^\ell_{M,t}$ are independent of the randomness of $\langle\cdot\rangle_{t,\sim}$. The Hamiltonian (3.30) that defines $\langle\cdot\rangle_{t,\sim}$ satisfies (3.40) and (3.41) provided $40\,\alpha\,D \le 1$. Thus we can apply Lemma 3.3.6 (with $v = 1$) to each of the terms created by Proposition 3.2.3. This argument is as in the proof of Theorem 3.3.7, but simpler, and we get

$$|\nu_t'(f)| \le L\alpha e^{LD}\left(\mathsf{E}\,\langle f\rangle_{t,\sim} + \exp\left(-\frac{N}{L}\right)\right) .$$

Using (3.91) we get

$$|\nu_t'(f)| \le L\alpha e^{LD}\left(\nu_t(f) + \exp\left(-\frac{N}{L}\right)\right) .$$

The result is then a consequence of the following lemma. □

Lemma 3.4.5. *If a function $\varphi \ge 0$ satisfies*

$$|\varphi'(t)| \le c_1\,\varphi(t) + c_2$$

for $0 < t < t_0$, where $c_1, c_2 \ge 0$ then

$$\varphi(t) \le \exp(c_1(t_0 - t))\left(\varphi(t_0) + \frac{c_2}{c_1}\right) . \tag{3.93}$$

Proof. We note that

$$\left|\left(\varphi(t) + \frac{c_2}{c_1}\right)'\right| \le c_1\left(\varphi(t) + \frac{c_2}{c_1}\right) ,$$

so that

$$\varphi(t) + \frac{c_2}{c_1} \le \exp(c_1(t_0 - t))\left(\varphi(t_0) + \frac{c_2}{c_1}\right) .$$

□

Corollary 3.4.6. *Under the conditions of Lemma 3.4.4 we have*

$$\mathsf{E}\,\langle f\rangle_{t,\sim} \le L\,e^{4D}\left(\nu(f) + \exp\left(-\frac{N}{L}\right)\right) . \tag{3.94}$$

Proof. Combine (3.91) and (3.92). □

Proof of Theorem 3.4.1. We use symmetry among sites to write

$$\nu((R_{1,2} - q)^2) = \nu(f) , \tag{3.95}$$

where $f = (\varepsilon_1\,\varepsilon_2 - q)(R_{1,2} - q)$, and we write

$$\nu(f) \le \nu_0(f) + \sup_{0<t<1} |\nu_t'(f)| . \tag{3.96}$$

Since $q = \mathsf{E}\,\mathrm{th}^2 Y$, we see from Lemma 3.2.2 that

$$\nu_0((\varepsilon_1 \varepsilon_2 - q)(R^-_{1,2} - q)) = (\mathsf{E}\,\mathrm{th}^2 Y - q)\,\nu_0(R^-_{1,2} - q) = 0\,,$$

and thus

$$\nu_0(f) = \frac{1}{N}\,\nu_0((\varepsilon_1 \varepsilon_2 - q)^2) = \frac{1}{N}\,(1 - q^2)\,. \tag{3.97}$$

To compute $\nu'_t(f)$ we use Proposition 3.2.3. We then approximate each term created by Proposition 3.2.3 using Theorem 3.3.7. The details will be given below but this idea is that relation (3.68) expresses each term created by Proposition 3.2.3 as sum of a "main contribution " given by (3.68) and an error term (bounded as in (3.69)). The key fact of the proof is that the sum of the main contributions created by the terms of Proposition 3.2.3 is 0.

We now detail the application of Theorem 3.3.7 to the terms given by Proposition 3.2.3. It follows exactly the same pattern for each of these terms, so we will detail only one case, that of

$$\mathrm{IV} = \alpha\,\nu_t(\varepsilon_1 \varepsilon_2\, f\, u'(S^1_{M,t})\, u'(S^2_{M,t})) = \alpha\,\mathsf{E}\,\frac{\langle \varepsilon_1 \varepsilon_2\, f\, U(S^1_{M,t}, S^2_{M,t})\rangle_{t,\sim}}{\langle V(S^1_{M,t}, S^2_{M,t})\rangle_{t,\sim}}\,,$$

where

$$V(x_1, x_2) = \exp(u(x_1) + u(x_2))\ ;\ U = \frac{\partial^2 V}{\partial x_1\,\partial x_2}\,,$$

and where of course

$$S^\ell_{M,t} = \frac{1}{\sqrt{N}} \sum_{i \le N-1} g_{i,M}\,\sigma^\ell_i + \sqrt{\frac{t}{N}}\, g_{N,M}\,\sigma^\ell_N\,.$$

We use (3.68) with $\tau_1 = \tau_2 = 2$, with base measure $\langle\cdot\rangle = \langle\cdot\rangle_{t,\sim}$ and with $D' = L^* e^{L^* D} \exp(N/250)$. Thus

$$D'^{\,4} \exp\left(-\frac{N}{40}\right) \le \frac{L}{N}\, e^{LD}\,.$$

We get, using (3.94) in the second line,

$$\begin{aligned}
&|\mathrm{IV} - C\ \ \mathsf{E}\,\langle \varepsilon_1 \varepsilon_2\, f\rangle_{t,\sim}| \\
&\quad \le L\alpha e^{LD}\left((\mathsf{E}\,\langle f^2\rangle_{t,\sim})^{1/2}(\mathsf{E}\,\langle (R_{1,2} - q)^2\rangle_{t,\sim})^{1/2} + \frac{1}{N}\right) \\
&\quad \le L\alpha e^{LD}\left(\nu_t(f^2)^{1/2}\,\nu_t((R_{1,2} - q)^2)^{1/2} + \frac{1}{N}\right)\,,
\end{aligned} \tag{3.98}$$

where

$$C = \alpha\,\mathsf{E}\,\frac{U(\theta_1, \theta_2)}{\mathsf{E}_\xi\, V(\theta_1, \theta_2)}\,. \tag{3.99}$$

Since ξ_1, ξ_2 are independent, we have

$$\mathsf{E}_\xi\, V(\theta_1, \theta_2) = (\mathsf{E}_\xi \exp u(\theta))^2$$

$$\mathsf{E}_\xi U(\theta_1, \theta_2) = (\mathsf{E}_\xi u'(\theta) \exp u(\theta))^2 ,$$

where $\theta = z\sqrt{q} + \xi\sqrt{1-q}$. Thus, using (3.11), we have

$$\frac{\mathsf{E}_\xi U(\theta_1, \theta_2)}{\mathsf{E}_\xi V(\theta_1, \theta_2)} = \Phi^2(z\sqrt{q}, \sqrt{1-q})$$

and using (3.13) we see that $C = r$.

Using now the same method with $U(x_1, x_2) = \exp(u(x_1) + u(x_2))$, we can bound

$$|r\, \nu_t(\varepsilon_1 \varepsilon_2 f) - r\, \mathsf{E}\, \langle \varepsilon_1 \varepsilon_2 f \rangle_{t,\sim}|$$

as in (3.98).

Proceeding in this manner shows that the main contribution of each term of (3.25) cancels out with the main contribution of the corresponding term of (3.26).

The main contribution arising from the term (3.24) is zero, because the main contributions of the three terms there are respectively

$$C\, \mathsf{E}\, \langle f \rangle_{t,\sim} ,\quad C\, \mathsf{E}\, \langle f \rangle_{t,\sim} ,\quad -2C\, \mathsf{E}\, \langle f \rangle_{t,\sim}$$

where now

$$C = \mathsf{E}\, \frac{\mathsf{E}_\xi (u'' + u'^2)(\theta) \exp u(\theta)}{\mathsf{E}_\xi \exp u(\theta)} .$$

In this manner we obtain that

$$|\nu_t'(f)| \le L \alpha e^{LD} \left((\mathsf{E}\, \langle f^2 \rangle_{t,\sim})^{1/2} (\mathsf{E}\, \langle (R_{1,2} - q)^2 \rangle_{t,\sim})^{1/2} + \frac{1}{N} \right) . \tag{3.100}$$

Using Corollary 3.4.6 and combining with (3.96) and (3.97) we get

$$\nu((R_{1,2} - q)^2) \le \frac{L}{N} + L \alpha e^{LD} \left(\nu((R_{1,2} - q)^2) + \frac{1}{N} \right) ,$$

and this proves Theorem 3.4.1. □

We now start the preparations for the proof of Theorem 3.4.2. We consider the following function of three variables:

$$\begin{aligned} F(q, r, \alpha) = & \frac{1}{2} r(q-1) + \mathsf{E} \log(2 \operatorname{ch}(z\sqrt{r})) \\ & + \alpha\, \mathsf{E} \log \mathsf{E}_\xi \exp u(z\sqrt{q} + \xi\sqrt{1-q}) . \end{aligned} \tag{3.101}$$

Lemma 3.4.7. *The conditions (3.12) and (3.13) mean respectively that* $\partial F / \partial r = 0$, $\partial F / \partial q = 0$.

Proof. We have, using integration by parts,

$$\frac{\partial F}{\partial r} = \frac{1}{2}\left(q - 1 + \frac{1}{\sqrt{r}} \mathsf{E}\, z \operatorname{th} z\sqrt{r}\right) = \frac{1}{2}\left(q - 1 + \mathsf{E}\,\frac{1}{\operatorname{ch}^2(z\sqrt{x})}\right)$$

so that $\partial F/\partial r = 0$ if

$$q = 1 - \mathsf{E}\,\frac{1}{\operatorname{ch}^2(z\sqrt{r})} = \mathsf{E}\operatorname{th}^2(z\sqrt{r}) \,.$$

Next, if

$$\theta = z\sqrt{q} + \xi\sqrt{1-q}\,,\ \theta' = z/\sqrt{q} - \xi/\sqrt{1-q}\,,$$

we have

$$\frac{\partial F}{\partial q} = \frac{r}{2} + \frac{\alpha}{2}\mathsf{E}\left(\theta' \frac{u'(\theta)\exp u(\theta)}{\mathsf{E}_\xi \exp u(\theta)}\right) . \tag{3.102}$$

To integrate by parts, we observe that $\mathsf{E}_\xi \exp u(\theta)$ does not depend on ξ, so it is a function $F_1(z)$ of z alone, and

$$\frac{\mathrm{d}F_1}{\mathrm{d}z} = \frac{\mathrm{d}}{\mathrm{d}z}\mathsf{E}_\xi \exp u(z\sqrt{q} + \xi\sqrt{1-q}) = \sqrt{q}\,\mathsf{E}_\xi\, u'(\theta)\,\exp u(\theta) \,.$$

We appeal to the formula (A.41) to find, since $\mathsf{E}(\theta'\theta) = 0$, $\mathsf{E}(\theta' z) = 1/\sqrt{q}$ that

$$\begin{aligned}\mathsf{E}\left(\theta' \frac{u'(\theta)\exp u(\theta)}{F_1(z)}\right) &= -\mathsf{E}\left(\frac{1}{F_1(z)^2} u'(\theta)\exp u(\theta)\,\mathsf{E}_\xi(u'(\theta)\exp u(\theta))\right)\\ &= -\mathsf{E}\,\frac{(\mathsf{E}_\xi u'(\theta)\exp u(\theta))^2}{(\mathsf{E}_\xi \exp u(\theta))^2}\end{aligned}$$

so that by (3.102), $\partial F/\partial q = 0$ if and only if (3.13) holds. □

If q, r are related by (3.12) and (3.13), the quantity (3.101) becomes a function $F(\alpha)$, and $F'(\alpha) = \mathsf{E}\log\mathsf{E}_\xi \exp u(\theta)$.

We will leave to the reader the (straightforward but) unrewarding task of proving that $F''(\alpha)$ remains bounded over the domain $L\alpha e^{LD} \le 1$, so that if $\alpha' \ge \alpha$, $L\alpha' e^{LD} \le 1$,

$$|F(\alpha') - F(\alpha) - (\alpha' - \alpha)\,\mathsf{E}\log\mathsf{E}_\xi \exp u(\theta)| \le L(\alpha' - \alpha)^2 \,. \tag{3.103}$$

Proof of Theorem 3.4.2. We observe the identity

$$Z_{N,M+1} = Z_{N,M}\left\langle \exp u\left(\frac{1}{\sqrt{N}}\sum_{i\le N} g_{i,M+1}\,\sigma_i\right)\right\rangle$$

so that

$$p_{N,M+1}(u) - p_{N,M}(u) = \frac{1}{N} \mathsf{E} \log \left\langle \exp u \left(\frac{1}{\sqrt{N}} \sum_{i \le N} g_{i,M+1} \sigma_i \right) \right\rangle . \quad (3.104)$$

Comparing with (3.103), and writing

$$\Delta_{N,M} = F\left(\frac{M}{N}\right) - p_{N,M}(u) ,$$

we get

$$|\Delta_{N,M+1} - \Delta_{N,M}| \le \frac{L}{N^2} + \frac{1}{N} \Bigg| \mathsf{E} \log \mathsf{E}_\xi \exp u(\theta) \quad (3.105)$$

$$- \mathsf{E} \log \left\langle \exp u \left(\frac{1}{\sqrt{N}} \sum_{i \le N} g_{i,M+1} \sigma_i \right) \right\rangle \Bigg| .$$

We will prove that

$$\left| \mathsf{E} \log \mathsf{E}_\xi \exp u(\theta) - \mathsf{E} \log \left\langle \exp u \left(\frac{1}{\sqrt{N}} \sum_{i \le N} g_{i,M+1} \sigma_i \right) \right\rangle \right| \le \frac{L e^{LD}}{N} . \quad (3.106)$$

Thus (3.105) implies

$$|\Delta_{N,M'+1} - \Delta_{N,M'}| \le \frac{L}{N^2} + \frac{L e^{LD}}{N^2} \quad (3.107)$$

when $LM'e^{LD}/N \le 1$. We observe that $\Delta_{N,0} = 0$, because $p_{N,0}(u) = F(0) = \log 2$. Thus, to prove Theorem 3.4.2, it suffices to sum the inequalities (3.107) for $M' = 0, \ldots, M-1$.

To prove (3.106) we consider

$$g_t(\boldsymbol{\sigma}) = \sqrt{t}\, g(\boldsymbol{\sigma}) + \sqrt{1-t}\, \theta ,$$

where

$$g(\boldsymbol{\sigma}) = \frac{1}{\sqrt{N}} \sum_{i \le N} g_{i,M+1} \sigma_i , \; \theta = z\sqrt{q} + \xi\sqrt{1-q}$$

and

$$\varphi(t) = \mathsf{E} \log \mathsf{E}_\xi \langle \exp u(g_t(\boldsymbol{\sigma})) \rangle .$$

The left-hand side of (3.106) is $|\varphi(1) - \varphi(0)|$. We compute

$$\varphi'(t) = \frac{1}{2} \mathsf{E} \frac{\left\langle \left(\frac{1}{\sqrt{t}} g(\boldsymbol{\sigma}) - \frac{1}{\sqrt{1-t}} \theta \right) u'(g_t(\boldsymbol{\sigma})) \exp u(g_t(\boldsymbol{\sigma})) \right\rangle}{\mathsf{E}_\xi \langle \exp u(g_t(\boldsymbol{\sigma})) \rangle} .$$

Integrating by parts in a computation that resembles that of (3.102) we get,

$$\varphi'(t) = -\frac{1}{2}\,\mathsf{E}\,\frac{\langle (R_{1,2}-q)\,u'(g_{t,1})\,u'(g_{t,2})\exp(u(g_{t,1})+u(g_{t,2}))\rangle}{(\mathsf{E}_\xi\langle \exp u(g_t(\boldsymbol{\sigma}))\rangle)^2} \tag{3.108}$$

where

$$g_{t,\ell} = \sqrt{t}\,g(\boldsymbol{\sigma}^\ell) + \sqrt{1-t}\,\theta_\ell\,;\; \theta_\ell = z\sqrt{q} + \xi_\ell\sqrt{1-q}\,.$$

In particular (see the computation (3.99))

$$\varphi'(0) = \frac{r}{2\alpha}\,\nu(R_{1,2}-q)\,.$$

We will show in the next section that $(r/\alpha)|\nu(R_{1,2}-q)| \le L/N$, so that, using the inequality

$$|\varphi(1)-\varphi(0)-\varphi'(0)| \le \sup_{0<t<1} |\varphi''(t)|\,,$$

we have to show that $|\varphi''(t)| \le LN^{-1}e^{LD}$. To understand the procedure, suppose that we instead need to bound $|\varphi'(t)|$. We simply appeal to Lemma 3.3.6 to see that

$$|\varphi'(t)| \le L\,e^{LD}\left(\nu(|R_{1,2}-q|) + \exp\left(-\frac{N}{L}\right)\right)$$

and by Theorem 3.4.1, this is at most $Le^{LD}/\sqrt{N}$. To handle the case of $\varphi''(t)$, we simply compute it from (3.108). The derivation "brings out one more factor $|R_{1,2}-q|$" and we apply Lemma 3.3.6 and Theorem 3.4.1 as in the case of $\varphi'(t)$. □

3.5 Exponential Inequalities

This section parallels Section 2.5. We first improve the control of second moments of $R_{1,2}-q$ into the control of higher moments; then we give first applications.

Starting with this section, matters get complicated at times in this chapter. In fact, the rest of this chapter contains several of the most technically challenging results of the book. The novice reader is thus advised to move to the easy Section 2.8, where (s)he will learn the essential technique of approximate integration by parts, and after this is done, to move to Chapter 5. While Chapter 5 is not exactly technically easy, it so happened that it was written first, when the author had the energy to explain everything in extreme detail.

Theorem 3.5.1. *There exists a number L such that if $L\alpha e^{LD} \le 1$, and $|u^{(\ell)}| \le L^* e^{L^* D}\exp(N/250)$ for $\ell \le 4$, we have*

$$\forall k \ge 0\,,\quad \nu((R_{1,2}-q)^{2k}) \le \left(\frac{Lk}{N}\right)^k\,. \tag{3.109}$$

Proof. The proof will be nearly identical to that of Theorem 2.5.1. As induction hypothesis we will use (2.239). Since $\nu((R_{1,2}-q)^{2k}) \le 4^k$, we can assume $L_0\, k/N \le 4$, i.e. $k \le 4N/L_0$. Since we can assume $L_0 \ge 4$, we can assume $k \le N-1$ and thus that (2.242) holds.

Using symmetry among the sites, we have

$$\nu((R_{1,2}-q)^{2k+2}) = \nu(f)\,, \tag{3.110}$$

where

$$f = (\varepsilon_1\,\varepsilon_2 - q)\,(R_{1,2}-q)^{2k+1}\,. \tag{3.111}$$

Thus

$$\nu((R_{1,2}-q)^{2k+2}) \le \nu_0(f) + \sup_t |\nu_t'(f)|\,. \tag{3.112}$$

We compute $\nu_t'(f)$ through Proposition 3.2.3, and we compute each resulting term through Theorem 3.3.7, as we did in the proof of Theorem 3.4.1, but now with $\tau_1 = (2k+2)/(2k+1)$ and $\tau_2 = 2k+2$. Since $|f| \le 2^{2k+2}$, we get

$$\begin{aligned} |\nu_t'(f)| \le L\,\alpha\, e^{LD}\Bigg(&\nu((R_{1,2}-q)^{2k+2}) \\ &+ \frac{1}{N}\nu\left((R_{1,2}-q)^{2k+2}\right)^{\frac{2k+1}{2k+2}} + 2^{2k}\exp\left(-\frac{N}{L}\right)\Bigg)\,. \end{aligned} \tag{3.113}$$

Since $\nu_0((\varepsilon_1\,\varepsilon_2-q)(R^-_{1,2}-q)^{2k+1}) = 0$, using the inequality $|x^{2k+1}-y^{2k+1}| \le (2k+1)\,|x-y|\,(x^{2k}+y^{2k})$, we get

$$\nu_0(f) \le \frac{2(2k+1)}{N}\,\nu_0((R^-_{1,2}-q)^{2k} + (R_{1,2}-q)^{2k})\,.$$

Using (3.92), (2.239) and (2.242) we get

$$\begin{aligned} \nu_0(f) &\le \frac{8(k+1)}{N}\left(\nu((R^-_{1,2}-q)^{2k} + (R_{1,2}-q)^{2k}) + 2^{2k}\exp\left(-\frac{N}{L}\right)\right) \\ &\le \frac{32(k+1)^{k+1}}{N^{k+1}}\,L_0^k + L\,2^{2k}\exp\left(-\frac{N}{L}\right)\,. \end{aligned} \tag{3.114}$$

We note that, since we have $xy \le x^{\tau_2} + y^{\tau_1}$, we have

$$\frac{1}{N}\nu\left((R_{1,2}-q)^{2k+1}\right)^{\frac{2k+1}{2k+2}} \le (\frac{1}{N})^{2k+2} + \nu\left((R_{1,2}-q)^{2k+1}\right)\,.$$

Combining with (3.112), (3.113) and (3.114), if $L\,\alpha\,e^{LD} \le 1/2$, we get

$$\nu((R_{1,2}-q)^{2k+2}) \le \frac{L}{L_0}\left(\frac{L_0(k+1)}{N}\right)^{k+1} + L\,2^{2k}\exp\left(-\frac{N}{L}\right)\,. \tag{3.115}$$

Since, for $x > 0$, we have $(ax)^x \ge \exp(-1/a)$, we have

$$\left(\frac{L_0(k+1)}{N}\right)^{k+1} \geq 2^{2(k+1)} \exp\left(-\frac{4N}{L_0}\right)$$

so that if L_0 is large enough the first term in the right-hand side of (3.115) is larger than the second term and (3.115) implies

$$\nu((R_{1,2}-q)^{2k+2}) \leq \frac{2L}{L_0}\left(\frac{L_0(k+1)}{N}\right)^{k+1} \leq \left(\frac{L_0(k+1)}{N}\right)^{k+1} . \tag{3.116}$$

This completes the induction and the proof of Theorem 3.5.1. □

Our program is now to learn how to make second moment computations, (as in Section 2.6) and higher moment computations (as in Section 2.7). The basic method is as follows. We write

$$\nu(f) = \nu_0(f) + \int_0^1 \nu_t'(f)\,\mathrm{d}t . \tag{3.117}$$

We express $\nu_t'(f)$ with Proposition 3.2.3. We then develop a "second order version" of Theorem 3.3.7 to evaluate each term occurring in Proposition 3.2.3; that is, we replace (3.73) by

$$|\varphi(1) - \varphi(0) - \varphi'(0)| \leq \sup_{0<v<1} |\varphi''(v)| . \tag{3.118}$$

Or, in other words, we will approximate $\varphi(1)$ by $\varphi(0)+\varphi'(0)$. The resulting approximation will not depend on t, and, when substituted in (3.117), will give a relation playing the role that (2.248) played for the SK model. We realize that we are going to have a hard time summing the terms $\varphi'(0)$ given by Proposition 3.3.8 for each of the terms of Proposition 3.2.3. This difficulty is of algebraic nature, and, once we adopt the right point of view, it will turn out to be manageable. We will worry about this later, and we first prove the "second-order" version of Theorem 3.3.7. We recall the notation (3.39).

Theorem 3.5.2. *There exists a constant L with the following property. Assume that $L\alpha e^{LD} \leq 1$, and that*

$$\forall \ell \leq 6, \quad |u^{(\ell)}| \leq L^* e^{L^* D} \exp\frac{N}{250} . \tag{3.119}$$

Consider an integer m, integers $s_1, \ldots, s_m$ with $s = \sum_{\ell\leq m} s_\ell \leq 2$. Consider the functions

$$V(x_1,\ldots,x_m) = \exp\sum_{\ell\leq m} u(x_\ell) \tag{3.120}$$

$$U(x_1,\ldots,x_m) = \frac{\partial^s V}{\partial x_1^{s_1}\ldots\partial x_m^{s_m}} . \tag{3.121}$$

Consider a function f on Σ_N^m with $|f|\leq 1$. Then we have

$$\mathsf{E}\,\frac{\langle f\,U(g(\boldsymbol{\sigma}^1),\ldots,g(\boldsymbol{\sigma}^m))\rangle_{t,\sim}}{\langle V(g(\boldsymbol{\sigma}^1),\ldots,g(\boldsymbol{\sigma}^m))\rangle_{t,\sim}} = \mathrm{I}+\mathrm{II}+\mathrm{III}+\mathrm{IV}+\mathcal{R}\,, \tag{3.122}$$

where, setting

$$Z = \mathsf{E}_\xi \exp u(\theta)\,, \tag{3.123}$$

we have

$$\mathrm{I} = \mathsf{E}\,\langle f\rangle_{t,\sim}\,\mathsf{E}\left(\frac{1}{Z^m}U(\theta_1,\ldots,\theta_m)\right) \tag{3.124}$$

$$\mathrm{II} = \sum_{\ell<\ell'\le m}\mathsf{E}\langle f(R_{\ell,\ell'}-q)\rangle_{t,\sim}\,\mathsf{E}\left(\frac{1}{Z^m}\frac{\partial^2 U}{\partial x_\ell\,\partial x_{\ell'}}(\theta_1,\ldots,\theta_m)\right) \tag{3.125}$$

$$\begin{aligned}\mathrm{III} = -m\sum_{\ell\le m}&\mathsf{E}\langle f(R_{\ell,m+1}-q)\rangle_{t,\sim}\\ &\times\mathsf{E}\left(\frac{1}{Z^{m+1}}\frac{\partial U}{\partial x_\ell}(\theta_1,\ldots,\theta_m)\,V_1(\theta_{m+1})\right)\end{aligned} \tag{3.126}$$

where $V_1(x) = u'(x)\exp u(x)$,

$$\begin{aligned}\mathrm{IV} = \frac{m(m+1)}{2}&\,\mathsf{E}\langle f(R_{m+1,m+2}-q)\rangle_{t,\sim}\\ &\times\mathsf{E}\left(\frac{1}{Z^{m+2}}U(\theta_1,\ldots,\theta_m)\,V_1(\theta_{m+1})\,V_1(\theta_{m+2})\right)\,,\end{aligned} \tag{3.127}$$

and where

$$|\mathcal{R}| \le K\left(\frac{\nu(f^2)^{1/2}}{N}+\exp\left(-\frac{N}{L}\right)\right)\,, \tag{3.128}$$

for a number K depending on D, m only.

We could get an explicit dependence of K on D and m, but this is not really needed. What is crucial is that K does not depend on u. Let us stress that the statement is true for all u and all N, so that one is permitted choose u depending on N. This is what we will do below, when u will be a suitable approximation, depending on N, of a given fixed function.

Proof. In this proof K depends on D, m only. We use (3.118) for the function φ given by (3.71) with $\langle\cdot\rangle = \langle\cdot\rangle_{t,\sim}$. The term I is $\varphi(0)$. To compute $\varphi'(0)$ we use Proposition 3.3.8. We first observe that the term (3.78) can be absorbed into $\mathcal{R}$ because of the factor $1/N$. Next we observe that when $v = 0$ the terms (3.79) to (3.81) would be the terms (3.125) to (3.127) respectively, if it were true that $\Delta_{\ell,\ell'} = R_{\ell,\ell'} - q$. This is not exactly true. The corresponding error is given by (3.75) and yields

$$|\varphi'(0) - (\mathrm{II}+\mathrm{III}+\mathrm{IV})| \le \frac{K}{N}\,\mathsf{E}\,\langle|f|\rangle_{t,\sim} \le \frac{K}{N}\left(\nu(f^2)^{1/2}+\exp\left(-\frac{N}{L}\right)\right)$$

using Corollary 3.4.6. Thus we have to prove that $|\varphi''(v)|$ can be bounded by the right-hand side of (3.128). Proposition 3.3.8 asserts that $\varphi'(v)$ is sum of terms

$$\pm \mathsf{E}\left(\frac{1}{Z_v^n}\langle f\, \Delta_{\ell,\ell'}\, U'(x_1,\ldots,x_n)\rangle_{t,\sim}\right) , \tag{3.129}$$

where $n = m, m+1$ or $m+2$, and

$$U' = \frac{\partial^s}{\partial x_1^{s_1}\ldots\partial x_n^{s_n}} \exp\left(\sum_{\ell\le n} u(x_\ell)\right) \tag{3.130}$$

for $s = s_1 + \cdots + s_n \le 4$, and of similar terms with $(t-1)/(2N)$ instead of $\Delta_{\ell,\ell'}$. Thus the second derivative will be a sum of terms

$$\pm \mathsf{E}\left(\frac{1}{Z_v^n}\langle f\, \Delta_{\ell,\ell'}\, \Delta_{\ell_1,\ell_2}\, U'(x_1,\ldots,x_n)\rangle_{t,\sim}\right) \tag{3.131}$$

where now $n \le m+4$ and U' is as in (3.131), but with $s \le 6$, and of similar terms where $\Delta_{\ell,\ell'}$ or $\Delta_{\ell,\ell'}$ and Δ_{ℓ_1,ℓ_2} are replaced by terms of the type $(t-1)/(2N)$. We use Lemma 3.3.6 with $B = (L^* e^{L^* D}\exp(N/250))^6 e^{(m+4)D}$ to obtain that the term (3.131) is bounded by

$$K\left(\mathsf{E}\,\langle |f|\,|\Delta_{\ell,\ell'}|\,|\Delta_{\ell_1,\ell_2}|\rangle_{t,\sim} + \exp\left(-\frac{N}{L}\right)\right) . \tag{3.132}$$

This is possible because $s \le 6$ in (3.130), and we control now $|u^{(\ell)}|$ for $\ell \le 6$ by (3.119). We use (3.75) to bound the term (3.132) by

$$K\left(\mathsf{E}\,\langle |f|\,|R_{\ell,\ell'}-q|\,|R_{\ell_1,\ell_2}-q|\rangle_{t,\sim} + \frac{1}{N}\,\mathsf{E}\,\langle |f|\rangle_{t,\sim} + \exp\left(-\frac{N}{L}\right)\right) ,$$

and then Corollary 3.4.6 to bound it by

$$K\left(\nu(|f|\,|R_{\ell,\ell'}-q|\,|R_{\ell_1,\ell_2}-q|) + \frac{1}{N}\,\nu(|f|) + \exp\left(-\frac{N}{L}\right)\right) .$$

We then use Hölder's inequality

$$\nu(|f|\,|R_{\ell,\ell'}-q|\,|R_{\ell_1,\ell_2}-q|) \le \nu(f^2)^{1/2}\,\nu((R_{1,2}-q)^4)^{1/2}$$

and Theorem 3.5.1 with $k=4$ to get the desired result. □

A drawback of Theorem 3.5.2 is that it involves the quantities $\mathsf{E}\,\langle f(R_{\ell,\ell'}-q)\rangle_{t,\sim}$ rather than the quantities $\nu(f(R_{\ell,\ell'}-q))$, so we have to relate these. This is made easier by the extra factor $R_{\ell,\ell'}-q$.

Lemma 3.5.3. *Under the conditions of Theorem 3.5.2 we have, given t, t' in $[0,1]$ that*

$$|\mathsf{E}\,\langle f(R_{\ell,\ell'}-q)\rangle_{t,\sim} - \nu_t(f(R_{\ell,\ell'}-q))| \le K\left(\frac{\nu(f^2)^{1/2}}{N} + \exp\left(-\frac{N}{L}\right)\right) \quad (3.133)$$

$$|\nu_t(f(R_{\ell,\ell'}-q)) - \nu_{t'}(f(R_{\ell,\ell'}-q))| \le K\left(\frac{\nu(f^2)^{1/2}}{N} + \exp\left(-\frac{N}{L}\right)\right) . \quad (3.134)$$

This will be used in the following form.

Corollary 3.5.4. *In Theorem 3.5.2 we can replace the terms $\mathsf{E}\,(\langle f(R_{\ell,\ell'} - q)\rangle_{t,\sim})$ either by $\nu_0(f(R_{\ell,\ell'} - q))$ or by $\nu_1(f(R_{\ell,\ell'} - q))$.*

Proof of Lemma 3.5.3. To prove (3.133), we write

$$\nu_t(f(R_{\ell,\ell'} - q)) = \mathsf{E}\,\frac{\langle f(R_{\ell,\ell'} - q)\, V(g(\boldsymbol{\sigma}^1), \ldots, g(\boldsymbol{\sigma}^n))\rangle_{t,\sim}}{\langle V(g(\boldsymbol{\sigma}^1), \ldots, g(\boldsymbol{\sigma}^n))\rangle_{t,\sim}}$$

where $n \le m+2$ and

$$V(x_1, \ldots, x_n) = \exp\left(\sum_{\ell \le n} u(x_\ell)\right) .$$

We then use Theorem 3.3.7 with $s = 0$, $\tau_1 = 4/3$ and $\tau_2 = 4$ to see that the left-hand side of (3.133) is bounded by

$$K\left(\mathsf{E}\,\langle |f(R_{\ell,\ell'} - q)|^{4/3}\rangle_{t,\sim}^{3/4}\, \mathsf{E}\,\langle |R_{1,2} - q|^4\rangle_{t,\sim}^{1/4} + \exp\left(-\frac{N}{L}\right)\right) .$$

Now since

$$\mathsf{E}\,\langle |f(R_{\ell,\ell'} - q)|^{4/3}\rangle_{t,\sim} \le \mathsf{E}\langle f^2\rangle_{t,\sim}^{2/3}\, \mathsf{E}\,\langle (R_{1,2} - q)^4\rangle_{t,\sim}^{1/3}$$

Corollary 3.4.6 and Theorem 3.5.1 (for $k = 4$) provide the required bound.

To prove (3.134), we prove that $\nu_t'(f(R_{\ell,\ell'} - q))$ is bounded by the right-hand side of (3.134). We use Proposition 3.3.8 and we compute each term with Theorem 3.3.7. The term $R_{\ell,\ell'} - q$ creates an extra factor $1/\sqrt{N}$, and Corollary 3.4.6 and Theorem 3.5.1 (for $k = 4$) provide the required bound. □

Proposition 3.5.5. *Under the conditions of Theorem 3.5.1, we have*

$$|\nu(R_{1,2} - q)| \le \frac{L}{N} .$$

The proof of this result will illustrate all the steps of the mechanisms that will be used in the next section to obtain very complete results, except the issue of the algebraic computations, an issue that will be bypassed here.

Proof. We use symmetry between sites to write

$$\nu(R_{1,2} - q) = \nu(f) \ ,$$

where $f = \varepsilon_1 \varepsilon_2 - q$. Thus by Lemma 3.2.2, we have $\nu_0(f) = \mathsf{E}\,\mathrm{th}^2 Y - q = 0$. We write

$$\nu(f) = \int_0^1 \nu_t'(f)\,\mathrm{d}t \ . \tag{3.135}$$

We compute $\nu_t'(f)$ from Proposition 3.2.3, and in each term we express $\langle\cdot\rangle_t$ as a function of $\langle\cdot\rangle_{t,\sim}$ as in (3.31). We use Corollary 3.5.4 to compute each term. We find

$$|\nu_t'(f) - A| \le \frac{K}{N} \tag{3.136}$$

where A is a sum of terms

$$C_{\ell,\ell'}\,\nu_0((\varepsilon_1 \varepsilon_2 - q)\,\varepsilon_\ell\,\varepsilon_{\ell'}\,(R_{\ell,\ell'} - q)) \ ,$$

where the coefficients $C_{\ell,\ell'}$ are of the type

$$\alpha\,\mathsf{E}\,\frac{1}{Z^m}\,\frac{\partial^s V}{\partial x_1^{s_1} \dots \partial x_m^{s_m}}(\theta_1, \dots, \theta_m) \ ,$$

where $Z = \mathsf{E}_\xi \exp u(\theta)$, $V(x_1, \dots, x_m) = \exp\left(\sum_{\ell \le m} u(x_\ell)\right)$ and $m \le 4$, $s = s_1 + \dots + s_m \le 5$. Thus, using Lemma 3.3.2 and the trick of (3.60) we get that $|C_{\ell,\ell'}| \le L\,\alpha\,e^{LD}$. By Lemma 3.2.2, we have

$$\begin{aligned}\nu_0((\varepsilon_1 \varepsilon_2 - q)\,\varepsilon_\ell\,\varepsilon_{\ell'}\,(R_{\ell,\ell'} - q)) &= \nu_0((\varepsilon_1 \varepsilon_2 - q)\,\varepsilon_\ell\,\varepsilon_{\ell'})\,\nu_0(R_{\ell,\ell'} - q)\\ &= \nu_0((\varepsilon_1 \varepsilon_2 - q)\,\varepsilon_\ell\,\varepsilon_{\ell'})\,\nu_0(R_{1,2} - q) \ .\end{aligned}$$

By (3.134) used for $f = 1$, we have

$$|\nu_0(R_{1,2} - q) - \nu(R_{1,2} - q)| \le \frac{L}{N} \ .$$

We then see from (3.136) that

$$|\nu_t'(f) - B\,\nu(R_{1,2} - q)| \le \frac{L}{N} \ ,$$

where $|B| \le L\,\alpha\,e^{LD}$. Using (3.135) we get

$$|\nu(R_{1,2} - q)(1 - B)| \le \frac{L}{N}$$

so that $|\nu(R_{1,2} - q)| \le L/N$ if $|B| \le 1/2$. □

A last objective is to prove the result corresponding to Theorem 2.4.12, by proving that if $L\alpha \exp LD \le 1$, given n, we can find independent standard normal r.v. $z_1, \ldots, z_n$ such that

$$\mathsf{E}\sum_{i\le n}(\langle\sigma_i\rangle - \mathrm{th}\,(z_i\sqrt{r}))^2 \le \frac{K}{N}\,. \tag{3.137}$$

The main obstacle is that this really requires (as Theorem 2.4.12 does) tools that are more general than those that we have developed. A clean proof would require to build these tools patiently, which space does not permit here. Our solution is to state this result as a research problem.

Research problem 3.5.6. (Level 1) Prove (3.137).

Since the approach most likely to succeed is not the one we used in Theorem 2.4.12, we will give a few hints. These hints do not provide a complete sketch of solution; in particular some extra ideas might be needed to reach the rate K/N rather than $K/\sqrt{N}$ in (3.137). The basic issue is to find the appropriate version of Lemma 2.4.13. Consider the Hamiltonian

$$-H_{N-1,M} = \sum_{k\le M} u(S_{k,0}) = \sum_{k\le M} u\left(\frac{1}{\sqrt{N}}\sum_{i\le N-1} g_{i,k}\,\sigma_i\right)\,,$$

and denote by $\langle\cdot\rangle_-$ the corresponding Gibbs' measure. The natural statement corresponding to Lemma 2.4.13 is that

$$\mathsf{E}\left(\langle\sigma_N\rangle - \mathrm{th}\left(\frac{1}{\sqrt{N}}\sum_{k\le N} g_{N,k}\langle u'(S_{k,0})\rangle_-\right)\right)^2 \le \frac{L}{N}\,. \tag{3.138}$$

To prove this, we suggest to consider the function

$$\varphi(t) = \mathsf{E}\left((\langle\sigma_N\rangle_t - \mathrm{th}\,Y_t)^2\right)\,,$$

where

$$Y_t = \sqrt{\frac{t}{N}}\sum_{k\le M} g_{N,k}\langle u'(S_{k,0})\rangle_- + \sqrt{1-t}\,Y\,,$$

so that $\varphi(0) = 0$, and $\varphi(1)$ is the left-hand side of (3.138). To prove (3.138) the natural approach would be to show that $|\varphi'(0)| \le L/N$ and $|\varphi''(t)| \le L/N$.

3.6 Central Limit Theorems

These theorems will be proved following the method of Section 2.7, and it seems necessary before the real work starts to review what the tools were.

A first important ingredient consists of (2.237) and (2.247). Combined with (2.248) these relations imply that when f is a product of k terms of the type $R^-_{\ell,\ell'} - q$, then

$$\nu(f) = \nu_0(f) + O(k+1) \tag{3.139}$$

$$\nu(f) = \nu_0(f) + \nu_0'(f) + O(k+2) \tag{3.140}$$

where $O(k)$ denotes a quantity such that $|O(k)| \le KN^{-k/2}$. Usually (3.139) is applied to $f(R^-_{\ell,\ell'} - q)$ rather than f, and then achieves the same accuracy as (3.140). The already proved relation (3.134) will perform the same task here.

The second pillar of Section 2.7 is (2.169), when used in the special case where $f = f^-(\varepsilon_x \varepsilon_y - q)$ (x, y integers), where f^- is a product of terms of the type $R^-_{\ell,\ell'} - q$. If f depends on n replicas, using Lemma 2.4.4, we see that

$$\begin{aligned}\nu_0'((\varepsilon_x \varepsilon_y - q) f^-) = & \sum_{\ell<\ell'\le n} b(\ell,\ell',x,y)\, \nu_0(f^-(R^-_{\ell,\ell'} - q)) \\ & - n \sum_{\ell\le n} b(\ell, n+1, x, y)\, \nu_0(f^-(R^-_{\ell,n+1} - q)) \\ & + \frac{n(n+1)}{2} b(n+1, n+2, x, y)\, \nu_0(f^-(R^-_{n+1,n+2} - q))\end{aligned} \tag{3.141}$$

where

$$b(\ell,\ell',x,y) = \beta^2 \nu_0((\varepsilon_x \varepsilon_y - q)\, \varepsilon_\ell \varepsilon_{\ell'}) \,. \tag{3.142}$$

We observe that

$$b(\ell,\ell',x,y) = b(\operatorname{card}(\{x,y\} \cap \{\ell,\ell'\}))\,, \tag{3.143}$$

where, setting $\widehat{q} = \mathsf{E}\,\mathrm{th}^4 Y$,

$$b(2) = \beta^2 \nu_0(1 - q\, \varepsilon_1 \varepsilon_2) = \beta^2(1-q^2) \tag{3.144}$$

$$b(1) = \beta^2 \nu_0((\varepsilon_1 \varepsilon_2 - q)\, \varepsilon_1 \varepsilon_3) = \beta^2(q - q^2) \tag{3.145}$$

$$b(0) = \beta^2 \nu_0((\varepsilon_1 \varepsilon_2 - q)\, \varepsilon_3 \varepsilon_4) = \beta^2(\widehat{q} - q^2) \,. \tag{3.146}$$

In particular, we observe that

$$b(n+1, n+2, x, y) = b(0) \,.$$

The specific values (3.144) to (3.146) are not so important. What seems to be essential is the relations (3.143) and (3.141), or more exactly the following consequence, (through (3.139))

$$\nu((\varepsilon_x\,\varepsilon_y - q)\,f^-) = \sum_{\ell<\ell'\le n} b(\ell,\ell',x,y)\,\nu(f^-(R_{\ell,\ell'}-q))$$
$$- n\sum_{\ell\le n} b(\ell,n+1,x,y)\,\nu(f^-(R_{\ell,n+1}-q))$$
$$+ \frac{n(n+1)}{2}\, b(0)\,\nu(f^-(R_{n+1,n+2}-q))$$
$$+ O(k+2)\,. \tag{3.147}$$

The story of this section is that this essential structure (3.143), (3.147) also holds in the case of the present model, but with more complicated values of $b(0)$, $b(1)$, $b(2)$. Once this has been proved, the proofs of Section 2.7 carry out verbatim, so that we will not repeat them and we will only compute the new values of A, B, C in Theorem 2.7.3. In order to convince the reader of the central importance of (3.143), and to make the point that a formula like (3.147) is rather manageable, we redo in this abstract setting the computation leading to (2.259). From (3.147) we have

$$\nu((\varepsilon_1-\varepsilon_2)(\varepsilon_3-\varepsilon_4)\,f^-) = \sum_{\ell<\ell'\le n} b'(\ell,\ell')\,\nu(f^-(R_{\ell,\ell'}-q))$$
$$- n\sum_{\ell\le n} b'(\ell,n+1)\,\nu(f^-(R_{\ell,n+1}-q))$$
$$+ \frac{n(n+1)}{2}\, b'(n+1,n+2)\,\nu(f^-(R_{n+1,n+2}-q))$$
$$+ O(k+2) \tag{3.148}$$

where

$$b'(\ell,\ell') = b(\ell,\ell',1,3) - b'(\ell,\ell',2,3) - b(\ell,\ell',1,4) + b(\ell,\ell',2,4)\,.$$

We claim that this is zero unless $\ell \in \{1,2\}$ and $\ell' \in \{3,4\}$. Indeed, if for example $\ell,\ell' \notin \{1,2\}$, then

$$b(\ell,\ell',1,3) = b(\ell,\ell',2,3)\,;\ b(\ell,\ell',1,4) = b(\ell,\ell',2,4)\,.$$

Thus all the numbers $b(\ell,\ell')$ are zero except

$$b(1,3) = b(2,4) = -b(2,3) = -b(2,4) = b(0) - 2b(1) + b(2)\,,$$

and we get from (3.148) that

$$\nu((\varepsilon_1-\varepsilon_2)(\varepsilon_3-\varepsilon_4)\,f^-) = (b(0)-2b(1)+b(2))$$
$$\times\nu(f^-(R_{1,3}-R_{2,3}-R_{1,4}+R_{2,4})) + O(k+2)\,. \tag{3.149}$$

Under (3.144) to (3.146), $b(0)-2b(1)+b(2)$ is the familiar quantity $\beta^2(1-2q+\widehat{q})$.

We now turn to the description of the values of $b(0)$, $b(1)$, $b(2)$ that occur in the present model. These values are complicated, although some underlying structure seems to be present. The good news is that, as explained, *all* the complication is contained there; the rest is as in Section 2.7. We consider the numbers

$$d(2) = 1 - q^2 \,;\; d(1) = q - q^2 \,;\; d(0) = \widehat{q} - q^2 \,, \tag{3.150}$$

where q is given by (3.12) and $\widehat{q} = \mathsf{E}\,\mathrm{th}^4(z\sqrt{r})$. Recalling that $\theta = z\sqrt{q} + \xi\sqrt{1-q}$, we consider the numbers

$$c(2) = \mathsf{E}\left(\frac{\mathsf{E}_\xi((u''(\theta) + u'^2(\theta))\exp u(\theta))}{\mathsf{E}_\xi \exp u(\theta)}\right)^2 \tag{3.151}$$

$$c(1) = \mathsf{E}\,\frac{\mathsf{E}_\xi((u''(\theta) + u'^2(\theta))\exp u(\theta))(\mathsf{E}_\xi(u'(\theta)\exp u(\theta)))^2}{(\mathsf{E}_\xi \exp u(\theta))^3} \tag{3.152}$$

$$c(0) = \mathsf{E}\left(\frac{\mathsf{E}_\xi(u'(\theta)\exp u(\theta))}{\mathsf{E}_\xi \exp u(\theta)}\right)^4 . \tag{3.153}$$

Maybe these formulas are easier to parse if we write $V(x) = \exp u(x)$, and

$$c(0) = \mathsf{E}\left(\frac{\mathsf{E}_\xi V'(\theta)}{\mathsf{E}_\xi V(\theta)}\right)^4 \;;\; c(1) = \mathsf{E}\,\frac{\mathsf{E}_\xi V''(\theta)(\mathsf{E}_\xi V'(\theta))^2}{(\mathsf{E}_\xi V(\theta))^3} \tag{3.154}$$

$$c(2) = \mathsf{E}\left(\frac{\mathsf{E}_\xi V''(\theta)}{\mathsf{E}_\xi V(\theta)}\right)^2 .$$

Consider the numbers

$$b_0(2) = \alpha\,(c(2)\,d(2) - 4\,c(1)\,d(1) + 3\,c(0)\,d(0)) \tag{3.155}$$

$$\begin{aligned} b_0(1) &= \alpha\,(c(1)\,d(2) + c(2)\,d(1) - 2\,c(1)\,d(1) - 3\,c(0)\,d(1) - 3\,c(1)\,d(0) \\ &\quad + 6\,c(0)\,d(0)) \end{aligned} \tag{3.156}$$

$$\begin{aligned} b_0(0) &= \alpha\,(c(2)\,d(0) + c(0)\,d(2) + 4\,c(1)\,d(1) - 8\,c(1)\,d(0) - 8\,c(0)\,d(1) \\ &\quad + 10\,c(0)\,d(0)) . \end{aligned} \tag{3.157}$$

Finally let

$$b(2) = b_0(2) - \frac{r^2}{\alpha}\,(d(2) - 4\,d(1) + 3\,d(0)) \tag{3.158}$$

$$b(1) = b_0(1) - \frac{r^2}{\alpha}\,(d(2) - 4\,d(1) + 3\,d(0)) \tag{3.159}$$

$$b(0) = b_0(0) - \frac{r^2}{\alpha}\,(d(2) - 4\,d(1) + 3\,d(0)) . \tag{3.160}$$

The term r^2/α looks absurd until one remembers that, with the notation of (3.154),

$$\frac{r^2}{\alpha} = \alpha \left(\mathsf{E} \left(\frac{\mathsf{E}_\xi V'(\theta)}{\mathsf{E}_\xi V(\theta)} \right)^2 \right)^2$$

stays bounded as $L\alpha e^{LD} \le 1$.

To make the point that (despite a negative first impression) there is some order in the expressions (3.155) and (3.157), we state the following, that is checked by simple algebra.

Lemma 3.6.1. *We have*

$$b(2) - 2\,b(1) + b(0) = \alpha\,(c(2) - 2\,c(1) + c(0))(d(2) - 2\,d(1) + d(0)) \quad (3.161)$$

$$b(2) - 4\,b(1) + 3\,b(0) = \alpha\,(c(2) - 4\,c(1) + 3\,c(0))(d(2) - 4\,d(1) + 3\,d(0))\,. \quad (3.162)$$

One should also mention that $d(2) - 4\,d(1) + 3\,d(0)$ is the familiar quantity $1 - 4\,q + 3\,\widehat{q}$.

Theorem 3.6.2. *There exists a number L with the following property. If $L\alpha \exp LD \le 1$, and if $|u^{(\ell)}| \le L^* e^{L^* D} \exp(N/250)$ for $\ell \le 6$, then, given a function f^- on Σ_{N-1}^n, that is a product of k terms of the type $R^-_{\ell,\ell'} - q$, $\ell, \ell' \le n$, and given $x, y \le n$, we have*

$$\begin{aligned}\nu((\varepsilon_x\,\varepsilon_y - q)\,f^-) = &\sum_{1 \le \ell < \ell' \le n} b(\ell, \ell', x, y)\,\nu(f^-(R_{\ell,\ell'} - q)) \\ &- n \sum_{\ell \le n} b(\ell, n+1, x, y)\,\nu(f^-(R_{\ell,n+1} - q)) \\ &+ \frac{n(n+1)}{2}\,b(0)\,\nu(f^-(R_{n+1,n+2} - q)) \\ &+ O(k+2)\,, \end{aligned} \quad (3.163)$$

where $b(\ell, \ell', x, y)$ satisfies (3.143) for the values (3.158) to (3.160).

Once this theorem is proved, (3.147) and Lemma 3.3.2 allow the proof of Theorem 2.7.3 to be copied in the present setting. Before we prove Theorem 3.6.2, we use it to compute the corresponding values of A, B, C.

From (3.149) and (3.161) we see that (2.282) will hold with

$$A^2 = \frac{1 - 2\,q + \widehat{q}}{N(1 - \alpha(1 - 2\,q + \widehat{q})(c(2) - 2\,c(1) + c(0)))}\,. \quad (3.164)$$

To see what corresponds to (2.282) we state the following.

Lemma 3.6.3. *Under the conditions of Theorem 3.6.2, and if f^- does not depend on the third replica, we have*

$$\begin{aligned}\nu((\varepsilon_1-\varepsilon_2)\,\varepsilon_3\,f^-) &= (b(2)-4\,b(1)+3\,b(0))\,\nu(f^-(R_{1,3}-R_{2,3}))\\ &\quad+(b(1)-b(0))\sum_{3\le\ell\le n}\nu(f^-(R_{1,\ell}-R_{2,\ell}-R_{1,n+1}+R_{2,n+1}))\\ &\quad+O(k+2)\,. \end{aligned}\tag{3.165}$$

Proof. We use (3.163) to see that

$$\begin{aligned}\nu((\varepsilon_1-\varepsilon_0)\,\varepsilon_3\,f^-) = &\sum_{\ell<\ell'\le n} b'(\ell,\ell')\,\nu(f^-(R_{\ell,\ell'}-q))\\ &-n\sum_{\ell\le n} b'(\ell,n)\,\nu(f^-(R_{\ell,n+1}-q))\\ &+\frac{n(n+1)}{2}\,b'(n+1,n+2)\,\nu(f^-(R_{n+1,n+2}-q))\\ &+O(k+2)\,, \end{aligned}\tag{3.166}$$

where

$$b'(\ell,\ell')=b(\ell,\ell',1,3)-b(\ell,\ell',2,3)\,.$$

As in the proof of (3.149) we see that $b'(\ell,\ell')=0$ if $\ell\notin\{1,2\}$, that $b'(1,2)=0$ and that

$$b'(1,3)=-b'(2,3)=b(2)-b(1)$$

$$\ell'\ge 4\Rightarrow b'(1,\ell')=-b'(2,\ell')=b(1)-b(0)\,.$$

Regrouping the terms for $\ell=1$ and $\ell=2$, (3.166) shows that

$$\begin{aligned}\nu((\varepsilon_1-\varepsilon_2)\,\varepsilon_3\,f^-) = &\,(b(2)-b(1))\,\nu(f^-(R_{1,3}-R_{2,3}))\\ &+\sum_{4\le\ell\le n}(b(1)-b(0))\,\nu(f^-(R_{1,\ell}-R_{2,\ell}))\\ &-n(b(1)-b(0))\,\nu(f^-(R_{1,n+1}-R_{2,n+1}))\\ &+O(k+2)\,. \end{aligned}\tag{3.167}$$

If f^- does not depend on the third replica, then

$$\nu(f^-(R_{1,n+1}-R_{2,n+1}))=\nu(f^-(R_{1,3}-R_{2,3}))\,.$$

□

If we examine the computations of Proposition 2.6.7 (and in particular (2.287)), we see that now the value of B is given by

$$B^2=\frac{1}{1-(b(2)-4\,b(1)+3\,b(0))}\left(\frac{1}{N}(q-\widehat{q})+(b(1)-b(0))\,A^2\right)\tag{3.168}$$

and if we repeat the computations of Proposition 2.6.9, we get that

$$C^2 = \frac{1}{1-(b(2)-4\,b(1)+3\,b(0))} \left(\frac{1}{N}(\widehat{q}-q^2)+b(0)\,A^2+(4\,b(1)-6\,b(0))\,B^2\right) . \qquad (3.169)$$

Theorem 3.6.4. *With the notations of Theorem 3.6.2, and if* $L\alpha e^{LD} \le 1$, $|u^{(\ell)}| \le L^* e^{L^* D} \exp(N/250)$ *for* $\ell \le 6$ *then (2.327) holds for the values of* A, B, C *given in (3.164), (3.168) and (3.169).*

We now turn to the proof of Theorem 3.6.2. This proof is not difficult but requires care and patience. Our method is to compute $\nu_t'((\varepsilon_x\,\varepsilon_y - q)\,f^-)$ through Proposition 3.2.3 and Theorem 3.5.2. The first step is to show that the term (3.24) does not contribute.

Lemma 3.6.5. *Under the conditions of Theorem 3.6.2, we have*

$$\sum_{\ell\le n} \nu_t((\varepsilon_x\,\varepsilon_y - q)\,f^-(u''+u'^2)(S^\ell_{M,t})) - n\,\nu_t((\varepsilon_x\,\varepsilon_y - q)\,f^-(u''+u'^2)(S^{n+1}_{M,t})) = O(k+2) . \qquad (3.170)$$

It is understood here and everywhere that the constant implicit in the quantity $O(k+2)$ does not depend on t.

Proof. We write $\langle\cdot\rangle_t$ as a function of $\langle\cdot\rangle_{t,\sim}$, and we apply Theorem 3.5.2 to each term (with $U = \partial^2 V/\partial x_\ell^2$ for the $\ell^{\rm th}$-term of the sum). The contributions of the terms I cancel out, so all we have to show is that

$$\mathsf{E}\,\langle(\varepsilon_x\,\varepsilon_y - q)\,f^-(R_{\ell,\ell'}-q)\rangle_{t,\sim} = O(k+2) .$$

Since $\sim$ means that we have replaced M by $M-1$, it suffices to show that

$$\nu_t((\varepsilon_x\,\varepsilon_y - q)\,f^-(R^-_{\ell,\ell'}-q)) = O(k+2) . \qquad (3.171)$$

Now, by Lemma 3.2.2, and since $\nu_0(\varepsilon_x\,\varepsilon_y - q) = 0$, (3.171) is true for $t=0$, so it is true for all t by (3.134). □

We develop now a tool to compute the terms (3.25). We keep in mind that to prove (3.163) in the right-hand side, we can replace ν by ν_0 and each term of the type R_{ℓ_1,ℓ_2} by $R^-_{\ell_1,\ell_2}$. All of our subsequent lemmas are formulated in term of these quantities. This makes the use of Lemma 3.2.2 easier.

Lemma 3.6.6. *Under the conditions of Theorem 3.6.2, if* f *depends on* n *replicas, if* $\ell,\ell' \le n$, *and if* $\nu(f^2) \le K/N^k$, *then*

$$\nu_t(f\,u'(S^\ell_{M,t})\,u'(S^{\ell'}_{M,t})) = \frac{r}{\alpha}\,\mathsf{E}\,\langle f\rangle_{t,\sim} + \sum_{\ell_1<\ell_2\le n} c(\ell,\ell',\ell_1,\ell_2)\,\nu_0(f(R^-_{\ell_1,\ell_2}-q))$$

$$
\begin{aligned}
&- n \sum_{\ell_1 \le n} c(\ell, \ell', \ell_1, n+1)\, \nu_0(f(R^-_{\ell_1, n+1} - q)) \\
&+ \frac{n(n+1)}{2} c(0)\, \nu_0(f(R^-_{n+1,n+2} - q)) \\
&+ O(k+2)\,, \qquad (3.172)
\end{aligned}
$$

where

$$c(\ell, \ell', \ell_1, \ell_2) = c(\mathrm{card}\{\ell, \ell'\} \cap \{\ell_1, \ell_2\})$$

and $c(2)$, $c(1)$ *and* $c(0)$ *are given by (3.151) to (3.153).*

This will be in particular applicable to the case where $f = (\varepsilon_x \varepsilon_y - q)\, \varepsilon_\ell \varepsilon_{\ell'} f^-$, where f^- is the product of k terms of the type $R^-_{\ell_1,\ell_2} - q$.

Proof. We express $\langle \cdot \rangle_t$ as a function of $\langle \cdot \rangle_{t,\sim}$, and we use Theorem 3.5.2 with $s_\ell = s_{\ell'} = 1$ and $s_{\ell''} = 0$ if $\ell'' \notin \{\ell, \ell'\}$. Now,

$$\mathsf{E}_\xi\, U(\theta_1, \ldots, \theta_n) = (\mathsf{E}_\xi\, u'(\theta) \exp u(\theta))^2\, (\mathsf{E}_\xi \exp u(\theta))^{n-2}$$

and hence

$$\mathsf{E}\left(\frac{U(\theta_1, \ldots, \theta_n)}{Z^n}\right) = \mathsf{E}\left(\frac{(\mathsf{E}_\xi (u'(\theta) \exp u(\theta)))^2}{Z^2}\right) = \frac{r}{\alpha}\,.$$

Similarly,

$$
\begin{aligned}
\mathsf{E}\left(\frac{1}{Z^{n+2}} U(\theta_1, \ldots, \theta_n)\, V_1(\theta_{n+1})\, V_1(\theta_{n+2})\right) &= \mathsf{E}\left(\frac{1}{Z^4} (\mathsf{E}_\xi (u'(\theta) \exp u(\theta)))^4\right) \\
&= c(0)\,,
\end{aligned}
$$

$$\mathsf{E}\left(\frac{1}{Z^{n+1}} \frac{\partial U}{\partial x_{\ell_1}} (\theta_1, \ldots, \theta_n)\, V_1(\theta_{n+1})\right) = c(\mathrm{card}(\{\ell, \ell'\} \cap \{\ell_1, n+1\}))\,,$$

$$\mathsf{E}\left(\frac{1}{Z^n} \frac{\partial^2 U}{\partial x_{\ell_1} \partial x_{\ell_2}} (\theta_1, \ldots, \theta_n)\right) = c(\mathrm{card}(\{\ell, \ell'\} \cap \{\ell_1, \ell_2\}))\,.$$

To conclude the proof we have to show that in the expression given by Theorem 3.5.2 we can replace each term $\mathsf{E}\langle f(R_{\ell,\ell'} - q)\rangle_{t,\sim}$ by $\nu_0(f(R^-_{\ell,\ell'} - q))$ within error $O(k+2)$. This follows from Lemma 3.5.3. □

In order to perform the algebra, it helps to introduce the quantity $e_i = \langle \sigma_i \rangle_0$, and

$$T^0_{\ell,\ell'} = \frac{1}{N} \sum_{i \le N-1} (\sigma_i^\ell - e_i)(\sigma_i^{\ell'} - e_i)$$

$$T^0_\ell = \frac{1}{N} \sum_{i \le N-1} (\sigma_i^\ell - e_i)\, e_i$$

$$T^0 = \frac{1}{N} \sum_{i \le N-1} e_i^2 - q \,.$$

A difference between these and the quantities $T, T_\ell, T_{\ell,\ell'}$ of Section 2.7 is that we recenter at $\langle \sigma_i \rangle_0$ rather than at $\langle \sigma_i \rangle$. Still we have the relation

$$R^-_{\ell_1,\ell_2} - q = T^0_{\ell_1,\ell_2} + T^0_{\ell_1} + T^0_{\ell_2} + T^0 \,, \tag{3.173}$$

that replaces (2.266)

Lemma 3.6.7. *Under the conditions of Lemma 3.6.6, we have*

$$\begin{aligned} \nu_t(f\, u'(S^\ell_{M,t})\, u'(S^{\ell'}_{M,t})) &= \frac{r}{\alpha} \mathsf{E}\, \langle f \rangle_{t,\sim} \\ &\quad + \sum_{\ell_1 < \ell_2} c(\ell,\ell',\ell_1,\ell_2)\, \nu_0(f\, T^0_{\ell_1,\ell_2}) \\ &\quad + \sum_{\ell_1} c(\ell_1;\ell,\ell')\, \nu_0(f\, T^0_{\ell_1}) \\ &\quad + (c(2) - 4\, c(1) + 3\, c(0))\, \nu_0(f\, T^0) \\ &\quad + O(k+2) \,, \end{aligned} \tag{3.174}$$

where

$$c(\ell_1;\ell,\ell') = \begin{cases} 2\, c(1) - 3\, c(0) & \text{if} \quad \ell_1 \notin \{\ell,\ell'\} \\ c(2) - 2\, c(1) & \text{if} \quad \ell_1 \in \{\ell,\ell'\} \,. \end{cases}$$

Proof. We simply substitute the relation (3.173) in the terms of Lemma 3.6.6, and we regroup the terms. This is where care and patience are required. The coefficient of $\nu_0(f\, T^0)$ is

$$\sum_{\ell_1 < \ell_2 \le n} c(\ell,\ell',\ell_1,\ell_2) - n \sum_{\ell_1 \le n} c(\ell,\ell',\ell_1,n+1) + \frac{n(n+1)}{2} c(0) \,. \tag{3.175}$$

In the first sum, one term exactly is $c(2)$. There are $(n-2)(n-3)/2$ terms for which $\{\ell_1,\ell_2\} \cap \{\ell,\ell'\} = \emptyset$, and which are equal to $c(0)$. Since the sum has $n(n-1)/2$ terms, there are exactly

$$\frac{n(n-1)}{2} - \frac{(n-2)(n-3)}{2} - 1 = 2n - 4$$

terms for which $\operatorname{card} \{\ell_1,\ell_2\} \cap \{\ell,\ell'\} = 1$, and which are equal to $c(1)$. The second sum of (3.175) is $2\, c(1) + (n-2)\, c(0)$ so that (3.175) is

$$\begin{aligned} &c(2) - 4\, c(1) + c(0) \left(\frac{(n-2)(n-3)}{2} - n(n-2) + \frac{n(n+1)}{2} \right) \\ &= c(2) - 4\, c(1) + 3\, c(0) \,. \end{aligned}$$

To compute the coefficient of $\nu_0(f\, T^0_{\ell_1})$, we observe that we can assume $\ell_1 \le n$, for otherwise $\nu_0(f\, T^0_{\ell_1}) = 0$. This coefficient is then

$$\sum_{\ell_2 \le n, \ell_2 \ne \ell_1} c(\ell, \ell', \ell_1, \ell_2) - n\, c(\ell, \ell', \ell_1, n+1)\ .$$

When $\ell_1 \notin \{\ell, \ell'\}$, this is

$$2\,c(1) + (n-3)\,c(0) - n\,c(0) = 2\,c(1) - 3\,c(0)\ ,$$

while if $\ell_1 \in \{\ell, \ell'\}$ this is

$$c(2) + (n-2)\,c(1) - n\,c(1) = c(2) - 2\,c(1)\ .$$

□

Lemma 3.6.8. *Under the conditions of Theorem 3.6.2, the contribution of the terms* II *of Proposition 3.2.3 applied to* $f = (\varepsilon_x\, \varepsilon_y - q)\, f^-$ *is* I+II+$O(k+2)$ *where*

$$\begin{aligned} \mathrm{I} = r\Bigg(&\sum_{\ell < \ell' \le n} \mathsf{E}\, \langle \varepsilon_\ell\, \varepsilon_{\ell'}\, (\varepsilon_x\, \varepsilon_y - q)\, f^- \rangle_{t,\sim} - n \sum_{\ell \le n} \mathsf{E}\, \langle \varepsilon_\ell\, \varepsilon_{n+1}\, (\varepsilon_x\, \varepsilon_y - q)\, f^- \rangle_{t,\sim} \\ &+ \frac{n(n+1)}{2}\, \mathsf{E}\, \langle \varepsilon_{n+1}\, \varepsilon_{n+2}\, (\varepsilon_x\, \varepsilon_y - q)\, f^- \rangle_{t,\sim} \Bigg)\ . \end{aligned} \tag{3.176}$$

$$\begin{aligned} \mathrm{II} = &\sum_{\ell < \ell' \le n} b_0(\ell, \ell', x, y)\, \nu_0(f^-(R^-_{\ell,\ell'} - q)) \\ &- n \sum_{\ell \le n} b_0(\ell, n+1, x, y)\, \nu_0(f^-(R^-_{\ell,n+1} - q)) \\ &+ \frac{n(n+1)}{2}\, b_0(0)\, \nu_0(f^-(R^-_{n+1,n+2} - q))\ , \end{aligned} \tag{3.177}$$

where

$$b_0(\ell, \ell', x, y) = b_0(\mathrm{card}(\{\ell, \ell'\} \cap \{x, y\}))$$

and where $b_0(2)$, $b_0(1)$, $b_0(0)$ *are given by (3.155) to (3.157).*

Proof. If we express $R^-_{\ell,\ell'}$ as in (3.173), the computation of Lemma 3.6.7 shows that the term II of (3.177) is equal to

$$\begin{aligned} \mathrm{III} = &\sum_{\ell_1 < \ell_2} b_0(\ell_1, \ell_2, x, y)\, \nu_0(f^-\, T^0_{\ell_1,\ell_2}) + \sum_{\ell_1} b_0(\ell_1; x, y)\, \nu_0(f^-\, T^0_{\ell_1}) \\ &+ (b_0(2) - 4\, b_0(1) + 3\, b_0(0))\, \nu_0(f^-\, T^0) \end{aligned} \tag{3.178}$$

where

$$b_0(\ell_1; x, y) = \begin{cases} 2\, b_0(1) - 3\, b_0(0) & \text{if } \ell_1 \notin \{x, y\} \\ b_0(2) - 2\, b_0(1) & \text{if } \ell_1 \in \{x, y\} \end{cases}$$

and we will show that the term II of Proposition 3.2.3 is equal to I + III, when this Proposition is applied to $f = f^-(\varepsilon_x \varepsilon_y - q)$. We use (3.172) (for $f = f^-(\varepsilon_x \varepsilon_y - q)\varepsilon_\ell \varepsilon_{\ell'}$) and Lemma 3.2.2 to see that we have

$$\begin{aligned} A_{\ell,\ell'} &:= \nu_t(f^-(\varepsilon_x \varepsilon_y - q)\varepsilon_\ell \varepsilon_{\ell'} u'(S^\ell_{M,t})\, u'(S^{\ell'}_{M,t})) \\ &= \frac{r}{\alpha} \mathsf{E}\, \langle f^-(\varepsilon_x \varepsilon_y - q)\varepsilon_\ell \varepsilon_{\ell'} \rangle_{t,\sim} \\ &+ \sum_{\ell_1<\ell_2} c(\ell,\ell',\ell_1,\ell_2)\, d(x,y,\ell,\ell')\, \nu_0(f^- T^0_{\ell_1,\ell_2}) \\ &+ \sum_{\ell_1} c(\ell_1;\ell,\ell')\, d(x,y,\ell,\ell')\, \nu_0(f^- T^0_{\ell_1}) \\ &+ (c(2) - 4c(1) + 3c(0))\, d(x,y,\ell,\ell')\, \nu_0(f^- T^0) + O(k+2)\;, \end{aligned}$$

where of course $d(x,y,\ell,\ell') = d(\mathrm{card}\,\{x,y\} \cap \{\ell,\ell'\})$. We have to show that

$$\alpha\left(\sum_{\ell<\ell'\le n} A_{\ell,\ell'} - n\sum_{\ell\le n} A_{\ell,n+1} + \frac{n(n+1)}{2} A_{n+1,n+2}\right) = \mathrm{I} + \mathrm{III} + O(k+2)\;,$$

and for this we will show that the coefficients of $\nu_0(f^- T^0_{\ell_1,\ell_2})$, $\nu_0(f^- T^0_{\ell_1})$, $\nu_0(f^- T^0)$ in the left-hand side and in the right-hand side coincide, i.e. that

$$\begin{aligned} &\sum_{\ell<\ell'\le n} c(\ell,\ell',\ell_1,\ell_2)\, d(x,y,\ell,\ell') - n\sum_{\ell\le n} c(\ell,n+1,\ell_1,\ell_2)\, d(x,y,\ell,n+1) \\ &+ \frac{n(n+1)}{2} c(n+1,n+2,\ell_1,\ell_2)\, d(x,y,n+1,n+2) \\ &= \frac{1}{\alpha} b_0(\ell_1,\ell_2,x,y) \end{aligned} \tag{3.179}$$

$$\begin{aligned} &\sum_{\ell<\ell'\le n} c(\ell_1;\ell,\ell')\, d(x,y,\ell,\ell') - n\sum_{\ell\le n} c(\ell_1;\ell,n+1)\, d(x,y,\ell,n+1) \\ &+ \frac{n(n+1)}{2} c(\ell_1;n+1,n+2)\, d(x,y,n+1,n+2) \\ &= \frac{1}{\alpha} b_0(\ell_1;x,y) \end{aligned} \tag{3.180}$$

$$\begin{aligned} &(c(2) - 4c(1) + 3c(0))\left(\sum_{\ell<\ell'\le n} d(x,y,\ell,\ell') - n\sum_{\ell\le n} d(x,y,\ell,n+1)\right. \\ &\left. + \frac{n(n+1)}{2} d(x,y,n+1,n+2)\right) = \frac{1}{\alpha}(b_0(2) - 4b_0(1) + 3b_0(0))\;. \end{aligned} \tag{3.181}$$

To prove these relations we can assume $n > \ell_1, \ell_2$. This is because in Proposition 3.2.3 we can increase n if we wish; the extra terms cancel out.

The proof of these identities is as straightforward as it is boring. We start with (3.179).

Case 1. $\{\ell_1, \ell_2\} = \{x, y\}$.

As we have shown in the proof of Lemma 3.6.7, there are $2n - 4$ choices of $\ell < \ell'$ such that $\mathrm{card}\,(\{\ell, \ell'\} \cap \{x, y\}) = 1$, so that the left-hand side of (3.179) is

$$\begin{aligned} & c(2)\,d(2) + c(1)\,d(1)\,(2n-4) + c(0)\,d(0)\left(\frac{(n-2)(n-3)}{2}\right) \\ & - n(2\,c(1)\,d(1) + (n-2)\,c(0)\,d(0)) + \frac{n(n+1)}{2}\,c(0)\,d(0) \\ & = c(2)\,d(2) - 4\,c(1)\,d(1) + 3\,c(0)\,d(0) \end{aligned}$$

as in Lemma 3.6.7.

Case 2. $\mathrm{card}\,(\{x, y\} \cap \{\ell_1, \ell_2\}) = 1$.

It is easier to perform the computation when $\ell_1 = x = 1$, $\ell_2 = 2$, $y = 3$, which does not decrease generality. The contribution of the sum $\sum_{\ell < \ell' \le n}$ is better computed by calculating the sum over ℓ' for a given ℓ. This contribution is

If $\ell = 1$:

$$c(2)\,d(1) + c(1)\,d(2) + (n-3)\,c(1)\,d(1)\,,$$

corresponding respectively to $\ell' = 2$, $\ell' = 3$, $\ell' \ge 4$.

If $\ell = 2$:

$$c(1)\,d(1) + (n-3)\,c(1)\,d(0)\,,$$

corresponding respectively to $\ell' = 3$, $\ell' \ge 4$.

If $\ell = 3$:

$$(n-3)\,c(0)\,d(1)\,.$$

Moreover, the sum of the contributions for $3 \le \ell \le n$ is

$$\frac{(n-3)(n-4)}{2}\,c(0)\,d(0)\,.$$

The contribution of the sum $\sum_{\ell \le n}$ is

$$c(1)\,d(1) + c(1)\,d(0) + c(0)\,d(1) + (n-3)\,c(0)\,d(0)\,.$$

Collecting the terms and using that

$$\frac{(n-3)(n-4)}{2} - n(n-3) + \frac{n(n+1)}{2} = 6$$

to compute the coefficient of $c(0)\,d(0)$, we get a total contribution of

$$c(2)\,d(1) + c(1)\,d(2) - 2\,c(1)\,d(1) - 3\,c(0)\,d(1) - 3\,c(1)\,d(0) + 6\,c(0)\,d(0)\,.$$

Case 3. $\{x,y\}\cap\{\ell_1,\ell_2\}=\emptyset$.

We observe that there are 6 choices of pairs (ℓ,ℓ') for which $\{\ell,\ell'\}\subset\{x,y,\ell_1,\ell_2\}$, so that there are

$$\frac{n(n-1)}{2}-\frac{(n-4)(n-5)}{2}-6=4n-16$$

choices of pairs (ℓ,ℓ'), $\ell<\ell'\le n$ for which

$$\operatorname{card}\{\ell,\ell'\}\cap\{x,y,\ell_1,\ell_2\}=1\,.$$

Thus the term (3.179) is

$$\begin{aligned}&c(2)\,d(0)+c(0)\,d(2)+4\,c(1)\,d(1)+(2n-8)(c(0)\,d(1)+c(1)\,d(0))\\&+\frac{(n-4)(n-5)}{2}c(0)\,d(0)-n(2\,c(1)\,d(0)+2\,c(0)\,d(1)+(n-4)\,c(0)\,d(0))\\&+\frac{n(n+1)}{2}c(0)\,d(0)\,.\end{aligned}$$

Using that

$$\frac{(n-4)(n-5)}{2}-n(n-4)+\frac{n(n+1)}{2}=10\,,$$

this is

$$c(2)\,d(0)+c(0)\,d(2)+4\,c(1)\,d(1)-8(c(0)\,d(1)+c(1)\,d(0))+10\,c(0)\,d(0)\,.$$

We have finished the proof of (3.179). We turn to the proof of (3.180), which is just as much fun.

We set $c'(1)=c(2)-2\,c(1)$, $c'(0)=2\,c(1)-3\,c(0)$, so that $c(\ell_1;\ell,\ell')=c'(\operatorname{card}(\{\ell_1\}\cap\{\ell,\ell'\}))$.

Case 1. $\ell_1\notin\{x,y\}$.

It is easier to write the proof when $\ell_1=1$, $x=2$, $y=3$, which does not decrease generality. We study the contribution of $\sum_{\ell<\ell'\le n}$ for the various values of ℓ:

If $\ell=1$:

$$2\,c'(1)\,d(1)+(n-3)\,c'(1)\,d(0)\,.$$

If $\ell=2$:

$$c'(0)\,d(2)+(n-3)\,c'(0)\,d(1)\,.$$

If $\ell=3$:

$$(n-3)\,c'(0)\,d(1)\,.$$

Moreover the sum of the contributions for $4\le\ell\le n$ is

$$\frac{(n-3)(n-4)}{2}c(0)\,d(0)\,.$$

The sum $\sum_{\ell\le n}$ is

$$c'(1)\,d(0)+2\,c'(0)\,d(1)+(n-3)\,c'(0)\,d(0)$$

and collecting the terms we find a total contribution of

$$2\,c'(2)\,d(2)+c'(0)\,d(2)-3\,c'(1)\,d(0)-6\,c'(0)\,d(1)+6\,c'(0)\,d(0)\,.$$

Substituting the values of $c'(1)$, $c'(0)$, algebra yields the following expression

$$\begin{aligned}&2(c(2)\,d(1)+c(1)\,d(2))-3(c(2)\,d(0)+c(0)\,d(2))-16\,c(1)\,d(1)\\&+18(c(1)\,d(0)+c(0)\,d(1))-18\,c(0)\,d(0)\end{aligned}$$

and this is indeed $\alpha^{-1}(2\,b_0(1)-3\,b_0(0))$.

Case 2. $\ell_1\in\{x,y\}$.

To write the proof it is easier to assume $\ell_1=x=1$, $y=2$. This is similar to the preceding case, but easier, and left to the reader.

We have proved (3.180). The proof of (3.181) is much easier (using (3.162)) and the computation has been done many times. □

Now we come to the last piece of the proof of Theorem 3.6.2.

Lemma 3.6.9. *Under the conditions of Theorem 3.6.2, the contribution of the term* III *of Proposition 3.2.3 when applied to* $f=(\varepsilon_x\,\varepsilon_y-q)\,f^-$ *is given by* $\mathrm{I}+\mathrm{IV}$, *where* I *is as in (3.176) and where*

$$\begin{aligned}\mathrm{IV}=\;&\frac{r^2}{\alpha}\,(d(2)-4\,d(1)+3\,d(0))\Bigg(\sum_{\ell<\ell'\le n}\nu_0(f^-(R^-_{\ell,\ell'}-q))\\&-n\sum_{\ell\le n}\nu_0(f^-(R^-_{\ell,n+1}-q))+\frac{n(n+1)}{2}\,\nu_0(f^-(R^-_{n+1,n+2}-q))\Bigg)\\&+O(k+2)\,.\end{aligned}$$

Proof. The basic formula is that, under the condition of Lemma 3.6.6, we have

$$\begin{aligned}\nu_t(f)=\mathsf{E}\,\langle f\rangle_{t,\sim}+\frac{r}{\alpha}\Bigg[&\sum_{\ell_1<\ell_2\le n}\nu_0(f(R^-_{\ell,\ell'}-q))-n\sum_{\ell_1\le n}\nu_0(f(R^-_{\ell,n+1}-q))\\&+\frac{n(n+1)}{2}\,\nu_0(f(R^-_{n+1,n+2}-q))\Bigg]+O(k+2)\,.\end{aligned}\tag{3.182}$$

To see this, we apply Theorem 3.5.2 to the case $U=V$, and we see that

$$\mathsf{E}\left(\frac{1}{Z^n}\,\frac{\partial^2U}{\partial x_\ell\,\partial x_{\ell'}}\,(\theta_1,\dots,\theta_n)\right)=\mathsf{E}\left(\left(\frac{\mathsf{E}_\xi\,u'(\theta)\exp u(\theta)}{\mathsf{E}_\xi\exp u(\theta)}\right)^2\right)=\frac{r}{\alpha}$$

and

$$\mathsf{E}\left(\frac{1}{Z^{n-1}}\frac{\partial U}{\partial x_\ell}(\theta_1,\dots,\theta_n)\,V_1(\theta_{n+1})\right)$$
$$= \mathsf{E}\left(\frac{1}{Z^{n+2}}\,U(\theta_1,\dots,\theta_n)\,V_1(\theta_{n+1})\,V_2(\theta_{n+2})\right) = \frac{r}{\alpha}\,.$$

At this point we realize that it is not necessary to do any work, because the proof is going to be identical to that of Lemma 3.6.8 except that we have to replace $c(2)$, $c(1)$ and $c(0)$ by r/α, and the coefficient α in (3.155) to (3.157) by r. But doing this replaces each of $b_0(2)$, $b_0(1)$, $b_0(0)$ by $(r^2/\alpha)(d(2) - 4\,d(1) + 3\,d(0))$.

□

3.7 An Approximation Procedure

In the previous sections we have worked under the assumption that we control the first six derivatives of u, i.e.

$$\forall\,\ell \le 6\,,\ |u^{(\ell)}| \le L^* e^{L^* D} \exp\left(\frac{N}{250}\right)\,. \tag{3.183}$$

In this section we will show that this hypothesis is not necessary. For this purpose, we will show that, given a function u, we can define a regularized version v of u, which satisfies (3.183) and such that the associated Gibbs measure, in some sense, does not differ too much from the Gibbs measure associated to u. Of course the choice of v will depend on N.

We consider a number $a > 0$, depending on N, that will be chosen later. This parameter controls how well v will approximate u.

It is elementary that there exists a function ζ supported by $[-a, a]$, $\zeta \ge 0$, such that

$$\int \zeta(x)\,\mathrm{d}x = 1 \tag{3.184}$$

$$\forall\,\ell \le 6\,,\ |\zeta^{(\ell)}| \le \frac{L}{a^{\ell+1}}\,. \tag{3.185}$$

We define the function v by

$$e^v = \zeta * e^u \tag{3.186}$$

where $*$ means convolution. Since $|u| \le D$, by (3.184) we have $e^v \le e^D$, so that $v \le D$, and, similarly, $|v| \le D$. We claim that for a certain number L^* we have

$$\forall\,\ell \le 6\,,\ |v^{(\ell)}| \le \frac{L^* e^{L^* D}}{a^\ell}\,. \tag{3.187}$$

This follows from (3.185) and computation of the derivatives of v. For example,

$$v' = \frac{\zeta' * e^u}{\zeta * e^u}$$

so that

$$|v'| \le |\zeta' * e^u|\, e^D \le e^{2D} \int |\zeta'(x)|\, dx = \frac{L\, e^{2D}}{a}$$

using (3.185).

Lemma 3.7.1. *If g is a centered Gaussian r.v. and b is a number, then*

$$|\mathsf{E} \exp u(g+b) - \mathsf{E} \exp v(g+b)| \le \frac{a\, e^D}{\sqrt{\mathsf{E}\, g^2}} . \tag{3.188}$$

Proof. Consider a function V such that $V'(x) = \exp u(x)$. Since $|V'(x)| \le e^D$, and since the support of ζ is contained in $[-a, a]$, we have

$$|V - \zeta * V| \le a\, e^D . \tag{3.189}$$

Now, since

$$(\zeta * V)' = \zeta * V' = \zeta * e^u = e^v ,$$

we have

$$(V - \zeta * V)' = e^u - e^v ,$$

and integration by parts shows that

$$\mathsf{E}\,(g(V(g+b) - \zeta * V(g+b))) = \mathsf{E}\, g^2\, \mathsf{E}(\exp u(g+b) - \exp v(g+b)) .$$

Using (3.189), we get

$$\mathsf{E}\, g^2 |\mathsf{E} \exp u(g+b) - \mathsf{E} \exp v(g+b)| \le a\, e^D (\mathsf{E}\, g^2)^{1/2} .$$

□

Lemma 3.7.2. *Consider a subset A of Σ_N. Then*

$$\mathsf{E}\left(\sum_{\sigma \in A} (\exp(-H_{N,M,u}(\boldsymbol{\sigma})) - \exp(-H_{N,M,v}(\boldsymbol{\sigma})))\right)^2$$
$$\le L\, e^{2MD} (\operatorname{card} A + a^2\, N\, M^2 (\operatorname{card} A)^2) , \tag{3.190}$$

where

$$-H_{N,M,u}(\boldsymbol{\sigma}) = \sum_{k \le M} u(S_k(\boldsymbol{\sigma})) , \tag{3.191}$$

$$S_k(\boldsymbol{\sigma}) = \frac{1}{\sqrt{N}} \sum_{i \le N} g_{i,k}\, \sigma_i \tag{3.192}$$

and $H_{N,M,v}(\boldsymbol{\sigma})$ is defined similarly.

Proof. The left-hand side of (3.190) is

$$\sum_{\boldsymbol{\sigma},\boldsymbol{\sigma}'\in A} \mathsf{E}\,(B(\boldsymbol{\sigma})\,B(\boldsymbol{\sigma}')) \tag{3.193}$$

where

$$B(\boldsymbol{\sigma}) = \exp(-H_{N,M,u}(\boldsymbol{\sigma})) - \exp(-H_{N,M,v}(\boldsymbol{\sigma}))\ .$$

In the sum (3.193), we bound separately the terms for which $\boldsymbol{\sigma} = \boldsymbol{\sigma}'$ or $\boldsymbol{\sigma} = -\boldsymbol{\sigma}'$. For these we use the trivial bound $|B(\boldsymbol{\sigma})| \le 2\,e^{MD}$ so that

$$\mathsf{E}\,|B(\boldsymbol{\sigma})\,B(\boldsymbol{\sigma}')| \le 4\,e^{2MD} \tag{3.194}$$

and thus

$$\sum_{\boldsymbol{\sigma}=\pm\boldsymbol{\sigma}',\boldsymbol{\sigma},\boldsymbol{\sigma}'\in A} \mathsf{E}\,(B(\boldsymbol{\sigma})\,B(\boldsymbol{\sigma}')) \le 8\,\mathrm{card}\,A\,e^{2MD}\ . \tag{3.195}$$

Next, we consider a pair $(\boldsymbol{\sigma},\boldsymbol{\sigma}')$, with $\boldsymbol{\sigma}' \ne \pm\boldsymbol{\sigma}$, and we find a better bound than (3.194). Since $\boldsymbol{\sigma} \ne -\boldsymbol{\sigma}'$, there exists a coordinate i with $\sigma_i \ne -\sigma'_i$, i.e. $\sigma_i = \sigma'_i$. Since $\boldsymbol{\sigma} \ne \boldsymbol{\sigma}'$, there exists a coordinate j with $\sigma_j = -\sigma'_j$. Without loss of generality, we assume that $\sigma_1 = \sigma'_1$, $\sigma_2 = -\sigma'_2$. Let us denote by E_0 integration in the variables $g_{i,k}$, $k \le M$, $i = 1,2$, at all other r.v. fixed.

The key observation is that all the variables of the type

$$\sigma_1\,g_{1,k} + \sigma_2\,g_{2,k} \text{ and } \sigma'_1\,g_{1,k} + \sigma'_2\,g_{2,k} = \sigma_1\,g_{1,k} - \sigma_2\,g_{2,k}$$

are independent, so that, for E_0, $B(\boldsymbol{\sigma})$ and $B(\boldsymbol{\sigma}')$ are independent, and

$$\mathsf{E}_0(B(\boldsymbol{\sigma})\,B(\boldsymbol{\sigma}')) = \mathsf{E}_0 B(\boldsymbol{\sigma})\,\mathsf{E}_0 B(\boldsymbol{\sigma}')\,.$$

Now

$$\mathsf{E}_0 B(\boldsymbol{\sigma}) = \prod_{k\le M} X_k - \prod_{k\le M} Y_k\ , \tag{3.196}$$

where

$$X_k = \mathsf{E}_0 \exp u(S_k(\boldsymbol{\sigma}))\,;\ Y_k = \mathsf{E}_0 \exp v(S_k(\boldsymbol{\sigma}))\ .$$

We use Lemma 3.7.1 with

$$g = \frac{1}{\sqrt{N}}\,(\sigma_1\,g_{1,k} + \sigma_2\,g_{2,k})$$

$$b = \frac{1}{\sqrt{N}} \sum_{i\ge 3} \sigma_i\,g_{i,k}$$

and we see that

$$|X_k - Y_k| \le a\sqrt{N}\ e^{D}\ . \tag{3.197}$$

Next, we use that for numbers $(x_k)_{k\le M}$, $(y_k)_{k\le M}$, if $|x_k|, |y_k| \le B$, then

$$\left|\prod_{k\le M} x_k - \prod_{k\le M} y_k\right| \le \left(\sum_{k\le M} |x_k - y_k|\right) B^{M-1}\,, \tag{3.198}$$

to deduce from (3.196) and (3.197) that

$$|\mathsf{E}_0 B(\boldsymbol{\sigma})| \le a\sqrt{N}\, M\, e^{MD}\,.$$

The same inequality holds for $\boldsymbol{\sigma}'$ rather than $\boldsymbol{\sigma}$, and thus

$$\sum \mathsf{E}\,(B(\boldsymbol{\sigma})\,B(\boldsymbol{\sigma}')) \le a^2\, N\, M^2 (\mathrm{card}\, A)^2\, e^{2MD}\,,$$

where the summation is over $\boldsymbol{\sigma} \ne \pm\boldsymbol{\sigma}'$, $\boldsymbol{\sigma}, \boldsymbol{\sigma}' \in A$. Combining with (3.195) finishes the proof. □

Using the Cauchy-Schwarz inequality, from (3.190) we obtain the following

Corollary 3.7.3. *With the notation of Lemma 3.7.2, we get*

$$\mathsf{E}\left(\left|\sum_{\boldsymbol{\sigma}\in A}(\exp(-H_{N,M,u}(\boldsymbol{\sigma})) - \exp(-H_{N,M,v}(\boldsymbol{\sigma}))\right|\right)$$
$$\le L\, e^{MD}(\sqrt{\mathrm{card}\, A} + a\sqrt{N}\, M\, \mathrm{card}\, A)\,. \tag{3.199}$$

We use the notation $\langle\cdot\rangle_u$ (resp. $\langle\cdot\rangle_v$) to indicate that we deal with the Gibbs' measure relative to $H_{N,M,u}$ (resp. $H_{N,M,v}$).

Corollary 3.7.4. *Given a subset I of $\{1,\dots,N\}$, we have*

$$\mathsf{E}\left|\left\langle\prod_{i\in I}\sigma_i\right\rangle_u - \left\langle\prod_{i\in I}\sigma_i\right\rangle_v\right| \le L\, e^{2MD}(2^{-N/2} + a\sqrt{N}\, M)\,. \tag{3.200}$$

Proof. Consider the set

$$A = \left\{\boldsymbol{\sigma};\, \prod_{i\in I}\sigma_i = 1\right\}$$

so that

$$\left\langle\prod_{i\in I}\sigma_i\right\rangle_u = \langle \mathbf{1}_A\rangle_u - \langle \mathbf{1}_{A^c}\rangle_u = 2\,\langle \mathbf{1}_A\rangle_u - 1$$

and

$$\left|\left\langle\prod_{i\in I}\sigma_i\right\rangle_u - \left\langle\prod_{i\in I}\sigma_i\right\rangle_v\right| \le 2\,|\langle \mathbf{1}_A\rangle_u - \langle \mathbf{1}_A\rangle_v|\,.$$

We define

$$S_u = \sum_{\sigma \in A} \exp(-H_{N,M,u}(\boldsymbol{\sigma}))$$

$$Z_u = \sum_{\sigma} \exp(-H_{N,M,u}(\boldsymbol{\sigma}))$$

(and similarly we define S_v, Z_v) so that $\langle \mathbf{1}_A \rangle_u = S_u/Z_u$, $\langle \mathbf{1}_A \rangle_v = S_v/Z_v$, and

$$\begin{aligned} |\langle \mathbf{1}_A \rangle_u - \langle \mathbf{1}_A \rangle_v| &= \left| \frac{S_u}{Z_u} - \frac{S_v}{Z_v} \right| = \left| \frac{S_u - S_v}{Z_u} + \frac{S_v(Z_v - Z_u)}{Z_u Z_v} \right| \\ &\le \frac{|S_u - S_v|}{Z_u} + \frac{|Z_u - Z_v|}{Z_u} \end{aligned} \tag{3.201}$$

since $S_v \le Z_v$. Since $Z_u \ge 2^N e^{-MD}$, we get the result from (3.199). □

Lemma 3.7.5. *Consider for $\ell < \ell' \le n$ integers $k(\ell, \ell')$. Then*

$$\begin{aligned} &\mathsf{E} \left| \left\langle \prod_{\ell < \ell' \le n} R_{\ell,\ell'}^{k(\ell,\ell')} \right\rangle_u - \left\langle \prod_{\ell < \ell' \le n} R_{\ell,\ell'}^{k(\ell,\ell')} \right\rangle_v \right| \\ &\le L n e^{2MD} (2^{-N/2} + a\sqrt{N} M) \,. \end{aligned} \tag{3.202}$$

Proof. We write

$$R_{\ell,\ell'} = \frac{1}{N} \sum_{i \le N} \sigma_i^\ell \sigma_i^{\ell'}$$

so that

$$R_{\ell,\ell'}^{k(\ell,\ell')} = \frac{1}{N^{k(\ell,\ell')}} \sum \prod_{k \le k(\ell,\ell')} \sigma_{i(\ell,\ell',k)}^{\ell} \sigma_{i(\ell,\ell',k)}^{\ell'}$$

where the summation is over all choices of $i(\ell, \ell', 1), \ldots, i(\ell, \ell', k(\ell, \ell'))$ of integers smaller than or equal to N. Thus

$$\prod_{\ell < \ell'} R_{\ell,\ell'}^{k(\ell,\ell')} = \frac{1}{N^{\overline{k}}} \sum \prod_{\ell < \ell'} \prod_{k \le k(\ell,\ell')} \sigma_{i(\ell,\ell',k)}^{\ell} \sigma_{i(\ell,\ell',k)}^{\ell'}$$

where $\overline{k} = \sum_{\ell < \ell' \le n} k(\ell, \ell')$ and where the sum is over the $N^{\overline{k}}$ choices of indices $i(\ell, \ell', k)$, $1 \le \ell < \ell' \le n$, $k \le k(\ell, \ell')$. If we rearrange the terms of the double product, we see that it is of the type $\prod_{\ell \le n} \prod_{i \in I(\ell)} \sigma_i^\ell$ (where $I(\ell)$ is a certain subset of $\{1, \ldots, N\}$) so that

$$\left\langle \prod_{\ell < \ell'} R_{\ell,\ell'}^{k(\ell,\ell')} \right\rangle_u = \frac{1}{N^{\overline{k}}} \sum \prod_{\ell \le n} \left\langle \prod_{i \in I(\ell)} \sigma_i^\ell \right\rangle_u ,$$

where the sum contains $N^{\overline{k}}$ terms. Thus to obtain (3.202) it suffices to bound the quantities

$$\mathsf{E}\left|\prod_{\ell\le n}\Big\langle\prod_{i\in I(\ell)}\sigma_i^\ell\Big\rangle_u - \prod_{\ell\le n}\Big\langle\prod_{i\in I(\ell)}\sigma_i^\ell\Big\rangle_v\right| .$$

This follows from (3.198), (with $B=1$) and (3.200).

Lemma 3.7.6. *If $L\alpha e^{LD}\le 1$, we have*

$$|q_u - q_v| \le a .$$

Proof. It follows from Lemma 3.7.1 that if $y>1/2$ we have

$$|\mathsf{E}\exp u(x+\xi y) - \mathsf{E}\exp v(x+\xi y)| \le L a e^D .$$

By a similar argument, using the relation

$$\mathsf{E}\,(g\,U'(g)) = \frac{1}{\mathsf{E}\,g^2}\,\mathsf{E}\,g^2\,U(g) - \mathsf{E}\,U(g)$$

we get

$$|\mathsf{E}\,\xi\exp u(x+\xi y) - \mathsf{E}\,\xi\exp v(x+\xi y)| \le L a e^D .$$

Elementary manipulations (as in the proof of (3.201)) then show that

$$|\Phi_u(x,y) - \Phi_v(x,y)| \le L a e^{LD} , \tag{3.203}$$

where Φ_u is given by (3.10). As in Section 3.2, substituting (3.12) into (3.13) we obtain r_u as solution of an equation $r_u = w_u(r_u)$ where $w_u' \le 1/2$ and r_v as solution of $r_v = w_v(r_v)$, where $|w_u - w_v| \le L\alpha a e^{LD}$. Thus

$$\begin{aligned}|r_u - r_v| &= |w_u(r_u) - w_v(r_v)|\\ &\le |w_u(r_u) - w_u(r_v)| + |w_u(r_v) - w_v(r_v)|\\ &\le \frac{1}{2}|r_u - r_v| + L\alpha a e^{LD}\end{aligned}$$

and hence $|r_u - r_v| \le L\alpha a e^{LD}$, so that

$$|q_u - q_v| \le L\alpha a e^{LD} .$$

□

Theorem 3.7.7. *There exists a constant L with the following property. If $|u|\le D$ is piecewise continuous and if $L\alpha e^{LD}\le 1$, then*

$$\forall k,\ \nu((R_{1,2}-q)^{2k}) \le \left(\frac{Lk}{N}\right)^k . \tag{3.204}$$

Proof. We see from (3.187) that if

$$a = \exp\left(-\frac{N}{1500}\right) , \tag{3.205}$$

then v satisfies the conditions of Theorem 3.5.1, so that

$$\mathsf{E}\,\langle (R_{1,2} - q_v)^{2k} \rangle_v \le \left(\frac{L\,k}{N}\right)^k .$$

Expanding the product and using Lemmas 3.7.5 and 3.7.6, if $L a D \le 1$ we have

$$\begin{aligned} \mathsf{E}\,\langle (R_{1,2} - q_u)^{2k} \rangle_u &\le \mathsf{E}\,\langle (R_{1,2} - q_v)^{2k} \rangle_v + 2^{2k}(a + L\, e^{2MD}(2^{-N/2} + a\sqrt{N}\,M)) \\ &\le \left(\frac{L\,k}{N}\right)^k + L\,2^{2k} \exp\left(-\frac{N}{2000}\right) . \end{aligned} \tag{3.206}$$

Since $(ax)^x \ge \exp(-1/a)$, we have

$$\left(\frac{L\,k}{N}\right)^k \ge 2^{2k} \exp\left(-\frac{4\,N}{L}\right)$$

so that if L is large enough, (3.206) implies (3.204). □

A similar (easier) argument shows that Theorem 3.6.4 still holds under the conditions of Theorem 3.7.7. It is left to the reader.

3.8 The Bernoulli Perceptron and Approximate Integration by Parts

The Hamiltonian (3.7) is very natural from the point of view of geometry, but from the point of view of neural networks it is more appropriate to replace the variables $(g_{i,k})$ by independent r.v. $\eta_{i,k}$, where $\mathsf{P}(\eta_{i,k} = \pm 1) = 1/2$. (These variables will be called Bernoulli r.v. or sometimes random signs.) That is, we will consider the Hamiltonian

$$-H_{N,M}(\boldsymbol{\sigma}) = \sum_{k \le M} u\left(\frac{1}{\sqrt{N}} \sum_{i \le N} \eta_{i,k}\,\sigma_i\right) . \tag{3.207}$$

The study of the Hamiltonian (3.207) is more challenging than the study of the "Gaussian" Hamiltonian (3.7) because we cannot use some special properties of Gaussian r.v. (such as integration by parts.).

Research problem 3.8.1. (Level 2) Decide if the results of Section 3.6 hold for the Hamiltonian (3.207).

In this section we will prove that we can still describe the system if u is smooth enough. For example, if u is 6 times differentiable, we will prove that (3.89) can be replaced by

$$N \geq N(u) \Rightarrow \nu((R_{1,2} - q)^2) \leq \frac{L}{N} , \tag{3.208}$$

where $N(u)$ depends on u through the size of its first 6 derivatives.

When u is not differentiable, the only result we know how to obtain (in the domain $L\alpha e^{LD} \leq 1$) is the computation of

$$\lim_{N\to\infty, M/N\to\alpha} p_{N,M}(u)$$

through an approximation argument of u by a smooth function. (We refer to [142] for this.) In particular the following special case of Problem 3.8.1 is open.

Research problem 3.8.2. (Level 2) If $u(x) = \mathbf{1}_{\{x\geq 0\}}$, is it true that for α_0 small enough we have

$$\lim_{N\to\infty, M/N\to\alpha_0} \nu((R_{1,2} - q)^2) = 0 \ ? \tag{3.209}$$

Is it true that

$$\sup_N N\,\nu((R_{1,2} - q)^2) < L$$

as $\alpha = M/N \leq \alpha_0$?

Here of course q is given by (3.12) and (3.13).

Our approach to the study of (3.207) will mimic our approach to the Gaussian case. We will replace integration by parts by "approximate integration by parts". This technique of "approximate integration by parts" will be of fundamental importance in the study of the Hopfield model in the next chapter.

The basic fact is that if v is a three times differentiable function on $\mathbb{R}$, then

$$v(1) - v(-1) = v'(1) + v'(-1) + \frac{1}{2}\int_{-1}^{1} (x^2 - 1)\, v'''(x)\, dx . \tag{3.210}$$

This is proved by integrating the last term by parts,

$$\int_{-1}^{1} \frac{1}{2}(x^2 - 1)\, v'''(x)\, dx = \frac{1}{2}(x^2 - 1)\, v''(x)\Big|_{x=-1}^{x=1} - \int_{-1}^{1} x v''(x)\, dx ,$$

$$\int_{-1}^{1} x v''(x)\,\mathrm{d}x = x v'(x)\Big|_{x=-1}^{x=1} - \int_{-1}^{1} v'(x)\,\mathrm{d}x$$
$$= v'(1) + v'(-1) - (v(1) - v(-1)) .$$

If η is a r.v. such that $\mathsf{P}(\eta = \pm 1) = 1/2$, then (3.210) implies

$$\mathsf{E}\,\eta v(\eta) = \mathsf{E}\,v'(\eta) + \frac{1}{4}\int_{-1}^{1} (x^2 - 1)\, v'''(x)\,\mathrm{d}x . \tag{3.211}$$

We will call $\mathsf{E}\,v'(\eta)$ the *main term* and the last term the *error term*. This term will have a tendency to be small because v will depend on η through $\eta/\sqrt{N}$, so the third derivative creates a factor $N^{-3/2}$. We will typically bound the error term through the crude inequality

$$\left|\frac{1}{4}\int_{-1}^{1} (x^2 - 1)\, v'''(x)\,\mathrm{d}x\right| \le \sup_{|x|\le 1} |v'''(x)| . \tag{3.212}$$

To define the smart path we will proceed as in (3.20) and (3.21), except that now we define

$$S_{k,t}(\boldsymbol{\sigma}) = \frac{1}{\sqrt{N}} \sum_{i\le N-1} \eta_{i,k}\,\sigma_i + \sqrt{\frac{t}{N}}\,\eta_{N,k}\,\sigma_N . \tag{3.213}$$

We define ν_t as in Section 3.2.

Proposition 3.8.3. *If f is a function on n replicas, we have*

$$\nu_t'(f) = \mathrm{I} + \mathrm{II} + \mathrm{III} + \mathcal{R} , \tag{3.214}$$

where I, II, III *are as in Proposition 3.2.3, and where*

$$|\mathcal{R}| \le \frac{\alpha\, K(n, D')}{N}\, \nu_t(|f|) \tag{3.215}$$

for

$$D' = \sup_{\ell\le 3} |u^{(\ell)}| . \tag{3.216}$$

Proof. The proof will repeat that of Proposition 3.2.3, replacing exact integration by parts with approximate integration by parts. Thus, rather than (3.29), we have

$$\nu_t'(f) = \mathrm{IV} + \mathrm{V} \tag{3.217}$$

where

$$\mathrm{IV} = \frac{\alpha}{2}\sqrt{\frac{N}{t}}\left(\sum_{\ell\le n} \mathsf{E}\,\eta_M \langle \varepsilon_\ell\, u'(S^\ell_{M,t})\, f\rangle_t - n\,\mathsf{E}\,\eta_M \langle \varepsilon_{n+1}\, u'(S^{n+1}_{M,t})\, f\rangle_t\right) ,$$

where V is as in the proof of Proposition 3.2.3, and where $\eta_M = \eta_{N,M}$. We will use approximate integration by parts to compute $\mathsf{E}\,\eta_M \langle \varepsilon_\ell\, u'(S^\ell_{M,t})\, f\rangle_t$. We consider the function $v_\ell(x)$ that is obtained by replacing each occurrence of η_M in the quantity $\langle \varepsilon_\ell\, u'(S^\ell_{M,t})\, f\rangle_t$ by x, and considering all the other r.v. $\eta_{i,k}$ as fixed. It should be obvious that if we add for the terms in IV the contributions of the "main terms" in the quantities $\mathsf{E}\, v'_\ell(\eta)$ we will get the terms I + II of Proposition 3.2.3, so that all we have to do is to take care of the error term. Since each occurrence of η_M in $\langle \varepsilon_\ell\, u'(S^\ell_{M,t})\, f\rangle_t$ has a factor $\sqrt{t/N}$, it should be obvious that

$$|v'''_\ell(x)| \le \frac{t^{3/2}}{N^{3/2}} K(n)\, D'^3 \langle |f| \rangle_{t,x} \,, \tag{3.218}$$

where $\langle \cdot \rangle_{t,x}$ means that each occurrence of η_M is replaced by x. We would like to relate $\langle |f| \rangle_{t,x}$ with $\langle |f| \rangle_t$. For this we simply observe that

$$\left| \frac{\mathrm{d}}{\mathrm{d}x} \langle |f| \rangle_{t,x} \right| \le \frac{K(n)\, D'}{\sqrt{N}} \langle |f| \rangle_{t,x}$$

so that, by integration

$$\sup_{|x|\le 1} \langle |f| \rangle_{t,x} \le K(n, D') \langle |f| \rangle_{t,1} \,.$$

Since expectation averages η_M over ± 1, we get

$$\mathsf{E} \sup_{|x|\le 1} \langle |f| \rangle_{t,x} \le K(n, D')\, \nu_t(|f|) \,. \tag{3.219}$$

Combining with (3.218) we get

$$\mathsf{E} \sup_{|x|\le 1} |v'''_\ell(x)| \le \frac{t^{3/2}}{N^{3/2}} K(n, D')\, \nu_t(|f|) \,.$$

Using (3.211) and (3.212) completes the proof. □

Remark. There is a lot of room in the proof of (3.219). It would suffice to know that $|\varepsilon_\ell| \le K\sqrt{N}$, or even, using Lemma 3.4.5, to know that $\nu_{t,x}(\{|\varepsilon_\ell| \ge K\sqrt{N}\})$ is very small. This will be useful in Chapter 4.

We also used integration by parts in the proof of Theorem 3.3.7. Rather than reproving a suitable version of this result it is easier to reduce to the Gaussian case by using the following principle.

Proposition 3.8.4. *Consider a function f on Σ_N^n, consider $\ell_1, \ell_2 \le n$, and consider the function*

$$V(x_1, \ldots, x_n) = \exp\left(\sum_{\ell \le n} u(x_\ell) \right)$$

and $U = \frac{\partial^2 V}{\partial x_{\ell_1} \partial x_{\ell_2}}$. *Consider* $0 < t < 1$ *and set*

$$g(\boldsymbol{\sigma}) = \frac{1}{\sqrt{N}} \sum_{i \le N-1} g_i \sigma_i + \sqrt{\frac{t}{N}} g_N \sigma_N$$

$$\eta(\boldsymbol{\sigma}) = \frac{1}{\sqrt{N}} \sum_{i \le N-1} \eta_i \sigma_i + \sqrt{\frac{t}{N}} \eta_N \sigma_N ,$$

where $(g_i)_{i \le N}$ *are independent standard Gaussian,* $(\eta_i)_{i \le N}$ *are independent,* $\mathsf{P}(\eta_i = \pm 1) = 1/2$. *Then*

$$\left| \mathsf{E} \frac{\langle f\, U(g(\boldsymbol{\sigma}^1), \ldots, g(\boldsymbol{\sigma}^n)) \rangle}{\langle V(g(\boldsymbol{\sigma}^1), \ldots, g(\boldsymbol{\sigma}^n)) \rangle} - \mathsf{E} \frac{\langle f\, U(\eta(\boldsymbol{\sigma}^1), \ldots, \eta(\boldsymbol{\sigma}^n)) \rangle}{\langle V(\eta(\boldsymbol{\sigma}^1), \ldots, \eta(\boldsymbol{\sigma}^n)) \rangle} \right| \le \frac{K(n, D')}{N} \mathsf{E}\langle |f| \rangle ,$$

where

$$D' = \sup_{0 \le \ell \le 6} |u^{(\ell)}| . \tag{3.220}$$

This allows to use Theorem 3.3.7 (in the case $s = 2$, but the case $s = 1$, $s = 0$ are similar) even in the case of the Hamiltonian (3.207). The philosophy is the same as that of Proposition 3.8.3. When $|u^{(\ell)}|$ is bounded for $\ell \le 6$, the Bernoulli case (3.207) can be handled as the Gaussian case, at the cost of an extra error term containing a factor $K(n, D')/N$, where D' is as in (3.220).

We will let the reader revisit the proofs of the results of the previous sections to investigate what can be recovered, and we will only state the result corresponding to Theorem 3.4.1 and Theorem 3.4.2

Theorem 3.8.5. *There exists a constant* L *with the following property. Let* u *be 6 times differentiable, and consider* D' *as in (3.220). Then if* $L \alpha e^{LD} \le 1$, *for* $N \ge K(D')$ *the inequalities (3.89) and (3.90) hold.*

Proposition 3.8.4 is a special case of an interesting method, called Trotter's method, that was invented to prove central limit theorems without using Fourier transform. We will prove only a simple version, that is sufficient to prove Proposition 3.8.4.

Proposition 3.8.6. *Consider a function* F *on* $\mathbb{R}^N$, *and assume that the absolute value of each partial derivative of* F *of order 4 is less than or equal to* A. *Then we have*

$$|\mathsf{E} F(g_1, \ldots, g_N) - \mathsf{E} F(\eta_1, \ldots, \eta_N)| \le LNA ,$$

where $g_1, \ldots, g_N, \eta_1, \ldots, \eta_N$ *are as in Proposition 3.8.4.*

Proof. We replace the variables η_i by the variables g_i one at a time, and we prove that

$$|\mathsf{E} F(g_1, \ldots, g_p, \eta_{p+1}, \ldots, \eta_N) - \mathsf{E} F(g_1, \ldots, g_{p-1}, \eta_p, \ldots, \eta_N)| \le LA .$$

Thus, working at $g_1, \ldots, g_{p-1}, \eta_{p+1}, \ldots, \eta_N$ fixed, we need only consider the case $N = 1$. We then write

$$F(x) = F(0) + x\,F'(0) + \frac{x^2}{2}\,F''(0) + \frac{x^3}{3!}\,F'''(0) + \mathcal{R}\,,$$

where $|\mathcal{R}| \le Ax^4/4!$. We substitute $x = g_1$ (resp. $x = \eta_1$), we take expectation, and we use that $\mathsf{E}\,g_1^\ell = \mathsf{E}\,\eta_1^\ell$ for $\ell \le 3$. □

Proof of Proposition 3.8.4. We fix the randomness in $\langle\cdot\rangle$ and we consider the function F on $\mathbb{R}^N$ such that

$$F(g_1, \ldots, g_N) = \frac{\langle f\,U(g(\boldsymbol{\sigma}^1), \ldots, g(\boldsymbol{\sigma}^n))\rangle}{\langle V(g(\boldsymbol{\sigma}^1), \ldots, g(\boldsymbol{\sigma}^n))\rangle}\,.$$

To end the argument, we crudely bound the partial derivatives of F of order 4 by $K(n, D')\,\langle|f|\rangle/N^2$. This is possible because each g_i occurs in $g(\boldsymbol{\sigma}^1)$ with a coefficient at most $1/\sqrt{N}$, so taking derivatives 4 times creates a factor $1/N^2$. □

3.9 Estimates for the Unbounded Case

In this section we learn how to work under weaker hypothesis on u than the condition $|u| \le D$. This is necessary in order to prove (3.4) for small α. From now on we will assume that

$$u \le 0 \tag{3.221}$$

and that for a certain number τ we have

$$x \ge \tau \Rightarrow u(x) = 0\,. \tag{3.222}$$

Only small changes would be needed to replace these conditions by the conditions $u \le D$ and $u(x) \ge -D$ for $x \ge \tau$, thereby obtaining a generalization of the case $|u| \le D$. We will not do this for simplicity.

It must be mentioned that the material from this point on until the end of next chapter was added after the author became aware of the remarkable results of [128], long after most of the rest of the book was written. There was less time to let this material mature, so it is probably even further from final form than the rest of the book. Moreover, at this stage, the fun of the entire project had long vanished, and the author is unsure as to whether he succeeded in investing as much effort as before into polishing details. Finally, the writing was greatly impaired by a lack of funds available for typing. It was unrealistic to even think of making significant changes to the previous sections of this chapter, that had already been typed. At several places we had no other route than arguing that modifications of previous arguments do the job. In short, this material is going to be more demanding, and the first time reader is urged to jump to Chapter 5.

The new difficulty when working under (3.222) rather than under the condition $|u| \le D$, is that it is much more difficult to control from below quantities such as

$$Z = \left\langle \exp u \left(\frac{\sum_{i \le N} g_{i,M}\, \sigma_i}{\sqrt{N}} \right) \right\rangle_{\sim} . \tag{3.223}$$

In the previous sections we used the bound $Z \ge e^{-D}$. Instead, we will now establish a kind of integrability property for $1/Z$, that is expressed in Theorem 3.9.1 below. The proof of this estimate occupies much of the section.

Theorem 3.9.1. *There exists a number* $L^{\sim}$ *with the following property. Consider a (non-random) probability measure* G *on* $\{-1,1\}^N$. *Consider jointly Gaussian r.v.* $(w(\boldsymbol{\sigma}))_{\boldsymbol{\sigma} \in \Sigma_N}$, *and assume that*

$$\forall \boldsymbol{\sigma}\,,\ \frac{3}{4} \le \mathsf{E}\, w^2(\boldsymbol{\sigma}) \le 1\,. \tag{3.224}$$

Let

$$A = G^{\otimes 2}\left(\left\{ (\boldsymbol{\sigma}^1, \boldsymbol{\sigma}^2)\,;\ \mathsf{E}\, w(\boldsymbol{\sigma}^1)\, w(\boldsymbol{\sigma}^2) \ge \frac{1}{L^{\sim}} \right\} \right) . \tag{3.225}$$

Then we have

$$\varepsilon \ge 2A^{1/L} \Rightarrow \mathsf{P}(G(\{\boldsymbol{\sigma}\,;\ w(\boldsymbol{\sigma}) \ge \tau\}) \le \varepsilon) \le L e^{L\tau^2} \varepsilon^{16}\,.$$

Of course the number 16 plays no special role here. To see the connection between Theorem 3.9.1 and the study of the quantity (3.223), we observe that if G is the probability measure $\langle \cdot \rangle_{\sim}$, then by (3.222) we have

$$Z \ge G(\{\boldsymbol{\sigma}\,;\ \frac{1}{\sqrt{N}} \sum_{i \le N} g_{i,M} \sigma_i \ge \tau\})\,.$$

Given $n \ge 1$, we consider the set

$$Q_n = \left\{ (\boldsymbol{\sigma}^1, \dots, \boldsymbol{\sigma}^n) \in \Sigma_N^n\,;\ \forall \ell \ne \ell'\,, \mathsf{E}\, w(\boldsymbol{\sigma}^\ell)\, w(\boldsymbol{\sigma}^{\ell'}) \le \frac{1}{L^{\sim}} \right\} .$$

We think of the quantity $L^{\sim}$ as a parameter, whose value will be fixed toward the end of the proof of Theorem 3.9.1. Until this is done, the value of the constant L may vary at each occurrence but does NOT depend on the value of $L^{\sim}$.

For simplicity we will write w_ℓ rather that $w(\boldsymbol{\sigma}^\ell)$.

Given $\boldsymbol{\sigma}^1, \dots, \boldsymbol{\sigma}^n$ in Σ_N, and given a number $\varepsilon > 0$, we define the event $\Omega_\varepsilon(\boldsymbol{\sigma}^1, \dots, \boldsymbol{\sigma}^n)$ as the event

$$\operatorname{card}\{\ell \le n\,;\ w_\ell \ge \tau\} \le 2\varepsilon n\,.$$

First, we reduce the proof of Theorem 3.9.1 to the proof of the following statement.

Proposition 3.9.2. *If $L^\sim$ is large enough the following occurs. Whenever $(\boldsymbol{\sigma}^1, \ldots, \boldsymbol{\sigma}^n) \in Q_n$, we have*

$$\varepsilon \geq n^{-1/L} \Rightarrow \mathsf{P}(\Omega_\varepsilon(\boldsymbol{\sigma}^1, \ldots, \boldsymbol{\sigma}^n)) \leq L\varepsilon^{16} e^{L\tau^2} . \tag{3.226}$$

Proof of Theorem 3.9.1. Assume that

$$G(\{\boldsymbol{\sigma}\,;\ w(\boldsymbol{\sigma}) \geq \tau\}) \leq \varepsilon . \tag{3.227}$$

Then

$$\int \operatorname{card}\{\ell \leq n\,;\ w_\ell \geq \tau\}\, \mathrm{d}G(\boldsymbol{\sigma}^1) \ldots \mathrm{d}G(\boldsymbol{\sigma}^n) \leq \varepsilon n ,$$

and thus, by Markov's inequality

$$G^{\otimes n}(\{(\boldsymbol{\sigma}^1, \ldots, \boldsymbol{\sigma}^n)\,;\ \operatorname{card}\{\ell \leq n\,;\ w_\ell \geq \tau\} \leq 2\varepsilon n\}) \geq \frac{1}{2} .$$

Now, we have $G^{\otimes n}(Q_n) \geq 1 - n(n-1)A/2$, so that if $n^2 A \leq 1/2$ we have $G^{\otimes n}(Q_n) \geq 3/4$, and thus

$$\begin{aligned}\frac{1}{4} &\leq G^{\otimes n}(\{(\boldsymbol{\sigma}^1, \ldots, \boldsymbol{\sigma}^n) \in Q_n\,;\ \operatorname{card}\{\ell \leq n\,;\ w_\ell \geq \tau\} \leq 2\varepsilon n\}) \\ &= \int_{Q_n} \mathbf{1}_{\Omega_\varepsilon(\boldsymbol{\sigma}^1, \ldots, \boldsymbol{\sigma}^n)}\, \mathrm{d}G(\boldsymbol{\sigma}^1) \ldots \mathrm{d}G(\boldsymbol{\sigma}^n) := Y .\end{aligned}$$

Thus (3.227) implies that $Y \geq 1/4$, so that

$$\mathsf{P}(G(\{\boldsymbol{\sigma}\,;\ w(\boldsymbol{\sigma}) \geq \tau\}) \leq \varepsilon) \leq \mathsf{P}\left(Y \geq \frac{1}{4}\right) \leq 4\,\mathsf{E}Y .$$

Now, by Proposition 3.9.2, if $\varepsilon \geq n^{-1/L}$

$$\mathsf{E}Y = \int_{Q_n} \mathsf{P}(\Omega_\varepsilon(\boldsymbol{\sigma}^1, \ldots, \boldsymbol{\sigma}^n))\, \mathrm{d}G(\boldsymbol{\sigma}^1) \ldots \mathrm{d}G(\boldsymbol{\sigma}^n) \leq L\varepsilon^{16} e^{L\tau^2} .$$

To conclude, we take for n the largest integer with $n^2 A \leq 1/2$. □

Without loss of generality, we now assume that $L^\sim \geq 4$. To prove Proposition 3.9.2 we will prove the following.

Proposition 3.9.3. *Consider jointly Gaussian r.v. $(w_\ell)_{\ell \leq n}$. Assume that*

$$\forall \ell \leq n,\ \frac{3}{4} \leq \mathsf{E} w_\ell^2 \leq 1\,;\ \forall \ell \neq \ell',\ \mathsf{E} w_\ell w_{\ell'} \leq \frac{1}{L^\sim} . \tag{3.228}$$

Then for $L \leq s \leq \sqrt{\log n}/L$, we have

$$\mathsf{P}(\operatorname{card}\{\ell \leq n\,;\ w_\ell \geq s\} \leq n \exp(-Ls^2)) \leq L \exp\left(-\frac{L^\sim s^2}{L}\right) . \tag{3.229}$$

Proof of Proposition 3.9.2. It follows from (3.229) that

$$\max(L,\tau) \le s \le \sqrt{\log n}/L \Rightarrow \mathsf{P}(\operatorname{card}\{\ell \le n\,;\; w_\ell \ge \tau\} \\ \le n\exp(-Ls^2)) \le L\exp\left(-\frac{L^\sim s^2}{L}\right).$$

Setting $\varepsilon = \exp(-Ls^2)$, this means that

$$n^{-1/L} \le \varepsilon \le \frac{1}{L}\exp(-L\tau^2) \Rightarrow \mathsf{P}(\operatorname{card}\{\ell \le n\,;\; w_\ell \ge \tau\} \le n\varepsilon) \\ \le L\varepsilon^{L^\sim/L}. \tag{3.230}$$

We now take $L^\sim$ large enough that $L^\sim/L \ge 16$. Then (3.230) implies (3.226) because for $\varepsilon \ge L^{-1}\exp(-L\tau^2)$ the right-hand side of (3.226) is at least 1. □

Lemma 3.9.4. *Consider* $(\boldsymbol{\sigma}^1,\ldots,\boldsymbol{\sigma}^n) \in Q_n$. *Then*

$$L \le s \le \sqrt{\log n} \Rightarrow \mathsf{E}\log\left(1 + \frac{1}{n}\sum_{\ell\le n}\exp s\,w_\ell\right) \ge \frac{s^2}{64}.$$

Proof. We consider the function F on $\mathbb{R}^n$ given by

$$F(x_1,\ldots,x_n) = \log\left(1 + \frac{1}{n}\sum_{\ell\le n}\exp s\,x_\ell\right). \tag{3.231}$$

Thus, if $D = 1 + n^{-1}\sum_{\ell\le n}\exp s\,x_\ell$, we have

$$\frac{\partial F}{\partial x_\ell} = \frac{s\exp s\,x_\ell}{nD}\,;\quad \frac{\partial^2 F}{\partial x_\ell^2} = \frac{s^2}{n}\left(\frac{\exp s\,x_\ell}{D} - \frac{1}{n}\left(\frac{\exp s\,x_\ell}{D}\right)^2\right) \ge 0\,,$$

and, if $\ell \ne \ell'$

$$\frac{\partial^2 F}{\partial x_\ell\,\partial x_{\ell'}} = -\frac{s^2}{n^2D^2}\exp(s\,x_\ell + s\,x_{\ell'}) \le 0\,.$$

We consider standard normal r.v. z and $\xi_\ell, \ell \le n$, all independent, and independent of the r.v. w_ℓ. We consider the r.v.

$$g_\ell = \frac{z}{\sqrt{L^\sim}} + \xi_\ell\sqrt{\frac{3}{4} - \frac{1}{L^\sim}}\,.$$

Thus,

$$\mathsf{E}\,g_\ell^2 = \frac{3}{4} \le \mathsf{E}\,w_\ell^2$$

and, if $\ell \ne \ell'$

$$\mathsf{E}\, g_\ell\, g_{\ell'} = \frac{1}{L^\sim} \geq \mathsf{E}\, w_\ell\, w_{\ell'}\ ,$$

by hypothesis. We use Slepian's Lemma (Proposition 2.4.5 and the following remark) to see that

$$\mathsf{E}\, F(w_1, \ldots, w_n) \geq \mathsf{E}\, F(g_1, \ldots, g_n)\ . \tag{3.232}$$

We now proceed to bound the right-hand side from below. For any $a > 0$ we have

$$\begin{aligned}\mathsf{E}\, F(g_1, \ldots, g_n) &\geq \mathsf{P}(z \geq 0)\, \mathsf{E} \log \left(1 + \frac{1}{n} \sum_{\ell \leq n} \exp\left(s\sqrt{\frac{3}{4} - \frac{1}{L^\sim}}\, \xi_\ell\right)\right) \\ &\geq \frac{1}{2} \log(1+a) \mathsf{P}\left(\sum_{\ell \leq n} \exp\left(s\sqrt{\frac{3}{4} - \frac{1}{L^\sim}}\, \xi_\ell\right) \geq na\right)\ . \end{aligned} \tag{3.233}$$

Consider $u > 0$, and observe that $3/4 - 1/L^\sim \geq 1/4$, so that

$$\sum_{\ell \leq n} \exp\left(s\sqrt{\frac{3}{4} - \frac{1}{L^\sim}}\, \xi_\ell\right) \geq \operatorname{card}\{\ell;\ \xi_\ell \geq u\} \exp \frac{su}{2}\ .$$

Combining with (3.233), we get

$$\begin{aligned}\mathsf{E}\, F(g_1, \ldots, g_n) \geq \frac{1}{2} \log(1+a)\, \mathsf{P}\Big(&\operatorname{card}\{\ell;\ \xi_\ell \geq u\} \\ &\geq na \exp\left(-\frac{su}{2}\right)\Big)\ . \end{aligned} \tag{3.234}$$

The r.v. $X_\ell = \mathbf{1}_{\{\xi_\ell \geq u\}}$ are independent and equidistributed, and

$$\mathsf{E}\, X_\ell = \delta := \mathsf{P}(\xi_\ell \geq u) \geq \frac{1}{L\sqrt{u}} \exp\left(-\frac{u^2}{2}\right)\ .$$

Now

$$\begin{aligned}\mathsf{E}\left(\sum_{\ell \leq n} X_\ell - n\delta\right)^2 &= \mathsf{E}\left(\sum_{\ell \leq n} X_\ell\right)^2 - n^2\,\delta^2 \\ &= \mathsf{E} \sum_{\ell \leq n} X_\ell^2 + \sum_{\ell \neq \ell'} \mathsf{E}\, X_\ell\, \mathsf{E}\, X_{\ell'} - n^2\,\delta^2 \\ &= n\delta + n(n-1)\delta^2 - n^2\delta^2 \leq n\delta \end{aligned}$$

so that, by Markov's inequality

$$\mathsf{P}\left(\sum_{\ell \leq n} X_\ell \geq \frac{n\delta}{2}\right) \geq 1 - \frac{4}{n\delta} \geq \frac{1}{2}$$

if $n\delta \geq 8$. Thus, if we choose a such that $a \exp(-su/2) = \delta/2$, we see from (3.234) that if $n\delta \geq 8$, we have

$$\mathsf{E}\, F(g_1, \ldots, g_n) \geq \frac{1}{4} \log(1+a) \; .$$

We take $u = s/2$ so that

$$a = \frac{\delta}{2} \exp \frac{su}{2} \geq \frac{1}{L\sqrt{s}} \exp \frac{s^2}{8}$$

and we have shown that

$$\frac{n}{L\sqrt{s}} \exp\left(-\frac{s^2}{8}\right) \geq 8 \Rightarrow \mathsf{E}\, F(g_1, \ldots, g_n) \geq \frac{1}{4} \log\left(1 + \frac{1}{L\sqrt{s}} \exp \frac{s^2}{8}\right) \; .$$

For s large enough ($s \geq L$) we have

$$s \leq \sqrt{\log n} \Rightarrow \frac{n}{L\sqrt{s}} \exp\left(-\frac{s^2}{8}\right) \geq 8$$

and

$$\log\left(1 + \frac{1}{L\sqrt{s}} \exp \frac{s^2}{8}\right) \geq \frac{s^2}{16} \; .$$

□

Lemma 3.9.5. *Assuming (3.228), for $L \leq s \leq \sqrt{\log n}/L$, we have*

$$\mathsf{P}(\operatorname{card}\{\ell \leq n \, ; \; w_\ell \geq s\} \geq n \exp(-Ls^2)) \geq \frac{1}{L} \; .$$

Proof. Since $\mathsf{E}\, w_\ell^2 \leq 1$ we have

$$\mathsf{E}\left(1 + \frac{1}{n} \sum_{\ell \leq n} \exp s\, w_\ell\right) \leq 1 + \exp\left(\frac{s^2}{2}\right) ,$$

so that by Markov's inequality for any number v we have

$$\mathsf{P}\left(1 + \frac{1}{n} \sum_{\ell \leq n} \exp s\, w_\ell \geq e^v \left(1 + \exp\left(\frac{s^2}{2}\right)\right)\right) \leq e^{-v}$$

and hence

$$\mathsf{P}\left(\log\left(1 + \frac{1}{n} \sum_{\ell \leq n} \exp s\, w_\ell\right) \geq v + \log\left(1 + \exp\left(\frac{s^2}{2}\right)\right)\right) \leq e^{-v} \; . \quad (3.235)$$

For $s \geq 1$ we have $\log(1+\exp(s^2/2)) \leq Ls^2$, and thus from (3.235) we see that

$$\mathsf{E}\left(\log\left(1+\frac{1}{n}\sum_{\ell\leq n}\exp s\,w_\ell\right)\right)^2 \leq Ls^4\,.$$

It follows from Lemma 3.9.4 and the second moment method (Proposition 2.2.3) that for $L \leq s \leq \sqrt{\log n}$ we have

$$\mathsf{P}\left(\log\left(1+\frac{1}{n}\sum_{\ell\leq n}\exp s\,w_\ell\right) \geq \frac{s^2}{128}\right) \geq \frac{1}{L}$$

and thus, since we assume $s \geq L$,

$$\mathsf{P}\left(\sum_{\ell\leq n}\exp s\,w_\ell \geq n\exp\left(\frac{s^2}{200}\right)\right) \geq \frac{1}{L}\,. \tag{3.236}$$

On the other hand, we have

$$\mathsf{E}\left(\sum_{\ell\leq n}\exp 2s\,w_\ell\right) \leq n\exp 2s^2\,,$$

so that, for $v>0$, by Markov's inequality,

$$\mathsf{P}\left(\sum_{\ell\leq n}\exp 2s\,w_\ell \geq nv\exp 2s^2\right) \leq \frac{1}{v}\,.$$

Taking $v = 2L$, where L occurs in (3.236), we see that if we define the event Ω by

$$\sum_{\ell\leq n}\exp s\,w_\ell \geq n\exp\left(\frac{s^2}{200}\right)\ ;\ \sum_{\ell\leq n}\exp 2s\,w_\ell \leq 2Ln\exp 2s^2 \tag{3.237}$$

then we have

$$\mathsf{P}(\Omega) \geq \frac{1}{2L}\,. \tag{3.238}$$

Since $\exp s\,w_\ell \leq \exp(-3s^2)\exp 2s\,w_\ell$ for $w_\ell \geq 3s$, we have

$$\sum\{\exp s\,w_\ell\,;\ w_\ell \geq 3s\} \leq \exp(-3s^2)\sum_{\ell\leq n}\exp 2s\,w_\ell\,.$$

Also,

$$\sum_{\ell\leq n}\left\{\exp s\,w_\ell\,;\ w_\ell \leq \frac{s}{300}\right\} \leq n\exp\left(\frac{s^2}{300}\right)\,.$$

Thus, on the event Ω, for $s \geq L$, we have

$$\sum_{\ell \leq n} \left\{ \exp s\, w_\ell\, ;\ \frac{s}{300} \leq w_\ell \leq 3s \right\}$$

$$\geq n \left(\exp\left(\frac{s^2}{200}\right) - \exp\left(\frac{s^2}{300}\right) - L\exp(-s^2) \right) \geq n\,.$$

Noting that

$$\sum_{\ell \leq n} \left\{ \exp s\, w_\ell\, ;\ \frac{s}{300} \leq w_\ell \leq 3s \right\} \leq \exp(3s^2)\mathrm{card}\left\{ \ell \leq n\, ;\ \frac{s}{300} \leq w_\ell \leq 3s \right\}$$

and changing s into $300s$ finishes the proof. □

Lemma 3.9.6. *Consider a number $a > 0$, consider vectors $(\mathbf{x}_\ell)_{\ell \leq n}$ of $\mathbb{R}^M$, and assume that $\|\mathbf{x}_\ell\| \leq 1$ and that $|\mathbf{x}_\ell \cdot \mathbf{x}_{\ell'}| \leq a$ whenever $\ell \neq \ell'$. Consider a vector $\mathbf{x}$ in $\mathbb{R}^M$. Then, for $a > 0$, we have*

$$\mathrm{card}\left\{ \ell\, ;\ \mathbf{x} \cdot \mathbf{x}_\ell \geq \|\mathbf{x}\|\, \sqrt{2a} \right\} \leq \frac{1}{a}\,.$$

Proof. Assume that $\mathbf{x} \cdot \mathbf{x}_\ell \geq \|\mathbf{x}\|\, \sqrt{2a}$ for $\ell \leq m$. Then

$$m\,\|\mathbf{x}\|\,\sqrt{2a} \leq \mathbf{x} \cdot \left(\sum_{\ell \leq m} \mathbf{x}_\ell \right) \leq \|\mathbf{x}\| \left\| \sum_{\ell \leq m} \mathbf{x}_\ell \right\|\,,$$

and

$$\left\| \sum_{\ell \leq m} \mathbf{x}_\ell \right\|^2 = \sum_{\ell \leq m} \|\mathbf{x}_\ell\|^2 + \sum_{\ell \neq \ell'} \mathbf{x}_\ell \cdot \mathbf{x}_{\ell'} \leq m + am(m-1)\,.$$

Thus

$$m\sqrt{2a} \leq \sqrt{m + am(m-1)} \leq \sqrt{m}\,\sqrt{1 + am}$$

so that $\sqrt{2a} \leq \sqrt{m^{-1} + a}$, i.e. $m \leq 1/a$. □

Proof of Proposition 3.9.3. We find vectors $(\mathbf{x}_\ell)_{\ell \leq n}$ in $\mathbb{R}^M$ (for some M) such that the joint law of the variables $(w_\ell)_{\ell \leq n}$ is the same as the joint law of the variables $(\mathbf{g} \cdot \mathbf{x}_\ell)$ where $\mathbf{g}$ is a standard Gaussian vector valued in $\mathbb{R}^M$. (We do not prove the existence of these vectors because it will be obvious for the variables w_ℓ we will need in the next section.) Thus it follows from Lemma 3.9.5 that if $L \leq s \leq L^{-1}\sqrt{\log n}$ we have $\mathsf{P}(\mathbf{g} \in A) \geq 1/L$ where

$$A = \{\mathbf{x}\, ;\ \mathrm{card}\{\ell \leq n, \mathbf{x} \cdot \mathbf{x}_\ell \geq s\} \geq n\exp(-Ls^2)\}\,.$$

For $v > 0$, consider the set

$$A_v = \{\mathbf{x};\ \mathrm{d}(\mathbf{x}, A) \le v\} ,$$

where d denotes the Euclidean distance in $\mathbb{R}^M$. Using Lemma 2.2.11 we see that

$$\mathsf{P}(\mathbf{g} \in A_v) \ge 1 - L\exp\left(-\frac{v^2}{L}\right) .$$

For $\mathbf{g} \in A_v$ we can find $\mathbf{g}'$ in A with $\|\mathbf{g}' - \mathbf{g}\| \le v$. Since $\mathbf{x}_\ell \cdot \mathbf{x}_{\ell'} = \mathsf{E}\, w_\ell w_{\ell'}$, we have $|\mathbf{x}_\ell \cdot \mathbf{x}_{\ell'}| \le 1/L^\sim$. Thus, by Lemma 3.9.6 (used for $a = 1/L^\sim$) we have

$$\mathrm{card}\left\{\ell \le n;\ (\mathbf{g} - \mathbf{g}') \cdot \mathbf{x}_\ell \ge v\sqrt{\frac{2}{L^\sim}}\right\} \le L^\sim .$$

Since $\mathbf{g} \in A$, we have

$$\mathrm{card}\{\ell \le n;\ \mathbf{g} \cdot \mathbf{x}_\ell \ge s\} \ge n\exp(-Ls^2) ,$$

so that

$$\mathrm{card}\left\{\ell \le n;\ \mathbf{g}' \cdot \mathbf{x}_\ell \ge s - v\sqrt{\frac{2}{L^\sim}}\right\} \ge n\exp(-Ls^2) - L^\sim .$$

Taking $v = s\sqrt{L^\sim/8}$, and using that $L \le s \le \sqrt{\log n}$, we have shown that for $\mathbf{g}' \in A_v$, we have, since $s \ge L$,

$$\mathrm{card}\left\{\ell \le n;\ \mathbf{g}' \cdot \mathbf{x}_\ell \ge \frac{s}{2}\right\} \ge n\exp(-Ls^2) .$$

Equivalently, this shows that for $L \le s \le \sqrt{\log n}/L$, we have

$$\mathsf{P}\left(\mathrm{card}\left\{\ell \le n;\ w_\ell \ge \frac{s}{2}\right\} \ge n\exp(-Ls^2)\right) \ge 1 - L\exp\left(-\frac{s^2 L^\sim}{L}\right) .$$

This proves (3.229). □

In order for Theorem 3.9.1 to be useful, we need that the quantity (3.225) be small. The following two results should look familiar.

Lemma 3.9.7. *Consider a Gibbs' measure G on Σ_N with Hamiltonian $H \ge 0$. Assume that for a certain number κ we have*

$$Z = \sum_{\boldsymbol{\sigma}} \exp(-H(\boldsymbol{\sigma})) \ge 2^N \exp(-N\kappa) .$$

Then

$$A = G^{\otimes 2}(\{R_{1,2} \ge 4\sqrt{\kappa}\}) \le \exp(-6N\kappa) .$$

Proof. We have, using the subgaussian inequality (A.3)

$$\begin{aligned} Z^2 A &= \sum \{\exp(-H(\boldsymbol{\sigma}^1) - H(\boldsymbol{\sigma}^2))\,;\; R_{1,2} \geq 4\sqrt{\kappa}\} \\ &\leq \operatorname{card}\{(\boldsymbol{\sigma}^1, \boldsymbol{\sigma}^2)\,;\; R_{1,2} \geq 4\sqrt{\kappa}\} \leq 2^{2N} \exp(-8N\kappa)\,. \end{aligned}$$

□

Corollary 3.9.8. *Consider a Gibbs' measure G on Σ_N with Hamiltonian $H \geq 0$. Assume that*

$$\sum_{\boldsymbol{\sigma}} \exp(-H(\boldsymbol{\sigma})) \geq 2^N \exp\left(-\frac{N}{16L^{\sim 2}}\right)\,.$$

Then, for independent standard Gaussian r.v. $(g_i)_{i \leq N}$ we have

$$\begin{aligned} &\mathsf{P}\left(G\left(\left\{\frac{1}{\sqrt{N}} \sum_{i \leq N} g_i\,\sigma_i \geq \tau\right\}\right) \leq \varepsilon\right) \\ &\leq L\varepsilon^{16}\, e^{L\tau^2} + L \exp\left(L\tau^2 - \frac{N}{L}\right)\,. \end{aligned} \tag{3.239}$$

Proof. We apply Lemma 3.9.7 with $4\sqrt{\kappa} = 1/L^{\sim}$, and then Theorem 3.9.1 with $w(\boldsymbol{\sigma}) = N^{-1/2} \sum_{i \leq N} g_i\,\sigma_i$. □

Remark. We will work under the condition $L\,\alpha\, e^{L\tau^2} = L\,M\, e^{L\tau^2}/N \leq 1$, so we will have $N \geq L\,M\, e^{L\tau^2}$, and the last term of (3.239) can then be replaced by $\exp(-N/L)$. This will be used many times.

The following lemma is designed to replace Lemma 3.9.7 when we deal with our interpolating Hamiltonians.

Lemma 3.9.9. *Consider a Gibbs' measure G on Σ_N with Hamiltonian H such that*

$$-H(\boldsymbol{\sigma}) = -H_0(\boldsymbol{\sigma}) + \sigma_N\, y \tag{3.240}$$

where $-H_0(\boldsymbol{\sigma}) \leq 0$ and, for $\varepsilon = 1$ or $\varepsilon = -1$

$$\sum_{\sigma_N = \varepsilon} \exp(-H_0(\boldsymbol{\sigma})) \geq 2^N \exp\left(-\frac{N}{32L^{\sim 2}}\right)\,. \tag{3.241}$$

Then, we have

$$A = G^{\otimes 2}\left(\left\{(\boldsymbol{\sigma}^1, \boldsymbol{\sigma}^2)\,;\; R_{1,2} \geq \frac{1}{2L^{\sim}}\right\}\right) \leq \exp\left(-\frac{N}{16L^{\sim 2}}\right)\,. \tag{3.242}$$

Proof. Using (3.241) we see that

$$\begin{aligned}Z := \sum_{\boldsymbol{\sigma}} \exp(-H(\boldsymbol{\sigma})) &= \sum_{\varepsilon} \sum_{\sigma_N = \varepsilon} \exp(-H_0(\boldsymbol{\sigma}) + \sigma_N\, y) \\ &\geq 2^N \sum_{\varepsilon} \exp \varepsilon\, y \exp\left(-\frac{N}{32 L^{\sim 2}}\right) \\ &= 2^{N+1} \operatorname{ch} y \exp\left(-\frac{N}{8L^{\sim 2}}\right) .\end{aligned}$$

Also,

$$\begin{aligned}Z^2 A &= \sum \left\{ \exp(-H(\boldsymbol{\sigma}^1) - H(\boldsymbol{\sigma}^2))\, ;\ R_{1,2} \geq \frac{1}{2L^{\sim}} \right\} \\ &\leq \sum \left\{ \exp(\sigma_N^1\, y + \sigma_N^2\, y)\, ;\ R_{1,2} \geq \frac{1}{2L^{\sim}} \right\} \\ &\leq (2 \operatorname{ch} y)^2\, 2^{2N} \exp\left(-\frac{N}{8L^{\sim 2}}\right) ,\end{aligned}$$

since $\exp x \leq 2 \operatorname{ch} x$ and using the subgaussian inequality. The result follows. □

3.10 The Gardner Formula for Small α

Our first goal is to prove Theorem 3.10.1 below. This result resembles Theorem 3.4.2. We will then use it to prove (3.4) for small α.

Theorem 3.10.1. *There exists a constant L with the following property. Assume that the function u satisfies (3.221) and (3.222). Then if $L\alpha e^{L\tau^2} \leq 1$, the equations (3.12) and (3.11) have a solution. Moreover, if*

$$\forall \ell, \quad 0 \leq \ell \leq 6, \qquad |u^{(\ell)}| \leq \exp \frac{N}{L} , \tag{3.243}$$

then (3.89) and (3.90) hold.

Remark. The dependence given in τ is not sharp when $\tau < 0$. In that case minor modifications of our arguments show we can replace the condition $L\alpha e^{L\tau^2} \leq 1$ by $L\alpha \leq 1$. These are left to the reader.

The proof of Theorem 3.10.1 closely follows that of Theorem 3.4.2, except that "we have to find another way to control the denominators". The main idea is illustrated in the proof of the following simple result.

Lemma 3.10.2. *If u satisfies conditions (3.221), (3.222) and if $L\alpha e^{L\tau^2} \leq 1$, then the equations (3.12) and (3.13) have a solution.*

In this statement we allow "u to take the value $-\infty$". The definition of the function Φ of (3.16) depends only on the function $U(x) = \exp u(x)$, and we allow that U takes the value 0.

Proof. The proof follows that of Proposition 3.2.1. Rather than (3.15) we require that

$$|r| \le r_0 \Rightarrow q = \mathsf{E}\,\mathrm{th}^2(z\sqrt{r}) \le \frac{1}{20} \,.$$

From (3.16) we see that

$$\Phi(z\sqrt{q}, \sqrt{1-q}) \le \frac{\mathsf{E}\,|\xi|}{\sqrt{1-q}\mathsf{E}_\xi \exp u(z\sqrt{q} + \xi\sqrt{1-q})} \le \frac{L}{Z} \tag{3.244}$$

where

$$Z = \mathsf{E}_\xi \exp u(z\sqrt{q} + \xi\sqrt{1-q}) \ge P_\xi(z\sqrt{q} + \xi\sqrt{1-q} \ge \tau)$$

using (3.222). Thus

$$Z \ge P_\xi\left(\xi \ge \frac{\tau - z\sqrt{q}}{\sqrt{1-q}}\right) \ge \frac{1}{1 + L|z\sqrt{q} - \tau|} \exp\left(-\frac{(z\sqrt{q} - \tau)^2}{2(1-q)}\right) ,$$

using (A.38). Writing $(z\sqrt{q} - \tau)^2 \le 2qz^2 + 2\tau^2$, and since $q/(1-q) \le 1/19$, we have

$$\mathsf{E}\,\frac{1}{Z^8} \le Le^{L\tau^2} , \tag{3.245}$$

and in particular from (3.244) that

$$\mathsf{E}\,\alpha\,\Phi^2(z\sqrt{q}, \sqrt{1-q}) \le L\alpha e^{L\tau^2} \,.$$

To finish the proof it remains to show that the derivative of the function $q \to \mathsf{E}\,\Phi^2(z\sqrt{q}, \sqrt{1-q})$ is bounded by $Le^{L\tau^2}$ where $q \le 1/20$. To see this, we compute this derivative using (3.16), we use the Cauchy-Schwarz inequality, (3.245), and the fact that (as in Proposition 3.2.1) the derivatives of the functions (3.18) are bounded. □

The work of Section 3.2 carries out verbatim to the present setting. We keep the notation $g(\boldsymbol{\sigma})$ of (3.39).

Lemma 3.10.3. *For every Gibbs' measure G that satisfies the conditions of Lemma 3.9.9, (3.42) holds.*

Proof. The simple modifications of the arguments of Lemma 3.3.3 needed to do this should be obvious. □

We recall the notation $g_{v,\ell}$ of (3.53).

Proposition 3.10.4. *Consider a non-random Gibbs' measure on Σ_N that satisfies the conditions of Lemma 3.9.9, and denote by $\langle\cdot\rangle$ integration with respect to this measure. Consider an integer $m \le 6$, integers $s_1, \dots, s_m \ge 0$ with $s = s_1 + \cdots + s_m \le 2$. Consider the functions U and V given by (3.65) and (3.66), and assume that $|U| \le B$. Consider a function f on Σ_N^m with $|f| \le 1$. Then we have*

$$\left|\mathsf{E}\frac{\langle f\,U(g_{v,1},\dots,g_{v,m})\rangle}{(\mathsf{E}_\xi\langle\exp u(g_{v,1})\rangle)^m}\right| \le Le^{L\tau^2}\left(\langle|f|\rangle + B^4\exp\left(-\frac{N}{L}\right)\right) . \tag{3.246}$$

Proof. Let us write $Z = (\mathsf{E}_\xi\langle\exp u(g_{v,1})\rangle)^m$. Consider a number $a > 0$. We have

$$\left|\mathsf{E}\frac{\langle f\,U(g_{v,1},\dots,g_{v,m})\rangle}{Z}\right| \le \mathrm{I} + \mathrm{II} \tag{3.247}$$

$$\mathrm{I} = \left|\mathsf{E}\,\mathbf{1}_{\{Z<a\}}\frac{\langle f\,U(g_{v,1},\dots,g_{v,m})\rangle}{Z}\right|$$

$$\begin{aligned}\mathrm{II} &= \left|\mathsf{E}\,\mathbf{1}_{\{Z\ge a\}}\frac{\langle f\,U(g_{v,1},\dots,g_{v,m})\rangle}{Z}\right| \\ &\le \left(\mathsf{E}\,\mathbf{1}_{\{Z\ge a\}}\frac{1}{Z^2}\right)^{1/2}\left(\mathsf{E}\,\langle f\,U(g_{v,1},\dots,g_{v,m})\rangle^2\right)^{1/2} .\end{aligned}$$

Since we assume that $|U| \le B$ we have

$$|\langle f\,U(g_{v,1},\dots,g_{v,m})\rangle| \le B\left\langle\exp\sum_{\ell\le m}u(g_{v,\ell})\right\rangle = B\langle\exp u(g_{v,1})\rangle^m ,$$

and since $\mathbf{1}_{\{Z<a\}}$ is independent of the r.v. $\xi_1,\dots,\xi_m$, we see that $|\mathrm{I}| \le B\,\mathsf{P}(Z \le a)$.

We will leave it to the reader to make the small changes to the proof of Lemma 3.3.6 to see that

$$\mathsf{E}\,\langle f\,U(g_{v,1},\dots,g_{v,m})\rangle^2 \le L\langle|f|\rangle^2 + L\,B^2\exp\left(-\frac{N}{L}\right) .$$

Thus all we have to do is to show that we can find a such that

$$\mathsf{P}(Z \le a) \le L\exp\left(L\tau^2 - \frac{N}{L}\right) \;;\; \mathsf{E}\mathbf{1}_{\{Z\ge a\}}\frac{1}{Z^2} \le Le^{L\tau^2} . \tag{3.248}$$

Consider $Z' = \langle\exp u(g_{v,1})\rangle$ and observe that $Z = E_\xi(Z'^m)$. We have

$$\mathsf{E}(g_{v,1}\,g_{v,2}) = v\,\mathsf{E}\,g(\boldsymbol{\sigma}^1)\,g(\boldsymbol{\sigma}^2) + (1-v)\,q \le v\,|R_{1,2}| + (1-v)\,q + \frac{1}{N} .$$

Since we assume $\alpha\,e^{L\tau^2} \le L$ there is no loss of generality to assume that $|q| \le 1/(2L^{\sim})$. Thus, by Lemma 3.9.7, we see that

$$A = G^{\otimes 2}\left(\left\{(\boldsymbol{\sigma}^1,\boldsymbol{\sigma}^2)\,;\ \mathsf{E}\, g_{v,1}\, g_{v,2} \ge \frac{1}{L^{\sim}}\right\}\right) \le \exp\left(-\frac{N}{16\,L^{\sim 2}}\right) .$$

By Theorem 3.9.1, we see that

$$\varepsilon \ge \exp\left(-\frac{N}{L}\right) \Rightarrow \mathsf{P}(Z' \le \varepsilon) \le Le^{L\tau^2}\,\varepsilon^{16} .$$

It follows that

$$\forall\, \varepsilon \ge 0\,,\ \mathsf{P}\left(Z' + \exp\left(-\frac{N}{L}\right) \le \varepsilon\right) \le Le^{L\tau^2}\,\varepsilon^{16}$$

and thus, since $m \le 6$ we have

$$\mathsf{E}\,\frac{1}{(Z' + \exp(-N/L))^{2m}} \le Le^{L\tau^2} ,$$

so that

$$\mathsf{E}\,\frac{1}{(Z'^m + \exp(-N/L))^2} \le Le^{L\tau^2} . \tag{3.249}$$

By the Cauchy-Schwarz inequality we have $1 \le \mathsf{E}_\xi\, X\, \mathsf{E}_\xi(1/X)$ and using it again we have $1 \le (\mathsf{E}_\xi\, X)^2\, \mathsf{E}_\xi(1/X^2)$. Using this for $X = Z'^m + \exp(-N/L)$, and since $Z = \mathsf{E}_\xi\, Z'^m$, we get

$$\mathsf{E}\,\frac{1}{(Z + \exp(-N/L))^2} \le Le^{L\tau^2} .$$

Thus, taking $a = \exp(-N/L)$, where L is as above, this implies (3.248). □

We recall the notation $\theta_\ell = z\sqrt{q} + \xi_\ell\sqrt{1-q}$.

Theorem 3.10.5. *Consider a non-random Gibbs' measure G as in Proposition 3.10.4, but assume now $m \le 4$ and that $|u^\ell| \le D'$. Then we have*

$$\mathsf{E}\,\frac{\langle f\, U(g(\boldsymbol{\sigma}^1),\ldots,g(\boldsymbol{\sigma}^m))\rangle}{\mathsf{E}_\xi\langle \exp u(g(\boldsymbol{\sigma}^1))\rangle^m} = \langle f\rangle\, \mathsf{E}\,\frac{U(\theta_1,\ldots,\theta_m)}{(\mathsf{E}_\xi \exp u(\theta_1))^m} + \mathcal{R} ,$$

where

$$|\mathcal{R}| \le Le^{L\tau^2}\left(\langle f^2\rangle^{1/2}\left(\langle (R_{1,2}-q)^2\rangle^{1/2} + \frac{1}{N}\right) + D'^4 \exp\left(-\frac{N}{L}\right)\right) .$$

Proof. It follows the proof of Theorem 3.3.7, using now Proposition 3.10.4 to bound the derivatives of the function φ of (3.71). □

It is interesting to observe that our argument does not allow to control all values of m. This raises the following issue.

Research problem 3.10.6. (Level 2) Prove (or disprove!) Theorem 3.6.4 under the hypothesis of Theorem 3.10.1.

Lemma 3.10.7. *If f is a non-negative function on Σ_N^n, $n \le 4$, we have* $\mathsf{E}\,\langle f\rangle_{t,\sim} \le L e^{L\tau^2}\,\nu_t(f)$.

Proof. Since we assume $u \le 0$, we have

$$\langle f\rangle_t = \frac{\left\langle f \exp \sum_{\ell\le n} u(S_{M,t}^\ell)\right\rangle_{t,\sim}}{\left\langle \exp \sum_{\ell\le n} u(S_{M,t}^\ell)\right\rangle_{t,\sim}} \ge \left\langle f \exp \sum_{\ell\le n} u(S_{M,t}^\ell)\right\rangle_{t,\sim} .$$

We then take expectation. In the right-hand side, we can take expectation in the variables $S_{M,t}^\ell = g(\boldsymbol{\sigma}^\ell)$ inside the bracket, and using (3.222) we get

$$\nu_t(f) \ge \frac{1}{L}\, e^{-L\tau^2}\,\mathsf{E}\,\langle f\rangle_{t,\sim} .$$

□

To prove Theorem 3.10.1, we want to follow the proof of Theorem 3.4.1. This requires applying Theorem 3.10.5 to the random measure $\langle\cdot\rangle_{t,\sim}$. This random measure may not always satisfy condition (3.241), but, as the following lemma proves, it does so with overwhelming probability (that is, the probability of failure is at most $K\exp(-N/K)$).

Lemma 3.10.8. *If $L\alpha e^{L\tau^2} \le 1$, the measure $\langle\cdot\rangle_{t,\sim}$ satisfies condition (3.241) with overwhelming probability.*

Proof. By (3.30) we see that (3.240) holds where

$$-H_0(\boldsymbol{\sigma}) = \sum_{k\le M-1} u\left(\frac{1}{\sqrt{N}}\sum_{i\le N-1} g_{i,k}\,\sigma_i + \sqrt{\frac{t}{N}}\, g_{N,k}\,\sigma_N\right) .$$

Thus to satisfy (3.241) it suffices that the following two conditions occur

$$\forall\, k \le M\,, \qquad |g_{N,k}| \le \sqrt{N} \tag{3.250}$$

$$\operatorname{card}\{\boldsymbol{\sigma} \in \Sigma_N\,;\ \forall\, k\le M\,,\ S_k \ge 2+\tau\} \ge 2^N \exp\left(-\frac{N}{8L^{\sim 2}}\right) \tag{3.251}$$

where

$$S_k = N^{-1/2} \sum_{k\le N-1} g_{i,k}\,\sigma_i\,. \tag{3.252}$$

Certainly (3.250) occurs with overwhelming probability. To study (3.251) let us write $\tau' = \tau + 2$,

$$V_M = 2^{-N}\,\mathrm{card}\,\{\boldsymbol{\sigma} \in \Sigma_N\,;\ \forall k \le M, S_k \ge \tau'\}$$

and consider the event Ω_M given by $V_M \ge \exp(-N/8L^{\sim 2})$. We want to show that this event occurs with overwhelming probability. We observe that $V_{M+1}/V_M = \mu(\{S_{M+1} \ge \tau'\})$ where μ is the uniform measure on the set $\{\boldsymbol{\sigma} \in \Sigma_N\,\forall k \le M, S_k \ge \tau'\}$, so that by Corollary 3.9.8 if we denote by E_{M+1} expectation in the r.v. $g_{i,M+1}$ only, we have

$$\mathbf{1}_{\Omega_M}\,\mathsf{E}_{M+1}\left(\mathbf{1}_{\Omega'_{M+1}}\frac{V_M}{V_{M+1}}\right) \le Le^{L\tau^2}\,. \tag{3.253}$$

Here $\Omega'_{M+1} = \{V_{M+1}/V_M \ge \exp(-N/L)\}$ and $\mathsf{P}\,(\Omega_M\backslash\Omega'_{M+1}) \le \exp(-N/L)$. Let us define by induction the events $\Omega''_{M+1} = \Omega''_M \cap \Omega'_{M+1} \cap \Omega_{M+1}$. We prove by induction over M that we have

$$\mathsf{E}\left(\mathbf{1}_{\Omega''_M}\frac{1}{V_M}\right) \le (Le^{L\tau^2})^M\,. \tag{3.254}$$

In fact, since $\Omega''_M \subset \Omega_M$, (3.254) and (3.253) imply that

$$\begin{aligned}\mathsf{E}\left(\mathbf{1}_{\Omega''_M\cap\Omega'_{M+1}}\frac{1}{V_{M+1}}\right) &= \mathsf{E}\left(\mathbf{1}_{\Omega''_M\cap\Omega'_{M+1}}\frac{V_M}{V_{M+1}}\frac{1}{V_M}\right) \\ &= \mathsf{E}\left(\mathbf{1}_{\Omega''_M}\frac{1}{V_M}\,\mathsf{E}_{M+1}\mathbf{1}_{\Omega'_{M+1}}\frac{V_M}{V_{M+1}}\right) \le (Le^{L\tau^2})^{M+1}\,,\end{aligned} \tag{3.255}$$

so that, by Markov's inequality we have

$$\mathsf{P}\,(\Omega''_M \cap \Omega'_{M+1}\backslash\Omega_{M+1}) \le (Le^{L\tau^2})^{M+1}\exp\left(-\frac{N}{L}\right)\,.$$

Since

$$\mathsf{P}(\Omega''_{M+1}) \ge \mathsf{P}(\Omega''_M) - \mathsf{P}(\Omega''_M \setminus \Omega'_{M+1}) - \mathsf{P}(\Omega''_M \cap \Omega'_{M+1} \setminus \Omega_{M+1})\,,$$

we see by induction over M that

$$\mathsf{P}(\Omega''_M) \ge 1 - M(Le^{L\tau^2})^M\exp\left(-\frac{N}{L}\right)\,,$$

so that since $\mathsf{P}(\Omega_M) \ge \mathsf{P}(\Omega''_M)$, for $ML\tau^2 \le N$, Ω_M occurs with overwhelming probability. □

Proof of Theorem 3.10.1. First we follows the proof of Theorem 3.4.1 to prove that $\nu((R_{1,2} - q)^2) \le L/N$. To take care of the exceptional case where the measure $\langle\cdot\rangle_{t,\sim}$ does not satisfy (3.241) (and where we cannot use Theorem 3.10.5 conditionally on this measure) we proceed as in the bounds of the term I of (3.247) to show that this case does not change the nature of the error term.

There is no obstacle either to adapt Theorems 3.4.2 and 3.5.2 (the latter is why we need to control the first 6 derivatives of u). $\square$

Theorem 3.10.1, together with an approximation procedure, will let us handle the limiting case where $\exp u(x) = \mathbf{1}_{\{x \geq \tau\}}$. In this case, the function Φ of (3.10) becomes

$$\Phi(x,y) = \frac{1}{y} \frac{\mathsf{E}(\xi \mathbf{1}_{\{x+\xi y \geq \tau\}})}{\mathsf{P}(x+\xi y \geq \tau)} = \frac{1}{y} A\left(\frac{\tau - x}{y}\right)$$

where

$$A(x) = \frac{1}{\sqrt{2\pi}} \frac{e^{-x^2/2}}{\mathsf{P}(\xi \geq x)} \tag{3.256}$$

and the equations (3.12), (3.13) become

$$q = \mathsf{E}\, \mathrm{th}^2(z\sqrt{r})\,;\ r = \frac{\alpha}{1-q} \mathsf{E}\, A^2\left(\frac{\tau - z\sqrt{q}}{\sqrt{1-q}}\right). \tag{3.257}$$

We use the notation $U_k = \{S_k \geq \tau\}$.

Theorem 3.10.9. *There exists a constant L with the following property. Consider a number τ, and α with $L\alpha e^{L\tau^2} \leq 1$. Then the equations (3.257) have at least one solution. If $M/N = \alpha$ and $t \leq 1/(1+\tau^2)$, we have*

$$\mathsf{P}\left(\left|\frac{1}{N}\log\left(2^{-N}\mathrm{card}\bigcap_{k\leq M} U_k\right) - RS(\alpha)\right| \geq t\right) \leq \exp\left(-\frac{Nt^2}{L}\right) \tag{3.258}$$

where

$$\begin{aligned} RS(\alpha) &= -\frac{1}{2} r(1-q) + \mathsf{E}\log \mathrm{ch}(z\sqrt{r}) \\ &\quad + \alpha\, \mathsf{E} \log \mathsf{P}_\xi\left(\xi \geq \frac{\tau - z\sqrt{q}}{\sqrt{1-q}}\right). \end{aligned} \tag{3.259}$$

and where q, r are a solution of the equations (3.257).

As pointed out, the equations (3.257) do have a unique solution, even though we did not prove it.

Research problem 3.10.10. (Level 2) Investigate the case where the uniform measure on Σ_N is replaced by the uniform measure on the cube $[-1,1]^N$, or, in other words, compute the typical volume of the intersection of this cube with random half-spaces.

This research problem is closely connected to the material of next chapter.

The proof of Theorem 3.10.9 relies on an approximation procedure. This procedure succeeds because the conditions on the derivatives of u in Theorem 3.10.1 are very weak.

Consider $\tau_1 < \tau < \tau_2$, and a number $C \geq 0$. We consider two functions u_1 and u_2 with the following properties

$$-C \leq u_1 \leq 0\,;\ x \geq \tau \Rightarrow u_1(x) = 0\,;\ x \leq \tau_1 \Rightarrow u_1(x) = -C \tag{3.260}$$

$$-2C \leq u_2 \leq 0\,;\ x \geq \tau_2 \Rightarrow u_2(x) = 0\,;\ x \leq \tau \Rightarrow u_2(x) = -2C\,. \tag{3.261}$$

Recalling (3.252) and writing $Z_j = \sum_{\boldsymbol{\sigma}} \exp\left(\sum_{k\leq M} u_j(S_k)\right)$, we observe that, since $u_1(x) = 0$ for $x \geq \tau$,

$$\frac{1}{N}\log\operatorname{card}\bigcap_{k\leq M} U_k = \frac{1}{N}\log\operatorname{card}\{\boldsymbol{\sigma}\in\Sigma_N\,;\ \forall k\leq M, S_k\geq\tau\} \leq \frac{1}{N}\log Z_1\,. \tag{3.262}$$

Since $u_2(x) \leq u_1(x)$ and $u_2(x) \leq u_1(x) - C$ for $x \leq \tau$, we have

$$\sum\left\{\exp\sum_{k\leq M} u_2(S_k)\,;\ \exists k\,,\ S_k\leq\tau\right\} \leq e^{-C} Z_1$$

and thus

$$\begin{aligned}\operatorname{card}\{\boldsymbol{\sigma}\,;\ \forall k\leq M\,,\ S_k\geq\tau\} &\geq \sum\left\{\exp\sum_{k\leq M} u_2(S_k)\,;\ \forall k\leq M\,,\ S_k\geq\tau\right\}\\ &\geq Z_2 - e^{-C} Z_1\,.\end{aligned} \tag{3.263}$$

We choose u_1 and u_2 such that

$$\forall \ell \leq 6\,,\ |u_1^{(\ell)}|,\ |u_2^{(\ell)}| \leq \exp\left(\frac{N}{L}\right)\,,$$

where L is the constant occurring in Theorem 3.10.1, and where $|\tau_2 - \tau_1| \leq \exp(-N/L)$, $C = N/L$. For $j = 1, 2$, we denote by q_j, r_j solutions to the equations (3.12) and (3.13) corresponding to $u = u_j$. As in Lemma 3.7.6, we see that

$$|q_j - q| \leq L\exp\left(-\frac{N}{L}\right)\,;\ |r_j - r| \leq L\exp\left(-\frac{N}{L}\right)$$

and

$$\begin{aligned}&\Big|-\frac{r_j}{2}(1-q_j) + \mathsf{E}\log\operatorname{ch} z\sqrt{r_j} + \alpha\,\mathsf{E}\,\log\,\mathsf{E}_\xi\exp u_j(z\sqrt{q}+\xi\sqrt{1-q})\\ &\quad - RS(\alpha)\Big| \leq L\exp\left(-\frac{N}{L}\right)\end{aligned} \tag{3.264}$$

where $RS(\alpha)$ is given by (3.259). Using Theorem 3.10.1 for u_1 and u_2 we then see that

$$\left|\frac{1}{N}\mathsf{E}\log Z_1 - \frac{1}{N}\mathsf{E}\log Z_2\right| \le \frac{L}{N}\,. \tag{3.265}$$

Lemma 3.10.11. *We have*

$$\left|\mathsf{E}\frac{1}{N}\log(2^{-N}Z_1) - RS(\alpha)\right| \le \frac{L}{N}\,, \tag{3.266}$$

and with probability at least 3/4 we have

$$\left|\frac{1}{N}\log\operatorname{card}\bigcap_{k\le M}U_k - \frac{1}{N}\log Z_1\right| \le \frac{L}{N}\,. \tag{3.267}$$

Proof. First (3.266) follows from Theorem 3.10.1 and (3.264). By (3.265) we have $\mathsf{E}\log(Z_1/Z_2) \le L$, and since $Z_2 \le Z_1$, we get that $\mathsf{P}(Z_1 \le LZ_2) \ge 3/4$. When $Z_1 \le LZ_2$, we have $Z_2 - e^{-C}Z_1 \ge Z_1/L$ (for N large enough) and (3.267) follows from (3.262) and (3.263). □

Given a number $A \ge 0$ we set

$$\log_A(x) = \max(\log x, -A)$$

(so $\log_A 0 = -A$).

The following strongly resembles Theorem 2.2.1.

Proposition 3.10.12. *Set $A = N/(16L^{\sim 2})$. Then for each $t > 0$ we have*

$$\mathsf{P}\left(\left|\frac{1}{N}\log_A\left(2^{-N}\operatorname{card}\bigcap_{k\le M}U_k\right) - \frac{1}{N}\mathsf{E}\log_A\left(2^{-N}\operatorname{card}\bigcap_{k\le M}U_k\right)\right| \ge t\right)$$
$$\le 2\exp\left(-\frac{1}{L}\min\left(\frac{N^2t^2}{M(1+\tau^2)}, \frac{Nt}{1+\tau^2}\right)\right)\,. \tag{3.268}$$

$$\mathsf{P}\left(\left|\frac{1}{N}\log_A(2^{-N}Z_1) - \frac{1}{N}\mathsf{E}\log_A(2^{-N}Z_1)\right| \ge t\right)$$
$$\le 2\exp\left(-\frac{1}{L}\min\left(\frac{N^2t^2}{M(1+\tau^2)}, \frac{Nt}{1+\tau^2}\right)\right)\,. \tag{3.269}$$

Proof of Theorem 3.10.9. We first observe that by (3.266)

$$\frac{1}{N}\mathsf{E}\log_A(2^{-N}Z_1) \ge \frac{1}{N}\mathsf{E}\log(2^{-N}Z_1) \ge RS(\alpha) - \frac{L}{N}\,.$$

Since we assume $L\alpha e^{L\tau^2} \le 1$, we can assume $RS(\alpha) \ge -A/2N$. Then (3.269) used for $t = A/2N$ shows that

$$\mathsf{P}(\log(2^{-N} Z_1) \le -A) \le L \exp\left(-\frac{N}{L}\right) . \tag{3.270}$$

Since

$$Z_1 \ge \exp(M \min u_1) \ge \exp(-NM/L) \ge \exp(-N^2)$$

it follows that

$$|\mathsf{E} \log(2^{-N} Z_1) - \mathsf{E} \log_A(2^{-N} Z_1)| \le \frac{L}{N}$$

and (3.266) implies

$$\frac{1}{N}|\mathsf{E} \log_A(2^{-N} Z_1) - RS(\alpha)| \le \frac{L}{N} ,$$

so that by (3.269) we have

$$\mathsf{P}\left(\left|\frac{1}{N} \log_A(2^{-N} Z_1) - RS(\alpha)\right| \ge t + \frac{L}{N}\right)$$
$$\le 2 \exp\left(-\frac{1}{L} \min\left(\frac{N^2 t^2}{M(1+\tau^2)}, \frac{Nt}{1+\tau^2}\right)\right) .$$

Since $N \ge M e^{L\tau^2}$, for $t = L/\sqrt{N}$ the right-hand side is $\le 1/4$. Thus with probability at least 1/4, we have that (3.267) occurs together with

$$\left|\frac{1}{N} \log_A(2^{-N} Z_1) - RS(\alpha)\right| \le \frac{L}{\sqrt{N}} \tag{3.271}$$

$$\left|\frac{1}{N} \log_A\left(2^{-N} \operatorname{card} \bigcap_{k \le M} U_k\right) - \frac{1}{N} \mathsf{E} \log_A\left(2^{-N} \operatorname{card} \bigcap_{k \le M} U_k\right)\right| \le \frac{L}{\sqrt{N}} .$$

We note that (3.267) implies

$$\left|\frac{1}{N} \log_A(2^{-N} Z_1) - \frac{1}{N} \log_A\left(2^{-N} \operatorname{card} \bigcap_{k \le M} U_k\right)\right| \le \frac{L}{N} .$$

Combining these inequalities yields

$$\left|\frac{1}{N} \mathsf{E} \log_A\left(2^{-N} \operatorname{card} \bigcap_{k \le M} U_k\right) - RS(\alpha)\right| \le \frac{L}{\sqrt{N}} ,$$

and (3.268) shows that

$$\mathsf{P}\left(\left|\frac{1}{N} \log_A\left(2^{-N} \operatorname{card} \bigcap_{k \le M} U_k\right) - RS(\alpha)\right| \ge t + \frac{L}{\sqrt{N}}\right)$$
$$\le L \exp\left(-\frac{1}{L} \min\left(\frac{N^2 t^2}{M(1+\tau^2)}, \frac{Nt}{1+\tau^2}\right)\right) .$$

Since $M(1+\tau^2) \le N$, for $t \le 1/(1+\tau^2)$ the right-hand side is at most $L\exp(-Nt^2/L)$ and this yields the result. □

The idea of the proof of Proposition 3.10.12 is to apply (A.20) to a suitable martingale difference sequence.

Lemma 3.10.13. *If $x, y \le 1$, we have*

$$|\log_A x - \log_A xy| \le |\log_A y| \,. \tag{3.272}$$

Proof. We have

$$0 \le \log_A x - \log_A xy \le -\log_A xy \le A$$

and

$$|\log_A x - \log_A xy| \le |\log x - \log xy| \le |\log y| \,,$$

so that

$$|\log_A x - \log_A xy| \le \min(A, |\log y|) = |\log_A y| \,.$$

□

We denote by Ξ_m the σ-algebra generated by the r.v. $(g_{i,k})$, for $i \le N$, $k \le m$. We denote by E_m conditional expectation given Ξ_m, so that $\mathsf{E} = \mathsf{E}_0$. We set $W = 2^{-N} \operatorname{card} \bigcap_{k \le M} U_k$, and we write

$$\frac{1}{N} \log_A W - \frac{1}{N} \mathsf{E} \log_A W = \sum_{m=1}^{M} X_m \,,$$

where

$$X_m = \mathsf{E}_m \left(\frac{1}{N} \log_A W \right) - \mathsf{E}_{m-1} \left(\frac{1}{N} \log_A W \right) . \tag{3.273}$$

Thus X_m is Ξ_m-measurable, and $\mathsf{E}_{m-1} X_m = 0$. (In other words, X_m is a martingale difference sequence.) We write

$$C_m = \bigcap_{k \ne m, k \le M} U_k \,, \; W_m = 2^{-N} \operatorname{card} C_m \ge W \,.$$

Using (3.272) for $x = W_m$ and $y = W/W_m$, we get

$$|\log_A W - \log_A W_m| \le \left| \log_A \left(\frac{W}{W_m} \right) \right| \,.$$

The left-hand side is 0 for $W_m \le e^{-A}$. Thus we have

$$|\log_A W - \log_A W_m| \le Y := \mathbf{1}_{\{W_m \ge \exp(-A)\}} \left| \log_A \left(\frac{W}{W_m} \right) \right| \,. \tag{3.274}$$

The idea of introducing W_m is that this quantity does not depend on the r.v. $g_{i,m}$, $i \le N$, so that $\mathsf{E}_m \log_A W_m = \mathsf{E}_{m-1} \log_A W_m$ and from (3.273) and (3.274), we get

$$N\,|X_m| \le \mathsf{E}_m\, Y + \mathsf{E}_{m-1}\, Y\,.$$

Given $\lambda > 0$, we have, using the Jensen and Cauchy-Schwarz inequalities,

$$\begin{aligned}\mathsf{E}_{m-1} \exp \lambda N\,|X_m| &\le \mathsf{E}_{m-1}(\exp \lambda\, \mathsf{E}_m\, Y\, \exp \lambda\, \mathsf{E}_{m-1}\, Y)\\ &\le (\mathsf{E}_{m-1} \exp 2\lambda\, \mathsf{E}_m\, Y)^{1/2} (\mathsf{E}_{m-1} \exp 2\lambda\, \mathsf{E}_{m-1}\, Y)^{1/2}\\ &\le \mathsf{E}_{m-1} \exp 2\lambda Y = \mathsf{E}_{m-1}\, \mathsf{E}'_m \exp 2\lambda\, Y\,, \end{aligned}\tag{3.275}$$

where E'_m denotes integration in the r.v. $(g_{i,m})_{i\le N}$ only, so that $\mathsf{E}_{m-1} = \mathsf{E}_{m-1}\,\mathsf{E}'_m$.

Lemma 3.10.14. *If $A = N/(16L^{\sim 2})$, we have, if $\lambda \le 1/L$*

$$\mathsf{E}'_m \exp 2\lambda\, Y \le L e^{L\tau^2}\,. \tag{3.276}$$

Proof. In this proof we work at $(g_{i,k})$, $k \ne m$ fixed and we write E for E'_m and P the corresponding probability. We can assume $W_m \ge e^{-A}$, for otherwise $Y = 0$. Using Corollary 3.9.8, we see that

$$\mathsf{P}\left(\frac{W}{W_m} \le \varepsilon\right) \le L\varepsilon e^{L\tau^2} + L\exp\left(-\frac{N}{L}\right)\,. \tag{3.277}$$

(In the right-hand side we could put ε^{16} rather than ε; but we try to make the point that any exponent works here.) Now,

$$\mathsf{E} \exp 2\lambda\, Y = \mathsf{E} \min\left(e^A, \frac{W_m}{W}\right)^{2\lambda} = \int_0^{e^A} 2\lambda t^{2\lambda-1}\, \mathsf{P}\left(\frac{W_m}{W} \ge t\right) \mathrm{d}t\,.$$

Using the bound $\mathsf{P}(W_m \ge tW) \le 1$ if $t \le 1$ and (3.277) for $\varepsilon = 1/t$ otherwise, we get, assuming $\lambda \le 1/L$,

$$\begin{aligned}\mathsf{E} \exp 2\lambda\, Y &\le L + L\int_1^{e^A} 2\lambda\left(t^{2\lambda-2}\, e^{L\tau^2} + t^{2\lambda-1} \exp\left(-\frac{N}{L}\right)\right) \mathrm{d}t\\ &\le L\, e^{L\tau^2} + \exp\left(2\lambda\, A - \frac{N}{L}\right) \le L\, e^{L\tau^2}\end{aligned}$$

if $\lambda \le 1/4L$. □

We are now in position to prove (3.268). By (3.275), (3.276) and Hölder's inequality we see that we have

$$\mathsf{E}_{m-1} \exp \frac{N|X_m|}{L(1+\tau^2)} \le 2$$

so that, using Berstein's inequality for martingale difference sequences (A.20) we get (3.268).

The proof of (3.269) is very similar, except that the ratio W/W_m has to be replaced by the quantity $\langle \exp u_1(S_m)\rangle$, where $\langle\cdot\rangle$ is the Gibbs' measure with Hamiltonian $\sum_{k\le M, k\neq m} \exp u(S_k)$. □

As a consequence of (3.268) and Theorem 3.10.9, we prove the following.

Corollary 3.10.15. *There exists $\alpha_0 < 1$ such that if $M \ge \alpha_0 N$ the set (3.1) is empty with overwhelming probability.*

It is possible to give a more direct (yet non-trivial) proof of this fact [85], [138].

Proof. First, we find $\alpha > 0$ and $\varepsilon > 0$ such that if $M_1 = \lfloor \alpha N \rfloor$, with overwhelming probability we have

$$\log \operatorname{card}\left(\Sigma_N \cap \bigcap_{k\le M} U_k\right) \le N(\log 2 - \alpha - 2\varepsilon)\,. \tag{3.278}$$

This follows from (3.278) and the fact that $RS(\alpha) < (1-\alpha)\log 2$ for $\alpha > 0$. This is because if $F(\alpha, r, q)$ denotes the right-hand side of (3.259), we have $RS(\alpha) = \sup_r \inf_q F(\alpha, r, q)$ and $(1-\alpha)\log 2 = \sup_r F(r, 0)$. Consider $M' = \lfloor (1-\varepsilon)\,N \rfloor$ and denote by E_0 expectation in the r.v. $g_{i,k}$, $M \le k \le M'$ only. Then

$$\mathsf{E}_0 \operatorname{card}\left(\Sigma_N \cap \bigcap_{k\le M'} U_k\right) = 2^{M'-M} \operatorname{card}\left(\Sigma_M \cap \bigcap_{k\le M} U_k\right) \le 2^{-N\varepsilon}$$

under (3.278). □

3.11 The Gardner Formula in the Bernoulli Case

In this section we investigate how to extend the previous results to the Bernoulli case, where the r.v. $g_{i,k}$ are replaced by variables $\eta_{i,k}$ with $\mathsf{P}(\eta_{i,k} = \pm 1) = 1/2$. Thus from this point on $H_{N,M}$ refers to the Hamiltonian (3.207).

Theorem 3.11.1. *There exists a constant L with the following property. Consider a function u that satisfies (3.221) and (3.222), that is 6 times differentiable and such that $|u^{(\ell)}| \le D'$ for $\ell \le 6$. Then if $LMe^{L\tau^2} \le N$, (3.89) and (3.90) hold for $N \ge K(D')$.*

The proof is identical to that of Theorem 3.10.1 except that rather than using integration by parts one now uses approximate integration by parts.

This creates extra error terms that become irrelevant for $N \geq N(D')$. This is rather straightforward. The *essential* fact is that thanks to Proposition 3.8.4 it is *not* necessary to find a version of Theorem 3.9.1 for the Bernoulli case. □

We recall the notation (3.259).

Theorem 3.11.2. *There exists a constant L with the following property. Consider a number τ and solutions q, r to the equations (3.257). Then given $0 < \varepsilon \leq 1/L(1+\tau^2)$, there exists an integer $N(\varepsilon, \tau)$ such that if $N \geq N(\varepsilon, \tau)$, $LMe^{L\tau^2} \leq N$, we have*

$$\mathsf{P}\left(\left|\frac{1}{N}\log\left(2^{-N}\mathrm{card}\bigcap_{k\leq M} U_k\right) - RS(\alpha)\right| \geq \varepsilon\right) \leq L\exp\left(-\frac{N\varepsilon^2}{L}\right).$$

Here of course $U_k = \{S_k \geq \tau\}$, $S_k = N^{-1/2}\sum_{i\leq N}\eta_{i,k}\sigma_i$ and $\alpha = M/N$.

Research problem 3.11.3. (Level 2) Find the rate at which the median of $N^{-1}\log\mathrm{card}\bigcap_{k\leq M} U_k$ converges to its limit.

It seems that our arguments provide a rate in $1/\log N$. Proving a rate in $N^{-\delta}$ ($\delta > 0$) would be of interest and might be easy. But of course finding the true rate is another matter.

One of the steps necessary to prove Theorem 3.11.2 is to prove that (3.268) remains valid. It suffices for this to know a result similar to Theorem 3.9.1, but where the power ε^{16} is replaced by $\varepsilon^{1/L}$.

The following resembles Proposition 3.9.3.

Lemma 3.11.4. *Consider numbers $(x_{i,\ell})$, $i \leq N$, $\ell \leq n$. Assume that for a certain number d, the following holds:*

$$\forall \ell \leq n\,,\ \sum_{i\leq N} x_{i,\ell}^2 = 1\,;\ \forall \ell \leq n\,,\ \forall i \leq N\,,\ |x_{i,\ell}| \leq d \tag{3.279}$$

$$\forall \ell < \ell'\,,\ \sum_{i\leq N} x_{i,\ell}x_{i,\ell'} \leq \frac{1}{2}\,. \tag{3.280}$$

Then if $d^4 \leq 1/(LN\log n)$ we have

$$\mathsf{P}\left(\forall \ell \leq n\,,\ \sum_{i\leq N}\eta_{i,\ell}x_{i,\ell} \leq \frac{\sqrt{\log n}}{L}\right) \leq Ln^{-1/L}\,.$$

The ideas behind this proof are well worth developing (see e.g. [151]) but we will give a self-contained arguments.

Proof of Lemma 3.11.4. We set $w_\ell = \sum_{i \le N} \eta_{i,\ell}\, x_{i,\ell}$. Given a number $s \ge 0$ we consider $F(x_1, \ldots, x_n)$ given by (3.231). If z, ξ_ℓ are independent standard Gaussian r.v. we set $g_\ell = z/\sqrt{2} + \xi_\ell/\sqrt{2}$, so that

$$\mathsf{E}\, g_\ell^2 = \mathsf{E}\, w_\ell^2\,;\ \ell \ne \ell' \Rightarrow \mathsf{E}\, g_\ell\, g_{\ell'} = \frac{1}{2} \ge \mathsf{E}\, w_\ell\, w_{\ell'}\ .$$

We have shown in the proof of Lemma 3.9.4 that

$$L \le s \le \frac{1}{L}\sqrt{\log n} \Rightarrow \mathsf{E}\, F(g_1, \ldots, g_n) \ge \frac{s^2}{L}\ . \tag{3.281}$$

Also,

$$F(w_1, \ldots, w_n) \le \log(1 + \exp(s \max_{\ell \le n} w_\ell)) \le 1 + s \max(0, \max_{\ell \le n} w_\ell)\ .$$

Since $x = \max(0, x) + \min(0, x)$, we get

$$\begin{aligned}\max_{\ell \le n} w_\ell &\ge \frac{1}{s}(F(w_1, \ldots, w_n) - 1) + \min(0, \max w_\ell) \\ &\ge \frac{1}{s}(F(w_1, \ldots, w_n) - 1) - |w_1| \end{aligned} \tag{3.282}$$

because $\min(0, \max_{\ell \le n} w_\ell) \ge \min(0, w_1) \ge -|w_1|$. We would like to use Slepian's lemma to say that $\mathsf{E}\, F(w_1, \ldots, w_n) \ge \mathsf{E}\, F(g_1, \ldots, g_n)$. Unfortunately, the r.v. w_ℓ are not Gaussian, and the proof of Slepian's lemma involved integration by parts. So, we have to revisit this proof, using approximate integration by parts, and to estimate the contribution of the error terms. It is straightforward to see that this contribution is at most $LN(ds)^4$, but since seeing this does require rewriting the proof of Slepian's lemma we will leave it to the reader (details of sharper estimates are given in [151]). Thus we get

$$\mathsf{E}\, F(w_1, \ldots, w_n) \ge \mathsf{E}\, F(g_1, \ldots, g_n) - LN(ds)^4$$

and combining with (3.281) and (3.282) we get

$$\mathsf{E} \max_{\ell \le n} w_\ell \ge \frac{s}{L} - L - LNd^4 s^3\ .$$

We take $s = \sqrt{\log n}/L$ and we see that if $d^4 < 1/LN \log n$ we have

$$\mathsf{E} \max_{\ell \le n} w_\ell \ge \frac{\sqrt{\log n}}{L}\ .$$

To finish the proof, we will appeal to the following known inequality

$$\mathsf{P}\left(\left|\sup_{\ell\le n} w_\ell - \mathsf{E}\sup_{\ell\le n} w_\ell\right| \ge t\right) \le L\exp\left(-\frac{t^2}{L}\right) . \tag{3.283}$$

This inequality is based on a kind of generalization of Theorem 2.2.4. The proof is not long, and we believe that it should be profitable for the reader to learn in depth the underlying principle. This is better done by studying ([133], [132], [90]). □

Proposition 3.11.5. *There exists a constant L with the following property. Consider a Gibbs' measure G with Hamiltonian H such that $-H \le 0$. Assume that*

$$\sum \exp(-H(\boldsymbol{\sigma})) \ge 2^N \exp\left(-\frac{N}{64}\right) .$$

Then, given independent r.v. η_i with $\mathsf{P}(\eta_i = \pm 1) = 1/2$, we have for each $\varepsilon > 0$ that

$$\mathsf{P}\left(G\left(\left\{\boldsymbol{\sigma} ; \frac{1}{\sqrt{N}}\sum_{i\le N}\eta_i\sigma_i \ge \tau\right\}\right) \le \varepsilon\right) \le L\varepsilon^{1/L}e^{L\tau^2} + L\exp\left(L\tau^2 - \frac{N}{L}\right) .$$

Proof. We recall Lemma 3.3.6; we consider Lemma 3.11.4 in the case $d = 1/\sqrt{N}$, so the condition $d^4 \le 1/(LN\log n)$ means $\log n \le N/L$. It should then be obvious how to adapt the proof of Theorem 3.9.1 to reach the desired conclusion. □

Throughout the rest of this section $A = N/64$ and

$$S_k = \frac{1}{\sqrt{N}}\sum_{i\le N}\eta_{i,k}\sigma_i ; \; U_k = \{S_k \ge \tau\} .$$

The proof of Theorem 3.11.2 relies on the following approximation procedure.

Proposition 3.11.6. *Consider $\varepsilon > 0$. Then there is a number ε' with the following property. Consider a function u that satisfies (3.221), (3.222), and*

$$x \le \tau - \varepsilon' \Rightarrow u(x) \le -\frac{1}{\varepsilon'} . \tag{3.284}$$

Then if N is large enough, whenever $LMe^{L\tau^2} \le N$, we have

$$\left|\mathsf{E}\frac{1}{N}\log_A\left(2^{-N}\sum_{\boldsymbol{\sigma}}\exp\left(\sum_{k\le M}u(S_k)\right)\right) - \mathsf{E}\frac{1}{N}\log_A\left(2^{-N}\mathrm{card}\bigcap_{k\le M}U_k\right)\right| \le \varepsilon . \tag{3.285}$$

Proof of Theorem 3.11.2. We recall the quantity $RS(\alpha)$ given by (3.259), where q,r are solutions of (3.257). We write $RS(\alpha,u)$ the quantity (3.259) when q,r are solutions of (3.12), (3.13). Given $\varepsilon>0$, we can find $\varepsilon'>0$ such that if u satisfies (3.221), (3.222) and (3.284) then $|RS(\alpha)-RS(\alpha,u)|\leq\varepsilon$. We fix such a function u with $|u^{(\ell)}|\leq K(\varepsilon')$ for $\ell\leq 6$. Next, we see from Proposition 3.11.5 that (3.268) remains valid (where now $A=N/64$) and also (3.269) if we take $Z_1=\sum_\sigma \exp\left(\sum_{k\leq M} u(S_k)\right)$. Thus by Theorem 3.11.1, for $N\geq N(\varepsilon,\tau)$, we have

$$\left|\frac{1}{N}\mathsf{E}\log 2^{-N} Z_1 - RS(\alpha,u)\right| \leq \varepsilon\,. \tag{3.286}$$

Since $L\alpha e^{L\tau^2}$, we can assume by taking L large enough that $RS(\alpha,u)\geq -A/2N$ $(=-1/128)$ and as in the proof of Theorem 3.10.9 we see that (3.286) implies

$$\left|\frac{1}{N}\mathsf{E}\log_A 2^{-N} Z_1 - RS(\alpha,u)\right| \leq 2\,\varepsilon$$

and, using (3.285), that

$$\left|\mathsf{E}\frac{1}{N}\log_A\left(2^{-N}\mathrm{card}\bigcap_{k\leq M} U_k\right) - RS(\alpha)\right| \leq 4\,\varepsilon\,.$$

If $5\varepsilon\leq A/2N$ and $RS(\alpha)\geq -A/2N$, then

$$\left|\frac{1}{N}\log\left(2^{-N}\mathrm{card}\bigcap_{k\leq M} U_k\right) - RS(\alpha)\right| \geq 5\,\varepsilon$$

$$\Rightarrow \left|\frac{1}{N}\log_A\left(2^{-N}\mathrm{card}\bigcap_{k\leq M} U_k\right) - \mathsf{E}\frac{1}{N}\log_A\left(2^{-N}\mathrm{card}\bigcap_{k\leq M} U_k\right)\right| \geq \varepsilon$$

and (3.268) finishes the proof. □

Lemma 3.11.7. *If $x,y,z\leq 1$, we have*

$$|\log_A xz - \log_A yz| \leq |\log_A x - \log_A y|\,\mathbf{1}_{\{z\geq \exp(-A)\}}\,.$$

Proof. The result is obvious if $z<\exp(-A)$, since both sides are 0. So we assume $z\geq\exp(-A)$.

If $\log_A x=-A$, then $\log_A xz=-A$, and the result follows since $-A\leq \log_A yz\leq \log_A y$. The same argument works if $\log_A y=-A$. And if $\log_A x = \log x$, $\log_A y=\log y$, then

$$|\log_A xy - \log_A yz| \leq |\log xz - \log yz| = |\log x - \log y|\,.$$

□

Lemma 3.11.8. *If* $0 < y \le x \le 1$, *then, for any* $c > 0$,

$$|\log_A x - \log_A y| \le \log_A y|\, \mathbf{1}_{\{y \le c\}} + \frac{|x-y|}{c} \, .$$

Proof. Since $\log_A y \le \log_A x \le 0$, we have $|\log_A x - \log_A y| \le |\log_A y|$. If $y > c$, we have $|\log x - \log y| \le |x - y|/c$. □

We start the proof of Proposition 3.11.6. Given $m \le M$, we consider the quantity

$$V_m = \sum \exp\left(\sum_{m \le k \le M} u(S_k) \right)$$

where the summation is over $\boldsymbol{\sigma}$ in $\bigcap_{k<m} U_k$. Thus, the left-hand side of (3.285) is

$$\frac{1}{N} |\mathsf{E} \log_A 2^{-N} V_1 - \mathsf{E} \log_A 2^{-N} V_{M+1}|$$
$$\le \frac{1}{N} \sum_{1 \le m \le M} |\mathsf{E} \log_A 2^{-N} V_{m+1} - \mathsf{E} \log_A 2^{-N} V_m| \, .$$

We proceed to bound the terms of the right-hand side. We consider the Gibbs' measure G given by

$$G(B) = \frac{1}{Z_m} \sum \exp\left(\sum_{m+1 \le k \le M} u(S_k) \right)$$

where the summation is over the set $B \cap \bigcap_{k<m} U_k$, and where Z_m is the normalizing factor. (It can happen that $\bigcap_{k<m} U_k$ is empty in which case G is not defined, and we set $Z_m = 0$. The rest of the proof will show that there is nothing to do in that case, or even in the case where $\log_A Z_m = -A$.)

We denote by $\langle \cdot \rangle$ an integration with respect to G, and we observe the identities

$$V_m = Z_m \langle \exp u(S_m) \rangle$$
$$V_{m+1} = Z_m \langle \mathbf{1}_{U_m} \rangle \, .$$

We note that $Y := \langle \mathbf{1}_{U_m} \rangle \le X := \langle \exp u(S_m) \rangle$ because $u(x) = 0$ if $x \ge \tau$. Thus, using Lemmas 3.11.7 and 3.11.8, we get that for any $c > 0$ we have

$$\frac{1}{N} |\log_A 2^{-N} V_{m+1} - \log_A 2^{-N} V_m| \le \mathbf{1}_{\{2^{-N} Z_m \ge \exp(-A)\}} \times \left(|\log_A Y|\, \mathbf{1}_{\{Y \le c\}} + \frac{1}{c} |X - Y| \right) . \tag{3.287}$$

We first take expectation only in the r.v. $(\eta_{i,m})_{i \le N}$ and we denote by P the corresponding probability. We can assume $Z_m \ge \exp(-A)$, for there is nothing to do otherwise. By (an obvious extension of) Proposition 3.11.5 we have for $t > 0$ that

$$\mathsf{P}\,(Y \le t) \le L\,t^{1/L}\,e^{L\tau^2} + L\exp\left(-\frac{N}{L}\right)$$

and it should then be obvious that given ε we can find c such that for N large enough (depending only on ε, τ) we have $\mathsf{E}\,(|\log_A Y|\,\mathbf{1}_{\{Y \le c\}}) \le \varepsilon/2$.

Thus, to prove Proposition 3.11.6, it suffices to prove the following

Lemma 3.11.9. *Given $\kappa > 0$, if ε' is small enough and if u satisfies (3.284), we have $\mathsf{E}\,|X - Y| \le \kappa$.*

Proof. Since $u(x) \le -1/\varepsilon'$ for $x \le \tau - \varepsilon'$, it suffices to show that $\mathsf{E}\,\langle \mathbf{1}_{\{\tau-\varepsilon' \le S_m \le \tau\}}\rangle \le \kappa$. Let us write $S^1 = S_m(\boldsymbol{\sigma}^1)$, $S^2 = S_m(\boldsymbol{\sigma}^2)$, $S^\sim = S^1 - S^2$, $I = [\tau - \varepsilon', \tau]$. We observe that

$$\mathsf{E}\,\langle \mathbf{1}_{\{S_m \in I\}}\rangle \le (\mathsf{E}\,\langle \mathbf{1}_{\{S_m \in I\}}\rangle^2)^{1/2} \tag{3.288}$$

and

$$\langle \mathbf{1}_{\{S_m \in I\}}\rangle^2 = \langle \mathbf{1}_{\{S^1 \in I\}}\,\mathbf{1}_{\{S^2 \in I\}}\rangle \le \langle \mathbf{1}_{\{|S^\sim| \le \varepsilon'\}}\rangle\,. \tag{3.289}$$

By (an obvious extension of) Lemma 3.9.7, since we assume $2^{-N} Z_m \ge \exp(-N/64)$, we have $G^{\otimes 2}(\{R_{1,2} \ge 1/2\}) \le \exp(-N/L)$, so that

$$G^{\otimes 2}\left(\left\{(\boldsymbol{\sigma}^1, \boldsymbol{\sigma}^2)\,;\ \mathrm{card}\,\{i \le N\,;\ \sigma_i^1 = \sigma_i^2\} \ge \frac{3N}{4}\right\}\right) \le \exp\left(-\frac{N}{L}\right)\,.$$

Since $\sigma_i^\sim = \sigma_i^1 - \sigma_i^2$ satisfies $|\sigma_i^\sim| \in \{0, 2\}$, it suffices to prove from (3.288) and (3.289) that if ε' is small enough we have (for N large enough)

$$\mathsf{P}\left(\left|\frac{1}{\sqrt{N}} \sum_{i \le N} \eta_{i,m}\,\sigma_i^\sim\right| \le \varepsilon'\right) \le \kappa$$

as soon as $\mathrm{card}\,\{i \le N\,;\ |\sigma_i^\sim| = 2\} \ge N/4$. This is obvious from the central limit theorem, or, if one wishes to use the methods in favor in this book, we consider a function φ with $\varphi(x) = 0$ for $|x| \ge 3\,\varepsilon'$, $\varphi(x) = 1$ for $|x| \le 2\,\varepsilon'$, and we compare $\mathsf{E}\,\varphi\left(N^{-1/2} \sum_{i \le N} \eta_{i,m}\,\sigma_i^\sim\right)$ and $\mathsf{E}\,\varphi\left(N^{-1/2} \sum_{i \le N} g_i\,\sigma_i^\sim\right)$ using Trotter's method. □

Research problem 3.11.10. (Level 2) Consider a distinguished point in Σ_N, for example the point $\mathbf{1} = (1, \ldots, 1)$. Find the typical cardinality of the set

$$\{\boldsymbol{\sigma}\,;\ \forall\, k \le M\,,\ S_k(\mathbf{1})\,S_k(\boldsymbol{\sigma}) \ge 0\}\,.$$

In geometric terms, each sum $N^{-1/2}\sum \eta_{i,k}\,\sigma_i$ determines two half-spaces, the half-space $\sum_{i\le N}\eta_{i,k}\,\sigma_i \ge 0$ and the half-space $\sum_{i\le N}\eta_{i,k}\,\sigma_i \le 0$. We choose the one of these two half-spaces that contains the distinguished point, and we intersect these as k varies (when the distinguished point lies on the boundary we do nothing).

This motivates the consideration of Hamiltonians of the type

$$\sum_{k\le M} u(S_k(\mathbf{1})\,S_k(\boldsymbol{\sigma}))\,.$$

Let us observe that

$$S_k(\mathbf{1})\,S_k(\boldsymbol{\sigma}) = \frac{1}{\sqrt{N}}\sum \xi_{i,k}\,\sigma_i$$

where $\xi_{i,k} = S_k(\mathbf{1})\,\eta_{i,k}$. The variables $(\xi_{i,k})_{i\le N}$ are however not independent. Still, one can hope that a suitable extension of the methods we have developed will allow progress on this question.

3.12 Notes and Comments

The problems considered in this chapter are studied in [59] and [49].

It took a long time to prove Theorem 3.4.1. The weaker methods developed previously [142] for this model or for the SK and the Hopfield models just would not work. During this struggle, it became clear that the smart path method as used here was a better way to go for these three models.

The results of Section 3.9 and after were added after I saw the paper [128]. It actually took only hours to discover the crucial estimates of Section 3.9. I simply had not believed that Theorem 3.10.1 was yet accessible.

The general idea of the interpolation method of Section 3.10 comes from [128], but the reasons that make it work here are different. I also learned in this paper about the truncations $\log_A$. While this idea of truncation is very natural, it takes some time to realize how efficient and helpful it is.

The question of the fluctuations of the size of $\bigcap_{k\le M} U_k$ is studied in [137]. This paper is another striking example of how devastating prejudice can be. I was so convinced that the isoperimetric methods of [132] are far superior to the method of martingale difference sequences that I did not check whether this most obvious approach (the one used in Proposition 3.10.12 here) works. (The results of [137] have more complicated proofs *and* a parasitic logarithm term...) Incidentally, it is overall true that the method of martingale difference sequences cannot compete with the more advanced tools. We urge the reader to study the book [90] to get convinced of this.

It is predicted in [88] that the replica-symmetric solution holds up to α^*, so Problem 3.1.1 amounts to controlling the entire replica-symmetric region, typically a very difficult task.

Problem 3.11.2 was shown to me by B. Derrida (see [50], model B).

4. Capacity of the Perceptron: The Gaussian and the Spherical Case

4.1 The Shcherbina and Tirozzi Model

The purpose of this chapter is to solve Problem 3.1.1 when the configuration space Σ_N is replaced by the sphere $S_N = \left\{\boldsymbol{\sigma} \in \mathbb{R}^N\,;\ \sum_{i \le N} \sigma_i^2 = N\right\}$. So, this chapter is an exception to our policy of studying only Ising models, that is, models with configuration space Σ_N.

In a first stage we will allow all of $\mathbb{R}^N$ as configuration space. We will consider the Hamiltonian

$$-H_{N,M}(\boldsymbol{\sigma}) = \sum_{k \le M} u\left(\frac{1}{\sqrt{N}} \sum_{i \le N} \eta_{i,k}\, \sigma_i\right) + h \sum_{i \le N} g_i\, \sigma_i - \kappa\, \|\boldsymbol{\sigma}\|^2 . \tag{4.1}$$

Here $\|\boldsymbol{\sigma}\|^2 = \sum_{i \le N} \sigma_i^2$, and $(g_i)_{i \le N}$ are independent standard Gaussian and independent of the random signs $\eta_{i,k}$. (As usual $\mathsf{P}(\eta_{i,k} = \pm 1) = 1/2$.) One can also consider the case where the r.v. $\eta_{i,k}$ are replaced by standard Gaussian r.v. $g_{i,k}$. This is easier, and will not be detailed.

We will always assume

$$u \le 0\,;\ u \text{ concave} . \tag{4.2}$$

Thus $H_{N,M}$ is a convex function of $\boldsymbol{\sigma}$. In fact,

$$\frac{1}{2} H_{N,M}(\boldsymbol{\sigma}) + \frac{1}{2} H_{N,M}(\boldsymbol{\sigma}') - H_{N,M}\left(\frac{\boldsymbol{\sigma} + \boldsymbol{\sigma}'}{2}\right) \ge \kappa \left\|\frac{\boldsymbol{\sigma} - \boldsymbol{\sigma}'}{2}\right\|^2 . \tag{4.3}$$

We will consider the Gibbs' measure G on $\mathbb{R}^N$ with Hamiltonian $H_{N,M}$, that is, for a subset B of $\mathbb{R}^N$,

$$G(B) = \frac{1}{Z_{N,M}} \int_B \exp(-H_{N,M}(\boldsymbol{\sigma}))\, d\boldsymbol{\sigma} ,$$

where $d\boldsymbol{\sigma}$ denotes Lebesgue's measure and $Z_{N,M} = \int \exp(-H_{N,M}(\boldsymbol{\sigma}))\, d\boldsymbol{\sigma}$ is the normalization factor.

The beauty of the present approach (introduced in [128]) is that it allows the use of tools from convexity. These tools are now well understood, and, while very powerful, they have simple self-contained proofs. Rather than reproducing these proofs, we will refer the reader to the delightful book [90] by M. Ledoux. The following ([90] Theorem 2.13) is a functional version of the Brunn-Minkowski theorem.

Theorem 4.1.1. *Consider non-negative functions U, V, W on $\mathbb{R}^N$, $0 < s < 1$, and assume that for all $\mathbf{x}, \mathbf{y}$ in $\mathbb{R}^N$ we have*

$$W(s\,\mathbf{x} + (1-s)\,\mathbf{y}) \geq U(\mathbf{x})^s\, V(\mathbf{y})^{1-s} . \tag{4.4}$$

Then

$$\int W(\mathbf{x})\mathrm{d}\mathbf{x} \geq \left(\int U(\mathbf{x})\,\mathrm{d}\mathbf{x}\right)^s \left(\int V(\mathbf{x})\,\mathrm{d}\mathbf{x}\right)^{1-s} . \tag{4.5}$$

In particular, if A, B are convex sets, and $\mathrm{Vol}A = \int_A \mathrm{d}\mathbf{x}$, we have

$$\mathrm{Vol}\,(s\,A + (1-s)\,B) \geq (\mathrm{Vol}\,A)^s\,(\mathrm{Vol}\,B)^{1-s} , \tag{4.6}$$

the Brunn-Minkowski inequality.

Proposition 4.1.2. *Consider a function V on $\mathbb{R}^N$, and assume that for some number $\kappa > 0$ we have*

$$\forall \mathbf{x}, \mathbf{y} \in \mathbb{R}^N ,\ \frac{1}{2}(V(\mathbf{x}) + V(\mathbf{y})) - V\left(\frac{\mathbf{x}+\mathbf{y}}{2}\right) \geq \kappa \left\|\frac{\mathbf{x}-\mathbf{y}}{2}\right\|^2 .$$

Consider the probability measure μ on $\mathbb{R}^N$ of density with respect to Lebesgue's measure proportional to $\exp(-V(\mathbf{x}))$. Consider a function f on $\mathbb{R}^N$ such that, for a certain number A we have

$$\forall \mathbf{x}, \mathbf{y} \in \mathbb{R} ,\ |f(\mathbf{x}) - f(\mathbf{y})| \leq A\,\|\mathbf{x} - \mathbf{y}\| . \tag{4.7}$$

Then for $v > 0$ we have

$$\mu\left(\left\{\mathbf{x};\ \left|f(\mathbf{x}) - \int f\mathrm{d}\mu\right| \geq v\right\}\right) \leq L \exp\left(-\frac{\kappa\, v^2}{8\,A^2}\right) . \tag{4.8}$$

This is a very nice generalization of Theorem 2.2.4.

Proof. We consider a median m of f, that is $\mu(\{f \geq m\}) \geq 1/2$ and $\mu(\{f \leq m\}) \geq 1/2$. Then Theorem 2.1.5 of [90] asserts that

$$\forall v \geq 0 ,\ \mu(\{|f - m| \geq v\}) \leq 4 \exp\left(-\frac{\kappa\, v^2}{4\,A^2}\right) .$$

Thus we have $\left|m - \int f \mathrm{d}\mu\right| \leq LA/\sqrt{\kappa}$, and hence

$$\mu\left(\left\{\left|f - \int f\mathrm{d}\mu\right| \geq v - LA/\sqrt{\kappa}\right\}\right) \leq 4 \exp\left(-\frac{\kappa\, v^2}{4\,A^2}\right)$$

from which (4.8) follows. □

Remark. We will hardly use Theorem 4.1.1. We quote it mainly because it is a central fact about convexity. In particular Theorem 2.15 of [90], on which the previous result is based, follows rather easily from it.

Using (4.8) for G, we get the following

Corollary 4.1.3. *We have*

$$\left\langle \exp \frac{\kappa}{L}(\sigma_N - \langle \sigma_N \rangle)^2 \right\rangle \le 2 \,.$$

Lemma 4.1.4. *Consider a probability* μ *on* $\mathbb{R}^N$ *such that*

$$\mu(B) = \frac{1}{Z}\int_B \exp\left(U(\boldsymbol{\sigma}) - \kappa \|\boldsymbol{\sigma}\|^2 + \sum_{i\le N} a_i\,\sigma_i \right) \mathrm{d}\boldsymbol{\sigma} \,,$$

where $U \le 0$ *and where* Z *is the normalizing factor. Then we have*

$$\begin{aligned} & v \ge \frac{2}{\kappa}\sqrt{\sum_{i\le N} a_i^2}\ ;\ \left(\frac{4\pi}{\kappa}\right)^{N/2} \frac{1}{Z}\exp\left(-\frac{\kappa v^2}{8}\right) \le 1 \\ & \Rightarrow \mu(\{\|\boldsymbol{\sigma}\| \ge v\}) \le \exp\left(-\frac{\kappa\, v^2}{8}\right) \,. \end{aligned} \tag{4.9}$$

Proof. We have

$$\begin{aligned} \mu(\{\|\boldsymbol{\sigma}\| \ge v\}) &= \frac{1}{Z}\int_{\{\|\boldsymbol{\sigma}\|\ge v\}} \exp\left(U(\boldsymbol{\sigma}) - \kappa\,\|\boldsymbol{\sigma}\|^2 + \sum_{i\le N} a_i\,\sigma_i \right) \mathrm{d}\boldsymbol{\sigma} \\ &\le \frac{1}{Z}\int_{\{\|\boldsymbol{\sigma}\|\ge v\}} \exp\left(-\kappa\,\|\boldsymbol{\sigma}\|^2 + \sum_{i\le N} a_i\,\sigma_i \right) \mathrm{d}\boldsymbol{\sigma} \,. \end{aligned}$$

If $v \ge 2\sqrt{\sum_{i\le N} a_i^2}/\kappa$, for $\|\boldsymbol{\sigma}\| \ge v$ we have $-\kappa\,\|\boldsymbol{\sigma}\|^2 + \sum_{i\le N} a_i\,\sigma_i \le -\kappa\,\|\boldsymbol{\sigma}\|^2/2$, so that

$$\begin{aligned} \mu(\{\|\boldsymbol{\sigma}\| \ge v\}) &\le \frac{1}{Z}\int_{\{\|\boldsymbol{\sigma}\|\ge v\}} \exp\left(-\frac{\kappa}{2}\,\|\boldsymbol{\sigma}\|^2\right) \mathrm{d}\boldsymbol{\sigma} \\ &\le \frac{1}{Z}\exp\left(-\frac{\kappa}{4}\,v^2\right)\int \exp\left(-\frac{\kappa}{4}\,\|\boldsymbol{\sigma}\|^2\right) \mathrm{d}\boldsymbol{\sigma} \,. \end{aligned}$$

In this chapter and the next, we will use many times the fact that

$$\int \exp(-t\,\|\boldsymbol{\sigma}\|^2)\,\mathrm{d}\boldsymbol{\sigma} = \left(\int \exp(-tx^2)\,\mathrm{d}x\right)^N = \left(\frac{\pi}{t}\right)^{N/2} \,. \tag{4.10}$$

Thus

$$\mu(\{\|\boldsymbol{\sigma}\| \geq v\}) \leq \left(\frac{4\pi}{\kappa}\right)^{N/2} \frac{1}{Z} \exp\left(-\frac{\kappa}{4} v^2\right) ,$$

from which the result follows. □

Lemma 4.1.5. *Assume that for a certain number D, the function u satisfies*

$$\forall x\,,\ u(x) \geq -D(1+|x|)\ . \tag{4.11}$$

Then we have

$$\frac{1}{Z_{N,M}} \leq \left(\frac{\kappa}{\pi}\right)^{N/2} \exp\left(MD\left(1+\frac{L}{\sqrt{\kappa}}\right)\right)\ . \tag{4.12}$$

Proof. Consider the Gaussian measure γ on $\mathbb{R}^N$ of density $W\exp(-\kappa\|\boldsymbol{\sigma}\|^2)$ with respect to Lebesgue's measure. By (4.10) we have $W = (\kappa/\pi)^{N/2}$. Let $\mathbf{x}_k = (N^{-1/2}\eta_{i,k})_{i\leq N}$, so that $\|\mathbf{x}_k\| = 1$, and hence

$$\int |\mathbf{x}_k \cdot \boldsymbol{\sigma}|\, \mathrm{d}\gamma(\boldsymbol{\sigma}) \leq \frac{L}{\sqrt{\kappa}}\ . \tag{4.13}$$

Now,

$$\begin{aligned} Z_{N,M} &= \int \exp\left(\sum_{k\leq M} u(\mathbf{x}_k\cdot\boldsymbol{\sigma}) - \kappa\|\boldsymbol{\sigma}\|^2 + h\sum_{i\leq N} g_i\,\sigma_i\right) \mathrm{d}\boldsymbol{\sigma} \\ &= W^{-1}\int \exp\left(\sum_{k\leq M} u(\mathbf{x}_k\cdot\boldsymbol{\sigma}) + h\sum_{i\leq N} g_i\,\sigma_i\right)\mathrm{d}\gamma(\boldsymbol{\sigma}) \\ &\geq W^{-1}\exp\left(\sum_{k\leq M}\int u(\mathbf{x}_k\cdot\boldsymbol{\sigma})\,\mathrm{d}\gamma(\boldsymbol{\sigma})\right) \end{aligned}$$

using Jensen's inequality. The result follows from (4.11) and (4.13). □

We will often assume that

$$\kappa \geq \kappa_0\,,\ h \leq h_0\,,\ M \leq 10N \tag{4.14}$$

where κ_0 and h_0 are given numbers. The condition $M \leq 10N$ is simply to avoid trivial complications; the most interesting case is $M \leq 2N$.

Until further notice we make the convention that K denotes a number that might depend on κ_0, h_0, D but that does not depend on M and N. This number need not be the same at each occurrence.

Corollary 4.1.6. *Under (4.11) and (4.14) we have*

$$v \geq K\left(\sqrt{N} + \sqrt{\sum_{i\leq N} g_i^2}\right) \Rightarrow G(\{\|\boldsymbol{\sigma}\| \geq v\}) \leq \exp\left(-\frac{\kappa v^2}{8}\right) . \tag{4.15}$$

$$v \geq K\sqrt{N} \Rightarrow \mathsf{E}G(\{\|\boldsymbol{\sigma}\| \geq v\}) \leq \exp\left(-\frac{v^2}{K}\right) \tag{4.16}$$

Proof. The first assertion follows from Lemmas 4.1.4 and 4.1.5. The second assertion follows from the first and the fact that for $t \geq 0$ we have $\mathsf{P}(\sum_{i\leq N} g_i^2 \geq tN) \leq L\exp(-t^2/L)$. □

We now turn to a crucial ingredient.

Theorem 4.1.7. *Under (4.11) and (4.14) we have*

$$\mathsf{E}\langle(R_{1,2} - \langle R_{1,2}\rangle)^2\rangle \leq \frac{K}{N} . \tag{4.17}$$

In fact, we could even control exponential moments of $R_{1,2} - \langle R_{1,2}\rangle$, but this will not be needed.

Proof. We write $\mathbf{b} = \langle\boldsymbol{\sigma}\rangle$, so that $\langle R_{1,2}\rangle = \|\mathbf{b}\|^2/N$, and

$$|R_{1,2} - \langle R_{1,2}\rangle| \leq \left|\frac{\boldsymbol{\sigma}^1\cdot\boldsymbol{\sigma}^2}{N} - \frac{\boldsymbol{\sigma}^1\cdot\mathbf{b}}{N}\right| + \left|\frac{\boldsymbol{\sigma}^1\cdot\mathbf{b}}{N} - \frac{\mathbf{b}\cdot\mathbf{b}}{N}\right| . \tag{4.18}$$

If we fix $\boldsymbol{\sigma}^1$, the map $\mathbf{x} \mapsto \boldsymbol{\sigma}^1\cdot\mathbf{x}/N$ satisfies (4.7) with $A = \|\boldsymbol{\sigma}^1\|/N$, so that by (4.8) used for $\mu = G$ we have

$$\int\left(\frac{\boldsymbol{\sigma}^1\cdot\boldsymbol{\sigma}^2}{N} - \frac{\boldsymbol{\sigma}^1\cdot\mathbf{b}}{N}\right)^2 \mathrm{d}G(\boldsymbol{\sigma}^2) \leq \frac{L\,\|\boldsymbol{\sigma}^1\|^2}{N^2\,\kappa} .$$

Using (4.16) we conclude that

$$\mathsf{E}\left\langle\left(R_{1,2} - \frac{\boldsymbol{\sigma}^1\cdot\mathbf{b}}{N}\right)^2\right\rangle \leq \frac{K}{N} .$$

We proceed in a similar manner for the second term on the right-hand side of (4.18). □

We will now deduce from Theorem 4.1.7 and the Ghirlanda-Guerra identities that $\langle R_{1,2}\rangle$ is essentially non-random. To simplify notation we denote by δ a quantity depending on u, κ, h, M, N and such that given $\kappa_0 < \kappa_1$ and $h_0 > 0$, we have

$$\lim_{N\to\infty}\sup_{M\leq 10N}\int_{\kappa_0}^{\kappa_1}\int_0^{h_0}|\delta|\,\mathrm{d}h\,\mathrm{d}\kappa = 0 . \tag{4.19}$$

Theorem 4.1.8. *Under (4.11) and (4.14), we have*

$$\mathsf{E}\left\langle (R_{1,2} - \mathsf{E}\langle R_{1,2}\rangle)^2 \right\rangle = \delta \tag{4.20}$$

$$\mathsf{E}\left\langle (R_{1,1} - \mathsf{E}\langle R_{1,1}\rangle)^2 \right\rangle = \delta \,. \tag{4.21}$$

Here and below, $R_{1,1} = \|\boldsymbol{\sigma}^1\|^2/N$. We write

$$F_{N,M} = \frac{1}{N}\log \int \exp(-H_{N,M}(\boldsymbol{\sigma}))\,\mathrm{d}\boldsymbol{\sigma} \,.$$

Thus, the dependence in u, κ, h is kept implicit.

Lemma 4.1.9. *If u satisfies (4.11), and under (4.14), we have*

$$\mathsf{E}\left((F_{N,M} - \mathsf{E}\,F_{N,M})^2\right) \le \frac{K}{N} \,. \tag{4.22}$$

Proof. We could in fact prove an exponential inequality as in Proposition 3.10.12, but (4.22) suffices here. For simplicity we will assume that $h = 0$, since the influence of this term can be controlled through Theorem 2.2.4. As in Proposition 3.10.12 we write

$$X_m = \mathsf{E}_m\, F_{N,M} - \mathsf{E}_{m-1}\, F_{N,M}$$

so that X_m is a martingale difference sequence, and we will use that

$$\mathsf{E}\left(\sum_{m\le M} X_m\right)^2 = \sum_{m\le M} \mathsf{E}\, X_m^2$$

because if $m < m'$, then $\mathsf{E}\, X_m\, X_{m'} = \mathsf{E}\,(X_m\, \mathsf{E}_m\, X_{m'}) = 0$. Let us denote by $\langle\cdot\rangle_m$ an average with respect to the Gibbs' measure with Hamiltonian

$$-\sum_{k\ne m, k\le M} u(S_k) + \kappa\,\|\boldsymbol{\sigma}\|^2$$

where of course $S_k = N^{-1/2}\sum_{i\le N} \eta_{i,k}\,\sigma_i$. Thus

$$N\,F_{N,M} = B + \log\langle \exp u(S_m)\rangle_m$$

where

$$B = \log \int \exp\left(\sum_{k\ne m} u(S_k) - \kappa\,\|\boldsymbol{\sigma}\|^2\right)\mathrm{d}\boldsymbol{\sigma}$$

does not depend on the variables $\eta_{i,m}$, so that $\mathsf{E}_m\, B = \mathsf{E}_{m-1}\, B$, and

$$N\,X_m = \mathsf{E}_m \log\langle \exp u(S_m)\rangle_m - \mathsf{E}_{m-1}\log\langle \exp u(S_m)\rangle_m \,.$$

Since $u \le 0$, we have $1 \ge \langle \exp u(S_m)\rangle_m \ge \exp\langle u(S_m)\rangle_m$ and

$$N\,|X_m| \le \mathsf{E}_m\,|\langle u(S_m)\rangle_m| + \mathsf{E}_{m-1}\,|\langle u(S_m)\rangle_m|$$

and thus

$$N^2\,\mathsf{E}\,X_m^2 \le 4\,\mathsf{E}\,\langle u(S_m)\rangle_m^2 \le LD^2\mathsf{E}\,\langle 1 + S_m^2\rangle_m \le LD^2\mathsf{E}\,\langle 1 + \frac{\|\boldsymbol{\sigma}\|^2}{N}\rangle_m\,,$$

where in the last inequality we have first integrated in the variables $\eta_{i,m}$. Using (4.16) we see that $N^2\mathsf{E}X_m^2 \le K$ and this proves (4.22). □

To lighten notation we write $\nu(f) = \mathsf{E}\,\langle f\rangle$. We still assume (4.14).

Lemma 4.1.10. *We have*

$$\nu\left(\left(\frac{1}{N}\sum_{i\le N} g_i\,\sigma_i - \frac{1}{N}\nu\left(\sum_{i\le N} g_i\,\sigma_i\right)\right)^2\right) = \delta \tag{4.23}$$

$$\nu((R_{1,1} - \nu(R_{1,1}))^2) = \delta\,. \tag{4.24}$$

Proof. One observes the identities

$$\frac{1}{N}\left\langle\sum_{i\le N} g_i\,\sigma_i\right\rangle = \frac{\partial}{\partial h}F_{N,M}\,;\ \langle -R_{1,1}\rangle = \frac{\partial}{\partial\kappa}F_{N,M}$$

and one copies the arguments of Theorem 2.12.1. □

From Lemma 4.1.10 we get a suitable version of the Ghirlanda-Guerra identities. If f is a function of Σ_N^n, such that $\nu(f^2)$ is bounded independently of N, then (4.23) and the Cauchy-Schwarz inequality imply that

$$\nu\left(\frac{1}{N}\left(\sum_{i\le N} g_i\,\sigma_i^1\right) f\right) = \nu\left(\frac{1}{N}\sum_{i\le N} g_i\,\sigma_i\right)\nu(f) + \delta\,. \tag{4.25}$$

To compute these we integrate by parts as in Section 2.12 (keeping in mind that it is no longer true that $\sigma_i^2 = 1$), and we get

$$\nu\left(\frac{1}{N}\sum_{i\le N} g_i\,\sigma_i\right) = h(\nu(R_{1,1}) - \nu(R_{1,2}))$$

$$\nu\left(\frac{1}{N}\left(\sum_{i\le N} g_i\,\sigma_i\right) f\right) = h\left(\sum_{\ell\le n}\nu(R_{1,\ell}\,f) - n\nu(R_{1,n+1}\,f)\right)\,,$$

and by (4.25)

$$n\nu(R_{1,n+1}\, f) = \nu(R_{1,1}\, f) - \nu(R_{1,1})\,\nu(f) + \nu(R_{1,2})\,\nu(f) \\ + \sum_{2\le\ell\le n} \nu(R_{1,\ell}\, f) + \delta\,. \tag{4.26}$$

Now, by (4.24) and the Cauchy-Schwarz inequality we have $\nu(R_{1,1}\, f) - \nu(R_{1,1})\,\nu(f) = \delta$, so that the Ghirlanda-Guerra identities take the same form as in Section 2.12.

Proof of Theorem 4.1.8. By (4.17) we have $\nu((R_{1,2} - \langle R_{1,2}\rangle)^2) = \delta$, and this implies (4.20) by (the proof of) (2.453). □

4.2 The Replica-Symmetric Equations

In this section we consider a concave function u, $u \le 0$, we assume (4.14) and that for a certain number D, we have

$$|u(x)| \le D(1+|x|)\,;\ \forall \ell \le 6\,,\ |u^{(\ell)}(x)| \le D\,. \tag{4.27}$$

Our objective is to show that the numbers

$$q = \mathsf{E}\,\langle R_{1,2}\rangle\,;\ \rho = \mathsf{E}\,\langle R_{1,1}\rangle \tag{4.28}$$

(nearly) satisfy certain equations (4.31), (4.45) and (4.46) below. The notation, that is similar to that of the previous chapters, should not hide the fact that the procedure is different. The numbers q and ρ are not defined through the replica-symmetric equations, but through the system. They depend on N, M (and of course on h, κ, u).

Consider independent standard Gaussian r.v. ξ, z. Consider numbers $0 \le x \le y$, the r.v. $\theta = z\sqrt{x} + \xi\sqrt{y-x}$, and define

$$\Psi(x,y) = \alpha\,\mathsf{E}\left(\frac{\mathsf{E}_\xi\,(u'(\theta)\exp u(\theta))}{\mathsf{E}_\xi \exp u(\theta)}\right)^2 \\ = \frac{\alpha}{y-x}\,\mathsf{E}\left(\frac{\mathsf{E}_\xi\,(\xi\exp u(\theta))}{\mathsf{E}_\xi \exp u(\theta)}\right)^2 \tag{4.29}$$

using integration by parts (of course E_ξ denotes averaging in ξ only). We also define

$$\overline{\Psi}(x,y) = \alpha\,\mathsf{E}\,\frac{\mathsf{E}_\xi\,((u''(\theta)+u'^2(\theta))\exp u(\theta))}{\mathsf{E}_\xi \exp u(\theta)} \\ = \frac{\alpha}{y-x}\,\mathsf{E}\,\frac{\mathsf{E}_\xi\,((\xi^2-1)\exp u(\theta))}{\mathsf{E}_\xi \exp u(\theta)}\,, \tag{4.30}$$

integrating by parts twice. We set

$$r = \Psi(q,\rho)\,;\ \overline{r} = \overline{\Psi}(q,\rho)\,. \tag{4.31}$$

This makes sense because by the Cauchy-Schwarz inequality $R_{1,2} \le R_{1,1}^{1/2} R_{2,2}^{1/2}$ and thus $q \le \rho$.

Lemma 4.2.1. *We have* $\overline{r} \le r$.

Proof. Consider a concave function w and assume that for a certain number $c > 0$ we have $w'' \le -c$. Consider the unique point x^* where $w'(x^*) = 0$, so that $w'(x) \ge -c(x - x^*)$ for $x \le x^*$ and $w'(x) \le -c(x-x^*)$ for $x \ge x^*$. Thus $c(x-x^*)^2 \le -w'(x)(x-x^*)$ and

$$\begin{aligned} c\int (x-x^*)^2 \exp w(x)\,\mathrm{d}x &\le \int -w'(x)(x-x^*)\exp w(x)\,\mathrm{d}x \\ &= \int \exp w(x)\,\mathrm{d}x \end{aligned} \tag{4.32}$$

by integration by parts, and assuming $\lim_{x\to\pm\infty} x \exp w(x) = 0$. Taking $w(x) = u(x) - x^2/2$ where u is concave gives

$$\mathsf{E}\,((\xi - x^*)^2 \exp u(\xi)) \le \mathsf{E} \exp u(\xi)\,.$$

Minimization of the left-hand side over x^* yields

$$\mathsf{E}\,(\xi^2 \exp u(\xi)) - \frac{(\mathsf{E}\,(\xi \exp u(\xi)))^2}{\mathsf{E}\exp u(\xi)} \le \mathsf{E}\exp u(\xi)$$

from which the result follows using (4.29) and (4.30). □

Lemma 4.2.2. *Under (4.14) we have*

$$\mathsf{E}\,\langle \sigma_N^8 \rangle \le K\,.$$

Proof. Using that $\langle \sigma_N^8 \rangle \le L(\langle \sigma_N\rangle^8 + \langle (\sigma_N - \langle \sigma_N \rangle)^8\rangle)$ we see from Corollary 4.1.3 that it suffices to prove that $\mathsf{E}\,\langle\sigma_N\rangle^8 \le K$. The core of the proof is the fact that if $\boldsymbol{\rho} = (\sigma_1,\cdots,\sigma_{N-1})$ and if $F(\boldsymbol{\rho},\sigma_N)$ is a concave function on $\mathbb{R}^N$, then the function $f(\sigma_N) = \log\int \exp F(\boldsymbol{\rho},\sigma_N)\,\mathrm{d}\boldsymbol{\rho}$ is concave. To see this, considering $x_1, x_2 \in \mathbb{R}$, and $0 < s < 1$, we apply (4.5) to $W(\boldsymbol{\rho}) = \exp F(\boldsymbol{\rho}, sx_1 + (1-s)\,x_2)$, $U(\boldsymbol{\rho}) = \exp F(\boldsymbol{\rho}, x_1)$, $V(\boldsymbol{\rho}) = \exp F(\boldsymbol{\rho}, x_2)$. Thus, the function

$$f(\sigma_N) = \log \int \exp\left(\sum_{k\le M} u(S_k) - \kappa \sum_{i\le N-1} \sigma_i^2 + h \sum_{i \le N-1} g_i\,\sigma_i\right) \mathrm{d}\boldsymbol{\rho}$$

is a concave function of σ_N. Now, using the definition of Gibbs' measure and Fubini's theorem, we see that

$$\langle \sigma_N \rangle = \frac{1}{A} \int x \exp(f(x) - \kappa x^2 + h\, g_N\, x)\, \mathrm{d}x$$

where $A = \int \exp(f(x) - \kappa x^2 + h\, g_N\, x)\, \mathrm{d}x$. Consider the point x^* where the concave function $w(x) = f(x) - \kappa x^2 + h\, g_N\, x$ is maximum. It follows from (4.32) that $\langle (\sigma_N - x^*)^2 \rangle \le 1/2\,\kappa_0$, so that $|\langle \sigma_N \rangle - x^*| \le L/\sqrt{\kappa_0}$. Hence all we have to show is that $\mathsf{E}\, x^{*8} \le K$. Since f is concave, we have $|x^*| \le |x_1|$ where $x_1 = (f'(0) + h\, g_N)/2\kappa$. (To see this, without loss of generality assume $f'(0) > 0$. Then since $f'(x_1) \le f'(0)$ we have $f'(x_1) - 2\kappa\, x_1 + h\, g_N \le 0$.) Thus it suffices to show that $\mathsf{E}\, f'(0)^8 \le K$. Now it is straightforward to see that

$$f'(0) = \frac{1}{\sqrt{N}} \sum_{k \le M} \eta_{N,k} \langle u'(S_{k,0}) \rangle_0$$

where $\langle \cdot \rangle_0$ is a certain Gibbs' measure on $\mathbb{R}^{N-1}$ that does not depend on the r.v. $\eta_{N,k}$, and $S_{k,0}$ is obtained from S_k by replacing σ_N by 0. Thus $\langle u'(S_{k,0}) \rangle_0$ does not depend on the r.v. $\eta_{N,k}$, and $|\langle u'(S_{k,0}) \rangle_0| \le D$. The result follows from (A.3). □

Consider a centered Gaussian r.v. Y, independent of all the other r.v. already considered, with $\mathsf{E}\, Y^2 = r$, and consider $0 < t < 1$. We write

$$S_{k,t} = \frac{1}{\sqrt{N}} \sum_{i \le N-1} \eta_{i,k}\, \sigma_i + \sqrt{\frac{t}{N}}\, \eta_{N,k}\, \sigma_N\ ,$$

and we consider the Hamiltonian

$$-H_{N,M,t}(\boldsymbol{\sigma}) = \sum_{k \le M} u(S_{k,t}(\boldsymbol{\sigma})) - \kappa\, \|\boldsymbol{\sigma}\|^2 + h \sum_{i \le N} g_i\, \sigma_i$$
$$+ \sigma_N \sqrt{1-t}\, Y - (1-t)(r - \overline{r})\, \frac{\sigma_N^2}{2}\ .$$

We denote by $\langle \cdot \rangle_t$ the Gibbs' measure with Hamiltonian $H_{N,M,t}$. We write $\nu_t(f) = \mathsf{E}\, \langle f \rangle_t$. It should be obvious that Corollary 4.1.6 and Lemma 4.2.2 remain true when ν is replaced by ν_t.

Proposition 4.2.3. *We have, uniformly in t,*

$$\nu_t((R_{1,2} - q)^2) = \delta\ ,\ \nu_t((R_{1,1} - \rho)^2) = \delta\ . \tag{4.33}$$

The uniformity in t refers of course to the convergence in (4.19)

Proof. The proof of Theorem 4.1.8 shows that, uniformly in t we have $\nu_t((R_{1,2} - q_t)^2) = \delta$ and $\nu_t((R_{1,1} - \rho_t)^2) = \delta$ where $q_t = \nu_t(R_{1,2})$ and

$\rho_t = \nu_t(R_{1,1})$. Thus all we have to do is to show that, uniformly in t, we have $q = q_t + \delta$ and $\rho = \rho_t + \delta$. Let us write

$$p_t = \frac{1}{N}\mathsf{E}\log\int\exp(-H_{N,M,t})\,\mathrm{d}\boldsymbol{\sigma}\;.$$

We have $\partial p_t/\partial\kappa = -\rho_t$, and by integration by parts we have as usual $\partial p_t/\partial h = h(\rho_t - q_t)$. So it suffices to show that $\partial p_t/\partial\kappa = \partial p_1/\partial\kappa + \delta$ and $\partial p_t/\partial h = \partial p_1/\partial h + \delta$. Since p_t is a convex function of h and κ, it suffices to show that $p_t = p_1 + \delta$. (The elementary arguments to see this have been explained in the proof of Theorem 2.4.19.) In fact we have $|\partial p_t/\partial t| \le K/N$. This is seen by computing the derivative using approximate integration by parts in the r.v. $\eta_{N,k}$. The only non-trivial point is the control of the error terms in this approximate integration by parts. It is achieved by noticing that the functionals $\nu_{t,x}$ of the proof of Proposition 3.8.3 still satisfy Corollary 4.1.6, so that it is very rare for these functionals that $|\sigma_N| \ge KN$, and one can use that remark following this Proposition to perform the step corresponding to (3.219). One can then control these error terms using Lemma 4.2.2. □

Lemma 4.2.4. *If f is a function on $(\mathbb{R}^N)^n$ and if $n \le 4$, then for $\alpha = M/N$ we have*

$$\nu_t'(f) = \mathrm{I} + \mathrm{II} + \mathrm{III} + \mathrm{IV} + +\mathrm{V} + \mathcal{R}\;,$$

where

$$\begin{aligned}\mathrm{I} = \frac{\alpha}{2}\Bigg(&\sum_{\ell\le n}\nu_t\left(\varepsilon_\ell^2(u'^2(S_{M,t}^\ell) + u''(S_{M,t}^\ell))\,f\right)\\ &- n\,\nu_t\left(\varepsilon_{n+1}^2(u'^2(S_{M,t}^{n+1}) + u''(S_{M,t}^{n+1}))\;f\right)\Bigg)\end{aligned} \tag{4.34}$$

$$\begin{aligned}\mathrm{II} = \alpha\Bigg(&\sum_{1\le\ell<\ell'\le n}\nu_t\left(\varepsilon_\ell\,\varepsilon_{\ell'}\,u'(S_{M,t}^\ell)\,u'(S_{M,t}^{\ell'})\,f\right)\\ &- n\sum_{\ell\le n}\nu_t\left(\varepsilon_\ell\,\varepsilon_{n+1}\,u'(S_{M,t}^\ell)\,u'(S_{M,t}^{n+1})\,f\right)\\ &+\frac{n(n+1)}{2}\,\nu_t\left(\varepsilon_{n+1}\,\varepsilon_{n+2}\,u'(S_{M,t}^{n+1})\,u'(S_{M,t}^{n+2})\,f\right)\Bigg)\end{aligned} \tag{4.35}$$

$$\begin{aligned}\mathrm{III} = -r\Bigg(&\sum_{1\le\ell<\ell'\le n}\nu_t(\varepsilon_\ell\,\varepsilon_{\ell'}\,f) - n\sum_{\ell\le n}\nu_t(\varepsilon_\ell\,\varepsilon_{n+1}\,f)\\ &+\frac{n(n+1)}{2}\,\nu_t(\varepsilon_{n+1}\,\varepsilon_{n+2}\,f)\Bigg)\end{aligned} \tag{4.36}$$

$$\mathrm{IV} = \frac{-r}{2}\left(\sum_{\ell \le n} \nu_t(\varepsilon_\ell^2 f) - n\,\nu_t(\varepsilon_{n+1}^2 f)\right)$$

$$\mathrm{V} = \frac{1}{2}(r - \overline{r})\left(\sum_{\ell \le n} \nu_t(\varepsilon_\ell^2 f) - n\,\nu_t(\varepsilon_{n+1}^2 f)\right)$$

$$|\mathcal{R}| \le \frac{K}{N}\,(\nu_t(f^2))^{1/2}\,. \tag{4.37}$$

The restriction $n \le 4$ is inessential, and motivated by our convention that K should not depend on n. In fact we will use only the case $n = 2$, but we stated the general formulas because these should be easier for the reader to recognize.

Proof. This is proved like Proposition 3.2.3, using appropriate integration by parts. The term V is created by the term $(1-t)(\overline{r}-r)\,\sigma_N^2$ of the Hamiltonian $H_{N,M,t}$. The only issue is the control of the error terms due to approximate integration by parts, which is done exactly as in the case of Proposition 4.2.3. □

The idea now is that the terms II and III nearly cancel out, while the term I nearly cancels out with IV + V.

Lemma 4.2.5. *Consider an integer $n \le 4$. Then, for a function f on $\mathbb{R}^{Nn}$, such that $\nu(f^2) \le 1$ we have*

$$\alpha\nu_t(f(u'^2(S_{M,t}^1) + u''(S_{M,t}^1))) = \overline{r}\,\nu_t(f) + \mathcal{R}$$

$$\alpha\nu_t(f\,u'(S_{M,t}^1)\,u'(S_{M,t}^2)) = r\,\nu_t(f) + \mathcal{R}$$

where

$$|\mathcal{R}| \le K\left((\nu_t((R_{1,2}-q)^2))^{1/2} + (\nu_t((R_{1,1}-\rho)^2))^{1/2} + \frac{1}{N}\right)\,.$$

Even though this formally looks like the estimates of Section 3.3, this is a very different (and much easier) game because we allow a possibly very large constant K in the above bound for $\mathcal{R}$.

Proof. We bring out the dependence of $\langle\cdot\rangle_t$ in $S_{M,t}^\ell$ by writing it in term of $\langle\cdot\rangle_{t,\sim}$ as in (3.31). We then introduce a parameter v, and the family of Gibbs' measures indexed by v obtained when $S_{M,t}^\ell$ is replaced by $g_{v,\ell} = \sqrt{v}\,S_{M,t}^\ell + \sqrt{1-v}\,\theta_\ell$, as in (3.53). This defines a functional $\nu_{t,v}$. These functionals still satisfy Corollary 4.1.6 and Lemma 4.2.2. The basic step is the inequality

$$\left|\frac{\partial}{\partial v}\nu_{t,v}(fB)\right| \le K\Bigg(\sum_{\ell \le n+1} \nu_{t,v}(|f|\,|R_{\ell,\ell}-\rho|) + \sum_{1<\ell<\ell' \le n+2} \nu_{t,v}(|f|\,|R_{\ell,\ell'}-q|) + \frac{1}{N}\Bigg)\,, \tag{4.38}$$

where either $B = 1$, $B = u'^2(S^1_{M,t}) + u''(S^1_{M,t})$ or $B = u'(S^1_{M,t})\, u'(S^2_{M,t})$. To this aim, we proceed as in Proposition 3.3.8. We have an extra error term due to approximate integration by parts, which is handled as in Proposition 4.2.3. Once (4.38) is proved, we use Corollary 4.1.6 and Lemma 3.4.5 to see that if f is non-negative,

$$\nu_{t,v}(f) \le K\left(\nu_t(f) + \frac{1}{N}\right)$$

and hence we have

$$|\nu_{t,0}(fB) - \nu_t(fB)| \le \mathcal{R}$$

where $\mathcal{R}$ is as in (4.37). For $B = 1$, we have $\nu_{t,0}(f) = \mathsf{E}\,\langle f\rangle_{t,\sim}$. For $B = u'^2(S^1_{M,t}) + u''(S^1_{M,t})$, then $\alpha\nu_{t,0}(f) = \overline{r}\,\mathsf{E}\,\langle f\rangle_{t,\sim}$ and similarly in the last case. □

If we combine Proposition 4.2.3, Lemma 4.2.4 and Lemma 4.2.5 as in Section 3.4 we have proved the following.

Proposition 4.2.6. *If either $f = \varepsilon_1\,\varepsilon_2$ or $f = \varepsilon_1^2$, we have $\nu_0(f) = \nu(f) + \delta$.*

We now observe that, by definition of ν_0, and by Fubini's theorem, we have

$$\nu_0(\varepsilon_1^2) = \mathsf{E}\frac{1}{Z}\int \varepsilon^2 \exp\left(\varepsilon(Y + hg_N) - \frac{\varepsilon^2}{2}(2\kappa + r - \overline{r})\right) \mathrm{d}\varepsilon \tag{4.39}$$

$$\nu_0(\varepsilon_1\,\varepsilon_2) = \mathsf{E}\left(\frac{1}{Z}\int \varepsilon \exp\left(\varepsilon(Y + hg_N) - \frac{\varepsilon^2}{2}(2\kappa + r - \overline{r})\right) \mathrm{d}\varepsilon\right)^2, \tag{4.40}$$

where

$$Z = \int \exp\left(\varepsilon(Y + hg_N) - \frac{\varepsilon^2}{2}(2\kappa + r - \overline{r})\right) \mathrm{d}\varepsilon\,.$$

We compute these Gaussian integrals as follows. If z is centered Gaussian, and d is a number, writing $z^2\, e^{dz} = z(z\, e^{dz})$, by integration by parts we get

$$\mathsf{E}\, z^2\, e^{dz} = \mathsf{E}\, z^2(\mathsf{E}\, e^{dz} + d\,\mathsf{E}\, z\, e^{dz})$$

$$\mathsf{E}\, z\, e^{dz} = d\,\mathsf{E}\, z^2\,\mathsf{E}\, e^{dz}\,.$$

Thus

$$\frac{\mathsf{E}\, z^2\, e^{dz}}{\mathsf{E}\, e^{dz}} = \mathsf{E}\, z^2 + d^2(\mathsf{E}\, z^2)^2\,.$$

Using this for $d = Y + hg_N$, $\mathsf{E}\, z^2 = 1/(2\kappa + r - \overline{r})$ and taking expectation, we get

$$\nu_0(\varepsilon_1^2) = \frac{1}{2\kappa + r - \overline{r}} + \frac{r + h^2}{(2\kappa + r - \overline{r})^2}\,. \tag{4.41}$$

Since $\nu(\varepsilon_1^2) = \rho$ by symmetry among sites, combining with Proposition 4.2.6 we have proved the first assertion of the following result. The second is similar.

Theorem 4.2.7. *If we define $r, \overline{r}$ by (4.31) we have*

$$\rho = \frac{1}{2\kappa + r - \overline{r}} + \frac{r + h^2}{(2\kappa + r - \overline{r})^2} + \delta \tag{4.42}$$

$$q = \frac{r + h^2}{2\kappa + r - \overline{r}} + \delta\,. \tag{4.43}$$

Let us consider the function

$$\begin{aligned} F(q, \rho) &= \alpha\, \mathsf{E} \log \mathsf{E}_\xi \exp u(z\sqrt{q} + \xi\sqrt{\rho - q}) \\ &+ \frac{1}{2}\frac{q}{\rho - q} + \frac{1}{2}\log(\rho - q) - \kappa\,\rho + \frac{h^2}{2}\,(\rho - q)\,, \end{aligned} \tag{4.44}$$

which is defined for $0 \le \rho < q$. It is elementary (calculus and integration by parts) to show that the conditions $\partial F/\partial \rho = 0 = \partial F/\partial q$ mean that

$$\rho = \frac{1}{2\kappa + r - \overline{r}} + \frac{r + h^2}{(2\kappa + r - \overline{r})^2} \tag{4.45}$$

$$q = \frac{r + h^2}{(2\kappa + r - \overline{r})^2}\,, \tag{4.46}$$

where $r, \overline{r}$ satisfy (4.31). Interestingly, it does not seem obvious at all that these equations have a solution. But it follows from Theorem 4.2.7 and a cluster point argument that indeed these solutions exist when $\kappa > 0$, u is concave and satisfies (4.11).

Research problem 4.2.8. (Level 2) Find general conditions under which the replica symmetric equations (4.31), (4.45) and (4.46) have a unique solution. Given u, for which values of α, κ, h is it the case?

4.3 Approximations of the Limiting Case

In the setting of Section 4.2, we would like to show that, if $F(\rho, q)$ is given by (4.44), and if $M/N \to \alpha$, then

$$F(\rho, q) = \lim_{N\to\infty} \mathsf{E} \frac{1}{N} \log \int \exp(-H_{N,M}(\boldsymbol{\sigma}))\, \mathrm{d}\boldsymbol{\sigma}\,.$$

The main obstacle to do this is that we do not know that the equations (4.31), (4.45) and (4.46) have a unique solution.

Fortunately, this is known in a significant domain of parameters in the "limiting" case where $\exp u(x) = \mathbf{1}_{\{x \ge \tau\}}$. In that case the function (4.44) takes the form

$$F(q,\rho) = \alpha \,\mathsf{E} \log \mathsf{P}_\xi(z\sqrt{q} + \xi\sqrt{\rho - q} \geq \tau) + \frac{1}{2}\frac{q}{\rho - q}$$
$$+ \frac{1}{2}\log(\rho - q) - \kappa\,\rho + \frac{h^2}{2}(\rho - q)\,. \tag{4.47}$$

Equivalently,

$$\mathsf{P}_\xi(z\sqrt{q} + \xi\sqrt{\rho - q} \geq \tau) = B\left(\frac{\tau - z\sqrt{q}}{\sqrt{\rho - q}}\right)$$

where $B(x) = \mathsf{P}(\xi \geq x)$.

We now fix once and for all $\tau \geq 0$.

Theorem 4.3.1. a) *If $\alpha < 2$ there is a unique solution $q_0 = q_0(\alpha, \kappa, h)$, $\rho_0 = \rho_0(\alpha, \kappa, h)$ to the equations*

$$\frac{\partial F}{\partial q}(q_0, \rho_0) = 0 = \frac{\partial F}{\partial \rho}(q_0, \rho_0)\,. \tag{4.48}$$

b) *Given κ_0, $h_0 > 0$, $\alpha_0 < 2$ and $\varepsilon > 0$, there exists ε' with the following property. Consider a concave function u that satisfies*

$$x \geq \tau \Rightarrow u(x) = 0\,;\; x \leq \tau - \varepsilon' \Rightarrow u(x) \leq -\frac{1}{\varepsilon'} \tag{4.49}$$

$$\forall x\,,\; |u(x)| \leq D(1 + |x|)\,;\; \forall \ell \leq 6\,,\; |u^{(\ell)}| \leq D\,. \tag{4.50}$$

Then we can find a number $\varepsilon'' > 0$ with the following property. Given $\kappa \geq \kappa_0$, $h \leq h_0$, $\alpha \leq \alpha_0$, and any numbers $0 \leq q \leq \rho$, consider $r = \Psi(q, \rho)$, $\overline{r} = \overline{\Psi}(q, \rho)$, where Ψ and $\overline{\Psi}$ are given by (4.29) and (4.30). Assume that

$$\left|q - \frac{r + h^2}{(2\kappa + r - \overline{r})^2}\right| \leq \varepsilon''\,;\; \left|\rho - q - \frac{1}{2\kappa + r - \overline{r}}\right| \leq \varepsilon''\,. \tag{4.51}$$

Then we have

$$|q - q_0(\alpha, \kappa, h)| \leq \varepsilon\,;\; |\rho - \rho_0(\alpha, \kappa, h)| \leq \varepsilon\,. \tag{4.52}$$

In words: if $\exp u$ approaches $\mathbf{1}_{\{x \geq \tau\}}$, any approximate solution of (4.31), (4.45) and (4.46) is close to (q_0, ρ_0).

We make the change of variable $x = q/(\rho - q)$, so that $q = x\rho/(1 + x)$, $\rho - q = \rho/(1 + x)$, and

$$F(q,\rho) = G(x,\rho) := \alpha \,\mathsf{E} \log B\left(\frac{\tau\sqrt{1 + x}}{\sqrt{\rho}} - z\sqrt{x}\right) + \frac{x}{2}$$
$$+ \frac{1}{2}\log\rho - \frac{1}{2}\log(1 + x) - \kappa\,\rho + \frac{h^2\,\rho}{2(1 + x)}\,. \tag{4.53}$$

Proposition 4.3.2. *For all $\alpha \ge 0$, the equations $\partial G/\partial\rho = 0$ and $\partial G/\partial x = 0$ have at most a solution.*

This will follow from the next result.

Proposition 4.3.3. *We have*

$$\frac{\partial^2 G}{\partial \rho^2} < 0\,, \quad \frac{\partial}{\partial x}\left(\frac{x+1}{x}\frac{\partial G}{\partial x}\right) > 0\,. \tag{4.54}$$

Proof of Proposition 4.3.2. Suppose that we have $\partial G/\partial x = 0$ and $\partial G/\partial\rho = 0$ at the points (x_1, ρ_1) and (x_2, ρ_2). Then since $\partial^2 G/\partial\rho^2 < 0$, we have $G(x_1, \rho_2) < G(x_1, \rho_1)$ unless $\rho_2 = \rho_1$. Since $\partial G/\partial x(x_2, \rho_2) = 0$, by the second part of (4.54) we have $\partial G/\partial x(x, \rho_2) < 0$ if $x < x_2$ and > 0 if $x > x_2$. Thus $G(x_1, \rho_2) > G(x_2, \rho_2)$ unless $x_1 = x_2$. So $G(x_2, \rho_2) < G(x_1, \rho_1)$ unless $(x_1, \rho_1) = (x_2, \rho_2)$. Reversing the argument shows that $(x_1, \rho_1) = (x_2, \rho_2)$. □

The proof of Proposition 4.3.3 will not be difficult to follow line by line. But unfortunately the author cannot explain why it works and how it could ever be found. (These questions should be directed to the authors of [128].)

We recall the function

$$A(x) = -\frac{\mathrm{d}}{\mathrm{d}x}\log B(x) = \frac{1}{\sqrt{2\pi}}\frac{e^{-x^2/2}}{B(x)}\,.$$

Lemma 4.3.4. *We have*

$$A(v) \ge v \tag{4.55}$$

$$A'(v) = A^2(v) - vA(v) \ge 0 \tag{4.56}$$

$$vA(v)\,A'(v) \le A^2(v) \tag{4.57}$$

$$A(v) \le v + \frac{1}{v}\,. \tag{4.58}$$

Proof. To prove (4.55) we can assume $v \ge 0$. Then

$$v\int_v^\infty \exp\left(-\frac{t^2}{2}\right)\mathrm{d}t \le \int_v^\infty t\exp\left(-\frac{t^2}{2}\right)\mathrm{d}t = \exp\left(-\frac{v^2}{2}\right)$$

which is (4.55). The equality in (4.56) is straightforward, and the inequality follows from (4.55) since $A(v) \ge 0$.

Using (4.56), we see that

$$vA(v)\,A'(v) = A^2(v)(vA(v) - v^2)$$

so that (4.57) is equivalent to (4.58).

Now, by integration by parts,

$$\int_v^\infty t^2 \exp\left(-\frac{t^2}{2}\right) dt = v \exp\left(-\frac{v^2}{2}\right) + \sqrt{2\pi}\, B(v)$$

$$\int_v^\infty t \exp\left(-\frac{t^2}{2}\right) dt = \exp\left(-\frac{v^2}{2}\right)$$

so that

$$0 \le \int_v^\infty (t-v)^2 \exp\left(-\frac{t^2}{2}\right) dt = (1+v^2)\sqrt{2\pi}\, B(v) - v \exp\left(-\frac{v^2}{2}\right) .$$

This proves (4.58). □

We write $W = \tau\sqrt{1+x}/\sqrt{\rho} - z\sqrt{x}$.

Lemma 4.3.5. *We have*

$$2\,\frac{\partial G}{\partial \rho} = \alpha\,\tau\sqrt{1+x}\,\rho^{-3/2}\,\mathsf{E}\,A(W) + \frac{1}{\rho} - 2\kappa + \frac{h^2}{1+x} \tag{4.59}$$

$$2\left(\frac{x+1}{x}\right)\frac{\partial G}{\partial x} = -\frac{\alpha}{x}\,\mathsf{E}(A^2(W)) + 1 - \frac{h^2\rho}{x(1+x)} . \tag{4.60}$$

Proof. The proof of (4.59) is straightforward. To prove (4.60) we start with the simple fact that

$$2\,\frac{\partial G}{\partial x} = -\alpha\,\mathsf{E}\left(\left(-\frac{z}{\sqrt{x}} + \frac{\tau}{\sqrt{1+x}\,\sqrt{\rho}}\right) A(W)\right) + 1 - \frac{1}{1+x} - \frac{h^2\rho}{(1+x)^2} . \tag{4.61}$$

Now, by integration by parts and (4.56)

$$\begin{aligned}\mathsf{E}\left(\frac{z}{\sqrt{x}} A(W)\right) &= -\mathsf{E}\,(A'(W)) = \mathsf{E}\,(W A(W)) - \mathsf{E}\,(A^2(W)) \\ &= \mathsf{E}\left(\left(\frac{\tau\sqrt{1+x}}{\sqrt{\rho}} - z\sqrt{x}\right) A(W)\right) - \mathsf{E}\,(A^2(W))\end{aligned}$$

and thus

$$(1+x)\,\mathsf{E}\left(\frac{z}{\sqrt{x}} A(W)\right) = \frac{\tau\sqrt{1+x}}{\sqrt{\rho}}\,\mathsf{E}\,A(W) - \mathsf{E}\,A^2(W) . \tag{4.62}$$

Plugging back this value in (4.61) yields (4.60). □

Proof of Proposition 4.3.3. It follows from (4.59) that

$$2\frac{\partial^2 G}{\partial \rho^2} = -\frac{5}{2}\frac{\alpha\tau\sqrt{1+x}}{\rho^{5/2}}\,\mathsf{E}\,A(W) - \alpha\left(\frac{\tau\sqrt{1+x}}{\rho^{3/2}}\right)^2 \mathsf{E}\,A'(W) - \frac{1}{\rho^2}$$

and this is ≤ 0 since $A' \geq 0$. By (4.60) we get

$$2\frac{\partial}{\partial x}\left(\frac{1+x}{x}\frac{\partial G}{\partial x}\right) = -\frac{\alpha}{x}\,\mathsf{E}\left(\left(\frac{\tau}{\sqrt{1+x}\sqrt{\rho}} - \frac{z}{\sqrt{x}}\right) A(W)\,A'(W)\right) + \frac{\alpha}{x^2}\,\mathsf{E}\,A^2(W) + \frac{h^2\rho}{x^2} - \frac{h^2\rho}{(1+x)^2}\,. \tag{4.63}$$

Now, we observe the identity

$$\frac{\tau}{\sqrt{1+x}\sqrt{\rho}} - \frac{z}{\sqrt{x}} = \frac{W}{x} - \frac{\tau}{x\sqrt{1+x}\sqrt{\rho}}\,,$$

so that (4.63) yields

$$2\frac{\partial}{\partial x}\left(\frac{1+x}{x}G\right) = \frac{\alpha\tau}{x^2\sqrt{1+x}\sqrt{\rho}}\,\mathsf{E}\,(A(W)\,A'(W)) + \frac{\alpha}{x^2}\,\mathsf{E}\,(A^2(W) - W A(W)\,A'(W)) + h^2\rho\left(\frac{1}{x^2} - \frac{1}{(1+x)^2}\right) \tag{4.64}$$

and all the terms are ≥ 0 by (4.57) and since $x \geq 0$. □

Lemma 4.3.6. *Consider a concave function u with $u(x) = 0$ for $x \geq \tau$. Set $\theta = z\sqrt{q} + \xi\sqrt{\rho - q}$, $Y = (\tau - z\sqrt{q})/\sqrt{\rho - q}$. Then*

$$\left(\frac{\mathsf{E}_\xi\,\xi \exp u(\theta)}{\mathsf{E}_\xi \exp u(\theta)}\right)^2 \leq \begin{cases} Y^2 + L & \textit{if} \quad Y \geq 0 \\ L & \textit{if} \quad Y \leq 0\,. \end{cases} \tag{4.65}$$

Proof. We observe that, since $u' \geq 0$,

$$\mathsf{E}_\xi\,\xi \exp u(\theta) = \mathsf{E}_\xi \sqrt{\rho - q}\,u'(\theta) \exp u(\theta) \geq 0\,. \tag{4.66}$$

Consider first the case where $Y \geq 0$. Let $U = \mathsf{E}_\xi(\exp u(\theta)\,\mathbf{1}_{\{\xi \leq Y\}})$. Then

$$\mathsf{E}_\xi\,\xi \exp u(\theta) \leq YU + \mathsf{E}_\xi(\xi\,\mathbf{1}_{\{\xi \geq Y\}}) = YU + \frac{1}{\sqrt{2\pi}}\exp\left(-\frac{Y^2}{2}\right)\,.$$

Since $u(\theta) = 0$ for $\xi \geq Y$ we have

$$\mathsf{E}_\xi \exp(u(\theta)\,\mathbf{1}_{\{\xi \geq Y\}}) \geq \mathsf{P}_\xi(\xi \geq Y) \geq \frac{Y}{1+Y^2}\frac{1}{\sqrt{2\pi}}\exp\left(-\frac{Y^2}{2}\right) \tag{4.67}$$

by (4.58), and thus

$$0 \le \frac{\mathsf{E}_\xi\, \xi \exp u(\theta)}{\mathsf{E}_\xi \exp u(\theta)} \le \frac{\sqrt{2\pi}\, YU + \exp\left(-\frac{Y^2}{2}\right)}{\sqrt{2\pi}\, U + \frac{Y}{1+Y^2}\exp\left(-\frac{Y^2}{2}\right)} \le Y + \frac{1}{Y}\,. \tag{4.68}$$

Taking squares proves the first part of (4.65) when $Y \ge 1$. When $Y \le 1$ we can replace the term $Y/(1+Y^2)$ in (4.67) by $1/L$ and in (4.68) we get a bound L. Taking squares this completes the proof of the first part of (4.65). Assume now $Y \le 0$. Then, since $u \le 0$ we have

$$\mathsf{E}_\xi\, \xi \exp u(\theta) \le \mathsf{E}_\xi\, |\xi| = \sqrt{\frac{2}{\pi}}$$

and the result since $\mathsf{E}_\xi \exp u(\theta) \ge 1/2$ by the first inequality of (4.67). □.

Lemma 4.3.7. *Consider $\kappa_0 > 0$, $\alpha_0 < 2$, $h_0 > 0$. Then we can find a number C depending only on κ_0, α_0, h_0, such that if $\kappa \ge \kappa_0$, $\alpha \le \alpha_0$, $h \le h_0$, and if the concave function u satisfies*

$$x \ge \tau \Rightarrow u(x) = 0\,,$$

then whenever q, ρ satisfy the equations (4.31), (4.45) and (4.46) we have $q, \rho \le C$.

Proof. Since

$$(\rho - q)(2\kappa + r - \overline{r}) = 1\,, \tag{4.69}$$

and since $\overline{r} \le r$, we have $\rho - q \le 1/2\kappa$, so the issue is to control the size of q. Since $r = \Psi(\rho, q)$, by (4.29) and (4.69) we have

$$q = \frac{r + h^2}{(2\kappa + r - \overline{r})^2} \le \frac{h_0^2}{(2\kappa_0)^2} + \alpha(\rho - q)\,\mathsf{E}\left(\frac{\mathsf{E}_\xi\, \xi \exp u(\theta)}{\mathsf{E}_\xi \exp u(\theta)}\right)^2\,,$$

where $\theta = z\sqrt{q} + \xi\sqrt{\rho - q}$. By Lemma 4.3.6, we have

$$q \le \frac{h_0^2}{(2\kappa_0)^2} + \frac{L}{\kappa_0} + \alpha(\rho - q)\,\mathsf{E}\, Y^2 \mathbf{1}_{\{Y \ge 0\}} \tag{4.70}$$

where $Y = (\tau - z\sqrt{q})/\sqrt{\rho - q}$. We have

$$\mathsf{E}\, Y^2 \mathbf{1}_{\{Y \ge 0\}} = \frac{q}{\rho - q}\,\mathsf{E}\left(\left(z - \frac{\tau}{\sqrt{q}}\right)^2 \mathbf{1}_{\{z \ge \tau/\sqrt{q}\}}\right)\,.$$

Consider $\theta > 1/2$ such that $\theta\alpha_0 < 1$. Then there is a number $q(\tau, \theta)$ such that if $q \ge q(\tau, \theta)$ we have

$$(\rho - q)\,\mathsf{E}\, Y^2 \mathbf{1}_{\{Y \ge 0\}} \le \theta q\,,$$

and combining with (4.70) this completes the proof. □

Corollary 4.3.8. *Consider* $\kappa_0 > 0$, $\alpha_0 < 2$, $h_0 > 0$. *Then we can find a number* ε' *depending only on* κ_0, α_0, h_0 *such that if* $\alpha \le \alpha_0$, $h \le h_0$, $\kappa \ge \kappa_0$ *and if the concave function* u *satisfies (4.49) then whenever* q, ρ *satisfy (4.31), (4.45) and (4.46) we have* $|q - q_0| \le \varepsilon$, $|\rho - \rho_0| \le \varepsilon$.

Proof. Knowing Proposition 4.3.2 and Lemma 4.3.7, this is a simple compactness argument. □

Proof of Theorem 4.3.1. To finish the proof of part 2 of that theorem we have to show that given u satisfying conditions (4.49) and (4.50), if q, ρ are approximate solutions of the equations (4.31), (4.45) and (4.46), they are close to a true solution of these equations. This is again a compactness argument, using that $|\overline{r}|$, $|r| \le LD^2$ from (4.29), (4.30) and (4.31). □

For $\alpha < 2$, the equations (4.48) have a unique solution q_0, ρ_0, and we define

$$RS_0(\alpha) = RS_0(\alpha, \kappa, h) = \alpha \, \mathsf{E} \log B\left(\frac{\tau - z\sqrt{q_0}}{\sqrt{\rho_0 - q_0}}\right) + \frac{1}{2}\frac{q_0}{\rho_0 - q_0} + \frac{1}{2}\log(\rho_0 - q_0) - \kappa\,\rho_0 + \frac{h^2}{2}(\rho_0 - q_0)\,, \quad (4.71)$$

where $B(x) = \mathsf{P}(z \ge x)$. (The reader will recognize that this is $F(q_0, \rho_0)$, where F is defined in (4.47).) The value of κ, h, will be kept implicit. (We recall that τ has been fixed once and for all.)

Theorem 4.3.9. *Consider* $\alpha_0 < 2$, $0 < \kappa_0 < \kappa_1$, $h_0 > 0$, $\varepsilon > 0$. *Then we can find* ε' *with the following property. If* u *is a concave function satisfying (4.49) and (4.50), then, for* N *large enough, we have*

$$\left|\mathsf{E}\frac{1}{N}\log\int \exp(-H_{N,M}(\boldsymbol{\sigma}))\,\mathrm{d}\boldsymbol{\sigma} - \left(RS_0\left(\frac{M}{N}\right) + \frac{1}{2}\log(2e\pi)\right)\right| \le \varepsilon \quad (4.72)$$

(where $H_{N,M}(\boldsymbol{\sigma})$ *is the Hamiltonian (4.1)) whenever* $\kappa_0 \le \kappa \le \kappa_1$, $h \le h_0$, $M/N \le \alpha_0$.

Let us write

$$p_{N,M} = \mathsf{E}\frac{1}{N}\log\int \exp(-H_{N,M}(\boldsymbol{\sigma}))\,\mathrm{d}\boldsymbol{\sigma} \quad (4.73)$$

and let us first consider the case $M = 0$. In that case

$$\int \exp\left(-\kappa\,\|\boldsymbol{\sigma}\|^2 + h\sum_{i\le N} g_i\,\sigma_i\right)\mathrm{d}\boldsymbol{\sigma} = \left(\frac{\pi}{\kappa}\right)^{N/2}\exp\frac{h^2}{4\kappa}\sum_{i\le N} g_i^2$$

and

$$p_{N,M} = \frac{1}{2}\log\left(\frac{\pi}{\kappa}\right) + \frac{h^2}{4\kappa}\,.$$

When $\alpha = 0$, we have $r = \overline{r} = 0$, so $\rho - q = 1/(2\kappa)$, $q = h^2/(4\kappa^2)$, and a straightforward computation gives

$$RS_0(0) = \frac{1}{2}\log\left(\frac{1}{2\kappa}\right) - \frac{1}{2} + \frac{h^2}{4\kappa}\,,$$

so that (4.72) holds in that case.

The proof of the following is now straightforward if one handles the error terms occurring from approximate integration by parts in the smart path method as in Proposition 4.2.3.

Lemma 4.3.10. *Assume that u satisfies (4.50), and consider independent random signs $(\eta_i)_{i\le N}$ independent of the randomness of $H_{N,M}$. Then we have*

$$\mathsf{E}\log\left\langle \exp u\left(\frac{1}{\sqrt{N}}\sum_{i\le N}\eta_i\,\sigma_i\right)\right\rangle = \mathsf{E}\log\mathsf{E}_\xi\exp u(z\sqrt{q}+\xi\sqrt{\rho-q})+\mathcal{R} \quad (4.74)$$

where

$$|\mathcal{R}| \le K(\kappa_0,h_0,D)\left(\mathsf{E}\left\langle|R_{1,2}-q|\right\rangle + \mathsf{E}\left\langle|R_{1,1}-\rho|\right\rangle + \frac{1}{N}\right)\,.$$

Proof of Theorem 4.3.9. Using integration by parts, we observe that

$$\frac{\partial p_{N,M}}{\partial h} = h(\mathsf{E}\left\langle R_{1,1}\right\rangle - \mathsf{E}\langle R_{1,2}\rangle)\,;\quad \frac{\partial p_{N,M}}{\partial\kappa} = -\mathsf{E}\left\langle R_{1,1}^2\right\rangle$$

and that, by (4.15) these quantities stay bounded for $\kappa\ge\kappa_0$, $h\le h_0$, (and given D such that (4.50) holds). Thus, to prove (4.72), it suffices to prove that given a domain

$$\Delta = \{(\kappa,h)\,;\ \kappa_1\le\kappa_2\le\kappa_0\,;\ h_1\le h\le h_2\}$$

then, for N large enough, whenever $M\le\alpha_0 N$, we have

$$\left|\int\!\!\int_\Delta\left(p_{N,M} - \left(RS_0\left(\frac{M}{N}\right) + \frac{1}{2}\log(2e\pi)\right)\right)\mathrm{d}\kappa\,\mathrm{d}h\right| \le 10\,\varepsilon\,\mathrm{Area}\,\Delta\,. \quad (4.75)$$

We observe that this is true for $M = 0$, as shown after the statement of Theorem 4.3.9.

Next we observe (as shown in the proof of Theorem 4.3.1) that as $\alpha\le\alpha_0<2$, $\kappa\ge\kappa_0$, the solutions q_0,ρ_0 of the equations (4.48) remain bounded by a certain number K_0. We then find a number $\varepsilon_1 > 0$ and a number $\varepsilon' > 0$ such that if $0 \le q_0 \le \rho_0 \le K_0$, then, if u satisfies (4.49) we have

$$|q - q_0| \le \varepsilon_1 \,,\ |\rho - \rho_0| \le \varepsilon_1 \tag{4.76}$$
$$\Rightarrow |\mathsf{E}\log \mathsf{E}_\xi\, u(z\sqrt{q} + \xi\sqrt{\rho - q}) - \mathsf{E}\log \mathsf{P}_\xi(z\sqrt{q_0} + \xi\sqrt{\rho_0 - q_0} \ge \tau)| \le \varepsilon \,.$$

Moreover we can further reduce ε' to ensure that part b of Theorem 4.3.1 (condition (4.52)) holds for $\varepsilon = \varepsilon_1$.

Combining (4.74), (4.52) and (4.76) we get that for N large enough we have

$$\sup_{M \le \alpha_0 N} \left| \int\int_\Delta \left(\mathsf{E}\left\langle \exp u\left(\frac{1}{\sqrt{N}}\sum_{i\le N} \eta_{i,M+1}\,\sigma_i\right)\right\rangle \right.\right.$$
$$\left.\left. - \mathsf{E}\log \mathsf{P}_\xi(z\sqrt{q_0} + \xi\sqrt{\rho_0 - q_0} \ge \tau)\right) \mathrm{d}\kappa\, \mathrm{d}h \right| \le 2\,\varepsilon\, \mathrm{Area}\,\Delta$$

from which (4.72) follows by summation over M, as in Section 3.4. □

We now write

$$U_k = \{S_k \ge \tau\}\,;\ S_k = \frac{1}{\sqrt{N}}\sum_{i\le N} \eta_{i,k}\,\sigma_i$$

and we state the main result of this chapter.

Theorem 4.3.11. *Consider $\kappa > 0$, $\alpha_0 < 2$, $\varepsilon > 0$. Then there exists $K = K(\kappa, \alpha_0, \varepsilon)$ such that if N is large enough, than for each $M \le \alpha_0 N$, we have*

$$\mathsf{P}\left(\left|\frac{1}{N}\log \int_{\cap_{k\le M} U_k} \exp\left(-\kappa\,\|\boldsymbol{\sigma}\|^2 + h\sum_{i\le N} g_i\,\sigma_i\right) \mathrm{d}\boldsymbol{\sigma}\right.\right.$$
$$\left.\left. - \left(RS_0\left(\frac{M}{N}\right) + \frac{1}{2}\log(2e\pi)\right)\right| \ge \varepsilon\right) \le K\exp\left(-\frac{N}{K}\right) . \tag{4.77}$$

Consider the (random) measure γ of density

$$W\exp\left(-\kappa\,\|\boldsymbol{\sigma}\|^2 + h\sum_{i\le N} g_i\,\sigma_i\right) \tag{4.78}$$

with respect to Lebesgue's measure, where

$$W = \left(\frac{\kappa}{\pi}\right)^{N/2} \exp\left(-h^2\sum_{i\le N}\frac{g_i^2}{4\kappa}\right) .$$

Rather than (4.77), we can prove that

$$\mathsf{P}\left(\left|\frac{1}{N}\log\gamma\left(\bigcap_{k\le M}U_k\right)-\left(RS_0\left(\frac{M}{N}\right)+\frac{1}{2}\log(2e\kappa)+\frac{h^2}{4\kappa}\right)\right|\ge\varepsilon\right)$$
$$\le K\exp\left(-\frac{N}{K}\right)\,. \tag{4.79}$$

A crucial step will be the following, where we recall the notation $\log_A x = \max(\log x, -A)$.

Proposition 4.3.12. *Given $a>0$, $\varepsilon>0$, there exists $\varepsilon'>0$ such that if u satisfies (4.49) we have*

$$\left|\frac{1}{N}\mathsf{E}\log_{Na}\gamma\left(\bigcap_{k\le M}U_k\right)-\frac{1}{N}\mathsf{E}\log_{Na}\int\exp\left(-\sum_{k\le M}u(S_k)\right)\mathrm{d}\gamma(\boldsymbol{\sigma})\right|\le\varepsilon \tag{4.80}$$

for N large enough.

The approach is similar to that of Proposition 3.11.6. We write

$$V_m=\int\exp\left(-\sum_{k\ge m}u(S_k)\right)\mathrm{d}\gamma(\boldsymbol{\sigma})\,,$$

where the integral is over the set $\bigcap_{k<m}U_k$; and we bound the left-hand side of (4.80) by

$$\sum_{1\le m\le M}\frac{1}{N}\left|\mathsf{E}\log_{Na}V_{m+1}-\mathsf{E}\log_{Na}V_m\right|\,. \tag{4.81}$$

Consider the Gibbs' measure G given by

$$G(B)=\frac{1}{Z_m}\int\exp\left(-\sum_{k>m}u(S_k)\right)\mathrm{d}\gamma$$

where the integral is over $B\cap\bigcap_{k<m}U_k$, and where Z_m is the normalizing factor. Thus

$$V_m=Z_m\langle u(S_m)\rangle\,;\;V_{m+1}=Z_m\langle\mathbf{1}_{U_m}\rangle$$

and, as in (3.287), for $c>0$ we have

$$|\log_{Na}V_{m+1}-\log_{Na}V_m|\le\mathbf{1}_{\{Z_m\ge\exp(-Na)\}}\left(\log_{Na}Y\,\mathbf{1}_{\{Y\le c\}}+\frac{|X-Y|}{c}\right)$$

where $Y=\langle\mathbf{1}_{U_m}\rangle$ and $X=\langle\exp u(U_m)\rangle$.

We will proceed exactly as in Section 3.11, but for this we need a generalization of Proposition 3.11.5, the key step of which is the following.

Lemma 4.3.13. *Consider numbers* $(x_{i,\ell})$, $i \le N$, $\ell \le n$. *Assume that for certain numbers* b, d, *we have*

$$\ell \ne \ell' \Rightarrow \sum_{i \le N} \min(|x_{i,\ell} - x_{i,\ell'}|^2, d) \ge b^2 \; . \tag{4.82}$$

Then we have

$$\mathsf{E} \max_{\ell \le n} \sum_{i \le N} x_{i,\ell}\, \eta_i \ge \frac{1}{L} \min\left(b\sqrt{\log n}\,,\ \frac{b^2}{d} \right) \; . \tag{4.83}$$

Here of course $(\eta_i)_{i \le N}$ are random signs.

Proof. This is a simple consequence of [131], Theorem 2.3. Some of the ingredients of this proof are related to the present ideas (see [151]) but there are other ingredients, and there is no point to reproduce them here. □

We need to construct large families that satisfy (4.82). This will be done through the following.

Lemma 4.3.14. *Recall the measure (4.78), consider a function* $-H \le 0$ *and the probability* G *of density* $Z^{-1}\exp(-H(\boldsymbol{\sigma}))$ *with respect to* γ, *where* Z *is the normalizing factor. Then given* $a > 0$ *there exists a number* $b > 0$ *depending only on* a, κ *such that*

$$Z \ge \exp(-Na) \Rightarrow G^{\otimes 2}\left(\left\{ (\boldsymbol{\sigma}^1, \boldsymbol{\sigma}^2)\, ;\ \sum_{i \le N} \min(|\sigma_i^1 - \sigma_i^2|^2, 1) \le N b^2 \right\} \right) \\ \le \exp(-Na) \; . \tag{4.84}$$

Proof. Proceeding as usual, it suffices to show that

$$\gamma^{\otimes 2}\left(\left\{ (\boldsymbol{\sigma}^1, \boldsymbol{\sigma}^2)\, ;\ \sum_{i \le N} \min(|\sigma_i^1 - \sigma_i^2|^2, 1) \le N b^2 \right\} \right) \le \exp(-3Na) \; .$$

The image of $\gamma^{\otimes 2}$ under the map $(\boldsymbol{\sigma}^1, \boldsymbol{\sigma}^2) \to \boldsymbol{\sigma}^{\sim} = \boldsymbol{\sigma}^1 - \boldsymbol{\sigma}^2$ is the Gaussian measure γ' of density proportional to $\exp(-\kappa \|\boldsymbol{\sigma}^{\sim}\|^2/2)$. Thus it suffices to prove that

$$\gamma'\left(\sum_{i \le N} \min((\sigma_i^{\sim})^2, 1) \le N b^2 \right) \le \exp(-3Na) \; . \tag{4.85}$$

This is an exercise about independent r.v. If $\lambda > 0$, the left-hand side of (4.85) is at most

$$(\exp \lambda b^2\, \mathsf{E} \exp(-\lambda \min(g^2, 1)))^N \; ,$$

where g is centered Gaussian with $\mathsf{E}\, g^2 = 1/\kappa$. Thus $\mathsf{E} \exp(-\lambda \min(g^2, 1)) = \exp(-\varphi(\lambda))$ where $\lim_{\lambda \to \infty} \varphi(\lambda) = \infty$. Thus if b is small enough we can find λ with $\lambda b^2 - \varphi(\lambda) \le -3a$. □

Proposition 4.3.15. *Consider the Gibbs' measure G as in Lemma 4.3.14. Consider independent random signs $(\eta_i)_{i\le N}$. Then, for some number $K = K(a,\kappa)$, we have*

$$\mathsf{P}\left(G\left(\left\{\boldsymbol{\sigma}\,;\ \sum_{i\le N}\eta_i\,\sigma_i \ge \tau\right\}\right)\le\varepsilon\right)\le K\,\varepsilon^{1/K}\,e^{K\tau^2}+K\exp\left(-\frac{N}{K}\right)\,.$$

Proof. This is a straightforward modification of the proof of Proposition 3.11.6. □

From this point on, the arguments of Section 3.11 carry out in a straightforward manner.

The reader might wonder what happens for $\alpha > 2$. In that case, it is known [79] that $\bigcap_{k\le M} U_k$ is empty with overwhelming probability.

Research problem 4.3.16. (Level 2) When $\tau < 0$, Theorem 4.3.11 remains valid if M/N is small enough. Find the largest value $\alpha_c(\tau)$ for which this is true.

Research problem 4.3.17. (Level 2) When the random signs $\eta_{i,k}$ are replaced by Gaussian r.v. $g_{i,k}$ prove a rate of convergence in $K/\sqrt{N}$ in Theorem 4.3.11.

For small α, this is likely to be easy. For larger α, this is a very interesting program. The first objective is to get a proper rate of convergence in the case studied in Section 4.2. Assuming that the replica-symmetric equations of Theorem 4.3.1 have a unique solution (q_0, ρ_0) one would like to prove that

$$\mathsf{E}\,\langle(R_{1,2}-q_0)^2\rangle\le K/N\,,\ \ \mathsf{E}\,\langle(R_{1,1}-\rho_0)^2\rangle\le K/N\,. \tag{4.86}$$

It is probably rather easy, based on the present technique to prove that $R_{1,2}$ essentially belongs to $[q_0-\varepsilon, q_0+\varepsilon]$ and $R_{1,1}$ to $[\rho_0-\varepsilon,\rho_0+\varepsilon]$. One is then in a situation similar to that of Section 2.8, and the relations (4.86) could be attacked with the cavity method. The program cannot even start without a better understanding of the replica-symmetric equations.

4.4 The Gardner Formula for the Sphere

We consider the uniform probability μ_N on the sphere

$$S_N = \{\boldsymbol{\sigma}\in\mathbb{R}^N\,;\ \|\boldsymbol{\sigma}\|^2 = N\}\,,$$

and we keep the notation $U_k = \left\{\boldsymbol{\sigma}\,;\ N^{-1/2}\sum_{i\le N}\eta_{i,k}\,\sigma_i\ge\tau\right\}$.

Theorem 4.4.1. *Consider $\tau \geq 0$, and define*

$$\alpha(\tau) = \frac{1}{\mathsf{E}\max^2(z+\tau,0)} \tag{4.87}$$

where z is standard Gaussian. Then if $\alpha_0 < \alpha(\tau)$, given $\varepsilon > 0$ we have for N large enough that

$$\frac{M}{N} \leq \alpha_0 \Rightarrow \mathsf{P}\left(\left|\frac{1}{N}\log \mu_N\left(\bigcap_{k\leq M} U_k\right) - RS\left(\frac{M}{N}\right)\right| \geq \varepsilon\right) \leq \exp\left(-\frac{N}{K}\right) \tag{4.88}$$

where

$$RS(\alpha) = \min_q \left(\alpha \mathsf{E}\log B\left(\frac{\tau - z\sqrt{q}}{\sqrt{1-q}}\right) + \frac{1}{2}\frac{q}{1-q} + \frac{1}{2}\log(1-q)\right). \tag{4.89}$$

Moreover given any number a we have, for N large enough

$$\frac{M}{N} \geq \alpha(\tau) \Rightarrow \mathsf{P}\left(\frac{1}{N}\log \mu_N\left(\bigcap_{k\leq M} U_k\right) \geq a\right) \leq \exp\left(-\frac{N}{K}\right). \tag{4.90}$$

Here as usual $B(x) = \mathsf{P}(z \geq x)$.

Research problem 4.4.2. (Level 1) Extend (4.90) to the case $\tau < 0$, α small.

Conjecture 4.4.3. (Level 2) If $M/N \geq \alpha_0 > \alpha(\tau)$, for large N the set $S_N \cap \bigcap_{k\leq M} U_k$ is empty with probability at least $1 - \exp(-N/K)$.

Another obvious question of interest is to find rates of convergence in Theorem 4.4.1.

Let us denote by $\mathrm{Vol}\, A$ the N-dimensional volume of a subset A of $\mathbb{R}^N$ and by Area A the $(N-1)$-dimensional area of a subset of A of vS_N, for any $v > 0$. Thus, if $A \subset S_N$, we have

$$\mu_N(A) = \frac{\text{Area}\, A}{\text{Area}\, S_N}$$

and

$$\frac{1}{N}\log \mu_N(A) = \frac{1}{N}\log \text{Area}\, A - \frac{1}{N}\log \text{Area}\, S_N \tag{4.91}$$

and (classically)

$$\frac{1}{N}\log \text{Area}\, S_N \simeq \frac{1}{2}\log(2e\pi)\,. \tag{4.92}$$

We write

$$C = C_{N,M} = \bigcap_{k \le M} U_k$$

and

$$F_{N,M}(\kappa) = \frac{1}{N} \log \int_{C_{N,M}} \exp(-\kappa \, \|\boldsymbol{\sigma}\|^2) \, \mathrm{d}\boldsymbol{\sigma} \; , \tag{4.93}$$

a quantity we have studied in Theorem 4.3.11.

A simple observation is that, since $\kappa \ge 0$, if $v > 1$ we have $vC \subset C$.Thus

$$v \ge 1 \Rightarrow \mathrm{Area}\,(C \cap vS_N) \ge v^{N-1} \mathrm{Area}\,(C \cap S_N) \tag{4.94}$$

$$v \le 1 \Rightarrow \mathrm{Area}\,(C \cap vS_N) \le v^{N-1} \mathrm{Area}\,(C \cap S_N) \; . \tag{4.95}$$

Lemma 4.4.4. *We have*

$$\frac{1}{N} \log \mathrm{Area}\,(C \cap S_N) \le F_{N,M}(\kappa) + \kappa + \frac{\log(\kappa\sqrt{N})}{N} + \frac{2}{N} + \frac{1}{\kappa N^2} \; . \tag{4.96}$$

Proof. Consider $w > 1$, and observe that

$$\int_C \exp(-\kappa \, \|\boldsymbol{\sigma}\|^2) \, \mathrm{d}\boldsymbol{\sigma} \ge \exp(-\kappa \, N \, w^2) \, \mathrm{Vol}(C \cap \{\boldsymbol{\sigma}; \; N \le \|\boldsymbol{\sigma}\|^2 \le N w^2\}) \; .$$

Using (4.94) we get

$$\begin{aligned} \mathrm{Vol}(C \cap \{\boldsymbol{\sigma}; \; N \le \|\boldsymbol{\sigma}\|^2 \le N w^2\}) &= \sqrt{N} \int_1^w \mathrm{Area}(C \cap vS_N) \, \mathrm{d}v \\ &\ge \frac{1}{\sqrt{N}} (w^N - 1) \, \mathrm{Area}(C \cap S_N) \; . \end{aligned}$$

Now we take $w = 1 + 1/\kappa N$, so $w^N - 1 \ge 1/\kappa$. Taking logarithms yields the result. □

What will occur is that there is near equality in (4.96) if we take the infimum over κ in the right-hand side. To prepare for this we need the following. We recall the solutions $q_0 = q_0(\alpha, \kappa), \rho_0 = \rho_0(\alpha, \kappa)$ of the equations (4.48). (Since $h = 0$ we omit it from the notation.)

Proposition 4.4.5. *We have*

$$\frac{\partial \rho_0}{\partial \kappa} < 0 \tag{4.97}$$

$$\lim_{\kappa \to \infty} \rho_0(\alpha, \kappa) = \rho(\alpha) := \inf \left\{ \rho; \; \alpha \, \mathsf{E} \max{}^2 \left(z + \frac{\tau}{\sqrt{\rho}}, 0 \right) \le 1 \right\} \; . \tag{4.98}$$

Proof. We recall the function G of (4.53), so that $x_0 = q_0/(\rho_0 - q_0)$ and ρ_0 are also solutions of the equations $\partial G/\partial\rho = 0$ and $\partial G/\partial x = 0$. We write $\rho' = \partial\rho_0/\partial\kappa$ and $x' = \partial x_0/\partial\kappa$. Since $\partial^2 G/\partial\kappa\,\partial\rho = -1$, if we take the derivative with respect to κ of the relation $\partial G/\partial\rho = 0$, we obtain the relation

$$\rho' \frac{\partial^2 G}{\partial\rho^2} + x' \frac{\partial^2 G}{\partial x\,\partial\rho} = 1\ .$$

It is understood here and in this proof that unless specified otherwise all the functions are computed at the arguments $x = x_0$, $\rho = \rho_0$. Since $\partial G/\partial x = 0$, we have similarly

$$\rho' \frac{\partial^2 G}{\partial\rho\,\partial x} + x' \frac{\partial^2 G}{\partial x^2} = 0$$

and thus

$$\rho' = \rho'^2 \frac{\partial^2 G}{\partial\rho^2} - x'^2 \frac{\partial^2 G}{\partial x^2}\ .$$

Now, by (4.54) we have $\partial^2 G/\partial\rho^2 < 0$ and

$$0 < \frac{\partial}{\partial x}\left(\frac{x+1}{x}\frac{\partial G}{\partial x}\right) = \frac{x+1}{x}\frac{\partial^2 G}{\partial x^2}$$

since $\partial G/\partial x = 0$. This proves (4.97).

By (4.54) we have $\partial G/\partial x(x,\rho_0) > 0$ for $x > x_0$. Thus, by (4.60), we have

$$\frac{\alpha}{x}\mathsf{E}\,A^2\left(\frac{\tau\sqrt{1+x}}{\sqrt{\rho_0}} - z\sqrt{x}\right) < 1\ .$$

Using that $A > 0$ and $A(v) \geq v$ by (4.55) we have

$$\frac{\alpha}{x}\mathsf{E}\max{}^2\left(\frac{\tau\sqrt{1+x}}{\sqrt{\rho_0}} - z\sqrt{x}, 0\right) < 1\ .$$

Letting $x \to \infty$, and changing z into $-z$ we get $\alpha\,\mathsf{E}\max^2(z+\tau/\sqrt{\rho_0}, 0) \leq 1$ i.e. $\rho_0 \geq \rho(\alpha)$.

Finally, we set $\rho_\infty(\alpha) = \lim_{\kappa\to\infty}\rho_0(\kappa,\alpha)$, and we assume for contradiction that $\rho_\infty(\alpha) > \rho(\alpha)$. We observe that since $\alpha < 2$ we have $\rho(\alpha) \geq 0$. By (4.59) we have

$$\alpha\frac{\tau\sqrt{1+x}}{\rho^{3/2}}\mathsf{E}\,A\left(\frac{\tau\sqrt{1+x}}{\sqrt{\rho}} - z\sqrt{x}\right) + \frac{1}{\rho} = 2\kappa$$

where $x = x_0$, $\rho = \rho_0$. If $\rho_0 \geq \rho_\infty(\alpha) > \rho(\alpha) \geq 0$ the only way that the left-hand side can become large as $\kappa \to \infty$ is that $x_0 \to \infty$. But by (4.60) we have

$$1 = \frac{\alpha}{x}\mathsf{E}\left(A^2\left(\frac{\tau\sqrt{1+x}}{\sqrt{\rho}} - z\sqrt{x}\right)\right)\ . \tag{4.99}$$

By (4.55) and (4.58) we have $A(v) \sim v$ for large v, so letting $\kappa \to \infty$, and thus $x \to \infty$, $\rho \to \rho_\infty(\alpha)$ we get from (4.99) that

$$1 = \alpha \mathsf{E}\left(\max^2\left(z + \frac{\tau}{\sqrt{\rho_\infty(\alpha)}}, 0\right)\right)$$

and this contradicts the fact that $\rho_\infty(\alpha) > \rho(\alpha)$. □

We recall the quantity $RS_0(\alpha,\kappa)$ of (4.71) (again, the notation omits to mention that $h = 0$).

Lemma 4.4.6. a) *If $\alpha < \alpha(\tau)$ (where $\alpha(\tau)$ is given by (4.87)) then the function $f(\kappa) = \kappa + RS_0(\alpha,\kappa)$ attains its minimum at a unique value $\kappa(\alpha)$, and this minimum is the quantity $RS(\alpha)$ of (4.89).*
b) *We have*

$$\lim_{\alpha \to \alpha(\tau)} RS(\alpha) = -\infty \,. \tag{4.100}$$

Proof. We have $f'(\kappa) = 1 - \rho_0(\alpha,\kappa)$. So $f'(\kappa)$ increases by Proposition 4.4.5 and $\lim_{\kappa\to\infty} f'(\kappa) = 1 - \rho(\alpha)$. When $\alpha < \alpha(\tau)$ we have $\rho(\alpha) < 1$. Since by (4.59) we have $\rho_0(\alpha,\kappa) \ge 1/(2\kappa)$, we have $f'(\kappa) < 0$ for small κ; thus there is a unique point $\kappa(\alpha)$ where $\rho_0 = \rho_0(\alpha,\kappa(\alpha)) = 1$ and $f' = 0$. Hence by (4.71) we have

$$RS_0(\alpha,\kappa(\alpha)) + \kappa(\alpha) = \alpha \mathsf{E} \log B\left(\frac{\tau - z\sqrt{q_0}}{\sqrt{1-q_0}}\right) + \frac{1}{2}\frac{q_0}{1-q_0} + \frac{1}{2}\log(1-q_0) \,.$$

Using the second part of (4.54), and that $x = q/(1-q)$ is an increasing function of q, we see that q_0 is the value that minimizes the right-hand side. To prove b) we observe that, since $B(x) = \mathsf{P}(\xi \ge x) \le \exp(-\max^2(0,x)/2)$, we have, given any $q < 1$

$$\begin{aligned}
\mathsf{E}\log B\left(\frac{\tau - z\sqrt{q}}{\sqrt{1-q}}\right) &\le -\frac{1}{2(1-q)}\mathsf{E}\left(\max^2(\tau - z\sqrt{q}, 0)\right) \\
&\le -\frac{q}{2(1-q)}\mathsf{E}\left(\max^2\left(z + \frac{\tau}{\sqrt{q}}, 0\right)\right) \\
&\le -\frac{q}{2(1-q)}\mathsf{E}\left(\max^2(z+\tau, 0)\right) \\
&= -\frac{q}{2\alpha(\tau)(1-q)}
\end{aligned}$$

so that

$$\begin{aligned}
&\alpha \mathsf{E}\log B\left(\frac{\tau - z\sqrt{q}}{\sqrt{1-q}}\right) + \frac{1}{2}\frac{q}{1-q} + \frac{1}{2}\log(1-q) \\
&\le \frac{1}{2(1-q)}\left(1 - \frac{\alpha}{\alpha(\tau)} q\right) + \frac{1}{2}\log(1-q) \,.
\end{aligned}$$

As $\alpha \to \alpha(\tau)$, this tends to $(1+\log(1-q))/2$, and this proves the result. □

Proof of Theorem 4.4.1. We observe that the case $\alpha > \alpha(\tau)$ follows from the case $\alpha < \alpha(\tau)$ and (4.100).

To prove (4.88) we prove that given $\alpha < \alpha(\tau)$, given $\varepsilon > 0$, then if $N \to \infty$, $M = \lfloor \alpha N \rfloor$, for N large enough we have

$$\mathsf{P}\left(\left|\frac{1}{N}\log \mu_N\left(\bigcap_{k\le M} U_k\right) - RS(\alpha)\right| \ge \varepsilon\right) \le \exp\left(-\frac{N}{K}\right) . \tag{4.101}$$

The claim in (4.90) that the same value of N achieves (4.88) for all $M \le \alpha_0 N$ follows by a simple argument from the fact that $\mu_N\left(\bigcap_{k\le M} U_k\right)$ decreases as M increases.

By Theorem 4.3.11, given $\varepsilon > 0$ and $\kappa > 0$, for N large enough (and $M = \lfloor \alpha N \rfloor$) we have

$$\mathsf{P}\left(\left|F_{N,M}(\kappa) - \left(RS_0(\alpha,\kappa) + \frac{1}{2}\log 2e\pi\right)\right| \ge \varepsilon\right) \le \exp\left(-\frac{N}{K}\right) \tag{4.102}$$

so that by Lemma 4.4.4 and (4.91), choosing κ such that $RS_0(\alpha,\kappa) + \kappa = RS(\alpha)$, and combining with Lemma 4.4.4 and (4.91), this proves that for N large we have

$$\mathsf{P}\left(\frac{1}{N}\log \mu_N\left(\bigcap_{k\le N} U_k\right) \ge RS(\alpha) + \varepsilon\right) \le \exp\left(-\frac{N}{K}\right) .$$

To prove the lower bound, we consider the number κ' at which the function $f(\kappa) = RS_0(\alpha,\kappa) + \kappa$ is minimum. Considering $v < 1$, for δ small enough we have

$$f(\kappa') < f(\kappa' + \delta)\,;\; f(\kappa'+\delta) > f(\kappa' + 2\delta) - (1-v^2)\,\delta\,,$$

the second statement resulting from the fact that $\lim_{\delta\to 0} f(\kappa'+\delta) - f(\kappa')/\delta = 0$. Thus, setting $\kappa = \kappa' + \delta$, we have

$$RS_0(\alpha,\kappa-\delta) - \delta < RS_0(\alpha,\kappa)\,;\; RS_0(\alpha,\kappa+\delta) + \delta v^2 < RS_0(\alpha,\kappa)\,. \tag{4.103}$$

We observe the inequalities

$$\begin{aligned}\int_{C\cap\{\|\boldsymbol{\sigma}\|\ge\sqrt{N}\}} &\exp(-\kappa\,\|\boldsymbol{\sigma}\|^2)\,\mathrm{d}\boldsymbol{\sigma} \le I_1 \\ := \exp(-N\delta) \int_C &\exp(-(\kappa-\delta)\,\|\boldsymbol{\sigma}\|^2)\,\mathrm{d}\boldsymbol{\sigma}\end{aligned}$$

$$\int_{C\cap\{\|\boldsymbol{\sigma}\|\le v\sqrt{N}\}} \exp(-\kappa\|\boldsymbol{\sigma}\|^2)\,\mathrm{d}\boldsymbol{\sigma} \le I_2$$
$$:= \exp(N\delta v^2)\int_C \exp(-(\kappa+\delta)\|\boldsymbol{\sigma}\|^2)\,\mathrm{d}\boldsymbol{\sigma}$$

and thus

$$\int_{C\cap\{v\sqrt{N}\le\|\boldsymbol{\sigma}\|\le\sqrt{N}\}} \exp(-\kappa\|\boldsymbol{\sigma}\|^2)\,\mathrm{d}\boldsymbol{\sigma} \ge \int_C \exp(-\kappa\|\boldsymbol{\sigma}\|^2)\,\mathrm{d}\boldsymbol{\sigma} - I_1 - I_2 \,.$$

The left-hand side is at most

$$\exp(-\kappa\, v^2\, N)\,\mathrm{Vol}(C\cap\{\|\boldsymbol{\sigma}\|\,;\ \|\boldsymbol{\sigma}\|\le\sqrt{N}\})\,.$$

Now, using (4.95),

$$\begin{aligned}\mathrm{Vol}(C\cap\{\|\boldsymbol{\sigma}\|\,;\ \|\boldsymbol{\sigma}\|\le\sqrt{N}\}) &= \sqrt{N}\int_0^1 \mathrm{Area}(C\cap v\,S_N)\,\mathrm{d}v\\ &\le \sqrt{N}\int_0^1 v^{N-1}\mathrm{Area}(C\cap S_N)\,\mathrm{d}v\\ &= \frac{1}{\sqrt{N}}\,\mathrm{Area}(C\cap S_N)\,.\end{aligned}$$

Thus we have the bound (recalling (4.93))

$$\begin{aligned}\mathrm{Area}(C\cap S_N) \ge \exp(N\,\kappa\, v^2)\,\big(&\exp N\,F_{M,N}(\kappa)\\ &- \exp N(-\delta + F_{M,N}(\kappa+\delta)) - \exp N(\delta\, v^2 + F_{M,N}(\kappa+\delta))\big)\,,\end{aligned} \tag{4.104}$$

and (4.103), (4.102) imply that, with overwhelming probability

$$\begin{aligned}\frac{1}{N}\log\mathrm{Area}(C\cap S_N) &\ge \kappa\, v^2 + RS_0(\alpha,\kappa) - \varepsilon\\ &\ge \kappa + RS_0(\alpha,\kappa) - \varepsilon - (1-v^2)\,\kappa\\ &\ge RS(\alpha) - \varepsilon - (1-v^2)\,\kappa\,.\end{aligned}$$

The result follows. □

4.5 Notes and Comments

The results of this chapter are essentially proved in [128]. (The difference being that [128] does not reach the exponential bound in (4.77) but a weaker bound.) In this paper I learned the general approach, the idea of the approximation procedure of Section 4.3, Proposition 4.3.3 and Lemma 4.2.2. The way [128] obtains the replica-symmetric equation seems genuinely different from what I do. It would be nice to make explicit the rationale behind their approach.

Rather than basing the approach on Theorem 2.15 of [90] one can also use the Brascamp-Lieb inequalities [39]. In 1 dimension, the inequality is stated in (4.106) below. In more dimensions, it is convenient to use a simplified form of these inequalities as follows. Consider a measure μ on $\mathbb{R}^N$ as in Proposition 4.1.2. Consider a function f on $\mathbb{R}^N$ (that need not satisfy (4.7)). Then, if $\boldsymbol{\nabla} f$ denotes the gradient of f,

$$\int f^2 \,\mathrm{d}\mu - \left(\int f \mathrm{d}\mu\right)^2 \le \frac{1}{\kappa}\int \|\boldsymbol{\nabla} f\|^2 \,\mathrm{d}\mu \,. \tag{4.105}$$

This inequality can be iterated. If f satisfies (4.7), then $\|\boldsymbol{\nabla} f\| \le A$. Using (4.105) for $\exp(\lambda f/2)$ rather than f yields

$$\int \exp \lambda f \,\mathrm{d}\mu - \left(\int \exp\left(\frac{\lambda f}{2}\right) \mathrm{d}\mu\right)^2 \le \frac{\lambda^2 A^2}{\kappa}\int \exp \lambda f \,\mathrm{d}\mu$$

so that if $\lambda^2 A^2 \le \kappa$,

$$\int \exp \lambda f \,\mathrm{d}\mu \le \frac{1}{1-\frac{\lambda^2 A^2}{\kappa}} \left(\int \exp \frac{\lambda f}{2} \,\mathrm{d}\mu\right)^2 .$$

By iteration we get

$$\int \exp \lambda f \,\mathrm{d}\mu \le \prod_{1\le \ell\le k} \left(\frac{1}{1-\frac{\lambda^2 A^2}{\kappa\, 2^{2(k-1)}}}\right)^{2^{k-1}} \left(\int \exp\left(\frac{\lambda f}{2^k}\right) \mathrm{d}\mu\right)^{2^k}$$

so that if $\lambda^2 A^2 \le \kappa/2$ we have

$$\int \exp \lambda f \,\mathrm{d}\mu \le L \left(\int \exp\left(\frac{\lambda f}{2^k}\right) \mathrm{d}\mu\right)^{2^k} .$$

Now the inequality $|e^x - x - 1| \le x^2\, e^x$ shows that if $\int f \,\mathrm{d}\mu = 0$, the right-hand side goes to 1 as $k \to \infty$, so that if f satisfies (4.7), we have

$$\int \exp\left(-\lambda\left(f - \int f \,\mathrm{d}\mu\right)\right) \mathrm{d}\mu \le L$$

when $|\lambda| = \sqrt{\kappa}/2A$, a result that is not far from (4.8), at least for our purposes here. The reader should consult [10] for more on the relations between these different inequalities.

Inequality (4.105) follows from a nice general result, namely that

$$\int f^2 \,\mathrm{d}\mu - \left(\int f \,\mathrm{d}\mu\right)^2 \le \int \frac{f'^2}{V''} \,\mathrm{d}\mu \,. \tag{4.106}$$

(The Brascamp-Lieb inequality in dimension 1.) For $f(x) = x$, $-V(x) = w(x) - x^2/2$ this implies Lemma 4.4.4; the proof we give is just a simplified proof of (4.106) in the special case we need.

The results of the present chapter allow one to prove a (good) upper bound for the quantity α^* of Problem 3.1.1 using "domination of the Ising model by the spherical model" as follows. If μ_N denotes the uniform measure on S_N, and if it is very likely that

$$\frac{1}{N} \log \mu_N \left(\bigcap_{k \le M} U_k \right) \le -\log 2 - \varepsilon \tag{4.107}$$

where ε does not depend on N and where $U_k = \left\{ N^{-1/2} \sum_{i \le N} g_{i,k}\, \sigma_i \ge \tau \right\}$, then it is very likely that $\Sigma_N \cap \bigcap_{k \le M} U_k \ne \emptyset$. This is because if V is a linear isometry of $\mathbb{R}^N$, the probability that $V(\Sigma_N) \cap \bigcap_{k \le M} U_k \ne \emptyset$ is independent of V by rotational invariance of Gaussian measures. Averaging over V, using Fubini's Theorem and (4.107) then yields the previous claim. This argument unfortunately does not work if one replaces the Gaussian r.v. $g_{i,k}$ by random signs $\eta_{i,k}$.

5. The Hopfield Model

5.1 Introduction: The Curie-Weiss Model

The Curie-Weiss model is the "canonical" model for mean-field (deterministic) ferromagnetic interaction, i.e. interaction where the spins tend to align with each other. For each pair of spins we want a term $\sigma_i \sigma_j$ in the Hamiltonian, so that this Hamiltonian will be (proportional to) $\sum_{i<j} \sigma_i \sigma_j$. Equivalently, we consider

$$-H_N(\boldsymbol{\sigma}) = \frac{\beta N}{2} \left(\frac{1}{N} \sum_{i \le N} \sigma_i \right)^2 - \frac{\beta}{2N} \left(\sum_{i \le N} \sigma_i \right)^2 . \tag{5.1}$$

This is a simple, almost trivial model that can be studied in considerable detail (see [51]). It is not our purpose to do this, but we will explain the basic facts that are relevant to this chapter. The partition function is given by

$$Z_N(\beta) = \sum_{\boldsymbol{\sigma}} \exp(-H_N(\boldsymbol{\sigma})) = \sum_k A_k \exp\left(\frac{\beta}{2N} k^2 \right) , \tag{5.2}$$

where

$$A_k = \operatorname{card} \left\{ \boldsymbol{\sigma} \in \Sigma_N \, ; \, \sum_{i \le N} \sigma_i = k \right\} .$$

Considering the function

$$\mathcal{I}(t) = \frac{1}{2} \left((1+t) \log(1+t) + (1-t) \log(1-t) \right) , \tag{5.3}$$

we recall (see (A.10)) that

$$\frac{2^N}{L\sqrt{N}} \exp\left(-N\mathcal{I}\left(\frac{k}{N} \right) \right) \le A_k \le 2^N \exp\left(-N\mathcal{I}\left(\frac{k}{N} \right) \right) , \tag{5.4}$$

so by (5.2) we have

$$\frac{2^N}{L\sqrt{N}} \exp\left(N \max_k \left(\frac{\beta}{2}\left(\frac{k}{N}\right)^2 - \mathcal{I}\left(\frac{k}{N}\right)\right)\right)$$
$$\le Z_N(\beta) \le (2N+1)\, 2^N \exp\left(N \max_t \left(\frac{\beta t^2}{2} - \mathcal{I}(t)\right)\right) ,$$

and thus

$$\frac{1}{N} \log Z_N(\beta) = \log 2 + \max_t \left(\frac{\beta t^2}{2} - \mathcal{I}(t)\right) + o(1) , \tag{5.5}$$

where $o(1)$ is a quantity such that $o(1) \to 0$ as $N \to \infty$. The function $\beta t^2/2 - \mathcal{I}(t)$ attains its maximum at a point t such that

$$\beta t = \mathcal{I}'(t) = \frac{1}{2} \log\left(\frac{1+t}{1-t}\right) \tag{5.6}$$

or, equivalently,

$$\frac{1+t}{1-t} = \exp(2\beta t) ,$$

i.e.

$$t = \frac{\exp(2\beta t) - 1}{\exp(2\beta t) + 1} = \operatorname{th} \beta t . \tag{5.7}$$

If $\beta \le 1$, the only root of (5.7) is $t = 0$. For $\beta > 1$, there is a unique root $m^* > 0$. That is, $m^* = m^*(\beta)$ satisfies

$$\operatorname{th} \beta m^* = m^* . \tag{5.8}$$

Since $\operatorname{th} x = x - x^3/3 + x^3\, o(x)$, where $o(x) \to 0$ as $x \to 0$, we get from (5.8) that

$$\beta m^* - \frac{\beta^3 m^{*3}}{3} + \beta^3 m^{*3}\, o(\beta m^*) = m^* ,$$

so that

$$m^*(\beta) \sim \sqrt{3(\beta - 1)} \quad \text{as} \quad \beta \to 1_+ . \tag{5.9}$$

We write

$$b^* = \frac{\beta m^{*2}}{2} - \mathcal{I}(m^*) \tag{5.10}$$

so (5.5) reads

$$\frac{1}{N} \log Z_N(\beta) = \log 2 + b^* + o(1) . \tag{5.11}$$

When $\beta > 1$, as $N \to \infty$, Gibbs' measure is essentially supported by the set of configurations $\boldsymbol{\sigma}$ for which $N^{-1} \sum_{i \le N} \sigma_i \simeq \pm m^*$. This is because for a subset U of $\mathbb{R}$,

$$G_N\left(\left\{\boldsymbol{\sigma} ; \frac{1}{N} \sum_{i \le N} \sigma_i \in U\right\}\right) = Z_N^{-1}(\beta) \sum \exp(-H_N(\boldsymbol{\sigma})) ,$$

where the summation is over all sequences for which $N^{-1}\sum_{i\leq N}\sigma_i \in U$. Thus

$$G_N\left(\left\{\boldsymbol{\sigma}; \frac{1}{N}\sum_{i\leq N}\sigma_i \in U\right\}\right) \leq \frac{1}{Z_N(\beta)}\sum_{k/N\in U} A_k \exp\frac{\beta N}{2}\left(\frac{k}{N}\right)^2 \tag{5.12}$$
$$\leq \frac{2^N}{Z_N(\beta)}(2N+1)\exp\left(N\sup_{t\in U}\left(\frac{\beta t^2}{2}-\mathcal{I}(t)\right)\right)$$

using (5.4), and bounding the sum in the second line by $(2N+1)$ times a bound for the largest term. If we take

$$U = \{t;\ |t \pm m^*| \geq \varepsilon\}$$

where ε is given (independent of N) then (5.5) shows that the right-hand side of (5.12) goes to zero as $N \to \infty$.

Thus, (when weighted with Gibbs' measure), the set of configurations is made of two different pieces: the configurations for which $N^{-1}\sum_{i\leq N}\sigma_i \simeq m^*$ and those for which $N^{-1}\sum_{i\leq N}\sigma_i \simeq -m^*$. The global invariance of Gibbs' measure by the transformation $\boldsymbol{\sigma} \to -\boldsymbol{\sigma}$ shows that these two pieces have the same weight. The system "spontaneously breaks down in two states".

This situation changes drastically if one adds an external field, i.e. one considers

$$-H_N(\boldsymbol{\sigma}) = \frac{\beta N}{2}\left(\frac{1}{N}\sum_{i\leq N}\sigma_i\right)^2 + h\sum_{i\leq N}\sigma_i\ . \tag{5.13}$$

To see where Gibbs' measure lies, one should now maximize

$$f(t) = \frac{\beta}{2}t^2 - \mathcal{I}(t) + th\ .$$

This maximum is attained at a point $t > 0$ because $f(t) > f(-t)$ for $t > 0$; this point t must satisfy $\beta t + h = \mathcal{I}'(t)$, i.e.

$$t = \mathrm{th}(\beta t + h)\ , \tag{5.14}$$

and there is a unique positive root to this equation. The external field "breaks the symmetry between the two states".

Consider now a random sequence $(\eta_i)_{i\leq N}$, $\eta_i = \pm 1$, and the random Hamiltonian

$$-H_N(\boldsymbol{\sigma}) = \frac{\beta}{2N}\left(\sum_{i\leq N}\eta_i\,\sigma_i\right)^2\ . \tag{5.15}$$

The randomness is not intrinsic, and can be removed by setting $\sigma'_i = \eta_i\,\sigma_i$. We can describe the Hamiltonian (5.15) by saying that it tends to align the spins $(\sigma_i)_{i\leq N}$ with the sequence $(\eta_i)_{i\leq N}$ or the sequence $(-\eta_i)_{i\leq N}$ rather than with the sequences $(1,\dots,1)$ or $(-1,\dots,-1)$.

The situation is much more interesting if we put in the Hamiltonian several terms that "pull in different directions". Consider numbers $(\eta_{i,k})_{i\le N,k\le M}$, $\eta_{i,k} = \pm 1$, and the Hamiltonian

$$-H_{N,M}(\boldsymbol{\sigma}) = \frac{\beta N}{2} \sum_{k\le M} \left(\frac{1}{N} \sum_{i\le N} \eta_{i,k}\,\sigma_i \right)^2 . \tag{5.16}$$

We will always write

$$m_k = m_k(\boldsymbol{\sigma}) = \frac{1}{N} \sum_{i\le N} \eta_{i,k}\,\sigma_i . \tag{5.17}$$

The effect of the term $\beta N m_k^2/2$ of (5.16) is (for $\beta > 1$) to tend to align the sequence $\boldsymbol{\sigma}$ with the sequence $(\eta_{i,k})_{i\le N}$ or the sequence $(-\eta_{i,k})_{i\le N}$. If the sequences $(\eta_{i,k})_{i\le N}$ are really different as k varies, this creates conflict.

The Hopfield model is the system with random Hamiltonian (5.16), when the numbers $\eta_{i,k}$ are independent r.v. with $\mathsf{P}(\eta_{i,k} = \pm 1) = 1/2$. It simplifies notation to observe that, equivalently, one can assume that

$$\begin{cases} \eta_{i,1} = 1 \quad \forall i \le N \text{ ; the numbers } (\eta_{i,k})_{i\le N, 2\le k\le N} \\ \text{are independent r.v. with } \mathsf{P}(\eta_{i,k} = \pm 1) = 1/2 . \end{cases} \tag{5.18}$$

The Hopfield model is already of interest if we fix M and let $N \to \infty$. We will however focus on the more challenging case where $N \to \infty$, $M \to \infty$, $M/N \to \alpha, \alpha > 0$.

The Hopfield model has a "high temperature phase" somewhat similar to the phase $\beta < 1$, $h = 0$ of the SK model. This phase occurs in the region

$$\beta(1 + \sqrt{\alpha}) < 1 \tag{5.19}$$

and it is quite interesting to see how this condition occurs. We will refer the reader to Section 2 of [135] for this, because this study does not use the cavity method and is somewhat distinct from the main theme we pursue here.

Another topic of interest is the "temperature zero" problem, i.e. the study of the (random) function

$$\boldsymbol{\sigma} \to -H_N(\boldsymbol{\sigma})$$

on Σ_N. We will not study this topic either because we feel that the current results in this direction are too far from their optimal form. We refer the reader to [104], [93], [54] for increasingly more sophisticated results.

Compared with the SK model, the Hopfield model brings two kinds of new features. One is the ferromagnetic interaction (5.1). For β close to one, this interaction creates difficulties that arise from the fact that the root of the equation $m^* = \operatorname{th} \beta m^*$ is "not so stable", in the sense that the slope of the tangent to the function $t \mapsto \operatorname{th} \beta t$ at $t = m^*$ gets close to 1 as $\beta \to 1$. This simple fact creates many of the technical difficulties inherent to the Hopfield model.

Another difference between the Hopfield model and the SK model is that the nature of the disorder is not exactly the same. In fact the Hopfield model is more similar to the "Bernoulli perceptron" model of Section 3.8 (but again with significant twists).

It would be more pedagogical, before attacking the Hopfield model, to study a disordered system that presents the difficulties due to the ferromagnetic interaction, but with a familiar disorder. Such a model exists. It is the SK model with ferromagnetic interaction. The Hamiltonian is

$$-H_N(\boldsymbol{\sigma}) = \frac{\beta_1 N}{2}\left(\frac{1}{N}\sum_{i\le N}\sigma_i\right)^2 + \frac{\beta_2}{\sqrt{N}}\sum_{i<j} g_{ij}\,\sigma_i\,\sigma_j + h\sum_{i\le N}\sigma_i\,. \tag{5.20}$$

Space (and energy!) limitation do not allow the study of this model here.

Research problem 5.1.1. (Level 1) Extend the results of Chapter 2 to the Hamiltonian (5.20).

What is really interesting is not to study this model for β_1, β_2 small, but, given β_1 (possibly large) to study the system for β_2 as large as possible. The "replica-symmetric" equations for this model are given in (5.361) and (5.362).

Throughout the chapter we will consider the Hopfield model with external field, so that the Hamiltonian is

$$\begin{aligned}-H_{N,M}(\boldsymbol{\sigma}) &= \frac{N\beta}{2}\sum_{k\le M}\left(\frac{1}{N}\sum_{i\le N}\eta_{i,k}\,\sigma_i\right)^2 + h\sum_{i\le N}\sigma_i\\ &= \frac{N\beta}{2}\sum_{k\le M} m_k^2 + Nhm_1\,.\end{aligned} \tag{5.21}$$

(It is of course important that we have chosen $\eta_{i,1} = 1$, so that the external field "pulls in the same direction as m_1".) Throughout the chapter, we denote by $m^* = m^*(\beta, h)$ the solution to (5.14), i.e. we have

$$m^* = \operatorname{th}(\beta m^* + h)\,. \tag{5.22}$$

We set

$$b^* = \log\operatorname{ch}(\beta m^* + h) - \frac{\beta^2}{2}m^{*2}\,. \tag{5.23}$$

Lemma 5.1.2. *We have*

$$|\log Z_{N,1} - N(b^* + \log 2)| \le K(\beta, h)\,.$$

Of course $Z_{N,M} = Z_{N,M}(\beta, h)$ denotes the partition function of the Hamiltonian (5.21). Thus $Z_{N,1}$ is the partition function of the Curie-Weiss model with external field. The proof serves as an introduction to the method of Section 5.3. It is much more effective and accurate than the more natural method leading to (5.11).

Proof. We start with the identity (see (A.36))

$$\mathsf{E} \exp ag = \exp \frac{a^2}{2}$$

if g is standard Gaussian, so that

$$Z_{N,1} = \sum_{\sigma} \mathsf{E} \exp(\sqrt{N\beta}\, g m_1 + N h m_1) = 2^N \mathsf{E} \operatorname{ch}\left(\sqrt{\frac{\beta}{N}}\, g + h\right)^N$$

because $m_1 = N^{-1} \sum_{i \le N} \sigma_i$. Thus

$$\begin{aligned} Z_{N,1} &= 2^N \frac{1}{\sqrt{2\pi}} \int_{\mathbb{R}} \exp\left(N \log \operatorname{ch}\left(\sqrt{\frac{\beta}{N}}\, t + h\right) - \frac{t^2}{2}\right) \mathrm{d}t \\ &= 2^N \sqrt{\frac{N\beta}{2\pi}} \int_{\mathbb{R}} \exp N \left(\log \operatorname{ch}(\beta z + h) - \frac{\beta z^2}{2}\right) \mathrm{d}z \end{aligned}$$

with the change of variable $t = \sqrt{N\beta}\, z$. The function $z \to \log \operatorname{ch}(\beta z + h) - \beta z^2/2$ has its maximum at the point z such that $\operatorname{th}(\beta z + h) = z$, i.e. $z = m^*$, and this maximum is b^*. Thus

$$Z_{N,1} = 2^N \exp(N b^*)\, A_N$$

where

$$A_N = \sqrt{\frac{N\beta}{2\pi}} \int_{\mathbb{R}} \exp N\psi(z)\, \mathrm{d}z \ ,$$

for

$$\psi(z) = \log \operatorname{ch}(\beta z + h) - \frac{\beta z^2}{2} - b^* \ .$$

It is elementary to see that there is a number K such that $\psi(z) \le -(z - m^*)^2/K$; making the change of variable $z = m^* + x/\sqrt{N}$ then implies easily that $\log A_N$ stays bounded as $N \to \infty$. □

Interestingly, it is not obvious that the value (5.23) is the same as the value (5.10), which, in the presence of the external field, is

$$\frac{\beta m^{*2}}{2} + m^* h - \mathcal{I}(m^*) \ . \tag{5.24}$$

To see the equality of (5.23) and (5.24), we observe that from (A.6) we have

$$\mathcal{I}(x) = \max_{\lambda}(\lambda x - \log \operatorname{ch} \lambda)$$

and that the maximum is obtained for $\operatorname{th} \lambda = x$, so that, if $x = m^*$, λ is $m^* \beta + h$ and hence $\mathcal{I}(m^*) = m^{*2}\beta + m^* h - \log \operatorname{ch}(m^*\beta + h)$.

5.2 The Replica-Symmetric Equations

Consider as usual a standard Gaussian r.v. z. For three numbers q, μ, r, consider the equations

$$q = \mathsf{E}\,\mathrm{th}^2(\beta z\sqrt{r} + \beta\mu + h) \tag{5.25}$$

$$\mu = \mathsf{E}\,\mathrm{th}(\beta z\sqrt{r} + \beta\mu + h) \tag{5.26}$$

$$r = \frac{\alpha q}{(1-\beta(1-q))^2} \,. \tag{5.27}$$

The reader who is disturbed by these formulas because they differ from those used by the physicists is referred to Section 5.12.

Our goal in Section 5.9 is to show (in a certain domain of parameters) that, as $N \to \infty$ and $M/N \to \alpha$, we have $R_{1,2} \simeq q$, $m_1 = m_1(\boldsymbol{\sigma}) \simeq \mu$ and $\sum_{2\le k\le M} m_k(\boldsymbol{\sigma}^1)\, m_k(\boldsymbol{\sigma}^2) \simeq r$. So we must at some point prove that the equations (5.25) to (5.27) do have a solution. This, of course, is not going to be enjoyable, but the interesting point is that this does not seem to be so easy. Moreover, the difficulties we will meet in solving these equations will surface again later in a more complicated form. Given a number L_0, we define the *admissible region* of parameters as the domain

$$\left\{(\alpha,\beta,h); h \ge 0, 1 \le \beta \le 2, \alpha \le \frac{m^{*4}}{L_0}\right\}$$
$$\cup \left\{(\alpha,\beta,h); h \ge 0, \beta \ge 2, \alpha \le \frac{1}{L_0 \log\beta}\right\}$$

where m^* is given by (5.22). It is useful to think to L_0 as a parameter that we will take as large as convenient. Our main results will hold for values of parameters in the admissible region, provided L_0 has been taken large enough. We see that for a result of this type, the choice of the number 2 in the definition of the admissible region is pretty much irrelevant. For $h = 0$, the admissible region seems to have the optimal shape (that is, we reach values of α that are proportional to the optimal values). Larger values of h seem to improve matters, but we decided not to clutter the exposition by trying to find the optimal shape in that case. Let us note that we can assume without loss of generality that $\alpha \le 1$ in the admissible region. We should also stress that we consider the number L_0 defining the admissible region as a parameter; this means in particular that when we write "L" for a certain constant, this means that the value of L has been determined through calculations that do *not* depend on the value of L_0. All dependences on L_0 are always explicit. At the end of every computation we then see how large L_0 has to be taken to ensure the properties we desire.

Theorem 5.2.1. *There exists a number L with the following property. If the number L_0 is large enough, then for (α, β, h) in the admissible region, the equations (5.25) to (5.27) have a unique solution such that*

$$|q - m^{*2}| \le \frac{m^{*2}}{L\beta^2}\,;\ |\mu - m^*| \le \frac{m^*}{L}\,. \tag{5.28}$$

For simplicity we write

$$Y = \beta z\sqrt{r} + \beta\mu + h\,. \tag{5.29}$$

We consider the functions

$$\Phi(r,\mu) = \mathsf{E}\,\mathrm{th}^2\,Y \tag{5.30}$$

$$\Psi(r,\mu) = \mathsf{E}\,\mathrm{th}\,Y \tag{5.31}$$

$$r(q) = \frac{\alpha q}{(1-\beta(1-q))^2}\,. \tag{5.32}$$

The principle of the proof is, given (α, β, h) in the admissible region, to find a domain D such that the map

$$T : (q,\mu) \to (\Phi(r(q),\mu), \Psi(r(q),\mu)) \tag{5.33}$$

sends D into D, and, for a certain distance on D, is a contraction, so that it has a unique fixed point. The main difficulty occurs when β is close to 1 and $h = 0$, because then $\partial\Psi/\partial\mu$ gets dangerously close to 1 (and we must be rather cautious). This is a consequence of the fact that

$$(\mathrm{th}(\beta t + h))' = \beta/\mathrm{ch}^2(\beta t + h) = \beta(1 - \mathrm{th}^2(\beta t + h))$$

is equal to

$$\beta(1 - \mathrm{th}^2(\beta m^* + h)) = \beta(1 - m^{*2})$$

when $t = m^*$, and that this gets close to 1 if β is close to 1 and h is close to zero.

Lemma 5.2.2. *We have*

$$\beta \ge 2 \Rightarrow m^* = m^*(\beta, h) \ge m^*(2,0);\ \beta(1 - m^{*2}) \le L\exp(-\beta/L) \tag{5.34}$$

$$\frac{m^{*2}}{L} \le a^* := 1 - \beta(1 - m^{*2}) \le 2m^{*2}\,. \tag{5.35}$$

Proof. The first part of (5.34) is obvious. To prove the second part we observe that

$$m^* = \mathrm{th}(\beta m^* + h) \ge \mathrm{th}(\beta m^*(2,0))$$

and

$$\beta(1 - m^{*2}) \le \frac{\beta}{\mathrm{ch}^2(\beta m^*(2,0))}\,.$$

The right-hand side inequality of (5.35) is obvious for $\beta \ge 2$ since then $1 - \beta(1 - m^{*2}) \le 1 - 2(1 - m^{*2}) \le 2m^{*2}$. If $\beta \le 2$, it follows from the fact that $1-\beta(1-m^{*2}) \le \beta m^{*2}$. To prove the left-hand side inequality of (5.35), we observe that, since this inequality is equivalent to $\beta - 1 \le (\beta - 1/L)\, m^{*2}$, and since $m^*(\beta, h)$ obviously increases with h, we can assume $h = 0$. We then observe that for $x > 0$ we have

$$1 - \frac{x}{\operatorname{th} x \operatorname{ch}^2 x} = \frac{\operatorname{th} x \operatorname{ch}^2 x - x}{\operatorname{th} x \operatorname{ch}^2 x} \ge \frac{1}{L} \frac{x^2}{1+x^2}$$

as is seen by studying the behavior at $x \to 0$ and $x \to \infty$. Taking $x = \beta m^*$ where $m^* = m^*(\beta, 0)$, we get, since $m^* = \operatorname{th} \beta m^*$,

$$\begin{aligned} 1 - \beta(1-m^{*2}) &= 1 - \frac{\beta}{\operatorname{ch}^2 \beta m^*} = 1 - \frac{\beta m^*}{m^* \operatorname{ch}^2 \beta m^*} \\ &= 1 - \frac{x}{\operatorname{th} x \operatorname{ch}^2 x} \ge \frac{1}{L} \frac{x^2}{1+x^2} \ge \frac{1}{L} \frac{\beta^2 m^{*2}}{(1+\beta^2)} \ge \frac{1}{L} m^{*2} , \end{aligned}$$

using that $1 + x^2 \le 1 + \beta^2$. □

As a consequence of (5.34), we observe that in the admissible region we have

$$\alpha \le \frac{L m^{*4}}{L_0} . \tag{5.36}$$

We recall the formula

$$\frac{\mathrm{d}}{\mathrm{d}x} \mathsf{E} f(z\sqrt{x}) = \frac{1}{2} \mathsf{E} f''(z\sqrt{x})$$

and we obtain the relations

$$\frac{\partial \Phi}{\partial r} = \beta^2 \mathsf{E} \frac{1 - 2\operatorname{sh}^2 Y}{\operatorname{ch}^4 Y} \tag{5.37}$$

$$\frac{\partial \Phi}{\partial \mu} = \beta \mathsf{E} \frac{2 \operatorname{th} Y}{\operatorname{ch}^2 Y} \tag{5.38}$$

$$\frac{\partial \Psi}{\partial r} = -\beta^2 \mathsf{E} \frac{\operatorname{th} Y}{\operatorname{ch}^2 Y} \tag{5.39}$$

$$\frac{\partial \Psi}{\partial \mu} = \beta \mathsf{E} \frac{1}{\operatorname{ch}^2 Y} . \tag{5.40}$$

We will consider the domain

$$D = \left\{ (q, \mu);\, |q - m^{*2}| \le \frac{m^{*2}}{B\beta^2};\, |\mu - m^*| \le \frac{m^*}{A} \right\} \tag{5.41}$$

where the value of the parameters A and B will be adjusted later.

Lemma 5.2.3. *There exists a number L_1 such that if $B \ge L_1$ and $(q, \mu) \in D$ we have*

$$r(q) \le \frac{L_2\, \alpha}{m^{*2}} \tag{5.42}$$

$$\left| \frac{\mathrm{d}r}{\mathrm{d}q}(q) \right| \le \frac{L\alpha\beta}{m^{*4}}\ . \tag{5.43}$$

Proof. We have

$$1 - \beta(1-q) = 1 - \beta(1-m^{*2}) + \beta(q - m^{*2}) \ge \frac{m^{*2}}{L} - \frac{m^{*2}}{B}$$

using (5.35) and (5.41). Thus if $B \ge 2L$, we get

$$1 - \beta(1-q) \ge m^{*2}/L\ , \tag{5.44}$$

and

$$r = \frac{\alpha q}{(1-\beta(1-q))^2} \le \frac{L\alpha}{m^{*2}}$$

because by (5.41), we have $q \le Lm^{*2}$. Since

$$\frac{\mathrm{d}r}{\mathrm{d}q} = \alpha \frac{1-\beta(1+q)}{(1-\beta(1-q))^3}$$

we get

$$\left| \frac{\mathrm{d}r}{\mathrm{d}q} \right| \le \frac{L\alpha((\beta-1)+q\beta)}{m^{*6}}\ .$$

Now, $q \le Lm^{*2}$, and by (5.35) we have $1 - \beta(1-m^{*2}) \ge 0$, and thus $\beta - 1 \le \beta m^{*2}$. This proves (5.43). □

We now fix $B = L_1$ where L_1 is the constant of Lemma 5.2.3.

Lemma 5.2.4. *If we assume*

$$\beta \ge 2,\ |\mu - m^*| \le \frac{m^*}{2},\ r \le \frac{L_2\, \alpha}{m^{*2}} \tag{5.45}$$

then if L_0 is large enough, and (α, β, h) belongs to the admissible region we have

$$\left|\frac{\partial \Phi}{\partial r}\right|, \left|\frac{\partial \Phi}{\partial \mu}\right|, \left|\frac{\partial \Psi}{\partial r}\right|, \left|\frac{\partial \Psi}{\partial \mu}\right| \le \frac{L}{\beta^3} \tag{5.46}$$

over D.

Thus Theorem 5.2.1 is rather easy for large β, because T is nearly constant.

Proof. If f is a function on $\mathbb{R}$, $|f| \le 1$, we have

$$\begin{aligned}\mathsf{E} f(Y) &\leq \mathsf{E}\,(\mathbf{1}_{\{z\leq -\mu/(2\sqrt{r})\}}\, f(Y)) + \mathsf{E}(\mathbf{1}_{\{z\geq -\mu/(2\sqrt{r})\}}\, f(Y)) \\ &\leq \mathsf{P}\left(z \leq -\frac{\mu}{2\sqrt{r}}\right) + \sup\left\{f(x); x \geq \frac{\beta\mu}{2}\right\} ,\end{aligned} \tag{5.47}$$

because

$$z \geq -\frac{\mu}{2\sqrt{r}} \Rightarrow \beta z\sqrt{r} + \beta\mu + h \geq \frac{\beta\mu}{2} .$$

By (5.45), we have $r \leq L_2\,\alpha/m^{*2} \leq L\alpha$ since $m^* \geq 1/L$ for $\beta \geq 2$. Since $|\mu - m^*| \leq m^*/2$, we have $\mu \geq m^*/2 \geq 1/L$, and thus

$$\begin{aligned}\mathsf{P}\left(\left\{z \leq -\frac{\mu}{2\sqrt{r}}\right\}\right) &\leq \exp\left(-\frac{1}{L\alpha}\right) \\ &\leq \exp\left(-\frac{L_0}{L}\log\beta\right) = \beta^{-L_0/L} \leq \frac{L}{\beta^3}\end{aligned} \tag{5.48}$$

if L_0 is large enough, and since $\alpha \leq 1/(L_0 \log\beta)$ in the admissible region.

To bound $\partial\Phi/\partial r$, (5.37) shows that we can use (5.47) in the case $f(x) = |1 - 2\,\mathrm{sh}^2 x|/\mathrm{ch}^4 x$, so that $f(x) \leq L\exp(-x)$, and (5.47), (5.48) show that $\mathsf{E} f(Y) \leq L/\beta^3$. The other cases are similar. □

Lemma 5.2.5. *If the segment between the points (x_0, y_0) and (x_1, y_1) is contained in D, then for any function $F(x,y)$ on D we have*

$$|F(x_1,y_1) - F(x_0,y_0)| \leq |x_1 - x_0| \sup_D \left|\frac{\partial F}{\partial x}\right| + |y_1 - y_0| \sup_D \left|\frac{\partial F}{\partial y}\right| . \tag{5.49}$$

Proof. We use the fact that

$$F(x_1,y_1) - F(x_0,y_0) = \int_0^1 \frac{\mathrm{d}}{\mathrm{d}t} F(x_0 + t(x_1 - x_0), y_0 + t(y_1 - y_0))\,\mathrm{d}t$$

and trivial bounds. □

We now turn to the case of large β, the easiest.

Lemma 5.2.6. *There exists β_0 such that the conclusion of Theorem 5.2.1 holds for $\beta \geq \beta_0$.*

Proof. We consider the domain D of (5.41) with $A = 1/2$ (and $B = L_1$, where L_1 is determined in Lemma 5.2.3). Then (5.45) holds for $r = r(q)$ by (5.42), so that (5.46) holds. Since

$$(\Phi(0, m^*), \Psi(0, m^*)) = (m^{*2}, m^*) , \tag{5.50}$$

we see by (5.42) and (5.49) that if $(q, \mu) \in D$ then

$$|\Phi(r(q), \mu) - m^{*2}| \leq \frac{L}{\beta^3} \tag{5.51}$$

$$|\Psi(r(q),\mu) - m^*| \le \frac{L}{\beta^3} \, . \tag{5.52}$$

Thus $T(D) \subset D$ if $\beta \ge \beta_0$. Also, by (5.49), T is a contraction on D if β_0 is large enough and (5.28) follows from (5.51) and (5.52). □

We now turn to the case $1 \le \beta < \beta_0$, a case that is significantly harder, because T will hardly be a contraction.

Lemma 5.2.7. *If $A \ge L_3$, $L_0 \ge L_3$, if $\beta \le \beta_0$, $r \le L_2\alpha/m^{*2}$, and if $|\mu - m^*| \le m^*/A$, then if (α,β,h) belongs to the admissible region we have*

$$\left|\frac{\partial \Psi}{\partial \mu}(r,\mu)\right| \le 1 - \frac{m^{*2}}{L} \tag{5.53}$$

$$\left|\frac{\partial \Psi}{\partial r}(r,\mu)\right| \le L\,m^* \tag{5.54}$$

$$\left|\frac{\partial \Phi}{\partial \mu}(r,\mu)\right| \le L\,m^* \tag{5.55}$$

$$\left|\frac{\partial \Phi}{\partial r}(r,\mu)\right| \le L \, . \tag{5.56}$$

Proof. It is condition (5.53) that creates the headache. This condition requires looking at higher order derivatives. First we observe that, as $\beta \le \beta_0$, $\alpha \le 1$, the partial derivatives of Ψ, Φ of order less than or equal to 3 remain bounded. Next, from (5.40), we have

$$\frac{\partial^2 \Psi}{\partial \mu^2} = -2\beta^2 \, \mathsf{E} \frac{\operatorname{th} Y}{\operatorname{ch}^2 Y}$$

so that

$$\frac{\partial^2 \Psi}{\partial \mu^2}(0,m^*) = -2\beta^2 \frac{\operatorname{th} \beta m^*}{\operatorname{ch}^2 \beta m^*} = -2\beta^2 m^*(1 - m^{*2})$$

and

$$\left|\frac{\partial^2 \Psi}{\partial \mu^2}(0,m^*)\right| \le L\,m^* \, . \tag{5.57}$$

Using (5.49), we get

$$\left|\frac{\partial^2 \Psi}{\partial \mu^2}(r,\mu) - \frac{\partial^2 \Psi}{\partial \mu^2}(0,m^*)\right| \le L(r + |\mu - m^*|) \, .$$

We note that $|\mu - m^*| \le m^*/2$, so that $\mu \le 2m^*$. Also

$$r \le \frac{L_2\,\alpha}{m^{*2}} \le Lm^{*2} \, ,$$

because, in the admissible region, we have $\alpha \le Lm^{*4}$ for $\beta \le \beta_0$. Combining with (5.57), we get

$$\left| \frac{\partial^2 \Psi}{\partial \mu^2}(r,\mu) \right| \le L m^* \, . \tag{5.58}$$

From (5.40) and (5.35)we have

$$\frac{\partial \Psi}{d\mu}(0,m^*) = \beta(1-m^{*2}) \le 1 - \frac{m^{*2}}{L} \tag{5.59}$$

and, using (5.49), (5.58)

$$\left| \frac{\partial \Psi}{\partial \mu}(r,\mu) - \frac{\partial \Psi}{\partial \mu}(0,m^*) \right| \le L\, r + L m^* \, |\mu - m^*| \, .$$

Now $r \le \alpha L / m^{*2}$ and $|\mu - m^*| \le m^*/A$. Since $\alpha \le L m^{*4}/L_0$ in the admissible region, we have using (5.59) that

$$\left| \frac{\partial \Psi}{\partial \mu}(r,\mu) \right| \le 1 - \frac{m^{*2}}{L} + \frac{L m^{*2}}{L_0} + \frac{L m^{*2}}{A} \le 1 - \frac{m^{*2}}{2L}$$

if L_0 and A are large enough. This proves (5.53).

The proof of (5.54) and (5.55) is much easier. We use (5.49) and the fact that, from (5.39)

$$\frac{\partial \Psi}{\partial r}(0,m^*) = -\beta^2 \, m^*(1-m^{*2})$$

so that

$$\left| \frac{\partial \Psi}{\partial r}(r,\mu) \right| \le L m^* \, ,$$

and, similarly, $\left| \frac{\partial \Phi}{\partial \mu}(r,\mu) \right| \le L m^*$ from (5.38). □

Proof of Theorem 5.2.1. We assume $\beta \le \beta_0$, and $A \ge L_3$, $L_0 \ge L_3$ so that Lemma 5.2.7 holds. Then, for (q,μ) in D, we have

$$\begin{aligned} |\Psi(r(q),\mu) - m^*| &= |\Psi(r(q),\mu) - \Psi(0,m^*)| \\ &\le L \frac{\alpha}{m^{*2}} m^* + \frac{m^*}{A}\left(1 - \frac{m^{*2}}{L}\right) \end{aligned} \tag{5.60}$$

using (5.42), (5.49), (5.53) and (5.54). Since $\alpha \le L m^{*4}/L_0$ this is at most

$$\frac{m^*}{A} + m^{*3}\left[\frac{L}{L_0} - \frac{1}{LA}\right] \le \frac{m^*}{A} \tag{5.61}$$

if $L_0 \ge LA$. Also, using (5.49), (5.55) and (5.56)

$$\begin{aligned} |\Phi(r(q),\mu) - m^{*2}| &= |\Phi(r(q),\mu) - \Phi(0,m^*)| \\ &\le \frac{L\alpha}{m^{*2}} + \frac{m^*}{A} L m^* \le \frac{m^{*2}}{L_1 \beta_0^2} \end{aligned} \tag{5.62}$$

since $\alpha \leq Lm^{*4}/L_0$, and provided A and L_0 are large enough. It follows from (5.60) to (5.62) that $T(D) \subset D$ if A, L_0 have been chosen as explained.

Next, we show that for a certain distance d on D, T is a contraction (provided again that L_0 is large enough). This will finish the proof. Given (q,μ), (q',μ') in D, we have, as in (5.60), and using (5.43) to see that

$$|r(q) - r(q')| \leq L\,\alpha\,|q - q'|/m^{*4}\,, \tag{5.63}$$

$$\begin{aligned}|\Psi(r(q),\mu) - \Psi(r(q'),\mu')| &\leq \frac{L\alpha}{m^{*4}}\,m^*|q-q'| + \left(1 - \frac{m^{*2}}{L}\right)|\mu - \mu'| \\ &\leq \frac{L}{L_0}\,m^*|q-q'| + \left(1 - \frac{m^{*2}}{L}\right)|\mu - \mu'|. \end{aligned} \tag{5.64}$$

As in (5.62), we have, using (5.63) again,

$$\begin{aligned}|\Phi(r(q),\mu) - \Phi(r(q'),\mu')| &\leq \frac{L\alpha}{m^{*4}}\,|q-q'| + L\,m^*|\mu - \mu'| \\ &\leq \frac{L}{L_0}\,|q-q'| + L\,m^*|\mu - \mu'|\,. \end{aligned} \tag{5.65}$$

Let us consider the distance d on D given by

$$d((q,\mu),(q',\mu')) = a\,|q-q'| + |\mu - \mu'|\,,$$

where a will be chosen later. Then (5.64) and (5.65) show that

$$\begin{aligned}d(T(q,\mu),T(q',\mu')) &\leq \frac{L}{L_0}(a + m^*)\,|q-q'| + \left(1 - \frac{m^{*2}}{L} + L\,a\,m^*\right)|\mu - \mu'| \\ &\leq \max\left(1 - \frac{m^{*2}}{L} + a\,L\,m^*, \frac{L}{L_0}\left(1 + \frac{m^*}{a}\right)\right) \\ &\quad\times (a|q-q'| + |\mu - \mu'|)\,. \end{aligned}$$

We take $a = m^*/L'$ where L' is large enough, and if L_0 is large enough we obtain

$$d(T(q,\mu),T(q',\mu')) \leq \left(1 - \frac{m^{*2}}{L}\right) d((q,\mu),(q',\mu'))\,.$$

□

For (α,β,h) in the admissible region, we will from now on always denote by q,μ,r the solutions of the equations (5.25) to (5.27) that satisfy (5.28). While we are at it, we gather some more information that will be used later.

Lemma 5.2.8. *In the admissible region we have (if L_0 is large enough)*

$$|\mu - m^*| \leq \frac{L}{L_0}\,m^* \tag{5.66}$$

$$|\mu - m^*| \le \frac{L}{\beta^{10}} \tag{5.67}$$

$$|q - m^{*2}| \le \frac{L}{L_0} m^{*2} \tag{5.68}$$

$$|r - \alpha| \le \frac{L}{\beta^{10}}\,,\; |r - \alpha\, m^*| \le \frac{L}{\beta^{10}} \tag{5.69}$$

$$\frac{q}{1-\beta(1-q)} \le L \tag{5.70}$$

$$\beta(1 - 2\mu^2 + q) \le 1 - \frac{m^{*2}}{L}\,. \tag{5.71}$$

Proof. To prove (5.66), we write

$$|\mu - m^*| = |\Psi(r,\mu) - \Psi(0,m^*)| \le Lr\, m^* + \left(1 - \frac{m^{*2}}{L}\right) |\mu - m^*|$$

using (5.49), (5.54) and (5.53). The conclusion then follows from (5.42).

To prove (5.67) we write $|\mu - m^*| \le 1 - \mu + 1 - m^*$ and we prove that

$$1 - \mu \le L\,\beta^{-10}\,. \tag{5.72}$$

For this, we follow the proof of (5.46). We write

$$\begin{aligned}
\mu &= \mathsf{E}\,\mathrm{th}\,(\beta z\sqrt{r} + \beta\mu + h) \\
&\ge \mathsf{P}\left(z \ge -\frac{\mu}{2\sqrt{r}}\right) \mathrm{th}\left(\frac{\beta\mu}{2}\right) - \mathsf{P}\left(z \le -\frac{\mu}{2\sqrt{r}}\right) \\
&= \mathrm{th}\left(\frac{\beta\mu}{2}\right) - \mathsf{P}\left(z < -\frac{\mu}{2\sqrt{r}}\right)\left(1 + \mathrm{th}\left(\frac{\beta\mu}{2}\right)\right) \\
&\ge \mathrm{th}\left(\frac{\beta\mu}{2}\right) - 2\exp\left(-\frac{\mu^2}{2r}\right)\,,
\end{aligned} \tag{5.73}$$

and we proceed as in (5.48) to obtain (5.72) if L_0 is large enough.

To prove (5.68) we write

$$|q - m^{*2}| = |\Phi(r,\mu) - \Phi(0,m^*)| \le Lr + L\,m^*|\mu - m^*|$$

using (5.55), (5.56) and (5.49). Now, by (5.27) and (5.43), we get

$$r \le \frac{L\,\alpha\, q}{m^{*2}} \le \frac{Lm^{*2}}{L_0}$$

and (5.68) follows from (5.66).

To prove (5.69), it suffices to prove the first part since $|1 - m^*| \le L\beta^{-11}$. We write

$$|r-\alpha| = \alpha\left|\frac{q}{(1-\beta(1-q))^2} - 1\right| \leq \frac{|q-1|}{(1-\beta(1-q))^2} + \frac{|1-(1-\beta(1-q))^2|}{(1-\beta(1-q))^2}\,.$$

Using that $(1-x)^2 \geq 1-2x$ we see that $|1-(1-\beta(1-q))^2| \leq 2\beta(1-q)$ so that using (5.44) we get $r \leq L\beta(1-q)$. Thus it suffices to prove that $1-q \leq L\beta^{-11}$, which is obtained exactly as (5.72).

To prove (5.70) we combine (5.44) with the inequality $q \leq Lm^{*2}$. To prove (5.71) we combine (5.66) with (5.68). □

Having had to work rather hard just to prove the existence of the solutions to (5.25) to (5.27), the reader might start to worry that the study of the Hopfield model is going to be technically challenging. This is unfortunately true, but a part of the difficulty is that we want to control all the values of parameters in the admissible region, and that for $h = 0$ we have to work hard to get the correct behavior $\alpha \leq (\beta-1)^2/2$ as $\beta \to 1_+$, $\alpha \leq 1/L \log \beta$ as $\beta \to \infty$. Getting this correct behavior is an interesting point, but it is a side story. The main story is that, given $\beta \neq 1$, there exists $\alpha(\beta) > 0$ such that we can understand the Hopfield model for $\alpha \leq \alpha(\beta)$. Simpler arguments suffice in that case, as the reader can verify easily in the case of Theorem 5.2.1.

5.3 The Bovier-Gayrard Localization Theorem

Understanding the Hopfield model requires to pass a number of technical obstacles. One of them is that it does not seem possible to use the smart path method unless one already has some information about Gibbs' measure. The purpose of this section and of the next is to gather such information.

If we rewrite the Hamiltonian as

$$-H_{N,M}(\boldsymbol{\sigma}) = \frac{N\beta}{2}\sum_{k\leq M} m_k^2(\boldsymbol{\sigma}) + Nhm_1(\boldsymbol{\sigma}) \tag{5.74}$$

we see that it is defined entirely as a function of the quantities $(m_k(\boldsymbol{\sigma}))_{k\leq M}$. Therefore these could be important objects.

Consider now the image G' of G under the random map $\boldsymbol{\sigma} \to (m_k(\boldsymbol{\sigma}))_{k\leq M}$. This is a random probability measure on $\mathbb{R}^M$. The purpose of this section is to show that if

$$\alpha = \alpha(N,M) = \frac{M}{N}$$

is not too large, we have some precious information on "where G' lives". Before we state the result, we introduce some terminology.

Definition 5.3.1. *We say that a set A of $\mathbb{R}^M$ is* negligible *if*

$$\mathsf{E}\,G'(A) \leq K\exp\left(-\frac{N}{K}\right) \tag{5.75}$$

where K does not depend on N, M. We say that G' is essentially supported *by A if $\mathbb{R}^M \backslash A$ is negligible.*

We say that an event Ω occurs with overwhelming probability *if $\mathsf{P}(\Omega^c) \leq K \exp(-N/K)$ when K does not depend on N, M.*

Of course the set A also depends on M, N, so it would be more formal to say that "a family $A_{N,M}$ of sets if negligible if $\mathsf{E}\, G(A_{N,M}) \leq K \exp(-N/K)$, where K does not depend on M, N".

In a similar manner, we will say that a subset A of Σ_N is negligible if $\mathsf{E}\, G(A) \leq K \exp(-N/K)$, where K does not depend on N, M. We denote by $(\mathbf{e}_k)_{k \leq M}$ the canonical basis of $\mathbb{R}^M$.

Theorem 5.3.2. *(The Bovier-Gayrard localization theorem.) Consider $\beta > 1$, $h \geq 0$, $\rho_0 \leq m^*/2$. If $\alpha \leq m^{*2} \rho_0^2/L$, then G' is essentially supported by the union of the $2M$ balls in $\mathbb{R}^M$ of radius ρ_0 centered at the points $\pm m^* \mathbf{e}_k$, $k \leq M$.*

It is useful to note that the balls of this theorem are disjoint since $\rho_0 \leq m^*/2$.

This is a good place to point out a subtle point. Even though intuitively we want to study the case $M/N \to \alpha$, where α is a given number, many of our results are stated to be true for *all* values of M, N, α being *defined* as M/N. Let us also stress that we insist in Definition 5.3.1 that K be independent of N *and* M (a requirement that is essential in the proof of Theorem 5.9.4). Thus with our definitions it is *not* true that G' is essentially supported by the balls of center $\pm m^* \mathbf{e}_k$ and radius $L\sqrt{\alpha}/m^*$ (which is the usual formulation of the Bovier-Gayrard theorem). This is shown by taking $M = 1$. On the other hand, Theorem 5.3.2 says that this is true if α remains bounded below independently of N. The same feature occurs in a number of subsequent results, and motivates their formulations.

It is intuitive that something of the type of Theorem 5.3.2 should happen when $h = 0$. (The case $h > 0$ is discussed in Section 5.5.) Each of the terms $m_k(\boldsymbol{\sigma})$ in the Hamiltonian attempts to create a Curie-Weiss model "in the direction of $(\eta_{i,k})_{i \leq N}$"; and in such a model $m_k(\boldsymbol{\sigma}) \simeq \pm m^*$. What Theorem 5.3.2 says is that if there are not too many such terms, for (nearly) each configuration, one of them wins over the others. For one k (depending on the configuration) we have $m_k(\boldsymbol{\sigma}) \simeq \pm m^*$, and for $k' \neq k$, $m_k(\boldsymbol{\sigma})$ is smaller. What is not intuitive is how large α can be taken. It is a separate problem to know whether the same k "wins" independently of the configuration.

The Bovier-Gayrard localization theorem is a deep fact, that will require significant work. The methods are of interest, but they are not related to the main theme of the book (the cavity method) and will be used only in this section and the next.

As an auxiliary tool, we consider the probability measure γ on $\mathbb{R}^M$, of density $W \exp(-\beta N \|\mathbf{z}\|^2/2)$ with respect to Lebesgue measure on $\mathbb{R}^M$, where $\|\mathbf{z}\|$ is the Euclidean norm of $\mathbf{z}$ and W is the normalizing factor $W = (N\beta/2\pi)^{M/2}$.

It will be useful to replace G' by its convolution $\overline{G} = G' * \gamma$ with γ. This method is called the Hubbard-Stratonovitch transform. It is an elaboration of the trick used in Lemma 5.1.2.

The first observation is that $\overline{G}$ can be seen as a small perturbation of G', because if $\alpha \leq \beta\rho^2/4$, then γ is essentially supported by the ball centered at the origin, of radius ρ. To see this, we observe that, by change of variable,

$$\int_{\mathbb{R}^M} \exp\left(\frac{\beta N}{4} \|z\|^2\right) \mathrm{d}\gamma(\mathbf{z}) = W \int_{\mathbb{R}^M} \exp\left(-\frac{\beta N}{4} \|\mathbf{z}\|^2\right) \mathrm{d}\mathbf{z} = 2^{M/2} .$$

Thus,

$$\gamma(\{\|\mathbf{z}\|^2 \geq \rho^2\}) \exp\left(\frac{N\beta}{4}\rho^2\right) \leq 2^{M/2} \leq \exp\left(\frac{\alpha N}{2}\right)$$

and, since $\alpha \leq \beta\rho^2/4$, we have

$$\gamma(\{\|\mathbf{z}\|^2 \geq \rho^2\}) \leq \exp\left(-\frac{N}{4}(\beta\rho^2 - 2\alpha)\right) \leq \exp\left(-\frac{N\beta\rho^2}{8}\right) . \tag{5.76}$$

This inequality shows in particular that if

$$B = \left\{\mathbf{z};\ \|\mathbf{z}\| \leq 2\sqrt{\frac{\alpha}{\beta}}\right\} \tag{5.77}$$

we have $\gamma(B) \geq 1/L$ (observe that $\alpha N = M \geq 1$). Thus, given any subset A of $\mathbb{R}^M$, we have

$$\overline{G}(A+B) = G' \otimes \gamma(\{(\mathbf{x},\mathbf{y}); \mathbf{x}+\mathbf{y} \in A+B\}) \geq G'(A)\,\gamma(B) ,$$

and thus

$$G'(A) \leq L\overline{G}(A+B) . \tag{5.78}$$

To prove that a set A is negligible for G' it hence suffices to prove that $A+B$ is negligible for $\overline{G}$. Thus it suffices to prove Theorem 5.3.2 for $\overline{G}$ rather than for G' (because if $\alpha \leq m^{*2}\,\rho_0^2/4$, then $2\sqrt{\alpha/\beta} \leq \rho_0$).

The reason why this is more convenient is that $\overline{G}$ has a simple density with respect to Lebesgue's measure. We consider the vector

$$\boldsymbol{\eta}_i = (\eta_{i,k})_{k \leq M}$$

of $\mathbb{R}^M$, and the (random) function ψ on $\mathbb{R}^M$ given by

$$\psi(\mathbf{z}) = -\frac{N\beta}{2}\|\mathbf{z}\|^2 + \sum_{i \leq N} \log \mathrm{ch}(\beta\,\boldsymbol{\eta}_i \cdot \mathbf{z} + h) , \tag{5.79}$$

where of course $\|\mathbf{z}\|^2 = \sum_{k \leq M} z_k^2$ and $\boldsymbol{\eta}_i \cdot \mathbf{z} = \sum_{k \leq M} \eta_{i,k}\, z_k$.

Lemma 5.3.3. *The probability $\overline{G}$ has a density*

$$W\, 2^N\, Z_{N,M}^{-1} \exp \psi(\mathbf{z})$$

with respect to Lebesgue's measure, where

$$Z_{N,M} = \sum_{\boldsymbol{\sigma}} \exp(-H_{N,M}(\boldsymbol{\sigma})) \; .$$

Proof. If we consider the positive measure δ consisting of a mass a at a point $\mathbf{x}$, the density of $\delta * \gamma$ at a point $\mathbf{z}$ is given by

$$a\, W \exp\left(-\frac{\beta N}{2} \|\mathbf{z} - \mathbf{x}\|^2\right) \; .$$

Also, G' gives mass $Z_{N,M}^{-1} \exp(-H_{N,M}(\boldsymbol{\sigma}))$ to the points $(m_k(\boldsymbol{\sigma}))_{k \le M}$. Thus, if we write $\mathbf{m}(\boldsymbol{\sigma}) = (m_k(\boldsymbol{\sigma}))_{k \le M}$, the density at $\mathbf{z}$ of $G' * \gamma$ is

$$\frac{1}{Z_{N,M}} \sum_{\boldsymbol{\sigma}} W \exp\left(\frac{N\beta}{2} \|\mathbf{m}(\boldsymbol{\sigma})\|^2 - \frac{N\beta}{2} \|\mathbf{z} - \mathbf{m}(\boldsymbol{\sigma})\|^2 + Nhm_1(\boldsymbol{\sigma})\right)$$

because $-H_{N,M}(\boldsymbol{\sigma}) = N\beta \|\mathbf{m}(\boldsymbol{\sigma})\|^2/2 + Nhm_1(\boldsymbol{\sigma})$. This is

$$\frac{W}{Z_{N,M}} \exp\left(-\frac{N\beta}{2} \|\mathbf{z}\|^2\right) \sum_{\boldsymbol{\sigma}} \exp(N\beta\, \mathbf{z} \cdot \mathbf{m}(\boldsymbol{\sigma}) + Nhm_1(\boldsymbol{\sigma})) \; .$$

Now

$$\begin{aligned} N\beta\, \mathbf{z} \cdot \mathbf{m}(\boldsymbol{\sigma}) + Nhm_1(\boldsymbol{\sigma}) &= \beta \sum_{k \le M} z_k \left(\sum_{i \le N} \eta_{i,k}\, \sigma_i\right) + h \sum_{i \le N} \sigma_i \\ &= \sum_{i \le N} \sigma_i\, \beta \sum_{k \le M} z_k\, \eta_{i,k} + h \sum_{i \le N} \sigma_i \\ &= \sum_{i \le N} \sigma_i (\beta\, \boldsymbol{\eta}_i \cdot \mathbf{z} + h) \; , \end{aligned}$$

and

$$\begin{aligned} \sum_{\boldsymbol{\sigma}} \exp(N\beta\, \mathbf{z} \cdot \mathbf{m}(\boldsymbol{\sigma}) + Nhm_1(\boldsymbol{\sigma})) &= 2^N \prod_{i \le N} \operatorname{ch}(\beta\, \boldsymbol{\eta}_i \cdot \mathbf{z} + h) \\ &= 2^N \exp \sum_{i \le N} \log \operatorname{ch}(\beta\, \boldsymbol{\eta}_i \cdot \mathbf{z} + h) \; . \end{aligned}$$

□

As a consequence, for a subset A of $\mathbb{R}_N$, we have

$$\overline{G}(A) = \frac{W \int_A \exp \psi(\mathbf{z})\, d\mathbf{z}}{2^{-N} Z_{N,M}} \qquad (5.80)$$

and the strategy to prove that A is negligible is simply to prove that typically the numerator in (5.80) is much smaller than the denominator. For this it certainly helps to bound the denominator from below.

Proposition 5.3.4. *If $\beta \le 2$ and $\alpha \le m^{*4}/L$, we have*

$$2^{-N} Z_{N,M} \ge \left(\frac{1}{La^*}\right)^{M/2} \exp Nb^* \tag{5.81}$$

where

$$b^* = \log \operatorname{ch}(\beta m^* + h) - \beta m^{*2}/2, a^* = 1 - \beta(1 - m^{*2}) .$$

This bound is true for any realization of the randomness. It somewhat resembles Lemma 5.1.2.

Before we start the proof, we mention the following elementary fact. The function $\xi(x) = \log \operatorname{ch} x$ satisfies

$$\xi'(x) = \operatorname{th} x;\ \xi''(x) = \frac{1}{\operatorname{ch}^2 x};\ \xi'''(x) = -2\frac{\operatorname{th} x}{\operatorname{ch}^2 x};\ |\xi^{(4)}(x)| \le 4 . \tag{5.82}$$

Proof of Proposition 5.3.4. Since $\overline{G}$ is a probability, Lemma 5.3.3 shows that

$$2^{-N} Z_{N,M} = W \int \exp \psi(\mathbf{z})\, d\mathbf{z} ;$$

so that

$$2^{-N} Z_{N,M} = W \int \exp \psi^{\sim}(\mathbf{v})\, d\mathbf{v} , \tag{5.83}$$

where

$$\begin{aligned}\psi^{\sim}(\mathbf{v}) &= \psi(m^* \mathbf{e}_1 + \mathbf{v}) \\ &= -\frac{N\beta}{2} \|m^* \mathbf{e}_1 + \mathbf{v}\|^2 + \sum_{i \le N} \log \operatorname{ch}(\beta \boldsymbol{\eta}_i \cdot (m^* \mathbf{e}_1 + \mathbf{v}) + h) .\end{aligned} \tag{5.84}$$

Since we assume $\eta_{i,1} = 1$, we have

$$\boldsymbol{\eta}_i \cdot (m^* \mathbf{e}_1 + \mathbf{v}) = m^* + \boldsymbol{\eta}_i \cdot \mathbf{v} .$$

We make an order 4 Taylor expansion of log ch around $b = \beta m^* + h$. We get

$$\begin{aligned}\psi^{\sim}(\mathbf{v}) = &-\frac{N\beta}{2} m^{*2} + N \log \operatorname{ch} b - \frac{N\beta}{2}\|\mathbf{v}\|^2 - N\beta m^* \mathbf{v} \cdot \mathbf{e}_1 \\ &+ \beta \operatorname{th} b \sum_{i \le N} \boldsymbol{\eta}_i \cdot \mathbf{v} + \frac{\beta^2}{2\operatorname{ch}^2 b} \sum_{i \le N} (\boldsymbol{\eta}_i \cdot \mathbf{v})^2 \\ &- \frac{\beta^3}{3} \frac{\operatorname{th} b}{\operatorname{ch}^2 b} \sum_{i \le N} (\boldsymbol{\eta}_i \cdot \mathbf{v})^3 + \frac{\beta^4}{6} \sum_{i \le N} R_i(\mathbf{v}) (\boldsymbol{\eta}_i \cdot \mathbf{v})^4\end{aligned} \tag{5.85}$$

where $|R_i(\mathbf{v})| \le 1$.

The idea of the proof is to simplify (5.83) by averaging over rotations. If U denotes a rotation of $\mathbb{R}^M$, the invariance of Lebesgue's measure by rotation shows from (5.83) that

$$2^{-N} Z_{N,M} = W \int \exp \psi^{\sim}(U(\mathbf{v}))\, \mathrm{d}\mathbf{v} \tag{5.86}$$

so that, if $\mathrm{d}U$ denotes Haar measure on the group of rotations, we have

$$\begin{aligned} 2^{-N} Z_{N,M} &= W \int \exp \psi^{\sim}(U(\mathbf{v}))\, \mathrm{d}U\, \mathrm{d}\mathbf{v} \\ &\geq W \int \exp\left(\int \psi^{\sim}(U(\mathbf{v}))\, \mathrm{d}U \right) \mathrm{d}\mathbf{v} \end{aligned} \tag{5.87}$$

by Jensen's inequality.

Given a vector $\mathbf{x}$ of $\mathbb{R}^M$, we have for a certain constant c_p that

$$\int (\mathbf{x} \cdot U(\mathbf{v}))^p\, \mathrm{d}U = c_p \, \|\mathbf{x}\|^p \, \|\mathbf{v}\|^p \tag{5.88}$$

because the left-hand side depends only on $\|\mathbf{x}\|$, $\|\mathbf{v}\|$.

To compute the quantity c_p, we consider a vector $\mathbf{g} = (g_k)_{k \leq M}$ where (g_k) are independent standard Gaussian. We apply (5.88) to $\mathbf{x} = \mathbf{g}$ and we take expectation. We observe that

$$\mathsf{E}((\mathbf{g} \cdot U(\mathbf{v}))^p) = \|U(\mathbf{v})\|^p \, \mathsf{E}\, g^p = \|\mathbf{v}\|^p \, \mathsf{E}\, g^p \,,$$

where g is standard Gaussian. Thus

$$c_p = \frac{\mathsf{E}\, g^p}{\mathsf{E}\, \|\mathbf{g}\|^p} \,.$$

In particular, $c_p = 0$ when p is odd, $c_2 = 1/M$, and, since

$$\mathsf{E}\, \|\mathbf{g}\|^4 \geq (\mathsf{E}\, \|\mathbf{g}\|^2)^2 = M^2 \,,$$

we get $c_4 \leq 3/M^2$.

We observe that $\|\boldsymbol{\eta}_i\|^2 = M$, so that (5.85) implies

$$\begin{aligned} \int \psi^{\sim}(U(\mathbf{v}))\, \mathrm{d}U \geq N &\left(\log \operatorname{ch} b - \frac{\beta m^{*2}}{2} \right) \\ &- \frac{N\beta}{2} \left(1 - \frac{\beta}{\operatorname{ch}^2 b} \right) \|\mathbf{v}\|^2 - \frac{N\beta^4}{2} \|\mathbf{v}\|^4 \,. \end{aligned} \tag{5.89}$$

Since $b = \beta m^* + h$, we have

$$1 - \frac{\beta}{\operatorname{ch}^2 b} = 1 - \beta(1 - \operatorname{th}^2 b) = 1 - \beta(1 - m^{*2}) = a^*$$

and thus

$$2^{-N} Z_{N,M} \geq (\exp Nb^*)\, W \int \exp\left(-\frac{N\beta}{2}\, a^* \|\mathbf{v}\|^2 - \frac{N\beta^4}{2}\, \|\mathbf{v}\|^4\right) \mathrm{d}\mathbf{v}$$
$$= \left(\frac{1}{a^*}\right)^{M/2} (\exp Nb^*)\, W \int \exp\left(-\frac{N\beta}{2}\, \|\mathbf{v}\|^2 - \frac{N\beta^4 \|\mathbf{v}\|^4}{2a^{*2}}\right) \mathrm{d}\mathbf{v}$$

by change of variable. Thus

$$2^{-N} Z_{N,M} \geq \left(\frac{1}{a^*}\right)^{M/2} (\exp Nb^*) \int \exp\left(-\frac{N\beta^4}{2} \frac{\|\mathbf{v}\|^4}{a^{*2}}\right) \mathrm{d}\gamma(\mathbf{v}) .$$

If we recall the set B of (5.77), we have

$$\int \exp\left(-\frac{N\beta^4}{2} \frac{\|\mathbf{v}\|^4}{a^{*2}}\right) \mathrm{d}\gamma(\mathbf{v}) \geq \gamma(B) \exp\left(-\frac{N\beta^4}{2a^{*2}} \left(2\sqrt{\frac{\alpha}{\beta}}\right)^4\right)$$
$$\geq \frac{1}{L} \exp\left(-\frac{LN\beta^2\alpha^2}{a^{*2}}\right)$$
$$= \frac{1}{L} \exp\left(-\frac{LM\beta^2\alpha}{a^{*2}}\right) .$$

Thus, since we assume $\beta \leq 2$, we have, using (5.35)

$$\alpha \leq m^{*4} \leq La^{*2} \Rightarrow 2^{-N} Z_{N,M} \geq \left(\frac{1}{La^*}\right)^{M/2} \exp Nb^* . \tag{5.90}$$

□

When $\beta \geq 2$, we will use a different bound. We will use the vector $\boldsymbol{\theta}$ given by

$$\boldsymbol{\theta} = \frac{m^*}{N} \sum_{i \leq N} (\boldsymbol{\eta}_i - \mathbf{e}_1) \tag{5.91}$$

so that $\theta_1 = 0$, and for $2 \leq k \leq M$ we have

$$\theta_k = m^* \left(\frac{1}{N} \sum_{1 \leq i \leq N} \eta_{i,k}\right) .$$

Proposition 5.3.5. *We have*

$$2^{-N} Z_{N,M} \geq \exp\left(Nb^* + \frac{N\beta}{2}\, \|\boldsymbol{\theta}\|^2\right) . \tag{5.92}$$

This bound is true for any realization of the disorder.

Proof. We start with (5.83). Using the convexity of the function log ch, and using that $\operatorname{th} b = m^*$, we get from (5.84) that

$$\begin{aligned}\psi^{\sim}(\mathbf{v}) &\geq -\frac{N\beta}{2}\|m^*\mathbf{e}_1+\mathbf{v}\|^2+N\log\operatorname{ch} b+(\operatorname{th} b)\left(\sum_{i\leq N}\beta\boldsymbol{\eta}_i\cdot\mathbf{v}\right)\\&=Nb^*-\frac{N\beta}{2}\|\mathbf{v}\|^2+\beta m^*\left(\sum_{i\leq N}(\boldsymbol{\eta}_i-\mathbf{e}_1)\cdot\mathbf{v}\right)\\&=Nb^*-\frac{N\beta}{2}\|\mathbf{v}\|^2+N\beta\,\boldsymbol{\theta}\cdot\mathbf{v}\\&=Nb^*+N\beta\|\boldsymbol{\theta}\|^2-\frac{N\beta}{2}\|\mathbf{v}-\boldsymbol{\theta}\|^2\,.\end{aligned}$$

Thus

$$\begin{aligned}W\int\exp\psi^{\sim}(\mathbf{v})\,d\mathbf{v}&\geq\exp\left(Nb^*+\frac{N\beta}{2}\|\boldsymbol{\theta}\|^2\right)W\int\exp\left(-\frac{N\beta}{2}\|\mathbf{v}-\boldsymbol{\theta}\|^2\right)d\mathbf{v}\\&=\exp\left(Nb^*+\frac{N\beta}{2}\|\boldsymbol{\theta}\|^2\right)\,.\end{aligned}$$

□

Since $a^*\geq 1/L$ for $\beta\geq 2$, we have the following convenient consequence of Propositions 5.3.4 and 5.3.5:

$$2^{-N}Z_{N,M}\geq\left(\frac{1}{La^*}\right)^{M/2}\exp b^*N\,. \tag{5.93}$$

The following elementary fact will play an important role.

Lemma 5.3.6. *The function*

$$\varphi(x)=\log\operatorname{ch}(\beta\sqrt{x}+h) \tag{5.94}$$

is concave. Moreover, if $x\leq 2$ then

$$\varphi''(x)\leq-\frac{\beta}{L}\,. \tag{5.95}$$

Proof. Setting $y=\beta\sqrt{x}+h$, computation shows that

$$\varphi''(x)=\frac{\beta}{4x^{3/2}}\left(\frac{\beta\sqrt{x}}{\operatorname{ch}^2 y}-\operatorname{th} y\right)\leq-\frac{\beta}{8x^{3/2}}\left(\frac{\operatorname{sh}2y-2y}{\operatorname{ch}^2 y}\right)\leq 0\,.$$

Moreover, distinguishing the cases $y\leq 1$ and $y\geq 1$, we see that

$$\frac{\operatorname{sh}2y-2y}{\operatorname{ch}^2 y}\geq\frac{1}{L}\min(1,y^3)\geq\frac{1}{L}\min(1,x^{3/2})\,.$$

The result follows. □

We are now ready to prove our first localization result.

Proposition 5.3.7. *If $\alpha \le m^{*4}/L$, the set*

$$A = \{\mathbf{z} \in \mathbb{R}^M\,;\ \|\mathbf{z}\| \ge 2m^*\}$$

is negligible for $\overline{G}$, that is

$$\mathsf{E}\,\overline{G}(A) \le K \exp\left(-\frac{N}{K}\right),$$

where K depends only on β, h.

Proof. We write

$$\sum_{i\le N} \log \mathrm{ch}\,(\beta\,\boldsymbol{\eta}_i\cdot\mathbf{z}+h) \le \sum_{i\le N}\varphi((\boldsymbol{\eta}_i\cdot\mathbf{z})^2) \le N\varphi\left(\frac{1}{N}\sum_{i\le N}(\boldsymbol{\eta}_i\cdot\mathbf{z})^2\right) \tag{5.96}$$

by concavity of φ.

Using Corollary A.4.4 we see that provided

$$\alpha \le \frac{a^{*2}}{L} \tag{5.97}$$

then the event

$$\forall \mathbf{z}\in\mathbb{R}^M, \qquad \frac{1}{N}\sum_{i\le N}(\boldsymbol{\eta}_i\cdot\mathbf{z})^2 \le \left(1+\frac{a^*}{8}\right)\|\mathbf{z}\| \tag{5.98}$$

occurs with overwhelming probability (that is, the probability of failure is at most $K\exp(-N/K)$, where K depends on β, h only). Under (5.98) we get from (5.96) that

$$\psi(\mathbf{z}) \le -\frac{N\beta}{2}\|\mathbf{z}\|^2 + N\varphi\left(\left(1+\frac{a^*}{8}\right)\|\mathbf{z}\|^2\right). \tag{5.99}$$

The rest of the proof requires struggling with calculus. The second derivative of the function

$$f(t) = \log \mathrm{ch}\,(\beta t + h) - \frac{\beta t^2}{2}$$

is $f''(t) = \beta^2 \mathrm{ch}^{-2}(\beta t + h) - \beta$, which decreases as t increases from 0. We have

$$f''(m^*) = -\beta\left(1 - \frac{\beta}{\mathrm{ch}^2(\beta m^* + h)}\right) = -\beta a^*,$$

where $a^* = 1 - \beta(1 - m^{*2})$. For $t \ge m^*$, we have $f''(t) \le f''(m^*)$, and since $f(m^*) = b^*$, $f'(m^*) = 0$, for $t \ge m^*$ we have

$$f(t) \le b^* - \frac{\beta}{2}a^*(t-m^*)^2.$$

If $t \geq 2m^*$, we have $t - m^* \geq t/2$, so that

$$\log \operatorname{ch}(\beta t + h) \leq b^* + \frac{\beta}{2}\left(1 - \frac{a^*}{4}\right) t^2$$

and thus, if $t \geq 2m^*$;

$$\begin{aligned}\varphi\left(\left(1 + \frac{a^*}{8}\right) t^2\right) &= \log \operatorname{ch}\left(\beta t\left(1 + \frac{a^*}{8}\right)\right) - \frac{\beta t^2}{2} \\ &\leq b^* - \frac{\beta t^2}{2}\left(1 - \left(1 - \frac{a^*}{4}\right)\left(1 + \frac{a^*}{8}\right)\right) \\ &\leq b^* - \frac{\beta t^2 a^*}{8}\end{aligned}$$

since we have $(1 - a^*/4)(1 + a^*/8) \leq 1 - a^*/8$.

Using (5.97), we see that under (5.98) we have

$$\begin{aligned}\int_A \psi(\mathbf{z})\, d\mathbf{z} &\leq \exp N b^* \int_A \exp\left(-\frac{N\beta}{8}\, a^* \|\mathbf{z}\|^2\right) d\mathbf{z} \qquad (5.100)\\ &\leq \exp\left(N\left(b^* - \frac{\beta a^{*2} m^{*2}}{4}\right)\right) \int_A \exp\left(-\frac{N\beta}{16}\, a^* \|\mathbf{z}\|^2\right) d\mathbf{z}\end{aligned}$$

because $\|\mathbf{z}\|^2 \geq 4m^{*2}$ on A. Now, by change of variable,

$$\int \exp\left(-\frac{N\beta}{16}\, a^* \|\mathbf{z}\|^2\right) d\mathbf{z} = \left(\frac{8}{a^*}\right)^{M/2} W^{-1} ,$$

so that

$$\begin{aligned}\int_A \psi(\mathbf{z})\, d\mathbf{z} &\leq \exp\left(N\left(b^* - \frac{\beta a^* m^{*2}}{8}\right)\right)\left(\frac{8}{a^*}\right)^{M/2} W^{-1} \\ &\leq \exp\left(N\left(b^* - \frac{\beta m^{*4}}{L}\right)\right)\left(\frac{L}{a^*}\right)^{M/2} W^{-1}\end{aligned}$$

using Lemma 5.2.2. Combining with (5.93), we get

$$\overline{G}(A) \leq L^M \exp\left(-N\,\frac{\beta m^{*4}}{L}\right) = \exp N\left(\alpha L - \frac{\beta m^{*4}}{L}\right) \leq \exp\left(-N\frac{\beta m^{*4}}{L}\right)$$

provided $\alpha \leq m^{*4}/L$. □

Our further results will be based on the following upper bound for ψ.

Lemma 5.3.8. *We have*

$$\begin{aligned}\psi(\mathbf{z}) \leq N b^* &+ \frac{\beta}{2}\left(\sum_{i \leq N} (\boldsymbol{\eta}_i \cdot \mathbf{z})^2 - N\,\|\mathbf{z}\|^2\right) \\ &- \frac{\beta}{L} \sum_{i \leq N} \min(1, ((\boldsymbol{\eta}_i \cdot \mathbf{z})^2 - m^{*2})^2) . \qquad (5.101)\end{aligned}$$

Proof. From Lemma 5.3.6, and assuming for definiteness that $x \geq m^{*2}$, we see that

$$\begin{aligned}\varphi(x) &= \varphi(m^{*2}) + \varphi'(m^{*2})(x - m^{*2}) + \int_{m^{*2}}^{x} (x-t)\,\varphi''(t)\,\mathrm{d}t \\ &\leq \varphi(m^{*2}) + \varphi'(m^{*2})(x - m^{*2}) - \frac{\beta}{L}\int_{m^{*2}}^{\min(x,2)} (x-t)\,\mathrm{d}t \\ &\leq \varphi(m^{*2}) + \varphi'(m^{*2})(x - m^{*2}) - \frac{\beta}{L}\min(1,(x-m^{*2})^2)\,. \quad (5.102)\end{aligned}$$

Now,

$$\varphi'(m^{*2}) = \frac{\beta}{2m^*}\,\mathrm{th}\,(\beta m^* + h) = \frac{\beta}{2}$$

and

$$\varphi(m^{*2}) - \frac{\beta}{2}m^{*2} = \log\mathrm{ch}\,(\beta m^* + h) - \frac{\beta}{2}m^{*2} = b^*\,,$$

so that (5.102) implies

$$\varphi(x) \leq b^* + \frac{\beta}{2}x - \frac{\beta}{L}\min(1,(x-m^{*2})^2)\,.$$

Using this for $x = (\boldsymbol{\eta}_i \cdot \mathbf{z})^2$ and summing over $i \leq N$ yields the result. □

The last term in (5.101) has a crucial influence. To use it conveniently, we need to learn to control it uniformly over certain sets. It will be convenient to introduce a parameter $d \leq 1$ and to write

$$R_d(\mathbf{z}) = \mathsf{E}\min(d,((\boldsymbol{\eta}_i \cdot \mathbf{z})^2 - m^{*2})^2)\,, \tag{5.103}$$

a quantity which is of course independent of i.

Proposition 5.3.9. *Consider a ball B of $\mathbb{R}^M$, of radius ρ, and assume that*

$$B \subset \{\mathbf{z};\|\mathbf{z}\| \leq 2m^*\}\,. \tag{5.104}$$

Then, for each $\varepsilon > 0$, with overwhelming probability, we have

$$\begin{aligned}\forall \mathbf{z} \in B, \quad &\sum_{i\leq N}\min(d,((\boldsymbol{\eta}_i \cdot \mathbf{z})^2 - m^{*2})^2) \\ &\geq \frac{N}{4}\left(R_d(\mathbf{z}) - 4\alpha\log\left(1+\frac{\rho}{\varepsilon}\right) - L\,\varepsilon\, m^*\sqrt{d}\right)\,. \quad (5.105)\end{aligned}$$

The proof has two steps. In the first we will show that the left-hand side of (5.105) can be controlled uniformly over many points. In the second we will show that this implies uniform control over B.

Lemma 5.3.10. *Consider a finite subset A of $\mathbb{R}^M$ and $C > 1$. Then, with probability at least $1 - 1/C$ we have*

$$\forall \mathbf{z} \in A, \sum_{i\leq N}\min(d,((\boldsymbol{\eta}_i \cdot \mathbf{z})^2 - m^{*2})^2) \geq \frac{N}{4}R_d(\mathbf{z}) - \log\mathrm{card}\,A - \log C\,. \tag{5.106}$$

Proof. Let

$$A_1 = \{\mathbf{z} \in A; \frac{N}{4} R_d(\mathbf{z}) \geq \log \operatorname{card} A + \log C\} .$$

To prove (5.106), it suffices to achieve control over $\mathbf{z} \in A_1$. Let us fix $\mathbf{z}$ in A_1. The r.v.

$$X_i = \min(d, ((\boldsymbol{\eta}_i \cdot \mathbf{z})^2 - m^{*2})^2)$$

are i.i.d., $0 \leq X_i \leq 1$, $\mathsf{E} X_i = R_d(\mathbf{z})$. We prove an elementary exponential inequality about these variables. Since $\exp(-x) \leq 1 - x/2$ for $x \leq 1$, we have

$$\mathsf{E}\exp(-X_i) \leq 1 - \frac{\mathsf{E} X_i}{2} \leq \exp\left(-\frac{\mathsf{E} X_i}{2}\right) = \exp\left(-\frac{R_d(\mathbf{z})}{2}\right)$$

and thus

$$\mathsf{E}\exp\left(-\sum_{i\leq N} X_i\right) \leq \exp\left(-\frac{NR_d(\mathbf{z})}{2}\right) ,$$

so that

$$\mathsf{P}\left(\sum_{i\leq N} X_i \leq \frac{NR_d(\mathbf{z})}{4}\right) \exp\left(-\frac{NR_d(\mathbf{z})}{4}\right) \leq \exp\left(-\frac{NR_d(\mathbf{z})}{2}\right) ,$$

and

$$\mathsf{P}\left(\sum_{i\leq N} X_i \leq \frac{NR_d(\mathbf{z})}{4}\right) \leq \exp\left(-\frac{NR_d(\mathbf{z})}{4}\right) \leq \frac{1}{C \operatorname{card} A}$$

since $z \in A_1$. Thus, with probability at least $1 - 1/C$, we have

$$\forall \mathbf{z} \in A_1, \sum_{i\leq N} \min(1, ((\boldsymbol{\eta}_i \cdot \mathbf{z})^2 - m^{*2})^2) \geq \frac{NR_d(\mathbf{z})}{4} .$$

□

Lemma 5.3.11. *On the event Ω of Corollary A.4.4 we have*

$$\begin{aligned} \forall \mathbf{z}_1, \mathbf{z}_2 \in \mathbb{R}^M, \quad & \Bigg|\sum_{i\leq N} \min(d, ((\boldsymbol{\eta}_i \cdot \mathbf{z}_1)^2 - m^{*2})^2) \\ & - \sum_{i\leq N} \min(d, ((\boldsymbol{\eta}_i \cdot \mathbf{z}_2)^2 - m^{*2})^2)\Bigg| \\ & \leq NL\sqrt{d}\, \|\mathbf{z}_1 - \mathbf{z}_2\| \left(\|\mathbf{z}_1\| + \|\mathbf{z}_2\|\right) . \end{aligned} \tag{5.107}$$

Moreover

$$|R_d(\mathbf{z}_1) - R_d(\mathbf{z}_2)| \leq L\sqrt{d}\, \|\mathbf{z}_1 - \mathbf{z}_2\| \left(\|\mathbf{z}_1\| + \|\mathbf{z}_2\|\right) . \tag{5.108}$$

Proof. We start with the observation that, since $d \le 1$, we have

$$\begin{aligned} |\min(d, x^2) - \min(d, y^2)| &= |\min{}^2(\sqrt{d}, |x|) - \min{}^2(\sqrt{d}, |y|)| \\ &= |\min(\sqrt{d}, |x|) - \min(\sqrt{d}, |y|)| \\ &\quad \times (\min(\sqrt{d}, |x|) + \min(\sqrt{d}, |y|)) \\ &\le 2\sqrt{d}\,|x - y| \,, \end{aligned} \tag{5.109}$$

and thus the left-hand side of (5.107) is bounded by

$$\begin{aligned} 2\sqrt{d} \sum_{i \le N} |(\boldsymbol{\eta}_i \cdot \mathbf{z}_1)^2 - (\boldsymbol{\eta}_i \cdot \mathbf{z}_2)^2| &\le 2\sqrt{d} \sum_{i \le N} |\boldsymbol{\eta}_i \cdot (\mathbf{z}_1 - \mathbf{z}_2)|(|\boldsymbol{\eta}_i \cdot \mathbf{z}_1| + |\boldsymbol{\eta}_i \cdot \mathbf{z}_2|) \\ &\le 2\sqrt{d} \left(\sum_{i \le N} (\boldsymbol{\eta}_i \cdot (\mathbf{z}_1 - \mathbf{z}_2))^2 \right)^{1/2} \\ &\quad \times \left[\left(\sum_{i \le N} (\boldsymbol{\eta}_i \cdot \mathbf{z}_1)^2 \right)^{1/2} + \left(\sum_{i \le N} (\boldsymbol{\eta}_i \cdot \mathbf{z}_2)^2 \right)^{1/2} \right] \\ &\le L\sqrt{d}\, N \|\mathbf{z}_1 - \mathbf{z}_2\| \, (\|\mathbf{z}_1\| + \|\mathbf{z}_2\|) \end{aligned} \tag{5.110}$$

on the event Ω of Corollary A.4.4. Relation (5.108) is a simple consequence of (5.109) and (5.110), taking expectations. □

Proof of Proposition 5.3.9. It is shown in Proposition A.3.1 that we can find a subset A of B such that

$$\operatorname{card} A \le \left(1 + \frac{\rho}{\varepsilon}\right)^M$$

and such that each point of B is within distance 2ε of A. We apply Lemma 5.3.10 with $C = \exp(-N\varepsilon m^* \sqrt{d})$. We observe that given $\mathbf{z}_2$ on B, there exists $\mathbf{z}_1$ in A with $\|\mathbf{z}_2 - \mathbf{z}_1\| \le 2\varepsilon$, and $\|\mathbf{z}_1\|, \|\mathbf{z}_2\| \le 2m^*$. We then apply (5.107) and (5.108) to obtain the result. □

To use efficiently (5.105), we need to understand the geometric nature of $R_d(\mathbf{z})$, and this is what we start now. We consider the r.v.

$$X = ((\boldsymbol{\eta}_i \cdot \mathbf{z})^2 - m^{*2})^2 \,, \tag{5.111}$$

so that by Proposition 2.2.3 we have

$$\mathsf{P}\left(X \ge \frac{1}{2} \mathsf{E} X \right) \ge \frac{1}{4} \frac{(\mathsf{E}\, X)^2}{\mathsf{E}\,(X^2)} \,, \tag{5.112}$$

and thus

$$\begin{aligned}\mathsf{E}\min(d,X) &\geq \min\left(d,\frac{\mathsf{E}\,X}{2}\right)\mathsf{P}\left(X\geq\frac{\mathsf{E}\,X}{2}\right)\\ &\geq \min\left(d,\frac{\mathsf{E}\,X}{2}\right)\frac{(\mathsf{E}\,X)^2}{4\,\mathsf{E}\,(X^2)}\,. \end{aligned} \tag{5.113}$$

We have

$$X=(U+a)^2\,, \tag{5.114}$$

where $a=\|\mathbf{z}\|^2-m^{*2}$,

$$U=\sum_{k\neq\ell}\eta_{i,k}\,\eta_{i,\ell}\,z_k\,z_\ell\,,$$

so that $\mathsf{E}\,U=0$, $\mathsf{E}\,U^2=\sum_{k\neq\ell}z_k^2\,z_\ell^2$ and thus

$$\mathsf{E}\,X=\sum_{k\neq\ell}z_k^2\,z_\ell^2+a^2\,. \tag{5.115}$$

It can be checked simply by force (expansion) that

$$\mathsf{E}\,U^4\leq L(\mathsf{E}\,U^2)^2\,, \tag{5.116}$$

but there are much nicer arguments to do this [13]. By (5.113), we have

$$\begin{aligned}\mathsf{E}\,X^2 &= \mathsf{E}\,U^4+4\,a\,\mathsf{E}\,U^3+6\,a^2\,\mathsf{E}\,U^2+4\,a^3\,\mathsf{E}\,U+a^4\\ &\leq \mathsf{E}\,U^4+4\,a(\mathsf{E}\,U^4)^{3/4}+6\,a^2\,\mathsf{E}\,U^2+a^4\\ &\leq L((\mathsf{E}\,U^2)^2+a(\mathsf{E}\,U^2)^{3/2}+a^2\,\mathsf{E}\,U^2+a^4)\end{aligned}$$

using (5.116). Using that $ab\leq a^4+b^{4/3}$, we get $\mathsf{E}\,X^2\leq L(\mathsf{E}\,U^2+a^2)^2=L(\mathsf{E}\,X)^2$ and (5.113) implies that

$$\mathsf{E}\min(d,X)\geq\frac{1}{L}\min\left(d,\frac{\mathsf{E}\,X}{2}\right)\,.$$

We have proved the following.

Lemma 5.3.12. *We have*

$$R_d(\mathbf{z})\geq\frac{1}{L}\min\left(d,(\|\mathbf{z}\|^2-m^{*2})^2+\sum_{k\neq\ell}z_k^2\,z_\ell^2\right)\,. \tag{5.117}$$

How to use concretely this formula is shown by the following.

Lemma 5.3.13. *Consider a number* $0\leq\xi\leq1$. *Assume that*

$$\forall k\in M,\quad \|\mathbf{z}-m^*\,\mathbf{e}_k\|\geq\xi\,m^*\,. \tag{5.118}$$

Then if

$$d=\xi^2\,m^{*4}\,, \tag{5.119}$$

we have

$$R_d(\mathbf{z})\geq\frac{\xi^2}{L}\,m^{*4}\,. \tag{5.120}$$

Proof. It is enough to prove that if

$$(\|\mathbf{z}\|^2 - m^{*2})^2 + \sum_{k\neq\ell} z_k^2\, z_\ell^2 \le \frac{\xi^2\, m^{*4}}{16} \tag{5.121}$$

then we can find $k \le M$ and $\varepsilon = \pm 1$ such that

$$\|\mathbf{z} - \varepsilon\, m^*\, \mathbf{e}_k\| \le \xi\, m^* \ . \tag{5.122}$$

First, we observe from (5.121) that

$$|\|\mathbf{z}\|^2 - m^{*2}| \le \frac{\xi\, m^{*2}}{4} \tag{5.123}$$

so that

$$|\|\mathbf{z}\| - m^*| = \frac{|\|\mathbf{z}\|^2 - m^{*2}|}{\|\mathbf{z}\| + m^*} \le \frac{\xi\, m^*}{4} \ . \tag{5.124}$$

Next, from (5.121) we also have that

$$\begin{aligned} \frac{\xi^2\, m^{*4}}{16} \ge \sum_{k\neq\ell} z_k^2\, z_\ell^2 = \|\mathbf{z}\|^4 - \sum_{\ell\le M} z_\ell^4 & \\ \ge \|\mathbf{z}\|^4 - (\max_{\ell\le M} z_\ell^2)\left(\sum_{\ell\le M} z_\ell^2\right) & \\ = \|\mathbf{z}\|^2(\|\mathbf{z}\|^2 - \max_{\ell\le M} z_\ell^2) \ . & \end{aligned} \tag{5.125}$$

Consider k such that $z_k^2 = \max_{\ell\le M} z_\ell^2$. Then, since $\|\mathbf{z}\|^2 \ge 3\, m^{*4}/4$ by (5.123), we have

$$\|\mathbf{z}\|^2 - z_k^2 \le \frac{\xi^2\, m^{*4}}{16\, \|\mathbf{z}\|^2} \le \frac{\xi^2\, m^{*2}}{12} \ .$$

Now $\|\mathbf{z}\|^2 - z_k^2 = \|\mathbf{z} - z_k\, \mathbf{e}_k\|^2$, so that

$$\|\mathbf{z} - z_k\, \mathbf{e}_k\| \le \frac{\xi\, m^*}{3} \tag{5.126}$$

and consequently

$$|\|\mathbf{z}\| - |z_k|| \le \frac{\xi\, m^*}{3} \ . \tag{5.127}$$

Moreover, if $\varepsilon = \operatorname{sign} z_k$, we have

$$\begin{aligned} \|z_k\, \mathbf{e}_k - \varepsilon\, m^*\, \mathbf{e}_k\| = |z_k - m^*\, \varepsilon| = ||z_k| - m^*| & \\ \le |\|\mathbf{z}\| - m^*| + |\|\mathbf{z}\| - |z_k|| \le \left(\frac{1}{4} + \frac{1}{3}\right) \xi\, m^* & \end{aligned} \tag{5.128}$$

using (5.124) and (5.127). Combining with (5.126) proves (5.122). □

We now put together the different pieces, and we state our main tool for the proof of Theorem 5.3.2.

Proposition 5.3.14. *Assume that $\alpha \le m^{*4}/L$. Consider a ball B of $\mathbb{R}^M$ of radius ρ, and assume that*

$$B \subset \{\|\mathbf{z}\|; \|\mathbf{z}\| \le 2m^*\} .$$

Consider a subset A of B and assume that for some number $\xi < 1$, we have

$$\mathbf{z} \in A \Rightarrow \forall k \le M, \|\mathbf{z} \pm m^* \mathbf{e}_k\| \ge \xi m^* .$$

Then, with overwhelming probability, we have

$$\overline{G}(A) \le (L a^*)^{M/2} W \int_A \exp\biggl(\frac{\beta}{2}\Bigl(\sum_{i\le N}(\boldsymbol{\eta}_i \cdot \mathbf{z})^2 - N\|\mathbf{z}\|^2\Bigr) - \frac{N\beta}{L}\Bigl(\xi^2 m^{*4} - L\alpha \log\Bigl(1+\frac{L\rho}{\xi m^*}\Bigr)\Bigr)\biggr) \mathrm{d}\mathbf{z} . \qquad (5.129)$$

Proof. We take $d = \xi^2 m^{*4}$. We consider Proposition 5.3.9 and Lemma 5.3.13 to see that, with overwhelming probability, if $\varepsilon > 0$, we have,

$$\forall \mathbf{z} \in A, \psi(\mathbf{z}) \le Nb^* + \frac{\beta}{2}\Bigl(\sum_{i\le N}(\boldsymbol{\eta}_i \cdot \mathbf{z})^2 - N\|\mathbf{z}\|^2\Bigr) - \frac{N\beta}{L}\Bigl(\xi^2 m^{*4} - L\alpha \log\Bigl(1+\frac{\rho}{\varepsilon}\Bigr) - L\,\varepsilon\,\xi\, m^{*3}\Bigr) .$$

We choose $\varepsilon = \xi m^*/L$ to get that, with overwhelming probability,

$$\forall \mathbf{z} \in A, \psi(\mathbf{z}) \le Nb^* + \frac{\beta}{2}\Bigl(\sum_{i\le N}(\boldsymbol{\eta}_i \cdot \mathbf{z})^2 - N\|\mathbf{z}\|^2\Bigr) - \frac{N\beta}{L}\Bigl(\xi^2 m^{*4} - L\alpha \log\Bigl(1+\frac{L\rho}{\xi m^*}\Bigr)\Bigr) .$$

The result then follows from (5.80) and (5.93). □

Proof of Theorem 5.3.2. First we consider the case

$$B = \{\mathbf{z}; \|\mathbf{z}\| \le 2m^*\}$$

$$A = \Bigl\{\mathbf{z}; \|\mathbf{z}\| \le 2m^*; \forall k \le M, \|\mathbf{z} \pm m^* \mathbf{e}_k\| \ge \frac{1}{2} m^*\Bigr\} . \qquad (5.130)$$

Thus, we can use (5.129) with $\rho = 2m^*$, $\xi = 1/2$. Consider a number $0 < c < 1$. Using Corollary A.4.4 we see that if $\alpha \le c^2/L$, with overwhelming probability we have

$$\overline{G}(A) \le (La^*)^{M/2} W \int_A \exp N\beta\Bigl(c\,\|\mathbf{z}\|^2 - \frac{m^{*4}}{L} + L\alpha\Bigr) \mathrm{d}\mathbf{z} . \qquad (5.131)$$

If $c = m^{*2}/L'$ where L' is large enough, then

$$\|\mathbf{z}\| \le 2m^* \Rightarrow c\,\|\mathbf{z}\|^2 - \frac{m^{*4}}{L} + L\alpha \le -\frac{m^{*4}}{4L} - \frac{m^{*2}}{4L}\,\|\mathbf{z}\|^2$$

so that (5.131) yields that if $\alpha \le m^{*4}/L$, with overwhelming probability we have

$$\begin{aligned}\overline{G}(A) &\le (La^*)^{M/2}\, W \exp\left(-\frac{N\beta m^{*4}}{L}\right)\int \exp\left(-\frac{N\beta m^{*2}}{L}\,\|\mathbf{z}\|^2\right) \mathrm{d}\mathbf{z}\\ &\le \left(\frac{La^*}{m^{*2}}\right)^{M/2} \exp\left(-\frac{N\beta m^{*4}}{L}\right) \le \exp\left(-\frac{N\beta m^{*4}}{L}\right) \end{aligned} \tag{5.132}$$

if $\alpha \le m^{*4}/L$, and using that $a^* \le Lm^{*2}$ by (5.35). Since (5.132) holds with overwhelming probability, and since (because $\overline{G}(A) \le 1$), for any $\varepsilon > 0$, we have

$$\mathsf{E}\,\overline{G}(A) \le \mathsf{P}(\overline{G}(A) \ge \varepsilon) + \varepsilon\,,$$

it follows that $\mathsf{E}\,\overline{G}(A) \le K\exp(-N/K)$. (Thus, to show that a set A is negligible, it suffices to show that with overwhelming probability we have $\overline{G}(A) \le K\exp(-N/K)$. This will be used many times.)

Next, given an integer $\ell \ge 1$, given $\varepsilon = \pm 1$, given $k \le M$, we want to apply (5.129) to the sets

$$B = \{\mathbf{z};\, \|\mathbf{z} - \varepsilon\, m^*\, \mathbf{e}_k\| \le 2^{-\ell}\, m^*\}\,,$$

$$A = \{\mathbf{z}; 2^{-\ell-1}\, m^* \le \|\mathbf{z} - \varepsilon\, m^*\, \mathbf{e}_k\| \le 2^{-\ell}\, m^*\}\,. \tag{5.133}$$

Thus, we can use (5.129) with $\xi = 2^{-\ell-1}$, $\rho = 2^{-\ell}\, m^*$. We set $\mathbf{v} = \mathbf{z} - \varepsilon\, m^*\, \mathbf{e}_k$, so that

$$\begin{aligned}\sum_{i\le N} (\boldsymbol{\eta}_i \cdot \mathbf{z})^2 - N\,\|\mathbf{z}\|^2 &= \sum_{i\le N} (\boldsymbol{\eta}_i \cdot \mathbf{v})^2 - N\,\|\mathbf{v}\|^2 \\ &\quad + 2\varepsilon\, m^* \sum_{i\le N} ((\boldsymbol{\eta}_i \cdot \mathbf{v})(\boldsymbol{\eta}_i \cdot \mathbf{e}_k) - N\mathbf{v}\cdot\mathbf{e}_k)\,,\end{aligned} \tag{5.134}$$

where we have used that $|\boldsymbol{\eta}_i \cdot \mathbf{e}_k| = 1$. Consider now a parameter c to be chosen later (depending only on ℓ, β, h). Using Corollary A.4.4, we see that if $\alpha \le c^2/L$, with overwhelming probability the quantity (5.134) is at most

$$N\, c\,(\|\mathbf{v}\|^2 + m^*\,\|\mathbf{v}\|)\,,$$

and (5.129) implies

$$\overline{G}(A) \le (La^*)^{M/2}\, W \int_{\{\|\mathbf{v}\| \le 2^{-\ell} m^*\}} \exp N\,\frac{\beta}{2}\, C(\mathbf{v})\, \mathrm{d}\mathbf{v}\,, \tag{5.135}$$

where

$$C(\mathbf{v}) = c\,(\|\mathbf{v}\|^2 + m^* \,\|\mathbf{v}\|) - \frac{m^{*4}\, 2^{-2\ell}}{L} + L\alpha \,.$$

Since $\alpha \le c^2$ and $\|\mathbf{v}\| \le 2^{-\ell}\, m^*$, we get

$$C(\mathbf{v}) \le Lc\,(2^{-\ell}\, m^{*2}) - \frac{m^{*4} 2^{-2\ell}}{L} + Lc^2 \,.$$

We see that it is a good idea to take $c = 2^{-\ell}\, m^{*2}/L$ because then, for $\|\mathbf{v}\| \le 2^{-\ell} m^*$ we have

$$C(\mathbf{v}) \le -\frac{m^{*4}\, 2^{-2\ell}}{L} - \frac{m^{*2}}{L}\,\|\mathbf{v}\|^2 \,,$$

and, since $a^* \le L\, m^{*2}$, (5.135) gives

$$\begin{aligned} \overline{G}(A) &\le (La^*)^{M/2}\, W \exp\left(-\frac{N\beta\, m^{*4}\, 2^{-2\ell}}{L}\right) \int \exp\left(-\frac{N\beta\, m^{*2}}{L}\,\|\mathbf{v}\|^2\right) d\mathbf{v} \\ &\le\; L^{M/2} \exp\left(-\frac{N\beta\, 2^{-2\ell}\, m^{*4}}{L}\right) \le K \exp\left(-\frac{N}{K}\right) \end{aligned} \tag{5.136}$$

provided $\alpha \le 2^{-2\ell}\, m^{*4}/L$.

As in the case of the set (5.130), it follows that the set (5.133) is negligible. Thus, if $\alpha \le m^{*2}\rho_0^2/L$, the set (5.133) is negligible whenever $\ell \ge 0$ and $m^*\, 2^{-\ell} \ge \rho_0$. Combining with Proposition 5.3.7, we have proved Theorem 5.3.2. □

It might be useful here to point out a small technical point. In the last few lines of the previous proof, we use the fact that if k does not depend on N, M, and if $A_1, \ldots, A_k$ are negligible sets, (in one of the meanings of Definition 5.3.1), then $\cup_{\ell \le k} A_\ell$ is negligible. One should however refrain to use this argument when the number of sets is not independent of N and M, because we insist in Definition 5.3.1 that K be independent of M and N.

For reasons that will be apparent in Section 5.9, we need to extend our results to the case where the Hamiltonian is

$$\begin{aligned} -H_{N,M,\beta'}(\boldsymbol{\sigma}) &= \frac{N\beta}{2} \sum_{1 \le k \le M-1} m_k(\boldsymbol{\sigma})^2 \\ &\quad + \frac{N\beta'}{2}\, m_M^2(\boldsymbol{\sigma}) + Nhm_1(\boldsymbol{\sigma}) \,, \end{aligned} \tag{5.137}$$

where $0 \le \beta' \le \beta$. It is intuitively obvious that decreasing the coefficient of m_M can only improve matters, but this still has to be checked. Doing this is unreasonably long and tedious and only the expert should read the rest of this section. A first observation is that, with obvious notation, we have $Z_{N,M-1} \le Z'_{N,M}$ and thus

$$\langle \mathbf{1}_A \rangle' \le \langle \mathbf{1}_A \exp \frac{N\beta'}{2}\, m_M^2 \rangle_\sim \tag{5.138}$$

where $\langle\cdot\rangle_\sim$ denotes an average with respect to Gibbs' measure relative to $H_{N,M-1}$, and $\langle\cdot\rangle'$ denotes an average with respect to Gibbs' measure relative to the Hamiltonian (5.137). Since the randomness in m_N is independent of the randomness in $\langle\cdot\rangle_\sim$, we have

$$\mathsf{E}\,\langle \mathbf{1}_A\rangle' \le \mathsf{E}\left\langle \mathbf{1}_A\,\mathsf{E}\exp\frac{N\beta'}{2}\,m_N^2\right\rangle_\sim .$$

For $\beta' \le 1/2$, we have $\mathsf{E}\exp(N\beta'\,m_N^2/2) \le L$ so that the localization results for the Gibbs' measure relative to $H'_{N,M}$ follow from the localization results for the Gibbs' measure relative to $H_{N,M-1}$. Thus we can assume $\beta' \ge 1/2$. In that case, we attempt to modify the proof of Theorem 5.3.2.

The function ψ of (5.79) has to be replaced by

$$\psi'(\mathbf{z}) = -\frac{N\beta}{2}\sum_{k\le M-1} z_k^2 - \frac{N\beta'}{2}\,z_M^2 + \sum_{i\le N}\log\operatorname{ch}\left(\sum_{k\le M-1}\beta\,\eta_{i,k}\,z_k + \beta'\eta_{i,M}\,z_M + h\right) . \tag{5.139}$$

If we define $T(\mathbf{z})$ by $T(\mathbf{z})_k = z_k$ for $k \le M-1$ and $T(\mathbf{z})_M = \beta'/\beta\, z_M$, then (5.139) is

$$\psi'(\mathbf{z}) = \frac{N\beta}{2}\left(1-\frac{\beta}{\beta'}\right)T^2(\mathbf{z})_M + \psi(T(\mathbf{z})) . \tag{5.140}$$

Thus, if A is a subset of $\mathbb{R}^M$,

$$\begin{aligned}\int_{T(A)}\exp\psi'(\mathbf{z})\,d\mathbf{z} &= \int_{T(A)}\exp\left(\psi(T(\mathbf{z})) + \frac{N\beta}{2}\left(1-\frac{\beta}{\beta'}\right)T(\mathbf{z})_M^2\right)d\mathbf{z}\\ &= \frac{\beta}{\beta'}\int_A \exp\left(\psi(\mathbf{z}) + \frac{N\beta}{2}\left(1-\frac{\beta}{\beta'}\right)z_M^2\right)d\mathbf{z}\\ &\le \frac{\beta}{\beta'}\int_A \exp\psi(\mathbf{z})\,d\mathbf{z} .\end{aligned} \tag{5.141}$$

Since we assume $\beta' \ge 1/2$, the proof of Theorem 5.3.2 shows that if $\alpha \le m^{*2}\rho_0^2/L$, then G' is essentially supported by the union of the sets $T(B_{k,\varepsilon})$, where $B_{k,\varepsilon}$ is the ball centered at $\varepsilon\, m^*\,\mathbf{e}_k$ of radius ρ_0.

It is true, as one should expect, that $T(B_{M,\varepsilon})$ is negligible whenever $\beta' < \beta$, but we will be able to prove this only after considerable work.

Lemma 5.3.15. *If $\beta/\beta' - 1 \ge L\,\rho_0^2$, $\rho_0 \le m^*/2$ and $\alpha \le m^{*2}\rho_0^2/L$, then $T(B_{M,\varepsilon})$ is negligible.*

Corollary 5.3.16. *If ρ_0 is small enough and if $\alpha \le m^{*2}\rho_0^2/L$, then G' is essentially supported by the union of the balls of center $\pm\varepsilon\, m^*\,\mathbf{e}_k$ and radius $2\rho_0$.*

Proof. It suffices to prove that if $T(B_{M,\varepsilon})$ is not negligible, then, it is contained in the ball of center $\varepsilon\, m^* \mathbf{e}_M$ of radius $2\rho_0$. For this it suffices to show that $|m^*(\beta'/\beta - 1)| \le \rho_0$. By Lemma 5.3.15, if $T(B_{M,\varepsilon})$ is not negligible, then $\beta' \ge \beta/(1 + L\rho_0)$, so that

$$\left| m^* \left(\frac{\beta'}{\beta} - 1 \right) \right| = m^* \left(1 - \frac{\beta'}{\beta} \right) \le m^* \left(1 - \frac{1}{1 + L\rho_0^2} \right) \le L m^* \rho_0^2 \le \rho_0$$

if ρ_0 is small enough. □

As a consequence of Corollary 5.3.16, Theorem 5.3.2 remains true for the Hamiltonian (5.137).

Proof of Lemma 5.3.15. We observe that for $\mathbf{z}$ in $B_{M,\varepsilon}$, we have $\mathbf{z}_M^2 \ge m^{*2}/4$ (since $\rho_0 \le m^*/2$), and thus

$$\frac{N\beta}{2} \left(1 - \frac{\beta}{\beta'} \right) z_M^2 \le -\frac{N m^{*2} \beta}{L} \left(\frac{\beta}{\beta'} - 1 \right)$$

and by (5.141)

$$\int_{T(B_{M,\varepsilon})} \exp \psi'(\mathbf{z})\, d\mathbf{z} \le \frac{\beta}{\beta'} \exp \left(-\frac{N m^{*2} \beta}{L} \left(\frac{\beta}{\beta'} - 1 \right) \right) \int_{B_{M,\varepsilon}} \exp \psi(\mathbf{z})\, d\mathbf{z} \,. \tag{5.142}$$

To bound the last term, we repeat, in a simpler manner, some of our previous computations. Using (5.101) we have

$$\psi(\mathbf{z}) \le N b^* + \frac{\beta}{2} \left(\sum_{i \le N} (\boldsymbol{\eta}_i \cdot \mathbf{z})^2 - N \|\mathbf{z}\|^2 \right) .$$

Proceeding as in the case of (5.134), we see that for $\alpha \le m^{*2} \rho_0^2 / L$, then with overwhelming probability we have

$$\mathbf{z} \in B_{M,\varepsilon} \Rightarrow \sum_{i \le N} (\boldsymbol{\eta}_i \cdot \mathbf{z})^2 - N \|\mathbf{z}\|^2 \le m^{*2} \rho_0^2 \,,$$

so that

$$\begin{aligned} W \int_{B_{M,\varepsilon}} \exp \psi(\mathbf{z})\, d\mathbf{z} &\le \exp N \left(b^* + \frac{\beta}{2} m^{*2} \rho_0^2 \right) W \int_{\{\|\mathbf{z}\| \le \rho_0\}} d\mathbf{z} \\ &\le \exp N (b^* + \beta m^{*2} \rho_0^2)\, W \int \exp \left(-\frac{N}{2} \beta m^{*2} \|\mathbf{z}\|^2 \right) d\mathbf{z} \\ &\le \left(\frac{1}{m^{*2}} \right)^{M/2} \exp N (b^* + \beta m^{*2} \rho_0^2) \,. \end{aligned}$$

We now use (5.142), (5.93), and the fact that $Z'_{N,M} \geq Z_{N,M-1}$ to see that with overwhelming probability

$$G(T(B_{M,\varepsilon})) \leq \frac{\beta}{m^* \beta'} L^M \exp N \left(\beta \, m^{*2} \, \rho_0^2 - \frac{\beta m^{*2}}{L} \left(\frac{\beta}{\beta'} - 1 \right) \right)$$

and thus $T(B_{M,\varepsilon})$ is negligible if

$$\frac{\beta}{\beta'} - 1 \geq L \, m^* \, \rho_0^2 , \alpha \leq m^{*2} \, \rho_0^2 / L \, .$$

□

5.4 Localization on Balls with Random Centers

Theorem 5.3.2 is in a sense optimal. But when β is large we can however get more precise results if we allow balls with random centers. These results will be crucial in controlling the model, when β is large, for values of α up to $1/L \log \beta$, but will be used only for this purpose. (Theorem 5.3.2 would allow to reach values of α of up to $1/L\beta$.) The arguments are of the same order of difficulty as those of Section 5.3, and are not related to any subsequent material.

We recall the vector $\boldsymbol{\theta}$ of Proposition 5.3.5.

Theorem 5.4.1. *If $\beta \geq 2$ and if $\alpha \leq 1/L_0 \log \beta$ for L_0 large enough, the set*

$$A = \left\{ \mathbf{z} \, ; \frac{L}{\beta^{10}} \leq \| \mathbf{z} - m^* \, \mathbf{e}_1 - \boldsymbol{\theta} \| \leq \frac{1}{2} \, m^* \right\} \tag{5.143}$$

is negligible for G', that is $\mathsf{E}\, G'(A) \leq K \exp(-N/K)$.

Equivalently, this means that the set

$$\left\{ \boldsymbol{\sigma} \, ; \frac{L}{\beta^{10}} \leq \| \mathbf{m}(\boldsymbol{\sigma}) - m^* \, \mathbf{e}_1 - \boldsymbol{\theta} \| \leq \frac{1}{2} \, m^* \right\}$$

is negligible for G. Of course the exponent 10 plays no special role. In some sense the content of Theorem 5.3.2 is that, given α, we can take ρ_0 as small as $L\sqrt{\alpha}$. Theorem 5.4.1 means that this radius can be reduced to $L\beta^{-10}$ if we accept that $\alpha \leq 1/L \log \beta$, and $m^* \, \mathbf{e}_1 + \boldsymbol{\theta}$ as the center of the ball rather than $m^* \, \mathbf{e}_1$. By the law of large numbers, $\|\boldsymbol{\theta}\|^2$ is typically very close to $m^{*2}\alpha$ so that the radius $L\sqrt{\alpha}$ of Theorem 5.3.2 is of optimal order, since $m^* \geq 1/L$ for $\beta \geq 2$.

The value $L\beta^{-10}$ in (5.143) is not optimal, but rather what we need at this stage. (The tools developed later would allow to compute the true value with extreme accuracy.)

To prove Theorem 5.4.1, we can assume without loss of generality that β is large enough so that $m^* \geq 3/4$. We write $y^+ = \max(y, 0)$, and we consider the function

$$\xi(x) = \frac{\beta^2 x^2}{\mathrm{ch}^2(\beta/4)} + 2\beta\left(|x| - \frac{1}{2}\right)^+ . \tag{5.144}$$

Here is a simple fact.

Lemma 5.4.2. *We have*

$$\xi_1(x) := \log \mathrm{ch}(\beta x + \beta m^* + h) - \log \mathrm{ch}(\beta m^* + h) - \beta x m^* \leq \xi(x) . \tag{5.145}$$

Proof. For $|x| \leq 1/2$, we have $\beta(m^* + x) \geq \beta(3/4 - 1/2) = \beta/4$, so that

$$\xi_1''(x) = \frac{\beta^2}{\mathrm{ch}^2(\beta x + \beta m^* + h)} \leq \frac{\beta^2}{\mathrm{ch}^2(\beta/4)} .$$

Since $\xi_1(0) = \xi_1'(0) = 0$, we have

$$\xi_1(x) \leq \frac{x^2}{2} \frac{\beta^2}{\mathrm{ch}^2(\beta/4)} \leq \xi(x) .$$

Moreover,

$$|\xi_1'(x)| = \beta\,|\mathrm{th}(\beta x + \beta m^* + h) - m^*| \leq 2\beta ,$$

so that for every real numbers x, y we have

$$|\xi_1(x) - \xi_1(y)| \leq 2\beta\,|x - y| .$$

Thus, assuming e.g. $x > 0$, if $|x| \geq 1/2$, we have

$$\xi_1(x) \leq \xi_1\left(\frac{1}{2}\right) + 2\beta\left(x - \frac{1}{2}\right) \leq \xi\left(\frac{1}{2}\right) + 2\beta\left(|x| - \frac{1}{2}\right) \leq \xi(x) .$$

□

The approach to Theorem 5.4.1 starts with the following.

Lemma 5.4.3. *For a subset A of $\mathbb{R}^M$, we have*

$$\overline{G}(A + m^*\mathbf{e}_1 + \boldsymbol{\theta}) \leq W \int_A \exp\left(-\frac{N\beta}{2}\|\mathbf{w}\|^2 + \sum_{i\leq N} \xi(\boldsymbol{\eta}_i \cdot (\mathbf{w} + \boldsymbol{\theta}))\right) \mathrm{d}\mathbf{w} . \tag{5.146}$$

Proof. If $\mathbf{z} = m^*\mathbf{e}_1 + \mathbf{v}$, then using (5.145) we have

$$\begin{aligned}
\psi(\mathbf{z}) &= -\frac{N\beta}{2}\|\mathbf{z}\|^2 + \sum_{i\leq N} \log \mathrm{ch}(\beta(m^* + \boldsymbol{\eta}_i \cdot \mathbf{v}) + h) \\
&\leq -\frac{N\beta}{2}\|m^*\mathbf{e}_1 + \mathbf{v}\|^2 + N \log \mathrm{ch}(\beta m^* + h) \\
&+ \beta m^* \sum_{i\leq N} \boldsymbol{\eta}_i \cdot \mathbf{v} + \sum_{i\leq N} \xi(\boldsymbol{\eta}_i \cdot \mathbf{v}) .
\end{aligned}$$

Since $\boldsymbol{\theta} = (m^*/N)\sum_{i\le N}(\boldsymbol{\eta}_i - \mathbf{e}_1)$, we get

$$\begin{aligned}\psi(\mathbf{z}) &\le Nb^* - \frac{N\beta}{2}\|\mathbf{v}\|^2 + N\beta\,\boldsymbol{\theta}\cdot\mathbf{v} + \sum_{i\le N}\xi(\boldsymbol{\eta}_i\cdot\mathbf{v})\\ &\le Nb^* + \frac{N\beta}{2}\|\boldsymbol{\theta}\|^2 - \frac{N\beta}{2}\|\mathbf{v}-\boldsymbol{\theta}\|^2 + \sum_{i\le N}\xi(\boldsymbol{\eta}_i\cdot\mathbf{v})\,.\end{aligned}$$

Setting $\mathbf{w} = \boldsymbol{\theta} - \mathbf{v}$, the result follows from (5.80) and (5.92). □

To use (5.146), we note that by convexity of ξ we have

$$\xi(x+y) \le \frac{1}{2}\left(\xi(2x) + \xi(2y)\right)\,,$$

so that

$$\xi(\boldsymbol{\eta}_i\cdot(\mathbf{w}+\boldsymbol{\theta})) \le \frac{1}{2}(\xi(2\boldsymbol{\eta}_i\cdot\boldsymbol{\theta}) + \xi(2\boldsymbol{\eta}_i\cdot\mathbf{w}))\,. \tag{5.147}$$

We can think of ξ as being made of two parts. The part $\beta^2x^2/\mathrm{ch}^2(\beta/4)$ is not dangerous because the coefficient of x^2 is very small for large β. Controlling the part $\beta(|x|-1/2)^+$, on the other hand, will require a real effort.

Lemma 5.4.4. *If* $t>0$ *and* $16t\,\|\mathbf{w}\|^2 \le 1$, *then*

$$\mathsf{E}\exp t\sum_{i\le N}\left(|\boldsymbol{\eta}_i\cdot\mathbf{w}| - \frac{1}{4}\right)^+ \le \exp\left(2N\exp\left(-\frac{1}{128\|\mathbf{w}\|^2}\right)\right)\,. \tag{5.148}$$

Proof. We observe that

$$\exp t\,y^+ \le 1 + \mathbf{1}_{\{y>0\}}\exp t\,y$$

so that

$$\begin{aligned}\mathsf{E}\exp t\left(|\boldsymbol{\eta}_i\cdot\mathbf{w}| - \frac{1}{4}\right)^+ &\le 1 + \mathsf{E}\left(\mathbf{1}_{\{|\boldsymbol{\eta}_i\cdot\mathbf{w}|\ge\frac{1}{4}\}}\exp t\,|\boldsymbol{\eta}_i\cdot\mathbf{w}|\right) \qquad (5.149)\\ &\le 1 + \mathsf{P}\left(|\boldsymbol{\eta}_i\cdot\mathbf{w}| \ge \frac{1}{4}\right)^{1/2}\left(\mathsf{E}\exp 2t\,|\boldsymbol{\eta}_i\cdot\mathbf{w}|\right)^{1/2},\end{aligned}$$

from the Cauchy-Schwarz inequality. Now, by the sub-Gaussian inequality (A.3) we have

$$\mathsf{E}\exp 2t\,\boldsymbol{\eta}_i\cdot\mathbf{w} \le \exp 2t^2\|\mathbf{w}\|^2 \tag{5.150}$$

so that

$$\mathsf{E}\exp 2t\,|\boldsymbol{\eta}_i\cdot\mathbf{w}| \le 2\exp 2t^2\|\mathbf{w}\|^2\,. \tag{5.151}$$

By Chebyshev's inequality, for $u>0$ we have

$$\mathsf{P}\left(|\boldsymbol{\eta}_i\cdot\mathbf{w}| \ge \frac{1}{4}\right) \le 2\exp\left(2u^2\|\mathbf{w}\|^2 - \frac{u}{2}\right)\,.$$

Taking $u = 1/(8\,\|\mathbf{w}\|^2)$, we get

$$\mathsf{P}\left(|\boldsymbol{\eta}_i \cdot \mathbf{w}| \geq \frac{1}{4}\right) \leq 2\exp\left(-\frac{1}{32\|\mathbf{w}\|^2}\right) ,$$

and combining with (5.150) and (5.149) gives

$$\begin{aligned}\mathsf{E}\exp t\left(|\boldsymbol{\eta}_i \cdot \mathbf{v}| - \frac{1}{4}\right)^+ &\leq 1 + 2\exp\left(t^2\,\|\mathbf{w}\|^2 - \frac{1}{64\|\mathbf{w}\|^2}\right)\\ &\leq 1 + 2\exp\left(-\frac{1}{128\|\mathbf{w}\|^2}\right)\end{aligned}$$

since $16t\,\|\mathbf{w}\|^2 \leq 1$. Thus

$$\mathsf{E}\exp t\left(|\boldsymbol{\eta}_i \cdot \mathbf{v}| - \frac{1}{4}\right)^+ \leq \exp\left(2\exp\left(-\frac{1}{128\|\mathbf{w}\|^2}\right)\right) ,$$

and (5.148) follows by independence. □

Lemma 5.4.5. *Given $a > 0$, with overwhelming probability, we have*

$$\sup_{\|\mathbf{w}\| \leq a} \sum_{i \leq N} \left(|\boldsymbol{\eta}_i \cdot \mathbf{w}| - \frac{1}{4}\right)^+ \leq LNa^2\left(\alpha + \exp\left(-\frac{1}{La^2}\right)\right) . \qquad (5.152)$$

Proof. It is proved in Proposition A.3.1 that we can find a subset A of $\mathbb{R}^M$, of cardinality at most 5^M, consisting of vectors of length $\leq 2a$, and whose convex hull contains every vector of length at most a. Then, since $\sum_{i \leq N}(|\boldsymbol{\eta}_i \cdot \mathbf{w}| - 1/4)^+$ is a convex function of $\mathbf{w}$, we have

$$\sup_{\|\mathbf{w}\| \leq a} \sum_{i \leq N} \left(|\boldsymbol{\eta}_i \cdot \mathbf{w}| - \frac{1}{4}\right)^+ \leq \sup_{\mathbf{w} \in A} \sum_{i \leq N} \left(|\boldsymbol{\eta}_i \cdot \mathbf{w}| - \frac{1}{4}\right)^+ .$$

Now, for $\mathbf{w} \in A$, we can use (5.148) for $t = 1/64a^2$, and by Chebyshev's inequality we get, for $y > 0$

$$\mathsf{P}\left(\sum_{i \leq N} \left(|\boldsymbol{\eta}_i \cdot \mathbf{w}| - \frac{1}{4}\right)^+ \geq Ny\right) \leq \exp N\left(-\frac{y}{64\,a^2} + 2\exp\left(-\frac{1}{La^2}\right)\right) .$$

Thus

$$\begin{aligned}&\mathsf{P}\left(\sup_{\mathbf{w} \in A} \sum_{i \leq N} \left(|\boldsymbol{\eta}_i \cdot \mathbf{w}| - \frac{1}{4}\right)^+ \geq Ny\right)\\ &\leq 5^M \exp N\left(-\frac{y}{64\,a^2} + 2\exp\left(-\frac{1}{La^2}\right)\right) . \qquad (5.153)\end{aligned}$$

Since $5^M = \exp N\alpha \log 5$, this proves the result if we take

$$y = 64\,a^2\left(2\,\alpha\log 5 + 4\exp\left(-\frac{1}{La^2}\right)\right)\,.$$

□

Lemma 5.4.6. *Consider* $0 < c \le 10^{-10}$. *Then, with overwhelming probability, we have*

$$\alpha \le c \Rightarrow \sum_{i\le N}\left(|\boldsymbol{\eta}_i\cdot\boldsymbol{\theta}| - \frac{1}{4}\right)^+ \le LN\exp\left(-\frac{1}{Lc}\right)\,. \tag{5.154}$$

Since $\boldsymbol{\theta}$ is of length of order $\sqrt{\alpha}$, this should be compared to (5.152) for $a = \sqrt{c}$. In (5.154), we do not have the term $LNa^2\alpha$. This term was created by the need to control all vectors $\mathbf{w}$ in (5.152); in (5.154) we have only $\boldsymbol{\theta}$ to control; unfortunately, this vector is random. Still one feels that one would be extremely unlucky for $\boldsymbol{\theta}$ to be one of the few vectors $\mathbf{w}$ for which $\sum_{i\le N}(|\boldsymbol{\eta}_i\cdot\mathbf{w}| - 1/4)^+$ might be large. Yet we see no simple proof of this fact.

Before going to the main argument, we state a simple fact.

Lemma 5.4.7. *Consider numbers* $(\rho_i)_{i\le N}$. *Then*

$$\mathsf{P}\left(\sum_{k\le M}\left(\sum_{i\le N}\eta_{i,k}\,\rho_i\right)^2 \ge t\right) \le 2^{M/2}\exp\left(-\frac{t}{4\sum_{i\le N}\rho_i^2}\right)\,. \tag{5.155}$$

Proof. We use (A.5) to see that

$$\mathsf{E}\exp\frac{\left(\sum_{i\le N}\eta_{i,k}\,\rho_i\right)^2}{4\sum_{i\le N}\rho_i^2} \le \sqrt{2}$$

so that, by independence,

$$\mathsf{E}\exp\frac{\sum_{k\le M}\left(\sum_{i\le N}\eta_{i,k}\,\rho_i\right)^2}{4\sum_{i\le N}\rho_i^2} \le 2^{M/2}\,,$$

and the result by Chebyshev's inequality. □

Proof of Lemma 5.4.6. The proof of this rather technical point is almost absurdly hard. It should be skipped by all except the experts.

Consider a subset I of $\{1,\ldots,N\}$, and

$$\boldsymbol{\eta}_I = \sum_{i\notin I}(\boldsymbol{\eta}_i - \mathbf{e}_1) \tag{5.156}$$

so that

$$\|\boldsymbol{\eta}_I\|^2 = \sum_{k\geq 2}\left(\sum_{i\notin I}\eta_{i,k}\right)^2 ,$$

and, by (5.155), (used for $\rho_i = 1$ if $i \notin I$, $\rho_i = 0$ otherwise)

$$\mathsf{P}(\Omega_I^c) \leq 2^{-cN} , \tag{5.157}$$

where

$$\Omega_I = \{\|\boldsymbol{\eta}_I\|^2 \leq 8\,c\,N^2\} .$$

We claim that

$$\mathsf{E}\left(\mathbf{1}_{\Omega_I}\exp\sum_{i\in I}\left(\frac{1}{N}|\boldsymbol{\eta}_i\cdot\boldsymbol{\eta}_I| - \frac{1}{32}\right)^+\right) \leq \exp\left(2N\exp\left(-\frac{1}{Lc}\right)\right) . \tag{5.158}$$

To see this, we observe that the random vectors $(\boldsymbol{\eta}_i)_{i\in I}$ are independent of the vector $\boldsymbol{\eta}_I$. Thus, by taking first in (5.158) expectation in the r.v. $(\eta_{i,k})$, $i \in I$, it suffices to prove that, setting $\mathbf{w} = \boldsymbol{\eta}_I/N$

$$\begin{aligned}\|\mathbf{w}\|^2 \leq 8\,c \Rightarrow \mathsf{E}&\left(\exp\sum_{i\in I}\left(|\boldsymbol{\eta}_i\cdot\mathbf{w}| - \frac{1}{32}\right)^+\right)\\ &\leq \exp\left(2N\exp\left(-\frac{1}{Lc}\right)\right) .\end{aligned} \tag{5.159}$$

Now

$$\left(|\boldsymbol{\eta}_i\cdot\mathbf{w}| - \frac{1}{32}\right)^+ = \frac{1}{8}\left(|\boldsymbol{\eta}_i\cdot 8\mathbf{w}| - \frac{1}{4}\right)^+$$

so that (5.159) follows from (5.148) with $t = 1/8$, and $8\mathbf{w}$ rather than $\mathbf{w}$. The condition $16\,t\,\|8\mathbf{w}\|^2 \leq 1$ is satisfied for $\|\mathbf{w}\|^2 \leq 8c$ if $c \leq 10^{-10}$.

We have proved (5.158). Since for $X \geq 0$ we have

$$\mathsf{E}\exp\mathbf{1}_{\Omega_I}X \leq 1 + \mathsf{E}\left(\mathbf{1}_{\Omega_I}\exp X\right) ,$$

we have

$$\mathsf{E}\exp\left(\mathbf{1}_{\Omega_I}\sum_{i\in I}\left(\frac{1}{N}|\boldsymbol{\eta}_i\cdot\boldsymbol{\eta}_I| - \frac{1}{32}\right)^+\right) \leq 2\exp\left(2N\exp\left(-\frac{1}{Lc}\right)\right) . \tag{5.160}$$

Let us denote by Av average over all subsets I of $\{1,\ldots,N\}$. By Jensen's inequality and (5.160), we have

$$\mathsf{E}\exp\left(\mathrm{Av}\,\mathbf{1}_{\Omega_I}\sum_{i\in I}\left(\frac{1}{N}|\boldsymbol{\eta}_i\cdot\boldsymbol{\eta}_I| - \frac{1}{32}\right)^+\right) \leq 2\exp\left(2N\exp\left(-\frac{1}{Lc}\right)\right) , \tag{5.161}$$

so that, with overwhelming probability

$$\mathrm{Av}\,\mathbf{1}_{\Omega_I}\sum_{i\in I}\left(\frac{1}{N}|\boldsymbol{\eta}_i\cdot\boldsymbol{\eta}_I|-\frac{1}{32}\right)^+\le 3N\exp\left(-\frac{1}{Lc}\right)\,. \tag{5.162}$$

Now

$$\mathrm{Av}\sum_{i\in I}\left(\frac{1}{N}|\boldsymbol{\eta}_i\cdot\boldsymbol{\eta}_I|-\frac{1}{32}\right)^+=\mathrm{Av}\,\mathbf{1}_{\Omega_I}\sum_{i\in I}\left(\frac{1}{N}|\boldsymbol{\eta}_i\cdot\boldsymbol{\eta}_I|-\frac{1}{32}\right)^++\mathcal{R} \tag{5.163}$$

where

$$\mathcal{R}=\mathrm{Av}\,\mathbf{1}_{\Omega_I^c}\sum_{i\in I}\left(\frac{1}{N}|\boldsymbol{\eta}_i\cdot\boldsymbol{\eta}_I|-\frac{1}{32}\right)^+\,.$$

Since $\|\boldsymbol{\eta}_i\|\le\sqrt{M}$ and $\|\boldsymbol{\eta}_I\|\le N\sqrt{M}$, we have $\mathsf{E}\,\mathcal{R}\le MN\,2^{-cN}$ by (5.157). Thus, with overwhelming probability we have $\mathcal{R}\le N\exp(-1/Lc)$ and combining (5.162) with (5.163) we see that with overwhelming probability we have

$$\mathrm{Av}\sum_{i\in I}\left(\frac{1}{N}|\boldsymbol{\eta}_i\cdot\boldsymbol{\eta}_I|-\frac{1}{32}\right)^+\le 4N\exp\left(-\frac{1}{Lc}\right)\,. \tag{5.164}$$

Now, given $i\le N$ we have

$$\mathrm{Av}\,(\mathbf{1}_{\{i\in I\}}\,\boldsymbol{\eta}_i\cdot\boldsymbol{\eta}_I)=\mathrm{Av}\left(\mathbf{1}_{\{i\in I\}}\,\boldsymbol{\eta}_i\cdot\sum_{j\notin I}(\boldsymbol{\eta}_j-\mathbf{e}_1)\right)=\frac{1}{4}\boldsymbol{\eta}_i\cdot\left(\sum_{j\ne i}(\boldsymbol{\eta}_j-\mathbf{e}_1)\right)$$

because if $j\ne i$, there are exactly 2^{N-2} choices of I for which $i\in I$ and $j\notin I$. Since the function $x\to(|x|-1)^+$ is convex, by Jensen's inequality we have

$$\begin{aligned}\left(\frac{1}{N}\left|\frac{1}{4}\boldsymbol{\eta}_i\cdot\left(\sum_{j\ne i}(\boldsymbol{\eta}_j-\mathbf{e}_1)\right)\right|-\frac{1}{32}\right)^+&\le\mathrm{Av}\left(\frac{1}{N}\left|\mathbf{1}_{\{i\in I\}}\,\boldsymbol{\eta}_i\cdot\boldsymbol{\eta}_I\right|-\frac{1}{32}\right)^+\\&=\mathrm{Av}\left(\mathbf{1}_{\{i\in I\}}\left(\frac{1}{N}|\boldsymbol{\eta}_i\cdot\boldsymbol{\eta}_I|-\frac{1}{32}\right)^+\right)\end{aligned}$$

and summation over i gives

$$\sum_{i\le N}\frac{1}{4}\left(\frac{1}{N}\left|\boldsymbol{\eta}_i\cdot\left(\sum_{j\ne i}(\boldsymbol{\eta}_j-\mathbf{e}_1)\right)\right|-\frac{1}{8}\right)^+\le\mathrm{Av}\sum_{i\in I}\left(\frac{1}{N}|\boldsymbol{\eta}_i\cdot\boldsymbol{\eta}_I|-\frac{1}{32}\right)^+,$$

so that, by (5.164), with overwhelming probability, we have

$$\sum_{i\le N}\left(\frac{1}{N}\left|\boldsymbol{\eta}_i\cdot\left(\sum_{j\ne i}(\boldsymbol{\eta}_j-\mathbf{e}_1)\right)\right|-\frac{1}{8}\right)^+\le 16N\exp\left(-\frac{1}{Lc}\right).$$

To finish the proof, we show that if

$$a_i=\frac{1}{N}\left|\boldsymbol{\eta}_i\cdot\left(\sum_{j\ne i}(\boldsymbol{\eta}_j-\mathbf{e}_1)\right)\right|\,,\quad b_i=\frac{1}{N}\left|\boldsymbol{\eta}_i\cdot\boldsymbol{\theta}\right|$$

we have

$$\left(b_i-\frac{1}{4}\right)^+\le 2\left(a_i-\frac{1}{8}\right)^+.$$

To see this, we note that if $\alpha\le 1/16$ we have

$$|a_i-b_i|\le\frac{1}{N}\left|\boldsymbol{\eta}_i\cdot(\boldsymbol{\eta}_i-\mathbf{e}_1)\right|=\frac{M-1}{N}\le\frac{1}{16}$$

so that, if $\left(b_i-\frac{1}{4}\right)^+>0$, we have $b_i\ge 1/4$, and hence $a_i\ge 3/16$, so that

$$\left(b_i-\frac{1}{4}\right)^+\le\left(a_i-\frac{1}{4}\right)^++\frac{1}{16}\le\left(a_i-\frac{1}{4}\right)^++\frac{a_i}{3}\le 2\left(a_i-\frac{1}{8}\right)^+$$

because $a_i/3\le a_i-1/8$. □

We are finally ready to take a first step towards Theorem 5.4.1.

Proposition 5.4.8. *If $\beta\ge\beta_0$ and $\alpha\le c:=1/L_0\log\beta$, where β_0 and L_0 are large enough, the set*

$$\left\{\mathbf{z}\,;\ \frac{1}{4\sqrt{\beta}}\le\|\mathbf{z}-m^*\mathbf{e}_1-\boldsymbol{\theta}\|\le\frac{1}{L}\right\}\tag{5.165}$$

is negligible for $\overline{G}$.

Proof. Consider $a>0$ and the set

$$A=\{\mathbf{w}\,;\ a\le\|\mathbf{w}\|\le 2a\}\,.$$

We will show that for suitable values of a the set $A+m^*\mathbf{e}_1+\boldsymbol{\theta}$ is negligible. Using (5.147), Lemmas 5.4.5 and 5.4.6, and the fact that $(|2x|-1/2)^+=2(|x|-1/4)^+$, we have shown that with overwhelming probability, if $0<c<10^{-10}$ and $\alpha\le c$, for each $\mathbf{w}$ in A, we have

$$\sum_{i\le N}\xi(\boldsymbol{\eta}_i\cdot(\mathbf{w}+\boldsymbol{\theta}))\le\frac{4\beta^2}{\mathrm{ch}^2(\beta/4)}\left(\sum_{i\le N}((\boldsymbol{\eta}_i\cdot\mathbf{w})^2+(\boldsymbol{\eta}_i\cdot\boldsymbol{\theta})^2)\right)\tag{5.166}$$
$$+L\beta N\left(a^2\left(\alpha+\exp\left(-\frac{1}{La^2}\right)\right)+\exp\left(-\frac{1}{Lc}\right)\right).$$

Corollary A.4.4 shows that with overwhelming probability, for every $\mathbf{w}$ in $\mathbb{R}^M$ we have $\sum_{i\le N}(\boldsymbol{\eta}_i\cdot\mathbf{w})^2 \le LN\|\mathbf{w}\|^2$. In particular, if $\mathbf{w}\in A$ we have

$$\sum_{i\le N}(\boldsymbol{\eta}_i\cdot\mathbf{w})^2 \le LNa^2 .$$

We also have

$$\sum_{i\le N}(\boldsymbol{\eta}_i\cdot\boldsymbol{\theta})^2 \le LN\|\boldsymbol{\theta}\|^2 .$$

With overwhelming probability (using (5.155)) we have $\|\boldsymbol{\theta}\|^2 \le Lc$ if $\alpha\le c$, and thus $\sum_{i\le N}(\boldsymbol{\eta}_i\cdot\boldsymbol{\theta})^2 \le LNc$. Thus, with overwhelming probability, for each $\mathbf{w}$ in A we have

$$\sum_{i\le N}\xi(\boldsymbol{\eta}_i\cdot(\mathbf{w}+\boldsymbol{\theta})) \le LN\delta$$

where

$$\delta = \frac{(a^2+c)\beta^2}{\mathrm{ch}^2(\beta/4)} + \beta\left(a^2\left(c+\exp\left(-\frac{1}{La^2}\right)\right)+\exp\left(-\frac{1}{Lc}\right)\right) .$$

From (5.146) we get that, with overwhelming probability, we have

$$\overline{G}(A+m^*\mathbf{e}_1+\boldsymbol{\theta}) \le \exp(LN\delta)W\int_A \exp\left(-\frac{N\beta}{2}\|\mathbf{w}\|^2\right)d\mathbf{w}$$

and since $\|\mathbf{z}\|\ge a$ on A,

$$\begin{aligned}W\int_A \exp\left(-\frac{\beta N}{2}\|\mathbf{w}\|^2\right)d\mathbf{w} &\le W\exp\left(-\frac{N\beta a^2}{4}\right)\int\exp\left(-\frac{N\beta}{4}\|\mathbf{w}\|^2\right)d\mathbf{w}\\ &= 2^{M/2}\exp\left(-\frac{N\beta a^2}{4}\right) .\end{aligned}$$

Thus with overwhelming probability we have

$$\begin{aligned}\overline{G}(A+m^*\mathbf{e}_1+\boldsymbol{\theta}) &\le \exp\left(N\left(L\alpha+L\delta-\frac{\beta a^2}{4}\right)\right)\\ &\le \exp N\left(L\delta-\frac{\beta a^2}{8}\right)\end{aligned} \tag{5.167}$$

if we assume $a^2\beta\ge 1/64$ and $\alpha\le c\le 1/L$.

If we denote by L_1 the constant in the last term of (5.167), we see that $A+m^*\mathbf{e}_1+\boldsymbol{\theta}$ is negligible provided $\delta\le\beta a^2/16L_1$, or, equivalently, provided that

$$\frac{a^2+c}{\mathrm{ch}^2(\beta/4)}\beta + a^2\left(c+\exp\left(-\frac{1}{La^2}\right)\right)+L\exp\left(-\frac{1}{Lc}\right) \le \frac{a^2}{16L_1} . \tag{5.168}$$

Since we assumed $a^2 \geq 1/64\beta$, we see that this is satisfied if $\beta \geq \beta_0$, $c \leq 1/L_0 \log \beta$ and $a \leq a_0$, where β_0, L_0 are large enough, a_0 is small enough. To prove (5.165) we then use that the set $A + m^* \mathbf{e}_1 + \boldsymbol{\theta}$ is negligible where $a = 2^{-\ell}$, $2^{-\ell} \leq a_0$, $2^{-\ell} \geq 1/8\sqrt{\beta}$. □

Here is a second step toward Theorem 5.4.1.

Proposition 5.4.9. *If $\beta \geq \beta_0$, $\alpha \leq 1/L_0 \log \beta$, where β_0 and L_0 are large enough the set*

$$\left\{ \mathbf{z} \,;\, \|\mathbf{z} - m^* \mathbf{e}_1 - \boldsymbol{\theta}\| \leq \frac{1}{L}, z_1 \leq m^* - \frac{1}{\beta^{20}} \right\} \tag{5.169}$$

is negligible for $\overline{G}$.

There is of course nothing special about the power β^{20} that could be replaced by a larger power.

Proof. Recalling that the first component θ_1 of $\boldsymbol{\theta}$ is zero, it suffices to prove from Proposition 5.4.8 that if

$$A = \left\{ \mathbf{w};\, \|\mathbf{w}\| \leq \frac{1}{4\sqrt{\beta}}, w_1 \leq -\beta^{-20} \right\}$$

then $A + m^* \mathbf{e}_1 + \boldsymbol{\theta}$ is negligible for $\overline{G}$. We proceed as in Proposition 5.4.8, but without using Lemma 5.4.5, to get that with overwhelming probability

$$\overline{G}(A + m^* \mathbf{e}_1 + \boldsymbol{\theta}) \leq \exp\left(N \left(\frac{L\beta^2(\beta^{-1} + c)}{\operatorname{ch}^2(\beta/4)} + \log\left(-\frac{1}{Lc} \right) \right) \right) U \,, \tag{5.170}$$

where

$$\begin{aligned} U &= W \int_A \exp\left(-\frac{N\beta}{2} \|\mathbf{w}\|^2 \right) \exp\left(\frac{\beta}{2} \sum_{i \leq N} \left(|2\,\boldsymbol{\eta}_i \cdot \mathbf{w}| - \frac{1}{4} \right)^+ \right) d\mathbf{w} \\ &= \int_A \exp\left(\frac{\beta}{2} \sum_{i \leq N} \left(2|\boldsymbol{\eta}_i \cdot \mathbf{w}| - \frac{1}{4} \right)^+ \right) d\gamma(\mathbf{w}) \,. \end{aligned}$$

For $\mathbf{w}$ in A we have $\|\mathbf{w}\|^2 \leq 1/16\beta$. Thus we can use (5.148) with $t = \beta/2$ to see that

$$\mathsf{E} U \leq \gamma(A) \exp\left(2N \exp\left(-\frac{\beta}{L} \right) \right) \,,$$

and (5.170) shows that with overwhelming probability we have

$$G(A + m^* \mathbf{e}_1 + \boldsymbol{\theta}) \leq \gamma(A) \exp N\delta \,, \tag{5.171}$$

where now

$$\delta = 3\exp\left(-\frac{\beta}{L}\right) + L\exp\left(-\frac{1}{Lc}\right) + \frac{L\beta^2(\beta^{-1}+c)}{\mathrm{ch}^2(\beta/4)} .$$

If g is standard Gaussian, we have

$$\begin{aligned}\gamma(A) &\le \mathsf{P}\left(\sqrt{\frac{1}{\beta N}}\, g \le -\beta^{-20}\right) = \mathsf{P}(g \le -\sqrt{N}\,\beta^{-39/2}) \\ &\le \exp\left(-\frac{N\beta^{-39}}{2}\right) . \end{aligned} \tag{5.172}$$

The result follows from the fact that $\delta \le \beta^{-39}/4$ if $c = 1/L_0 \log\beta$ for L_0 large enough, $\beta \ge \beta_0$. □

Proposition 5.4.10. *Under the conditions of Proposition 5.4.9, the set*

$$A = \left\{\mathbf{z}\,;\, \|\mathbf{z} - m^*\mathbf{e}_1 - \boldsymbol{\theta}\| \le \frac{1}{L}, z_1 \le m^* - \frac{2}{\beta^{20}}\right\} \tag{5.173}$$

is negligible for G', that is $\mathsf{E}G'(A) \le K\exp(-N/K)$.

Proof. This is the same statement as Proposition 5.4.9, except that we replace $\overline{G}$ by G' (and β^{-20} by $2\beta^{-20}$).

In general, replacing G' by $\overline{G}$ allows one to study where G' lives "within distance of about $\sqrt{\alpha/\beta}$". The reason why we can do much better here is that "the boundary of A is almost 1-dimensional". Consider the positive measure ν on $\mathbb{R}^M$ defined by $\nu(D) = G'(D \cap A)$. Consider $\overline{\nu} = \nu * \gamma$. Consider the set B defined in (5.77), so that $\gamma(B) \ge 1/L$, and thus

$$\overline{\nu}(A+B) \ge \nu(A)\,\gamma(B) \ge \frac{1}{L}\,\nu(A) = \frac{1}{L}\,G'(A) ,$$

and hence

$$\begin{aligned} G'(A) \le L\overline{\nu}(A+B) &\le L\overline{\nu}\left((A+B) \cap \left\{\mathbf{z}; z_1 \le -m^* \frac{1}{\beta^{20}}\right\}\right) \\ &\quad + L\overline{\nu}\left(\left\{\mathbf{z}; z_1 \ge m^* - \frac{1}{\beta^{20}}\right\}\right) . \end{aligned} \tag{5.174}$$

Now

$$A + B \subset \left\{\mathbf{z};\, \|\mathbf{z} - m^*\mathbf{e}_1 - \boldsymbol{\theta}\| \le 2\sqrt{\frac{\alpha}{\beta}} + \frac{1}{L}\right\} ,$$

so that if the constant L in (5.173) is large enough, and since $\overline{\nu} \le \overline{G}$, we get from Proposition 5.4.9 that

$$\mathsf{E}G'(A) \le K\exp\left(-\frac{N}{K}\right) + L\,\mathsf{E}\overline{\nu}\left(\left\{\mathbf{x}; x_1 \ge -\frac{1}{\beta^{20}}\right\}\right) . \tag{5.175}$$

Now,

$$\overline{\nu}\left(\left\{\mathbf{x}; x_1 \geq m^* - \frac{1}{\beta^{20}}\right\}\right) = \nu \otimes \gamma\left(\left\{(\mathbf{z}, \mathbf{y}); z_1 + y_1 \geq m^* - \frac{1}{\beta^{20}}\right\}\right)$$
$$= G' \otimes \gamma\left(\left\{(\mathbf{z}, \mathbf{y}); z_1 + y_1 \geq m^* - \frac{1}{\beta^{20}}, \mathbf{z} \in A\right\}\right).$$

If $\mathbf{z} \in A$, then $z_1 \leq m^* - 2/\beta^{20}$, so that if $z_1 + y_1 \geq m^* - 1/\beta^{20}$, we must have $y_1 \geq 1/\beta^{20}$. Thus

$$\overline{\nu}\left(\left\{\mathbf{w}; w_1 \geq -\frac{1}{\beta^{20}}\right\}\right) \leq \gamma\left(\left\{\mathbf{y}; y_1 \geq \frac{1}{\beta^{20}}\right\}\right) \leq \exp\left(-\frac{N\beta^{-39}}{2}\right),$$

where the last inequality is proved as in (5.172). Combining with (5.175) finishes the proof. □

Proof of Theorem 5.4.1. We consider the vector

$$\mathbf{c} = \left(\frac{1}{N}\sum_{i \leq N} \eta_{i,k}\right)_{k \leq M}$$

so that $m^*\mathbf{c} = m^*\mathbf{e}_1 + \boldsymbol{\theta}$. We have

$$\|\mathbf{m}(\boldsymbol{\sigma}) - m^*\mathbf{c}\| \leq |m^* - m_1(\boldsymbol{\sigma})|\|\mathbf{c}\| + \|\mathbf{m}(\boldsymbol{\sigma}) - m_1(\boldsymbol{\sigma})\mathbf{c}\|, \tag{5.176}$$

where $m_1(\boldsymbol{\sigma}) = N^{-1}\sum_{i \leq N} \sigma_i$. For $k \geq 2$, the kth component of $\mathbf{m}(\boldsymbol{\sigma}) - m_1(\boldsymbol{\sigma})\mathbf{c}$ is

$$\frac{1}{N}\sum_{i \leq N} \eta_{i,k}(\sigma_i - m_1(\boldsymbol{\sigma})) = \frac{1}{N}\sum_{i \leq N} \eta_{i,k}\,\rho_i \tag{5.177}$$

where $\rho_i = \rho_i(\boldsymbol{\sigma}) = \sigma_i - m_1(\boldsymbol{\sigma})$. We note that

$$\sum_{i \leq N} \rho_i^2 = \sum_{i \leq N} (1 - 2\,\sigma_i\, m_1(\boldsymbol{\sigma}) + m_1(\boldsymbol{\sigma})^2) = N(1 - m_1^2(\boldsymbol{\sigma}))\,. \tag{5.178}$$

Assume now that $m_1(\boldsymbol{\sigma}) \geq m^* - 2\beta^{-20}$. Since $m^* = m^*(\beta, h) \geq m^*(\beta, h) = \text{th}(\beta\, m^*(2, 0))$, we have $m^* \geq 1 - L\beta^{-20}$, so that $m_1(\boldsymbol{\sigma}) \geq 1 - L\beta^{-20}$ and, by (5.178), we have

$$\sum_{i \leq N} \rho_i^2 \leq NL\beta^{-20},$$

so that by (5.155) we have

$$\mathsf{P}\left(\sum_{k \leq M}\left(\sum_{i \leq N} \eta_{i,k}\,\rho_i(\boldsymbol{\sigma})\right)^2 \geq t\right) \leq 2^{M/2}\exp\left(-\frac{t\beta^{20}}{LN}\right).$$

Thus, if $U = \{\boldsymbol{\sigma}; m_1(\boldsymbol{\sigma}) \geq m^* - 2\beta^{-20}\}$, we have

$$\mathsf{P}\left(\sup_{\boldsymbol{\sigma}\in U} \sum_{k\leq M}\left(\sum_{i\leq N} \eta_{i,k}\,\rho_i(\boldsymbol{\sigma})\right)^2 \geq t\right) \leq 2^{N+M/2}\exp\left(-\frac{t\beta^{20}}{LN}\right)$$

and thus with overwhelming probability we have

$$\begin{aligned} m_1(\boldsymbol{\sigma}) \geq m^* - 2\beta^{-20} &\Rightarrow \sum_{k\leq M}\left(\sum_{i\leq N}\eta_{i,k}\,\rho_i(\boldsymbol{\sigma})\right)^2 \leq \frac{LN^2}{\beta^{20}} \\ &\Rightarrow \|\mathbf{m}(\boldsymbol{\sigma}) - m_1(\boldsymbol{\sigma})\mathbf{c}\| \leq \frac{L}{\beta^{10}}\,. \end{aligned} \tag{5.179}$$

By (5.155) we have $\|\mathbf{c}\|^2 \leq L$ with overwhelming probability. Moreover $m^* - m_1(\boldsymbol{\sigma}) \geq m^* - 1$ and since $m_1(\boldsymbol{\sigma}) \geq m^* - 2\beta^{-20}$, we have

$$|m^* - m_1(\boldsymbol{\sigma})| \leq \max(1 - m^*, 2\beta^{-20}) \leq \frac{L}{\beta^{20}}$$

and (5.35), (5.176), (5.179) show that, with overwhelming probability,

$$m_1(\boldsymbol{\sigma}) \geq m^* - 2\beta^{-20} \Rightarrow \|\mathbf{m}(\boldsymbol{\sigma}) - m^*\mathbf{e}_1 - \boldsymbol{\theta}\| \leq \frac{L}{\beta^{10}}\,,$$

so that, with overwhelming probability, if A is as in (5.143), we have $G'(A) \leq G'(A_1)$ where

$$A_1 = \left\{\mathbf{z}\,;\|\mathbf{z} - m^*\mathbf{e}_1 - \boldsymbol{\theta}\| \leq \frac{m^*}{2}, z_1 \leq m^* - \frac{2}{\beta^{20}}\right\}\,.$$

Now, by Proposition 5.4.10, if $\beta \geq \beta_0$, we have

$$G'(A_1) \leq G'(A_2) + K\exp\left(-\frac{N}{K}\right)\,,$$

where

$$A_2 = \left\{\mathbf{z}\,; \frac{1}{L} \leq \|\mathbf{z} - m^*\mathbf{e}_1 - \boldsymbol{\theta}\| \leq \frac{m^*}{2}\right\}\,.$$

In the admissible region we have $\alpha \leq c := 1/(L_0 \log 2)$. Thus, with overwhelming probability we have

$$\|\boldsymbol{\theta}\| \leq 2m^*\sqrt{c}\,, \tag{5.180}$$

and hence

$$A_2 \subset A_3 = \left\{\mathbf{z}\,;\ \frac{1}{L} - L\sqrt{c} \leq \|\mathbf{z} - m^*\mathbf{e}_1\| \leq \frac{m^*}{2} + L\sqrt{c}\right\}\,.$$

On the other hand, by Theorem 5.3.2 (and since $\beta \geq \beta_0$), G' is essentially supported by the set

$$A_4 = \bigcup_{k \leq M} \{\mathbf{z}\,;\ \|\mathbf{z} \pm m^* \mathbf{e}_k\| \leq L\sqrt{c}\}\,.$$

If L_0 is large enough, we have $A_3 \cap A_4 = \emptyset$, so that A_3 is negligible.

As for the case $\beta \leq \beta_0$, there is nothing to prove if the constant L in (5.143) is large enough. □

Theorem 5.4.1 remains true for the Hamiltonian (5.137). The small changes needed to see this are better left to the reader.

5.5 Selecting a State with an External Field

A consequence of Theorem 5.3.2 is as follows. Consider, for $k \leq M$, the set

$$C_k = \left\{\boldsymbol{\sigma};\, \|\mathbf{m}(\boldsymbol{\sigma}) \pm m^* \mathbf{e}_k\| \leq \frac{m^*}{4}\right\}\,. \tag{5.181}$$

Then, if $\alpha \leq m^{*4}/L$, Gibbs' measure is essentially supported by the union of the sets C_k, as $k \leq M$.

Conjecture 5.5.1. (Level 2) Assume that $h = 0$. If M is large, prove that for the typical disorder there is a $k \leq M$ such that $G(C_k)$ is nearly 1.

The paper [36] contains some results relevant to this conjecture.

Research problem 5.5.2. (Level 2) More generally, understand precisely the properties of the random sequence $(G(C_k))_{k \leq M}$, when $h = 0$.

The goal of this section is the following

Theorem 5.5.3. *If $\beta > 1$, $h > 0$, then for $\alpha \leq m^{*4}/L$, the set*

$$A = \left\{\boldsymbol{\sigma};\, \|\mathbf{m}(\boldsymbol{\sigma}) - m^* \mathbf{e}_1\| \geq \frac{m^*}{4}\right\} \tag{5.182}$$

is negligible.

This means of course that $\mathsf{E}G(A) \leq K \exp(-N/K)$ where K depends only on β, h.

The content of Theorem 5.5.3 is as follows. We know from Theorem 5.3.2 that Gibbs' measure is essentially supported by the union of $2M$ balls centered at $\pm m^* \mathbf{e}_k$ with a certain radius. When $h \neq 0$, only the ball centered at $m^* \mathbf{e}_1$ matters.

Corollary 5.5.4. *If* (α, β, h) *belongs to the admissible region, if* $h > 0$, *and if the constant* L_0 *of the admissible region is large enough, then the set*

$$A = \left\{ \boldsymbol{\sigma}; \|\mathbf{m}(\boldsymbol{\sigma}) - m^* \mathbf{e}_1 - \boldsymbol{\theta}\| \geq \frac{L}{\beta^{10}} \right\} \tag{5.183}$$

is negligible.

Proof. Assume first $\beta \geq 2$. By (5.143) the set

$$A_1 = \left\{ \boldsymbol{\sigma}; \frac{L}{\beta^{10}} \leq \|\mathbf{m}(\boldsymbol{\sigma}) - m^* \mathbf{e}_1 - \boldsymbol{\theta}\| \leq \frac{m^*}{2} \right\}$$

is negligible, and by (5.182) the set

$$A_2 = \left\{ \boldsymbol{\sigma}; \|\mathbf{m}(\boldsymbol{\sigma}) - m^* \mathbf{e}_1\| \geq \frac{m^*}{4} \right\}$$

is negligible. Using (5.180) we see that with overwhelming probability we have $\|\boldsymbol{\theta}\| \leq 2m^* \sqrt{c} \leq m^*/4$ if L_0 is large enough, and thus

$$\|\mathbf{m}(\boldsymbol{\sigma}) - m^* \mathbf{e}_1 - \boldsymbol{\theta}\| > \frac{m^*}{2} \Rightarrow \|\mathbf{m}(\boldsymbol{\sigma}) - m^* \mathbf{e}_1\| \geq \frac{m^*}{4} ,$$

so that $A \subset A_1 \cup A_2$, and hence A is negligible. If $\beta \leq 2$ and if the constant L of (5.183) is large enough, then $A \subset \{\boldsymbol{\sigma}; \|\mathbf{m}(\boldsymbol{\sigma})\| \geq 2m^*\}$ is negligible by (5.182). □

We consider the sets

$$B_{k,\varepsilon} = \left\{ \boldsymbol{\sigma}; \|\mathbf{m}(\boldsymbol{\sigma}) - \varepsilon\, m^* \mathbf{e}_k\| \leq \frac{m^*}{4} \right\} \tag{5.184}$$

$$C_{k,\varepsilon} = \left\{ \boldsymbol{\sigma}; \varepsilon\, m_k(\boldsymbol{\sigma}) \geq \frac{3m^*}{4} \right\} . \tag{5.185}$$

Let us denote by $H^0_{N,M}(\boldsymbol{\sigma})$ the Hamiltonian (5.16), which corresponds to the case $h = 0$. We set

$$S(k, \varepsilon) = \sum \exp(-H^0_{N,M}(\boldsymbol{\sigma})) , \tag{5.186}$$

where the summation is over $\boldsymbol{\sigma}$ in $C_{k,\varepsilon}$. The crucial property is as follows.

Lemma 5.5.5. *There exists a number* a *such that for each* $k \leq M$, *each* $\varepsilon = \pm 1$, *we have*

$$0 \leq u \leq 1 \Rightarrow \mathsf{P}\left(\left| \frac{1}{N} \log S(k, \varepsilon) - a \right| \geq u \right) \leq K \exp\left(-\frac{Nu^2}{K} \right) . \tag{5.187}$$

It suffices to prove this for $k = \varepsilon = 1$, because the r.v. $S(k,\varepsilon)$ all have the same distribution. Thus (5.187) resembles Corollary 2.2.5. It relies on a principle that is somewhat similar to Theorem 2.2.4. This principle, that has received numerous applications, is explained in great detail in Section 6 of [133], and Lemma 5.5.5 is proved exactly as Theorem 6.8 there. The author believes that learning properly this principle is well worth the effort, and that this is better done by reading [133] than by reading only the proof of Lemma 5.5.5. Thus, Lemma 5.5.5 will be one of our few exceptions to our policy of giving complete self-contained proofs.

Proof of Theorem 5.5.3. We have

$$G(C_{1,1}) = \frac{1}{Z_{N,M}} \sum_{\sigma \in C_{1,1}} \exp(-H_{N,M}(\boldsymbol{\sigma})) \,.$$

For $\boldsymbol{\sigma}$ in $C_{1,1}$, we have $h\,m_1(\boldsymbol{\sigma}) \geq 3\,h\,m^*/4$, so that $-H_{N,M}(\boldsymbol{\sigma}) \geq -H^0_{N,M}(\boldsymbol{\sigma}) +3Nh\,m^*/4$, and

$$1 \geq G(C_{1,1}) \geq \exp\left(\frac{3}{4}\,Nh\,m^*\right) \frac{S(1,1)}{Z_{N,M}} \,. \tag{5.188}$$

If $(k,\varepsilon) \neq (1,1)$, we have $h\,m_1(\boldsymbol{\sigma}) \leq h\,m^*/4$ for $\boldsymbol{\sigma}$ in $B_{k,\varepsilon}$ so that

$$\begin{aligned} \sum_{\sigma \in B_{k,\varepsilon}} \exp(-H_{N,M}(\boldsymbol{\sigma})) &\leq \exp\left(\frac{1}{4}\,Nh\,m^*\right) \sum_{\sigma \in B_{k,\varepsilon}} \exp(-H^0_{N,M}(\boldsymbol{\sigma})) \\ &\leq \exp\left(\frac{1}{4}\,Nh\,m^*\right) S(k,\varepsilon) \end{aligned}$$

and thus

$$G(B_{k,\varepsilon}) \leq \exp\left(\frac{1}{4}\,Nh\,m^*\right) \frac{S(k,\varepsilon)}{Z_{N,M}} \,. \tag{5.189}$$

Taking $u = \min(1, h\,m^*/8)$ in Lemma 5.5.5 we see that with overwhelming probability we have

$$S(1,1) \geq \exp\left(Na - \frac{1}{8}\,Nh\,m^*\right)$$

$$\forall\, k,\varepsilon, \quad S(k,\varepsilon) \leq \exp\left(Na + \frac{1}{8}\,Nh\,m^*\right)$$

and thus

$$\forall\, k,\varepsilon,\; S(k,\varepsilon) \leq \exp\left(\frac{1}{4}\,Nh\,m^*\right) S(1,1) \,.$$

Combining with (5.188) and (5.189) we see that, with overwhelming probability we have

$$(k, \varepsilon) \neq (1,1) \Rightarrow G(B_{k,\varepsilon}) \leq \exp\left(-\frac{1}{4} N h\, m^*\right)$$

so that $B_{k,\varepsilon}$ is negligible. Combining with Theorem 5.3.2 finishes the proof. □

In the rest of this section we prove that Theorem 5.5.3 remains true for the Hamiltonian (5.137). The proof of this intuitively obvious fact is complicated and tedious, and is (again) reserved for the expert.

We set

$$R(k, \varepsilon) = \sum \exp(-H^0_{N,M}(\boldsymbol{\sigma}))$$

where the summation is over $\boldsymbol{\sigma}$ in $B(k, \varepsilon)$.

Lemma 5.5.6. *There exists a number a such that, given $u > 0$, there exists $K(u)$ for which, whenever $k \leq M$ and $\varepsilon = \pm 1$ we have*

$$\mathsf{P}\left(\left|\frac{1}{N} \log R(k, \varepsilon) - a\right| \geq u\right) \leq K(u) \exp\left(-\frac{N}{K(u)}\right) . \tag{5.190}$$

Proof. The number a will be the same as in (5.187).

We denote by Z_0 the partition function of the Hamiltonian (5.16), that is $Z_0 = \sum \exp(-H^0_{N,M}(\boldsymbol{\sigma}))$.

Let

$$U = \sum \exp(-H^0_{N,M}(\boldsymbol{\sigma}))$$

where the summation is over $\boldsymbol{\sigma} \notin \cup B_{k,\varepsilon}$, where the union is over $k \leq M$ and $\varepsilon = \pm 1$. It follows from Theorem 5.3.2 that for some number K_0 we have

$$\mathsf{P}\left(\frac{U}{Z_0} \geq \exp\left(-\frac{N}{K_0}\right)\right) \leq K_0 \exp\left(-\frac{N}{K_0}\right) . \tag{5.191}$$

From (5.187) we see that if $0 < u < 1$, then

$$\forall\, k, \varepsilon \quad \mathsf{P}(S(k, \varepsilon) \leq \exp N(a + u)) \geq 1 - K \exp\left(-\frac{N u^2}{K}\right) .$$

Now

$$Z_0 \leq \sum_{k,\varepsilon} S(k, \varepsilon) + U$$

so that, by (5.191)

$$\mathsf{P}\left(Z_0 \leq 2N \exp N(a + u) + Z_0 \exp\left(-\frac{N}{K}\right)\right)$$
$$\geq 1 - KN \exp\left(-\frac{N u^2}{K}\right) - K \exp\left(-\frac{N}{K}\right) . \tag{5.192}$$

For large N we have

$$Z_0 \le 2N \exp N(a+u) + Z_0 \exp\left(-\frac{N}{K}\right) \Rightarrow Z_0 \le 4N \exp N(a+u) .$$

We have

$$\begin{aligned}
&\mathsf{P}\left(U \ge 4N \exp N\left(a - \frac{1}{2K_0}\right)\right) \\
&\le \mathsf{P}(Z_0 \ge 4\,N \exp N\left(a + \frac{1}{2K_0}\right) + \mathsf{P}\left(\frac{U}{Z_0} \ge \exp\left(-\frac{N}{K_0}\right)\right) \\
&\le K \exp\left(-\frac{N}{K}\right) ,
\end{aligned} \tag{5.193}$$

where we use (5.191) and (5.192) in the last line, with $u = 1/2K_0$. Now

$$R(k,\varepsilon) \le S(k,\varepsilon) \le R(k,\varepsilon) + U$$

and by (5.187)

$$\mathsf{P}\left(S(k,\varepsilon) \le \exp N\left(a - \frac{u}{2}\right)\right) \le K \exp\left(-\frac{Nu^2}{K}\right) .$$

If $u < 1/2K_0$, for N large, we have

$$\exp N\left(a - \frac{u}{2}\right) \ge 4N \exp N\left(a - \frac{1}{2K_0}\right) + \exp N(a-u)$$

and thus, for N large,

$$\begin{aligned}
&\mathsf{P}(R(k,\varepsilon) \le \exp N(a-u)) \\
&\le \mathsf{P}\left(S(k,\varepsilon) \le \exp N\left(a - \frac{u}{2}\right)\right) + \mathsf{P}\left(U \ge 4\,N \exp N\left(a - \frac{1}{2K_0}\right)\right) \\
&\le K \exp\left(-\frac{Nu^2}{K}\right) .
\end{aligned}$$

The other half of (5.190), namely the inequality

$$\mathsf{P}(R(k,\varepsilon) \ge \exp N(a+u)) \le K \exp\left(-\frac{Nu^2}{K}\right)$$

follows from (5.187). □

Corollary 5.5.7. *For the Hamiltonian (5.137) if either $h > 0$ or $\beta' < \beta$, then $B_{M,\varepsilon}$ is negligible.*

Proof. The basic fact is that

$$\boldsymbol{\sigma} \in B_{1,1} \Rightarrow |m_M(\boldsymbol{\sigma})| \le \frac{m^*}{4}$$

$$\boldsymbol{\sigma} \in B_{M,\varepsilon} \Rightarrow |m_M(\boldsymbol{\sigma})| \geq \frac{3m^*}{4} .$$

If G denotes Gibbs' measure with Hamiltonian (5.137), we have, as in (5.188) that

$$1 \geq G(B_{1,1}) \geq \exp\left(\frac{3}{4}Nhm^* - \left(\frac{m^*}{4}\right)^2 (\beta - \beta')\right) \frac{R(1,1)}{Z_{N,M}}$$

while

$$G(B_{M,\varepsilon}) \leq \exp\left(\frac{1}{4}Nhm^* - \left(\frac{3m^*}{4}\right)^2 (\beta - \beta')\right) \frac{R(M,\varepsilon)}{Z_{N,M}}$$

and one then proceeds as in the proof of Theorem 5.5.3. □

To prove that Theorem 5.5.3 remains true for the Hamiltonian (5.137), one also has to show that if $h > 0$ the sets $B_{k,\varepsilon}$, $k \leq M-1$, $(k,\varepsilon) \neq (1,1)$ are negligible. This requires another argument, which is a direct adaptation of the proof of Theorem 5.5.3, establishing (5.187) for $k \leq M-1$, where in (5.186) $H^0_{N,M}$ now denotes the Hamiltonian (5.137) for $h = 0$.

From this point on, dealing with the Hamiltonian (5.137) requires only trivial adaptations of our main arguments. These will not be carried out.

5.6 Controlling $m_k(\boldsymbol{\sigma})$, $k \geq 2$

In this section we prove that if (α, β, h) belongs to the admissible region and if $h > 0$ then for $k \geq 2$, $m_k(\boldsymbol{\sigma})$ is typically small, about $1/\sqrt{N}$. This technical fact will be essential in Section 5.8, in controlling the error terms created by the "approximate integration by parts" of Section 3.8.

Theorem 5.6.1. *If (α, β, h) belongs to the admissible region and $h > 0$ (and if L_0 is large enough) then there exists a constant K depending only on β, h such that, for all $k \geq 2$, we have*

$$\mathsf{E}\left\langle \exp \frac{N m_k^2}{K} \right\rangle \leq K . \tag{5.194}$$

In particular for each p we have $\mathsf{E}\langle m_k^{2p}\rangle \leq KN^{-p}$, where K now depends also on p. To prove Theorem 5.6.1 we can assume by symmetry that $k = M$. The proof uses a "cavity upon M" argument. We denote by $\langle\cdot\rangle_\sim$ Gibbs' measure relative to the Hamiltonian

$$-H_{N,M-1}(\boldsymbol{\sigma}) = \frac{\beta}{2} \sum_{k \leq M-1} N\, m_k^2(\boldsymbol{\sigma}) + N\, h\, m_1(\boldsymbol{\sigma}) . \tag{5.195}$$

It should be obvious that for a function f on Σ_N we have

$$\langle f \rangle = \frac{\left\langle f \exp \frac{N\beta}{2} m_M^2 \right\rangle_\sim}{\left\langle \exp \frac{N\beta}{2} m_M^2 \right\rangle_\sim} . \tag{5.196}$$

To prove Theorem 5.6.1, we will consider the set

$$U = \{\boldsymbol{\sigma}; |m_1(\boldsymbol{\sigma}) - m^*| \leq \rho\} , \tag{5.197}$$

where ρ will be suitably chosen. The main step of the proof is as follows.

Proposition 5.6.2. *Assume that*

$$\beta(1 - m^{*2} + 2\rho m^*) < 1 . \tag{5.198}$$

Then we can find $\beta_1 > \beta$ *and* $s > 0$ *such that*

$$\mathsf{E}\left(\left\langle \mathbf{1}_U \exp \frac{N\beta_1}{2} m_M^2(\boldsymbol{\sigma}) \right\rangle_\sim^s\right) \leq K . \tag{5.199}$$

Proof. Let us write η_i rather than $\eta_{i,M}$. We have

$$m_M(\boldsymbol{\sigma}) = \frac{1}{N}\left(\sum_{i\leq N} \eta_i m^* + \sum_{i\leq N} \eta_i(\sigma_i - m^*)\right) . \tag{5.200}$$

Given $t > 0$, we have

$$(a+b)^2 \leq (1+t)\, a^2 + \left(1 + \frac{1}{t}\right) b^2 ,$$

and applying this to (5.200) we get

$$N m_M(\boldsymbol{\sigma})^2 \leq \frac{1+t}{N}\left(\sum_{i\leq N} \eta_i m^*\right)^2 + \frac{1+t}{Nt}\left(\sum_{i\leq N} \eta_i(\sigma_i - m^*)\right)^2 \tag{5.201}$$

and thus

$$\begin{aligned} &\left\langle \mathbf{1}_U \exp \frac{N\beta_1}{2} m_M^2(\boldsymbol{\sigma}) \right\rangle_\sim^s \\ &\leq \exp\left(s(1+t)\,\frac{\beta_1}{2N}\, m^{*2}\left(\sum_{i\leq N} \eta_i\right)^2\right) \\ &\times \left\langle \mathbf{1}_U \exp\left(\frac{(1+t)\beta_1}{2Nt}\left(\sum_{i\leq N} \eta_i(\sigma_i - m^*)\right)^2\right)\right\rangle_\sim^s . \end{aligned} \tag{5.202}$$

Thus, by the Cauchy-Schwarz inequality applied to (5.202) we have

$$\begin{aligned}&\left(\mathsf{E}\left\langle \mathbf{1}_U \exp \frac{N\beta_1}{2}\, m_M^2(\boldsymbol{\sigma})\right\rangle_{\sim}^{s}\right)^2\\ &\leq \mathsf{E}\exp\left(s(1+t)\,\frac{\beta_1}{N}\, m^{*2}\left(\sum_{i\leq N}\eta_i\right)^2\right)\\ &\times \mathsf{E}\left\langle \mathbf{1}_U \exp\left(\frac{(1+t)\beta_1}{2Nt}\left(\sum_{i\leq N}\eta_i(\sigma_i - m^*)\right)^2\right)\right\rangle_{\sim}^{2s} . \qquad (5.203)\end{aligned}$$

There is no loss of generality to assume that $2s \leq 1$. Thus the last term of (5.203) is at most

$$\left(\mathsf{E}\left\langle \mathbf{1}_U \exp\left(\frac{(1+t)\beta_1}{2Nt}\left(\sum_{i\leq N}\eta_i(\sigma_i - m^*)\right)^2\right)\right\rangle_{\sim}\right)^{2s} . \qquad (5.204)$$

Now,

$$\begin{aligned}\sum_{i\leq N}(\sigma_i - m^*)^2 &= N(1 - 2\,m_1(\boldsymbol{\sigma})\,m^* + m^{*2})\\ &= N(1 - m^{*2} + 2(m^* - m_1(\boldsymbol{\sigma}))\,m^*)\\ &\leq N(1 - m^{*2} + 2\rho\, m^*)\end{aligned}$$

if $\boldsymbol{\sigma} \in U$. Using (5.198), let us choose $\beta_1 > \beta$ and t large enough that

$$\frac{1+t}{t}\,\beta_1(1 - m^{*2} + 2\rho\, m^*) < 1 . \qquad (5.205)$$

Let us denote by E_0 integration in the r.v. $(\eta_i)_{i\leq N}$ alone. Then (5.205) and (A.5) show that for $\boldsymbol{\sigma}$ in U, we have

$$\mathsf{E}_0 \exp\left(\frac{(1+t)\beta_1}{2Nt}\left(\sum_{i\leq N}\eta_i(\sigma_i - m^*)\right)^2\right) \leq K ,$$

where K is independent of N. Since the disorder in $\langle\cdot\rangle_{\sim}$ is independent of the r.v. η_i, when taking expectation in (5.204), we can first take expectation E_0 inside the bracket $\langle\cdot\rangle_{\sim}$. Then the quantity (5.204) is bounded independently of N. To finish the proof, we choose s small enough that

$$s(1+t)\,\beta_1\, m^{*2} < \frac{1}{2} .$$

Then the first term on the right-hand side of (5.203) is bounded independently of N, again by (A.5). □

Lemma 5.6.3. *If the constant L_0 in the definition of the admissible region has been chosen large enough, at each point (α, β, h), $h > 0$, of the admissible region we can find ρ such that*

$$\beta(1 - m^{*2} + 2\rho m^*) < 1 \tag{5.206}$$

and K depending only on β, h such that

$$\mathsf{E}\, G(\{\boldsymbol{\sigma}; |m_1(\boldsymbol{\sigma}) - m^*| > \rho\}) \leq K \exp\left(-\frac{N}{K}\right) . \tag{5.207}$$

Proof. We take

$$\rho = \frac{1 - \beta(1 - m^{*2})}{4\beta m^*}$$

and this ensures (5.206) because $1 - \beta(1 - m^{*2}) > 0$ and hence $2\rho\beta m^* < 1 - \beta(1 - m^{*2})$. Using Lemma 5.2.2, we have

$$\rho \geq \rho' := \frac{m^*}{L\beta} .$$

Since the first component of $\boldsymbol{\theta}$ is zero, we have

$$\|\mathbf{m}(\boldsymbol{\sigma}) - m^* \mathbf{e}_1 - \boldsymbol{\theta}\| \geq |m_1(\boldsymbol{\sigma}) - m^*|$$

and

$$\{\boldsymbol{\sigma}; |m_1(\boldsymbol{\sigma}) - m^*| \geq \rho\} \subset \{\boldsymbol{\sigma}; \|\mathbf{m}(\boldsymbol{\sigma}) - m^* \mathbf{e}_1 - \boldsymbol{\theta}\| \geq \rho'\} . \tag{5.208}$$

Given $L_1 > 0$, there exists β_0 such that $\rho' \geq L_1/\beta^2$ for $\beta \geq \beta_0$, and Corollary 5.5.4, Theorems 5.5.3, 5.4.1 and 5.3.2 show that then the sets (5.207) are negligible. For $\beta \leq \beta_0$, we have $\rho' \geq \rho_0 = m^*/L_2$, and then

$$\{\boldsymbol{\sigma}; |m_1(\boldsymbol{\sigma}) - m^*| \geq \rho_0\} \subset \{\boldsymbol{\sigma}; \|\mathbf{m}(\boldsymbol{\sigma}) - m^* \mathbf{e}_1\| \geq \rho_0\} . \tag{5.209}$$

Theorems 5.5.3 and 5.3.2 show that these sets are negligible if the constant L_0 is large enough. □

Proof of Theorem 5.6.1. If U is given by (5.197), where ρ is as in (5.206), for any number A we have

$$\mathsf{E}\left\langle \exp \frac{N}{A} m_M^2(\boldsymbol{\sigma}) \right\rangle = \mathrm{I} + \mathrm{II}$$

where

$$\mathrm{I} = \mathsf{E}\left\langle \mathbf{1}_{U^c} \exp \frac{N}{A}\, m_M^2(\boldsymbol{\sigma}) \right\rangle$$

$$\mathrm{II} = \mathsf{E}\left\langle \mathbf{1}_{U} \exp \frac{N}{A}\, m_M^2(\boldsymbol{\sigma}) \right\rangle .$$

Now $|m_M| \le 1$, so that

$$\mathrm{I} \le \exp \frac{N}{A}\, \mathsf{E}\langle \mathbf{1}_{U^c} \rangle \le K \exp N \left(\frac{1}{A} - \frac{1}{K} \right) \tag{5.210}$$

by (5.207).

Given a number $u > 1$, by Hölder's inequality we have

$$\mathrm{II} \le \mathsf{E}\left(\left\langle \mathbf{1}_{U} \exp \frac{Nu}{A}\, m_M^2(\boldsymbol{\sigma}) \right\rangle^{1/u} \right) .$$

By (5.196) we have

$$\left\langle \mathbf{1}_{U} \exp \frac{Nu}{A}\, m_M^2(\boldsymbol{\sigma}) \right\rangle \le \left\langle \mathbf{1}_{U} \exp N \left(\frac{u}{A} + \beta \right) m_M^2(\boldsymbol{\sigma}) \right\rangle_{\sim} . \tag{5.211}$$

Using (5.206), Proposition 5.6.2 provides $s > 0$ and $\beta_1 > \beta$ such that (5.199) holds. We take $u = 1/s$, and then A large enough that $\beta + u/A \le \beta_1$ and $A \ge 2K$, where K occurs in (5.210). From (5.211) and (5.199) we get $\mathrm{II} \le K$ and we get $\mathrm{I} \le K$ by (5.210). The result follows. □

5.7 The Smart Path

Given $\boldsymbol{\sigma} = (\sigma_1, \ldots, \sigma_N)$ in Σ_N, we write $\boldsymbol{\rho} = (\sigma_1, \ldots, \sigma_{N-1}) \in \Sigma_{N-1}$. We write

$$n_k(\boldsymbol{\rho}) = \frac{1}{N} \sum_{i \le N-1} \eta_{i,k}\, \sigma_i \tag{5.212}$$

and we define

$$-H_{N-1,M}(\boldsymbol{\rho}) = \frac{N\beta}{2} \sum_{1 \le k \le M} n_k^2(\boldsymbol{\rho}) + Nh\, n_1(\boldsymbol{\rho}) . \tag{5.213}$$

For simplicity, we will write η_k rather than $\eta_{N,k}$. Thus

$$m_k(\boldsymbol{\sigma}) = n_k(\boldsymbol{\rho}) + \frac{\eta_k\, \sigma_N}{N} \tag{5.214}$$

and, by simple algebra

$$-H_{N,M}(\boldsymbol{\sigma}) = -H_{N-1,M}(\boldsymbol{\rho}) + \beta \sigma_N \sum_{1 \le k \le M} \eta_k\, n_k(\boldsymbol{\rho}) + \sigma_N h , \tag{5.215}$$

ignoring the constant $\beta M/(2N)$ that does not change Gibbs' measure. We assume throughout this section that (α, β, h) belongs to the admissible region and that $h > 0$. We consider the solution (q, μ, r) of (5.25) to (5.27). We consider a parameter $0 \le t \le 1$ and the Hamiltonian

$$-H_{N,M,t}(\boldsymbol{\sigma}) = -H_{N-1,M}(\boldsymbol{\rho}) + \beta\, \sigma_N \biggl(\sqrt{t} \sum_{2 \le k \le M} \eta_k\, n_k(\boldsymbol{\rho}) + \sqrt{1-t}\, z \sqrt{r} + t\, n_1(\boldsymbol{\rho}) + (1-t)\mu \biggr) + h\, \sigma_N \,. \tag{5.216}$$

Thus, if

$$Y = \beta\, z \sqrt{r} + \beta \mu h \tag{5.217}$$

then

$$-H_{N,M,1}(\boldsymbol{\sigma}) = -H_{N,M}(\boldsymbol{\sigma}) \tag{5.218}$$

$$-H_{N,M,0}(\boldsymbol{\sigma}) = -H_{N-1,M}(\boldsymbol{\rho}) + \sigma_N Y \,. \tag{5.219}$$

We observe that $H_{N,M,0}$ depends on σ_N only through $\sigma_N Y$. This is necessary to obtain the (familiar) Lemma 5.7.1 below; this requires replacing $n_1(\boldsymbol{\rho})$ by μ, the new feature of the smart path method here. We denote by $\langle \cdot \rangle_t$ the average with respect to Gibbs' measure (and its products) relative to the Hamiltonian (5.216) and we write $\nu_t(f) = \mathsf{E}\, \langle f \rangle_t$. We recall that as always $\varepsilon_\ell = \sigma_N^\ell$.

Lemma 5.7.1. *Given a function f^- on Σ_{N-1}^n and a subset I of $\{1, \dots, n\}$, we have*

$$\nu_0 \Bigl(f^- \prod_{\ell \in I} \varepsilon_\ell \Bigr) = \nu_0 \Bigl(\prod_{\ell \in I} \varepsilon_\ell \Bigr) \nu_0(f^-) = \mathsf{E}(\operatorname{th} Y)^{\operatorname{card} I} \nu_0(f^-) \,. \tag{5.220}$$

Proof. See Lemma 2.4.4. □

An essential tool will be the computation of $\nu_t'(f)$. This is made slightly more cumbersome than usual by the need to use approximate integration by parts.

To lighten notation we write

$$n_k^\ell = n_k(\boldsymbol{\rho}^\ell)\,;\ S_{\ell,\ell'} = \sum_{2 \le k \le M} n_k^\ell\, n_k^{\ell'} \,. \tag{5.221}$$

Proposition 5.7.2. *Consider a random function f on Σ_N^n. This function f can depend on the r.v. $(\eta_{i,k})_{i \le N-1, k \le M}$, but must not depend on the r.v. $\eta_k = \eta_{N,k}$. Then, if (α, β, h) belongs to the admissible region, $h > 0$, we have*

$$\nu_t'(f) = \mathrm{I} + \mathrm{II} + \mathrm{III} + \mathcal{R} \tag{5.222}$$

where

$$\mathrm{I} = \beta \left(\sum_{\ell \le n} \nu_t(f\, \varepsilon_\ell (n_1^\ell - \mu)) - n\nu_t(f\, \varepsilon_{n+1}(n_1^{n+1} - \mu)) \right) \tag{5.223}$$

$$\mathrm{II} = \frac{\beta^2}{2} \left(\sum_{\ell \le n} \nu_t(f\, S_{\ell,\ell}) - n\nu_t(f\, S_{n+1,n+1}) \right) \tag{5.224}$$

$$\begin{aligned}\mathrm{III} = \beta^2 \Big(&\sum_{1 \le \ell < \ell' \le n} \nu_t(f(S_{\ell,\ell'} - r)\, \varepsilon_\ell\, \varepsilon_{\ell'}) - n \sum_{\ell \le n} \nu_t(f(S_{\ell,n+1} - r)\, \varepsilon_\ell\, \varepsilon_{n+1}) \\ &+ \frac{n(n+1)}{2} \nu_t(f(S_{n+1,n+2} - r)\, \varepsilon_{n+1}\, \varepsilon_{n+2}) \Big) \end{aligned} \tag{5.225}$$

and

$$|\mathcal{R}| \le K \sum_{2 \le k \le M} \sum_{\ell \le n+1} \nu_t(|n_k^\ell|^4\, |f|) \tag{5.226}$$

where K depends on β, n, h only.

This resembles Proposition 3.2.3 when $u(x) = \beta x^2/2$. The term I here is a new feature, and so is the error term $\mathcal{R}$. One main difference with Proposition 3.2.3 is that we have *not* used the symmetry between the terms $k = 2, \dots, M$ in the sum $\sum_{2 \le k \le M} n_k^\ell n_k^{\ell'}$ to replace $S_{\ell,\ell'}$ by $n_M^\ell n_M^{\ell'}$. The reason for this is that we (somewhat) control $S_{\ell,\ell'}$ through the results of Sections 5.3 and 5.4.

Lemma 5.7.3. *If f is a random function on Σ_N^n, as in Proposition 5.7.2, we have $\nu_t'(f) = \mathrm{I} + \mathrm{IV} + \mathrm{V}$, where I is as in (5.223),*

$$\begin{aligned}\mathrm{IV} = \frac{\beta}{2\sqrt{t}} \Bigg(&\sum_{\ell \le n} \nu_t \Bigg(f \sum_{2 \le k \le M} \eta_k\, \varepsilon_\ell\, n_k^\ell \Bigg) \\ &- n\, \nu_t \Bigg(f \sum_{2 \le k \le M} \eta_k\, \varepsilon_{n+1}\, n_k^{n+1} \Bigg) \Bigg) \end{aligned} \tag{5.227}$$

$$\mathrm{V} = -\frac{\beta\sqrt{r}}{2\sqrt{1-t}} \left(\sum_{\ell \le n} \nu_t(f\, z\, \varepsilon_\ell) - n\, \nu_t(f\, z\, \varepsilon_{n+1}) \right) . \tag{5.228}$$

Proof. This follows from (3.27). □

Proof of Proposition 5.7.2. The proof will follow the same scheme as that of Proposition 3.8.3. We will use integration by parts to compute the term (5.228) and approximate integration by parts as in (3.211) to compute the term (5.227). The computation of the term (5.228) is straightforward:

$$\mathrm{V} = -\beta^2 r\Bigg(\sum_{\ell<\ell'} \nu_t(\varepsilon_\ell\,\varepsilon_{\ell'}\,f) - n\sum_{\ell\le n}\nu_t(\varepsilon_\ell\,\varepsilon_{n+1}\,f)$$
$$+\frac{n(n+1)}{2}\,\nu_t(\varepsilon_{n+1}\,\varepsilon_{n+2}\,f)\Bigg)\,. \tag{5.229}$$

To compute IV we write, for $\ell \le n+1$:

$$\nu_t(f\,\eta_k\,\varepsilon_\ell\,n_k^\ell) = \mathsf{E}\,(\eta_k\,\langle f\,\varepsilon_\ell\,n_k^\ell\rangle_t)\,.$$

To compute the right-hand side, we will fix k, and we will use (3.211) at all the r.v. except η_k being fixed. We consider the function $v(x) = \langle f\,\varepsilon_\ell\,n_k^\ell\rangle_{t,x}$, where the bracket $\langle\cdot\rangle_{t,x}$ means that all the occurences of η_k in the expression of $\langle f\,\varepsilon_\ell\,n_k^\ell\rangle_t$ have been replaced by x. Thus we have, when $\ell \le n$,

$$v'(x) = \beta\sqrt{t}\sum_{\ell'\le n}\langle\varepsilon_\ell\,\varepsilon_{\ell'}\,n_k^\ell\,n_k^{\ell'}\,f\rangle_{t,x} - \beta\sqrt{t}\,n\,\langle\varepsilon_\ell\,\varepsilon_{n+1}\,n_k^\ell\,n_k^{n+1}\,f\rangle_{t,x}$$

and a similar formula when $\ell = n+1$. If we regroup the contributions for $2\le k\le M$ of the "main terms" arising from (3.211), we see that it is

$$\frac{\beta^2}{2}\Bigg(\sum_{\ell\le n}\nu_t(S_{\ell,\ell}\,f) - n\,\nu_t(S_{n+1,n+1}\,f)\Bigg)$$
$$+\beta^2\Bigg(\sum_{\ell<\ell'}\nu_t(\varepsilon_\ell\,\varepsilon_{\ell'}\,S_{\ell,\ell'}\,f) - n\sum_{\ell\le n}\nu_t(\varepsilon_\ell\,\varepsilon_{n+1}\,S_{\ell,n+1}\,f)$$
$$+\frac{n(n+1)}{2}\,\nu_t(\varepsilon_{n+1}\,\varepsilon_{n+2}\,S_{n+1,n+2}\,f)\Bigg)\,.$$

In other words, to prove Proposition 5.7.2 it is enough to control the error terms arising from (3.211) and (3.212). It should be obvious, using the inequality $\prod_{\ell\le 4} a_\ell \le \sum_{\ell\le 4} a_\ell^4$ that

$$|v^{(3)}(x)| \le K(n,\beta)\sum_{\ell\le n+1}\langle |n_k^\ell|^4\,|f|\rangle_{t,x}\,. \tag{5.230}$$

On the other hand, given a function f' on $n+1$ replicas, $f'\ge 0$, we have

$$\frac{\mathrm{d}}{\mathrm{d}x}\,\langle f'\rangle_{t,x} \le K(n,\beta)\,\langle f'\rangle_{t,x}\,.$$

This follows from the formula

$$\frac{\mathrm{d}}{\mathrm{d}x}\,\langle f'\rangle_{t,x} = \beta\sqrt{t}\Bigg(\sum_{\ell\le n}\langle\varepsilon_\ell\,n_k^\ell\,f'\rangle_{t,x} - n\,\langle\varepsilon_{n+1}\,n_k^{n+1}\,f'\rangle_{t,x}\Bigg)$$

and the fact that $|n_k^\ell|\le 1$. Thus we have (as in Lemma 3.4.5)

$$\langle f' \rangle_{t,x} \le K(n,\beta) \langle f' \rangle_{t,1} \le K(n,\beta)(\langle f' \rangle_{t,1} + \langle f' \rangle_{t,-1})$$

and thus

$$\mathsf{E}\, \langle f' \rangle_{t,x} \le K(n,\beta)\, \nu_t(f') \; .$$

Using this for $f' = |n_k^\ell|^4\, |f|$ and combining with (5.230) finishes the proof. □

Proposition 5.7.4. *If f is as in Proposition 5.7.2, and $0 \le f \le 1$ then*

$$\nu_t(f) \le K\, \nu(f) + K \exp\left(-\frac{N}{K}\right) \tag{5.231}$$

where K depends only on n, β, h.

Proof. We use Proposition 5.7.2. We recall from Corollary A.4.4 that, with overwhelming probability, we have

$$\forall \boldsymbol{\rho}, \; \sum_{2 \le k \le M} n_k^2(\boldsymbol{\rho}) \le L \; .$$

Also, we always have $\sum_{2 \le k \le M} n_k^2(\boldsymbol{\rho}) \le M$. Using that r stays bounded in the admissible region, and that $n_k^4(\boldsymbol{\rho}) \le n_k^2(\boldsymbol{\rho})$ to control the term (5.226), we get

$$\nu_t'(f) \le K\, \nu_t(f) + K \exp\left(-\frac{N}{K}\right)$$

and we apply Lemma 3.4.5. □

Corollary 5.7.5. *Assume that for a certain number ρ $(= \rho(\beta,h))$, $\rho \le 1$, the sets*

$$\{|S_{1,1} - r| \ge \rho^2\}\, ; \; \{|S_{1,2} - r| \ge \rho^2\}\, ; \; \{|n_1(\boldsymbol{\rho}) - \mu| \ge \rho\} \tag{5.232}$$

are negligible. Then if f is as in Proposition 5.7.2, we have, if $|f| \le 1$,

$$\nu_t(|f|) \le \exp(L\, n^2\, \beta^2\, \rho)\, \nu(|f|) + K \exp\left(-\frac{N}{K}\right) , \tag{5.233}$$

$$|\nu_t(f) - \nu_0(f)| \le L\, n^2\, \beta^2\, \rho \exp(L\, n^2\, \beta^2\, \rho)\, \nu(|f|) + K \exp\left(-\frac{N}{K}\right) , \tag{5.234}$$

where K depends only on n, β, h.

Proof. Let us consider $\nu_t'(f)$ as given by Proposition 5.7.2. We observe that

$$\mathrm{II} = \frac{\beta^2}{2} \left(\sum_{\ell \le n} \nu_t((S_{\ell,\ell} - r)\, f) - n\, \nu_t((S_{n+1,n+1} - r)\, f) \right) .$$

Thus, using Proposition 5.7.4 and the hypothesis that the sets (5.232) are negligible, we get that

$$|\mathrm{I}+\mathrm{II}+\mathrm{III}| \le Ln^2\beta^2\rho\,\nu_t(|f|) + K\exp\left(-\frac{N}{K}\right) . \tag{5.235}$$

Note that we *cannot* put ρ^2 here instead of ρ because the last of the sets (5.232) is $\{|n_1 - \mu| \ge \rho\}$.

Given $a > 0$, it follows from Theorem 5.6.1 and Proposition 5.7.4 that the sets $U_{k,\ell} = \{|n_k^\ell| \ge a\}$ satisfy

$$\nu_t(\mathbf{1}_{U_{k,\ell}}) \le K\exp\left(-\frac{N}{K}\right)$$

where K depends only on β, h, a. Thus the quantity $\mathcal{R}$ of (5.222) satisfies

$$|\mathcal{R}| \le K_0\, a^2\, \nu_t\left(\sum_{\ell\le n}\sum_{k\le M}(n_k^\ell)^2\,|f|\right) + K\exp\left(-\frac{N}{K}\right) ,$$

where K_0 is the constant of (5.226). (We used here that $\sum_{\ell\le n}\sum_{k\le M}(n_k^\ell)^4 \le nM$.) We know by Corollary A.4.4 that with overwhelming probability we have $\sum_{k\le M}(n_k(\boldsymbol{\rho}))^2 \le L$ for each $\boldsymbol{\rho}$, so that the appropriate choice of a yields

$$|\mathcal{R}| \le n^2\,\beta^2\,\rho\,\nu_t(|f|) + K\exp\left(-\frac{N}{K}\right) , \tag{5.236}$$

where K depends only on β, h, n, ρ. Combining with (5.235) we get

$$|\nu_t'(f)| \le L\,n^2\,\beta^2\,\rho\,\nu_t(|f|) + K\exp\left(-\frac{N}{K}\right) \tag{5.237}$$

and combining with Lemma 3.4.5 this proves (5.233). Using (5.237) and (5.233) we obtain (5.234). □

Proposition 5.7.6. *If (α, β, h) belongs to the admissible region, if $h > 0$, and if the constant L_0 has been chosen large enough, then the sets (5.232) are negligible for*

$$\rho = L\min\left(\frac{m^*}{\beta^{10}}, \frac{m^*}{\sqrt{L_0}}\right) . \tag{5.238}$$

Thus, given n, the potentially devastating factor $\exp L\,n^2\,\beta^2\,\rho$ in (5.234) remains bounded independently of β.

Corollary 5.7.7. *If (α, β, h) belongs to the admissible region, $h > 0$, the constant L_0 has been chosen large enough, the function f is as in Proposition 5.7.2, $|f| \le N$, and the sets (5.232) are negligible, then*

$$\nu_t(|f|) \le L^{n^2}\, \nu(|f|) + K \exp\left(-\frac{N}{K}\right) , \tag{5.239}$$

$$|\nu_t(f) - \nu_0(f)| \le L^{n^2}\, \beta^2\, \rho\, \nu(|f|) + K \exp\left(-\frac{N}{K}\right) , \tag{5.240}$$

where K depends only on β, h, n.

Proof. If ρ is such that the sets (5.232) are negligible, we can assume by Proposition 5.7.2 that ρ is less than the quantity (5.238), so that in (5.234) we have $\exp L\, n^2\, \beta^2\, \rho \le L^{n^2}$. Observe also that the condition $|f| \le 1$ of Corollary 5.7.5 has been relaxed to $|f| \le N$, but $N \exp(-N/K) \le K' \exp(-N/K')$. □

Proof of Proposition 5.7.6. We first prove that we can take $\rho = Lm^*/\sqrt{L_0}$. In Theorem 5.3.2 we can take $\rho_0^2 = Lm^{*2}/L_0$. If we combine with Theorem 5.5.3, we see that the set

$$A = \left\{\boldsymbol{\sigma}\, ;\ \|\mathbf{m}(\boldsymbol{\sigma}) - m^* \mathbf{e}_1\| \ge \frac{Lm^*}{\sqrt{L_0}}\right\}$$

is negligible. On the complement of A we have

$$|m_1(\boldsymbol{\sigma}) - m^*| \le \frac{L}{\sqrt{L_0}} m^*$$

$$\sum_{k \ge 2} m_k^2 \le \frac{L}{L_0} m^{*2} \,.$$

Since $n_k(\boldsymbol{\rho}) = m_k(\boldsymbol{\sigma}) - \eta_{N,k}\, \sigma_N / N$, we see that for N large we have

$$|n_1(\boldsymbol{\rho}) - m^*| \le \frac{L}{\sqrt{L_0}} m^*$$

$$\sum_{2 \le k \le M} (n_k(\boldsymbol{\rho}))^2 \le \frac{L}{L_0} m^{*2}$$

whenever $\boldsymbol{\sigma} \notin A$. By the Cauchy-Schwarz inequality, for $\boldsymbol{\sigma}^\ell, \boldsymbol{\sigma}^{\ell'} \notin A$ we have $|S_{\ell,\ell'}| \le Lm^{*2}/L_0$. So all we have to do is to prove that

$$|\mu - m^*| \le \frac{L}{\sqrt{L_0}} m^*\, ;\ |r| \le \frac{Lm^{*2}}{L_0} \,.$$

This follows from (5.66) and (5.42).

Now we prove that we can take $\rho = Lm^*/\beta^{10}$. Since we have shown that we can take $\rho = Lm^*/\sqrt{L_0}$, we can assume $\beta \ge 2$, and it suffices to show that we can take $\rho = L/\beta^{10}$. We use (5.183) so that we know that the set

$$A = \left\{ \|\mathbf{m}(\boldsymbol{\sigma}) - m^* \mathbf{e}_1 - \boldsymbol{\theta}\| \geq \frac{L}{\beta^{10}} \right\}$$

is negligible. We recall that $\theta_1 = 0$, $\theta_k = N^{-1} m^* \left(\sum_{i \leq N} \eta_{i,k}\right)$. As previously, if $\boldsymbol{\sigma}$ does not belong to A, we have

$$|n_1(\boldsymbol{\rho}) - m^*| \leq \frac{L}{\beta^{10}} \,, \tag{5.241}$$

$$\sum_{2 \leq k \leq M} (n_k(\boldsymbol{\rho}) - \theta_k)^2 \leq \frac{L}{\beta^{20}} \,.$$

Now,

$$\begin{aligned} \left| S_{\ell,\ell'} - \sum_{2 \leq k \leq M} \theta_k^2 \right| &= \left| \sum_{2 \leq k \leq M} (n_k^\ell \, n_k^{\ell'} - \theta_k^2) \right| \\ &= \left| \sum_{2 \leq k \leq M} ((n_k^\ell - \theta_k)\, n_k^{\ell'} + \theta_k (n_k^{\ell'} - \theta_k)) \right| \\ &\leq \left(\sum_{2 \leq k \leq M} (n_k^\ell - \theta_k)^2 \sum_{2 \leq k \leq M} (n_k^{\ell'})^2 \right)^{1/2} \\ &\quad + \left(\sum_{2 \leq k \leq M} (n_k^{\ell'} - \theta_k)^2 \sum_{2 \leq k \leq M} \theta_k^2 \right)^{1/2} \,. \end{aligned} \tag{5.242}$$

With overwhelming probability, we have

$$\left| \sum_{2 \leq k \leq M} \theta_k^2 - \alpha \, m^* \right| \leq \frac{L}{\beta^{20}} \,,$$

so that, by (5.242), with overwhelming probability, whenever $\boldsymbol{\sigma}^\ell, \boldsymbol{\sigma}^{\ell'} \notin A$ we have

$$|S_{\ell,\ell'} - \alpha \, m^*| \leq \frac{L}{\beta^{10}} \,.$$

To complete the proof we appeal to (5.67) and (5.69). □

We set

$$s = \alpha \, \frac{1 - \beta(1-q)^2}{(1 - \beta(1-q))^2} \tag{5.243}$$

and we state the main result of this section.

Theorem 5.7.8. *If the constant L_0 in the definition of the admissible region is large enough, if (α, β, h) belongs to the admissible region and if $h > 0$, the following occurs. Given numbers $\tau_1, \tau_2 > 1$ with $1/\tau_1 + 1/\tau_2 = 1$, and a function f on Σ_N^n with $|f| \leq 1$, we have*

$$\nu(f) = \nu_0(f) + \mathrm{VI} + \mathrm{VII} + \mathrm{VIII} + \mathcal{R} \tag{5.244}$$

where

$$\mathrm{VI} = \beta\left(\sum_{\ell \leq n} \nu_0(f\,\varepsilon_\ell(n_1^\ell - \mu)) - n\,\nu_0(f\,\varepsilon_{n+1}(n_1^{n+1} - \mu))\right) \tag{5.245}$$

$$\mathrm{VII} = \frac{\beta^2}{2}\left(\sum_{\ell \leq n} \nu_0(f(S_{\ell,\ell} - s)) - n\,\nu_0(f(S_{n+1,n+1} - s))\right) \tag{5.246}$$

$$\begin{aligned}\mathrm{VIII} = \beta^2\Bigg(&\sum_{\ell < \ell' \leq n} \nu_0(f\,\varepsilon_\ell\,\varepsilon_{\ell'}(S_{\ell,\ell'} - r)) - n\sum_{\ell \leq n} \nu_0(f\,\varepsilon_\ell\,\varepsilon_{n+1}(S_{\ell,n+1} - r))\\ &+ \frac{n(n+1)}{2}\nu_0(f\,\varepsilon_{n+1}\,\varepsilon_{n+2}(S_{n+1,n+2} - r))\Bigg)\end{aligned} \tag{5.247}$$

and

$$\begin{aligned}|\mathcal{R}| \leq L^{n^2}\,\beta^4\,\nu(|f|^{\tau_1})^{1/\tau_1}\Bigg(&\nu(|S_{1,1} - s|^{2\tau_2})^{1/\tau_2}\\ &+ \nu(|S_{1,2} - r|^{2\tau_2})^{1/\tau_2} + \nu(|n_1 - \mu|^{2\tau_2})^{1/\tau_2} + \frac{K}{N}\Bigg)\\ &+ K\exp\left(-\frac{N}{K}\right).\end{aligned} \tag{5.248}$$

The constant K in (5.248) does not depend on τ_1, τ_2. Moreover, if the sets (5.232) are negligible we have

$$\nu(f) = \nu_0(f) + \mathrm{VI} + \mathcal{R}' \tag{5.249}$$

where

$$|\mathcal{R}'| \leq L^{n^2}\,\beta^4\,\rho^2\,\nu(|f|) + K\exp\left(-\frac{N}{K}\right). \tag{5.250}$$

The point of (5.249) is that there is a factor ρ^2 (and not ρ) in (5.250). This requires to keep the term VI in (5.249).

Proof. We write

$$\nu(f) = \nu_0(f) + \int_0^1 \nu_t'(f)\,\mathrm{d}t\,. \tag{5.251}$$

We denote by $A(t)$ the sum of the terms I to III of Proposition 5.7.2, and by $\mathcal{R}(t)$ the last term of (5.222). From (5.251) we get

$$\nu(f) = \nu_0(f) + A(0) + \int_0^1 (A(t) - A(0) + \mathcal{R}(t))\,\mathrm{d}t\,, \tag{5.252}$$

and we observe that $A(0) = \mathrm{VI} + \mathrm{VII} + \mathrm{VIII}$. Using (5.226) and Hölder's inequality we get

$$\begin{aligned} |\mathcal{R}(t)| &\le K\,M\,\nu_t((n_2^1)^{4\tau_2})^{1/\tau_2}\,\nu_t(|f|^{\tau_1})^{1/\tau_1} \\ &\le K\,M\,\nu((n_2^1)^{4\tau_2})^{1/\tau_2}\,\nu(|f|^{\tau_1})^{1/\tau_1} + K\exp\left(-\frac{N}{K}\right) \end{aligned} \tag{5.253}$$

using (5.239) in the second line. Thus, by Theorem 5.6.1, we have

$$|\mathcal{R}(t)| \le \frac{K}{N}\,\nu(|f|^{\tau_1})^{1/\tau_1} + K\exp\left(-\frac{N}{K}\right)\,. \tag{5.254}$$

Next we bound $A'(t)$ by computing it through Proposition 5.7.2. The remainder is handled as in (5.254). In the other terms, beside f, we have always *two* factors of the type $(S_{\ell,\ell} - s)$ or $(S_{\ell,\ell'} - r)$ or $(n_1^\ell - \mu)$; we then use Hölder's inequality as in

$$\nu_t(|f\,U_1\,U_2|) \le \nu_t(|f|^{\tau_1})^{1/\tau_1}(\nu_t(|U_1|^{2\tau_2})^{1/\tau_2} + \nu_t(|U_2|^{2\tau_2})^{1/\tau_2})$$

and we replace ν_t by ν as previously using (5.239). In this manner we bound $|A'(t)|$ by the right-hand side of (5.248). Combining with (5.252), this proves (5.244).

We turn to the proof of (5.249). Instead of using (5.253) to control $|\mathcal{R}(t)|$, we proceed as in (5.236) to obtain

$$|\mathcal{R}(t)| \le n^2\,\beta^2\,\rho^2\,\nu_t(|f|) + K\exp\left(-\frac{N}{K}\right)\,.$$

(We could in fact put any power of ρ here.) The sets (5.232) are negligible for ν_t by Proposition 5.7.4, so that

$$|A'(t)| \le L\,n^4\,\beta^4\,\rho^2\,\nu_t(|f|) + K\exp\left(-\frac{N}{K}\right)$$

$$|\mathrm{VI} + \mathrm{VIII}| \le L\,n^2\,\beta^2\,\rho^2\,\nu_t(|f|) + K\exp\left(-\frac{N}{K}\right)\,,$$

since we can replace s by r in VI. The result follows using (5.239). □

5.8 Integration by Parts

In the previous section occurred quantities such as $\nu(f\,\varepsilon_\ell\,\varepsilon_{\ell'}\,S_{\ell,\ell'})$. Since $S_{\ell,\ell'} = \sum_{2\le k\le M} n_k^\ell\, n_k^{\ell'}$, by symmetry we have

$$\nu(f\,\varepsilon_\ell\,\varepsilon_{\ell'}\,S_{\ell,\ell'}) = (M-1)\,\mathsf{E}\,\langle n_M^\ell\, n_M^{\ell'}\, f\rangle\;. \tag{5.255}$$

We have to learn how to evaluate these. We faced a similar problem in Chapter 3. There, it was solved through the use of the smart path method, in Section 3.3. The use of the smart path method is however more difficult in the present case because it is hard to control $\varphi'(v)$, when $\varphi(v)$ is given by (3.71). The only way we know how to do this requires integration by parts, and, as will be shown, integration by parts allows direct calculation of the quantities (5.255).

We write

$$D = 1-\beta(1-q)\,;\; r' = \frac{r}{\alpha} = \frac{q}{D^2}\,;\; s' = \frac{s}{\alpha} = \frac{1-\beta(1-q)^2}{D^2} \tag{5.256}$$

and we observe the relation

$$s'-r' = \frac{1-q}{D}\;. \tag{5.257}$$

Theorem 5.8.1. *If (α,β,h) belongs to the admissible region, $h>0$, and if the constant L_0 that defines this region has been chosen large enough, then for a function f on Σ_N^{n-1} (i.e. f depends on $n-1$ replicas) and $\tau>1$, we have*

$$N\nu((n_M^1)^2\, f) = s'\,\nu(f) + C_1 + \mathcal{R}_1 \tag{5.258}$$

$$N\nu(n_M^1\, n_M^2\, f) = r'\,\nu(f) + C_2 + \mathcal{R}_2 \tag{5.259}$$

where

$$C_k = \frac{V_k}{D} + \frac{\beta q}{D^2}\left(\sum_{p\le n} V_p - n\,V_n\right) \tag{5.260}$$

$$V_p = (1-\delta_{1,p})\,\nu((R_{1,p}-q)\,f) + \beta N \sum_{\ell\ne p,\ell\le n} \nu((R_{\ell,p}-q)\,n_M^1\, n_M^\ell\, f)$$
$$-\,\beta\,N\,n\,\nu((R_{p,n+1}-q)\,n_M^1\, n_M^{n+1}\, f) \tag{5.261}$$

and

$$|\mathcal{R}_1|, |\mathcal{R}_2| \le \frac{K}{N}\,\nu(|f|^\tau)^{1/\tau}\;. \tag{5.262}$$

Here $\delta_{1,p}=1$ if $p=1$ and $\delta_{1,p}=0$ otherwise. Of course $\sum_{p\le n} V_p - nV_n = \sum_{p\le n-1} V_p - (n-1)\,V_n$, but the present formulation is more convenient.

What Theorem 5.8.1 means is that (within the small error terms $\mathcal{R}_1$, $\mathcal{R}_2$) we have a kind of expansion

$$N\,\nu(n_M^1\,n_M^2\,f) = r'\,\nu(f) + C_2$$

where C_2 is made up of terms that have a tendency to be small because they contain the factor $R_{p,\ell} - q$. Some of these terms contain factors $n_M^1\,n_M^k$, and iterating the process will yield higher order expansions.

We will use the notation

$$A_k = N\,\nu(n_M^1\,n_M^k\,f)\;. \tag{5.263}$$

Lemma 5.8.2. *We have*

$$A_k = \nu(R_{1,k}\,f) + N\beta\left(\sum_{\ell\le n}\nu(R_{k,\ell}\,n_M^1\,n_M^\ell\,f) - n\nu(R_{k,n+1}\,n_M^1\,n_M^{n+1}\,f)\right) + \mathcal{R}_k \tag{5.264}$$

where $\mathcal{R}_k$ is bounded as in (5.262).

Proof. Writing for simplicity η_i rather than $\eta_{i,M}$, we have

$$A_k = \sum_{i\le N-1}\mathsf{E}\,(\eta_i\langle n_M^1\,\sigma_i^k\,f\rangle)\;. \tag{5.265}$$

To make the dependence of $\langle\cdot\rangle$ upon η_i explicit, we bring in the bracket $\langle\cdot\rangle_\sim$ of Section 5.6, and we write

$$\mathsf{E}\,(\eta_i\langle n_M^1\,\sigma_i^k\,f\rangle) = \mathsf{E}\left(\eta_i\frac{\left\langle n_M^1\,\sigma_i^k\,f\exp\left(\frac{N\beta}{2}\sum_{\ell\le n}(m_M^\ell)^2\right)\right\rangle_\sim}{\left\langle\exp\frac{N\beta}{2}\,(m_M^1)^2\right\rangle_\sim^n}\right)\;. \tag{5.266}$$

The randomness in $\langle\cdot\rangle_\sim$ is independent of η_i. We then use "approximate integration by parts" as in Propositions 3.8.3 and 5.7.2. Since (abusing notation in an obvious manner) $\partial\,m_M^\ell/\partial\,\eta_i = \sigma_i^\ell/N$ and $\partial\,n_M^1/\partial\,\eta_i = \sigma_i^1/N$, the main term obtained from (5.266) is

$$\begin{aligned}\frac{1}{N}\,\nu(\sigma_i^1\,\sigma_i^k\,f) &+ \beta\sum_{\ell\le n}\nu(n_M^1\,m_M^\ell\,\sigma_i^k\,\sigma_i^\ell\,f)\\ &- n\,\beta\,\nu(n_M^1\,m_M^{n+1}\,\sigma_i^k\,\sigma_i^{n+1}\,f)\;,\end{aligned} \tag{5.267}$$

and the remainder $\mathcal{R}_i$ satisfies

$$|\mathcal{R}_i| \le K\,\nu\left(|n_M^1||f|\sum_{\ell\le n+1}|m_M^\ell|^3\right) \le \frac{K}{N^2}\,\nu(|f|^\tau)^{1/t}\;,$$

using Theorem 5.6.1. (The constant K is independent of i.) We write (5.267) as

$$\frac{1}{N}\,\nu(\sigma_i^1\,\sigma_i^k\,f)+\beta\sum_{\ell\le n}\nu(n_M^1\,n_M^\ell\,\sigma_i^k\,\sigma_i^\ell\,f)-\beta\,n\,\nu(n_M^1\,n_M^{n+1}\,\sigma_i^k\,\sigma_i^{n+1}\,f)+\mathcal{R}_i' \tag{5.268}$$

where

$$\mathcal{R}_i'=\frac{\beta}{N}\left(\sum_{\ell\le n}\nu(n_M^1\,\eta_N\,\sigma_N^k\,\sigma_i^k\,\sigma_i^\ell\,f)-n\,\nu(n_M^1\,\eta_N\,\sigma_N^{n+1}\,\sigma_i^k\,\sigma_i^{n+1}\,f)\right). \tag{5.269}$$

Using again approximate integration by parts, and Theorem 5.6.1 we see that

$$|\mathcal{R}_i'|\le\frac{K}{N^2}\,\nu(|f|^\tau)^{1/\tau}\,.$$

Summation of the quantities (5.266) over $i\le N-1$ then yields

$$\begin{aligned}A_k&=\nu(R_{1,k}^-\,f)+\beta\sum_{\ell\le n}N\,\nu(R_{k,\ell}^-\,n_M^1\,n_M^\ell\,f)\\&\quad-\beta\,n\sum_{\ell\le n}N\,\nu(R_{k,n+1}^-\,n_M^1\,n_M^{n+1}\,f)+\mathcal{R}_k\,,\end{aligned}$$

where $R_{k,\ell}^-$ is given by (2.80) and where $\mathcal{R}_k$ is bounded as in (5.262). To replace $R_{k,\ell}^-$ by $R_{k,\ell}$ without changing the nature of the error term, we simply use Hölder's inequality and Theorem 5.6.1. □

Lemma 5.8.3. *For $k\le n$ we have*

$$DA_k=\nu(f)(q+\delta_{1,k}(1-q))+\beta q\left(\sum_{\ell\le n}A_\ell-nA_n\right)+V_k+\mathcal{R}_k \tag{5.270}$$

where V_k is given by (5.261) and $\mathcal{R}_k$ is bounded as in (5.262).

Proof. Since $R_{1,1}=1$, we have

$$\nu(R_{1,k}\,f)=(1-\delta_{1,k})\,\nu((R_{1,k}-q)\,f)+\nu(f)(q+\delta_{1,k}(1-q))\,.$$

This is seen by distinguishing the cases $k=1$ and $k\ne 1$. When $\ell\ne k$, $\ell\le n$, we have

$$\nu(R_{k,\ell}\,n_M^1\,n_M^\ell\,f)=\frac{q\,A_\ell}{N}+\nu((R_{k,\ell}-q)\,n_M^1\,n_M^\ell\,f)\,.$$

When $\ell=n+1$, we use the fact that by hypothesis f does not depend on the nth replica to write

$$\nu(R_{k,n+1}\,n_M^1\,n_M^{n+1}\,f)=\frac{q\,A_n}{N}+\nu((R_{k,n+1}-q)\,n_M^1\,n_M^{n+1}\,f)$$

because $A_n=A_{n+1}$. We also note that $A_k=N\nu(R_{k,k}\,n_M^1\,n_M^k\,f)$.

Substituting these relations in (5.264) yields the result. □

Lemma 5.8.4. *The system of equations*

$$Dx_k = b_k + \beta q \left(\sum_{\ell \le n} x_\ell - n\, x_n \right) \tag{5.271}$$

where $(x_k)_{k \le n}$ *are unknown and* $(b_k)_{k \le n}$ *are numbers has solutions*

$$x_k = \frac{b_k}{D} + \frac{\beta q}{D^2} \left(\sum_{\ell \le n} b_\ell - n\, b_n \right) . \tag{5.272}$$

Proof. Adding all the equations, and subtracting n times the last one, we get

$$D \left(\sum_{\ell \le n} x_\ell - n\, x_n \right) = \sum_{\ell \le n} b_\ell - n\, b_n \ ,$$

from which the result is obvious. □

Proof of Theorem 5.8.1. Lemma 5.8.3 means that $x_k = A_k$ satisfies the relation (5.271) for

$$b_k = \nu(f)(q + \delta_{1,k}(1-q)) + V_k + \mathcal{R}_k \ . \tag{5.273}$$

The formulae (5.272) are linear in $(b_k)_{k \le n}$. The contribution of the term (V_k) in (5.273) is the term C_k in (5.258) and (5.259). If now $b_k = q + \delta_{1,k}(1-q)$, then

$$\sum_{\ell \le n} b_\ell - n\, b_n = 1 - q$$

so that

$$x_1 = \frac{1}{D} + \frac{\beta q(1-q)}{D^2} = \frac{1 - \beta(1-q) + \beta q(1-q)}{D^2} = s'$$

and for $k \ge 1$

$$x_k = \frac{q}{D} + \frac{\beta q(1-q)}{D^2} = \frac{q}{D^2} = r' \ .$$

□

In Theorem 5.8.1 we can replace n_M by n_k for any $2 \le k \le M$. By summation over $2 \le k \le M$, we then get the following.

Theorem 5.8.5. *Under the hypothesis of Theorem 5.8.1, we have*

$$\nu(S_{1,1}\, f) = s\, \nu(f) + C_1' + \mathcal{R}_1 \tag{5.274}$$

$$\nu(S_{1,2}\, f) = r\, \nu(f) + C_2' + \mathcal{R}_2 \tag{5.275}$$

where

$$C'_k = \frac{U_k}{D} + \frac{\beta q}{D^2}\left(\sum_{p\le n} U_p - n\,U_n\right) \tag{5.276}$$

$$U_p = \alpha(1-\delta_{1,p})\,\nu((R_{1,p}-q)\,f) + \beta \sum_{\ell\ne p,\ell\le n} \nu((R_{\ell,p}-q)\,S_{1,\ell}\,f)$$
$$- \beta\, n\, \nu((R_{p,n+1}-q)\,S_{1,n+1}\,f)\,, \tag{5.277}$$

and where $\mathcal{R}_1$ *and* $\mathcal{R}_2$ *are as in (5.262).*

5.9 The Replica-Symmetric Solution

Our first goal is to prove the following.

Theorem 5.9.1. *If* (α,β,h) *belongs to the admissible region, and* $h>0$*, and if the constant* L_0 *that defines this region is large enough, then*

$$Q_N := \nu((R_{1,2}-q)^2) \le \frac{K}{N} \tag{5.278}$$

$$M_N := \nu((m_1-\mu)^2) \le \frac{K}{N} \tag{5.279}$$

$$S_N := \nu\left(\left(\sum_{2\le k\le M} (n_k^1)^2 - s\right)^2\right) \le \frac{K}{N} \tag{5.280}$$

$$R_N := \nu\left(\left(\sum_{2\le k\le M} n_k^1\, n_k^2 - r\right)^2\right) \le \frac{K}{N}\,. \tag{5.281}$$

The quantities $Q_N, R_N, \ldots$ also depend on M. We keep the dependence implicit to lighten notation. Of course, one could also define M_N by replacing m_1 by n_1.

In a later effort, we will then compute $p_{N,M}(\beta,h)$.

As usual the number K in (5.278) to (5.281) depends on β, h only. The region of validity of Theorem 5.9.1 excludes the case $h=0$. It should be obvious that (5.279) cannot hold if $h=0$. (In that case, one of the quantities $\pm m_k$ is close to μ, but by symmetry this cannot always be m_1.)

Research problem 5.9.2. (Level 2) Prove that (5.278), (5.280) and (5.281) hold even if $h=0$, or better that when (α,β,h) belongs to admissible region, they hold with a constant K that depends only on β.

The proof of Theorem 5.9.1 is rather delicate. The main difficulties already arose in the proof of the existence of the solutions to the equations (5.25) to (5.27).

The idea of the proof is to establish the following three relations

$$Q_N \le L \frac{m^*}{\beta^2} Q_N^{1/2} M_N^{1/2} + \frac{L}{\beta^2} Q_N^{1/2}(R_N^{1/2} + S_N^{1/2}) + \frac{K}{N} \tag{5.282}$$

$$R_N + S_N \le C \beta^4 Q_N + \frac{K}{N} \tag{5.283}$$

$$M_N \le \left(1 - \frac{m^{*2}}{L}\right) M_N + L \frac{m^*}{\beta^2} M_N^{1/2}(R_N^{1/2} + S_N^{1/2}) + \frac{K}{N} \,. \tag{5.284}$$

(One could in fact put larger powers of β in the denominators. There is all the room in the world for large β.) These relations hold in the admissible region (if $h > 0$) and K depends only on β, h. The number C in (5.283) can be made arbitrarily small by taking the constant L_0 of the admissible region suitably large. From (5.284) we get

$$m^* M_N \le \frac{L}{\beta^2} M_N^{1/2}(R_N^{1/2} + S_N^{1/2}) + \frac{K}{N} \,. \tag{5.285}$$

We observe the relation

$$2ab \le t\, a^2 + \frac{1}{t}\, b^2 \tag{5.286}$$

and using this in (5.285) for $t = \beta^2 m^*/L$, we get

$$m^* M_N \le \frac{1}{2} m^* M_N + \frac{L}{m^* \beta^4}(R_N + S_N) + \frac{K}{N} \,,$$

so that

$$M_N \le \frac{L}{m^{*2}\beta^4}(R_N + S_N) + \frac{K}{N} \le \frac{LC}{m^{*2}} Q_N + \frac{K}{N} \,, \tag{5.287}$$

using (5.283) in the second inequality. If we substitute (5.287) and (5.286) in (5.282) we get

$$Q_N \le L\sqrt{C}\, Q_N + \frac{K}{N}$$

so that by making L_0 large enough that $L\sqrt{C} \le 1/2$ we get $Q_N \le K/N$ and using (5.283), (5.287) this proves Theorem 5.9.1.

We set

$$\widehat{\mu} = \mathsf{E}\,\mathrm{th}^3 Y = \nu_0(\varepsilon_1 \varepsilon_2 \varepsilon_3) \,; \;\; \widehat{q} = \mathsf{E}\,\mathrm{th}^4 Y = \nu_0(\varepsilon_1 \varepsilon_2 \varepsilon_3 \varepsilon_4) \,, \tag{5.288}$$

using lemma 5.7.1.

We start with a simple fact.

Lemma 5.9.3. *We have*

$$\beta^4(1-q^2) \le L \,, \tag{5.289}$$

$$\beta^4|\mu(1-q)| \le Lm^*, \ \beta^4|\widehat{\mu}-\mu q| \le Lm^* \,. \tag{5.290}$$

Proof. We have seen in (5.72) that in the admissible region we have $|1-\mu| \le L\beta^{-4}$, and similarly one shows that $|1-q| \le L\beta^{-4}$, $|1-\widehat{\mu}| \le L\beta^{-4}$, $|1-\widehat{q}| \le L\beta^{-4}$ and this implies that all the quantities considered in (5.289) and (5.290) stay bounded. To prove (5.290) one can assume $\beta \le 2$. We have then seen in Theorem 5.2.1 that $\mu \le Lm^*$, $q \le Lm^{*2} \le Lm^*$, and of course $\widehat{\mu} = \mathsf{E}\,\mathrm{th}^3 Y \le q = \mathsf{E}\,\mathrm{th}^2 Y$. □

Proof of (5.282). We write as usual

$$Q_N = \nu((R_{1,2}-q)^2) = \nu((\varepsilon_1\,\varepsilon_2 - q)\, f) \tag{5.291}$$

where $f = R_{1,2} - q$. It certainly would not work to use on the last term an "order one expansion" as we did in Chapters 2 and 3. Rather, we will use an "order two expansion" as provided by Theorem 5.7.8. We use this result with $n = 2$, $f(\varepsilon_1\,\varepsilon_2 - q)$ to obtain that

$$Q_N = \nu_0((\varepsilon_1\,\varepsilon_2 - q)\, f) + \mathrm{I} + \mathrm{II} + \mathrm{III} + \mathcal{R} \tag{5.292}$$

where

$$\begin{aligned} \mathrm{I} = \beta\Big(&\sum_{\ell\le 2} \nu_0(\varepsilon_\ell(\varepsilon_1\,\varepsilon_2 - q)(n_1^\ell - \mu)\, f) \\ &- 2\,\nu_0((\varepsilon_1\,\varepsilon_2 - q)\,\varepsilon_3(n_1^3 - \mu)\, f)\Big) \end{aligned} \tag{5.293}$$

$$\begin{aligned} \mathrm{II} = \beta^2\Big(&\sum_{\ell\le 2} \nu_0((\varepsilon_1\,\varepsilon_2 - q)(S_{\ell,\ell} - s)\, f) \\ &- 2\,\nu_0((\varepsilon_1\,\varepsilon_2 - q)(S_{3,3} - s)\, f)\Big) \end{aligned} \tag{5.294}$$

$$\begin{aligned} \mathrm{III} = \beta^2(&\nu_0(\varepsilon_1\,\varepsilon_2(\varepsilon_1\,\varepsilon_2 - q)(S_{1,2} - r)\, f) \\ &- 2\sum_{\ell\le 2} \nu_0(\varepsilon_\ell\,\varepsilon_3(\varepsilon_1\,\varepsilon_2 - q)(S_{\ell,3} - r)\, f) \\ &+ 3\,\nu_0(\varepsilon_3\,\varepsilon_4(\varepsilon_1\,\varepsilon_2 - q)(S_{3,4} - r)\, f))\,, \end{aligned} \tag{5.295}$$

and where $\mathcal{R}$ is as in (5.248), with $\tau_1 = \tau_2 = 2$.

Let us observe that, since $q = \mathsf{E}\,\mathrm{th}^2 Y = \nu_0(\varepsilon_1\,\varepsilon_2)$, by Lemma 5.7.1 we have

$$\nu_0((\varepsilon_1 \varepsilon_2 - q) f) = \nu_0\left(\frac{1}{N}(\varepsilon_1 \varepsilon_2 - q)^2\right) \le \frac{4}{N}\,. \tag{5.296}$$

We consider a number ρ such that the sets (5.232) are negligible. By Proposition 5.7.6 we can assume that $\beta^6 \rho \le Lm^*$. Then the quantity $\mathcal{R}$ of (5.248) satisfies

$$|\mathcal{R}| \le L\beta^4 \rho\, Q_N^{1/2}(R_N^{1/2} + S_N^{1/2} + M_N^{1/2}) + \frac{K}{N}\,. \tag{5.297}$$

To see this, we simply use that if $|n_1^1 - \mu| \le \rho$, then

$$(m_1^1 - \mu)^4 \le (m_1^1 - \mu)^2\left(\rho + \frac{1}{N}\right)^2 \le 2\rho^2(m_1^1 - \mu)^2 + \frac{8}{N^2}$$

to see that $\nu((n_1^1 - \mu)^4) \le 2\rho^2 \nu((m_1^1 - \mu)^2) + K/N^2$, and we proceed similarly for the other terms. Thus

$$|\mathcal{R}| \le \frac{Lm^*}{\beta^2}\, Q_N^{1/2}\, M_N^{1/2} + \frac{L}{\beta^2}\, Q_N^{1/2}(S_N^{1/2} + R_N^{1/2}) + \frac{K}{N}\,. \tag{5.298}$$

To handle the term I we first observe that $\varepsilon_1(\varepsilon_1 \varepsilon_2 - q) = \varepsilon_2 - \varepsilon_1 q$ and that, from Lemma 5.7.1

$$|\nu_0((\varepsilon_2 - q\,\varepsilon_1)\, f(n_1^1 - \mu)) - \mu(1-q)\, \nu_0(f(n_1^1 - \mu))| \le \frac{K}{N} \tag{5.299}$$

since $\mu = \mathsf{E}\,\mathrm{th}\, Y = \nu_0(\varepsilon_1)$ and $|f - f^-| \le 1/N$, where $f^- = R_{1,2}^- - q$.

Next, we use that from (5.240), for a function f' on Σ_N^3, $|f'| \le 1$, we have

$$|\nu(f') - \nu_0(f')| \le L\beta^2 \rho \nu(|f'|) + K \exp\left(-\frac{N}{K}\right)\,.$$

We use this for $f' = f(n_1^1 - \mu)$ and (5.299) to get

$$\beta|\nu_0((\varepsilon_2 - q\,\varepsilon_1)\, f(n_1^1 - \mu))| \le \frac{K}{N} + L\beta^3 \rho \nu(|f||n_1^1 - \mu|) + \beta|\mu(1-q)||\nu(f(n_1^1 - \mu))|\,.$$

Now $\beta^5 \rho \le Lm^*$ by (5.238) and $\beta^3|\mu(1-q)| \le Lm^*$ by (5.290). By the Cauchy-Schwarz inequality we get

$$\beta|\nu_0((\varepsilon_2 - q\,\varepsilon_1)\, f(n_1^1 - \mu))| \le \frac{K}{N} + L\frac{m^*}{\beta^2}\, Q_N^{1/2}\, M_N^{1/2}\,,$$

and in the same manner we get

$$|\mathrm{I}| \le \frac{K}{N} + L\frac{m^*}{\beta^2}\, Q_N^{1/2}\, M_N^{1/2}\,,$$

using that $\beta|\widehat{\mu} - \mu q| \le Lm^*/\beta^2$.

We now turn to the terms II, III, for which we will use a closely related technique. Since the various terms of II, III are handled through the same method, we will consider only the case of

$$\nu_0(\varepsilon_1\,\varepsilon_2(\varepsilon_1\,\varepsilon_2 - q)(S_{1,2} - r)\,f)\,. \tag{5.300}$$

We define

$$S^{\sim}_{\ell,\ell'} = \sum_{2\le k\le M-1} n_k^\ell\, n_k^{\ell'}\,, \tag{5.301}$$

so that this quantity does not depend on the r.v. (η_k). Moreover, by Theorem 5.6.1 we have

$$|\nu_0(\varepsilon_1\,\varepsilon_2(\varepsilon_1\,\varepsilon_2 - q)(S_{1,2} - r)\,f) - \nu_0(\varepsilon_1\,\varepsilon_2(\varepsilon_1\,\varepsilon_2 - q)(S^{\sim}_{1,2} - r)\,f)| \le \frac{K}{N}\,.$$

Since $\nu_0(\varepsilon_1\,\varepsilon_2(\varepsilon_1\,\varepsilon_2 - q)) = 1 - q^2$, and since $|f - f^-| \le 1/N$, where $f^- = R^-_{1,2} - q$, we get

$$|\nu_0(\varepsilon_1\,\varepsilon_2(\varepsilon_1\,\varepsilon_2 - q)(S^{\sim}_{1,2} - r)\,f) - (1-q^2)\,\nu_0((S^{\sim}_{1,2} - r)\,f)| \le \frac{K}{N}\,.$$

The random function $(S^{\sim}_{1,2} - r)\,f$ does not depend on the r.v. (η_k), so that we can use (5.240) and the Cauchy-Schwarz inequality to get

$$|\nu((S^{\sim}_{1,2}-r)\,f) - \nu_0((S^{\sim}_{1,2}-r)\,f)| \le L\beta^2\rho\, Q_N^{1/2}\nu((S^{\sim}_{1,2}-r)^2)^{1/2} + \frac{K}{N}\,. \tag{5.302}$$

Since $S^{\sim}_{1,2} = S_{1,2} - n_M^1\, n_M^2$, we have $(S^{\sim}_{1,2} - r)^2 \le 2(S_{1,2} - r)^2 + 2(n_M^1\, n_M^2)^2$, and Theorem 5.6.1 shows that $\nu((S^{\sim}_{1,2} - r)^2) \le 2S_N + K/N^2$. The left-hand side of (5.302) is thus bounded by $L\beta^2\rho\, Q_N^{1/2}\, R_N^{1/2} + K/N$. Hence we get

$$|\nu_0((S^{\sim}_{1,2} - r)\,f) - \nu((S_{1,2} - r)\,f)| \le \frac{K}{N} + L\beta^2\rho\, Q_N^{1/2}\, R_N^{1/2}\,,$$

and

$$\begin{aligned}\beta^2|\nu_0(\varepsilon_1\,\varepsilon_2(\varepsilon_1\,\varepsilon_2 - q)(S_{1,2} - r)\,f)| &\le \frac{K}{N} + L\beta^4\rho\, Q_N^{1/2}\, R_N^{1/2} \\ &\quad + \beta^2(1-q^2)|\nu((S_{1,2} - r)\,f)| \\ &\le \frac{K}{N} + \frac{L}{\beta^2}\, Q_N^{1/2}\, R_N^{1/2}\end{aligned}$$

by (5.289) and the Cauchy-Schwarz inequality. We have proved (5.282).

Proof of (5.283). We will prove (5.283) with

$$C = \frac{L}{L_0^2}\,. \tag{5.303}$$

Using symmetry, we write

$$S_N = \nu((S_{1,1} - s)^2) = \nu(((M-1)(n_M^1)^2 - s)(S_{1,1} - s)) \tag{5.304}$$

and using Theorem 5.6.1,

$$\begin{aligned} S_N &\le \nu((M(n_M^1)^2 - s)(S^{\sim}_{1,1} - s)) + \frac{K}{N} \\ &= \alpha N \, \nu(((n_M^1)^2 - \frac{s'}{N})(S^{\sim}_{1,1} - s)) + \frac{K}{N} \,. \end{aligned} \tag{5.305}$$

We now use (5.258) with $n = 3$ to see that

$$N \, \nu(((n_M^1)^2 - \frac{s'}{N})(S^{\sim}_{1,1} - s)) = \frac{V_1}{D} + \frac{\beta q}{D^2} (V_1 + V_2 - 2V_3) + \mathcal{R}_1 \tag{5.306}$$

where

$$|\mathcal{R}_1| \le \frac{K}{N} \nu((S^{\sim}_{1,2} - s)^2)^{1/2} \le \frac{K}{N} \,,$$

and where V_p is given by (5.261) for $n = 3$ and $f = S^{\sim}_{1,1} - s$. We then see from (5.305) and (5.306) that

$$S_N \le \alpha \left(\frac{V_1}{D} + \frac{\beta q}{D^2} (V_1 + V_2 - 2V_3) \right) + \frac{K}{N} \,. \tag{5.307}$$

By Theorem 5.6.1 this remains true if in the expression of V_p we take now $f = S_{1,1} - s$. We then argue as in the proof of Theorem 5.8.5. If we denote by V_p^k the same quantity as V_p, but where each occurrence of a quantity n_M^ℓ is replaced by n_k^ℓ, we see from (5.307) that

$$S_N \le \sum_{2 \le k \le M} \frac{1}{N} \left(\frac{V_1^k}{D} + \frac{\beta q}{D^2} (V_1^k + V_2^k - 2V_3^k) \right) + \frac{K}{N} \,,$$

and thus

$$S_N \le \frac{U_1}{D} + \frac{\beta q}{D^2} (U_1 + U_2 - U_3) + \frac{K}{N} \tag{5.308}$$

where

$$\begin{aligned} U_p &= \alpha(1 - \delta_{1,p}) \, \nu((R_{1,p} - q) \, f) \\ &+ \beta \left(\sum_{\ell \ne p, \ell \le 3} \nu((R_{\ell,p} - q) \, S_{1,\ell} \, f) - 3\nu((R_{p,4} - q) \, S_{1,4} \, f) \right) \,, \end{aligned} \tag{5.309}$$

for $f = S_{1,1} - r$. Since we assume that the sets (5.232) are negligible, and since $r \le s$, this is also the case of the sets

$$\{|S_{1,\ell}| \ge s + \rho^2\} \,.$$

Thus

$$|U_p| \le \alpha(1-\delta_{1,p})\, Q_N^{1/2}\, S_N^{1/2} + 4\beta(s+\rho^2)\, Q_N^{1/2}\, S_N^{1/2} + \frac{K}{N}$$

and we deduce from (5.308) that

$$S_N \le \frac{K}{N} + L\left(\frac{\beta(s+\rho^2)}{D} + \frac{\alpha\beta q}{D^2} + \frac{\beta^2 q(s+\rho^2)}{D^2}\right) Q_N^{1/2}\, S_N^{1/2} .$$

By (5.70) we have $q \le LD$ in the admissible region, so that

$$\frac{\beta(s+\rho^2)}{D} + \frac{\alpha\beta q}{D^2} + \frac{\beta^2 q(s+\rho^2)}{D^2} \le L\beta^2\left(\frac{s+\rho^2+\alpha}{D}\right) .$$

Now, since $r = \alpha q/D^2$ and by (5.70) we have $r \le L\alpha/D$. Since, by (5.257), we have $s - r \le \alpha/D$, we have $s \le L\alpha/D$. Since $D \ge m^{*2}/L$ by (5.28) and (5.34) we have

$$L\beta^2\left(\frac{s+\rho^2+\alpha}{D}\right) \le L\beta^2\left(\frac{\alpha}{m^{*4}} + \frac{\rho^2}{m^{*2}}\right) \le \frac{L\beta^2}{L_0} ,$$

since $\alpha \le m^{*4}/L_0$, and since we can take $\rho^2 \le Lm^{*2}/L_0$ by Proposition 5.7.6. Thus we get

$$S_N \le \frac{K}{N} + \frac{L\beta^2}{L_0}\, S_N^{1/2}\, Q_N^{1/2}$$

and using (5.286) with $t = L_0/2L\beta^2$ we get

$$S_N \le \frac{K}{N} + \frac{L\beta^4}{L_0^2}\, Q_N .$$

The same argument works for R_N. □

Proof of (5.284). We have

$$\begin{aligned} M_N &= \nu\left(\left(\frac{1}{N}\sum_{i\le N-1}\sigma_i - \mu\right)^2\right) = \nu\left(\left(\frac{1}{N}\sum_{2\le i\le N}\sigma_i - \mu\right)^2\right) \\ &= \nu\left(\left(\frac{N-1}{N}\varepsilon_1 - \mu\right)\left(\frac{1}{N}\sum_{2\le i\le N}\sigma_i - \mu\right)\right) \\ &\le \nu((\varepsilon_1-\mu)\, f) + \frac{K}{N} , \end{aligned}$$

where $f = n_1^1 - \mu$. We use Theorem 5.7.8 for $n = 1$, and we get

$$M_N \le \nu_0((\varepsilon_1 - \mu)\, f) + \mathrm{IV} + \mathrm{V} + \mathrm{VI} + \mathcal{R} \tag{5.310}$$

where

$$\mathrm{IV} = \beta(\nu_0(\varepsilon_1(\varepsilon_1-\mu)(n_1^1-\mu)\,f) - \nu_0(\varepsilon_2(\varepsilon_1-\mu)(n_1^2-\mu)\,f)) \tag{5.311}$$

$$\mathrm{V} = \frac{\beta^2}{2}(\nu_0((\varepsilon_1-\mu)(S_{1,1}-s)\,f) - \nu_0((\varepsilon_1-\mu)(S_{2,2}-s)\,f)) \tag{5.312}$$

$$\mathrm{VI} = \beta^2(-\nu_0(\varepsilon_1\,\varepsilon_2(\varepsilon_1-\mu)(S_{1,2}-r)\,f)+\nu_0(\varepsilon_2\,\varepsilon_3(\varepsilon_1-\mu)(S_{2,3}-r)\,f))\ , \tag{5.313}$$

and where, as in the case of (5.282), we have

$$|\mathcal{R}| \le \frac{K}{N} + \beta^2\rho\, M_N^{1/2}(R_N^{1/2}+Q_N^{1/2})\ . \tag{5.314}$$

First, we use Lemma 5.7.1 to see that V $=0$,

$$\mathrm{IV} = \beta((1-\mu^2)\,\nu_0((n_1^1-\mu)\,f) - (q-\mu^2)\,\nu_0((n_1^2-\mu)\,f)) \tag{5.315}$$

$$\mathrm{VI} = \beta^2(-\mu(1-q)\,\nu_0((S_{1,2}-r)\,f) + (\widehat{\mu}-\mu q)\,\nu_0((S_{1,3}-r)\,f))\ . \tag{5.316}$$

The difficulty to study IV is that it does not work to use (5.240) to relate $\nu_0(f(n_1^1-\mu)) = \nu_0((n_1^1-\mu)^2)$ and $\nu((n_1^1-\mu)^2) = M_N$. We would get an error term $L\beta^2\rho M_N$, that is devastating (we need an extra factor m^*, that is NOT provided by the coefficient $1-\mu^2$). Rather than (5.240) we use (5.249) to get

$$\begin{aligned}\Big|\nu((n_1^1-\mu)^2) - \nu_0((n_1^1-\mu)^2) - \beta\Big(\sum_{\ell\le 2}\nu_0(\varepsilon_\ell(n_1^1-\mu)^2(n_1^\ell-\mu))\\ - 2\nu_0(\varepsilon_3(n_1^1-\mu)^2(n_1^3-\mu))\Big)\Big|\\ \le \frac{K}{N} + L\beta^4\rho^2 M_N\ .\end{aligned} \tag{5.317}$$

Now, since the set $\{|n_1^\ell-\mu|\ge\rho\}$ is negligible, we get

$$\begin{aligned}|\nu_0(\varepsilon_\ell(n_1^1-\mu)^2(n_1^\ell-\mu)| &= \mu|\nu_0((n_1^1-\mu)^2(n_1^\ell-\mu))|\\ &\le \frac{K}{N} + \mu\,\rho\,\nu_0((n_1^1-\mu)^2) \le \frac{K}{N} + L\mu\rho M_N\ ,\end{aligned}$$

using (5.239) in the last inequality to replace ν_0 bt ν. In this manner we get from (5.317) that

$$|\nu((n_1^1-\mu)^2) - \nu_0((n_1^1-\mu)^2)| \le \frac{K}{N} + L\beta^4(\rho^2+\mu\rho)\,M_N\ . \tag{5.318}$$

In a similar manner we get

$$|\nu((n_1^1-\mu)(n_1^2-\mu)) - \nu_0((n_1^1-\mu)(n_1^2-\mu))| \le \frac{K}{N} + L\beta^4(\rho^2+\mu\rho)\,M_N\ . \tag{5.319}$$

(Thanks to the coefficient $q-\mu^2$ of this term in (5.315), it would suffice here to use (5.240).) Thus, using the Cauchy-Schwarz inequality,

$$|\nu_0((n_1^1-\mu)(n_1^2-\mu))| \le M_N + \frac{K}{N} + L\beta^4(\rho^2+\mu\rho)\, M_N\ .$$

From (5.315), (5.316), and (5.319), and observing that $q = \mathsf{E}\,\mathrm{th}^2 Y \ge \mathsf{E}\,\mathrm{th}\, Y = \mu$, we then get

$$\begin{aligned}|\mathrm{IV}| &\le \beta(1+q-2\mu^2)\, M_N + L\beta^4(\rho^2+\rho\mu)\, M_N\\ &\le \beta(1+q-2\mu^2)\, M_N + L\beta^4\rho\, m^* M_N\end{aligned}$$

since $\mu \le Lm^*$ and $\rho \le Lm^*$. We then use (5.71) to get

$$|\mathrm{IV}| \le \left(1-\frac{m^{*2}}{L}+L\beta^4\rho\, m^*\right) M_N + \frac{K}{N}\ ,$$

and Proposition 5.7.6 then shows that if L_0 is large enough we have

$$|\mathrm{IV}| \le \left(1-\frac{m^{*2}}{L}\right) M_N + \frac{K}{N}\ .$$

To prove (5.284), using (5.290), it suffices to show that for $\ell = 2,3$ we have

$$|\nu_0((S_{1,\ell}-r)\, f)| \le \frac{K}{N} + L\, R_N^{1/2}\, M_N^{1/2}\ .$$

Replacing $S_{1,\ell}$ by $S_{1,\ell}^{\sim}$ as in (5.49), this follows from (5.240).

We have proved Theorem 5.9.1. □

Theorem 5.9.4. *If (α,β,h) belongs to the admissible region and $h>0$, we have*

$$|p_{N,M}(\beta,h) - F(\alpha,\beta,h,r,\mu,q)| \le \frac{K}{N} \tag{5.320}$$

where

$$\begin{aligned}F(\alpha,\beta,h,r,\mu,q) = \log 2 &- \frac{\mu^2\beta}{2} + \frac{\alpha}{2}\left(\frac{\beta q}{1-\beta(1-q)} - \log(1-\beta(1-q))\right)\\ &- \frac{\beta^2 r}{2}(1-q) + \mathsf{E}\log\mathrm{ch}(\beta z\sqrt{r}+\beta\mu+h)\ .\end{aligned} \tag{5.321}$$

Here, r,μ,q are given by (5.25) to (5.27), and K depends only on β,h.

Lemma 5.9.5. *If we consider α,β,h,r,μ,q as independent variables, the conditions (5.25) to (5.27) mean that $\partial F/\partial\mu = \partial F/\partial r = \partial F/\partial q = 0$.*

Proof. This is of course straightforward. Writing as usual $Y = \beta z\sqrt{r}+\beta\mu+h$, we have

$$\frac{\partial F}{\partial \mu} = -\mu\beta + \beta \mathsf{E}\,\mathrm{th}\, Y$$

$$\frac{\partial F}{\partial r} = -\frac{\beta^2}{2}(1-q) + \frac{\beta^2}{2}\mathsf{E}\frac{1}{\mathrm{ch}^2 Y}$$

$$\frac{\partial F}{\partial q} = \frac{\alpha}{2}\left(\frac{\beta}{D} - \frac{\beta^2 q}{D^2} - \frac{\beta}{D}\right) + \frac{\beta^2 r}{2} = \frac{\beta^2}{2}\left(\frac{\alpha q}{D^2} - r\right),$$

where $D = 1 - \beta(1-q)$. □

Thus, if we consider the quantity (5.321) as a function of (α, β, h) only, we have

$$\frac{\partial F}{\partial \alpha} = f(\alpha,\beta,h) := \frac{1}{2}\left(\frac{\beta q}{1-\beta(1-q)} - \log(1-\beta(1-q))\right). \tag{5.322}$$

Lemma 5.9.6. *We have*

$$\left| f(\alpha,\beta,h) - \mathsf{E}\log\left\langle \exp\frac{N\beta}{2} m_M^2\right\rangle_{\sim}\right| \le \frac{K}{N}. \tag{5.323}$$

Proof. We fix M, N, β, h. We consider

$$\varphi(u) = f(\alpha, u, h);\, \psi(u) = \mathsf{E}\log\left\langle \exp\frac{Nu}{2} m_M^2\right\rangle_{\sim}$$

so that $\varphi(0) = \psi(0) = 0$,

$$\psi'(u) = \frac{N}{2}\mathsf{E}\frac{\langle m_M^2 \exp\frac{Nu}{2} m_M^2\rangle_{\sim}}{\langle \exp\frac{Nu}{2} m_M^2\rangle_{\sim}}$$

and

$$\begin{aligned}\varphi'(u) &= \frac{1}{2}\left(\frac{q}{1-u(1-q)} + \frac{uq(1-q)}{(1-u(1-q))^2} + \frac{1-q}{1-u(1-q)}\right)\\ &= \frac{1}{2}\frac{1-u(1-q)^2}{(1-u(1-q))^2}.\end{aligned}$$

To prove (5.323), it suffices to prove that if $0 \le u \le \beta$, we have

$$\left| N\mathsf{E}\langle m_M^2\rangle_u - \frac{1-u(1-q)^2}{(1-u(1-q))^2}\right| \le \frac{K}{N}, \tag{5.324}$$

where $\langle\cdot\rangle_u$ denotes average for Gibbs' measure corresponding to the Hamiltonian (5.137)

$$-H_{N,M,u}(\boldsymbol{\sigma}) = \frac{N\beta}{2} \sum_{k \le M-1} m_k^2(\boldsymbol{\sigma}) + \frac{Nu}{2} m_M^2(\boldsymbol{\sigma}) + hNm_1(\boldsymbol{\sigma}) ,$$

and where K depends only on β, h, but not on M or u.

It is for the purpose of proving (5.324) that we have been interested in the Hamiltonian (5.137). As was mentioned at the end of Section 5.5, our results extend to this Hamiltonian, and in particular it follows from Theorem 5.9.1 that

$$\mathsf{E}\,\langle (R_{1,2} - q)^2 \rangle_u \le \frac{K}{N} . \tag{5.325}$$

We will show in Theorem 5.10.1 that $\nu((R_{1,2} - q)^4) \le K/N^2$ and in Lemma 5.10.2 that $|\nu(R_{1,2} - q)| \le K/N$. The same arguments show that

$$\mathsf{E}\,\langle (R_{1,2} - q)^4 \rangle_u \le \frac{K}{N^2} \tag{5.326}$$

$$|\mathsf{E}\,\langle (R_{1,2} - q) \rangle_u| \le \frac{K}{N} . \tag{5.327}$$

We can repeat the proof of Theorem 5.8.1 for $\langle \cdot \rangle_u$, if we replace β by u in the expressions of r', s'. Using (5.194), (5.326) and Hölder's inequality we get that

$$|N\mathsf{E}\langle (R_{\ell,p} - q)\, n_M^1\, n_M^\ell (R_{a,b} - q) \rangle_u| \le \frac{K}{N} .$$

Using (5.258) and (5.259) for $f = R_{\ell,p} - q$, we deduce that

$$|N\mathsf{E}\langle (R_{\ell,p} - q)\, n_M^1\, n_M^\ell \rangle_u| \le \frac{K}{N}$$

and using (5.258) for $f = 1$, we get that

$$\left| N\mathsf{E}\langle (n_M^1)^2 \rangle_u - \frac{1 - u(1-q)^2}{(1 - u(1-q))^2} \right| \le \frac{K}{N}$$

and this proves (5.324) and (5.323). □

The proof of Theorem 5.9.4 is then completed as the proof of Theorem 3.4.2. □

Research problem 5.9.7. (Level 1) Given (α, β, h) in the admissible region and an integer n, prove that there exist independent standard normal r.v. $(z_i)_{i \le n}$ such that

$$\forall i \le n, \quad \mathsf{E}(\langle \sigma_i \rangle - \mathrm{th}(\beta z_i \sqrt{r} + \beta\mu + h))^2 \le \frac{K}{N} . \tag{5.328}$$

The approach most likely to succeed is that of Theorem 2.4.12. Along that route, the following result, interesting in its own right, is likely to be a key step.

Proposition 5.9.8. *If (α, β, h) belongs to the admissible region and $h > 0$, then*

$$\mathsf{E}\left(\langle\sigma_N\rangle - \operatorname{th}\left(\beta \sum_{2\le k\le M} \eta_k \langle n_k\rangle_- + \beta\mu + h\right)\right)^2 \le \frac{K}{N} \tag{5.329}$$

where $\langle\cdot\rangle_-$ denotes Gibbs' measure relative to the Hamiltonian (5.213).

Even after this is proved, to solve Problem 4.9.2 there remain obstacles and in particular the need to replace the variables $\sum_{2\le k\le M} \eta_k \langle n_k\rangle_-$ (at $\langle n_k\rangle_-$ fixed) by actual Gaussian r.v., and it is not obvious how to do this while preserving the correct rate in (5.328) (so, let us point out that the whole point of Research problem 5.9.7 is to get the rate K/N).

Proof. This proof requires to know that

$$|\nu(m_1^1 - \mu)| \le \frac{K}{N} \tag{5.330}$$

$$|\nu(S_{1,2} - r)| \le \frac{K}{N} \tag{5.331}$$

and these will be proved in the next section. To prove (5.329) we consider

$$\varphi(t) = \mathsf{E}((\langle\sigma_N\rangle_t - \operatorname{th} Y_t)^2)$$

where $\langle\cdot\rangle_t$ is Gibbs' measure for the Hamiltonian (5.216) and where

$$Y_t = \beta\sqrt{t} \sum_{2\le k\le M} \eta_k \langle n_k\rangle_- + \sqrt{1-t}\,\beta z\sqrt{r} + \beta\mu + h$$

(so that $\varphi(0) = 0$ by Lemma 5.7.1).

We study separately the functions

$$\varphi_1(t) = \mathsf{E}(\langle\sigma_N\rangle_t^2) = \mathsf{E}(\langle\sigma_N^1\,\sigma_N^2\rangle_t) = \nu_t(\varepsilon_1\,\varepsilon_2)$$

$$\varphi_2(t) = \mathsf{E}(\langle\varepsilon_1\rangle_t \operatorname{th} Y_t)$$

$$\varphi_3(t) = \mathsf{E}(\operatorname{th}^2 Y_t)\ .$$

If we denote by $S_{\ell,-}$ the sum $\sum_{2\le k\le M} n_k^\ell \langle n_k\rangle_-$, computing using approximate integration by parts as in the proof of Theorem 5.7.8 we find

$$\varphi_2'(t) = \mathrm{VII} + \mathrm{VIII} + \mathrm{IX} + \mathrm{X} + \mathrm{XI} + \mathcal{R}$$

where

$$\mathrm{VII} = \beta\,\mathsf{E}((\langle n_1^1 - \mu\rangle_t - \langle \varepsilon_1\,\varepsilon_2(n_1^2 - \mu)\rangle_t)\,\mathrm{th}\,Y_t)$$

$$\mathrm{VIII} = \frac{\beta^2}{2}\,\mathsf{E}((\langle \varepsilon_1(S_{1,1} - s)\rangle_t - \langle \varepsilon_1(S_{2,2} - s)\rangle_t)\,\mathrm{th}\,Y_t)$$

$$\mathrm{IX} = \beta^2\,\mathsf{E}((-\langle \varepsilon_2(S_{1,2} - r)\rangle_t + \langle \varepsilon_1\,\varepsilon_2\,\varepsilon_3(S_{2,3} - r)\rangle_t)\,\mathrm{th}\,Y_t)$$

$$\mathrm{X} = \frac{\beta^2}{2}\,\mathsf{E}\left(\langle \varepsilon_1\rangle_t(\langle S_{1,2}\rangle_- - r)\,\frac{2\,\mathrm{th}\,Y_t}{\mathrm{ch}^2\,Y_t}\right)$$

$$\mathrm{XI} = \beta^2\,\mathsf{E}\left((\langle S_{1,-} - r\rangle_t - \langle \varepsilon_1\,\varepsilon_2(S_{2,-} - r)\rangle_t)\,\frac{1}{\mathrm{ch}^2\,Y_t}\right)$$

and

$$|\mathcal{R}| \le K/N\ .$$

We write

$$|\varphi_2(1) - \varphi_2(0)| = \left|\int_0^1 \varphi_2'(t)\,\mathrm{d}t\right| \le \frac{K}{N} + \left|\int_0^1 \psi(t)\,\mathrm{d}t\right| \tag{5.332}$$

where $\psi(t) = \mathrm{VII} + \mathrm{VIII} + \mathrm{IX} + \mathrm{X} + \mathrm{XI}$.

Next, we prove that $|\psi(0)| \le K/N$. It suffices to show that

$$|\mathsf{E}\,\langle n_1 - \mu\rangle_-| \le \frac{K}{N} \tag{5.333}$$

$$|\mathsf{E}\,\langle S_{1,2} - r\rangle_-| \le \frac{K}{N}\ . \tag{5.334}$$

To prove (5.333) we use that (5.330) means

$$|\mathsf{E}\,\langle m_1^1 - \mu\rangle| \le \frac{K}{N}$$

and using this for $N-1$ rather than N we get $|\mathsf{E}\,\langle n_1^1 - \mu_-\rangle_-| \le K/N$ where μ_-, q_-, r_- are given by the equations (5.25) to (5.27) when α is replaced by $\alpha_- = M/(N-1)$. It is simple to see that $|\mu_- - \mu| \le K/N$, by computing $\partial\mu/\partial\alpha$. In a similar manner, we have $|r - r_-| \le K/N$, and to prove (5.334) it suffices to prove that $\left|\mathsf{E}\left\langle \sum_{2\le k\le M} m_k^1\, m_k^2 - r\right\rangle\right| \le K/N$ and to change N into $N-1$. Since $m_k^1 = n_k^1 + \eta_{N,k}\,\sigma_N/N$, it suffices to show that

$$\left|\mathsf{E}\sum_{2\le k\le M} \eta_{N,k}\langle \sigma_N^1\, n_k^2\rangle\right| \le K\ ,$$

a simple matter using approximate integration by parts.

To control the right-hand side of (5.332), we write

$$\left|\int_0^1 \psi(t)\,\mathrm{d}t\right| \le |\psi(0)| + \sup|\psi'(t)| \le \frac{K}{N} + \sup|\psi'(t)|\ .$$

We claim that

$$|\psi'(t)| \le \frac{K}{N}, \tag{5.335}$$

so that

$$|\varphi_2(1) - \varphi_2(0)| \le \frac{K}{N}. \tag{5.336}$$

To prove (5.335), we compute $\psi'(t)$ in the same manner we computed φ_2', and we see that it is enough to show that

$$\mathsf{E}\,\langle (S_{1,2} - r)^2\rangle_- \le \frac{K}{N} \tag{5.337}$$

$$\mathsf{E}\,\langle (S_{1,2} - r)^2\rangle_t \le \frac{K}{N} \tag{5.338}$$

$$\mathsf{E}\,\langle (S_{1,-} - r)^2\rangle_t \le \frac{K}{N} \tag{5.339}$$

$$\nu_t((S_{1,1} - s)^2) \le \frac{K}{N}\,;\ \nu_t((n_1^1 - \mu)^2) \le \frac{K}{N}. \tag{5.340}$$

Even though (5.233) was proved for a non-random function f, it still holds, with the same proof for a random function that does not depend on the r.v. $(\eta_i)_{i\le N}$ (e.g. for $f = (S_{1,-} - r)^2$ or $f = (n_1^1 - \mu)^2$). Thus (5.340) follows from (5.233) and Theorem 5.9.1. In the right-hand side of (5.233) we can replace $\nu(|f|)$ by $\nu_0(|f|)$ (with the same proof). Moreover, for a function f^- on Σ_{N-1}^p, we have $\langle f\rangle_0 = \langle f\rangle_-$, so that (5.338) follows from (5.337), and so does (5.339), using Jensen's inequality. As just explained, to prove (5.337) it suffices to show that $\nu_0((S_{1,2} - r)^2) \le K/N$, and this follows from (5.233) and Theorem 5.9.1.

To finish the proof, one then establishes by similar methods the result corresponding to (5.336) for φ_1 and φ_3. □

5.10 Higher Moments and Central Limit Theorems

Our first result concerns higher moments.

Theorem 5.10.1. *Under the conditions of Theorem 5.9.1, given an integer k we have*

$$\nu((R_{1,2} - q)^{2k}) \le \frac{K(k)}{N^k} \tag{5.341}$$

$$\nu((n_1^1 - \mu)^{2k}) \le \frac{K(k)}{N^k} \tag{5.342}$$

$$\nu((S_{1,1} - s)^{2k}) = \nu\left(\left(\sum_{2\le k\le M} (n_k^1)^2 - s\right)^{2k}\right) \le \frac{K(k)}{N^k} \tag{5.343}$$

$$\nu((S_{1,2}-r)^{2k}) = \nu\left(\left(\sum_{2\le k\le M} n_k^1 n_k^2 - r\right)^{2k}\right) \le \frac{K(k)}{N^k} . \tag{5.344}$$

As in Theorem 5.9.1, we can replace the quantities n_k^ℓ by m_k^ℓ if we wish. If one is a little bit cautious, one can also show that

$$K(k) = (Kk)^k , \tag{5.345}$$

where K depends on β, h only, a result corresponding to the "exponential inequalities" of Chapters 2, 3.

The reader familiar with Section 2.5 and Section 3.5 should be able to adapt the proof of Theorem 5.9.1 in order to prove Theorem 5.10.1 by induction over k, and it does not seem appropriate to rewrite that proof. The reason why we did not state Theorem 5.10.1 in the stronger version (5.345) is that we will not need it, and that it is probably easier for the reader not wishing to write the proof of Theorem 5.10.1 himself to feel that it is quite obvious that this should be true.

The "second order expansion" needed for actual computations of moments has already been stated in Theorem 5.7.8 and we will (as in Chapters 2 and 3) replace (5.248) by the simpler bound

$$|\mathcal{R}| \le K\frac{\nu(f^2)^{1/2}}{N} + K\sup|f|\exp\left(-\frac{N}{K}\right) . \tag{5.346}$$

This will be used only in cases where $|f| \le N^K$ (e.g. $f = S_{1,2} - r$), and then $\sup|f|\exp(-N/K) \le K\exp(-N/K)$.

Another useful consequence of (5.244) (and (5.233), Theorem 5.9.1) is that

$$|\nu(f) - \nu_0(f)| \le \frac{K\nu(f^2)^{1/2}}{\sqrt{N}} + K\sup|f|\exp\left(-\frac{N}{K}\right) . \tag{5.347}$$

The reader, having mastered Sections 2.6 and 3.6, surely knows what methods we will use. Still, it might be useful to first cover a simple case. The following was used in the proof of Proposition 5.9.8. We recall that we denote by $O(k)$ a quantity U such that $|U| \le K/N^{k/2}$.

Lemma 5.10.2. *Under the conditions of Theorem 5.10.1, we have*

$$\nu(n_1 - \mu) = O(2) \tag{5.348}$$

$$\nu(S_{1,2} - r) = O(2) \tag{5.349}$$

$$\nu(R_{1,2} - q) = O(2) . \tag{5.350}$$

Proof. We write $f = \varepsilon_1 - \mu$, so that by Theorem 5.7.8 we have,

$$\nu(m_1 - \mu) = \nu(f) = \nu_0(f) + \mathrm{I} + \mathrm{II} + \mathrm{III} + O(2) \tag{5.351}$$

where

$$\mathrm{I} = \beta(\nu_0((\varepsilon_1 - \mu)\,\varepsilon_1(n_1^1 - \mu)) - \nu_0((\varepsilon_1 - \mu)\,\varepsilon_2(n_1^2 - \mu)))$$

$$\mathrm{II} = \frac{\beta^2}{2}(\nu_0((\varepsilon_1 - \mu)(S_{1,1} - s)) - \nu_0((\varepsilon_1 - \mu)(S_{2,2} - s)))$$

$$\mathrm{III} = \beta^2(-\nu_0(\varepsilon_1\,\varepsilon_2(\varepsilon_1 - \mu)(S_{1,2} - r)) + \nu_0(\varepsilon_2\,\varepsilon_3(\varepsilon_1 - \mu)(S_{2,3} - r)))\,.$$

Using Lemma 5.7.1, we have

$$\nu_0((\varepsilon_1 - \mu)\,\varepsilon_1(n_1^1 - \mu)) = (1 - \mu^2)\,\nu_0(n_1^1 - \mu)$$

and using (5.347)

$$\nu_0((\varepsilon_1 - \mu)\,\varepsilon_1(n_1^1 - \mu)) = (1 - \mu^2)\,\nu(n_1^1 - \mu) + O(2)\,.$$

In this manner we find, using the notation of (5.288), that

$$\mathrm{I} = \beta(1 - q)\,\nu(n_1 - \mu) + O(2)$$

$$\mathrm{II} = O(2)$$

$$\mathrm{III} = \beta^2(\widehat{\mu} - \mu)\,\nu(S_{1,2} - r) + O(2)\,.$$

Finally we get

$$(1 - \beta(1 - q))\,\nu(m_1 - \mu) = \beta^2(\widehat{\mu} - \mu)\,\nu(S_{1,2} - r) + O(2)\,. \tag{5.352}$$

We now use (5.275) with $f = 1$ to see that

$$\nu(S_{1,2}) = r + \frac{U_2}{D} + \frac{\beta q}{D^2}\left(\sum_{p \le 3} U_p - 3\,U_3\right) + O(2) \tag{5.353}$$

where

$$U_p = \alpha(1 - \delta_{1,p})\,\nu(R_{1,p} - q) + \beta \sum_{\ell \ne p, \ell \le 3} \nu((R_{\ell,p} - q)\,S_{1,\ell}) - 3\beta\nu((R_{p,4} - q)\,S_{1,4})\,.$$

Now we have

$$\nu((R_{\ell,p} - q)\,S_{1,1}) = s\,\nu(R_{\ell,p} - q) + O(2)$$

$$\nu((R_{\ell,p} - q)\,S_{1,\ell}) = r\,\nu(R_{\ell,p} - q) + O(2) \quad \text{if } \ell \ne 1\,.$$

Thus we have

$$U_1 = -\beta\,r\,\nu(R_{1,2} - q) + O(2)$$

$$U_2 = U_3 = (\alpha + \beta(s-2r))\,\nu(R_{1,2}-q) + O(2)$$
$$U_1 + U_2 + U_3 - 3U_3 = (-\alpha + \beta(r-s))\,\nu(R_{1,2}-q) + O(2)\,.$$

We observe that

$$-\alpha + \beta(r-s) = -\alpha(1+\beta(s'-r')) = -\alpha\left(1 + \frac{\beta(1-q)}{D}\right) = -\frac{\alpha}{D}$$

so that (5.353) gives

$$\nu(S_{1,2}-r) = \left(\frac{\alpha+\beta(s-2r)}{D} - \frac{\alpha\,\beta\,q}{D^3}\right)\nu(R_{1,2}-q) + O(2)\,.$$

Using again that $\alpha + \beta(s-r) = \alpha/D$ we get

$$\nu(S_{1,2}-r) = \alpha\left(\frac{1}{D^2} - \frac{2\,\beta\,q}{D^3}\right)\nu(R_{1,2}-q) + O(2)\,. \tag{5.354}$$

In a last effort, we use again Theorem 5.7.8 to write

$$\nu(R_{1,2}-q) = \nu(\varepsilon_1\,\varepsilon_2 - q) = \nu_0(\varepsilon_1\,\varepsilon_2 - q) + \mathrm{IV} + \mathrm{V} + \mathrm{VI} + O(2) \tag{5.355}$$

where

$$\mathrm{IV} = \beta\left(\sum_{\ell\le 2}\nu_0((\varepsilon_1\,\varepsilon_2 - q)\,\varepsilon_1(n_1^\ell - \mu)) - 2\nu_0((\varepsilon_1\,\varepsilon_2 - q)\,\varepsilon_3(n_1^3 - \mu))\right)$$

$$\mathrm{V} = \frac{\beta^2}{2}\left(\sum_{\ell\le 2}\nu_0((\varepsilon_1\,\varepsilon_2 - q)(S_{\ell,\ell} - s) - 2\nu_0((\varepsilon_1\,\varepsilon_2 - q)(S_{3,3} - s))\right)$$

$$\mathrm{VI} = \beta^2(\nu_0((\varepsilon_1\,\varepsilon_2 - q)\,\varepsilon_1\,\varepsilon_2(S_{1,2} - r)) - 2\sum_{\ell\le 2}\nu_0((\varepsilon_1\,\varepsilon_2 - q)\,\varepsilon_\ell\,\varepsilon_3(S_{\ell,3} - r)$$
$$+ 3\nu_0((\varepsilon_1\,\varepsilon_2 - q)\,\varepsilon_3\,\varepsilon_4(S_{3,4} - r)))\,.$$

Proceeding as in the proof of (5.352), and using the notation of (5.288), we get

$$\mathrm{IV} = 2\beta(\mu - \widehat{\mu})\,\nu(n_1^1 - \mu) + O(2)$$
$$\mathrm{V} = O(2)$$
$$\mathrm{VI} = \beta^2(1 - 4q + 3\widehat{q})\,\nu(S_{1,2} - r) + O(2)$$

so that (5.355) yields

$$\nu(R_{1,2}-q) = 2\beta(\mu-\widehat{\mu})\,\nu(m_1^1-\mu) + \beta^2(1-4q+3\widehat{q})\,\nu(S_{1,2}-r) + O(2)\,. \tag{5.356}$$

Finally the only task is to show that (5.352), (5.354) and (5.356) imply (5.348) to (5.350). Setting $A = \nu(m_1 - \mu)$, $B = \nu(S_{1,2} - r)$, $C = \nu(R_{1,2} - q)$, we have

$$(1-\beta(1-q))\,A=\beta(\widehat{\mu}-\mu)\,B+O(2) \tag{5.357}$$

$$B=\alpha\left(\frac{1}{D^2}-\frac{2\,\beta\,q}{D^3}\right)C+O(2) \tag{5.358}$$

$$C=2\beta(\mu-\widehat{\mu})\,A+\beta^2(1-4\,q+3\,\widehat{q})\,B+O(2)\,. \tag{5.359}$$

Substituting (5.358) into (5.359) we get

$$C=\frac{2\beta(\mu-\widehat{\mu})}{1-U}\,A+O(2) \tag{5.360}$$

where

$$U=\alpha\,\beta^2(1-4\,q+3\,\widehat{q})\left(\frac{1}{D^2}-\frac{2\,\beta\,q}{D^3}\right)$$

and using (5.358) and (5.360) we get from (5.357) that

$$(1-\beta(1-q))\,A=W A+O(2)$$

where

$$W=-\frac{2\,\beta^3\,\alpha(\mu-\widehat{\mu})^2}{1-U}\left(\frac{1}{D^2}-\frac{2\,\beta\,q}{D^3}\right)\,.$$

So it suffices to show that, in the admissible region, if L_0 is large enough, we have

$$U\le 1/2 \qquad 1-\beta(1-q)-W<1\,.$$

This is done as usual, distinguishing the cases $\beta\ge 2$, $\beta\le 2$. □

Even the simple Lemma 5.10.2 required significant work, so we must get organized if we want to go further. We must keep in mind however that the term VI in Theorem 5.7.8 creates a new situation compared to the models of Chapters 2 and 3. Such a term already occurs in the SK model with ferromagnetic interaction of (5.20), and this simpler model should be studied first. The "replica-symmetric" equations for this model are

$$\mu=\mathsf{E}\,\mathrm{th}(\beta_2 z\sqrt{q}+\beta_1\mu+h) \tag{5.361}$$

$$q=\mathsf{E}\,\mathrm{th}^2(\beta_2 z\sqrt{q}+\beta_1\mu+h)\,. \tag{5.362}$$

Research problem 5.10.3. (Level 1) For the SK model with ferromagnetic interaction, describe the joint asymptotic behavior of the quantities $m_1^\ell-\mu$ and $R_{\ell,\ell'}-q$.

To study this problem, it seems wise to consider the quantities $T_{\ell,\ell'}$, T_ℓ, T as in Sections 2.6, 3.6, and the quantities

$$U_\ell=m_1^\ell-\langle m_1\rangle\ ;\ U=\langle m_1\rangle-\mu\,.$$

It seems that the structure is as follows: The quantities $T_{\ell,\ell'}$ are (asymptotically) independent and independent of the quantities T_ℓ, T, U_ℓ, U. The pairs (T_ℓ, U_ℓ), as ℓ varies, are independent; however T_ℓ and U_ℓ *are* correlated, and this makes the computations more complicated. If we set $Y = \beta_2 z\sqrt{q}+\beta_1\mu+h$,

$$\widehat{\mu} = \mathsf{E}\,\mathrm{th}^3 Y\,, \qquad \widehat{q} = \mathsf{E}\,\mathrm{th}^4 Y\,,$$

and

$$A = \beta_2^2(1-4q+3\widehat{q})\,; \qquad C = \beta_1(1-q)$$
$$D = -2\beta_2^2(\mu-\widehat{\mu})\,; \qquad B = \beta_1^2(q-\widehat{q})$$

one can show (in a suitable domain of parameters) that

$$\nu(T_1^2) = \frac{1}{N}\frac{(1-C)(q-\widehat{q})+B(\mu-\widehat{\mu})}{(1-A)(1-C)-BD} + O(3)$$

$$\nu(U_1 T_1) = \frac{1}{N}\frac{D(q-\widehat{q})+(1-A)(\mu-\widehat{\mu})}{(1-A)(1-C)-BC} + O(3)$$

$$\nu(U_1^2) = \frac{1}{N}\frac{(1-A)(1-q)+D(\mu-\widehat{\mu})}{(1-A)(1-C)-BC} + O(3)\,,$$

but these formulas are not very encouraging.

Research problem 5.10.4. (Level 1) For the Hopfield model describe the joint asymptotic behavior of the quantities $m_1^\ell - \mu$ and $R_{\ell,\ell'} - q$.

One may also be more ambitious, and want to include the quantities $n_k^\ell n_k^{\ell'} - r'$, $(n_k^\ell)^2 - s'$, $S_{\ell,\ell'} - r$, etc. Since (roughly speaking) Problem 4.10.2 is to Problem 4.10.1 what Theorem 3.6.4 is to Theorem 2.7.3, one should certainly not attempt to solve Problem 4.10.2 before Problem 4.10.1.

We will content ourselves with the part of the program of Problem 4.10.2 that can be achieved as in Chapters 2 and 3.

Theorem 5.10.5. *In the admissible region formula (2.302) holds for*

$$A = \frac{1-2q+\widehat{q}}{N\left(1-\frac{\alpha\beta^2(1-2q+\widehat{q})}{(1-\beta(1-q))^2}\right)}\,. \tag{5.363}$$

If we follow the proof of (2.302), to prove Theorem 5.10.5, it suffices to prove the following

Proposition 5.10.6. *Consider a function f^- on Σ_{N-1}^n, and assume that f^- is the product of k terms of the type $T^-_{\ell,\ell'}$. Then we have*

$$\nu((\varepsilon_1-\varepsilon_2)(\varepsilon_3-\varepsilon_4)\,f^-) = \frac{\alpha\beta^2(1-2q+\widehat{q})}{D^2}\,\nu((T_{1,3}-T_{1,4}+T_{2,3}-T_{2,4})\,f^-)$$
$$+\,O(k+1)\,.$$

This is shown by combining the next two results.

Lemma 5.10.7. *We have*

$$\nu((\varepsilon_1-\varepsilon_2)(\varepsilon_3-\varepsilon_4)\,f^-) = \beta^2(1-2q+\widehat{q})\,\nu((S_{1,3}-S_{1,4}-S_{2,3}+S_{2,4})\,f^-) \\ + O(k+1)\,.$$

Proof. This follows the proof of (2.259), using now (5.244) rather than (2.248), and using (5.341) to see that the contribution of the term $\nu((f^-)^4)^{1/4}$ is $O(k-1)$. □

We write again $D = 1-\beta(1-q)$.

Lemma 5.10.8. *We have*

$$\nu((S_{1,3}-S_{1,4}-S_{2,3}+S_{2,4})\,f^-) = \frac{\alpha}{D^2}\,\nu((R_{1,3}-R_{1,4}-R_{2,3}+R_{2,4})\,f^-) \\ + O(k+1)\,. \tag{5.364}$$

Proof. We use Theorem 5.8.5, i.e.

$$\nu(S_{1,2}\,f^-) = r\nu(f^-) + \frac{U_2}{D} + \frac{\beta q}{D^2}\left(\sum_{p\le n} U_p - n\,U_n\right) + O(k+1) \tag{5.365}$$

for

$$U_p = (1-\delta_{1,p})\,\alpha\,\nu((R_{1,p}-q)\,f^-) + \beta\sum_{\ell\ne p}\nu((R_{\ell,p}-q)\,S_{1,\ell}\,f^-) \\ - n\,\beta\,\nu((R_{p,n+1}-q)\,S_{1,n+1}\,f^-)\,.$$

Using that $S_{1,\ell}$ is approximately $r+\delta_{1,\ell}(s-r)$ we have

$$\begin{aligned} U_p &= (1-\delta_{1,p})\,\alpha\,\nu((R_{1,p}-q)\,f^-) \\ &\quad + \beta\sum_{\ell\ne p}(r+\delta_{1,\ell}(s-r))\,\nu((R_{\ell,p}-q)\,f^-) \\ &\quad - n\,\beta\,r\,\nu((R_{p,n+1}-q)\,f^-) + O(k+1)\,. \end{aligned} \tag{5.366}$$

Let us now write $U_p(1,2)$ (etc.) rather than U_p to indicate that this quantity is relevant to the computation of $\nu(S_{1,2}\,f^-)$. Distinguishing the cases $p=1$ and $p\ge 2$, we see that

$$U_p(1,2) = U_p'(1,2) + U_p''(1,2) + O(k+1)$$

where

$$U_p'(1,2) = (1-\delta_{1,p})(\alpha+\beta(s-r))\,\nu((R_{1,p}-q)\,f^-)$$

$$U''_p(1,2) = \beta r \left(\sum_{\ell \le n} \nu((R_{\ell,p} - q)\, f^-) - n\, \nu((R_{p,n+1} - q)\, f^-) \right) .$$

We observe that

$$\alpha + \beta(s - r) = \alpha(1 + \beta(s' - r')) = \alpha \left(1 + \frac{\beta(1-q)}{D} \right) = \frac{\alpha}{D} \tag{5.367}$$

and (5.365) becomes

$$\nu(S_{1,2}\, f^-) = r\, \nu(f^-) + \frac{\alpha}{D^2}\, \nu((R_{1,2} - q)\, f^-) + \frac{\beta\, r}{D}\, U''_2(1,2) + \frac{\beta\, q}{D^2}\, W(1,2) + O(k+1)$$

where

$$W(1,2) = \sum_{p \le n} U_p(1,2) - n\, U_n(1,2) .$$

The point is that the quantity $W(1,2)$ does not depend on the index 2. This is apparent on the formulas (5.366). Thus

$$W(1,3) = W(1,4) \; ; \quad W(2,3) = W(2,4)$$

so

$$W(1,3) - W(1,4) + W(2,3) + W(3,4) = 0 .$$

Similarly, the quantity $U''_2(1,2)$ does not depend on the index 1, so it also cancels out when computing the left-hand side of (5.364), i.e.

$$U''_3(1,3) - U''_4(1,4) - U''_3(2,3) + U''_4(2,4) = 0 .$$

Thus, in computing the left-hand side of (5.364) through (5.365), only the contribution of the term $U'_2(1,2)$ remains, and this contribution is

$$\frac{1}{D}\, (U'_3(1,3) - U'_4(1,4) - U'_3(2,3) + U'_4(2,4)) = \frac{\alpha + \beta(s-r)}{D}\, \nu((R_{1,3} - R_{1,4} - R_{2,3} + R_{2,4})\, f^-) ,$$

and we use again (5.367) to conclude. □

We have proved Theorem 5.10.5.

5.11 The *p*-Spin Hopfield Model

Consider an integer $p \ge 3$. Keeping the notation $m_k(\boldsymbol{\sigma}) = N^{-1} \sum_{i \le N} \eta_{i,k}\, \sigma_i$, we consider the Hamiltonian

$$-H_{N,M}(\boldsymbol{\sigma}) = \frac{N\beta}{p} \sum_{k\le M} m_k^p(\boldsymbol{\sigma}) + Nh\,m_1(\boldsymbol{\sigma})\,. \tag{5.368}$$

The aim of this section is to show how to adapt part of the results of the previous sections to the present case. Despite superficial similarities, the case $p \ge 3$ is significantly different from the case $p = 2$. In particular, integration by parts will behave in a fundamentally different way. This makes more apparent what a remarkable and exceptional object the Hopfield model is.

The results of the present section are less accurate and less complete than those about the case $p = 2$.

Research problem 5.11.1. (Level 2) Study the p-spin Hopfield model for $p \ge 3$ with the same accuracy as was achieved in the previous sections for the case $p = 2$. In particular, find the correct shape of the admissible region.

Studying the Hamiltonian (5.368) requires doing again much of the work of Sections 5.1 to 5.10. For this reason, and because it is likely that readers who have reached this point and yet are still willing to read the details of the proofs are both skilled and motivated, we will depart from the general policy of the book and we will not provide all the details of the arguments when these are simple variations on previous themes. The style of writing of this section is closer to the usual style of a research paper than of a book. For this reason, the non-expert reader should directly jump to the last two pages of the section, where, after the proof of Proposition 5.11.13 we explain the nicest feature of this model, the different behavior of integration by parts.

For definiteness, and to avoid having to distinguish cases, we assume that p is *even*. We start with the case $M = 1$, and we assume as usual that $\eta_{i,1} = 1$. As in Section 5.1, we see that

$$\lim_{N\to\infty} p_N(\beta, h) = \log 2 + \sup_t \left(\frac{\beta\, t^p}{p} + ht - \mathcal{I}(t) \right) \tag{5.369}$$

and the supremum is obtained at a value such that $\beta\, t^{p-1} + h = \mathcal{I}'(t)$ i.e.

$$t = \operatorname{th}(\beta\, t^{p-1} + h)\,. \tag{5.370}$$

If h is small enough (the only case we consider), and β is small this equation has a unique solution. If β is large it (seems to) have three solutions. There is a value $\beta(h)$ such that if $\beta > \beta(h)$ the largest solution of (5.370) is the one that achieves the supremum in (5.369). (Of course $\beta(h)$ depends also on p, but the value of p is fixed once and for all.) This solution will be denoted by $m^* = m^*(\beta, h)$, so that

$$m^* = \operatorname{th}(\beta(m^*)^{p-1} + h)\,. \tag{5.371}$$

For $\mathbf{x}$ in $\mathbb{R}^M$, we write

$$|||\mathbf{x}||| = \left(\sum_{k\le M} x_k^{2(p-1)}\right)^{1/2(p-1)} . \tag{5.372}$$

Theorem 5.11.2. *Given $\beta > \beta(h)$, and given $\rho > 0$, there exists a number $\alpha_0 > 0$ such that if $M \le \alpha_0 N^{p-1}$, the set*

$$\{\boldsymbol{\sigma}\,;\ \forall k \le M\,,\ |||\mathbf{m}(\boldsymbol{\sigma}) - m^*\mathbf{e}_k||| \ge \rho\} \tag{5.373}$$

is negligible.

Despite a formal similarity with the case $p = 2$, the reasons why this holds are very different. We also observe that the most interesting case is when M is a proportion of N^{p-1}, so that we write

$$\alpha = \frac{M}{N^{p-1}} . \tag{5.374}$$

Without loss of generality we assume $\alpha \le 1$.

Lemma 5.11.3. *If*

$$2^{-j} \ge \left(\frac{8\log 2M}{N}\right)^{1/2} , \tag{5.375}$$

then with overwhelming probability, for each $\boldsymbol{\sigma}$ on Σ_N, we have

$$\operatorname{card}\{k \le M\,;\ |m_k(\boldsymbol{\sigma})| \ge 2^{-j}\} \le 2^{2j+3} . \tag{5.376}$$

Proof. Given $\boldsymbol{\sigma}$ and k, by the subgaussian inequality (A.3) we have

$$\mathsf{P}\,(|m_k(\boldsymbol{\sigma})| \ge 2^{-j}) \le 2\exp\left(-\frac{N2^{-2j}}{2}\right) .$$

Consider a subset I of $\{1,\dots,M\}$, with $n = \operatorname{card} I$. By independence, given $\boldsymbol{\sigma}$, we have

$$\mathsf{P}\,(\forall k \in I,\ |m_k(\boldsymbol{\sigma})| \ge 2^{-j}) \le 2^n \exp\left(-\frac{nN2^{-2j}}{2}\right) .$$

Thus, using that there are at most M^n choices for I, we see that

$$\mathsf{P}\,(\exists \boldsymbol{\sigma} \in \Sigma_N,\ \operatorname{card}\{k\,;\ |m_k(\boldsymbol{\sigma})| \ge 2^{-j}\} \ge n) \le 2^N M^n\, 2^n \exp\left(-\frac{nN2^{-2j}}{2}\right)$$

$$\le \exp\left(N + n\log(2M) - \frac{nN2^{-2j}}{2}\right) .$$

If $n = 2^{2j+2}$, $\log 2M \le 2^{-2j}N/8$ the exponent in the last term is at most

$$N + n\left(\log(2M) - \frac{N2^{-2j}}{2}\right) \le N - \frac{3nN2^{-2j}}{8} \le N - \frac{3N}{2} = -\frac{N}{2},$$

and this finishes the proof. □

We denote by j_0 the largest integer that satisfies

$$2^{-j_0} \ge \left(\frac{8\log 2N^{p-1}}{N}\right)^{1/2}. \tag{5.377}$$

Thus j_0 satisfies (5.375), since we assume $M \le N^{p-1}$.

Corollary 5.11.4. *With overwhelming probability, given $0 \le j \le j_0$, for each $\boldsymbol{\sigma}$ in Σ_N we have*

$$\sum_{k\le M} \{m_k(\boldsymbol{\sigma})^p\,;\ 2^{-j_0} \le |m_k(\boldsymbol{\sigma})| \le 2^{-j}\} \le L2^{-(p-2)j} \tag{5.378}$$

$$\sum_{k\le M} \{m_k(\boldsymbol{\sigma})^{2(p-1)}\,;\ 2^{-j_0} \le |m_k(\boldsymbol{\sigma})| \le 2^{-j}\} \le L2^{-2(p-2)j}\,. \tag{5.379}$$

Proof. This is immediate. By Lemma 5.11.3 if $j \le j_0 - 1$ we have, with overwhelming probability that

$$\sum_{k\le M} \{m_k(\boldsymbol{\sigma})^p\,;\ 2^{-j-1} \le |m_k(\boldsymbol{\sigma})| \le 2^{-j}\} \le (2^{-j})^p L 2^{2j}\,,$$

and summation over j yields (5.378). The same method implies (5.379). □

It is essential in Corollary 5.11.4 that $p \ge 3$. This corollary provides an excellent control of the values of $|m_k(\boldsymbol{\sigma})|$ that are not too small. We need another argument to control the small values. We set

$$d(p) = \mathsf{E}\, m_k(\boldsymbol{\sigma})^p\, \mathbf{1}_{\{|m_k(\boldsymbol{\sigma})| \le 2^{-j_0}\}}\,. \tag{5.380}$$

This quantity is of course independent of $\boldsymbol{\sigma}$. The introduction of $d(p)$ (and its subsequent use as a recentering constant) is motivated by the fact that, in contrast with the case $p = 2$, it is not true that

$$\sum_{2\le k\le M} \mathsf{E}\, m_k^p(\sigma)$$

stays bounded as $N \to \infty$ and $M \le \alpha N^{p-1}$.

Corollary 5.11.5. *Consider a number $c > 0$. Then if*

$$\alpha = \frac{M}{N^{p-1}} \le c \tag{5.381}$$

we have, with overwhelming probability that for each $\boldsymbol{\sigma}$ *in* Σ_N,

$$\left|\sum_{k\le M}(m_k(\boldsymbol{\sigma})^p\,\mathbf{1}_{\{|m_k(\boldsymbol{\sigma})|\le 2^{-j_0}\}}-d(p))\right|\le K(p)\sqrt{c} \tag{5.382}$$

$$\sum_{k\le M} m_k(\boldsymbol{\sigma})^{2(p-1)}\,\mathbf{1}_{\{|m_k(\boldsymbol{\sigma})|\le 2^{-j_0}\}}\le K(p)c\,. \tag{5.383}$$

Proof. Let us fix $\boldsymbol{\sigma}$ in Σ_N, and, for $k\le M$, let us consider the r.v.

$$X_k=m_k(\boldsymbol{\sigma})^p\,\mathbf{1}_{\{|m_k(\boldsymbol{\sigma})|\le 2^{-j_0}\}}-d(p)\,.$$

These r.v. are independent, centered, and satisfy

$$|X_k|\le 2^{-pj_0} \tag{5.384}$$

$$\mathsf{E}\,X_k^2\le L\,\mathsf{E}\,m_k(\boldsymbol{\sigma})^{2p}\le\frac{K(p)}{N^p}\,. \tag{5.385}$$

Thus, by Bernstein's inequality (A.18), we have, for $t>0$, that

$$\mathsf{P}\left(\left|\sum_{k\le M}X_k\right|\ge t\right)\le 2\exp\left(-\frac{1}{L}\min\left(\frac{N^p\,t^2}{K(p)M},\frac{t}{2^{-pj_0}}\right)\right)\,.$$

Since $p\ge 3$ and

$$2^{-j_0}\le K(p)\,N^{-1/3}\,, \tag{5.386}$$

we see that for $M\le c\,N^p$, we have

$$\mathsf{P}\left(\left|\sum_{k\le M}X_k\right|\ge K(p)\sqrt{c}\right)\le 2\exp(-N)\,,$$

and this proves that (5.382) holds with overwhelming probability. To prove (5.383), we define

$$Y_k=m_k(\boldsymbol{\sigma})^{2(p-1)}\,\mathbf{1}_{\{|m_k(\boldsymbol{\sigma})|\le 2^{-j_0}\}}$$

so that

$$\mathsf{E}\,Y_k\le\frac{K(p)}{N^{p-1}}\,.$$

If $X_k=Y_k-\mathsf{E}\,Y_k$, we have

$$|X_k|\le 2^{-2j_0(p-1)+1}$$

$$\mathsf{E}\,X_k^2\le\mathsf{E}\,Y_k^2\le 2^{-2j_0(p-1)}\,\mathsf{E}\,Y_k\,.$$

Thus, by Bernstein's inequality, we have, for $t>0$

$$\mathsf{P}\left(\left|\sum_{k\le M} X_k\right|\ge t\right)\le \exp\left(-\frac{1}{L}\min\left(\frac{t^2N^{p-1}2^{2j_0(p-1)}}{K(p)\,M}, t\,2^{2j_0(p-1)}\right)\right).$$

By definition of j_0, we have

$$2^{2j_0}\ge \frac{N}{Lp\log N}, \tag{5.387}$$

and we have $M\le N^{p-1}$, so that we see that given $t>0$ we have

$$\mathsf{P}\left(\left|\sum_{k\le M} X_k\right|\ge t\right)\le K(t,p)\exp(-N).$$

Now, for $M\le cN^{p-1}$, we have $\sum_{k\le M}\mathsf{E}\,Y_k\le K(p)\,c$, and this completes the proof. □

Lemma 5.11.6. *Consider $a,\varepsilon>0$. Then, with overwhelming probability, given any subset I of $\{1,\dots,M\}$ with*

$$n=\operatorname{card} I\le \frac{Na^2}{Lp\log N} \tag{5.388}$$

we have, for $w>0$, $\sqrt{N}(1-2\varepsilon)\,w\le(1+a)^{1/2}$, that

$$\begin{aligned}&\operatorname{card}\left\{\boldsymbol{\sigma}\,;\ \sum_{k\in I} m_k^2(\boldsymbol{\sigma})\ge w^2\right\}\\ &\le 2^N\left(1+\frac{1}{\varepsilon}\right)^n\exp\left(-N\mathcal{I}\left(\frac{(1-2\varepsilon)w}{(1+a)}\right)\right).\end{aligned} \tag{5.389}$$

Moreover, with overwhelming probability, we have

$$\operatorname{card}\left\{\boldsymbol{\sigma}\,;\ \sum_{k\in I} m_k^2(\boldsymbol{\sigma})\ge\left(\frac{1+a}{1-2\varepsilon}\right)^2\right\}=0. \tag{5.390}$$

We have already controlled the "intermediate" values of m_k and the "small values" of m_k. This is an attempt at controlling the sum of the squares of the few largest values.

Proof. We have

$$\left(\sum_{k\in I} m_k(\boldsymbol{\sigma})^2\right)^{1/2}=\sup\sum_{k\in I}\alpha_k\,m_k(\boldsymbol{\sigma}),$$

where the supremum is taken over all sequences $\boldsymbol{\alpha} = (\alpha_k)_{k\in I}$ such that $\sum_{k\in I} \alpha_k^2 \le 1$. It is shown in Proposition A.3.1 that we can find a set A of such sequences with $\operatorname{card} A \le (1+\varepsilon^{-1})^n$ and

$$(1-2\varepsilon)\sup_{\boldsymbol{\alpha}} \sum_{k\in I} \alpha_k\, m_k(\boldsymbol{\sigma}) \le \sup_{\boldsymbol{\alpha}\in A} \sum_{k\in I} \alpha_k\, m_k(\boldsymbol{\sigma})\,.$$

Thus

$$\operatorname{card}\left\{\boldsymbol{\sigma}\,;\ \sup_{\boldsymbol{\alpha}} \sum_{k\in I} \alpha_k\, m_k(\boldsymbol{\sigma}) \ge w\right\}$$
$$\le \left(1+\frac{1}{\varepsilon}\right)^n \max\operatorname{card}\left\{\boldsymbol{\sigma}\,;\ \sum_{k\in I} \alpha_k\, m_k(\boldsymbol{\sigma}) \ge w(1-2\varepsilon)\right\}$$

where the max is over $\boldsymbol{\alpha} = (\alpha_k)$ in A. Now

$$\sum_{k\in I} \alpha_k\, m_k(\boldsymbol{\sigma}) = \sum_{i\le N} a_i\, \sigma_i\ ,$$

where $a_i = \frac{1}{N}\sum_{k\in I} \eta_{i,k}\, \alpha_k$, and by (A.11),

$$\operatorname{card}\left\{\boldsymbol{\sigma}\,;\ \sum_{i\le N} a_i\,\sigma_i \ge w(1-2\varepsilon)\right\} = \operatorname{card}\left\{\boldsymbol{\sigma}\,;\ \sum_{i\le N} N\, a_i\,\sigma_i \ge Nw(1-2\varepsilon)\right\}$$
$$\le 2^N \exp\left(-NI\left(\frac{Nw(1-2\varepsilon)}{\left(N\sum_{i\le N} a_i^2\right)^{1/2}}\right)\right).$$

Now it is shown in Proposition A.4.3 that with overwhelming probability, for all sets I that satisfy (5.388) we have $N\sum_{i\le N} a_i^2 \le 1+a$. The last claim of the lemma follows from the inequality $\sum_{i\le N} a_i\,\sigma_i \le (N\sum_{i\le n} a_i^2)^{1/2}$. □

Proposition 5.11.7. *Consider a number $c > 0$, $c < 1$, and the smallest integer j such that*

$$2^{-(p-2)j} \le \sqrt{c}\,. \tag{5.391}$$

Then with overwhelming probability if $M/N^{p-1} \le c$ we have

$$\forall\,\boldsymbol{\sigma}\,,\ -H_{N,M}(\boldsymbol{\sigma}) \le Md(p) + \frac{N\beta}{p}\, s(\boldsymbol{\sigma})^p + Nhs(\boldsymbol{\sigma}) + NK(p)\sqrt{c}\,, \tag{5.392}$$

where

$$s(\boldsymbol{\sigma})^p = \sup_I \sum_{k\in I} m_k(\boldsymbol{\sigma})^p\ , \tag{5.393}$$

the supremum being taken over the subsets I of $\{1,\dots,M\}$ with $\operatorname{card} I \le 2^{2j+3}$.

In words, $s(\boldsymbol{\sigma})^p$ is the sum of the 2^{2j+3} largest terms of the family $(m_k(\boldsymbol{\sigma})^p)_{k\le M}$.

Proof. If we combine (5.378) and (5.382), we see that

$$\sum (m_k(\boldsymbol{\sigma})^p - d(p)) \le K(p)\sqrt{c}\,,$$

where the sum is over the values of k for which $|m_k(\boldsymbol{\sigma})| \le 2^{-j}$. Moreover, if $I = \{k; |m_k(\boldsymbol{\sigma})| \ge 2^{-j}\}$ then $\operatorname{card} I \le 2^{2j+3}$ by Lemma 5.11.3. Finally $m_1(\boldsymbol{\sigma}) \le s(\boldsymbol{\sigma})$. □

Quite naturally, we also require a lower bound for $Z_{N,M}$.

Lemma 5.11.8. *With overwhelming probability we have*

$$\begin{aligned} Z_{N,M} &= \sum_{\boldsymbol{\sigma}} \exp(-H_{N,M}(\boldsymbol{\sigma})) \\ &\ge \frac{2^N}{K\sqrt{N}} \exp\left(N\left(\frac{\beta}{p} m^{*p} + hm^* - \mathcal{I}(m^*) + Md(p)\right)\right) . \end{aligned} \tag{5.394}$$

Proof. Given an integer $q < N$, we have

$$\begin{aligned} Z_{N,M} &\ge \operatorname{card}\left\{\boldsymbol{\sigma}\,;\, \sum_{i\le N} \sigma_i = q\right\} \\ &\times \exp\left(N\frac{\beta}{p}\left(\frac{q}{N}\right)^p + hq + Mc(p) + \inf_{\boldsymbol{\sigma}} \sum_{2\le k\le M} (m_k(\boldsymbol{\sigma})^p - c(p))\right) , \end{aligned}$$

where

$$c(p) = \mathsf{E}\, m(\boldsymbol{\sigma})^p \ge d(p)\,.$$

The r.v. $c(p) - m_k(\boldsymbol{\sigma})^p$ are centered, independent, and smaller than or equal to $c(p)$, so we can apply an inequality of Hoeffding [80] implying that

$$\mathsf{P}\left(\sum_{2\le k\le M} c(p) - m_k(\boldsymbol{\sigma})^p \ge t\right) \le \exp\left(-\frac{2t^2}{(M-1)\,c(p)^2}\right) .$$

Since $c(p) \le K(p)\,N^{-p/2}$, and $M \le N^{p-1}$, we see that with overwhelming probability, for all $\boldsymbol{\sigma}$ we have

$$\sum_{2\le k\le M} m_k(\boldsymbol{\sigma})^p - c(p) \ge -K(p)\,, \tag{5.395}$$

and (5.395), (A.10) yield

$$Z_{N,M} \geq \frac{1}{K}\frac{2^N}{\sqrt{N}} \exp\left(N\left(\frac{\beta}{p}\left(\frac{q}{N}\right)^p + h\frac{q}{N} - \mathcal{I}\left(\frac{q}{N}\right)\right) + Md(p)\right) .$$

To conclude we choose q such that $|q/N - m^*| \leq 1/N$, (and we observe that the derivative of $t \to \beta t^p/p + ht - \mathcal{I}(t)$ is bounded in a neighborhood of m^*).

In the following lemmas, we keep the notation of (5.392).

Lemma 5.11.9. *Consider $\rho > 0$. Then, we can choose c small enough such that if $\alpha = M/N^{p-1} \leq c$, the set $A = \{\boldsymbol{\sigma}\,;\ s(\boldsymbol{\sigma}) \geq m^* + \rho\}$ is negligible. Here $s(\boldsymbol{\sigma})$ is defined from c as in Proposition 5.11.7.*

Proof. Consider the function $f(t) = \beta t^p/p + ht$. Then the function $t \to f(t) - \mathcal{I}(t)$ attains its maximum at $t = m^*$. We can then choose $\varepsilon > 0$ and $\delta > 0$ small enough that

$$m^* + \rho \leq t \leq (1-3\varepsilon)^{-1} \Rightarrow f(t) - \mathcal{I}((1-3\varepsilon)t) \leq f(m^*) - \mathcal{I}(m^*) - \delta . \quad (5.396)$$

We consider now $c > 0$, that will be determined later. We have

$$G_{N,M}(A) = \frac{S}{Z_{N,M}} \quad (5.397)$$

where

$$S = \sum_{\boldsymbol{\sigma}\in A} \exp(-H_{N,M}(\boldsymbol{\sigma})) .$$

By (5.392) we have

$$S \leq \exp(Md(p) + NK\sqrt{c}) \sum_{\boldsymbol{\sigma}\in A} \exp\left(N\frac{\beta}{p}\, s^p(\boldsymbol{\sigma}) + Nhs(\boldsymbol{\sigma})\right) . \quad (5.398)$$

Now, by (A.44) we have

$$\begin{aligned}
&\sum_{\boldsymbol{\sigma}\in A} \exp\left(\frac{N\beta}{p}\, s^p(\boldsymbol{\sigma}) + Nhs(\boldsymbol{\sigma})\right) \\
&= (\operatorname{card} A) \exp Nf(m^* + \rho) \\
&+ \int_{m^*+\rho}^{\infty} Nf'(t) \exp Nf(t) \operatorname{card}\{\boldsymbol{\sigma}\,;\ s(\boldsymbol{\sigma}) \geq t\}\, \mathrm{d}t . \quad (5.399)
\end{aligned}$$

Consider N large enough that (5.388) holds with $a = \varepsilon$ and $n = 2^{2j+3}$. We observe the fact that for numbers b_k, since $p \geq 2$, we have

$$\sum_{k\in I} b_k^{p/2} \leq \left(\sum_{k\in I} b_k\right)^{p/2}$$

so that

$$\Big(\sum_{k\in I} m_k^p\Big)^{1/p} \le \Big(\sum_{k\in I} m_k^2\Big)^{1/2}$$

and

$$\begin{aligned}\operatorname{card}\{\boldsymbol{\sigma}\,;\ s(\boldsymbol{\sigma})\ge t\} &\le \sum_I \operatorname{card}\Big\{\boldsymbol{\sigma}\,;\ \sum_{k\in I} m_k^2 \ge t^2\Big\}\\ &\le K\,N^K\,2^N\exp(-N\,\mathcal{I}((1-3\varepsilon)t))\ , \qquad (5.400)\end{aligned}$$

where K depends on c only. Here, we have used (5.389), the fact that there are at most $N^{2^{2j+3}}$ possible choices for I, and the inequality $(1-2\varepsilon)/(1+\varepsilon)\ge 1-3\varepsilon$. Also, by (5.390), if $(1-3\varepsilon)t>1$, there are no $\boldsymbol{\sigma}$ for which $s(\boldsymbol{\sigma})\ge t$. Thus, by (5.396) we see that with overwhelming probability we have

$$S \le K\,N^K\,2^N\exp(Md(p)+N(f(m^*)-\mathcal{I}(m^*)-\delta+K(p)c))\ .$$

Hence, if we have chosen c small enough, Lemma 5.11.8 shows that with overwhelming probability we have $G_{N,M}(A)\le K\,N^K\exp(-N\delta/2)$. □

Lemma 5.11.10. *Consider $\rho>0$. Then we can choose $c>0$ small enough that if $\alpha=M/N^{p-1}\le c$, with overwhelming probability we have*

$$\sup_k |m_k(\boldsymbol{\sigma})| \ge m^* - \rho\ .$$

Proof. Let $v=m^*-\rho$. Consider the function defined by $a(t)=t$ if $t\le v$, and $a(t)=t^{p/2}/v^{p/2-1}$ if $t\ge v$, so that $a(t)>t$ if $t\ge v=m^*-\rho$. Since the function $t\to f(t)-\mathcal{I}(t)$ attains its maximum at $t=m^*$, we can find $\varepsilon>0$ and $\delta>0$ such that

$$(1-3\varepsilon)\,a(t)\le 1 \Rightarrow f(t)-\mathcal{I}((1-3\varepsilon)\,a(t)) \le f(m^*)-\mathcal{I}(m^*)-\delta\ . \qquad (5.401)$$

Consider

$$A=\{\boldsymbol{\sigma}\,;\ \sup_k |m_k(\boldsymbol{\sigma})|\le v = m^*-\rho\}\ ,$$

and recall the formulas (5.397) and (5.398). Consider $c>0$, to be chosen later. Then

$$\sum_{\boldsymbol{\sigma}\in A}\exp Nf(s(\boldsymbol{\sigma})) = 1+\int_0^\infty Nf'(t)\exp(Nf(t))\operatorname{card}\{\boldsymbol{\sigma}\in A, s(\boldsymbol{\sigma})\ge t\}\,\mathrm{d}t\ .$$

If $t\le v$ we use the bound (5.400). If $t\ge v$, we observe that

$$\boldsymbol{\sigma}\in A \Rightarrow \sum_{k\in I} m_k^p(\boldsymbol{\sigma}) \le v^{p-2}\sum_{k\in I} m_k^2(\boldsymbol{\sigma})$$

so that, using (5.389) and (5.390), we have

$$\begin{aligned}\operatorname{card}\{\boldsymbol{\sigma}\in A;\ s(\boldsymbol{\sigma})\geq t\} &\leq \sum_I \operatorname{card}\left\{\boldsymbol{\sigma};\ \sum_{k\in I} m_k^2(\boldsymbol{\sigma})\geq \frac{t^p}{v^{p-2}}\right\}\\ &\leq KN^K 2^N \exp(-N\mathcal{I}((1-3\varepsilon)\,a(t))\end{aligned}$$

if $(1-\varepsilon)a(t)\leq 1$, while if $(1-\varepsilon)a(t)>1$ there are no $\boldsymbol{\sigma}$ in A with $s(\boldsymbol{\sigma})\geq t$. One then finishes the proof as in the case of Lemma 5.11.9. □

Proof of Theorem 5.11.2. Consider $\rho>0$ and c small enough that Lemmas 5.11.8 and 5.11.9 hold. Then $G_{N,M}$ is essentially supported by the set

$$A=\{\boldsymbol{\sigma};\ \sup_k |m_k(\boldsymbol{\sigma})|\geq m^*-\rho;\ s(\boldsymbol{\sigma})\leq m^*+\rho\}\ .$$

Consider $\boldsymbol{\sigma}$ in A, and the set I such that $\{m_k(\boldsymbol{\sigma});\ k\in I\}$ are the 2^{2j+3} largest values of the sequence $m_k(\boldsymbol{\sigma})$, and consider k_0 such that $|m_{k_0}(\boldsymbol{\sigma})|=\max_k |m_k(\boldsymbol{\sigma})|$. Then

$$(m^*-\rho)^p\leq m_{k_0}(\boldsymbol{\sigma})^p\leq \sum_{k\in I} m_k(\boldsymbol{\sigma})^p\leq (m^*+\rho)^p$$

and thus

$$\sum_{k\in I,k\neq k_0} m_k(\boldsymbol{\sigma})^p\leq (m^*+\rho)^p-(m^*-\rho)^p\ .$$

On the other hand, by (5.379) and (5.383) we have (with overwhelming probability)

$$\sum_{k\notin I} m_k^{2p}(\boldsymbol{\sigma})\leq K(p)\,c\ ,$$

and the result is then obvious. □

We will leave it to the reader to check, as in Section 5.5 that, when $h>0$, "the system chooses one of the possible states", as expressed in the following lemma.

Lemma 5.11.11. *Assume $h>0$. Then, given $\rho>0$ we can find $c>0$ such that if $\alpha\leq c$, the set $\{\boldsymbol{\sigma};\ |m_1(\boldsymbol{\sigma})-m^*|>\rho\}$ is negligible.*

We now turn to the control of the functions m_k, control that will be necessary to use approximate integration by parts. We consider the largest number β_0 such that we have

$$\sup_t\left(\frac{\beta_0\, t^p}{p}-\mathcal{I}(t)\right)=0\ . \tag{5.402}$$

It is of course likely that β_0 is exactly the number $\beta(0)$ previously considered, such that for $\beta>\beta(0)$ the maximum of the right-hand side of (5.369) is obtained at the largest root of the equation $t=\operatorname{th}(\beta t^p)$.

Lemma 5.11.12. *Consider numbers* $(a_i)_{i\le N}$. *Then if*

$$\left(\frac{1}{N}\sum_{i\le N} a_i^2\right)^{p/2} < \beta_0$$

we have

$$\mathsf{E}\exp\frac{N}{p}\left(\frac{1}{N}\sum_{i\le N}\eta_{i,M}\, a_i\right)^p \le K$$

where K *depends only on* $\sum_{i\le N} a_i^2$.

Proof. We have, by (A.6)

$$\begin{aligned}
&\mathsf{E}\exp\frac{N}{p}\left(\frac{1}{N}\sum_{i\le N}\eta_{i,M}\, a_i\right)^p \\
&\le 1+\int_0^\infty Nt^{p-1}\exp\frac{Nt^p}{p}\,\mathsf{P}\left(\sum_{i\le N}\eta_{i,M}\, a_i \ge tN\right)\mathrm{d}t \\
&\le 1+N\int_0^a t^{p-1}\exp N\left(\frac{t^p}{p}-\mathcal{I}\left(\frac{t}{a}\right)\right)\mathrm{d}t\ ,
\end{aligned}$$

using (A.11), and where $a=(N^{-1}\sum_{i\le N} a_i^2)^{1/2}$. By change of variables, this is

$$N\, a^p\int_0^1 t^{p-1}\exp N\left(\frac{a^p\, t^p}{p}-\mathcal{I}(t)\right)\mathrm{d}t\ .$$

We then use the fact that if $\beta<\beta_0$, by (5.402) we have $a^p\, t^p/p-\mathcal{I}(t)\le -(\beta_0-a^p)\, t^p/p$. □

Proposition 5.11.13. *Assume that*

$$\beta(1-m^*)^{p/2} < \beta_0\ . \tag{5.403}$$

Then there is $c>0$ *such that if* $\alpha\le c$, *we have*

$$\forall n \le \frac{N^{1-2/p}}{\log N}\ ,\ \mathsf{E}\,\langle m_M^{2n}\rangle \le \left(\frac{Kn}{N}\right)^n$$

where K *depends only on* β, h, p.

This provides a very good control of many moments of m_M. We do not know however if condition (5.403) always holds when $\beta>\beta_0$. It certainly holds when β is large enough because, as in the case $p=2$, m^* approaches 1 exponentially fast as $\beta\to\infty$.

Proof. We write

$$\mathsf{E}\,\langle m_M^{2n}\rangle \le \mathrm{I} + \mathrm{II} \tag{5.404}$$

where

$$\mathrm{I} = \mathsf{E}\,(\mathbf{1}_{\Omega_0}\,\langle m_M^{2n}\rangle)$$

$$\mathrm{II} = \mathsf{E}\,(\mathbf{1}_{\Omega_0^c}\,\langle m_M^{2n}\rangle)$$

for $\Omega_0 = \left\{\left|\sum_{i\le N}\eta_{i,M}\right| \le N^{1-1/p}\right\}$.
Thus, using the subgaussian inequality we have

$$\mathrm{II} \le \mathsf{P}\,(\Omega_0^c) \le \exp\left(-\frac{N^{1-2/p}}{2}\right) \le K\left(\frac{n}{N}\right)^n \tag{5.405}$$

provided $n \le N^{1-2/p}/\log N$.

Now we choose $b > 1$, $\tau_2 > 1$ and $\rho > 0$ such that

$$\beta\, b^{p-1}\,\tau_2(1 - m^{*2} + 2m^*\rho)^{p/2} < \beta_0\,. \tag{5.406}$$

This is possible by (5.403). We consider a number a with $1/a + 1/b = 1$, a number τ_1 with $1/\tau_1 + 1/\tau_2 = 1$. Given x, y, the convexity of the function $t \to t^p$ shows that

$$\left(\frac{x}{a} + \frac{y}{b}\right)^p \le \frac{x^p}{a} + \frac{y^p}{b}$$

so that

$$(x+y)^p \le a^{p-1}\,x^p + b^{p-1}\,y^p$$

and thus

$$\begin{aligned}\exp\frac{N\beta}{p}m_M^p \le \exp\frac{N\beta\, a^{p-1}}{p}&\left(\frac{1}{N}\sum_{i\le N}\eta_{i,M}\right)^p m^{*p}\\ \times\exp\frac{N\beta\, b^{p-1}}{p}&\left(\frac{1}{N}\sum_{i\le N}\eta_{i,M}(\sigma_i - m^*)\right)^p .\end{aligned} \tag{5.407}$$

Consider the set

$$A = \{\boldsymbol{\sigma}\,;\ |m_1(\boldsymbol{\sigma}) - m^*| \le \rho\}\,.$$

We write

$$\langle m_M^{2n}\rangle \le \langle m_M^{2n}\,\mathbf{1}_A\rangle + \langle m_M^{2n}\,\mathbf{1}_{A^c}\rangle \tag{5.408}$$

and thus

$$\mathsf{E}\,\mathbf{1}_{\Omega_0}\langle m_M^{2n}\rangle \le \mathsf{E}\,\mathbf{1}_{\Omega_0}\langle m_M^{2n}\,\mathbf{1}_A\rangle + \mathsf{E}\,\langle \mathbf{1}_{A^c}\rangle\,. \tag{5.409}$$

Now, using familiar notation

$$\langle m_M^{2n}\,\mathbf{1}_A\rangle = \frac{\langle m_M^{2n}\,\mathbf{1}_A\exp(N\beta\, m_M^p/p)\rangle_\sim}{\langle \exp(N\beta m_M^p/p)\rangle_\sim}\,.$$

Using that the denominator is at least 1, and since on Ω_0 we have $N\left(N^{-1}\sum_{i\le N}\eta_{i,M}\right)^p \le 1$, we get from (5.407) that

$$\begin{aligned}&\mathsf{E}\,(\mathbf{1}_{\Omega_0}\langle m_M^{2n}\,\mathbf{1}_A\rangle)\\&\le K\mathsf{E}\left\langle \mathbf{1}_A\, m_M^{2n}\exp\frac{N\beta\, b^{p-1}}{p}\left(\frac{1}{N}\sum_{i\le N}\eta_{i,M}(\sigma_i-m^*)\right)^p\right\rangle_\sim .\end{aligned} \tag{5.410}$$

If $\boldsymbol{\sigma}\in A$, we have

$$\sum_{i\le N}(\sigma_i-m^*)^2 = N-2m^*\left(\sum_{i\le N}\sigma_i\right)+Nm^{*2}\le N(1-m^{*2}+2m^*\rho)\ . \tag{5.411}$$

We denote by E_M expectation in $\eta_{i,M}$ only, and by Hölder's inequality we get

$$\mathsf{E}\,(\mathbf{1}_{\Omega_0}\langle m_M^{2n}\,\mathbf{1}_A\rangle)\le K\mathsf{E}\,\langle \mathbf{1}_A\,\mathsf{E}_M(m_M^{2n\tau_1})^{1/\tau_1}\,B^{1/\tau_2}\rangle \tag{5.412}$$

where

$$B=\mathsf{E}_M\exp\left(\frac{N\beta\, b^{p-1}\tau_2}{p}\left(\frac{1}{N}\sum_{i\le N}\eta_{i,M}(\sigma_i-m^*)\right)^p\right)\ .$$

By Lemma 5.11.12, we have $B\le K$, so that, since

$$\mathsf{E}_M(m_M^{2n\tau_1})^{1/\tau_1}\le\left(\frac{K(\tau_1)n}{N}\right)^n\ ,$$

we get from (5.409) that

$$\mathsf{E}\,\mathbf{1}_{\Omega_0}\langle m_M^{2n}\rangle\le\left(\frac{K(\tau_1)n}{N}\right)^n+\mathsf{E}\,\langle\mathbf{1}_{A^c}\rangle\ .$$

Lemma 5.11.11 shows that if c is small enough, and $\alpha\le c$, the set A^c is negligible, i.e. $\mathsf{E}\,\langle\mathbf{1}_{A^c}\rangle\le K\exp(-N/K)$. Since $\exp(-N/K)\le(Kn/N)^n$ for each n, this finishes the proof. □

Now we turn to the description of the "replica-symmetric equations". Considering two standard Gaussian r.v. g_1, g_2, with $a=\mathsf{E}\,g_1\,g_2$, we define

$$Q_p(a)=\mathsf{E}\,(g_1\,g_2)^p\ ,\ Q_p^-(a)=\mathsf{E}\,g_1^p\,g_2^{p-2}=\mathsf{E}\,g_2^p\,g_1^{p-2}\ .$$

If we write $(g_1\,g_2)^p=g_1\,g_1^{p-1}\,g_2^p$ and we use integration by parts we get

$$Q_p(a)=(p-1)\,Q_p^-(a)+p\,a\,Q_{p-1}(a)\ . \tag{5.413}$$

If we write $g_1^p\,g_2^{p-2}=g_1\,g_1^{p-1}\,g_2^{p-2}$, we get, if $p\ge 2$,

$$Q_p^-(a) = (p-1)\,Q_{p-2}(a) + (p-2)\,a\,Q_{p-1}^-(a)\,. \tag{5.414}$$

We can then compute $Q_p(a)$, $Q_p^-(a)$ by induction:

$$Q_0(a) = 1\,,\ Q_1(a) = a\,,\ Q_2^-(a) = 1\,,\ Q_2(a) = 1 + 2a^2\,,$$

$$Q_3^-(a) = 2Q_1(a) + a\,Q_2^-(a) = 2a + 1$$

$$Q_3(a) = 2Q_3^-(a) + 3aQ_2(a) = 4a + 1 + 3a(1 + 2a^2) = 6a^2 + 7a + 2\,.$$

We consider the system of equations

$$r = \alpha\, p\, Q_p(q) \tag{5.415}$$

$$\mu = \mathsf{E}\,\mathrm{th}(\beta z\sqrt{r} + \beta\mu^{p-1} + h) \tag{5.416}$$

$$q = \mathsf{E}\,\mathrm{th}^2(\beta z\sqrt{r} + \beta\mu^{p-1} + h)\,. \tag{5.417}$$

It is not difficult to show that given h and given $\beta > \beta(h)$ then if α is small enough these equations have a solution with μ close to m^* and q close to m^{*2}.

To define the smart path, we consider the Hamiltonian

$$-\,H_{N,M,t}(\sigma) = \frac{N\beta}{p}\sum_{2\le k\le M}\left(n_k(\boldsymbol{\sigma}) + \sigma_N\frac{\sqrt{t}}{N}\eta_{N,k}\right)^p + \frac{N\beta}{p}\,n_1^p(\boldsymbol{\sigma})$$
$$+\,\sigma_N(\beta z\sqrt{r(1-t)} + \beta(tn_1(\boldsymbol{\sigma}) + (1-t)\mu)^{p-1} + h)\,. \tag{5.418}$$

It is not exactly true that $H_{N,M,1} = H_{N,M}$ because

$$\frac{N\beta}{p}\left(n_1(\boldsymbol{\sigma}) + \frac{\sigma_N}{N}\right)^p \neq \frac{N\beta}{p}\,n_1^p(\boldsymbol{\sigma}) + \beta\,\sigma_N\,n_1^{p-1}(\boldsymbol{\sigma})$$

but the missing terms are of lower order. The presence of many lower order terms is a new feature compared to the case $p = 2$. For example, when we want to compute $\nu_t'(f)$, we have to replace in (5.224) and (5.225) the quantity $S_{\ell,\ell'}$ by

$$p\sum_{2\le k\le M}\left(n_k^\ell + \frac{\sqrt{t}}{N}\eta_{N,k}\,\sigma_N^\ell\right)^{p-1}\left(n_k^\ell + \frac{\sqrt{t}}{N}\eta_{N,k}\,\sigma_N^\ell\right)^{p-1}$$

and then, (to keep the expression manageable) we show that up to lower order terms this can be replaced by $p\sum_{2\le k\le M}(n_k^\ell\, n_k^{\ell'})^{p-1}$.

The real difference with the case $p = 2$ however lies in the integration by parts.

When we try to mimic the work of Section 5.8, and to compute

$$N\,\mathsf{E}\,\langle(n_M^1\, n_M^2)^{p-1}f\rangle = \sum_{i\le N}\mathsf{E}\,\eta_{i,M}\langle\sigma_i^1(n_M^1)^{p-2}(n_M^2)^{p-1}f\rangle \tag{5.419}$$

we first find two terms corresponding to the dependence of n_M^1 and n_M^2 on $\eta_{i,M}$, and they are

$$(p-2)\,\mathsf{E}\,\langle (n_M^1)^{p-3}(n_M^2)^{p-1}f\rangle + (p-1)\,\mathsf{E}\,\langle R_{1,2}(n_M^1\, n_M^2)^{p-2}f\rangle\,. \tag{5.420}$$

This should be compared with the right-hand side of (5.413) with $p-1$ instead of p. We then find terms occurring because of the dependence of $\langle\cdot\rangle$ on $\eta_{i,M}$. These terms are

$$N\beta\Bigg(\sum_{\ell\le n}\mathsf{E}\,\langle (n_M^1)^{p-2}(n_M^2)^{p-1}R_{1,\ell}(n_M^\ell)^{p-1}\rangle$$
$$-n\,\mathsf{E}\,\langle (n_M^1)^{p-2}(n_M^2)^{p-1}R_{1,n+1}(n_M^{n+1})^{p-1}\rangle\Bigg)\,. \tag{5.421}$$

In the domain of Proposition 5.11.13, "each factor n_M^ℓ contributes about like $1/\sqrt{N}$". There are $3p-4$ such factors in the terms of (5.421), and $2p-2$ factors in the left-hand side term of (5.419). For $p>2$, we have $3p-4>2p-4$, and the terms (5.421) are of lower order compared to those of (5.419).

We will let to the interested reader the task of adapting Theorem 5.8.1 and its proof to the values of β where (5.403) holds, and when α is small enough.

5.12 Notes and Comments

The Hopfield model was introduced in [111], but became popular only after Hopfield [81], [82] put it forward as a model of memory. For this aspect as a model of memory, it is the energy landscape, i.e. the function $\boldsymbol{\sigma}\mapsto\sum_{k\le m}m_k^2(\boldsymbol{\sigma})$ that matters. There are some rigorous results, [104], [93], [135], [127] but they are based on ad hoc methods, none of which deserves to appear in a book. A detailed study of the model from the physicists point of view appears in [7].

The first attempt at justifying the replica-symmetric equations can be found in [114]. The authors try to duplicate the results of [113] for the Hopfield model, i.e. to establish the replica-symmetric equations under the assumption that a certain quantity does not fluctuate with the disorder. This paper contains many interesting ideas, but one could of course wonder, among other things, how one could prove anything at all without addressing the question of uniqness of the solutions of these equations. There exists a remake of this article [115].

My notation differs from the traditional one as I call r what is traditionally called $r\alpha$. Thus, the replica-symmetric equations usually read

$$q=\mathsf{E}\,\mathrm{th}^2(\beta z\sqrt{r\alpha}+\beta\mu+h)$$

$$\mu = \mathsf{E}\,\text{th}(\beta z\sqrt{r\alpha} + \beta\mu + h)$$
$$r = \frac{q}{(1-\beta(1-q))^2}\,.$$

This might be natural when one derives these equations from the "replica trick". The reason for not following this tradition is that the entire approach starts with studying the sequence $(m_k(\boldsymbol{\sigma}))_{k\le M}$, and its global behavior (as in the Bovier-Gayrard localization theorem). Thus it is natural to incorporate the data about the length of this sequence (i.e. α) is the parameter r. (Maybe it is not such a good idea after all, but it is too late to change it anyway!)

The Bovier-Gayrard localization theorem is the culmination of a series of papers of these authors, sometimes with P. Picco. I am unsure as to whether the alternate approach I give here is better than the original one, but at least it is different. Bovier and Gayrard put forward the law of $(m_k(\boldsymbol{\sigma}))$ under σ as the central object. This greatly influenced the paper [135] where I first proved the validity of the replica-symmetric solution, using the cavity method. Very soon after seeing this paper, Bovier and Gayrard produced a much simpler proof [31], based on convexity properties of the function ψ of (4.79) (which they proved) and on the Brascamp-Lieb inequalities (see section 4.5). Interestingly these convexity properties do not hold in the entire admissible region, but only in the domain $\alpha \le L(\beta-1)^2$ if $\alpha \le L$ and $L\alpha\beta \le 1$ if $\alpha \ge 2$. For these reasons, I do not follow this approach, but I use the cavity method that allows to control the correct region. The proof I give is essentially that of [144] except that I no longer consider the sequence $(m_k(\boldsymbol{\sigma}))_{k\le M}$ as the central object, but, in line with the rest of the book, I work with Gibbs' measure itself.

In her paper [123], Shcherbina claims that her methods allow her to prove that the replica-symmetric solution holds on a large region. It would indeed be very nice to have a proof of the validity of the replica-symmetric solution that does not require to prove first something like the Bovier-Gayrard localization theorem. I am most certainly willing to believe that such a proof exists when I see it written down.

The paper [17] should not be missed.

The paper [31] motivated me to study the p-spin Hopfield model. A different aspect of this model is studied in [37].

The interesting paper [20] is also related to the present chapter.

6. The p-Spin Interaction Model at Low Temperature

6.1 A priori Estimates

The Hamiltonian of the p-spin interaction model replaces the 2-spin interaction of the SK model by a p-spin interaction. This Hamiltonian is given by

$$-H_N(\boldsymbol{\sigma}) = \beta \left(\frac{p!}{2N^{p-1}}\right)^{\frac{1}{2}} \sum g_{i_1,\ldots,i_p} \sigma_{i_1} \ldots \sigma_{i_p} \, . \tag{6.1}$$

The summation is over all possible choices of indices $1 \le i_1 < \cdots < i_p \le N$, and the r.v. $g_{i_1,\ldots,i_p}$ are independent standard Gaussian. The only purpose of the factor 2 in the denominator of (6.1) is to let this Hamiltonian coincide with (2.2) when $p = 2$, and otherwise it is a constant nuisance. As will be explained below, the case $p \ge 3$ is fundamentally different from the case $p = 2$.

To understand the normalization, we observe that for two configurations $\boldsymbol{\sigma}^1$ and $\boldsymbol{\sigma}^2$, we have

$$\begin{aligned}
\mathsf{E}(H_N(\boldsymbol{\sigma}^1)\, H_N(\boldsymbol{\sigma}^2)) &= \frac{\beta^2 p!}{2N^{p-1}} \sum_{i_1<\ldots<i_p} \sigma^1_{i_1} \ldots \sigma^1_{i_p}\, \sigma^2_{i_1} \ldots \sigma^2_{i_p} \\
&= \frac{\beta^2}{2N^{p-1}} \sum_d \sigma^1_{i_1}\, \sigma^2_{i_1} \ldots \sigma^1_{i_p}\, \sigma^2_{i_p} \, ,
\end{aligned}$$

where $\sum_d$ denotes the sum over all choices of indices that are all different. If $\sum_{nd}$ denotes the sum over all choices of indices $i_1, \ldots, i_p$ such that at least two of them coincide, we then have

$$\mathsf{E}(H_N(\boldsymbol{\sigma}^1)\, H_N(\boldsymbol{\sigma}^2)) = \frac{\beta^2 N}{2} R^p_{1,2} - \frac{\beta^2}{2N^{p-1}} \sum_{nd} \sigma^1_{i_1}\, \sigma^2_{i_1} \ldots \sigma^1_{i_p}\, \sigma^2_{i_p} \, . \tag{6.2}$$

Since there are at most KN^{p-1} terms in the sum $\sum_{nd}$, we get

$$\left| \mathsf{E}(H_N(\boldsymbol{\sigma}^1)\, H_N(\boldsymbol{\sigma}^2)) - \frac{\beta^2 N}{2} R^p_{1,2} \right| \le K \, . \tag{6.3}$$

When $\boldsymbol{\sigma}^1 \ne \pm\boldsymbol{\sigma}^2$, we have $R^p_{1,2} \to 0$ as $p \to \infty$, and the larger p, the more the p-spin interaction model should resemble the REM. This intuition will turn out to be correct, but it requires work to prove it, since we cannot exchange the limits $p \to \infty$ and $N \to \infty$.

Lemma 6.1.1. *We have*

$$\frac{\beta^2 N}{2} - K \le \mathsf{E}\, H_N^2(\boldsymbol{\sigma}) \le \frac{\beta^2 N}{2} \tag{6.4}$$

$$\mathsf{E}(H_N(\boldsymbol{\sigma}^1) + H_N(\boldsymbol{\sigma}^2))^2 \le \beta^2 N (1 + R_{1,2}^p) . \tag{6.5}$$

Proof. The relation (6.4) follows from (6.2), and (6.5) follows from (6.2) and the Cauchy-Schwarz inequality. □

In this chapter we will focus on low temperature results, but precious information will be gathered first at high temperature. We define the number γ_p by

$$\gamma_p^2 = 2 \inf_{0<t<1} (1 + t^{-p}) \mathcal{I}(t) \tag{6.6}$$

where $\mathcal{I}(t)$ is the function of (A.7).

Theorem 6.1.2. *If $\beta \le \gamma_p$ we have*

$$\lim_{N \to \infty} p_N(\beta) = \log 2 + \frac{\beta^2}{4} . \tag{6.7}$$

Since $\mathcal{I}'(0) = \mathcal{I}(0) = 0$ and $\mathcal{I}''(t) = 1/(1-t^2)$, we have $\mathcal{I}(t) \ge t^2/2$ and $\gamma_2 = 1$. Thus Theorem 6.1.2 recovers Theorem 2.2.1. For p large, after considerable work, we will be able to prove that (6.7) holds for $\beta \le \gamma_p'$ where $\gamma_p' > \gamma_p$. In view of (1.12), we only need a lower bound for $p_N(\beta)$ in order to prove (6.7).

Proof. Let us consider $\beta \le \gamma_p$, and the r.v.

$$X = \operatorname{card}\left\{\boldsymbol{\sigma}\,;\ -H_N(\boldsymbol{\sigma}) \ge \frac{N\beta^2}{2}\right\} .$$

We will estimate $\mathsf{E}X$ and $\mathsf{E}X^2$ in order to apply (2.18). We have $\mathsf{E}X = \sum_{\boldsymbol{\sigma}} P(-H_N(\boldsymbol{\sigma}) \ge N\beta^2/2)$, so that, using (A.36) and (6.4), we have

$$\mathsf{E}X \ge \frac{2^N}{K\sqrt{N}} \exp\left(-\frac{N^2\beta^2}{8\left(\frac{N}{2} - K\right)}\right) .$$

Now

$$\frac{N^2\beta^2}{8\left(\frac{N}{2} - K\right)} \le \frac{N\beta^2}{4\left(1 - \frac{K}{N}\right)} \le \frac{N\beta^2}{4} + K$$

so that

$$\mathsf{E}X \ge \frac{2^N}{K\sqrt{N}} \exp\left(-\frac{N\beta^2}{4}\right) . \tag{6.8}$$

If we write

$$X = \sum_{\sigma} \mathbf{1}_{\{-H_N(\boldsymbol{\sigma}) \geq N\beta^2/2\}}$$

we get

$$X^2 = \sum_{\boldsymbol{\sigma}^1, \boldsymbol{\sigma}^2} \mathbf{1}_{\{-H_N(\boldsymbol{\sigma}^1) \geq N\beta^2/2\}} \mathbf{1}_{\{-H_N(\boldsymbol{\sigma}^2) \geq N\beta^2/2\}}$$

and thus

$$\mathsf{E}X^2 = \sum_{\boldsymbol{\sigma}^1, \boldsymbol{\sigma}^2} A(\boldsymbol{\sigma}^1, \boldsymbol{\sigma}^2) , \tag{6.9}$$

where

$$\begin{aligned} A(\boldsymbol{\sigma}^1, \boldsymbol{\sigma}^2) &= \mathsf{P}\left(\left\{-H_N(\boldsymbol{\sigma}^1) \geq \frac{N\beta^2}{2}, -H_N(\boldsymbol{\sigma}^2) \geq \frac{N\beta^2}{2}\right\}\right) \\ &\leq \mathsf{P}(-(H_N(\boldsymbol{\sigma}^1) + H_N(\boldsymbol{\sigma}^2)) \geq N\beta^2) \\ &\leq \exp\left(-\frac{N^2\beta^4}{2\mathsf{E}((H_N(\boldsymbol{\sigma}^1) + H_N(\boldsymbol{\sigma}^2))^2)}\right) \end{aligned}$$

using (A.35). Using (6.5), we get

$$A(\boldsymbol{\sigma}^1, \boldsymbol{\sigma}^2) \leq \exp\left(-\frac{N\beta^2}{2(1 + R_{1,2}^p)}\right) \leq \exp\left(-\frac{N\beta^2}{2(1 + |R_{1,2}|^p)}\right) .$$

It is therefore natural to regroup the terms of (6.9) according to the value of $|R_{1,2}|$. Since $NR_{1,2}$ is an integer, we have

$$\begin{aligned} \sum_{(\boldsymbol{\sigma}^1, \boldsymbol{\sigma}^2)} A(\boldsymbol{\sigma}^1, \boldsymbol{\sigma}^2) \leq & \sum_{0 \leq k \leq N} \operatorname{card}\{(\boldsymbol{\sigma}^1, \boldsymbol{\sigma}^2); |R_{1,2}| = k/N\} \\ & \times \exp\left(-\frac{N\beta^2}{2(1 + (k/N)^p)}\right) . \end{aligned} \tag{6.10}$$

Now

$$\begin{aligned} \operatorname{card}\left\{(\boldsymbol{\sigma}^1, \boldsymbol{\sigma}^2); |R_{1,2}| = \frac{k}{N}\right\} &= 2^N \operatorname{card}\left\{\boldsymbol{\sigma} ; \left|\sum_{i \leq N} \sigma_i\right| = k\right\} \\ &\leq 2^{2N+1} \exp\left(-N\mathcal{I}\left(\frac{k}{N}\right)\right) . \end{aligned}$$

The condition $\beta^2 \leq \gamma_p^2$ implies that for $0 < t < 1$ we have

$$\beta^2 \leq 2(1 + t^{-p})\mathcal{I}(t)$$

and thus

$$\frac{\beta^2}{2} \leq \frac{\beta^2}{2(1 + t^p)} + \mathcal{I}(t)$$

so that (6.9) and (6.10) imply that

$$\mathsf{E}X^2 \le (N+1)\, 2^{2N+1} \exp\left(-\frac{N\beta^2}{2}\right).$$

By (2.18) we then have

$$\mathsf{P}\left(X \ge \frac{\mathsf{E}X}{2}\right) \ge \frac{1}{KN^2}.$$

Since

$$Z_N(\beta) \ge X \exp\left(\frac{N\beta^2}{2}\right)$$

we have

$$\mathsf{P}\left(Z_N(\beta) \ge \frac{2^N}{K\sqrt{N}} \exp\left(\frac{N\beta^2}{4}\right)\right) \ge \frac{1}{KN^2}$$

so that

$$\mathsf{P}\left(\frac{1}{N}\log Z_N(\beta) \ge \log 2 + \frac{\beta^2}{4} - K\frac{\log N}{N}\right) \ge \frac{1}{LN^2}.$$

The proof is then concluded as in Theorem 2.2.1. □

We will now perform a bit of calculus to gather information about γ_p.

Proposition 6.1.3. *If p is large enough we have*

$$\gamma_p \ge 2\sqrt{\log 2}\,(1 - 2^{-p}). \tag{6.11}$$

We observe that by Proposition 1.1.4, (6.7) cannot hold for $\beta > 2\sqrt{\log 2}$. It will be rather useful to know that as p increases, γ_p becomes close to $2\sqrt{\log 2}$ exponentially fast.

Lemma 6.1.4. *If p is large enough, the following holds. If $v \ge 2^{-p/2}$, then*

$$\inf_{0\le t\le 1-v} 2(1+t^{-p})\mathcal{I}(t) \ge 4\log 2 + \frac{pv}{4}. \tag{6.12}$$

Moreover

$$\gamma_p^2 \ge 4\log 2\left(1 - \frac{2^{-p}}{\log 2}\right) + \mathcal{R}_p \tag{6.13}$$

where $|\mathcal{R}_p| \le L p^2 2^{-2p}$.

Of course (6.11) follows from (6.13), since $(1-x)^{1/2} \le 1 - x/2$ and $2\log 2 \ge 1$.

Proof. Since $\mathcal{I}(t) \ge t^2/2$, we have

$$2(1+t^{-p})\mathcal{I}(t) \ge t^{2-p} = (1+(t-1))^{2-p} \ge 1 + (p-2)(1-t)$$

by convexity. If $1 - t \geq 3/p$, we then have

$$1 + (p-2)(1-t) = 1 + \frac{p}{4}(1-t) + \left(\frac{3p}{4} - 2\right)(1-t)$$
$$\geq 1 + \frac{p}{4}(1-t) + \frac{9}{4} - \frac{6}{p} \geq 4\log 2 + \frac{p}{4}(1-t)$$

if p is large. Thus we need only to consider values of t greater than $1 - 3/p$. Setting $u = 1 - t$, we observe that, by convexity

$$1 + t^{-p} = 1 + (1-u)^{-p} \geq 2 + pu$$

$$(1+t)\log(1+t) \geq 2\log 2 - (1 + \log 2)\, u$$

and thus

$$2\,\mathcal{I}(t) \geq 2\log 2 - (1 + \log 2)\, u + u \log u$$

so that

$$2(1 + t^{-p})\,\mathcal{I}(t) \geq \psi(u) := (2 + pu)(2\log 2 - (1 + \log 2)\, u + u\log u) \ .$$

Now, algebra shows that

$$\psi(u) = 4\log 2 + 2u(p-1)\log 2 - 2u + 2u\log u - pu^2(1 + \log 2) + pu^2 \log u \ . \tag{6.14}$$

Since $t \geq 1 - 3/p$ we have $u \leq 3/p$, and hence $pu^2 \leq 3u$,

$$u \log u \geq \frac{3}{p} \log \frac{3}{p} = -\frac{3}{p} \log \frac{p}{3} ,$$

so that $pu^2 \log u \geq -3u\log(p/3)$ and

$$\psi(u) \geq 4\log 2 + u\left(2(p-1)\log 2 - 2 + 2\log u - 3(1 + \log 2) - 3\log\frac{p}{3}\right) \ . \tag{6.15}$$

If we assume that $\log u \geq -\frac{p}{2} \log 2$, for p large enough we get

$$\psi(u) \geq 4\log 2 + \frac{pu}{4} ,$$

and this proves (6.12). Thus to prove (6.13) we can assume that $u \leq 2^{-p/2}$. In that case, proceeding as in (6.15) we get

$$\psi(u) \geq 4\log 2 + u(2(p-1)\log 2 - 2 + 2\log u - p2^{-p/2}(1 + \log 2) - p2^{-p/2}\log(2^{p/2})) \ .$$

If $u \geq 2^{-p+3}$, we have $2\log u \geq -2(p-3)\log 2$ and thus for p large enough we have $\psi(u) \geq 4\log 2$. For $u \leq 2^{-p+3}$, we have

$$|pu^2 \log u| \le Lp^2\, 2^{-2p}\,;\ |pu^2| \le Lp^2\, 2^{-2p}$$

so that (6.14) yields

$$\psi(u) \ge \varphi(u) + \mathcal{R}_p\,,$$

where $|\mathcal{R}_p| \le Lp^2\, 2^{-2p}$, and

$$\varphi(u) = 4\log 2 + 2u((p-1)\log 2 - 1) + 2u\log u\,.$$

Now $\varphi(u)$ has its minimum at $u = 2^{-p+1}$, and $\varphi(2^{-p+1}) = 4\log 2 - 2^{-p+2}$. □

We consider the function $\xi(\beta)$ given by

$$\xi(\beta) = \begin{cases} \log 2 + \dfrac{\beta^2}{4} & \text{if} \quad \beta \le 2\sqrt{\log 2} \\ \beta\sqrt{\log 2} & \text{if} \quad \beta \ge 2\sqrt{\log 2}\,. \end{cases} \tag{6.16}$$

This function occured in Proposition 1.1.5.

Corollary 6.1.5. *If p is sufficiently large, for all values of β we have*

$$\liminf_{N\to\infty} p_N(\beta) \ge \xi(\beta) - 2^{-p}\beta\,. \tag{6.17}$$

Proof. The function $\beta \to p_N(\beta)$ is convex, and by Theorem 6.1.2 for $\beta \le \gamma_p$ its limit is $\log 2 + \beta^2/4$. Elementary properties of convex functions imply that

$$\liminf_{N\to\infty} p'_N(\gamma_p) \ge \frac{\gamma_p}{2}$$

and using convexity again we get, for $\beta \ge \gamma_p$,

$$\liminf_{N\to\infty} p_N(\beta) \ge \log 2 + \frac{\gamma_p^2}{4} + \frac{1}{2}\gamma_p(\beta - \gamma_p)\,. \tag{6.18}$$

Using (6.11) we have

$$\begin{aligned} \frac{1}{2}\gamma_p(\beta-\gamma_p) &\ge \sqrt{\log 2}\,(\beta-\gamma_p) - 2^{-p}\sqrt{\log 2}\,(\beta-\gamma_p) \\ &\ge \sqrt{\log 2}\,(\beta-\gamma_p) - 2^{-p}(\beta-\gamma_p)\,. \end{aligned}$$

Combining this inequality with (6.18) and the fact that

$$\xi(\beta) \le \xi(\gamma_p) + (\beta-\gamma_p)\sqrt{\log 2}\,,$$

completes the proof. □

Corollary 6.1.5 expresses the fact that (at least as far as p_N is concerned) for p large, the p-spin interaction model approaches the REM.

We now turn to the main result of this chapter. In this result, as well as in the rest of the chapter, *we assume that p is odd.* (This enables us to avoid distinguishing some cases and to simplify some details in the sequel.)

We will say that a subset I of $\mathbb{R}$ is negligible if

$$\mathsf{E}\, G_N^{\otimes 2}(\{(\boldsymbol{\sigma}^1, \boldsymbol{\sigma}^2);\ R_{1,2} \in I\}) \le K \exp\left(-\frac{N}{K}\right),$$

where K depends on I, p, β, but not on N.

Theorem 6.1.6. *There exists an integer p_0 such that if $p \ge p_0$, for each β the sets $[-1, -2^{-p/4}]$ and $[2^{-p/4}, 1 - 2^{-p/2}]$ are negligible.*

To prepare for the proof, given u with $|u| \le 1$, we consider the quantity

$$A_N(u) = \sum \exp(-H_N(\boldsymbol{\sigma}^1) - H_N(\boldsymbol{\sigma}^2)) \tag{6.19}$$

where the summation is over all the pairs $(\boldsymbol{\sigma}^1, \boldsymbol{\sigma}^2)$ with $R_{1,2} = u$. (Of course we assume that such pairs exist.)

Lemma 6.1.7. *We have*

$$\frac{1}{N}\mathsf{E} \log A_N(u) \le \xi_u(\beta) \tag{6.20}$$

where

$$\xi_u(\beta) = \frac{\beta^2}{2}(1 + u^p) + 2\log 2 - \mathcal{I}(|u|) \tag{6.21}$$

if

$$\frac{\beta^2}{2}(1 + u^p) \le 2\log 2 - \mathcal{I}(|u|) \tag{6.22}$$

and

$$\xi_u(\beta) = \beta\sqrt{2(1 + u^p)(2\log 2 - \mathcal{I}(|u|))} \tag{6.23}$$

if (6.22) fails.

Proof. There are at most $2^{2N}\exp(-N\mathcal{I}(|u|))$ pairs in the summation (6.19). For each such pair,

$$\mathsf{E}\,((H_N(\boldsymbol{\sigma}^1) + H_N(\boldsymbol{\sigma}^2))^2) \le \beta^2 N(1 + u^p)$$

by (6.5). Thus the result follows from Proposition 1.1.3. □

Lemma 6.1.8. *If p is large enough, then if $u \in [-1, -2^{-p/4}]$ or $u \in [2^{-p/4}, 1 - 2^{-p/2}]$*

$$\xi_u(\beta) \le 2\xi(\beta) - \beta\, 2^{-\frac{p}{2}-3}\,. \tag{6.24}$$

Proof. The basic fact is that if u is as in the statement, we have

$$2u^p \log 2 - (1+u^p)\,\mathcal{I}(|u|) \le -2^{-\frac{p}{2}-2}\,. \tag{6.25}$$

To prove this suppose first $u < 0$, and observe that $1+u^p \ge 0$, so that (6.25) is obvious if $u^p \le -1/2$. If $u^p \ge -1/2$, since $u^p < 0$, the left-hand side of (6.25) is at most (since $\mathcal{I}(t) \ge t^2/2$)

$$-\frac{1}{2}\mathcal{I}(|u|) \le -\frac{u^2}{4} \le -2^{-\frac{p}{2}-2}$$

if $u \le -2^{-p/4}$. This proves (6.25) when $u < 0$. Suppose now $u > 0$. If $u^{p-2} \le 1/8$, then $u^p \le u^2/8$, and the left-hand side of (6.25) is at most

$$\frac{u^2}{4}\log 2 - \mathcal{I}(u) \le \frac{u^2}{4} - \frac{u^2}{2} \le -\frac{u^2}{4} \le -2^{-\frac{p}{2}-2}$$

if $u \ge 2^{-p/4}$. If $u^{p-2} \ge 1/8$, we use (6.12) with $v = 2^{-p/2}$ to see that

$$(1+u^p)\,\mathcal{I}(u) \ge \left(2\log 2 + \frac{p\,2^{-p/2}}{8}\right)u^p$$

and the left-hand side of (6.25) is at most

$$-\frac{p\,2^{-p/2}}{8}u^p \le -\frac{p\,2^{-p/2}}{8}\left(\frac{1}{8}\right)^{p/(p-2)} \le -2^{-\frac{p}{2}-2}$$

if p is large enough. This proves (6.25).

Now we write

$$\begin{aligned}2(1+u^p)(2\log 2 - \mathcal{I}(|u|)) &= 4\log 2 + 2(2u^p\log 2 - (1+u^p)\,\mathcal{I}(u))\\ &\le 4\log 2 - 2^{-\frac{p}{2}-1}\end{aligned}$$

by (6.25). Thus

$$\begin{aligned}\sqrt{2(1+u^p)(2\log 2 - \mathcal{I}(|u|))} &\le \sqrt{4\log 2 - 2^{-\frac{p}{2}-1}}\\ &= 2\sqrt{\log 2}\left(1 - \frac{2^{-\frac{p}{2}-3}}{\log 2}\right)^{1/2}\\ &\le 2\sqrt{\log 2} - 2^{-\frac{p}{2}-3}\end{aligned} \tag{6.26}$$

using the inequality $(1+x)^{1/2} \le 1 + x/2$. Thus, if (6.22) fails, by (6.23) we have

$$\xi_u(\beta) \le 2\beta\sqrt{\log 2} - \beta\,2^{-\frac{p}{2}-3}\,. \tag{6.27}$$

We observe that $2\beta\sqrt{\log 2} \le 2\xi(\beta)$, so that (6.27) implies (6.24). Suppose now that (6.22) holds, so that since $u > 0$ we have in particular that

$$\frac{\beta^2}{2} \le 2\log 2 - \mathcal{I}(|u|) \tag{6.28}$$

and

$$\xi_u(\beta) = \frac{\beta^2}{2} + 2\log 2 + \frac{\beta^2}{2}\,u^p - \mathcal{I}(|u|) = 2\xi(\beta) + \frac{\beta^2}{2}\,u^p - \mathcal{I}(|u|)\;, \tag{6.29}$$

since $\beta^2 \le 4\log 2$ by (6.28) and thus $\xi(\beta) = \log 2 + \beta^2/4$. Now, by (6.28) and (6.25)

$$\frac{\beta^2}{2}\,u^p - \mathcal{I}(|u|) \le 2u^p \log 2 - (1+u^p)\mathcal{I}(|u|) \le -2^{-\frac{p}{2}-2}\,. \tag{6.30}$$

Since $\beta \le 2$, (6.29) and (6.30) imply again (6.24). □

Proof of Theorem 6.1.6. If we consider the set

$$B = \{(\boldsymbol{\sigma}^1, \boldsymbol{\sigma}^2)\,;\; R_{1,2} \in [-1, -2^{-p/4}] \cup [2^{-p/4}, 1-2^{-p/2}]\}$$

then we have

$$G_N^{\otimes 2}(B) = \frac{\sum A_N(u)}{Z_N^2} \tag{6.31}$$

where the summation is for u in the intervals $[-1, -2^{-p/4}]$ and $[2^{-p/4}, 1 - 2^{-p/2}]$. As in Corollary 2.2.5 it follows from Theorem 2.2.4 and Lemma 6.1.7 that, with overwhelming probability, we have

$$\frac{1}{N}\log A_N(u) \le \xi_u(\beta) + \beta\, 2^{-p} \tag{6.32}$$

$$\frac{1}{N}\log Z_N \ge p_N(\beta) - \beta\, 2^{-p}\,. \tag{6.33}$$

Now, by (6.17), for N large enough, we have $p_N(\beta) \ge \xi(\beta) - \beta\, 2^{-p+1}$, and (6.24) implies $\xi_u(\beta) \le 2\xi(\beta) - \beta\, 2^{-\frac{p}{2}-3}$, so that (6.32) and (6.33) imply

$$\frac{1}{N}\log A_N(u) \le \frac{1}{N}\log Z_N^2 + 7\beta\, 2^{-p} - \beta\, 2^{-\frac{p}{2}-3} \le \frac{1}{N}\log Z_N^2 - \beta\, 2^{-\frac{p}{2}-4}$$

if p is large enough. Thus

$$A_N(u) \le Z_N^2 \exp(-N\beta\, 2^{-\frac{p}{2}-4})$$

and (6.31) implies

$$G_N^{\otimes 2}(B) \le (2N+1)\exp(-N\beta\, 2^{-\frac{p}{2}-4})\,.$$

This inequality holds with overwhelming probability, and this implies that $\mathsf{E}\, G_N^{\otimes 2}(B) \le K\exp(-N/K)$. □

Research problem 6.1.9. (Level 1) Prove a suitable version of the results of Sections 2.4 to 2.7 for the p-spin interaction model with external field at $\beta \le \beta_0$.

6.2 Construction and Separation of the Lumps

From Theorem 6.1.6 we know that, for a typical realization of the Gibbs measure, the overlap of two configurations is essentially always either close to zero or close to 1. It turns out that this simple property implies some interesting information about the structure of Gibbs' measure. This is shown by the following purely deterministic result.

Theorem 6.2.1. *Consider a probability measure μ on Σ_N. Consider two numbers a, a', $0 \le a' \le 1/2$, $0 < a < 1/28$ and set*

$$\varepsilon = \mu^{\otimes 2}(\{(\boldsymbol{\sigma}^1, \boldsymbol{\sigma}^2)\,;\ |R_{1,2}| \ge a'\,,\ R_{1,2} \le 1 - 2a\})\,. \tag{6.34}$$

Then we can find a partition $(C_\alpha)_{\alpha \ge 1}$ of Σ_N with the following properties

$$\boldsymbol{\sigma}^1, \boldsymbol{\sigma}^2 \in C_\alpha \Rightarrow R_{1,2} \ge 1 - 12a \tag{6.35}$$

$$\mu^{\otimes 2}\left(\{(\boldsymbol{\sigma}^1, \boldsymbol{\sigma}^2)\,;\ |R_{1,2}| \ge a'\} \setminus \bigcup_{\alpha \ge 1} C_\alpha^2\right) \le 3\,\varepsilon^{1/3}\,. \tag{6.36}$$

This will be used for ε very small. Of course there are at most 2^N non-empty subsets $C_\alpha \subset \Sigma_N$, but we prefer not to specify these non-empty subsets in order to keep simpler notation.

Theorem 6.1.6 means essentially the following: the only way that $|R_{1,2}| > a'$ (or, equivalently that $R_{1,2} \ge 1 - 2a$) is that $\boldsymbol{\sigma}^1$ and $\boldsymbol{\sigma}^2$ lie in the same set C_α. Conversely, if $\boldsymbol{\sigma}^1$ and $\boldsymbol{\sigma}^2$ belong to C_α, then $R_{1,2}$ is close to 1. Moreover,

$$\mu^{\otimes 2}(\{|R_{1,2}| \ge a'\}) \simeq \sum_{\alpha \ge 1} \mu(C_\alpha)^2\,. \tag{6.37}$$

It is *not* a consequence of the hypothesis of Theorem 6.2.1 that the quantity (6.37) is not very small. For example, the uniform probability on Σ_N satisfies

$$\mu^{\otimes 2}(\{|R_{1,2}| \ge t\}) \le 2 \exp\left(-\frac{Nt^2}{2}\right)$$

by (A.3). But, when the quantity (6.37) is not too small, some of the sets (C_α) must have a significant mass. These are small sets with significant mass, which we call *lumps*.

Proof. We recall that

$$\mathrm{d}\,(\boldsymbol{\sigma}^1, \boldsymbol{\sigma}^2) = \frac{1}{N}\,\mathrm{card}\,(\{i \le N\,;\ \sigma_i^1 \ne \sigma_i^2\}) = \frac{1}{2}\,(1 - R_{1,2}) \tag{6.38}$$

is a distance on Σ_N. We denote by $B(\boldsymbol{\sigma}, r)$ the ball of radius r and center $\boldsymbol{\sigma}$ for this distance. We set

$$B_0(\boldsymbol{\sigma}) = B(\boldsymbol{\sigma}, a)\,;\ B(\boldsymbol{\sigma}) = B(\boldsymbol{\sigma}, 3a)\,;\ D(\boldsymbol{\sigma}) = B(\boldsymbol{\sigma}, 6a)\backslash B(\boldsymbol{\sigma}, 2a)\,. \quad (6.39)$$

We observe that

$$\boldsymbol{\sigma}^1, \boldsymbol{\sigma}^2 \in B(\boldsymbol{\sigma}) \Rightarrow \mathrm{d}\,(\boldsymbol{\sigma}^1, \boldsymbol{\sigma}^2) \le 6a \Rightarrow R_{1,2} \ge 1 - 12a\,. \quad (6.40)$$

We will say that a set $B(\boldsymbol{\sigma})$ is a *lump* if $\mu(B_0(\boldsymbol{\sigma})) \ge \varepsilon^{1/3}$. We consider a maximal disjoint family $(C_\alpha)_{\alpha \le M}$ of lumps (this family might well be empty).

We will prove that (6.36) holds when replacing $\bigcup_{\alpha \ge 1} C_\alpha$ by $\bigcup_{\alpha \le M} C_\alpha$. By (6.40), the sets C_α satisfy (6.35), but they do not form a partition of Σ_N. To obtain a partition of Σ_N it suffices to decompose $\Sigma_N \backslash \bigcup_{\alpha \le M} C_\alpha$ into sets consisting of a single point.

To start, we prove that

$$B(\boldsymbol{\sigma}) \text{ is a lump } \Rightarrow \mu(D(\boldsymbol{\sigma})) \le \varepsilon^{2/3}\,. \quad (6.41)$$

This is because if $\boldsymbol{\sigma}^1 \in B_0(\boldsymbol{\sigma})$, $\boldsymbol{\sigma}^2 \in D(\boldsymbol{\sigma})$, then $a \le \mathrm{d}\,(\boldsymbol{\sigma}^1, \boldsymbol{\sigma}^2) \le 7a$, so that $1 - 14a \le R_{1,2} \le 1 - 2a$, and since $a < 1/28$, $a' \le 1/2$ we have

$$B_0(\boldsymbol{\sigma}) \times D(\boldsymbol{\sigma}) \subset \{a' < R_{1,2} \le 1 - 2a\}\,,$$

so that by (6.34)

$$\varepsilon^{1/3}\mu(D(\boldsymbol{\sigma})) \le \mu(B_0(\boldsymbol{\sigma}))\,\mu(D(\boldsymbol{\sigma})) = \mu^{\otimes 2}(B_0(\boldsymbol{\sigma}) \times D(\boldsymbol{\sigma})) \le \varepsilon\,.$$

This proves (6.41).

Since the sets C_α are disjoint and since $\mu(C_\alpha) \ge \varepsilon^{1/3}$ by definition of a lump, we have $M \le \varepsilon^{-1/3}$. For $\alpha \le M$, we consider $\boldsymbol{\sigma}_\alpha$ such that $C_\alpha = B(\boldsymbol{\sigma}_\alpha)$, and we set $D = \bigcup_{\alpha \le M} D(\boldsymbol{\sigma}_\alpha)$. By (6.41) we have $\mu(D(\boldsymbol{\sigma}_\alpha)) \le \varepsilon^{2/3}$, so that $\mu(D) \le \varepsilon^{1/3}$. Consider the set

$$A = \{(\boldsymbol{\sigma}^1, \boldsymbol{\sigma}^2)\,;\ \boldsymbol{\sigma}^2 \in B_0(\boldsymbol{\sigma}^1), \mu(B_0(\boldsymbol{\sigma}^1)) \le \varepsilon^{1/3}\}\,.$$

Then, by Fubini theorem, we have $\mu^{\otimes 2}(A) \le \varepsilon^{1/3}$. To prove (6.36), keeping (6.34) in mind, it suffices to prove that

$$R_{1,2} \ge 1 - 2a \Rightarrow (\boldsymbol{\sigma}^1, \boldsymbol{\sigma}^2) \in A \cup (D \times \Sigma_N) \cup \bigcup_{\alpha \le M} C_\alpha^2\,.$$

If $R_{1,2} \ge 1 - 2a$, then $\mathrm{d}\,(\boldsymbol{\sigma}^1, \boldsymbol{\sigma}^2) \le a$, so $\boldsymbol{\sigma}^2 \in B_0(\boldsymbol{\sigma}^1)$. If $\mu(B_0(\boldsymbol{\sigma}^1)) \le \varepsilon^{1/3}$, then $(\boldsymbol{\sigma}^1, \boldsymbol{\sigma}^2) \in A$ and we are done. Thus we can assume $\mu(B_0(\boldsymbol{\sigma}^1)) \ge \varepsilon^{1/3}$ and then $B(\boldsymbol{\sigma}^1)$ is a lump. Since the family $(C_\alpha)_{\alpha \le M}$ is a maximum disjoint family of lumps, the lump $B(\boldsymbol{\sigma}^1)$ is not disjoint from this family. That is, there exists $\alpha \le M$ with $B(\boldsymbol{\sigma}^1) \cap B(\boldsymbol{\sigma}_\alpha) \ne \emptyset$, so that $\mathrm{d}\,(\boldsymbol{\sigma}^1, \boldsymbol{\sigma}_\alpha) \le 6a$. If $\boldsymbol{\sigma}^1 \in D(\boldsymbol{\sigma}_\alpha)$ we are done because then $(\boldsymbol{\sigma}^1, \boldsymbol{\sigma}^2) \in D \times \Sigma_N$. Thus we can assume $\boldsymbol{\sigma}^1 \notin D(\boldsymbol{\sigma}_\alpha)$. Now $D(\boldsymbol{\sigma}_\alpha) = \{\boldsymbol{\sigma}\,;\ 2a \le \mathrm{d}\,(\boldsymbol{\sigma}_\alpha, \boldsymbol{\sigma}) \le 6a\}$, so that since $\mathrm{d}\,(\boldsymbol{\sigma}^1, \boldsymbol{\sigma}_\alpha) \le 6a$ we have $\mathrm{d}\,(\boldsymbol{\sigma}^1, \boldsymbol{\sigma}_\alpha) \le 2a$ and, since $\mathrm{d}\,(\boldsymbol{\sigma}^1, \boldsymbol{\sigma}^2) \le a$, we have $\mathrm{d}\,(\boldsymbol{\sigma}^2, \boldsymbol{\sigma}_\alpha) \le 3a$, which proves that $(\boldsymbol{\sigma}^1, \boldsymbol{\sigma}^2) \in C_\alpha^2$.

□

Theorem 6.2.2. *There exists a number p_0 with the following property. If $p \geq p_0$ is odd, then we can find a partition $(C_\alpha)_{\alpha\geq 1}$ of Σ_N (depending on the disorder) such that*

$$\boldsymbol{\sigma}^1, \boldsymbol{\sigma}^2 \in C_\alpha \Rightarrow R_{1,2} \geq 1 - 2^{-p/2+4} \tag{6.42}$$

$$\mathsf{E}\, G_N^{\otimes 2}\left(\{|R_{1,2}| \geq 2^{-p/4}\} \setminus \bigcup_{\alpha\geq 1} C_\alpha^2\right) \leq K \exp\left(-\frac{N}{K}\right) . \tag{6.43}$$

Proof. We use Theorem 6.2.1 at given disorder with $a' = 2^{-p/4}$, $a = 2^{-p/2}$ and Theorem 6.1.6. □

Let us stress again that this contains *no* information when $\mathsf{E}\, G_N^{\otimes 2}(\{|R_{1,2}| \geq 2^{-p/4}\}) \leq K \exp(-N/K)$.

For convenience, the sets $(C_\alpha)_{\alpha\geq 1}$ will be called the lumps (even though they can be of very small Gibbs' measure).

The content of (6.43) is that, if $\boldsymbol{\sigma}^1, \boldsymbol{\sigma}^2$ do not belong to the same lump then it is very likely that $|R_{1,2}| \leq 2^{-p/4}$. We will show that in fact in that case we have $R_{1,2} \simeq 0$. The lumps are "orthogonal".

Theorem 6.2.3. *There exists a number p_0 such that if $p \geq p_0$ and $\beta \leq 2^{p^2/10}$, we have*

$$\mathsf{E}\, \langle R_{1,2}^2 \mathbf{1}_{\{|R_{1,2}|\leq 1/2\}} \rangle \leq \frac{K}{N} . \tag{6.44}$$

Of course the value $1/2$ does not play a special role here.

Let us discuss a number of essential issues before proving Theorem 6.2.3. We have

$$p_N'(\beta) = \frac{1}{N}\left(\frac{p!}{2N^{p-1}}\right)^{1/2} \sum \mathsf{E}\, g_{i_1,\ldots,i_p} \langle \sigma_{i_1} \cdots \sigma_{i_p} \rangle ,$$

where the summation is over $1 \leq i_1 < \cdots < i_p \leq N$, and integration by parts as in the proof of (2.210) yields

$$p_N'(\beta) = \beta \frac{p!}{2N^p} \sum (1 - \mathsf{E} \langle \sigma_{i_1}^1 \cdots \sigma_{i_p}^1 \sigma_{i_1}^2 \cdots \sigma_{i_p}^2 \rangle) ,$$

where the summation is as before, and proceeding as in the proof of (6.3) yields

$$\left| p_N'(\beta) - \frac{\beta}{2}(1 - \mathsf{E}\, \langle R_{1,2}^p \rangle) \right| \leq \frac{K}{N} . \tag{6.45}$$

By (1.7), (6.4) and since $\beta N p_N'(\beta) = \mathsf{E}\, \langle -H_N(\boldsymbol{\sigma}) \rangle$, we have $p_N'(\beta) \leq \sqrt{\log 2}$, so that for $\beta > 2\sqrt{\log 2}$, we have

$$\liminf_{N\to\infty} \mathsf{E}\, \langle R_{1,2}^p \rangle \geq 1 - \frac{2\sqrt{\log 2}}{\beta} . \tag{6.46}$$

On the other hand, for $\beta < \gamma_p$, Theorem 6.1.2 (combined with elementary properties of convex functions) and (6.45) imply

$$\lim_{N\to\infty} \mathsf{E}\,\langle R_{1,2}^p\rangle = 0 \tag{6.47}$$

Since p is odd, this does not show that $R_{1,2}$ is typically small, but one may actually prove that

$$\mathsf{E}\,\langle R_{1,2}^2\rangle \le \frac{K}{N}\,. \tag{6.48}$$

The information contained in Theorem 6.2.3 is that the transition between the "high-temperature" behavior (6.48) and the "low-temperature" behavior (6.46) cannot be seen if one is looking only at the values taken by $R_{1,2}$ in the interval $[-1/2, 1/2]$. Rather, what happens is that $R_{1,2}$ starts taking values close to 1 (in fact, greater than $1 - 2^{-p/2}$). It is a "discontinuous transition". The central difficulty in controlling the entire high-temperature region of a model (such as in Chapter 3 or 5) is to show that a discontinuous transition does not occur inside this region. This is what Section 2.9 tries to achieve in the case of the SK model; as for the other models, the only indication that this should also be the case is numerical simulation.

The proof of Theorem 6.2.3 will use the smart path method. Given $\boldsymbol{\sigma} = (\sigma_1, \ldots, \sigma_N) \in \Sigma_N$, we consider

$$\boldsymbol{\rho} = (\sigma_1, \ldots, \sigma_{N-1}) \in \Sigma_{N-1}\,. \tag{6.49}$$

We consider the Hamiltonian

$$-H_{N-1}(\boldsymbol{\rho}) = \beta\left(\frac{p!}{2N^{p-1}}\right)^{1/2} \sum_{i_1<\cdots<i_p\le N-1} g_{i_1,\ldots,i_p}\,\sigma_{i_1}\cdots\sigma_{i_p} \tag{6.50}$$

so that

$$-H_N(\boldsymbol{\sigma}) = -H_{N-1}(\boldsymbol{\rho}) + \sigma_N\, g(\boldsymbol{\rho})\,, \tag{6.51}$$

where

$$g(\boldsymbol{\rho}) = \beta\left(\frac{p!}{2N^{p-1}}\right)^{1/2} \sum_{i_1<\cdots<i_{p-1}\le N} g_{i_1,\ldots,i_{p-1},N}\,\sigma_{i_1}\cdots\sigma_{i_{p-1}}\,. \tag{6.52}$$

The idea will be to replace the complicated process $(g(\boldsymbol{\rho}))$ by a simpler one $g'(\boldsymbol{\rho})$ and to move from one to the other using the Hamiltonian

$$-H_{N,t}(\boldsymbol{\sigma}) = -H_{N-1}(\boldsymbol{\rho}) + \sigma_N(\sqrt{t}\,g(\boldsymbol{\rho}) + \sqrt{1-t}\,g'(\boldsymbol{\rho}))\,. \tag{6.53}$$

The choice of the process $g'(\boldsymbol{\rho})$ will be guided by the following intuition. Theorem 6.2.2 says that when two configurations are in different lumps, they are "orthogonal". Orthogonality means independence for Gaussian r.v., so that we would like $g(\boldsymbol{\rho}^1)$ and $g(\boldsymbol{\rho}^2)$ to be independent when $\boldsymbol{\rho}^1$ and $\boldsymbol{\rho}^2$ lie in two different lumps. Since $\boldsymbol{\rho}^1, \boldsymbol{\rho}^2$ belong to Σ_{N-1}, "lump" here must mean "lump of Σ_{N-1} for the Hamiltonian H_{N-1}". The Hamiltonian $H_{N-1}(\boldsymbol{\rho})$ may be viewed as the Hamiltonian of the p-spin interaction model for an $(N-1)$-spin system, with a new value β^- of the inverse temperature parameter, β^- being related to β through

$$\frac{\beta^-}{(N-1)^{(p-1)/2}} = \frac{\beta}{N^{(p-1)/2}}$$

so that

$$\beta^- = \beta\left(1 - \frac{1}{N}\right)^{(p-1)/2} . \tag{6.54}$$

We define the process $g'(\boldsymbol{\rho})$ as follows.

Conditionally on the disorder of H_{N-1} (i.e. on the r.v. $g_{i_1,\ldots,i_p}$ where $1 \le i_1 < \cdots < i_p \le N-1$), the process $(g'(\boldsymbol{\rho}))$ is Gaussian. If $(C_\alpha)_{\alpha\ge1}$ denotes the decomposition of Σ_{N-1} into lumps relative to H_{N-1}, then

$$\boldsymbol{\rho}^1, \boldsymbol{\rho}^2 \in C_\alpha \Rightarrow \mathsf{E}_0(g'(\boldsymbol{\rho}^1)\, g'(\boldsymbol{\rho}^2)) = \mathsf{E}\,(g(\boldsymbol{\rho}^1)\, g(\boldsymbol{\rho}^2)) \tag{6.55}$$

$$\boldsymbol{\rho}^1 \in C_\alpha, \boldsymbol{\rho}^2 \in C_\gamma, \alpha \ne \gamma \Rightarrow \mathsf{E}_0(g'(\boldsymbol{\rho}^1)\, g'(\boldsymbol{\rho}^2)) = 0 , \tag{6.56}$$

where E_0 denotes conditional expectation given the randomness of H_{N-1}. The idea behind (6.55) is that we would like the process $(g'(\boldsymbol{\rho}))$ to be as close as we can to the process $(g(\boldsymbol{\rho}))$. In particular we are trying not to disturb anything within a given lump, so as not to be hampered by our lack of knowledge there. The existence of the process $(g'(\boldsymbol{\rho}))$ is obvious by "gluing together independent copies" of the processes $(g(\boldsymbol{\rho}))_{\boldsymbol{\rho}\in C_\alpha}$ for $\alpha \ge 1$.

We denote by $\langle\cdot\rangle_t$ the Gibbs' measure with Hamiltonian (6.53), and, for a function f on Σ_N^n, we write $\nu_t(f) = \mathsf{E}\,\langle f\rangle_t$, $\nu = \nu_1$.

We write, using symmetry between sites

$$\mathsf{E}\,\langle R_{1,2}^2\, \mathbf{1}_{\{|R_{1,2}|\le 1/2\}}\rangle = \nu(\varepsilon_1\, \varepsilon_2\, f) , \tag{6.57}$$

where

$$f = R_{1,2}\, \mathbf{1}_{\{|R_{1,2}|\le 1/2\}} . \tag{6.58}$$

Lemma 6.2.4. *If f is as in (6.58), we have*

$$|\nu_0(\varepsilon_1\, \varepsilon_2\, f)| \le \frac{K}{N} . \tag{6.59}$$

In this section, contrarily to what was happening previously, the use of the smart path method requires a control of ν_0 which is often nontrivial.

Proof. We observe that $|R_{1,2} - R_{1,2}^-| \le 1/N$. Thus

$$|\mathbf{1}_{\{|R_{1,2}|\le 1/2\}} - \mathbf{1}_{\{|R_{1,2}^-|\le 1/2\}}| \le \mathbf{1}_{\{|R_{1,2}^- - 1/2|\le 1/N\}} + \mathbf{1}_{\{|R_{1,2}^- + 1/2|\le 1/N\}} .$$

Hence, if we set

$$f^- = R^-_{1,2}\,\mathbf{1}_{\{|R^-_{1,2}|\le 1/2\}}\,,$$

we have

$$|f - f^-| \le \frac{1}{N} + f'$$

where

$$f' = \mathbf{1}_{\{|R^-_{1,2}-1/2|\le 1/N\}} + \mathbf{1}_{\{|R^-_{1,2}+1/2|\le 1/N\}}\,.$$

Thus

$$|\nu_0(f) - \nu_0(f^-)| \le \nu_0(|f - f^-|) \le \frac{1}{N} + \nu_0(f')\,. \tag{6.60}$$

We have

$$\nu_0(f') = \mathsf{E}\,\frac{\left\langle f' \operatorname{Av}\exp\sum_{\ell\le 2}\varepsilon_\ell\, g'(\boldsymbol{\rho}^\ell)\right\rangle_-}{Z^2}\,, \tag{6.61}$$

where $\langle\cdot\rangle_-$ denotes an average for Gibbs' measure relative to H_{N-1}, and where

$$Z = \langle \operatorname{Av}\exp\varepsilon\, g'(\boldsymbol{\rho})\rangle_- = \langle \operatorname{ch} g'(\boldsymbol{\rho})\rangle_-\,. \tag{6.62}$$

Since $Z \ge 1$, by taking expectation given the disorder of H_{N-1} in (6.61) we obtain

$$\nu_0(f') \le K\mathsf{E}\,\langle f'\rangle_-\,. \tag{6.63}$$

Using Theorem 6.1.6 for $\langle\cdot\rangle_-$ we get from (6.60) that

$$|\nu_0(\varepsilon_1\,\varepsilon_2\, f) - \nu_0(\varepsilon_1\,\varepsilon_2\, f^-)| \le \frac{K}{N}\,.$$

Now, since f^- is a function on Σ^2_{N-1}, we have

$$\begin{aligned}\nu_0(\varepsilon_1\,\varepsilon_2\, f^-) &= \mathsf{E}\,\frac{\left\langle \operatorname{Av}\varepsilon_1\,\varepsilon_2\, f^- \exp\left(\sum_{\ell\le 2}\varepsilon_\ell\, g'(\boldsymbol{\rho}^\ell)\right)\right\rangle_-}{Z^2}\\ &= \mathsf{E}\,\frac{\langle f^- \operatorname{sh} g'(\boldsymbol{\rho}^1)\operatorname{sh} g'(\boldsymbol{\rho}^2)\rangle_-}{Z^2}\,,\end{aligned}$$

where Z is as in (6.62). We observe that

$$\nu_0(\varepsilon_1\,\varepsilon_2\, f^-) = \sum_{\alpha,\gamma} W_{\alpha,\gamma} \tag{6.64}$$

where

$$W_{\alpha,\gamma} = \mathsf{E}\,\frac{\langle f^-\,\mathbf{1}_{\{\boldsymbol{\rho}^1\in C_\alpha,\boldsymbol{\rho}^2\in C_\gamma\}} \operatorname{sh} g'(\boldsymbol{\rho}^1)\operatorname{sh} g'(\boldsymbol{\rho}^2)\rangle_-}{Z^2}\,. \tag{6.65}$$

When $\alpha = \gamma$, we have $W_{\alpha,\gamma} = 0$ because $R^-_{1,2} > 1/2$ if $\boldsymbol{\rho}^1, \boldsymbol{\rho}^2 \in C_\alpha$ by (6.42), and because $f^- = 0$ when $R^-_{1,2} > 1/2$.

Finally we prove that if $\alpha \ne \gamma$ we have $W_{\alpha,\gamma} = 0$. This is a symmetry argument. The process $(g''(\boldsymbol{\rho}))$ defined by

$$g''(\boldsymbol{\rho}) = -g'(\boldsymbol{\rho}) \quad \text{if } \boldsymbol{\rho} \in C_\alpha$$

$$g''(\boldsymbol{\rho}) = g'(\boldsymbol{\rho}) \quad \text{if } \boldsymbol{\rho} \notin C_\alpha$$

has the same law as the process $(g'(\boldsymbol{\rho}))$ because of (6.56). Thus, replacing $g'(\boldsymbol{\rho})$ by $g''(\boldsymbol{\rho})$ does not change the value of $W_{\alpha,\gamma}$. On the other hand

$$\operatorname{ch} g''(\boldsymbol{\rho}) = \operatorname{ch} g'(\boldsymbol{\rho})$$

$$\boldsymbol{\rho}^1 \in C_\alpha \Rightarrow \operatorname{sh} g''(\boldsymbol{\rho}^1) = -\operatorname{sh} g'(\boldsymbol{\rho}^1)$$

$$\boldsymbol{\rho}^2 \in C_\gamma \Rightarrow \operatorname{sh} g''(\boldsymbol{\rho}^2) = \operatorname{sh} g'(\boldsymbol{\rho}^2)\,.$$

Thus (6.62) shows that replacing the process $g'(\boldsymbol{\rho})$ by the process $g''(\boldsymbol{\rho})$ results in a change of sign for $W_{\alpha,\gamma}$. Hence $W_{\alpha,\gamma} = -W_{\alpha,\gamma}$, i.e. $W_{\alpha,\gamma} = 0$. □

We write

$$\Delta_{\ell,\ell'} = \mathsf{E}_0\, g(\boldsymbol{\rho}^\ell)\, g(\boldsymbol{\rho}^{\ell'}) - \mathsf{E}_0\, g'(\boldsymbol{\rho}^\ell)\, g'(\boldsymbol{\rho}^{\ell'})\,. \tag{6.66}$$

Lemma 6.2.5. *If f is function on Σ_N^2, we have*

$$\nu_t'(f) := \frac{\mathrm{d}}{\mathrm{d}t}\nu_t(f) = \nu_t(\varepsilon_1\varepsilon_2 f\,\Delta_{1,2}) - 2\sum_{\ell=1,2}\nu_t(\varepsilon_\ell\varepsilon_3 f\,\Delta_{\ell,3}) + 3\,\nu_t(\varepsilon_3\varepsilon_4 f\,\Delta_{3,4})\,.$$

Proof. If we keep in mind that in similar results (starting with Proposition 2.3.1) the integrations by parts are always performed conditionally on the disorder of H_{N-1}, this is business as usual.

□

Corollary 6.2.6. *If $|f| \le 1$ we have*

$$|\nu_t'(f)| \le 6\,\nu_t(|\Delta_{1,2}|)\,. \tag{6.67}$$

Lemma 6.2.7. *We have*

$$\left| \mathsf{E}_0\, g(\boldsymbol{\rho}^1)\, g(\boldsymbol{\rho}^2) - \frac{\beta^2 p}{2}\,(R_{1,2}^-)^{p-1} \right| \le \frac{K}{N}\,. \tag{6.68}$$

Proof. This is almost identical to (6.3). □

Lemma 6.2.8. *We have*

$$|\Delta_{1,2}| \le \beta^2 p\, 2^{-(p-3)p/4}\,(R_{1,2}^-)^2\,\mathbf{1}_{\{|R_{1,2}^-| \le 1/2\}} + K\,\mathbf{1}_A \tag{6.69}$$

where

$$A = \left\{ (\boldsymbol{\sigma}^1, \boldsymbol{\sigma}^2)\,;\ |R_{1,2}^-| \ge 2^{-p/4},\, (\boldsymbol{\rho}^1, \boldsymbol{\rho}^2) \notin \bigcup_{\alpha \ge 1} C_\alpha^2 \right\}.$$

Proof. If $\boldsymbol{\rho}^1, \boldsymbol{\rho}^2$ belong to the same lump then $\Delta_{1,2} = 0$ by (6.56). If $|R^-_{1,2}| \le 2^{-p/4}$, then $|R^-_{1,2}|^{p-1} \le (R^-_{1,2})^2\, 2^{-(p-3)p/4}$. And if $|R^-_{1,2}| > 2^{-p/4}$, and $\boldsymbol{\rho}^1, \boldsymbol{\rho}^2$ do not belong to the same lump, then $(\boldsymbol{\sigma}^1, \boldsymbol{\sigma}^2) \in A$ and $|\Delta_{1,2}| \le K$. □

Lemma 6.2.9. *We have* $\nu_t(\mathbf{1}_A) \le K/N$*, where* K *does not depend on* t.

Proof. We proceed as in the proof of (6.63), writing

$$\begin{aligned}\nu_t(\mathbf{1}_A) &\le \mathsf{E}\left\langle \mathrm{Av}\,\mathbf{1}_A \exp\left(\sum_{\ell\le 2} \varepsilon_\ell(\sqrt{t}\,g(\boldsymbol{\rho}^\ell) + \sqrt{1-t}\,g'(\boldsymbol{\rho}^\ell))\right)\right\rangle_- \\ &\le K\,\mathsf{E}\,\langle \mathbf{1}_A\rangle_- \,,\end{aligned}$$

and we use (6.43) for H_{N-1}. □

Proof of Theorem 6.2.3. Let us assume that

$$\beta^2\, p\, 2^{-(p-3)p/4} \le \frac{1}{4}\,,$$

which occurs for $\beta \le 2^{p^2/10}$ and p large enough. We use (6.67) for $f' = (R^-_{1,2})^2\, \mathbf{1}_{\{|R^-_{1,2}|\le 1/2\}}$. Using (6.69) and Lemma 6.2.9, we get

$$|\nu'_t(f')| \le \frac{1}{4}\,\nu_t(f') + \frac{K}{N}$$

and by Lemma 3.4.5 we get

$$\nu_t(f') \le 2\nu(f') + \frac{K}{N}\,.$$

If we now use (6.67) and (6.69) for $\varepsilon_1\,\varepsilon_2\, f$, where f is as in (6.58), we get

$$|\nu'_t(\varepsilon_1\,\varepsilon_2\, f)| \le \frac{1}{4}\,\nu_t(f') + \frac{K}{N} \le \frac{1}{2}\,\nu(f') + \frac{K}{N}\,.$$

Combining with (6.57), (6.59) and the relation $|\nu(\varepsilon_1\,\varepsilon_2\, f) - \nu_0(\varepsilon_1\,\varepsilon_2\, f)| \le \sup_t |\nu'_t(\varepsilon_1\,\varepsilon_2\, f)|$, we obtain that

$$\begin{aligned}\nu(R^2_{1,2}\,\mathbf{1}_{\{|R_{1,2}|\le 1/2\}}) &\le \frac{1}{2}\,\nu((R^-_{1,2})^2\,\mathbf{1}_{\{|R_{1,2}|\le 1/2\}}) + \frac{K}{N} \\ &\le \frac{1}{2}\,\nu(R^2_{1,2}\,\mathbf{1}_{\{|R^-_{1,2}|\le 1/2\}}) + \frac{K}{N}\,.\end{aligned}$$

□

Conjecture 6.2.10. (Level 2) Prove that Theorem 6.2.2 is true for all values of β.

6.3 Pure States

The goal of this section is to prove that the lumps do not contain any further structure if β is not too large. We will prove, as the physicists say, that the lumps are "pure states" in the sense that the overlap of two configurations chosen randomly (according to their Gibbs' weights) inside the same lump C_α is essentially independent of these configurations. Thus this overlap will be close to a number q_α, which might depend on α as well as on the randomness. Of course it would be very nice if these numbers q_α were all practically equal to a certain non-random number (depending only on β and p), but this is a somewhat separate question which will be investigated later.

Theorem 6.3.1. *There exists a number p_0 such that if $p \ge p_0$ and $\beta p \le 2^{p/4-2}$, we have*

$$\mathsf{E}\left\langle \left(\frac{(\boldsymbol{\sigma}^1 - \boldsymbol{\sigma}^2)\cdot \boldsymbol{\sigma}^3}{N}\right)^2 \mathbf{1}_{\{R_{1,2}\ge 1/2\}} \right\rangle \le \frac{K}{N} \,. \tag{6.70}$$

It is not obvious to relate (6.70) to the statement that "the lumps are pure states". In order to understand the relation, we let

$$w_\alpha = G_N(C_\alpha) \tag{6.71}$$

denote the weights of the lumps; these random variables will play an essential part in the next section. Let $\langle\cdot\rangle_\alpha$ denote the Gibbs measure conditioned on C_α, so that

$$\langle f\rangle_\alpha = \frac{1}{w_\alpha}\langle f\,\mathbf{1}_{C_\alpha}\rangle \,.$$

We thus have

$$\sum_{\alpha\ge 1} w_\alpha^3 \left\langle \left(\frac{(\boldsymbol{\sigma}^1 - \boldsymbol{\sigma}^2)\cdot \boldsymbol{\sigma}^3}{N}\right)^2\right\rangle_\alpha = \left\langle \left(\frac{(\boldsymbol{\sigma}^1 - \boldsymbol{\sigma}^2)\cdot \boldsymbol{\sigma}^3}{N}\right)^2 f\right\rangle$$

where

$$f(\boldsymbol{\sigma}^1,\boldsymbol{\sigma}^2,\boldsymbol{\sigma}^3) = \sum_{\alpha\ge 1} \mathbf{1}_{C_\alpha}(\boldsymbol{\sigma}^1)\,\mathbf{1}_{C_\alpha}(\boldsymbol{\sigma}^2)\,\mathbf{1}_{C_\alpha}(\boldsymbol{\sigma}^3) \le \mathbf{1}_{\{R_{1,2}\ge 1/2\}} \,,$$

and (6.70) then yields

$$\mathsf{E}\sum_{\alpha\ge 1} w_\alpha^3 \left\langle \left(\frac{(\boldsymbol{\sigma}^1 - \boldsymbol{\sigma}^2)\cdot \boldsymbol{\sigma}^3}{N}\right)^2\right\rangle_\alpha \le \frac{K}{N} \,. \tag{6.72}$$

Observing that

$$\langle (R_{1,2}-R_{3,4})^2\rangle_\alpha \le 2\langle (R_{1,2}-R_{1,3})^2\rangle_\alpha + 2\langle (R_{1,3}-R_{3,4})^2\rangle_\alpha = 4\left\langle \left(\frac{(\boldsymbol{\sigma}^1-\boldsymbol{\sigma}^2)\cdot\boldsymbol{\sigma}^3}{N}\right)^2\right\rangle_\alpha$$

and that

$$\begin{aligned}\langle (R_{1,2}-R_{3,4})^2\rangle_\alpha &= \langle R_{1,2}^2 + R_{3,4}^2 - 2R_{1,2}\,R_{3,4}\rangle_\alpha \\ &= 2(\langle R_{1,2}^2\rangle_\alpha - \langle R_{1,2}\rangle_\alpha^2) \\ &= 2\langle (R_{1,2}-\langle R_{1,2}\rangle_\alpha)^2\rangle_\alpha ,\end{aligned}$$

one may then define $q_\alpha = \langle R_{1,2}\rangle_\alpha$, so that (6.72) then implies

$$\mathsf{E}\sum_{\alpha\ge 1} w_\alpha^3 \langle (R_{1,2}-q_\alpha)^2\rangle_\alpha \le \frac{K}{N}\,. \tag{6.73}$$

This means that if w_α is not too small, then $R_{1,2}$ is nearly equal to $q_\alpha = \langle R_{1,2}\rangle_\alpha$ for $\boldsymbol{\sigma}^1, \boldsymbol{\sigma}^2$ in C_α.

The proof of Theorem 6.3.1 is based on the smart path method, with a number of twists.

We define

$$V = \mathbf{1}_{\{R_{1,2}\ge 1-2^{-p/2+1}\}},\, V^- = \mathbf{1}_{\{R_{1,2}^-\ge 1-2^{-p/2+1}\}}\,. \tag{6.74}$$

By Theorem 6.1.6, we have

$$\begin{aligned} U_N := \mathsf{E}\left\langle \left(\frac{(\boldsymbol{\sigma}^1-\boldsymbol{\sigma}^2)\cdot\boldsymbol{\sigma}^3}{N}\right)^2 \mathbf{1}_{\{R_{1,2}\ge 1/2\}}\right\rangle \\ \le \mathsf{E}\left\langle \left(\frac{(\boldsymbol{\sigma}^1-\boldsymbol{\sigma}^2)\cdot\boldsymbol{\sigma}^3}{N}\right)^2 V\right\rangle + \frac{K}{N}\,.\end{aligned}$$

Using symmetry between sites yields

$$U_N \le \frac{K}{N} + \mathsf{E}\,\langle (\varepsilon_1-\varepsilon_2)\,\varepsilon_3\, f\rangle\,,$$

where

$$f = \frac{(\boldsymbol{\sigma}^1-\boldsymbol{\sigma}^2)\cdot\boldsymbol{\sigma}^3}{N} V = (R_{1,3}-R_{2,3})\,V\,. \tag{6.75}$$

Consider $f^- = (R_{1,3}^- - R_{2,3}^-)\,V^-$. As in the proof of Lemma 6.2.4 one shows that $\mathsf{E}\,\langle |f-f^-|\rangle \le K/N$, and thus

$$U_N \le \frac{K}{N} + \mathsf{E}\,\langle (\varepsilon_1-\varepsilon_2)\,\varepsilon_3\, f^-\rangle\,. \tag{6.76}$$

To lighten notation we write $\varepsilon^\sim = \varepsilon_1-\varepsilon_2$, $\boldsymbol{\sigma}^\sim = \boldsymbol{\sigma}^1-\boldsymbol{\sigma}^2$. Expressing the last bracket in (6.76) as a function of the Gibbs measure $\langle\cdot\rangle_-$ with Hamiltonian H_{N-1} of (6.50), one obtains, using the notations of Section 6.2

$$\begin{aligned} \mathsf{E}\langle \varepsilon^{\sim} \varepsilon_3 f^{-} \rangle &= \mathsf{E}\frac{1}{Z^3} \langle \mathrm{Av}\, \varepsilon^{\sim} \varepsilon_3 f^{-} \mathcal{E}_1 \mathcal{E}_2 \mathcal{E}_3 \rangle_{-} \\ &= \mathsf{E}\frac{1}{Z^3} \langle f^{-}\, \mathrm{Av}\, \varepsilon^{\sim} \varepsilon_3 \mathcal{E}_1 \mathcal{E}_2 \mathcal{E}_3 \rangle_{-} \ , \end{aligned} \tag{6.77}$$

where

$$\mathcal{E}_\ell = \exp \varepsilon_\ell\, g(\boldsymbol{\rho}^\ell)\ ;\ Z = \langle \mathrm{Av}\, \exp \varepsilon\, g(\boldsymbol{\rho}) \rangle_{-} \ .$$

Let us observe that

$$\begin{aligned} \varepsilon^{\sim} \mathcal{E}_1 \mathcal{E}_2 &= \varepsilon^{\sim} \exp(\varepsilon_1 g(\boldsymbol{\rho}^1) + \varepsilon_2 g(\boldsymbol{\rho}^2)) \\ &= \varepsilon^{\sim} \exp\left(\frac{1}{2} \varepsilon^{\sim} (g(\boldsymbol{\rho}^1) - g(\boldsymbol{\rho}^2))\right) \ . \end{aligned} \tag{6.78}$$

(This identity becomes obvious if one distinguishes the cases $\varepsilon_2 = \varepsilon_1$, $\varepsilon_2 = -\varepsilon_1$.) Let us further consider an independent copy $g'(\boldsymbol{\rho})$ of the r.v. $g(\boldsymbol{\rho})$; that is,

$$g'(\boldsymbol{\rho}) = \beta \left(\frac{p!}{2N^{p-1}}\right)^{1/2} \sum_{i_1 < \cdots < i_{p-1} \le N-1} g'_{i_1,\ldots,i_{p-1}} \sigma_{i_1} \cdots \sigma_{i_{p-1}}$$

where the r.v. $g'_{i_1,\ldots,i_{p-1}}$ are standard normal, independent of all the other r.v. involved. We write

$$g_\theta = g_\theta(\boldsymbol{\rho}^1, \boldsymbol{\rho}^2) = \sin\theta (g(\boldsymbol{\rho}^1) - g(\boldsymbol{\rho}^2)) + \cos\theta (g'(\boldsymbol{\rho}^1) - g'(\boldsymbol{\rho}^2)) \ , \tag{6.79}$$

$$\mathcal{E}_\theta = \exp\left(\frac{1}{2} \varepsilon^{\sim} g_\theta(\boldsymbol{\rho}^1, \boldsymbol{\rho}^2)\right) \ , \tag{6.80}$$

and we define

$$\varphi(\theta) = \mathsf{E}\frac{1}{Z^3} \langle f^{-}\, \mathrm{Av}\, \varepsilon^{\sim} \varepsilon_3 \mathcal{E}_\theta \mathcal{E}_3 \rangle_{-} \ . \tag{6.81}$$

We have thus created a path that has the effect of replacing the variables $g(\boldsymbol{\rho}^1) - g(\boldsymbol{\rho}^2)$ by independent copies. Using the relations (6.76) to (6.81), we see that

$$U_N \le \varphi\left(\frac{\pi}{2}\right) + \frac{K}{N} \ . \tag{6.82}$$

The following largely motivates the previous definitions.

Lemma 6.3.2. *We have*

$$|\varphi(0)| \le \frac{K}{N} \ . \tag{6.83}$$

Proof. Let E_0 denote expectation with respect to the r.v. $g'_{i_1,\ldots,i_{p-1}}$ only, all the other r.v. remaining fixed. One has

$$\mathsf{E}\frac{1}{Z^3} \langle f^{-}\mathrm{Av}\, \varepsilon^{\sim} \varepsilon_3 \mathcal{E}_0 \mathcal{E}_3 \rangle_{-} = \mathsf{E}\frac{1}{Z^3} \langle f^{-}\mathrm{Av}\, \varepsilon^{\sim} \varepsilon_3 \mathsf{E}_0 \mathcal{E}_0 \mathcal{E}_3 \rangle_{-} \tag{6.84}$$

because the r.v. $g'_{i_1,\ldots,i_{p-1}}$ appear in $\mathcal{E}_0$ only. Now

$$\mathrm{Av}\,\varepsilon^{\sim}\,\varepsilon_3\,\mathsf{E}_0\,\mathcal{E}_0\,\mathcal{E}_3 = \mathrm{Av}\,\varepsilon^{\sim}\,\mathsf{E}_0\,\mathcal{E}_0\,\mathrm{Av}\,\varepsilon_3\,\mathcal{E}_3$$

because $\varepsilon_1, \varepsilon_2$ appear in $\mathcal{E}_0$ only, whereas ε_3 appears in $\mathcal{E}_3$ only. But

$$\mathrm{Av}\,\varepsilon^{\sim}\,\mathsf{E}_0\,\mathcal{E}_0 = \mathrm{Av}\,\varepsilon^{\sim} \exp\left(\frac{1}{2}(\varepsilon^{\sim})^2\,\mathsf{E}((g'(\boldsymbol{\rho}^1) - g'(\boldsymbol{\rho}^2))^2)\right)$$

is 0. Thus the quantity (6.84) is also 0. □

We write

$$g'_\theta = \frac{\mathrm{d}}{\mathrm{d}\theta}\,g_\theta(\boldsymbol{\rho}^1, \boldsymbol{\rho}^2) = \cos\theta(g(\boldsymbol{\rho}^1) - g(\boldsymbol{\rho}^2)) - \sin\theta(g'(\boldsymbol{\rho}^1) - g'(\boldsymbol{\rho}^2))\,.$$

Lemma 6.3.3. *If $\ell = 3, 4$, we have*

$$|\mathsf{E}\,g'_\theta\,g(\boldsymbol{\rho}^\ell)| \le \frac{K}{N} + \frac{\beta^2 p^2}{2}\,|R^-_{1,\ell} - R^-_{2,\ell}|\,. \tag{6.85}$$

Proof. We have

$$\mathsf{E}\,g'_\theta\,g(\boldsymbol{\rho}^\ell) = \cos\theta(\mathsf{E}\,g(\boldsymbol{\rho}^1)\,g(\boldsymbol{\rho}^\ell) - \mathsf{E}\,g(\boldsymbol{\rho}^2)\,g(\boldsymbol{\rho}^\ell))\,.$$

Using Lemma 6.2.7, we get

$$|\mathsf{E}\,g'_\theta\,g(\boldsymbol{\rho}^\ell)| \le \frac{K}{N} + \frac{\beta^2 p}{2}\,|(R^-_{1,\ell})^{p-1} - (R^-_{2,\ell})^{p-1}|\,.$$

For $|x|, |y| \le 1$, we have $|x^{p-1} - y^{p-1}| \le (p-1)\,|x-y| \le p\,|x-y|$, and this implies (6.85). □

Lemma 6.3.4. *We have*

$$|\varphi'(\theta)| \le \frac{K}{N} + \beta^2 p^2\,\mathsf{E}\,\frac{1}{Z^3}\,\langle f'\mathrm{Av}(\varepsilon^{\sim})^2\,\mathcal{E}_\theta\,\mathcal{E}_3\rangle_-\,, \tag{6.86}$$

where

$$f' = (R^-_{1,3} - R^-_{2,3})^2\,V^-\,. \tag{6.87}$$

Proof. Since

$$\frac{\mathrm{d}}{\mathrm{d}\theta}\,\mathcal{E}_\theta = \frac{1}{2}\,\varepsilon^{\sim}\,g'_\theta\,\mathcal{E}_\theta\,,$$

we have

$$\varphi'(\theta) = \frac{1}{2}\,\mathsf{E}\,\frac{1}{Z^3}\,\langle f^-\mathrm{Av}(\varepsilon^{\sim})^2\,\varepsilon_3\,g'_\theta\,\mathcal{E}_\theta\,\mathcal{E}_3\rangle_-\,.$$

Integration by parts (as already performed many times), combined with the fact that $\mathsf{E}\,g'_\theta\,g_\theta = 0$ yields

$$\varphi'(\theta) = \frac{1}{2}\,\mathsf{E}\,\frac{1}{Z^3}\,\langle f^- \mathrm{Av}(\varepsilon^\sim)^2\,\mathsf{E}\,(g'_\theta\, g(\boldsymbol{\rho}^3))\,\mathcal{E}_\theta\,\mathcal{E}_3\rangle_- \\ - \frac{3}{2}\,\mathsf{E}\,\frac{1}{Z^4}\,\langle f^- \mathrm{Av}(\varepsilon^\sim)^2\,\varepsilon_3\,\varepsilon_4\,\mathsf{E}\,(g'_\theta\, g(\boldsymbol{\rho}^4))\,\mathcal{E}_\theta\,\mathcal{E}_3\,\mathcal{E}_4\rangle_- \,.$$

Using Lemma 6.3.3, we get

$$|\varphi'(\theta)| \le \frac{K}{N} + \frac{\beta^2\, p^2}{4}\,\mathsf{E}\,\frac{1}{Z^3}\,\langle |f^-||R^-_{1,3} - R^-_{2,3}|\,\mathrm{Av}(\varepsilon^\sim)^2\,\mathcal{E}_\theta\,\mathcal{E}_3\rangle_- \\ + \frac{3\,\beta^2\, p^2}{4}\,\mathsf{E}\,\frac{1}{Z^4}\,\langle |f^-||R^-_{1,4} - R^-_{2,4}|\,\mathrm{Av}(\varepsilon^\sim)^2\,\mathcal{E}_\theta\,\mathcal{E}_3\,\mathcal{E}_4\rangle_- \,. \tag{6.88}$$

We observe that

$$|f^-||R^-_{1,3} - R^-_{2,3}| = f' \,,$$

$$|f^-||R^-_{1,4} - R^-_{2,4}| = |R^-_{1,3} - R^-_{2,3}|\,|R^-_{1,4} - R^-_{2,4}|\,V^- \\ \le \frac{1}{2}\,((R^-_{1,3} - R^-_{2,3})^2\,V^- + (R^-_{1,4} - R^-_{2,4})^2\,V^-) \,.$$

Substituting this in the last term of (6.88) and using symmetry between replicas 3 and 4, we see that this last term is at most

$$\frac{3\,\beta^2\, p^2}{4}\,\mathsf{E}\,\frac{1}{Z^4}\,\langle (R^-_{1,3} - R^-_{2,3})^2\,V^- \mathrm{Av}(\varepsilon^\sim)^2\,\mathcal{E}_\theta\,\mathcal{E}_3\,\mathcal{E}_4\rangle_- \\ = \frac{3\,\beta^2\, p^2}{4}\,\mathsf{E}\,\frac{1}{Z^3}\,\langle f' \mathrm{Av}(\varepsilon^\sim)^2\,\mathcal{E}_\theta\,\mathcal{E}_3\rangle_- \,.$$

□

Lemma 6.3.5. *If $\beta\, p \le 2^{p/4-2}$ we have*

$$\mathsf{E}\,\frac{1}{Z^3}\,\langle f' \mathrm{Av}(\varepsilon^\sim)^2\,\mathcal{E}_\theta\,\mathcal{E}_3\rangle_- \le \frac{K}{N} + 2\,\mathsf{E}\,\langle f'(\varepsilon^\sim)^2\rangle \,. \tag{6.89}$$

Proof. Consider the function

$$\psi(\theta) = \mathsf{E}\,\frac{1}{Z^3}\,\langle f' \mathrm{Av}(\varepsilon^\sim)^2\,\mathcal{E}_\theta\,\mathcal{E}_3\rangle_- \,.$$

Proceeding as in the proof of Lemma 6.3.4, one obtains

$$|\psi'(\theta)| \le \frac{K}{N} + \beta^2\, p^2\,\mathsf{E}\,\frac{1}{Z^3}\,\langle |R^-_{1,3} - R^-_{2,3}|^3\,V^- \mathrm{Av}(\varepsilon^\sim)^2\,\mathcal{E}_\theta\,\mathcal{E}_3\rangle_- \,. \tag{6.90}$$

Now, using (6.38),

$$|R^-_{1,3} - R^-_{2,3}| = \frac{1}{N}\left|\sum_{i\le N-1} (\sigma^1_i - \sigma^2_i)\,\sigma^3_i\right| \\ \le \frac{2}{N}\,\mathrm{card}\,\{i \le N\,;\ \sigma^1_i \ne \sigma^2_i\} = 1 - R_{1,2} \,.$$

Thus, by (6.90),

$$\psi'(\theta) \leq \frac{K}{N} + \beta^2 p^2 \,\mathsf{E}\, \frac{1}{Z^3} \langle (R^-_{1,3} - R^-_{2,3})^2 V^-(1 - R^-_{1,2}) \,\mathrm{Av}(\varepsilon^\sim)^2 \,\mathcal{E}_\theta \,\mathcal{E}_3 \rangle_- \,. \quad (6.91)$$

Since $V^-(1 - R^-_{1,2}) \leq 2^{-p/2+1} V^-$, one then obtains

$$\psi'(\theta) \leq \frac{K}{N} + \beta^2 p^2 \, 2^{-p/2+1} \psi(\theta) \,,$$

which yields the announced inequality by Lemma 3.4.5.

Proof of Theorem 6.3.1. If we write

$$\psi\left(\frac{\pi}{2}\right) = \psi(0) + \int_0^{\pi/2} \psi'(\theta)\, d\theta$$

and combine (6.82), (6.83), (6.86) and (6.89), we get, recalling definition (6.87),

$$\begin{aligned} U_N &\leq \frac{K}{N} + 4\beta^2 p^2 \,\mathsf{E}\, \langle f'(\varepsilon^\sim)^2 \rangle \\ &\leq \frac{K}{N} + 4\beta^2 p^2 \,\mathsf{E} \left\langle \left(\frac{\boldsymbol{\sigma}^\sim \cdot \boldsymbol{\sigma}^3}{N} \right)^2 V(\varepsilon^\sim)^2 \right\rangle \,. \end{aligned} \quad (6.92)$$

Using symmetry between sites we can now replace $(\varepsilon^\sim)^2$ by

$$\frac{1}{N} \sum_{i \leq N} (\sigma_i^1 - \sigma_i^2)^2 = 2(1 - R_{1,2}) \,.$$

Since $V(1 - R_{1,2}) \leq V\, 2^{-p/2+1}$, (6.92) implies

$$U_N \leq \frac{K}{N} + \beta^2 p^2 \, 2^{-p/2+3} \,\mathsf{E} \left\langle \left(\frac{\boldsymbol{\sigma}^\sim \cdot \boldsymbol{\sigma}^3}{N} \right)^2 V \right\rangle \leq \frac{K}{N} + \beta^2 p^2 \, 2^{-p/2+3} U_N \,.$$

□

6.4 Perturbed Hamiltonian and Extended Ghirlanda-Guerra Identities

In the previous section we have shown "that the lumps are pure states". (So this is how we will call them from now on.) This means that the overlap of two configurations chosen in C_α randomly according to their Gibbs' weights is essentially always close to a number q_α. We would like to show that in fact all these numbers q_α are close to a given (non-random) number q. It seems difficult to achieve this unless one controls the Gibbs' weights (w_α) of the sets

C_α. The reason for this will be apparent later when we compute the value of q, a value which is closely connected to the properties of the sequence (w_α). The only way this can be done at the time of this writing is complicated [152], so we will take a shortcut and move to a slightly different problem. We will add a lower-order perturbation term to the Hamiltonian. In a manner that looks miraculous (because it is not really understood) the perturbation term will rather effortlessly provide a complete control of the sequence (w_α).

We consider independent standard Gaussian r.v. $g^s_{i_1,\ldots,i_s}$ where $1 \le i_1 < \ldots < i_s$ and where $s \ge 1$. We consider

$$g_{N,s}(\boldsymbol{\sigma}) = \left(\frac{s!}{N^{s-1}}\right)^{1/2} \sum g^s_{i_1,\ldots,i_s}\, \sigma_{i_1} \cdots \sigma_{i_s} \,, \tag{6.93}$$

where the summation is over $1 \le i_1 < \ldots < i_s \le N$. We set $g_{N,s}(\boldsymbol{\sigma}) = 0$ if $s > N$. Thus (6.93) is the Hamiltonian of an s-spin interaction model (without the parasitic factor 2 in the denominator). We consider a sequence $\xi(N)$, whose role is to adjust the strength of the perturbation term, and the Hamiltonian

$$\begin{aligned} -H_N(\boldsymbol{\sigma}) = \beta \left(\frac{p!}{2N^{p-1}}\right)^{1/2} & \sum_{i_1<\ldots<i_p} g_{i_1,\ldots,i_p}\, \sigma_{i_1} \cdots \sigma_{i_p} \\ & + \xi(N) \sum_{s\ge 1} \beta_s\, 2^{-s}\, g_{N,s}(\boldsymbol{\sigma}) \,. \end{aligned} \tag{6.94}$$

Here β_s are numbers, and the role of the factor 2^{-s} is to ensure convergence when $|\beta_s| \le 1$, which is the case we will consider. The Hamiltonian (6.94) depends on the parameters β and $\boldsymbol{\beta} = (\beta_s)_{s\ge1}$. We set

$$p_N(\beta, \boldsymbol{\beta}) = \frac{1}{N} \mathsf{E} \log Z_N \,,$$

where Z_N is the partition function $Z_N = \sum \exp(-H_N(\boldsymbol{\sigma}))$. The following expresses that the "perturbation term" $\xi(N) \sum_{s\ge1} \beta_s\, 2^{-s}\, g_{N,s}(\boldsymbol{\sigma})$ is a smaller order term.

Lemma 6.4.1. *We have*

$$p_N(\beta, \mathbf{0}) \le p_N(\beta, \boldsymbol{\beta}) \le p_N(\beta, \mathbf{0}) + \frac{\xi(N)^2}{2} \sum_{s\ge1} 2^{-2s} \beta_s^2 \,. \tag{6.95}$$

Proof. The right-hand side follows by Jensen's inequality, integrating in the r.v. $g^s_{i_1,\ldots,i_s}$ inside the log rather than outside. The left-hand side is obtained by observing that

$$p_N(\beta, \boldsymbol{\beta}) - p_N(\beta, \mathbf{0}) = \frac{1}{N} \mathsf{E} \log \langle \exp(-H_N^{\mathrm{per}}(\boldsymbol{\sigma})) \rangle \,,$$

where

$$-H_N^{\mathrm{per}}(\boldsymbol{\sigma}) = \xi(N) \sum_{s \geq 1} 2^{-s} \beta_s \, g_{N,s}(\boldsymbol{\sigma})$$

and where the bracket $\langle \cdot \rangle$ corresponds to the choice $\boldsymbol{\beta} = \mathbf{0}$. Now, by Jensen's inequality used for $\langle \cdot \rangle$, we have

$$\mathsf{E} \log \langle \exp(-H_N^{\mathrm{per}}(\boldsymbol{\sigma})) \rangle \geq \mathsf{E} \log \exp(\langle -H_N^{\mathrm{per}}(\boldsymbol{\sigma}) \rangle) = \mathsf{E} \langle -H_N^{\mathrm{per}}(\boldsymbol{\sigma}) \rangle = 0$$

as is seen by integrating first in the r.v. $g^s_{i_1,\ldots,i_s}$. □

One can use an upper bound similar to (6.95) when the summation over all values of $\boldsymbol{\sigma}$ is replaced by a summation over a subset. We then see that if $\xi(N) \to 0$, then Theorems 6.1.6 and 6.2.2 remain valid for the Hamiltonian (6.94). Despite this and (6.95), the system governed by the Hamiltonian (6.94) is genuinely different from the p-spin interaction model; and it is not clear how to relate the behavior of $R_{1,2}$ in these two models. These behaviors can actually be rather different. This can be shown in the case where p is even, where it has been proved (see [152]) that $R_{1,2}$ takes essentially 3 values (in a suitable range of β) for the standard p-spin interaction model, but only 2 values with the Hamiltonian (6.94), the extra term H_N^{per} destroying the global symmetry around the origin.

Lemma 6.4.2. *For each value of β, we have*

$$\int \mathsf{E}\left(\left| \left\langle \frac{g_{N,s}(\boldsymbol{\sigma})}{N} \right\rangle - \mathsf{E} \left\langle \frac{g_{N,s}(\boldsymbol{\sigma})}{N} \right\rangle \right| \right) \mathrm{d}\boldsymbol{\beta} \leq \frac{K(\beta)\, 2^s}{\xi(N) N^{1/4}} \,, \tag{6.96}$$

$\int \ldots d\boldsymbol{\beta}$ meaning here that each β_s is being integrated between -1 and 1.

Proof. We use Theorem 2.12.1 with $-H_2(\boldsymbol{\sigma}) = \xi(N)\, 2^{-s} g_{N,s}(\boldsymbol{\sigma})$ and $-H_1(\boldsymbol{\sigma}) = -H_N + \beta_s H_2(\boldsymbol{\sigma})$ (thus β_s plays here the rôle of the parameter x). This gives (6.96) when we integrate in β_s only from -1 to 1. We then use Fubini's theorem to conclude. □

It will turn out to be a good idea to assume

$$\xi(N) \to 0 \;; \quad \xi^2(N)\, N^{1/4} \to \infty \,, \tag{6.97}$$

an assumption that holds for the rest of this chapter.

We will denote by $\delta = \delta(\beta, \boldsymbol{\beta}, N)$ a quantity such that

$$\lim_{N \to \infty} \int |\delta| \, \mathrm{d}\boldsymbol{\beta} = 0 \,, \tag{6.98}$$

where the integral is as always over $-1 \leq \beta_s \leq 1, s \geq 1$.

Lemma 6.4.3. (The extended Ghirlanda-Guerra identities.) *Consider a function f on Σ_N^n with $|f| \le 1$ and a continuous function ψ on $\mathbb{R}$. Then*

$$\nu(\psi(R_{1,n+1})\, f) = \frac{1}{n}\,\nu(\psi(R_{1,2}))\,\nu(f) + \frac{1}{n}\sum_{2\le \ell\le n} \nu(\psi(R_{1,\ell})\, f) + \delta\,. \tag{6.99}$$

Here we revert to the compact notation $\nu(f) = \mathsf{E}\,\langle f\rangle$.

Proof. By approximation we can assume that $\psi(x)$ is a polynomial so that we can in fact assume that $\psi(x) = x^s$, for an integer $s \ge 1$. We then repeat the proof of (2.451) using (6.96) and (6.3) (this brings out another factor $\xi(N)$ and explains the square in (6.97)). □

We are now in a position to control the weights (w_α). Let us set

$$m = m_N(\beta, \boldsymbol{\beta}) = 1 - \nu(\mathbf{1}_{\{R_{1,2}\ge 1/2\}})\,,$$

and recall that the numbers $S^{(m)}(n_1, \dots, n_k)$ were defined in (1.44).

Theorem 6.4.4. *Given any value of $\boldsymbol{\beta}$ and integers $k, n_1, \dots, n_k$, we have*

$$\mathsf{E}\prod_{s\le k}\sum_{\alpha\ge 1} w_\alpha^{n_s} = S^{(m)}(n_1, \dots, n_k) + \delta\,. \tag{6.100}$$

Here of course $w_\alpha = G_N(C_\alpha)$. What this statement means is that if we order the lumps so that their weights are non-increasing the distribution of the sequence $(w_\alpha)_{\alpha\ge1}$ is nearly a Poisson Dirichlet distribution of parameter $m = m_N(\beta, \boldsymbol{\beta})$. This is because, as explained in Section 1.2, the numbers

$$S(n_1, \dots, n_k) = \mathsf{E}\prod_{s\le k}\sum_{\alpha\ge 1} w_\alpha^{n_s} \tag{6.101}$$

determine the distribution of the sequence w_α. The fact that we control the error δ of (6.100) only in average allows in principle for exceptional values of $\boldsymbol{\beta}$ where we have no information.

Proof. Let us use (6.99) with ψ such that $\psi(x) = 0$ if $x \le 1/2$ and $\psi(x) = 1$ if $x \ge 3/4$. If we use Theorem 6.1.6 (which, as we pointed out, remains valid for the perturbed Hamiltonian), we see that the contribution of configurations for which $R_{1,\ell} \in [1/2, 3/4]$ is exponentially small, so that (6.99) holds for $\psi = \mathbf{1}_{\{x\ge 3/4\}}$.

Consider now integers $n_1, \dots, n_k$, and $n = n_1 + \cdots + n_k$. Consider n replicas, and a partition $I_1, \dots, I_k$ of $\{1, \dots, n\}$ into disjoint sets with card $I_s = n_s$. Consider the function f given by

$$f(\boldsymbol{\sigma}^1, \dots, \boldsymbol{\sigma}^n) = \prod_{1\le s\le k}\ \prod_{\ell,\ell'\in I_s} \mathbf{1}_{\{R_{\ell,\ell'}\ge 3/4\}}\,.$$

Consider the set

$$A = \left\{ (\boldsymbol{\sigma}^1, \ldots, \boldsymbol{\sigma}^n); \exists \ell, \ell' \le n, R_{\ell,\ell'} \ge 3/4, (\boldsymbol{\sigma}^\ell, \boldsymbol{\sigma}^{\ell'}) \notin \bigcup_{\alpha \ge 1} C_\alpha^2 \right\} . \quad (6.102)$$

Then, by (6.43) we have $\nu(\mathbf{1}_A) \le K \exp(-N/K)$. If $(\boldsymbol{\sigma}^1, \ldots, \boldsymbol{\sigma}^n)$ does not belong to A and if $f(\boldsymbol{\sigma}^1, \ldots, \boldsymbol{\sigma}^n) = 1$, then for each $1 \le s \le k$ and for each $\ell, \ell' \in I_s$, there is an α for which $(\boldsymbol{\sigma}^\ell, \boldsymbol{\sigma}^{\ell'}) \in C_\alpha^2$. Thus there is an α_s such that $\boldsymbol{\sigma}^\ell \in C_{\alpha_s}$ whenever $\ell \in I_s$. Hence

$$\begin{aligned}(\boldsymbol{\sigma}^1, \ldots, \boldsymbol{\sigma}^n) \in B := \bigcup_{\alpha_1, \ldots, \alpha_k} \{ &(\boldsymbol{\sigma}^1, \ldots, \boldsymbol{\sigma}^n) ;\ \forall s \le k, \\ &\forall \ell \in I_s,\ \boldsymbol{\sigma}^\ell \in C_{\alpha_s} \} . \end{aligned} \quad (6.103)$$

Since by (6.42) we have $f = 1$ on B, we have

$$|\nu(f) - \nu(\mathbf{1}_B)| \le \nu(\mathbf{1}_A) \le K \exp\left(-\frac{N}{K}\right) . \quad (6.104)$$

Now

$$\langle \mathbf{1}_B \rangle = \sum_{\alpha_1, \ldots, \alpha_k} \prod_{s \le k} G_N(C_{\alpha_s})^{n_s} ,$$

because the union in (6.103) is a disjoint union. Thus, since $w_\alpha = G_N(C_\alpha)$, we get

$$\langle \mathbf{1}_B \rangle = \prod_{s \le k} \sum_{\alpha \ge 1} w_\alpha^{n_s} \quad (6.105)$$

and

$$|\nu(f) - S(n_1, \ldots, n_k)| \le K \exp\left(-\frac{N}{K}\right) . \quad (6.106)$$

Consider now $\ell \le n$, and assume without loss of generality that $1 \in I_1$. If $\ell \in I_1$, then $\mathbf{1}_{\{R_{1,\ell} \ge 3/4\}} f = f$. Otherwise assume that $\ell \in I_s$ where $s > 1$. To compute $\nu(\mathbf{1}_{\{R_{1,\ell} \ge 3/4\}} f)$, we observe that (outside A) the condition $R_{1,\ell} \ge 3/4$ means that $\boldsymbol{\sigma}^1, \boldsymbol{\sigma}^\ell$ must belong to the same set C_α. Condition (6.104) thus remains true when adding the requirement $\alpha_1 = \alpha_s$ to the definition of B in (6.103). Thus (6.105) has to be replaced by

$$\langle \mathbf{1}_B \rangle = \sum_{\alpha \ge 1} w_\alpha^{n_1 + n_s} \prod_{2 \le s' \le n, s' \ne s} \sum_{\alpha \ge 1} w_\alpha^{n_{s'}} , \quad (6.107)$$

and thus

$$|\nu(\mathbf{1}_{\{R_{1,\ell} \ge 3/4\}} f) - S(n_1 + n_s, \ldots, n_{s-1}, n_{s+1}, \ldots, n_k)| \le K \exp\left(-\frac{N}{K}\right) .$$

In a similar manner one shows that

$$|\nu(\mathbf{1}_{\{R_{1,n+1}\geq 3/4\}} f) - S(n_1+1, n_2, \ldots, n_k)| \leq K \exp\left(-\frac{N}{K}\right) .$$

Combining these relations with (6.99) we get

$$\begin{aligned} S(n_1+1, n_2, \ldots, n_k) &= \frac{n_1 - m}{n} S(n_1, n_2, \ldots, n_k) \\ &+ \sum_{2\leq s\leq k} \frac{n_s}{n} S(n_1+n_s, \ldots, n_k) + \delta . \end{aligned}$$

Since the quantities $S^{(m)}(n_1, \ldots, n_k)$ satisfy the same inductive relation, Theorem 6.4.4 follows by induction over n. □

There is a special property of Λ_m which will play an important role in the next section, and we now turn to this property. We consider a pair (U,V) of r.v. (that need not be independent) and assume

$$\mathsf{E} U^2 < \infty,\ \mathsf{E} V^2 < \infty,\ V \geq 1 . \tag{6.108}$$

We then let (U_α, V_α) denote i.i.d copies of this pair.

Theorem 6.4.5. *Assume that the sequence of weights $(v_\alpha)_{\alpha\geq 1}$ is distributed like Λ_m and is independent of the sequence (U_α, V_α). Then we have the formulas*

$$\mathsf{E} \frac{\sum_{\alpha\geq 1} v_\alpha U_\alpha}{\sum_{\alpha\geq 1} v_\alpha V_\alpha} = \frac{\mathsf{E}(UV^{m-1})}{\mathsf{E}V^m} \tag{6.109}$$

$$\mathsf{E} \frac{\sum_{\alpha\neq\beta} v_\alpha v_\beta U_\alpha U_\beta}{\left(\sum_{\alpha\geq 1} v_\alpha V_\alpha\right)^2} = m\left(\frac{\mathsf{E}(UV^{m-1})}{\mathsf{E}V^m}\right)^2 \tag{6.110}$$

$$\mathsf{E} \frac{\sum_{\alpha\geq 1} v_\alpha^2 U_\alpha^2}{\left(\sum_{\alpha\geq 1} v_\alpha V_\alpha\right)^2} = (1-m)\frac{\mathsf{E}(U^2V^{m-2})}{\mathsf{E}V^m} . \tag{6.111}$$

Lemma 6.4.6. *To prove Theorem 6.4.5, one can assume that U and V are bounded.*

Proof. This is a straightforward approximation argument. To explain the procedure, we explain how to reduce the proof of (6.110) to the case where U is bounded. We consider a number M and define

$$U_2 = U - M \quad \text{if } U \geq M ,$$

$$U_2 = U + M \quad \text{if } U \leq -M ,$$

whereas $U_2 = 0$ otherwise. Setting $U_1 = U - U_2$, we thus have to show that

$$\Delta_M = \mathsf{E}\left(D^{-2}\left(\sum_{\alpha\neq\beta} v_\alpha v_\beta U_\alpha U_\beta - \sum_{\alpha\neq\beta} v_\alpha v_\beta U_{1,\alpha} U_{1,\beta}\right)\right)$$

becomes small when M becomes large, where $D = \sum_\alpha v_\alpha V_\alpha \geq 1$. Now

$$\Delta_M \leq \mathsf{E} \sum_{\alpha\neq\beta} v_\alpha v_\beta |U_{1,\alpha} U_{2,\beta} + U_{2,\alpha} U_{1,\beta} + U_{2,\alpha} U_{2,\beta}| \, .$$

We then write, using successively the Cauchy-Schwarz inequality and the convexity of the function $x \mapsto x^2$, that

$$\begin{aligned}
\mathsf{E} \sum_{\alpha\neq\beta} v_\alpha v_\beta |U_{1,\alpha} U_{2,\beta}| &\leq \mathsf{E}\left(\sum_{\alpha\geq 1} v_\alpha |U_{1,\alpha}|\right)\left(\sum_{\alpha\geq 1} v_\alpha |U_{2,\alpha}|\right) \\
&\leq \left(\mathsf{E}\left(\sum_{\alpha\geq 1} v_\alpha |U_{1,\alpha}|\right)^2 \mathsf{E}\left(\sum_{\alpha\geq 1} v_\alpha |U_{2,\alpha}|\right)^2\right)^{1/2} \\
&\leq \left(\mathsf{E}\left(\sum_{\alpha\geq 1} v_\alpha U_{1,\alpha}^2\right) \mathsf{E}\left(\sum_{\alpha\geq 1} v_\alpha U_{2,\alpha}^2\right)\right)^{1/2} \\
&\leq (\mathsf{E}\, U_1^2\, \mathsf{E}\, U_2^2)^{1/2}
\end{aligned}$$

because $\sum_{\alpha\geq 1} v_\alpha = 1$ and the sequence U_α is independent of (v_α). The result is then obvious. □

Proof of Theorem 6.4.5. By Lemma 6.4.6, we can assume that U and V are bounded. Let us denote by μ the measure on $\mathbb{R}^+$ having density x^{-1-m} with respect to Lebesgue's measure and by ν the law of the pair (U, V). Consider a Poisson point process $(u_\alpha)_{\alpha\geq 1}$ of intensity measure μ. Consider an independent sequence $(U_\alpha, V_\alpha)_{\alpha\geq 1}$ of pairs of r.v., each pair having the law ν. This sequence is assumed to be independent of the process $(u_\alpha)_{\alpha\geq 1}$. The basic observation is that the set of triples $(u_\alpha, U_\alpha, V_\alpha)$ on $\mathbb{R}^3$ is a Poisson point process of intensity measure $\mu \otimes \nu$. The reason is that if C is a Borel set with $\mu(C) < \infty$, conditionally on the fact that there is a given number k of points u_α in C, the corresponding points $(u_\alpha, U_\alpha, V_\alpha)$ are distributed like a sample of k independent points chosen according to $\mu_C \otimes \nu$, where μ_C is the probability induced by μ on C; and this property is inherited by subsets of $C \times \mathbb{R}^2$.

To avoid convergence problems, we consider a parameter $a > 0$, and the measure μ_a induced by μ on $[a, 1/a]$. Those of the points $(u_\alpha, U_\alpha, V_\alpha)$ for which $a \leq u_\alpha \leq 1/a$ generate a Poisson point process of intensity measure $\mu_a \otimes \nu$. We denote by A the set of indices α for which $a \leq u_\alpha \leq 1/a$.

Consider a continuous function $h(x, y, z)$ on $\mathbb{R}^+$. The points $c_\alpha = h(u_\alpha, U_\alpha, V_\alpha)$, $\alpha \in A$, arise from a Poisson point process on $\mathbb{R}^+$, the intensity measure of which is the image ξ_a of $\mu_a \otimes \nu$ under the map $(x, y, z) \mapsto h(x, y, z)$. We observe that ξ_a has a finite mass and bounded support.

If f is a (Borel measurable) function taking only one value besides zero, we have

$$\mathsf{E} \exp \sum_{\alpha \in A} f(c_\alpha) = \exp\left(\int (\exp f(x) - 1)\, \mathrm{d}\xi_a(x)\right) . \tag{6.112}$$

This is because if $f = \lambda \mathbf{1}_C$, then $N_C = \operatorname{card}\{\alpha \in A;\ c_\alpha \in C\}$ is a Poisson r.v. of expectation $\xi_a(C)$, so that, since $\sum_{\alpha \in A} f(c_\alpha) = \lambda N_C$, we have

$$\mathsf{E} \exp \sum_{\alpha \in A} f(c_\alpha) = \mathsf{E} \exp \lambda N_C = \exp((e^\lambda - 1)\, \xi_a(C)) .$$

Since what happens for a Poisson point process on disjoint sets constitutes independent events, it follows that (6.112) holds for f taking only finitely many values; ξ_a has a finite mass and bounded support, so that a straightforward approximation procedure (using e.g. monotonicity) shows (6.112) holds true whenever f is continuous, in particular for $f(x) = x$. But ξ_a being the image of $\mu_a \otimes \nu$ under h, we have

$$\int (e^x - 1)\, \mathrm{d}\xi_a(x) = \int (\exp h(x, y, z) - 1)\, \mathrm{d}\mu_a(x)\, \mathrm{d}\nu(y, z) ,$$

and (6.112) becomes

$$\begin{aligned} &\mathsf{E} \exp\left(\sum_{\alpha \in A} h(u_\alpha, U_\alpha, V_\alpha)\right) \\ &= \exp\left(\int (\exp h(x, y, z) - 1)\, \mathrm{d}\mu_a(x)\, \mathrm{d}\nu(y, z)\right) . \end{aligned} \tag{6.113}$$

We now take

$$h(x, y, z) = s(xy)^k - txz ,$$

where s, t are parameters, $t > 0$. Differentiating both sides of the identity (6.113) at $s = 0$ one obtains

$$\mathsf{E} \sum_{\alpha \in A} (u_\alpha U_\alpha)^k \exp\left(-t \sum_{\alpha \in A} u_\alpha V_\alpha\right) := f_a(t) , \tag{6.114}$$

where

$$\begin{aligned} f_a(t) = &\int (xy)^k e^{-txz}\, \mathrm{d}\mu_a(x)\, \mathrm{d}\nu(y, z) \\ &\exp\left(\int (e^{-txz} - 1)\, \mathrm{d}\mu_a(x)\, \mathrm{d}\nu(y, z)\right) . \end{aligned} \tag{6.115}$$

Integrating (6.114) for t varying from 0 to ∞ yields

$$\mathsf{E}\,\frac{\sum_{\alpha\in A}(u_\alpha U_\alpha)^k}{\sum_{\alpha\in A} u_\alpha V_\alpha} = \int_0^\infty f_a(t)\,\mathrm{d}t\,. \tag{6.116}$$

To obtain (6.109) we will let $a \to 0$ in (6.116). Let us first examine the right-hand side; keeping in mind that $V \geq 1$, and that V is bounded we may assume that $1 \leq z \leq M$ in (6.115). We have

$$\int x^k\,\mathrm{e}^{-tx}\,\mathrm{d}\mu(x) < \infty$$

$$\int (1-\mathrm{e}^{-tMx})\,\mathrm{d}\mu(x) < \infty\,,$$

so that given t, we have $\lim_{a\to 0} f_a(t) = f(t)$, where f is defined as f_a, but with μ instead of μ_a.

In order to be able to exchange the limit $a \to 0$ and the integral in the right-hand side of (6.116) we investigate whether there is dominated convergence (the reader may skip this tedious point). Observe that

$$f_a(t) \leq M^k \int x^k\,\mathrm{e}^{-tx}\,\mathrm{d}\mu_a(x)\exp\left(\int(\mathrm{e}^{-tx}-1)\,\mathrm{d}\mu_a(x)\right)$$

where M is such that $|U| \leq M$. (This follows from the fact that $z \geq 1$, so that $\mathrm{e}^{-txz} - 1 \leq \mathrm{e}^{-tx} - 1$.) Thus, since $\mathrm{e}^{-tx} - 1 \leq -1/2$ for $x \geq 1/t$, we have

$$\begin{aligned} f_a(t) &\leq M^k \int_a^\infty x^{k-m-1}\,\mathrm{e}^{-tx}\,\mathrm{d}x \exp\left(-\frac{1}{2}\,\mu_a\left(\left[\frac{1}{t},\infty\right)\right)\right) \\ &\leq M^k\,t^{m-k}\int_{ta}^\infty x^{k-m-1}\,\mathrm{e}^{-x}\,\mathrm{d}x \exp\left(-\frac{1}{2}\,\mu_a\left(\left[\frac{1}{t},\infty\right)\right)\right)\,. \end{aligned}$$

Assuming $k \geq 1$, we have

$$\int_{ta}^\infty x^{k-m-1}\,\mathrm{e}^{-x}\,\mathrm{d}x \leq K(m)\exp\left(-\frac{ta}{K(m)}\right)\,,$$

where $K(m)$ depends on m only. Also,

$$\mu_a\left(\left[\frac{1}{t},\infty\right)\right) \geq \frac{1}{K(m)}\,\min\left(t^m,\frac{1}{a^m}\right)$$

and thus

$$f_a(t) \leq K(m)\,M^k\,t^{m-k}\exp\left(-\frac{1}{K(m)}\left(ta + \min\left(t^m,\frac{1}{a^m}\right)\right)\right)\,.$$

Now

$$ta + \min\left(t^m, \frac{1}{a^m}\right) = \min\left(ta + t^m, ta + \frac{1}{a^m}\right) .$$

Since $ta + a^{-m} \ge t^{m/(m+1)}$, we have

$$f_a(t) \le K(m)\, M^k\, t^{m-k} \exp\left(-\frac{1}{K(m)} \min(t^m, t^{\frac{m}{m+1}})\right) . \tag{6.117}$$

When $k = 1$, this shows that

$$\lim_{a\to 0} \int f_a(t)\, \mathrm{d}t = \int f(t)\, \mathrm{d}t .$$

Let us next compute $f(t)$; setting $x = w/tz$, we have

$$\begin{aligned} &\int (xy)^k\, e^{-txz}\, x^{-m-1}\, \mathrm{d}x\, \mathrm{d}\nu(y,z) \\ &= t^{m-k} \int w^{k-m-1}\, e^{-w}\, \mathrm{d}w \int y^k\, z^{m-k}\, \mathrm{d}\nu(y,z) . \end{aligned} \tag{6.118}$$

Since ν is the joint law of the pair (U, V) the last integral is $\mathsf{E}\, U^k\, V^{m-k}$. Thus, recalling the Γ function $\Gamma(c) = \int_0^\infty x^{c-1}\, e^{-x}\, \mathrm{d}x$, we get that the quantity (6.118) is

$$t^{m-k}\, \Gamma(k-m)\, \mathsf{E}\, U^k\, V^{m-k} . \tag{6.119}$$

Next, using the same change of variable,

$$\begin{aligned} \int (e^{-txz} - 1)\, x^{-1-m}\, \mathrm{d}x\, \mathrm{d}\nu(y,z) &= t^m\, \mathsf{E}(V^m) \int_0^\infty (e^{-w} - 1)\, w^{-1-m}\, \mathrm{d}w \\ &= -\frac{t^m}{m}\, \mathsf{E}(V^m)\, \Gamma(1-m) , \end{aligned} \tag{6.120}$$

where we have used integration by parts in the last line.

From (6.119) and (6.120), we find that when $k = 1$ we have $\int f(t)\, \mathrm{d}t = \mathsf{E}(U V^{m-1}) / \mathsf{E}\, V^m$.

Let us finally study the behavior of the left-hand side of (6.116) as $a \to 0$. We define

$$C = \sum_{\alpha\in A} u_\alpha\, U_\alpha\ ;\ C' = \sum_{\alpha\ge 1} u_\alpha\, U_\alpha$$

$$D = \sum_{\alpha\in A} u_\alpha\, V_\alpha\ ;\ D' = \sum_{\alpha\ge 1} u_\alpha\, V_\alpha$$

and observe that

$$\left|\frac{C}{D} - \frac{C'}{D'}\right| \le \frac{|C|\,|D' - D|}{DD'} + \frac{|C - C'|}{D'} .$$

If M is such that $|U_\alpha| \le M$, $1 \le V_\alpha \le M$, we get

$$\left|\mathsf{E}\,\frac{C}{D}-\mathsf{E}\,\frac{C'}{D'}\right|\le 2M^2\,\mathsf{E}\,\frac{\sum_{\alpha\notin A}u_\alpha}{\sum_{\alpha\ge 1}u_\alpha}\,. \tag{6.121}$$

We want to prove that the right-hand side of (6.121) goes to 0 as $a\to 0$. It suffices to show that $\sum_{\alpha\notin A}u_\alpha/\sum_{\alpha\ge 1}u_\alpha$ goes to zero in probability. This is very simple and left to the reader.

We have proved that taking the $a\to 0$ limit in (6.116) suffices to establish (6.109). To prove (6.111), we proceed as previously, fixing $k=2$, multiplying both sides of (6.115) by t before integrating, and using that $\Gamma(2-m)=(1-m)\,\Gamma(1-m)$.

In order to prove (6.110) we fix $k=1$ and differentiate twice 6.113) with respect to s at $s=0$. This yields

$$\mathsf{E}\left(\sum_{\alpha\in A}u_\alpha\,U_\alpha\right)^2\exp\left(-t\sum_{\alpha\in A}u_\alpha\,V_\alpha\right):=h_a(t) \tag{6.122}$$

where

$$\begin{aligned}h_a(t)=&\left[\left(\int xy\,e^{-txz}\,\mathrm{d}\mu_a(x)\,\mathrm{d}\nu(x,y)\right)^2+\int (xy)^2\,e^{-txz}\,\mathrm{d}\mu_a(x)\,\mathrm{d}\nu(x,y)\right]\\&\exp\left(\int (e^{-txz}-1)\,\mathrm{d}\mu_a(x)\,\mathrm{d}\nu(x,y)\right)\,.\end{aligned}$$

Multiplying both sides by t, and then integrating in t from 0 to ∞, we find as before that

$$\begin{aligned}h(t)=\lim_{a\to 0}h_a(t)=&\,[t^{2m-2}\,\Gamma(1-m)^2\,\mathsf{E}(UV^{m-1})^2+t^{m-2}\,\Gamma(2-m)\\&\times\mathsf{E}(U^2V^{m-2})]\exp\left(-\frac{t^m}{m}\,\mathsf{E}(V^m)\,\Gamma(1-m)\right)\,.\end{aligned}$$

Thus, setting $t^m=w$, using (6.122) and taking the $a\to 0$ limit yields

$$\begin{aligned}\int_0^\infty th(t)\,\mathrm{d}t&=m\left(\frac{\mathsf{E}(UV^{m-1})}{\mathsf{E}V^m}\right)^2+(1-m)\,\frac{\mathsf{E}(U^2V^{m-2})}{\mathsf{E}V^m}\\&=\mathsf{E}\left(\frac{\sum_{\alpha\ge 1}u_\alpha\,U_\alpha}{\sum_{\alpha\ge 1}u_\alpha\,V_\alpha}\right)^2\,.\end{aligned}$$

Combining this identity with (6.111) proves (6.110). □

The practical way to use Theorem 6.4.5 will be as follows.

Theorem 6.4.7. *Consider two numbers A, $\varepsilon'>0$. Then we can find a number n and a number $\varepsilon>0$ with the following property. Consider a random sequence $(w_\alpha)_{\alpha\ge 1}$ with $w_1\ge w_2\ge\cdots$ and $\sum_{\alpha\ge 1}w_\alpha=1$. Consider a number*

$0 \le m \le 1$, and assume that, given any integer k, any integers $n_1, \ldots, n_k$ with $n_s \ge 2$, $\sum_{s \le k} n_s \le n$, we have

$$\left| \mathsf{E}\left(\prod_{s \le k} \sum_{\alpha \ge 1} w_\alpha^{n_s} \right) - S^{(m)}(n_1, \ldots, n_k) \right| \le \varepsilon \,. \tag{6.123}$$

Consider a pair (U, V) of r.v. with $V \ge 1$, $\mathsf{E}U^2 \le A$, $\mathsf{E}V^2 \le A$. Consider an independent sequence of pairs $((U_\alpha, V_\alpha))_{\alpha \ge 1}$ that are distributed like the pair (U, V) and that are independent of the sequence (w_α). Then we have

$$\left| \mathsf{E} \frac{\sum_{\alpha \ge 1} w_\alpha U_\alpha}{\sum_{\alpha \ge 1} w_\alpha V_\alpha} - \frac{\mathsf{E}(U V^{m-1})}{\mathsf{E}V^m} \right| \le \varepsilon' \tag{6.124}$$

$$\left| \mathsf{E} \frac{\sum_{\alpha \ne \beta} w_\alpha w_\beta U_\alpha U_\beta}{\left(\sum_{\alpha \ge 1} w_\alpha V_\alpha \right)^2} - m \left(\frac{\mathsf{E}(U V^{m-1})}{\mathsf{E}V^m} \right)^2 \right| \le \varepsilon' \tag{6.125}$$

$$\left| \mathsf{E} \frac{\sum_{\alpha \ge 1} w_\alpha^2 U_\alpha^2}{\left(\sum_{\alpha \ge 1} w_\alpha V_\alpha \right)^2} - (1 - m) \frac{\mathsf{E}(U^2 V^{m-2})}{\mathsf{E}V^m} \right| \le \varepsilon' \,. \tag{6.126}$$

Proof. This is a compactness argument which is completely straightforward, so we do not give all the details. We consider only the case of (6.124). Assuming that the theorem fails, we can find A, ε', and for each integer $r \ge 1$ a random non-increasing sequence $(w_{r,\alpha})_{\alpha \ge 1}$, with $\sum_{\alpha \ge 1} w_{r,\alpha} = 1$, and we can find $0 \le m(r) \le 1$ such that, given any integers $n_1, \ldots, n_k$ with $n_s \ge 2$, we have

$$\left| \mathsf{E} \prod_{s \le k} \sum_{\alpha \ge 1} w_{r,\alpha}^{n_s} - S^{(m(r))}(n_1, \ldots, n_k) \right| \le \frac{1}{r} \,. \tag{6.127}$$

We can also find an independent identically distributed sequence $(U_{r,\alpha}, V_{r,\alpha})_{\alpha \ge 1}$ independent of the sequence $(w_{k,\alpha})$ with $\mathsf{E}U_{r,\alpha}^2 \le A$, $\mathsf{E}V_{r,\alpha}^2 \le A$, $V_{r,\alpha} \ge 1$, and

$$\left| \mathsf{E} \frac{\sum_{\alpha \ge 1} w_{r,\alpha} U_{r,\alpha}}{\sum_{\alpha \ge 1} w_{r,\alpha} V_{r,\alpha}} - \frac{\mathsf{E}(U_r V_r^{m(r)-1})}{\mathsf{E}V_r^{m(r)}} \right| \ge \varepsilon' \,. \tag{6.128}$$

We can assume without loss of generality that the sequence $m(r)$ converges. Let m be its limit.

First we show that we cannot have $m = 1$. Assuming, for contradiction, that $m = 1$, since $S^{(m)}(2) = 1 - m$, we see from (6.127) that $\lim_{r \to \infty} \mathsf{E}\left(\sum_{\alpha \ge 1} w_{r,\alpha}^2 \right) = 0$. Using relations such as

$$\begin{aligned}
\mathsf{E}\sum_{\alpha\geq 1} w_{r,\alpha} U_{r,\alpha} - \mathsf{E}\, U_r &= \mathsf{E}\sum_{\alpha\geq 1} w_{r,\alpha}(U_{r,\alpha} - \mathsf{E}\, U_r) \\
&\leq \left(\mathsf{E}\left(\sum_{\alpha\geq 1} w_{r,\alpha}(U_{r,\alpha} - \mathsf{E}\, U_{r,\alpha})\right)^2\right)^{1/2} \\
&= \left(\mathsf{E}\sum_{\alpha\geq 1} w_{r,\alpha}^2(\mathsf{E}(U_{r,\alpha}^2) - (\mathsf{E}\, U_{r,\alpha})^2)\right)^{1/2} \\
&\leq A\left(\mathsf{E}\sum_{\alpha\geq 1} w_{r,\alpha}^2\right)^{1/2} ,
\end{aligned}$$

we then see that

$$\left|\mathsf{E}\,\frac{\sum_{\alpha\geq 1} w_{r,\alpha} U_{r,\alpha}}{\sum_{\alpha\geq 1} w_{r,\alpha} V_{r,\alpha}} - \frac{\mathsf{E}\, U_r}{\mathsf{E}\, V_r}\right| \to 0$$

and this contradicts (6.128).

Next, we show that we cannot have $m = 0$. Indeed, if $m = 0$, by (6.127) we see that $\mathsf{E}\sum_{\alpha\geq 1} w_{r,\alpha}^2 \to 1$, and since $\sum_{\alpha=1} w_\alpha = 1$, this implies that $\mathsf{E}\,|1 - w_{r,1}| \to 0$, $\mathsf{E}\sum_{\alpha\geq 2} w_{r,\alpha} \to 0$. It is then easy to conclude that

$$\left|\mathsf{E}\,\frac{\sum_{\alpha\geq 1} w_{r,\alpha} U_{r,\alpha}}{\sum_{\alpha\geq 1} w_{r,\alpha} V_{r,\alpha}} - \mathsf{E}\,\frac{U_r}{V_r}\right| \to 0$$

and this again contradicts (6.128).

Thus we have $0 < m < 1$. Consider the law μ_r of the sequence $(w_{r,\alpha})$ on the compact set $\mathcal{S}$ of non-increasing sequences $(x_\alpha)_{\alpha\geq 1}$ with $\sum_{\alpha\geq 1} x_\alpha \leq 1$. We can assume that μ_r has a limit μ. Consider a sequence $(w_\alpha)_{\alpha\geq 1}$ of law μ. Then (since for $n \geq 2$ the maps $(x_\alpha)_{\alpha\geq 1} \to \sum_{\alpha\geq 1} x_\alpha^n$ are continuous on $\mathcal{S}$) for each $n_1, \ldots, n_k \geq 2$, we have

$$\mathsf{E}\left(\prod_{s\leq k}\sum_{\alpha\geq 1} w_\alpha^{n_s}\right) = S^{(m)}(n_1, \ldots, n_k) ,$$

and, as explained in Section 1.2, this implies that $\mu = \Lambda_m$ (and, in particular that $\sum_{\alpha\geq 1} w_\alpha = 1$ almost surely). We can assume that the joint law of the pair (U_r, V_r) has a limit. This limit is the law of a pair (U, V) such that $V \geq 1$, $\mathsf{E}\, U^k \leq A$, $\mathsf{E}\, V^k \leq A$, and by Theorem 6.4.5 we have

$$\mathsf{E}\,\frac{\sum_{\alpha\geq 1} v_\alpha U_\alpha}{\sum_{\alpha\geq 1} v_\alpha V_\alpha} = \frac{\mathsf{E}(U\, V^{m-1})}{\mathsf{E}(V^{m-1})} . \tag{6.129}$$

Thus to reach a contradiction it suffices to show that

$$\lim_{r\to\infty} \mathsf{E}\,\frac{\sum_{\alpha\ge 1} w_{r,\alpha}\,U_{r,\alpha}}{\sum_{\alpha\ge 1} w_{r,\alpha}\,V_{r,\alpha}} = \mathsf{E}\,\frac{\sum_{\alpha\ge 1} v_\alpha\,U_\alpha}{\sum_{\alpha\ge 1} v_\alpha\,V_\alpha}\,.$$

The same argument as in the proof of Theorem 6.4.5 enables us to assume additionally that $U, V \le M$. The function $(x, y) \mapsto x/y$ is continuous on the domain $y \ge 1$, so it suffices to show that for a continuous function f on $\mathbb{R}^2$, we have

$$\lim_{r\to\infty} \mathsf{E} f\left(\sum_{\alpha\ge 1} w_{r,\alpha}\,U_{r,\alpha}, \sum_{\alpha\ge 1} w_{r,\alpha}\,V_{r,\alpha}\right)$$
$$= \mathsf{E} f\left(\sum_{\alpha\ge 1} v_\alpha\,U_\alpha, \sum_{\alpha\ge 1} v_\alpha\,V_\alpha\right)\,. \tag{6.130}$$

Now, given $\varepsilon'' > 0$ we can find $n > 0$ such that $\mathsf{E}\sum_{\alpha\ge n} v_\alpha < \varepsilon''$, and thus, for large r, $\mathsf{E}\sum_{\alpha\ge n} w_{r,\alpha} \le \varepsilon''$. Thus it suffices to show that for each integer n we have

$$\lim_{r\to\infty} \mathsf{E} f\left(\sum_{\alpha\le n} w_{r,\alpha}\,U_{r,\alpha}, \sum_{\alpha\le n} w_{r,\alpha}\,V_{r,\alpha}\right)$$
$$= \mathsf{E} f\left(\sum_{\alpha\le n} v_\alpha\,U_\alpha, \sum_{\alpha\le n} v_\alpha\,V_\alpha\right)\,. \tag{6.131}$$

(The details of a similar argument appear in the proof of Theorem 1.2.1.) Since (6.131) is obvious, the proof is finished. □

6.5 One Step of Replica-Symmetry Breaking

In this section we complete the study of the structure of the p-spin interaction model with Hamiltonian (6.94). Physicists call this situation "one step of replica-symmetry breaking".

Given a number q, we consider the r.v.

$$Y = \beta\, z \sqrt{\frac{p\, q^{p-1}}{2}}$$

where z is standard Gaussian. We recall the notation

$$m = m_N(\beta, \boldsymbol{\beta}) = 1 - \mathsf{E}\left\langle \mathbf{1}_{\{R_{1,2}\ge 1/2\}}\right\rangle\,.$$

Theorem 6.5.1. *There exists a number p_0 and a number L such that if $p \ge p_0$, p is odd, and $3/2 \le \beta \le 2^{p/L}$, the equation*

$$q = \frac{\mathsf{E}\,\mathrm{th}^2 Y \,\mathrm{ch}^m Y}{\mathsf{E}\,\mathrm{ch}^m Y} \tag{6.132}$$

has a unique solution $q = q_N(\beta, \boldsymbol{\beta})$ *in the interval* $[1 - 2^{-p/2}, 1]$. *Given any* $\varepsilon > 0$ *we have*

$$\lim_{N \to \infty} \int \mathsf{E} G_N^{\otimes 2}(\{|R_{1,2}| \geq \varepsilon, |R_{1,2} - q| \geq \varepsilon\})\, \mathrm{d}\boldsymbol{\beta} = 0\,, \tag{6.133}$$

the integral being over $-1 \leq \beta_s \leq 1$ *for each* $s \geq 1$.

In words, for generic values of $\boldsymbol{\beta}$, $R_{1,2}$ can only take values that are either close to 0 or to $q = q_N(\beta, \boldsymbol{\beta})$. The value 0 is obtained when $\boldsymbol{\sigma}^1$ and $\boldsymbol{\sigma}^2$ belong to two different pure states, whereas the value q is obtained if they belong to the same pure state.

The restriction $\beta \geq 3/2$ is simply to ensure that (6.132) has a solution. We understand well the system when $\beta \leq 3/2$.

Research problem 6.5.2. (Level 2) Prove that Theorem 6.5.1 is true with the Hamiltonian (6.132), for each value of $\beta \leq 2^{p/L}$. Prove that for any $\varepsilon > 0$ we have

$$\mathsf{E}\, G_N^{\otimes 2}(\{|R_{1,2}| \geq \varepsilon, |R_{1,2} - q| \geq \varepsilon\}) \leq K \exp\left(-\frac{N}{K}\right)\,.$$

Research problem 6.5.3. (Level 2) Construct systems that exhibit a given number of steps of replica-symmetry breaking, that is such that the overlaps essentially take only a given number of values. Study such systems with the same level of detail as is achieved here.

It might be of some help to use the ideas of [141] for this last problem; such ideas can probably be transfered from the spherical model considered in [141] to the p-spin interaction model on Σ_N. (Of course Derrida's Generalized Random Energy model is not eligible here as an answer to our problem.)

We start the preparations for the proof of Theorem 6.5.1. This proof will use the smart path method, but it seems necessary to let the last two coordinates play a special rôle rather than by the last one only. Given $\boldsymbol{\sigma}$ in Σ_N we let $\boldsymbol{\rho}$ denote the sequence $(\sigma_1, \ldots, \sigma_{N-2})$ in Σ_{N-2} and consider the Hamiltonian

$$\begin{aligned} -H_{N-2}(\boldsymbol{\rho}) &= \beta \left(\frac{p!}{2N^{p-1}}\right)^{1/2} \sum_{1 \leq i_1 < \ldots < i_p \leq N-2} g_{i_1, \ldots, i_p} \sigma_{i_1} \cdots \sigma_{i_p} \\ &\quad + \xi(N) \sum_{s \geq 1} 2^{-s} \beta_s \left(\frac{s!}{N^{s-1}}\right)^{1/2} \sum_{1 \leq i_1 < \ldots < i_s \leq N-2} g^s_{i_1, \ldots, i_s} \sigma_{i_1} \cdots \sigma_{i_s}\,. \end{aligned} \tag{6.134}$$

We then have

$$-H_N(\boldsymbol{\sigma}) = -H_{N-2}(\boldsymbol{\rho}) + \sigma_{N-1}\, g'(\boldsymbol{\rho}) + \sigma_N\, g(\boldsymbol{\rho}) + \sigma_{N-1}\,\sigma_N\, g''(\boldsymbol{\rho}) \;, \quad (6.135)$$

the 4 terms appearing here on the right-hand side corresponding respectively to the terms of $-H_N(\boldsymbol{\sigma})$ which do not contain either σ_N or σ_{N-1} (resp. do not contain σ_N but contain σ_{N-1}; do not contain σ_{N-1} but contain σ_N; contain both σ_N and σ_{N-1}). Thus

$$\begin{aligned} g'(\boldsymbol{\rho}) = \beta\left(\frac{p!}{2N^{p-1}}\right)^{1/2} & \sum_{1\le i_1<\ldots<i_{p-1}\le N-2} g_{i_1,\ldots,i_{p-1},N-1}\,\sigma_{i_1}\cdots\sigma_{i_{p-1}} \\ & + \xi(N)\sum_{s\ge 1} 2^{-s}\,\beta_s\left(\frac{s!}{N^{s-1}}\right)^{1/2} \\ & \times \sum_{1\le i_1<\ldots<i_{s-1}\le N-2} g^s_{i_1,\ldots,i_{s-1},N-1}\,\sigma_{i_1}\cdots\sigma_{i_{s-1}} \end{aligned}$$

$$\begin{aligned} g(\boldsymbol{\rho}) = \beta\left(\frac{p!}{2N^{p-1}}\right)^{1/2} & \sum_{1\le i_1<\ldots<i_{p-1}\le N-2} g_{i_1,\ldots,i_{p-1},N}\,\sigma_{i_1}\cdots\sigma_{i_{p-1}} \\ & + \xi(N)\sum_{s\ge 1} 2^{-s}\,\beta_s\left(\frac{s!}{N^{s-1}}\right)^{1/2} \\ & \times \sum_{1\le i_1<\ldots<i_{s-1}\le N-2} g^s_{i_1,\ldots,i_{s-1},N}\,\sigma_{i_1}\cdots\sigma_{i_{s-1}} \end{aligned}$$

$$\begin{aligned} g''(\boldsymbol{\rho}) = \beta\left(\frac{p!}{2N^{p-1}}\right)^{1/2} & \sum_{1\le i_1<\ldots<i_{p-2}\le N-2} g_{i_1,\ldots,i_{p-2},N-1,N}\,\sigma_{i_1}\cdots\sigma_{i_{p-2}} \\ & + \xi(N)\sum_{s\ge 1} 2^{-s}\,\beta_s\left(\frac{s!}{N^{s-1}}\right)^{1/2} \\ & \times \sum_{1\le i_1<\ldots<i_{s-2}\le N-2} g^s_{i_1,\ldots,i_{s-2},N-1,N}\,\sigma_{i_1}\cdots\sigma_{i_{s-2}} \;. \end{aligned}$$

It is of some help to observe that the 4 terms in (6.135) are probabilistically independent, and that (just as in Lemma 6.2.7) we have

$$\left|\mathsf{E}\, g(\boldsymbol{\rho}^1)\, g(\boldsymbol{\rho}^2) - \frac{\beta^2 p}{2}\,(R^+_{1,2})^{p-1}\right| \le K\,\xi(N)^2 \quad (6.136)$$

$$\left|\mathsf{E}\, g'(\boldsymbol{\rho}^1)\, g'(\boldsymbol{\rho}^2) - \frac{\beta^2 p}{2}\,(R^+_{1,2})^{p-1}\right| \le K\,\xi(N)^2 \quad (6.137)$$

$$|\mathsf{E}\, g''(\boldsymbol{\rho})^2| \le \frac{K}{N}\,, \tag{6.138}$$

where

$$R^+_{1,2} = \frac{1}{N} \sum_{i \le N-2} \sigma_i^1 \sigma_i^2\,. \tag{6.139}$$

Here and below, it would be more consistent with our previous notation to write $R^{-,-}_{1,2}$ rather than $R^+_{1,2}$; however $R^+_{1,2}$ is easier to read.

If f is a function on Σ_N^n, we then have

$$\langle f \rangle = \frac{1}{Z^n} \langle \mathrm{Av}\, f\, \mathcal{E}_n \rangle_+\,, \tag{6.140}$$

where $\langle \cdot \rangle_+$ denotes average with respect to Gibbs' measure on Σ^n_{N-2} with Hamiltonian H_{N-2}; where Av means average over σ_N^ℓ, $\sigma_{N-1}^\ell = \pm 1$, where

$$\mathcal{E}_n = \exp \sum_{\ell \le n} (\sigma_N^\ell\, g(\boldsymbol{\rho}^\ell) + \sigma_{N-1}^\ell\, g'(\boldsymbol{\rho}^\ell) + \sigma_N^\ell\, \sigma_{N-1}^\ell\, g''(\boldsymbol{\rho}^\ell)) \tag{6.141}$$

and where

$$Z = \langle \mathrm{Av}\, \mathcal{E}_1 \rangle_+\,. \tag{6.142}$$

To lighten notation, we will as always write $\varepsilon_\ell = \sigma_N^\ell$, and in the same spirit we will write $\eta_\ell = \sigma_{N-1}^\ell$.

To construct the smart path, we consider the partition $(C_\alpha)_{\alpha \ge 1}$ of Σ_{N-2} produced by Theorem 6.2.1 applied to the Hamiltonian (6.134). We consider independent sequences $(z_\alpha)_{\alpha\ge1}$, $(z'_\alpha)_{\alpha\ge1}$ of standard Gaussian r.v. We define

$$Y(\boldsymbol{\rho}) = \beta \sqrt{\frac{p\, q^{p-1}}{2}}\, z_\alpha := Y_\alpha \tag{6.143}$$

where α is such that $\boldsymbol{\rho} \in C_\alpha$, and we define $Y'(\boldsymbol{\rho})$ similarly. Here of course $q = q_N(\beta, \boldsymbol{\beta})$ is the solution to (6.132). (The existence of this solution will be proved later.) We let

$$g_t(\boldsymbol{\rho}) = \sqrt{t}\, g(\boldsymbol{\rho}) + \sqrt{1-t}\, Y(\boldsymbol{\rho})\,.$$

and define $g'_t(\boldsymbol{\rho})$ similarly. We then let

$$\mathcal{E}_{n,t} = \exp \sum_{\ell \le n} (\varepsilon_\ell\, g_t(\boldsymbol{\rho}^\ell) + \eta_\ell\, g'_t(\boldsymbol{\rho}^\ell) + \sqrt{t}\, \varepsilon_\ell\, \eta_\ell\, g''(\boldsymbol{\rho}^\ell))$$

$$Z_t = \langle \mathrm{Av}\, \mathcal{E}_{1,t} \rangle_+$$

$$\langle f \rangle_t = \frac{\langle \mathrm{Av}\, f\, \mathcal{E}_{n,t} \rangle_+}{Z_t^n}$$

and finally $\nu_t(f) = \mathsf{E}\, \langle f \rangle_t$. This is the smart path.

Let us consider the quantities $q_{1,2} = q(\boldsymbol{\sigma}^1, \boldsymbol{\sigma}^2)$ given by

$$q_{1,2} = \begin{cases} q & \text{if} \quad R_{1,2} \geq 1/2 \\ 0 & \text{if} \quad R_{1,2} < 1/2 \end{cases}$$

and $q^+_{1,2} = q^+(\boldsymbol{\rho}^1, \boldsymbol{\rho}^2)$ given by

$$q^+_{1,2} = \begin{cases} q & \text{if} \quad R^+_{1,2} \geq 1/2 \\ 0 & \text{if} \quad R^+_{1,2} < 1/2 \,. \end{cases}$$

We next define

$$A_N(\beta, \boldsymbol{\beta}) = \mathsf{E}\,\langle (R_{1,2} - q_{1,2})^2 \rangle \,.$$

In order to prove Theorem 6.5.1, it suffices to establish that for any $\beta \leq 2^{p/L}$, we have

$$\lim_{N\to\infty} \int A_N(\beta, \boldsymbol{\beta})\, \mathrm{d}\boldsymbol{\beta} = 0 \,. \tag{6.144}$$

This is simply because $(R_{1,2} - q_{1,2})^2 \geq \varepsilon^2$ when $|R_{1,2}| \geq \varepsilon$ and $|R_{1,2} - q| \geq \varepsilon$.

We recall that we denote by δ $(= \delta_N(\beta, \boldsymbol{\beta}))$ a quantity such that $\lim_{N\to\infty} \int \delta_N(\beta, \boldsymbol{\beta})\, \mathrm{d}\boldsymbol{\beta} = 0$. This quantity need not be the same at each occurrence.

Thus, we can write (6.144) as

$$A_N(\beta, \boldsymbol{\beta}) = \delta \,.$$

If we expand the square $(R_{1,2} - q_{1,2})^2$, using that $R_{1,2} = N^{-1} \sum_{i\leq N} \sigma_i^1 \sigma_i^2$, and the symmetry between sites, we get

$$\begin{aligned} A_N = \nu((R_{1,2} - q_{1,2})^2) = \left(1 - \frac{1}{N}\right) \nu((\sigma_N^1 \sigma_N^2 - q_{1,2})(\sigma_{N-1}^1 \sigma_{N-1}^2 - q_{1,2})) \\ + \frac{1}{N} \nu((\sigma_N^1 \sigma_N^2 - q_{1,2})^2) \,. \end{aligned} \tag{6.145}$$

With the notation $\sigma_N^\ell = \varepsilon_\ell$, $\sigma_{N-1}^\ell = \eta_\ell$, this is

$$A_N = \left(1 - \frac{1}{N}\right) \nu((\varepsilon_1 \varepsilon_2 - q_{1,2})(\eta_1 \eta_2 - q_{1,2})) + \frac{1}{N} \nu((\varepsilon_1 \varepsilon_2 - q_{1,2})^2)$$

and thus

$$A_N \leq \nu((\varepsilon_1 \varepsilon_2 - q_{1,2})(\eta_1 \eta_2 - q_{1,2})) + \frac{L}{N} \,. \tag{6.146}$$

Using Theorem 6.1.6 as in the proof of Theorem 6.2.2 we can replace $q_{1,2}$ by $q^+_{1,2}$, so that

$$A_N = \nu(f) + \delta \,, \tag{6.147}$$

where

$$f = (\varepsilon_1 \varepsilon_2 - q^+_{1,2})(\eta_1 \eta_2 - q^+_{1,2}) \,. \tag{6.148}$$

Writing as usual $\nu_t(f) = \mathsf{E}\,\langle f\rangle_t$ and $\nu_t'(f) = \mathrm{d}\nu_t(f)/\mathrm{d}t$, we will use that

$$\nu(f) = \nu_0(f) + \nu_1'(f) - \frac{1}{2}\nu_1''(f) + \int_0^1 \frac{t^2}{2}\nu_t'''(f)\,\mathrm{d}t\,, \tag{6.149}$$

a formula proved by simple integration by parts.

Proposition 6.5.4. *We have*

$$\nu_0(f) = \delta\,. \tag{6.150}$$

Contrarily to what happened in the previous chapters, this is a non-trivial fact, and it is actually here that the work of Section 6.4 is really required.

Proof. We start by writing

$$Z_0 = \langle \mathrm{Av}\exp(\varepsilon_1 Y(\boldsymbol{\rho}^1) + \eta_1 Y'(\boldsymbol{\rho}^1))\rangle_+ = \langle \mathrm{ch}\,Y(\boldsymbol{\rho})\,\mathrm{ch}\,Y'(\boldsymbol{\rho})\rangle_+ \tag{6.151}$$

and

$$\langle f\rangle_0 = \frac{1}{Z_0^2}\left\langle \mathrm{Av}\,f\exp\left(\sum_{\ell\le 2}(\varepsilon_\ell Y(\boldsymbol{\rho}^\ell) + \eta_\ell Y'(\boldsymbol{\rho}^\ell))\right)\right\rangle_+ = \mathrm{I} + \mathrm{II} + \mathrm{III} + \mathrm{IV}$$

where

$$\mathrm{I} = \langle \varepsilon_1\varepsilon_2\eta_1\eta_2\rangle_0 = \frac{1}{Z_0^2}\langle \mathrm{sh}\,Y(\boldsymbol{\rho}^1)\,\mathrm{sh}\,Y(\boldsymbol{\rho}^2)\,\mathrm{sh}\,Y'(\boldsymbol{\rho}^1)\,\mathrm{sh}\,Y'(\boldsymbol{\rho}^2)\rangle_+$$

$$\mathrm{II} = -\langle \varepsilon_1\varepsilon_2 q_{1,2}^+\rangle_0 = -\frac{1}{Z_0^2}\langle q_{1,2}^+\,\mathrm{sh}\,Y(\boldsymbol{\rho}^1)\,\mathrm{sh}\,Y(\boldsymbol{\rho}^2)\,\mathrm{ch}\,Y'(\boldsymbol{\rho}^1)\,\mathrm{ch}\,Y'(\boldsymbol{\rho}^2)\rangle_+$$

$$\mathrm{III} = -\langle \eta_1\eta_2 q_{1,2}^+\rangle_0 = -\frac{1}{Z_0^2}\langle q_{1,2}^+\,\mathrm{ch}\,Y(\boldsymbol{\rho}^1)\,\mathrm{ch}\,Y(\boldsymbol{\rho}^2)\,\mathrm{sh}\,Y'(\boldsymbol{\rho}^1)\,\mathrm{sh}\,Y'(\boldsymbol{\rho}^2)\rangle_+$$

$$\mathrm{IV} = \frac{1}{Z_0^2}\langle (q_{1,2}^+)^2\,\mathrm{ch}\,Y(\boldsymbol{\rho}^1)\,\mathrm{ch}\,Y(\boldsymbol{\rho}^2)\,\mathrm{ch}\,Y'(\boldsymbol{\rho}^1)\,\mathrm{ch}\,Y'(\boldsymbol{\rho}^2)\rangle_+\,.$$

For $\boldsymbol{\rho}$ in C_α we have $Y(\boldsymbol{\rho}) = Y_\alpha$ and $Y'(\boldsymbol{\rho}) = Y'_\alpha$. Thus, if w_α is the weight of C_α for the Gibbs measure relative to the Hamiltonian (6.134) we have

$$Z_0 = \sum_{\alpha\ge 1} w_\alpha\,\mathrm{ch}\,Y_\alpha\,\mathrm{ch}\,Y'_\alpha$$

and, similarly,

$$\mathrm{I} = \frac{1}{Z_0^2}\sum_{\alpha,\beta\ge 1} w_\alpha w_\beta\,\mathrm{sh}\,Y_\alpha\,\mathrm{sh}\,Y'_\alpha\,\mathrm{sh}\,Y_\beta\,\mathrm{sh}\,Y'_\beta\,.$$

We observe that

$$\mathsf{E}\,\frac{1}{Z_0^2}\sum_{\alpha\neq\beta} w_\alpha\, w_\beta \operatorname{sh} Y_\alpha \operatorname{sh} Y'_\alpha \operatorname{sh} Y_\beta \operatorname{sh} Y'_\beta = 0\,.$$

This is seen by taking expectation in Y_α first, and using the fact that changing Y_α in $-Y_\alpha$ does not change Z_0 but changes $\operatorname{sh} Y_\alpha$ into $-\operatorname{sh} Y_\alpha$. Thus

$$\mathsf{E}\,\mathrm{I} = \mathsf{E}\,\frac{1}{Z_0^2}\sum_{\alpha\geq 1} w_\alpha^2 \operatorname{sh}^2 Y_\alpha \operatorname{sh}^2 Y'_\alpha = \mathsf{E}\,\frac{\sum_{\alpha\geq 1} w_\alpha^2\, U_\alpha^2}{\left(\sum_{\alpha\geq 1} w_\alpha\, V_\alpha\right)^2}$$

for $U_\alpha = \operatorname{sh} Y_\alpha \operatorname{sh} Y'_\alpha$, $V_\alpha = \operatorname{ch} Y_\alpha \operatorname{ch} Y'_\alpha$.

Let us consider

$$m^+ = 1 - \mathsf{E}\,\langle \mathbf{1}_{\{R^+_{1,2}\geq 1/2\}}\rangle_+\,.$$

Then we have shown in Theorem 6.4.4 that for any integers $n_1,\ldots,n_k$, we have

$$S(n_1,\ldots,n_k) = \mathsf{E}\prod_{s\leq k}\sum_{\alpha\geq 1} w_\alpha^{n_s} = S^{(m^+)}(n_1,\ldots,n_k) + \delta\,.$$

The purist will observe that this is not exactly the statement of Theorem 6.4.4 because in (6.134) the normalization factors involve $(s!/N^{s-1})^{1/2}$ rather than $(s!/(N-2)^{s-1})^{1/2}$. To remove this minor obstacle, one can either make the change of variable $\beta_s = \beta'_s(N/(N-2))^s$, or observe that the proof of Theorem 6.4.4 carries exactly to this case without any change.

It then follows from Theorem 6.4.7 that, defining U, V, Y, Y' in the obvious manner, we have

$$\mathsf{E}\,\frac{\sum_{\alpha\geq 1} w_\alpha^2\, U_\alpha^2}{\left(\sum_{\alpha\geq 1} w_\alpha\, V_\alpha\right)^2} = (1-m^+)\,\frac{\mathsf{E}(U^2\, V^{m^+-2})}{\mathsf{E}(V^{m^+})} + \delta$$

$$= (1-m^+)\left(\frac{\mathsf{E}(\operatorname{th}^2 Y \operatorname{ch}^{m^+} Y)}{\mathsf{E}\operatorname{ch}^{m^+} Y}\right)^2 + \delta\,,$$

since

$$\mathsf{E}\,U^2\, V^{m^+-2} = \mathsf{E}(\operatorname{th}^2 Y \operatorname{th}^2 Y' \operatorname{ch}^{m^+} Y \operatorname{ch}^{m^+} Y') = (\mathsf{E}(\operatorname{th}^2 Y \operatorname{ch}^{m^+} Y))^2$$

by independence.

To study $\mathsf{E}\,\mathrm{II}$, we observe by Theorem 6.2.2 that, since $q^+_{1,2} = 0$ when $R^+_{1,2} < 1/2$,

$$\begin{aligned}\mathsf{E}\,\mathrm{II} &= q\,\mathsf{E}\,\frac{1}{Z_0^2}\left\langle \bigcup_{\alpha\geq 1} \mathbf{1}_{C_\alpha\times C_\alpha}\,\mathrm{sh}\,Y(\boldsymbol{\rho}^1)\,\mathrm{sh}\,Y(\boldsymbol{\rho}^2)\,\mathrm{ch}\,Y'(\boldsymbol{\rho}^1)\,\mathrm{ch}\,Y'(\boldsymbol{\rho}^2)\right\rangle_+ + \delta \\ &= q\,\mathsf{E}\,\frac{1}{Z_0^2}\sum_{\alpha\geq 1} w_\alpha^2\,\mathrm{sh}^2\,Y_\alpha\,\mathrm{ch}^2\,Y'_\alpha + \delta\,.\end{aligned}$$

Proceeding as in the case of the term I, but with $U_\alpha = \mathrm{sh}\,Y_\alpha\,\mathrm{ch}\,Y'_\alpha$, we then get

$$\mathsf{E}\,\mathrm{II} = -q(1-m^+)\,\frac{\mathsf{E}\,\mathrm{th}^2\,Y\,\mathrm{ch}^{m^+}\,Y}{\mathsf{E}\,\mathrm{ch}^{m^+}\,Y} + \delta$$

and the same value for $\mathsf{E}\,\mathrm{III}$. We also find $\mathsf{E}\,\mathrm{III} = (1-m^+)\,q^2$, so that finally

$$\mathsf{E}\,\nu_0(f) = (1-m^+)\left(q - \frac{\mathsf{E}\,\mathrm{th}^2\,Y\,\mathrm{ch}^{m^+}\,Y}{\mathsf{E}\,\mathrm{ch}^{m^+}\,Y}\right)^2 + \delta\,.$$

Since q satisfies (6.132), it suffices to prove the next lemma. □

Lemma 6.5.5. *We have* $|m^+ - m| = \delta$.

Proof. If we use again Theorem 6.1.6, and the fact that the function $\mathbf{1}_{\{x\geq 1/2\}}$ can be uniformly approximated by polynomials on the set $[-1,1/4]\cup[3/4,1]$, we see that it suffices to show that for each integer $s\geq 1$, we have $|m_s^+ - m_s| = \delta$, where

$$m_s = \mathsf{E}\,\langle R_{1,2}^s\rangle\,,\quad m_s^+ = \mathsf{E}\,\langle (R_{1,2}^+)^s\rangle_+\,.$$

We observe that

$$\begin{aligned}\frac{\partial}{\partial\beta_s}\,p_N(\beta,\boldsymbol{\beta}) &= \frac{\xi(N)}{N}\,2^{-s}\left(\frac{s!}{N^{s-1}}\right)^{1/2}\mathsf{E}\left\langle \sum_{i_1<\ldots<i_s\leq N} g^s_{i_1,\ldots,i_s}\,\sigma_{i_1}\cdots\sigma_{i_s}\right\rangle \\ &= \xi(N)^2\,2^{-2s}\,\beta_s(1-\mathsf{E}\,\langle R_{1,2}^s\rangle) + \mathcal{R}\end{aligned} \tag{6.152}$$

where $|\mathcal{R}|\leq K\,\xi(N)^2/N$. This is obtained by mimicking the proof of (6.3).

Let us now define

$$p_N^+(\beta,\boldsymbol{\beta}) = \frac{1}{N}\,\mathsf{E}\log\sum\exp(-H_{N-2}(\boldsymbol{\rho}))\,,$$

where the summation is over all values of $\boldsymbol{\rho}$ and where $H_{N-2}(\boldsymbol{\rho})$ is given by (6.134). If we copy the proof of (6.152) we get that

$$\frac{\partial}{\partial\beta_s}\,p_N^+(\beta,\boldsymbol{\beta}) = \xi(N)^2\,\beta_s(1-\mathsf{E}\,\langle (R_{1,2}^+)^s\rangle_+) + \mathcal{R}\,, \tag{6.153}$$

where $|\mathcal{R}|\leq K\,\xi(N)^2/N$.

On the other hand, copying the proof of Lemma 6.4.1, we see that

$$p_N^+(\beta,\boldsymbol{\beta}) \le p_N(\beta,\boldsymbol{\beta}) \le p_N^+(\beta,\boldsymbol{\beta}) + \frac{K}{N} \,. \tag{6.154}$$

Both $p_N^+(\beta,\boldsymbol{\beta})$ and $p_N(\beta,\boldsymbol{\beta})$ are convex functions of β_s. For a convex function ψ and $h>0$ we have

$$\frac{\psi(x)-\psi(x-h)}{h} \le \psi'(x) \le \frac{\psi(x+h)-\psi(x)}{h} \,.$$

Using this fact for $h = N^{-1/2}$ together with (6.154) yields

$$\int_{-1}^{1} \left| \frac{\partial}{\partial\beta_s} p_N^+(\beta,\boldsymbol{\beta}) - \frac{\partial}{\partial\beta_s} p_N(\beta,\boldsymbol{\beta}) \right| \mathrm{d}\beta_s \le \frac{K}{\sqrt{N}} \,. \tag{6.155}$$

Combining with (6.152), (6.153), and using the fact that $\xi(N)^2 \ge N^{-1/4}$ enables us to conclude that $|m - m^+| = \delta$. □

We set

$$\Delta_{\ell,\ell'} = \mathsf{E}\, g(\boldsymbol{\rho}^\ell)\, g(\boldsymbol{\rho}^{\ell'}) - \mathsf{E}\, Y(\boldsymbol{\rho}^\ell)\, Y(\boldsymbol{\rho}^{\ell'}) = \mathsf{E}\, g'(\boldsymbol{\rho}^\ell)\, g'(\boldsymbol{\rho}^{\ell'}) - \mathsf{E}\, Y'(\boldsymbol{\rho}^\ell)\, Y'(\boldsymbol{\rho}^{\ell'}) \,,$$

$$a_{\ell,\ell'} = \mathsf{E}\, g''(\boldsymbol{\rho}^\ell)\, g''(\boldsymbol{\rho}^{\ell'}) \,.$$

Proposition 6.5.6. *Consider a function f on Σ_N^n. Then we have*

$$\begin{aligned} \nu_t'(f) = & \sum_{\ell<\ell'\le n} \nu_t((\varepsilon_\ell\, \varepsilon_{\ell'} + \eta_\ell\, \eta_{\ell'})\, \Delta_{\ell,\ell'}\, f) \\ & - n \sum_{\ell\le n} \nu_t((\varepsilon_\ell\, \varepsilon_{n+1} + \eta_\ell\, \eta_{n+1})\, \Delta_{\ell,n+1}\, f) \\ & + \frac{n(n+1)}{2} \nu_t((\varepsilon_{n+1}\, \varepsilon_{n+2} + \eta_{n+1}\, \eta_{n+2})\, \Delta_{n+1,n+2} f) + \mathcal{N}, \end{aligned} \tag{6.156}$$

where

$$\begin{aligned} \mathcal{N} = & \sum_{\ell<\ell'} \nu_t(\varepsilon_\ell\, \eta_\ell\, \varepsilon_{\ell'}\, \eta_{\ell'}\, a_{\ell,\ell'}\, f) - n \sum_{\ell\le n} \nu_t(\varepsilon_\ell\, \eta_\ell\, \varepsilon_{n+1}\, \eta_{n+1}\, a_{\ell,n+1}\, f) \\ & + \frac{n(n+1)}{2} \nu_t(\varepsilon_{n+1}\, \eta_{n+1}\, \varepsilon_{n+2}\, \eta_{n+2}\, a_{n+1,n+2}\, f) \,. \end{aligned}$$

Proof. This is integration by parts as usual, if one keeps in mind that $g_t(\boldsymbol{\rho})$ and $g_t'(\boldsymbol{\rho})$ are independent. The term $\mathcal{N}$ comes from $\sqrt{t}\, g''(\boldsymbol{\rho})$ and it is called so because its influence is negligible in the sense that it is at most K/N whenever $|f| \le 1$ (and the same uniform bound applies to all its derivatives in t).

Lemma 6.5.7. *Consider the sets*

$$B(\ell,\ell') = \left\{ (\boldsymbol{\sigma}^\ell, \boldsymbol{\sigma}^{\ell'});\ R_{\ell,\ell'} \ge \frac{1}{2},\ \exists \alpha \ne \gamma,\ \boldsymbol{\rho}^\ell \in C_\alpha,\ \boldsymbol{\rho}^{\ell'} \in C_\gamma \right\} ,$$

$$A(\ell,\ell') = \{(\boldsymbol{\sigma}^\ell, \boldsymbol{\sigma}^{\ell'})\, ;\, |R_{\ell,\ell'}| \geq 2^{-p/4}\,,\; R_{\ell,\ell'} \leq 1 - 2^{-p/2}\}\,.$$

Then we have

$$\left| \Delta_{\ell,\ell'} - \frac{\beta^2 p}{2}\,(R_{\ell,\ell'}^{p-1} - q_{\ell,\ell'}^{p-1}) \right| \leq K\,\xi(N)^2 + K\,\mathbf{1}_{B(\ell,\ell')} \tag{6.157}$$

$$|\Delta_{\ell,\ell'}| \leq \frac{\beta^2 p^2}{2}\,|R_{\ell,\ell'} - q_{\ell,\ell'}| + K\,\xi(N)^2 + K\,\mathbf{1}_{B(\ell,\ell')} \tag{6.158}$$

$$|R_{\ell,\ell'}^{p-1} - q_{\ell,\ell'}^{p-1}| \leq p\,2^{-p/2} + K\,\mathbf{1}_{A(\ell,\ell')} + K\,\mathbf{1}_{B(\ell,\ell')} \tag{6.159}$$

$$|\Delta_{\ell,\ell'}| \leq \beta^2\,p^2\,2^{-p/2} + K\,\xi(N)^2 + K\,\mathbf{1}_{B(\ell,\ell')} + K\,\mathbf{1}_{A(\ell,\ell')} \tag{6.160}$$

$$(1 - R_{\ell,\ell'})\,|R_{\ell,\ell'}^{p-1} - q_{\ell,\ell'}^{p-1}| \leq p\,2^{-p/2}|R_{\ell,\ell'} - q_{\ell,\ell'}| + K\,\mathbf{1}_{A(\ell,\ell')}\,. \tag{6.161}$$

Proof. From (6.136) we have

$$\left| \mathsf{E}\,g(\boldsymbol{\rho}^\ell)\,g(\boldsymbol{\rho}^{\ell'}) - \frac{\beta^2 p}{2}\,R_{\ell,\ell'}^{p-1} \right| \leq K\,\xi(N)^2\,.$$

It follows from the definition of $Y(\boldsymbol{\rho})$ that $\mathsf{E}\,Y(\boldsymbol{\rho}^\ell)\,Y(\boldsymbol{\rho}^{\ell'}) = \beta^2\,p\,q^{p-1}/2$ if $\boldsymbol{\rho}^\ell$ and $\boldsymbol{\rho}^{\ell'}$ belong to the same set C_α, while $\mathsf{E}\,Y(\boldsymbol{\rho}^\ell)\,Y(\boldsymbol{\rho}^{\ell'}) = 0$ otherwise. Thus $\mathsf{E}\,Y(\boldsymbol{\rho}^\ell)\,Y(\boldsymbol{\rho}^{\ell'}) = \beta^2\,p\,q_{\ell,\ell'}^{p-1}/2$ unless $R_{\ell,\ell'} \geq 1/2$ and $\boldsymbol{\rho}^\ell \in C_\alpha$, $\boldsymbol{\rho}^{\ell'} \in C_\gamma$ for $\alpha \neq \gamma$, i.e. unless $(\boldsymbol{\sigma}^\ell, \boldsymbol{\sigma}^{\ell'}) \in B(\ell,\ell')$. This proves (6.157).

To deduce (6.158) from (6.157) we simply use the fact that $|x^{p-1} - y^{p-1}| \leq p|x-y|$ if $|x|, |y| \leq 1$.

To prove (6.159) we note that if $(\boldsymbol{\sigma}^1, \boldsymbol{\sigma}^2) \notin A(\ell,\ell')$ and $(\boldsymbol{\sigma}^1, \boldsymbol{\sigma}^2) \notin B(\ell,\ell')$ then either $|R_{\ell,\ell'}| \leq 2^{-p/4}$, $q_{\ell,\ell'} = 0$, in which case

$$|R_{\ell,\ell'}^{p-1} - q_{\ell,\ell'}^{p-1}| \leq 2^{-(p-1)p/4}$$

or else $R_{\ell,\ell'} \geq 1 - 2^{-p/2}$, $q_{\ell,\ell'} = q \geq 1 - 2^{-p/2}$ and

$$|R_{\ell,\ell'}^{p-1} - q_{\ell,\ell'}^{p-1}| \leq p\,|R_{\ell,\ell'} - q_{\ell,\ell'}| \leq p\,2^{-p/2}\,.$$

Next, we observe that (6.160) follows from (6.157), (6.159).

Finally, to prove (6.161), we proceed as in the proof of (6.159). If $(\boldsymbol{\sigma}^1, \boldsymbol{\sigma}^2) \notin A(\ell,\ell')$ then either $R_{\ell,\ell'} \geq 1 - 2^{-p/2}$, in which case

$$(1 - R_{\ell,\ell'})\,|R_{\ell,\ell'}^{p-1} - q_{\ell,\ell'}^{p-1}| \leq p\,2^{-p/2}\,|R_{\ell,\ell'} - q_{\ell,\ell'}|\,,$$

or else $|R_{\ell,\ell'}| \leq 2^{-p/4}$, $q_{\ell,\ell'} = 0$, in which case

$$\begin{aligned}(1 - R_{\ell,\ell'})\,|R_{\ell,\ell'}^{p-1} - q_{\ell,\ell'}^{p-1}| &\leq 2\,|R_{\ell,\ell'}|^{p-1} \leq 2 \cdot 2^{-(p-2)p/4}\,|R_{\ell,\ell'}| \\ &\leq 2^{-p/2}\,|R_{\ell,\ell'} - q_{\ell,\ell'}|\,.\end{aligned}$$

□

The crucial fact in this section is that $|\Delta_{\ell,\ell'}|$ is essentially small (as shown by (6.160)).

Since $|\Delta_{\ell,\ell'}| \le K$, Lemma 3.4.5, Theorems 6.1.6, 6.2.2 and (6.156) show that

$$\nu_t(\mathbf{1}_{A(\ell,\ell')}) \le K \exp\left(-\frac{N}{K}\right) \; ; \; \nu_t(\mathbf{1}_{B(\ell,\ell')}) \le K \exp\left(-\frac{N}{K}\right) . \tag{6.162}$$

Corollary 6.5.8. *Consider a function* $f : \Sigma_N^2 \to [0,1]$. *Then whenever* $\beta^2 p^2 2^{-p/2} \le 1$, *we have*

$$\nu_t(f) \le L\,\nu(f) + K\,\xi(N)^2 .$$

Proof. We use (6.156), (6.160) and (6.162) to see that

$$|\nu_t'(f)| \le L\,\nu_t(f) + K\,\xi(N)^2 ,$$

and we conclude with Lemma 3.4.5. □

We are now ready to take care of the last term in (6.149).

Lemma 6.5.9. *If* f *is a function on* Σ_N^2 *such that* $|f| \le 2$, *then for* $\beta^2 p^2 2^{-p/2} \le 1$, *we have*

$$|\nu_t'''(f)| \le L\,\beta^6 p^6 2^{-p/2}\,\nu((R_{1,2} - q_{1,2})^2) + K\,\xi(N)^2 . \tag{6.163}$$

Proof. We iterate (6.156) three times in order to compute $\nu_t'''(f)$, and observe that any term that containing $a_{\ell,\ell'}$ will give a contribution of at most K/N. In the other terms, each derivation creates a factor $\Delta_{\ell,\ell'}$. Use of Hölder's inequality shows that

$$\nu_t'''(f) \le L\,\nu_t(|\Delta_{1,2}|^3) + \frac{K}{N} .$$

We apply (6.158) to two of the factors $\Delta_{1,2}$, and (6.160) to the third one. □

Lemma 6.5.10. *If* f *is as in (6.148), we have*

$$|\nu_1'(f)| \le L\,\beta^2 p^2 2^{-p/2}\,\nu((R_{1,2} - q_{1,2})^2) + K\,\xi(N)^2 . \tag{6.164}$$

Proof. We use (6.156). We have to bound terms of the type

$$\nu((\varepsilon_1\,\varepsilon_2 - q_{1,2}^{+})(\eta_1\,\eta_2 - q_{1,2}^{+})\,\varepsilon_\ell\,\varepsilon_{\ell'}\,\Delta_{\ell,\ell'}) . \tag{6.165}$$

First, using (6.157), we can replace the quantity (6.165) by

$$\frac{\beta^2 p}{2}\,\nu((\varepsilon_1\,\varepsilon_2 - q_{1,2})(\eta_1\,\eta_2 - q_{1,2})\,\varepsilon_\ell\,\varepsilon_{\ell'}\,(R_{\ell,\ell'}^{p-1} - q_{\ell,\ell'}^{p-1})) = \mathrm{V} + \mathrm{VI}$$

where

$$\mathrm{V} = \frac{\beta^2 p}{2}\,\nu((\varepsilon_1\,\varepsilon_2 - q_{1,2})(\eta_1\,\eta_2 - q_{1,2})(R_{\ell,\ell'}^{p-1} - q_{\ell,\ell'}^{p-1}))$$

$$\mathrm{VI} = -\frac{\beta^2 p}{2}\,\nu((1-\varepsilon_\ell\,\varepsilon_{\ell'})(\varepsilon_1\,\varepsilon_2 - q_{1,2})(\eta_1\,\eta_2 - q_{1,2})(R_{\ell,\ell'}^{p-1} - q_{\ell,\ell'}^{p-1}))\,.$$

As in the proof of (6.146), using symmetry between sites we see that

$$\begin{aligned}\mathrm{V} &\le \frac{\beta^2 p}{2}\,\nu((R_{1,2}-q_{1,2})^2(R_{\ell,\ell'}^{p-1} - q_{\ell,\ell'}^{p-1})) + \frac{K}{N}\\ &\le \frac{\beta^2 p^2}{2}\,2^{-p/2}\,\nu((R_{1,2}-q_{1,2})^2) + \frac{K}{N}\,,\end{aligned}$$

thanks to (6.159). This site-symmetry may also be used to show that

$$\begin{aligned}\mathrm{VI} &\le \frac{\beta^2 p}{2}\,\nu((1-\varepsilon_\ell\,\varepsilon_{\ell'})(\varepsilon_1\,\varepsilon_2 - q_{1,2})(R_{1,2} - q_{1,2})(R_{\ell,\ell'}^{p-1} - q_{\ell,\ell'}^{p-1})) + \frac{K}{N}\\ &\le \beta^2\,p\,\nu((1-\varepsilon_\ell\,\varepsilon_{\ell'})|R_{1,2}-q_{1,2}|\,|R_{\ell,\ell'}^{p-1} - q_{\ell,\ell'}^{p-1}|) + \frac{K}{N}\end{aligned}$$

because $1-\varepsilon_\ell\,\varepsilon_{\ell'} \ge 0$, as well as

$$\mathrm{VI} \le \beta^2\,p\,\nu((1-R_{\ell,\ell'})\,|R_{1,2}-q_{1,2}|\,|R_{\ell,\ell'}^{p-1} - q_{\ell,\ell'}^{p-1}|)\,.$$

we finally apply (6.161). □

The proof of the following is entirely similar.

Lemma 6.5.11. *If f is as in (6.148) then*

$$|\nu_1''(f)| \le L\,\beta^4\,p^4\,2^{-p/2}\,\nu((R_{1,2}-q_{1,2})^2) + K\,\xi(N)^2\,. \tag{6.166}$$

Proof of Theorem 6.5.1. We use (6.147), (6.149), (6.150), (6.163), (6.164) and (6.166) to get

$$A_N \le L\,\beta^6\,p^6\,2^{-p/2}\,A_N + \delta$$

so that if $\beta \le 2^{p/12}/Lp$, we have $A_N \le A_N/2 + \delta$. To complete the proof it remains only to prove that the equation (6.132) has a unique solution. This results from Proposition 6.5.13 below. □

Lemma 6.5.12. *If $3/2 \le \beta \le 2^{p/L}$ and if p is large enough, for N large enough we have $m^2\beta^2 \ge 17/8$.*

Of course the number 17/8 is just a convenient choice.

Proof. We have

$$m = 1 - \mathsf{E}\,\langle \mathbf{1}_{\{R_{1,2}\ge 1/2\}}\rangle \ge 1 - \mathsf{E}\,\langle \mathbf{1}_{\{R_{1,2}\ge 1-2^{-p/2}\}}\rangle - K\exp\left(-\frac{N}{K}\right)$$

by Theorem 6.1.6 and

$$(1-2^{-p/2})^p \mathsf{E}\langle \mathbf{1}_{\{R_{1,2}\geq 1-2^{-p/2}\}}\rangle \leq \mathsf{E}\langle R_{1,2}^p \mathbf{1}_{\{R_{1,2}\geq 1-2^{-p/2}\}}\rangle$$
$$\leq \mathsf{E}\langle R_{1,2}^p\rangle + 2^{-p^2/4} + \frac{K}{N},$$

since, according to Theorem 6.1.6

$$|\mathsf{E}\langle R_{1,2}^p \mathbf{1}_{\{R_{1,2}<1-2^{-p/2}\}}\rangle| \leq \mathsf{E}\langle |R_{1,2}^p| \mathbf{1}_{|R_{1,2}|\leq 2^{-p/4}}\rangle + \frac{K}{N}$$
$$\leq 2^{-p^2/4} + \frac{K}{N}.$$

Now, as usual, we have

$$\frac{\partial}{\partial \beta} p_N(\beta, \boldsymbol{\beta}) \leq \frac{\beta}{2}(1 - \mathsf{E}\langle R_{1,2}^p\rangle) + \frac{K}{N}$$

so that

$$\mathsf{E}\langle R_{1,2}^p\rangle \leq 1 - \frac{2}{\beta}\frac{\partial}{\partial\beta} p_N(\beta, \boldsymbol{\beta}) + \frac{K}{N}.$$

Combining these relations yields

$$m \geq 1 - (1-2^{-p/2})^{-p} - L 2^{-p^2/4} + \frac{2}{\beta}\frac{\partial p_N}{\partial \beta} - \frac{K}{N}. \tag{6.167}$$

We recall that the function $p_N(\beta, \boldsymbol{\beta})$ is convex in β so that $\partial p_N/\partial\beta$ increases in β. By (6.7) and (6.95) we have

$$\beta \leq \gamma_p \Rightarrow \lim_{N\to\infty} p_N(\beta, \boldsymbol{\beta}) = \frac{\beta^2}{4} + \log 2$$

and thus

$$\lim_{N\to\infty} \frac{\partial p_N}{\partial \beta}(\beta, \boldsymbol{\beta}) = \frac{\beta}{2}.$$

Since $\gamma_p > 3/2$ for large p, for $\beta > 3/2$ we have $\partial p_N/\partial\beta \geq 3/4$ for N large enough and (6.167) shows that for $\beta \leq 2^{p/4}$, p large enough, N large enough we have

$$m^2\beta^2 > 17/8.$$

□

Proposition 6.5.13. a) *If p is large enough and if $\beta \leq 2^{p/L}$, then for each number $0 < b < 2$ such that $b^2\beta^2 \geq 17/8$ the equation*

$$q = \frac{\mathsf{E}\,\mathrm{th}^2 Y\,\mathrm{ch}^b Y}{\mathsf{E}\,\mathrm{ch}^b Y}, \tag{6.168}$$

where $Y = \beta z\sqrt{pq^{p-1}/2}$ has a unique solution in the interval $[1-2^{-p/2}, 1]$. This solution is a continuous function of β, b.

b) *If* $0 < x \le 3/2$ *and* $x^2 \beta^2 \ge 17/8$, *then the system of equations (6.132) and*

$$x = 1 - q^p(1-m) \tag{6.169}$$

has a unique solution $q = A(x,\beta)$, $m = B(x,\beta)$ *and* $\partial B/\partial x > 0$.

Proof. To prove a), we consider the function

$$\Phi(q) = \frac{\mathsf{E}\,\mathrm{th}^2 Y \,\mathrm{ch}^b Y}{\mathsf{E}\,\mathrm{ch}^b Y} = 1 - \frac{\mathsf{E}\,\mathrm{ch}^{b-2} Y}{\mathsf{E}\,\mathrm{ch}^b Y}\,.$$

We will prove that

$$1 - 2^{-p/2} \le q \le 1 \Rightarrow 1 - 2^{-p/2} \le \Phi(q) \le 1 \tag{6.170}$$

$$1 - 2^{-p/2} \le q \le 1 \Rightarrow \Phi'(q) \le 1/2\,. \tag{6.171}$$

To prove (6.170) we simply use that $\mathrm{ch}\,Y \ge 1$ and $\mathrm{ch}\,Y \ge 2^{-1}\exp Y$ to write

$$\begin{aligned} 1 - \Phi(q) \le \frac{1}{\mathsf{E}\,\mathrm{ch}^b Y} \le \frac{2^b}{\mathsf{E}\exp bY} &= 2^b \exp\left(-\frac{b^2\beta^2 p}{4}\, q^{p-1}\right) \\ &\le 2^b \exp\left(-\frac{17}{16}\, pq^{p-1}\right) \end{aligned} \tag{6.172}$$

and this proves (6.170) for p large enough.

As for (6.171), it is a rather rough bound. Indeed, computing $\Phi'(q)$ via integration by parts and using the fact that $\mathrm{ch}^a Y \le 1$ when $a \le 0$ as well as (6.172), one obtains

$$|\Phi'(q)| \le L\beta^2 p\, q^{p-2} \frac{1}{\mathsf{E}\,\mathrm{ch}^b Y} \le L\beta^2 p\, 2^{-p/2}\,.$$

To prove b), we first observe that since

$$m = 1 - \frac{1-x}{q^p} = \frac{x}{q^p} + 1 - \frac{1}{q^p}$$

and since we assume $x^2\beta^2 \ge 17/8$, $q \ge 1 - 2^{-p/2}$ and p large, we have $m^2\beta^2 q^{p-1} \ge 2$ for $\beta \le 2^{p/L}$. It was then shown in the proof of a) that

$$q \in [1 - 2^{-p/2}, 1] \Rightarrow \Phi(q,m) = \frac{\mathsf{E}\,\mathrm{th}^2 Y\,\mathrm{ch}^m Y}{\mathsf{E}\,\mathrm{ch}^m Y} \in [1 - 2^{-p/2}, 1]\,.$$

Thus all we have to do is to show that the function

$$\Psi(q) = \Phi\left(q, 1 + \frac{x-1}{q^p}\right) \tag{6.173}$$

satisfies $|\Psi'(q)| \le 1/2$. This is because straightforward and brutal estimates show that $\partial\Phi/\partial m$ is very small if $m \le 2$, $m^2\beta^2 > 17/8$ (the case of $\partial\Phi/\partial q$ was considered in a)). Indeed, since $\Phi(q,m) = 1 - \mathsf{E}\,\mathrm{ch}^{m-2} Y/\mathsf{E}\,\mathrm{ch}^m Y$, we have

$$\frac{\partial \Phi}{\partial m} = -\frac{\mathsf{E}\log(\operatorname{ch} Y)\operatorname{ch}^{m-2} Y}{\mathsf{E}\operatorname{ch}^m Y} + \frac{\mathsf{E}\operatorname{ch}^{m-2} Y\,\mathsf{E}\log(\operatorname{ch} Y)\operatorname{ch}^m Y}{(\mathsf{E}\operatorname{ch}^m Y)^2}$$

and

$$\left|\frac{\partial \Phi}{\partial m}\right| \le \frac{\mathsf{E}\log \operatorname{ch} Y}{\mathsf{E}\operatorname{ch}^m Y} + \frac{\mathsf{E}\log(\operatorname{ch} Y)\operatorname{ch}^m Y}{(\mathsf{E}\operatorname{ch}^m Y)^2} . \tag{6.174}$$

Using the bound $\operatorname{ch} Y \le \exp|Y|$ we obtain

$$\mathsf{E}\log(\operatorname{ch} Y)\operatorname{ch}^m Y \le \mathsf{E}\,|Y|\operatorname{ch}^m Y = 2\,\mathsf{E}\,(\mathbf{1}_{\{Y\ge 0\}} Y \operatorname{ch}^m Y) .$$

Writing this as an integral and integrating by parts we obtain

$$\begin{aligned} 2\,\mathsf{E}\,(\mathbf{1}_{\{Y\ge 0\}} Y \operatorname{ch}^m Y) &= \beta^2 q^{p-1}(1 + m\,\mathsf{E}\,(\mathbf{1}_{\{Y\ge 0\}} \operatorname{th} Y \operatorname{ch}^m Y) \\ &\le \beta^2 q^{p-1}(1 + \mathsf{E}\operatorname{ch}^m Y) , \end{aligned}$$

so that using the bound of (6.172) the quantity $|\partial\Phi/\partial m|$ is indeed very small.

To prove the assertion that $\partial B/\partial x > 0$ (or, abusing notations, that $\partial m/\partial x > 0$) we derive from (6.169) that

$$1 = -pq^{p-1}(1-m)\frac{\partial q}{\partial x} + q^p \frac{\partial m}{\partial x}$$

so that it suffices to show that $\partial q/\partial x$ is very small. This follows from the previous estimates on the partial derivatives of Φ by differentiating the relation $q = \Phi(q, 1 + (x-1)/q^2)$ with respect to x. □

To complete Theorem 6.5.1, one would like to be able to compute $\lim_{N\to\infty} p_N(\beta, \boldsymbol{\beta})$. The following is a step in this direction.

Proposition 6.5.14. *Under the conditions of Theorem 6.5.1, we have*

$$(N+1)\,p_{N+1}(\beta, \boldsymbol{\beta}) - N\,p_N(\beta, \boldsymbol{\beta}) = F(\beta, q, m) + \delta' , \tag{6.175}$$

where

$$F(\beta, q, m) = \log 2 + \frac{\beta^2}{4}(1 - p\,q^{p-1}) + \frac{\beta^2}{4}(p-1)\,q^p(1-m) + \frac{1}{m}\log \mathsf{E}\operatorname{ch}^m Y \tag{6.176}$$

and where

$$\lim_{N\to\infty} \int_{3/2}^{2^{p/L}} \int |\delta'|\,\mathrm{d}\boldsymbol{\beta}\,\mathrm{d}\beta = 0 . \tag{6.177}$$

The proof of Proposition 6.5.14 requires yet another interesting property of the Poisson-Dirichlet distribution.

Proposition 6.5.15. *Consider a random sequence* $(v_\alpha)_{\alpha\geq 1}$ *that is distributed like* Λ_m. *Consider a r.v.* $U \geq 1$, $\mathsf{E}U^2 < \infty$ *and independent copies* $(U_\alpha)_{\alpha\geq 1}$. *Then*

$$\mathsf{E}\log\left(\sum_{\alpha\geq 1} v_\alpha U_\alpha\right) = \frac{1}{m}\log \mathsf{E}\, U^m \,. \tag{6.178}$$

Proof. By definition, we can write $v_\alpha = u_\alpha / \sum_{\gamma\geq 1} u_\gamma$, where the sequence $(u_\alpha)_{\alpha\geq 1}$ arises from a Poisson point process of intensity measure $\mu = x^{-m-1}\,\mathrm{d}x$. As in the proof of Theorem 6.4.5 we use the fact that the set $(u_\alpha, U_\alpha)_{\alpha\geq 1}$ generates a Poisson point process on $(\mathbb{R}^+)^2$ of intensity measure $\mu\otimes\nu$, where ν is the law of U. Consider the image measure η of $\mu\otimes\nu$ under the map $(x,y)\mapsto xy$. Then, for a Borel set of $\mathbb{R}^+$, we have

$$\eta(A) = \mu\otimes\nu(\{x,y\}\,;\ xy\in A) = \int \mu(\{x\,;\ xy\in A\})\,\mathrm{d}\nu(y)\,. \tag{6.179}$$

Now

$$\mu(\{x\,;\ xy\in A\}) = \int_{xy\in A} x^{-m-1}\,\mathrm{d}x = y^m\int_A z^{-m-1}\,\mathrm{d}z = y^m\,\mu(A)$$

by making the change of variable $z = xy$. Thus, from (6.179) we have

$$\eta(A) = \mu(A)\int y^m\,\mathrm{d}\nu(y) = \mu(A)\,\mathsf{E}\,U^m\,. \tag{6.180}$$

Consider a number $a > 0$, and the image μ_a of μ under the map $x\mapsto xa$. Then

$$\mu_a(A) = \int_{xa\in A} x^{-m-1}\,\mathrm{d}x = a^m\int_A z^{-m-1}\,\mathrm{d}z = a^m\,\mu(A) \tag{6.181}$$

with the change of variable $z = xa$. Comparing (6.180) with (6.181) we see that $\eta = \mu_a$ for $a = (\mathsf{E}\,U^m)^{1/m}$.

Since the family (u_α, U_α) is a Poisson point process of intensity measure $\eta\otimes\nu$, the family $(u_\alpha U_\alpha)_{\alpha\geq 1}$ is a Poisson point process of intensity measure $\nu = \mu_a$. Thus the law of this family is the law of $(a\,u_\alpha)_{\alpha\geq 1}$ and

$$\mathsf{E}\log\sum_{\alpha\geq 1} u_\alpha U_\alpha = \mathsf{E}\log a\sum_{\alpha\geq 1} u_\alpha$$

so that

$$\begin{aligned}\log a &= \mathsf{E}\log\sum_{\alpha\geq 1} u_\alpha U_\alpha - \mathsf{E}\log\sum_{\alpha\geq 1} u_\alpha \\ &= \mathsf{E}\log\frac{\sum_{\alpha\geq 1} u_\alpha U_\alpha}{\sum_{\gamma\geq 1} u_\gamma} = \mathsf{E}\log\left(\sum_{\alpha\geq 1} v_\alpha U_\alpha\right)\,.\end{aligned}$$

Here of course it is assumed that $\log \sum_{\alpha\geq 1} u_\alpha$ has an expectation. The proof of this fact is rather tedious, but straightforward, and we leave it to the reader. □

We consider

$$\beta' = \beta\left(\frac{N+1}{N}\right)^{\frac{p-1}{2}} \quad\text{and}\quad \boldsymbol{\beta}' = \left(\beta_s\left(\frac{N+1}{N}\right)^{\frac{s-1}{2}}\right)_{s\geq 1} .$$

Lemma 6.5.16. *We have*

$$(N+1)\, p_{N+1}(\beta', \boldsymbol{\beta}') - N\, p_N(\beta, \boldsymbol{\beta}) = \log 2 + \frac{\beta^2 p}{4}(1-q^{p-1}) + \frac{1}{m}\log \mathsf{E}\,\mathrm{ch}^m\, Y + \delta\, .$$

Proof. As in (2.217) we see that

$$(N+1)\, p_{N+1}(\beta', \boldsymbol{\beta}') - N\, p_N(\beta, \boldsymbol{\beta}) = \mathsf{E}\log 2\,\langle \mathrm{Av}\exp \sigma_{N+1}\, g(\boldsymbol{\sigma})\rangle \tag{6.182}$$

where $g(\boldsymbol{\sigma}) = g_1(\boldsymbol{\sigma}) + g_2(\boldsymbol{\sigma})$,

$$g_1(\boldsymbol{\sigma}) = \beta\left(\frac{p!}{2N^{p-1}}\right)^{p/2} \sum_{i_1<\ldots<i_{p-1}\leq N} g_{i_1,\ldots,i_{p-1},N+1}\,\sigma_{i_1}\cdots\sigma_{i_{p-1}}$$

$$g_2(\boldsymbol{\sigma}) = \xi(N)\sum_{s\geq 1} 2^{-s}\,\beta_s\left(\frac{s!}{N^{s-1}}\right)^{1/2} \sum_{i_1<\ldots<i_{s-1}\leq N} g^s_{i_1,\ldots,i_{s-1},N+1}\,\sigma_{i_1}\cdots\sigma_{i_{s-1}}\, .$$

We consider the decomposition $(C_\alpha)_{\alpha\geq 1}$ of Theorem 6.2.2 and set

$$Y(\boldsymbol{\sigma}) = \beta\sqrt{\frac{p\,q^{p-1}}{2}}\, z_\alpha = Y_\alpha$$

for $\boldsymbol{\sigma} \in C_\alpha$, where $(z_\alpha)_{\alpha\geq 1}$ is an independent sequence of standard Gaussian r.v. To estimate the right-hand side of (6.182), we use the smart path method and consider

$$\varphi(t) = \mathsf{E}\log\langle \mathrm{ch}\, g_t(\boldsymbol{\sigma})\rangle$$

where

$$g_t(\boldsymbol{\sigma}) = \sqrt{t}\, g(\boldsymbol{\sigma}) + \sqrt{1-t}\, Y(\boldsymbol{\sigma})\, .$$

Thus

$$g'_t(\boldsymbol{\sigma}) = \frac{\mathrm{d}}{\mathrm{d}t} g_t(\boldsymbol{\sigma}) = \frac{1}{2\sqrt{t}}\, g(\boldsymbol{\sigma}) - \frac{1}{2\sqrt{1-t}}\, Y(\boldsymbol{\sigma})\, .$$

We have

$$\varphi'(t) = \mathsf{E}\,\frac{\langle g'_t(\boldsymbol{\sigma})\,\mathrm{sh}\, g_t(\boldsymbol{\sigma})\rangle}{\langle \mathrm{ch}\, g_t(\boldsymbol{\sigma})\rangle}$$

and integration by parts shows that

$$\varphi'(t) = \mathsf{E}\,\frac{\langle \mathsf{E}(g_t'(\boldsymbol{\sigma})\, g_t(\boldsymbol{\sigma}))\,\mathrm{ch}\, g_t(\boldsymbol{\sigma})\rangle}{\langle \mathrm{ch}\, g_t(\boldsymbol{\sigma})\rangle} - \mathsf{E}\,\frac{1}{\langle \mathrm{ch}\, g_t(\boldsymbol{\sigma})\rangle^2} \langle \mathsf{E}(g_t'(\boldsymbol{\sigma}^1)\, g_t(\boldsymbol{\sigma}^2))\,\mathrm{sh}\, g_t(\boldsymbol{\sigma}^1)\,\mathrm{sh}\, g_t(\boldsymbol{\sigma}^2)\rangle\ . \quad (6.183)$$

We note that

$$\left|\mathsf{E}\, g_t'(\boldsymbol{\sigma})\, g_t(\boldsymbol{\sigma}) - \frac{\beta^2 p}{4}(1 - q^{p-1})\right| \le K\xi^2(N)$$

and

$$\left|\mathsf{E}\, g_t'(\boldsymbol{\sigma}^1)\, g_t(\boldsymbol{\sigma}^2) - \frac{\beta^2 p}{4}\,(R_{1,2}^{p-1} - q_{1,2}^{p-1})\right| \le K\,\xi^2(N) + \mathbf{1}_B$$

where $q_{1,2} = q\,\mathbf{1}_{\{R_{1,2} \ge 1/2\}}$ and

$$B = \{(\boldsymbol{\sigma}^1, \boldsymbol{\sigma}^2);\ R_{1,2} \ge 1/2, \exists\, \alpha \ne \gamma, \boldsymbol{\sigma}^1 \in C_\alpha, \boldsymbol{\sigma}^2 \in C_\gamma\}\,.$$

Thus (6.183) gives, since $\mathrm{ch}\, g_t(\boldsymbol{\sigma}) \ge 1$,

$$\left|\varphi'(t) - \frac{\beta^2 p}{4}\,(1 - q^{p-1})\right| \le K\,\mathsf{E}\langle |R_{1,2}^{p-1} - q_{1,2}^{p-1}|\,|\mathrm{sh}\, g_t(\boldsymbol{\sigma}^1)\,\mathrm{sh}\, g_t(\boldsymbol{\sigma}^2)|\rangle + K\,\xi^2(N)\,.$$

Integrating in $g_t(\boldsymbol{\sigma}^1)$, $g_t(\boldsymbol{\sigma}^2)$ inside the bracket rather than outside, and using Theorem 6.5.1 we obtain

$$\varphi'(t) = \frac{\beta^2 p}{4}\,(1 - q^{p-1}) + \delta$$

so that

$$\varphi(t) = \mathsf{E}\log\langle \mathrm{ch}\, g(\boldsymbol{\sigma})\rangle = \frac{\beta^2 p}{4}\,(1 - q^{p-1}) + \varphi(0) + \delta$$

$$= \frac{\beta^2 p}{4}\,(1 - q^{p-1}) + \mathsf{E}\log\left(\sum_{\alpha \ge 1} w_\alpha\,\mathrm{ch}\, Y_\alpha\right) + \delta$$

where $w_\alpha = G_N(C_\alpha)$.

As a last step, we need to deduce from (6.178) that

$$\mathsf{E}\log\left(\sum_{\alpha \ge 1} w_\alpha\,\mathrm{ch}\, Y_\alpha\right) = \frac{1}{m}\,\log \mathsf{E}\,\mathrm{ch}^m\, Y + \delta\,.$$

As in Theorem 6.4.7 (using now Proposition 6.5.15) it suffices to show that if the law of the random sequence $(w_{\alpha,r})_{\alpha \ge 1}$ converges to Λ_m then $\mathsf{E}\log\left(\sum_{\alpha \ge 1} w_\alpha\ \mathrm{ch}\, Y_\alpha\right)$ converges to $\mathsf{E}\log\left(\sum_{\alpha \ge 1} v_\alpha\,\mathrm{ch}\, Y_\alpha\right)$. This is easily done if we write

$$\log\left(\sum_{\alpha\geq 1} w_\alpha \operatorname{ch} Y_\alpha\right) = \log\left(1+\sum_{\alpha\geq 1} w_\alpha(\operatorname{ch} Y_\alpha - 1)\right)$$

and we use the fact that for $x, y \geq 0$, we have $|\log(1+x)-\log(1+y)| \leq |x-y|$, which shows that $\mathsf{E}\log\left(\sum_{\alpha\geq 1} w_\alpha \operatorname{ch} Y_\alpha\right)$ is essentially determined by the law of the few largest coefficients w_α. □

Proof of Proposition 6.5.14. It remains to prove that

$$\begin{aligned}&(N+1)\, p_{N+1}(\beta', \boldsymbol{\beta}') - (N+1)\, p_{N+1}(\beta, \boldsymbol{\beta}) \\ &= (p-1)\,\frac{\beta^2}{4}\,(1-(1-m)\, q^p) + \delta'\ . \end{aligned} \tag{6.184}$$

Here and below, δ' denotes a quantity as in (6.177), that needs not be the same at each occurrence. The proof of (6.184) is decomposed in several steps.

In a first step we prove that

$$(N+1)(p_{N+1}(\beta', \boldsymbol{\beta}') - p_{N+1}(\beta, \boldsymbol{\beta}')) = \frac{(p-1)\,\beta}{2}\,\frac{\partial p_{N+1}}{\partial \beta}\,(\beta', \boldsymbol{\beta}') + \delta'\ . \tag{6.185}$$

To see this, we define

$$\varphi(t) = p_{N+1}(t, \boldsymbol{\beta}')$$

so that φ is convex and

$$\begin{aligned}\varphi(\beta') &= \varphi(\beta) + \int_\beta^{\beta'} \varphi'(t)\,\mathrm{d}t \\ &= \varphi(\beta) + (\beta'-\beta)\,\varphi'(\beta') - \int_\beta^{\beta'} (t-\beta)\,\varphi''(t)\,\mathrm{d}t\ .\end{aligned} \tag{6.186}$$

Now

$$\beta' - \beta = \beta\left(\left(1+\frac{1}{N}\right)^{\frac{p-1}{2}} - 1\right) = \frac{(p-1)\,\beta}{2N} + \mathcal{R}$$

where $|\mathcal{R}| \leq K/N^2$. To control the last term of (6.186), since $\varphi''(t) > 0$ we can write

$$\left|\int_\beta^{\beta'} (t-\beta)\,\varphi''(t)\,\mathrm{d}t\right| \leq |\beta-\beta'| \int_\beta^{\beta'} \varphi''(t)\,\mathrm{d}t = |\beta-\beta'|\,(\varphi'(\beta') - \varphi'(\beta))$$

and, since φ' remains bounded, this quantity goes to zero as $N \to \infty$ uniformly as β varies over a fixed interval. Thus, multiplying (6.186) by $N+1$ proves (6.185). In a second stage we prove that

$$\frac{\partial p_{N+1}}{\partial \beta}\,(\beta', \boldsymbol{\beta}') - \frac{\partial p_N}{\partial \beta}\,(\beta, \boldsymbol{\beta}) = \delta'\ . \tag{6.187}$$

To see this we observe that

$$|p_{N+1}(\beta', \boldsymbol{\beta}') - p_N(\beta, \boldsymbol{\beta})| \le \frac{K}{N}$$

and we argue as in the proof of Lemma 6.5.5. To use (6.187) we observe that

$$\frac{\partial p_N}{\partial \beta}(\beta, \boldsymbol{\beta}) = \frac{\beta}{2}(1 - \mathsf{E}\langle R_{1,2}^p \rangle) + \mathcal{R} ,$$

where $|\mathcal{R}| \le K/N$ and that, according to Theorem 6.5.1, $R_{1,2}$ is essentially taking only the values 0 and q, so that

$$\frac{\partial p_N}{\partial \beta}(\beta, \boldsymbol{\beta}) = \frac{\beta}{2}(1 - q^p \, \mathsf{E}\langle \mathbf{1}_{\{R_{1,2} \ge 1/2\}} \rangle) + \delta' = \frac{\beta}{2}(1 - (1-m)\, q^p) + \delta' .$$

In a final step, we prove that

$$(N+1)\, p_{N+1}(\beta, \boldsymbol{\beta}') - (N+1)\, p_{N+1}(\beta, \boldsymbol{\beta}) = \delta' . \tag{6.188}$$

If $\beta'_s = \beta_s((N+1)/N)^{(s-1)/2}$, then

$$|\beta'_s - \beta_s| = \beta_s \left(\left(1 + \frac{1}{N}\right)^{(s-1)/2} - 1 \right) \le \beta_s \left(\exp\left(\frac{s-1}{2N}\right) - 1 \right) \le \frac{s}{N} ,$$

since $e^x - 1 \le 2x$ for $x \le 3/2$. Thus the left-hand side of (6.188) is bounded by

$$\frac{N+1}{N} \sum_{s \ge 1} s \sup \left| \frac{\partial p_N}{\partial \beta_s} \right| .$$

Now, computation of $\partial p_N / \partial \beta_s$ as usual gives

$$\frac{\partial p_N}{\partial \beta_s} = \xi(N)^2 \, \beta_s \, 2^{-2s} \Bigg(\frac{N(N-1)\dots(N-s+1)}{N^s} - \mathsf{E}\left\langle \frac{s!}{N^s} \sum_{i_1 < \dots < i_s} \sigma_{i_1}^1 \sigma_{i_1}^2 \dots \sigma_{i_p}^1 \sigma_{i_p}^2 \right\rangle \Bigg)$$

so that

$$\left| \sup \frac{\partial p_N}{\partial \beta_s} \right| \le 2\, \xi(N)^2 \, 2^{-2s}$$

and this proves (6.181). □

6.6 A Positivity Principle

Consider the function $f(\beta) = \inf F(\beta, q, m)$, where F is given by (6.176), and where the infimum is over $0 \le q \le 1$ and $0 \le m \le 1$. It is an obvious consequence of Proposition 6.5.14 that

$$\liminf_{N\to\infty} p_N(\beta,\mathbf{0}) \geq f(\beta)\,. \tag{6.189}$$

To find an upper bound for p_N, one can try to adapt the scheme of Section 2.11. It is not difficult to see that in this case, the quantities $(R_{1,2}-q_k)^2$ have to be replaced by $R_{1,2}^p - pR_{1,2}q_k^{p-1} + (p-1)q_k^p$. When p is *even* these quantities are non-negative, and one obtains that

$$\forall\, q, m\,,\ p_N(\beta) \leq F(\beta, q, m) + \frac{K}{N}$$

and, combined with (6.189) this yields $\lim_{N\to\infty} p_N(\beta) = f(\beta)$. Unfortunately, when p is odd, there is no obvious reason why these quantities should be non-negative. One can hope to be able to prove that, in some sense, $R_{1,2}$ is essentially non-negative; the same will then hold true for $R_{1,2}^p - pR_{1,2}q_k^{p-1} + (p-1)q_k^p$. Unfortunately, the random measures considered in Section 2.11 look like tricky objects, and it is not obvious how to extend to them the techniques we use elsewhere in this book. The (potentially important) program of understanding how to work with these objects is better left for future research. To compute $\lim_{N\to\infty} p_N(\beta,\mathbf{0})$ we will follow a route involving only familiar objects. The first step is the following

Theorem 6.6.1. *For each values of β the limit $\lim_{N\to\infty} p_N(\beta,\mathbf{0})$ exists.*

The obvious approach is to mimic the method of Section 2.10. Unfortunately, with the notation of the proof of Theorem 2.10.1 it is clear that

$$\left(\frac{N_1}{N}\,R' + \frac{N_2}{N}\,R''\right)^p \leq \frac{N_1}{N}\,(R')^p + \frac{N_2}{N}\,(R'')^p$$

only when p is even. When p is odd, this could be wrong unless $R' \geq 0$, $R'' > 0$. However there is a remarkably general principle which will be of much help to us, according to which $R_{1,2}$ is essentially non-negative whenever the extended Ghirlanda-Guerra identities hold.

Theorem 6.6.2. *Consider a Hamiltonian H_N depending on the parameter β such that the relations (6.99) hold. Then, for each $\varepsilon > 0$ we have*

$$\lim_{N\to\infty} \int \nu(1_{\{R_{1,2}\leq -\varepsilon\}})\,\mathrm{d}\beta = 0\,.$$

The basic principle is as follows.

Lemma 6.6.3. *Consider a probability measure G on Σ_N, $0<\varepsilon<1$ and $c<1$. Let*

$$U = \{\boldsymbol{\sigma}^1\,;\ G(\{\boldsymbol{\sigma}^2\,;\ R_{1,2} \leq -\varepsilon\}) > c\}\,.$$

Then we have

$$\varepsilon\, G(U) \leq 2(1-c)\,.$$

Proof. Set $a = G(U)$, and define the probability μ on Σ_N by $\mu(C) = G(C \cap U)/a$. Thus

$$\forall \boldsymbol{\sigma}^1 \in U\,,\ \mu(\{\boldsymbol{\sigma}^2\,;\ R_{1,2} > -\varepsilon\}) \le \frac{1-c}{a}\,,$$

and, by Fubini's theorem, if $i < j$ we have

$$\mu^{\otimes n}(\{R_{i,j} > -\varepsilon\}) \le \frac{1-c}{a}\,,$$

and thus, for any integer n,

$$\int \Bigl(\sum_{1 \le i < j \le n} \mathbf{1}_{\{R_{i,j} > -\varepsilon\}}\Bigr) \mathrm{d}\mu(\boldsymbol{\sigma}^1) \ldots \mathrm{d}\mu(\boldsymbol{\sigma}^n) \le \frac{1-c}{a} \frac{n(n-1)}{2}\,.$$

Thus we can find $\boldsymbol{\sigma}^1, \ldots, \boldsymbol{\sigma}^n$ in Σ_N such that

$$\operatorname{card} A \le \frac{1-c}{a} \frac{n(n-1)}{2}\,,$$

where $A = \{(i,j)\,;\ i < j\,,\ R_{i,j} > -\varepsilon\}$. Now

$$0 \le \Bigl\| \frac{1}{\sqrt{N}} \sum_{\ell \le n} \boldsymbol{\sigma}^\ell \Bigr\|^2 = n + 2 \sum_{i<j} R_{i,j} = n + 2 \sum_{(i,j) \in A} R_{i,j} + 2 \sum_{(i,j) \notin A} R_{i,j}\,.$$

Since $R_{i,j} \le 1$ when $(i,j) \in A$, and $R_{i,j} \le -\varepsilon$ when $(i,j) \notin A$, we get

$$0 \le n + n(n-1) \frac{1-c}{a} - \varepsilon\, n(n-1) \Bigl(1 - \frac{1-c}{a}\Bigr)\,,$$

and thus

$$\frac{1-c}{a}(1+\varepsilon) \ge \varepsilon - \frac{1}{n-1}$$

Since n is arbitrary, we have $an \le (1-c)(1+\varepsilon) \le 2(1-c)$. □

Consider $\eta > 0$, and fix n large enough so that $n \ge 1 + 2/\varepsilon$ and

$$\frac{\eta}{2n^{1-\eta}} \ge \exp\Bigl(-\frac{(n-1)\,\eta\varepsilon}{4n^{1-\eta}}\Bigr)\,. \tag{6.190}$$

Lemma 6.6.4. *Consider a probability measure G on Σ_N. Then*

$$G^{\otimes n}(\{(\boldsymbol{\sigma}^1, \ldots, \boldsymbol{\sigma}^n)\,;\ \forall \ell\,,\ 2 \le \ell \le n\,,\ R_{1,\ell} \le -\varepsilon\}) \le d := \frac{\eta}{n^{1-\eta}}\,. \tag{6.191}$$

Proof. We observe that

$$G^{\otimes n}(\{(\boldsymbol{\sigma}^1,\ldots,\boldsymbol{\sigma}^n)\,;\ \forall \ell\,,\ 2\le\ell\le n\,,\ R_{1,\ell}\le -\varepsilon\}) = \int f(\boldsymbol{\sigma})^{n-1}\mathrm{d}G(\boldsymbol{\sigma})$$

where

$$f(\boldsymbol{\sigma}^1) = G(\{\boldsymbol{\sigma}^2\,;\ R_{1,2}\le -\varepsilon\})\,.$$

Let us set

$$U = \left\{\boldsymbol{\sigma}\,;\ f(\boldsymbol{\sigma})\ge c := \left(\frac{d}{2}\right)^{\frac{1}{n-1}}\right\}\,.$$

so, since $f(\boldsymbol{\sigma})\le c$ outside U,

$$\int f(\boldsymbol{\sigma})^{n-1}\mathrm{d}G(\boldsymbol{\sigma}) \le G(U) + c^{n-1} \le G(U) + \frac{d}{2}\,.$$

Now, by (6.190) we have

$$c = \left(\frac{d}{2}\right)^{\frac{1}{n-1}} \ge \exp\left(-\frac{\eta\varepsilon}{4n^{1-\eta}}\right) \ge 1 - \frac{\eta\varepsilon}{4n^{1-\eta}}\,,$$

so that by Lemma 6.6.3 we have $G(U)\le 2(1-c)/\varepsilon\le d/2$. □

Proof of Theorem 6.6.2. We consider $\varepsilon>0$, $\eta>0$, and n as in (6.190). For $0\le r\le n$, we define

$$\varepsilon_r = 2\varepsilon\left(1-\frac{r}{2n}\right)\,,$$

so that $-2\varepsilon < -\varepsilon_r < -\varepsilon_{r+1} < -\varepsilon$. Define

$$C_r = \{(\boldsymbol{\sigma}^1,\ldots,\boldsymbol{\sigma}^r)\,;\ \forall \ell\,,\ 2\le\ell\le r\,,\ R_{1,\ell}\le -\varepsilon_r\}$$

$$C'_r = \{(\boldsymbol{\sigma}^1,\ldots,\boldsymbol{\sigma}^r)\,;\ \forall \ell\,,\ 2\le\ell\le r\,,\ R_{1,\ell}\le -\varepsilon_{r+1}\}\,.$$

For $r\le n$ consider a continuous function $\psi_r:\mathbb{R}\to[0,1]$ satisfying

$$x\le -\varepsilon_r \Rightarrow \psi_r(x)=1\,;\ x\ge -\varepsilon_{r+1}\Rightarrow \psi_r(x)=0\,. \tag{6.192}$$

Thus, for $\ell\le r$

$$\psi_r(R_{1,\ell})\,\mathbf{1}_{C'_r} \ge \mathbf{1}_{C_r} \tag{6.193}$$

because on C_r we have $R_{1,\ell}\le-\varepsilon_r$, so $\psi_r(R_{1,\ell})=1$, and because $-\varepsilon_r\le -\varepsilon_{r+1}$. In a similar manner if we see C'_r as a subset of Σ_N^{r+1}, we have

$$\mathbf{1}_{C_{r+1}} \ge \psi_r(R_{1,r+1})\,\mathbf{1}_{C'_r}\,. \tag{6.194}$$

We also observe that

$$\mathsf{E}\,\langle\psi_r(R_{1,2})\rangle \ge a := \mathsf{E}\,\langle\mathbf{1}_{\{R_{1,2}\le -2\varepsilon\}}\rangle\,.$$

Thus, using relation (6.99) together with (6.193) and (6.194) we have

$$\begin{aligned}\mathsf{E}\,\langle \mathbf{1}_{C_{r+1}}\rangle &\geq \mathsf{E}\,\langle \psi_r(R_{1,r+1})\,\mathbf{1}_{C'_r}\rangle \\ &\geq \frac{1}{r}\,a\,\mathsf{E}\,\langle \mathbf{1}_{C'_r}\rangle + \frac{1}{r}\sum_{2\leq \ell\leq r}\mathsf{E}\,\langle \psi_r(R_{1,\ell})\,\mathbf{1}_{C'_r}\rangle + \delta \\ &\geq \frac{r-1+a}{r}\,\mathsf{E}\,\langle \mathbf{1}_{C_r}\rangle + \delta\,. \end{aligned} \tag{6.195}$$

We note that $\log(1-x) \geq -x - x^2$ for $x \leq 1/2$, and thus, for $r \geq 2$ we have

$$\frac{r-1+a}{r} = 1 - \frac{(1-a)}{r} \geq \exp\left(-\frac{(1-a)}{r} - \frac{1}{r^2}\right)\,.$$

Now $1/2 + \cdots + 1/(n-1) \leq \log n$, so that iteration of (6.195) from $r = 2$ to $r = n-1$ yields

$$\mathsf{E}\,\langle \mathbf{1}_{C_n}\rangle \geq \frac{a}{Ln^{1-a}} + \delta\,. \tag{6.196}$$

Using (6.191), we see that $\langle \mathbf{1}_{C_n}\rangle \leq \eta/n^{1-\eta}$, so that (6.196) implies

$$\frac{a}{n^{1-a}} \leq \frac{L\eta}{n^{1-\eta}} + \delta$$

and thus (since $a \leq 1$) $a \leq \eta + \delta$. Since η is arbitrary, we have $a = \mathsf{E}\,\langle \mathbf{1}_{\{R_{1,2}\leq -2\varepsilon\}}\rangle = \delta$. □

Proof of Theorem 6.6.1. Consider integers N_1, N_2 and $N = N_1 + N_2$. We consider the following Hamiltonians

$$-H_1(\boldsymbol{\sigma}) = \beta\left(\frac{p!}{2N_1^{p-1}}\right)^{1/2}\sum_{1\leq i_1<\cdots<i_p\leq N_1} g'_{i_1\cdots i_p}\,\sigma_{i_1}\cdots\sigma_{i_p}$$

$$-H_2(\boldsymbol{\sigma}) = \xi(N_1)\sum_{s\geq 1}2^{-s}\,\beta'_s\left(\frac{s!}{N_1^{s-1}}\right)^{1/2}\sum_{1\leq i_1<\cdots<i_s\leq N_1} g'^{s}_{i_1\cdots i_s}\,\sigma_{i_1}\cdots\sigma_{i_s}$$

$$-H_3(\boldsymbol{\sigma}) = \beta\left(\frac{p!}{2N_2^{p-1}}\right)^{1/2}\sum_{N_1< i_1<\cdots<i_p\leq N} g''_{i_1\cdots i_p}\,\sigma_{i_1}\cdots\sigma_{i_p}$$

$$-H_4(\boldsymbol{\sigma}) = \xi(N_2)\sum_{s\geq 1}2^{-s}\,\beta''_s\left(\frac{s!}{N_2^{s-1}}\right)^{1/2}\sum_{N_1< i_1<\cdots<i_s\leq N} g''^{s}_{i_1\cdots i_s}\,\sigma_{i_1}\cdots\sigma_{i_s}$$

and the Hamiltonian $-H$ of (6.1). Here all the r.v. are standard normal and independent. Consider

$$H_t(\boldsymbol{\sigma}) = \sqrt{t}\,H(\boldsymbol{\sigma}) + \sqrt{1-t}(H_1(\boldsymbol{\sigma}) + H_3(\boldsymbol{\sigma})) + H_2(\boldsymbol{\sigma}) + H_4(\boldsymbol{\sigma})$$

and

$$\varphi(t) = \int\int \frac{1}{N} \log \sum_{\boldsymbol{\sigma}} \exp(-H_t(\boldsymbol{\sigma}))\, \mathrm{d}\boldsymbol{\beta}'\, \mathrm{d}\boldsymbol{\beta}'' ,$$

the integral being performed over $|\beta'_s| \le 1$ and $|\beta''_s| \le 1$, for $s \ge 1$. Setting $R' = N_1^{-1} \sum_{i \le N_1} \sigma_i^1 \sigma_i^2$ and $R'' = N_2^{-1} \sum_{i > N_1} \sigma_i^1 \sigma_i^2$, we then see as in Section 2.10 that

$$\begin{aligned}\varphi'(t) \ge \int\int \mathsf{E}\Big\langle -\Big(\frac{N_1}{N} R' + \frac{N_2}{N} R''\Big)^p + \frac{N_1}{N} R'^p + \frac{N_2}{N} R''^p \Big\rangle_t \mathrm{d}\boldsymbol{\beta}'\, \mathrm{d}\boldsymbol{\beta}'' \\ - \frac{K}{N_1} - \frac{K}{N_2} ,\end{aligned} \tag{6.197}$$

where K is independent of t. A minor variation on the argument of Theorem 6.6.2 shows that if the ratio N_1/N remains bounded from below, then for each $\varepsilon > 0$ we have

$$\lim_{N \to \infty} \int\int \mathsf{E}\langle \mathbf{1}_{\{R' \le -\varepsilon\}} \rangle_t\, \mathrm{d}\boldsymbol{\beta}'\, \mathrm{d}\boldsymbol{\beta}'' = 0$$

uniformly in t, and similarly for R''. Since (crudely)

$$x', x'' \ge -\varepsilon \Rightarrow \frac{N_1}{N} x'^p + \frac{N_2}{N} x''^p - \Big(\frac{N_1}{N} x' + \frac{N_2}{N} x''\Big)^p \ge -\varepsilon^p ,$$

we see from (6.197) that if $a > 0$, $N_1, N_2 \ge aN$, then $\varphi(1) - \varphi(0) \ge \eta_N(a)$, where $\eta_N(a) \to 0$. This implies in particular that for $N_1, N_2 \ge N/4$,

$$p_N(\beta, \mathbf{0}) \ge \frac{N_1}{N} p_{N_1}(\beta, \mathbf{0}) + \frac{N_2}{N} p_{N_2}(\beta, \mathbf{0}) - \varepsilon_N$$

where $\lim_{N \to \infty} \varepsilon_N = 0$. A variation on the standard sub-additive arguments (left to the reader) implies that $\lim_{N \to \infty} p_N(\beta, \mathbf{0})$ exists. □

6.7 Computing the Free Energy

We set

$$p_\infty(\beta) = \lim_{N \to \infty} p_N(\beta, \mathbf{0}) , \tag{6.198}$$

the existence of which has been established in Section 6.6. This is a convex function of β, so that it has a right-derivative at each point. We denote this right-derivative by $p'_\infty(\beta)$. We do not know (yet) that the function $p_\infty(\beta)$ is differentiable at every point, but we know that the set of points where it is not differentiable is at most countable.

Theorem 6.7.1. *There exists a number p_0 such that if $p \geq p_0$, if p is odd, and if $3/2 \leq \beta \leq 2^{p/L}$, then the system of equations (6.132) and*

$$p'_\infty(\beta) = \frac{\beta}{2}(1 - q^p(1-m)) \tag{6.199}$$

has a unique solution $q(\beta)$, $m(\beta)$, where $q(\beta) \in [1 - 2^{-p/2}, 1]$, $0 < m(\beta) \leq 2$, $\beta^2 m(\beta)^2 \geq 2$. If the function $p_\infty(\cdot)$ is differentiable at β, then for each $\varepsilon > 0$ we have

$$\lim_{N\to\infty} \int \mathsf{E}\, G_N^{\otimes 2}(\{|R_{1,2}| \geq \varepsilon\,,\ |R_{1,2} - q(\beta)| \geq \varepsilon\}) \mathrm{d}\boldsymbol{\beta} = 0\,. \tag{6.200}$$

The above integral depends on $\boldsymbol{\beta}$ only through Gibbs' measure $G_N^{\otimes 2}$ since $q(\beta)$ no longer depends on $\boldsymbol{\beta}$. Yet we do not know how to improve (6.200) into a convergence occurring for each value of $\boldsymbol{\beta}$, or even to prove the convergence for $\boldsymbol{\beta} = \mathbf{0}$ only.

Proof. Since $p'_N(\beta) \leq \sqrt{\log 2}$ by (1.7) and (6.4), we have

$$x := \frac{2p'_\infty(\beta)}{\beta} \leq \frac{2\sqrt{\log 2}}{3/2} \leq \frac{4}{3}\,.$$

Also, $p'_\infty(\beta) \geq 3/4$ (as seen in the proof of Lemma 6.5.12), so that $\beta^2 x^2 \geq 9/4$. The existence of $q(\beta)$ and $m(\beta)$ then follows from Proposition 6.5.13.

Let us now assume that the function $p_\infty(\cdot)$ is differentiable at β and show that

$$\lim_{N\to\infty} \frac{\partial}{\partial\beta} p_N(\beta, \boldsymbol{\beta}) = p'_\infty(\beta) \tag{6.201}$$

holds uniformly in $\boldsymbol{\beta}$ (where $|\beta_s| \leq 1$ for each s). This elementary fact is proved as follows: if $\beta < \beta_1$ we have, by convexity

$$\frac{\partial}{\partial\beta} p_N(\beta, \boldsymbol{\beta}) \leq \frac{p_N(\beta_1, \boldsymbol{\beta}) - p_N(\beta, \boldsymbol{\beta})}{\beta_1 - \beta}\,,$$

so that, using (6.198)

$$\limsup_{N\to\infty} \frac{\partial}{\partial\beta} p_N(\beta, \boldsymbol{\beta}) \leq \frac{p_\infty(\beta_1) - p_\infty(\beta)}{\beta_1 - \beta}\,,$$

and letting $\beta_1 \to \beta$ we get

$$\limsup_{N\to\infty} \frac{\partial}{\partial\beta} p_N(\beta, \boldsymbol{\beta}) \leq p'_\infty(\beta)\,,$$

whereas $\liminf_{N\to\infty} \partial p_N(\beta, \boldsymbol{\beta})/\partial\beta \geq p'_\infty(\beta)$ may be proved similarly.

We know that

$$\frac{\partial}{\partial\beta} p_N(\beta, \boldsymbol{\beta}) = \frac{\beta}{2}(1 - \mathsf{E}\langle R_{1,2}^p\rangle) + \delta\,,$$

where $|\delta| \le K/N$, and from Theorem 6.5.1 we have

$$\mathsf{E}\,\langle R_{1,2}^p\rangle = q^p\,\mathsf{E}\,\langle \mathbf{1}_{\{R_{1,2}\ge 1/2\}}\rangle + \delta = q^p(1-m) + \delta$$

where δ is as in (6.98). Thus

$$\frac{\partial}{\partial\beta}\,p_N(\beta,\boldsymbol{\beta}) = \frac{\beta}{2}\,(1 - q^p(1-m)) + \delta\,.$$

Since the system of equations (6.132), (6.169) has a unique solution, we then get from (6.201) that

$$m = m_N(\beta,\boldsymbol{\beta}) = m(\beta) + \delta\,,\ q = q_N(\beta,\boldsymbol{\beta}) = q(\beta) + \delta\,, \tag{6.202}$$

and the proof is now complete. □

Proposition 6.7.2. *For p large enough, p odd, $3/2 \le \beta \le 2^{p/L}$, we have*

$$p_\infty(\beta) = F(\beta, q(\beta), m(\beta)) \tag{6.203}$$

where $q(\beta)$ and $m(\beta)$ are defined in Theorem 6.7.1.

Since $q(\beta)$ and $m(\beta)$ are defined in terms of $p'_\infty(\beta)$, this is not the end of the story.

Proof. If we combine (6.175) and (6.202) we get

$$(N+1)\,p_{N+1}(\beta,\boldsymbol{\beta}) - N\,p_N(\beta,\boldsymbol{\beta}) = F(\beta, q(\beta), m(\beta)) + \delta'$$

where δ' is as in (6.177). Summation over N yields

$$p_N(\beta,\boldsymbol{\beta}) = F(\beta, q(\beta), m(\beta)) + \delta'$$

and (6.95) implies

$$p_\infty(\beta) = F(\beta, q(\beta), m(\beta)) + \delta'\,. \tag{6.204}$$

Still, there could be (a few) values of β for which $p_\infty(\beta) \ne F(\beta, q(\beta), m(\beta))$. But since the function $\beta \mapsto p'_\infty(\beta)$ is right-continuous, this is also the case of the functions $\beta \mapsto q(\beta)$ and $\beta \mapsto m(\beta)$ and thus (6.204) yields (6.203). □

For $\beta \ge 3/2$, we consider

$$\mathcal{D}_\beta = \{(q,m)\,;\ 1 - 2^{-p/2} \le q \le 1\,;\ 0 < m \le 2\,,\ \beta^2 m^2 \ge 2\}\,.$$

Proposition 6.7.3. *If p is large enough, for $3/2 \le \beta \le 2^{p/L}$, there is a unique point $(\overline{q}(\beta), \overline{m}(\beta)) \in \mathcal{D}_\beta$ such that*

$$F(\beta, \overline{q}(\beta), \overline{m}(\beta)) = \inf_{(q,m)\in\mathcal{D}_\beta} F(\beta, q, m)\,.$$

Moreover, $\overline{m}'(\beta) \le 0$ and $\overline{m}(3/2) = 1$. Furthermore, if $(q,m) \in \mathcal{D}_\beta$ and

$$\frac{\partial F}{\partial q}\,(\beta, q, m) = \frac{\partial F}{\partial m}\,(\beta, q, m) = 0$$

then $(q,m) = (\overline{q}(\beta), \overline{m}(\beta))$.

Theorem 6.7.4. *There exists* p_0 *such that if* $p \geq p_0$, p *is odd, then for* $3/2 \leq \beta \leq 2^{p/L}$ *we have*

$$q(\beta) = \overline{q}(\beta)\,; \; m(\beta) = \overline{m}(\beta) \tag{6.205}$$

$$p_\infty(\beta) = F(\beta, \overline{q}(\beta), \overline{m}(\beta))\,. \tag{6.206}$$

The idea of the proof is very simple. First, one should note the following magic properties of the function F:

$$\frac{\partial F}{\partial \beta} = \frac{\beta}{2}\,(1 - q^p(1-m)) \tag{6.207}$$

$$\frac{\partial F}{\partial q} = \frac{\beta^2}{4}\,pq^{p-2}(1-m)\left(q - \frac{\mathsf{E}\,\mathrm{th}^2 Y\,\mathrm{ch}^m Y}{\mathsf{E}\,\mathrm{ch}^m Y}\right) \tag{6.208}$$

so that

$$\frac{\partial F}{\partial q} = 0 \Leftrightarrow q = \frac{\mathsf{E}\,\mathrm{th}^2 Y\,\mathrm{ch}^m Y}{\mathsf{E}\,\mathrm{ch}^m Y}\,. \tag{6.209}$$

These are obtained by straightforward computation and integration by parts, and the reader having reached this point certainly does not need to be guided through this calculation.

Taking the (right) derivative of (6.203) with respect to β, we obtain

$$\begin{aligned} p'_\infty(\beta) = {} & \frac{\partial F}{\partial \beta}\,(\beta, q(\beta), m(\beta)) + q'(\beta)\,\frac{\partial F}{\partial q}\,(\beta, q(\beta), m(\beta)) \\ & + m'(\beta)\,\frac{\partial F}{\partial m}\,(\beta, q(\beta), m(\beta))\,. \end{aligned}$$

By (6.209) and (6.132) we have

$$\frac{\partial F}{\partial q}\,(\beta, q(\beta), m(\beta)) = 0\,. \tag{6.210}$$

Using (6.207) and (6.199), we obtain that

$$m'(\beta)\,\frac{\partial F}{\partial m}\,(\beta, q(\beta), m(\beta)) = 0\,.$$

The relation

$$\frac{\beta}{2}\,(1 - q(\beta)^p(1 - m(\beta))) = p'_\infty(\beta) \leq \sqrt{\log 2}\,,$$

that tends to let $m(\beta)$ decrease, shows that it cannot really happen that $m'(\beta) = 0$. This forces the relation $\partial F/\partial\beta\,(\beta, q(\beta), m(\beta)) = 0$. Bringing this together with (6.210) and Proposition 6.7.3 yields $(q(\beta), m(\beta)) = (\overline{q}(\beta), \overline{m}(\beta))$.

Unfortunately, filling in the details of this approach is not really gratifying and is reserved for the experts.

Lemma 6.7.5. *The function $\beta \mapsto m(\beta)$ is right-continuous and has left limits. Moreover we have $\lim_{\beta'\to\beta,\beta'<\beta} m(\beta') \le m(\beta)$.*

Proof. We recall the function B of Proposition 6.5.13. This function is continuous and $m(\beta) = B(2p'_\infty(\beta)/\beta, \beta)$. The function $\beta \mapsto p'_\infty(\beta)$ is right-continuous and has left limits, so this is also the case for the function $\beta \mapsto m(\beta)$. Moreover

$$\lim_{\beta'\to\beta,\beta'<\beta} m(\beta') = B\left(\frac{2}{\beta} \lim_{\beta'\to\beta,\beta'<\beta} p'_\infty(\beta'), \beta\right) \le B\left(\frac{2p'_\infty(\beta)}{\beta}, \beta\right) ,$$

because $\lim_{\beta'\to\beta,\beta'<\beta} p'_\infty(\beta') \le p'_\infty(\beta)$ and the function $x \mapsto B(x,\beta)$ is non-decreasing, as shown in Proposition 6.5.13. □

Lemma 6.7.6. *If*

$$C = \frac{\partial}{\partial m} F(\beta, q(\beta), m(\beta)) \ne 0 ,$$

then the function $\beta' \mapsto m(\beta')$ has a right derivative at β and this right derivative is zero.

Proof. Consider $0 < \varepsilon < 1$ and write

$$p_\infty(\beta+\varepsilon) - p_\infty(\beta) = F(\beta+\varepsilon, q(\beta+\varepsilon), m(\beta+\varepsilon)) - F(\beta, q(\beta), m(\beta)) .$$

Performing an order 2 expansion and using the fact that by 6.209) one has $\partial F(\beta, q(\beta), m(\beta))/\partial q = 0$ one obtains

$$\left|p_\infty(\beta+\varepsilon) - p_\infty(\varepsilon) - \varepsilon\frac{\partial F}{\partial \beta}(\beta, q(\beta), m(\beta)) - (m(\beta+\varepsilon) - m(\beta))\, C\right|$$
$$\le K(\varepsilon^2 + (m(\beta+\varepsilon) - m(\beta))^2 + (q(\beta+\varepsilon) - q(\beta))^2) .$$

Here and below, K is independent of ε. It follows that

$$|C| \left|\frac{m(\beta+\varepsilon) - m(\beta)}{\varepsilon}\right| \le \mathrm{I} + \mathrm{II} , \tag{6.211}$$

$$\mathrm{I} = \left|\frac{p_\infty(\beta+\varepsilon) - p_\infty(\beta)}{\varepsilon} - \frac{\partial F}{\partial \beta}(\beta, q(\beta), m(\beta))\right|$$

$$\mathrm{II} = K\left(\varepsilon + \frac{1}{\varepsilon}(m(\beta+\varepsilon) - m(\beta))^2 + \frac{1}{\varepsilon}(q(\beta+\varepsilon) - q(\beta))^2\right) .$$

We want to let $\varepsilon \to 0$ in (6.211). By (6.208), (6.199) we know that $\mathrm{I} \to 0$. The problem is to control II. Let us write

$$x = \frac{2p'_\infty(\beta)}{\beta} \, ; \, \eta = \frac{2p'_\infty(\beta+\varepsilon)}{\beta+\varepsilon} - x$$

so that $m(\beta+\varepsilon) = B(x+\eta, \beta+\varepsilon)$, and thus

$$|m(\beta+\varepsilon) - m(\beta)| \leq K(\varepsilon + |\eta|) ,$$

and similarly for q. Thus we have

$$\mathrm{II} \leq K\left(\varepsilon + \frac{\eta^2}{\varepsilon}\right) . \tag{6.212}$$

We will prove that the ratio $|\eta|/\varepsilon$ remains bounded as $\varepsilon \to 0$, and letting $\varepsilon \to 0$ in (6.211) will then finish the proof. Expanding the expression

$$m(\beta+\varepsilon) - m(\beta) = B(x+\eta, \beta+\varepsilon) - B(x,\beta)$$

to the second order again, we get

$$|m(\beta+\varepsilon) - m(\beta) - \eta D_1 - \varepsilon D_2| \leq K(\varepsilon^2 + \eta^2) ,$$

where $D_1 = \partial B/\partial x(x,\beta) > 0$, as was shown in Proposition 6.5.13. Combining with (6.211) and (6.212) we get

$$\left|\frac{\eta}{\varepsilon} D_1 + D_2\right| \leq K\left(\varepsilon + \frac{\eta^2}{\varepsilon}\right)$$

and hence

$$\frac{|\eta|}{\varepsilon}|D_1| \leq K\left(1 + |\eta|\frac{|\eta|}{\varepsilon}\right) .$$

Since $\lim_{\varepsilon\to 0} |\eta| = 0$, this shows that indeed the ratio $|\eta|/\varepsilon$ remains bounded as $\varepsilon \to 0$, and finishes the proof. □

Proof of Theorem 6.7.4. If $m(\beta) \neq \overline{m}(\beta)$, then by (6.132) and the last assertion of Proposition 6.7.3, we have

$$\frac{\partial F}{\partial m}(\beta, q(\beta), m(\beta)) \neq 0$$

so that, since $\overline{m}'(\beta) \leq 0$, by Lemma 6.7.6, the function $\beta \mapsto m(\beta) - \overline{m}(\beta)$ has a right-derivative at β that is non-negative. Combining with Lemma 6.7.5, we see that if $m(\beta_0) > \overline{m}(\beta_0)$, then for all $\beta_0 \leq \beta \leq \beta_1 := 2^{p/L_0}$, we have $m(\beta) > \overline{m}(\beta)$. Using again Lemmas 6.7.5 and 6.7.6, we see that $m(\beta_0) \leq m(\beta_1)$. Now for $\beta \geq 3/2$,

$$\frac{3}{4} \leq p'_\infty(\beta) = \frac{\beta}{2}(1 - q(\beta)^p(1-m(\beta))) \leq \sqrt{\log 2}$$

so that

$$1 - \frac{1}{q(\beta)^p} + \frac{1}{q(\beta)^p}\frac{3}{2\beta} \leq m(\beta) \leq 1 - \frac{1}{q(\beta)^p} + \frac{1}{q(\beta)^p}\frac{2\sqrt{\log 2}}{\beta}$$

and since $m(\beta_0) \le m(\beta_1)$ and $q(\beta) \in [1 - 2^{-p/2}, 1]$, one obtains

$$\frac{3}{2\beta_0} \le [1 - 2^{-p/2}]^{-p} \left(1 + \frac{2\sqrt{\log 2}}{\beta_1}\right) - 1 ,$$

which shows that $\beta_0 \ge \beta_1/2$ (for p large). Thus for $\beta \le \beta_0 \le 2^{p/L}$ we have $m(\beta) \le \overline{m}(\beta)$.

Should it happen that $m(\beta_2) < \overline{m}(\beta_2)$ for some $\beta_2 > 3/2$, then Lemmas 6.7.5 and 6.7.6 show that $m(3/2) < \overline{m}(3/2)$, which is impossible because $m(3/2) = \overline{m}(3/2) = 1$. □

Proof of Proposition 6.7.3. It would be nice to have a clean proof of this Proposition, but unfortunately our approach is very heavy. As shown in Proposition 6.5.13 (and (6.209)) the condition $\partial F/\partial q = 0$ defines a function $q = q(m)$. (We now think of β as fixed once and for all.) Moreover, the function

$$q \mapsto q - \frac{\mathsf{E}\,\mathrm{th}^2 Y\,\mathrm{ch}^m Y}{\mathsf{E}\,\mathrm{ch}^m Y}$$

increases over $[1-2^{-p/2}, 1]$, so that (6.209) shows that $\partial F/\partial q < 0$ if $q < q(m)$ and $\partial F/\partial q > 0$ if $q > q(m)$. Thus, if q, m is a minimum of F, we must have $q = q(m)$. Consider the function

$$\Psi(m) = F(\beta, q(m), m) .$$

We will prove that

$$\Psi''(m) > 0 \tag{6.213}$$

$$\Psi'\left(\sqrt{\frac{2}{\beta}}\right) < 0 , \tag{6.214}$$

so that the function Ψ has a unique minimum in the interval $[\sqrt{2/\beta}, 1]$.

The main idea of the proof is that in the range of parameters we are considering, we can nearly compute $\mathsf{E}\,\mathrm{ch}^m Y$. To see this we write

$$\mathrm{ch}\, Y = \frac{1}{2}\left(\exp|Y| + \exp(-|Y|)\right) ,$$

so that

$$\begin{aligned}\mathrm{ch}^m Y &= 2^{-m}(\exp|Y| + \exp(-|Y|))^m \\ &= 2^{-m}(\exp m|Y|)(1 + \exp(-2|Y|))^m .\end{aligned}$$

Since $(1+x)^m \le 1 + x$ for $x \ge 0$ and $m \le 1$, we get

$$\mathrm{ch}^m Y \le 2^{-m} \exp m|Y| + 1 ,$$

and thus

$$\mathsf{E}\,\mathrm{ch}^m Y \le 1 + 2^{1-m}\mathsf{E}\exp mY = 1 + 2^{1-m}\exp\frac{m^2\beta^2 q^{p-1}}{4}p\,. \tag{6.215}$$

Also,

$$\mathrm{ch}^m Y \ge 2^{-m}\exp m|Y|$$

and

$$\mathsf{E}\,\mathrm{ch}^m Y \ge 2^{1-m}\mathsf{E}(\mathbf{1}_{\{Y\ge 0\}}\exp mY) \ge 2^{1-m}\mathsf{E}(\exp mY - 1)\,;$$

so that, combining with (6.215), we have

$$\left|\mathsf{E}\,\mathrm{ch}^m Y - 2^{1-m}\exp\left(\frac{m^2\beta^2 q^p}{4}p\right)\right| \le 2\,. \tag{6.216}$$

The above inequality enables us to compute $\mathsf{E}\,\mathrm{ch}^m Y$ with a relative error that is exponentially small in p, since $m^2\beta^2 \ge 2$. So, using (6.216) one might expect that

$$f(m) = \frac{1}{m}\log \mathsf{E}\,\mathrm{ch}^m Y \simeq \frac{1-m}{m}\log 2 + \frac{m\beta^2 q^p}{4}p \tag{6.217}$$

has a second derivative which may be approximated by $m^{-3}2\log 2$. However this does not follow directly from (6.217). Indeed, one would rather have to compute $f''(m)$ and use estimates such as (6.216) for each term. This is long and tedious and since the reader has nothing to learn from this computation we will not do it. This proves (6.213) because the influence of $q(m)$ is small, as the derivative in m are exponentially small in p. The proof of (6.214), and of the fact that $\Psi'(1) < 0$ if $\beta = 3/4$, are obtained through a similar approximation, which also lead to the relation $\overline{m}'(\beta) < 0$. □

6.8 Notes and Comments

The p-spin interaction model, invented by B. Derrida, was studied in [66] and [58]. Interestingly, at the end of her paper, E. Gardner mentions that this model should be amenable to rigorous study.

A first rigorous progress was achieved in [140], in particular the key observation that the a priori estimates are the way to get started. This paper [140] fails to prove entirely that the lumps are pure states, and the attempt in that direction illustrates the curse of the cavity method. Among the many possible variations, it is difficult to spot the correct combinations. The proof given in Section 6.3 is almost easy; on the other hand, the attempt of [140], trying to relate the N-spin system with the $(N-1)$-spin system, runs into impossible difficulties.

In the rest of the book we have presented results that nearly represent what was known at the time of writing, whereas in this chapter, we have opted for a different strategy, choosing not to cover the most general results known to us. It is indeed possible to prove Theorem 6.5.1 when there is an external field, i.e. when a term

$$\beta h \sum_{i \le N} \sigma_i \tag{6.218}$$

is added to the Hamiltonian, where $|h| \le 1/L$. This does not require essential new ideas, but it is quite more complicated to prove the a priori estimates. In this case the two values q_0, q_1 taken by $R_{1,2}$ are given by a system of 2 equations with 2 unknown.

It is also possible to prove Theorem 6.5.1 without adding a perturbation term to the Hamiltonian; thus one can do *without* the extended Ghirlanda-Guerra identities, but the technical devices required to go around this difficulty are really complicated.

These two extensions of Theorem 6.5.1 can be found in [152].

Research problem 6.8.1. (Level 2) Extend Theorem 6.5.1 to the case where the external field (6.218) is added to the Hamiltonian but when one is *NOT* allowed the perturbation term of the Hamiltonian.

Fate had it that we decided to cover the case p odd, for the simple reason that when p is even the a priori estimates are a bit more cumbersome, since one cannot rule out the possibility that the overlaps take values close to -1. Much later we figured out the way to compute the "free energy" and we realized that this runs into complications that do not occur when p is even, where Guerra's bound provides a quick kill. On the positive side, this led us to discover the positivity principle of Section 6.6, and moreover one can always hope that the main idea of the cumbersome Section 6.7 (which shows "how to determine the last parameter in the free energy") will be of use somewhere else.

A few days before sending the final version of this work to the editor, we discovered the choice of parameters in Theorem 2.11.14 that leads to Lemmas 2.11.17 and 2.11.18. Theorem 2.11.14 has a version for the p-spin interaction model when p is even. We hope that this version, combined with the ideas of these lemmas and of Theorem 2.9.10 will allow (when p is even) striking progress on the study of the p-spin interaction model, and in particular the solution of Research problems 6.5.2, 6.5.3 and 6.8.1. We plan to investigate this approach in the near future.

7. The Diluted SK Model and the K-Sat Problem

7.1 Introduction

In the SK model, each individual (or spin) interacts with every other individual. For large N, this does not make physical sense. Rather, we would like that, as $N \to \infty$, a given individual typically interacts only with a bounded number of other individuals. This motivates the introduction of the diluted SK model. In this model, the Hamiltonian is given by

$$-H_N(\boldsymbol{\sigma}) = \beta \sum_{i<j} g_{ij}\, \gamma_{ij}\, \sigma_i\, \sigma_j \ . \tag{7.1}$$

As usual, $(g_{ij})_{i<j}$ are i.i.d. standard Gaussian. The r.v. γ_{ij} are independent from the variables g_{ij}, and i.i.d., with $\mathsf{P}(\gamma_{ij} = 1) = \gamma/N$, $\mathsf{P}(\gamma_{ij} = 0) = 1-\gamma/N$. Here $\gamma > 0$ is a parameter indicating "how diluted is the interaction". The Hamiltonian (7.1) contains an interaction term between σ_i and σ_j only when $\gamma_{ij} = 1$. Thus, the expected number of terms in (7.1) is $\gamma(N-1)/2$, and the expected number of terms that contain σ_i is about $\gamma/2$. One should observe that the usual normalizing factor $1/\sqrt{N}$ does not occur in (7.1).

If we draw an edge between i and j when $\gamma_{ij} = 1$, the resulting random graph is well understood [11]. When $\gamma < 1$, this graph has only small connected components, so there is no "global interaction" and the situation is not so interesting. In order to get a challenging model we must certainly allow the case $\gamma \gg 1$.

In an apparently unrelated direction, let us remind the reader that the motivation of Chapter 3 is the problem as to whether certain random subsets of $\{-1,1\}^N$ have a non-empty intersection. In Chapter 3, we considered "random half-spaces". These somehow "depend on all coordinates". What would happen if instead we considered sets depending only on a given number p of coordinates? For example the sets

$$\{\boldsymbol{\sigma}\,;\ (\sigma_{i_1}, \ldots, \sigma_{i_p}) \neq (\eta_1, \ldots, \eta_p)\} \tag{7.2}$$

where $1 \le i_1 < i_2 < \cdots < i_p \le N$, and $\eta_i = \pm 1$.

The question of knowing whether M random independent sets of the type (7.2) have a non-empty intersection is known as the (random) p-sat problem, and is of (considerable) interest in theoretical computer science. By a random subset of the type (7.2), we of course mean one that is chosen uniformly among all possible choices. This motivates the introduction of the Hamiltonian

$$-H_N(\boldsymbol{\sigma}) = \beta \sum_{k \le M} W_k(\boldsymbol{\sigma}) \tag{7.3}$$

where $W_k(\boldsymbol{\sigma}) = 1$ if $(\sigma_{i(k,1)}, \ldots, \sigma_{i(k,p)}) \ne (\eta_{k,1}, \ldots, \eta_{k,p})$, and $W_k(\boldsymbol{\sigma}) = 0$ otherwise. The indices $1 \le i(k,1) < i(k,2) < \cdots < i(k,p) \le N$ and the numbers $\eta_{k,i} = \pm 1$ are chosen randomly uniformly over all possible choices. The interesting case is when M is proportional to N.

If, rather than insisting in (7.3) that we want exactly M terms in the Hamiltonian, we decide at random, independently for each possible term, whether we want to include it or not in the sum, we obtain a model that is technically simpler and (probably) with the same behavior. To formalize this idea, let us denote by $[N]^p$ the set of subsets of $\{1, \ldots, N\}$ of cardinality p. Let us write

$$I = \{i(I,1), \ldots, i(I,p)\} ,$$

where $i(I,1) < \cdots < i(I,p)$. We consider i.i.d. r.v. $\eta_{I,\ell}$, $1 \le \ell \le p$, $I \in [N]^p$, $\mathsf{P}(\eta_{I,\ell} = \pm 1) = 1/2$, and we define

$$W_I(\boldsymbol{\sigma}) = 1 \quad \text{if} \quad (\sigma_{i(I,1)}, \ldots, \sigma_{i(I,p)}) \ne (\eta_{I,1}, \ldots, \eta_{I,p})$$

and $W_I(\boldsymbol{\sigma}) = 0$ otherwise. We consider i.i.d. r.v. (γ_I) with $\mathsf{P}(\gamma_I = 1) = \gamma/N^{p-1}$ and $\mathsf{P}(\gamma_I = 0) = 1 - \gamma/N^{p-1}$, and the Hamiltonian

$$-H_N(\boldsymbol{\sigma}) = \beta \sum_I \gamma_I \, W_I(\boldsymbol{\sigma}) . \tag{7.4}$$

The expected number of terms in the sum is

$$\frac{\gamma}{N^{p-1}} \binom{N}{p} = \frac{\gamma N}{p!} \frac{(N-1) \ldots (N-p+1)}{N^{p-1}} \sim \frac{\gamma N}{p!}$$

for large N, so that (7.4) resembles (7.3) when $M = \gamma N / p!$.

To cover simultaneously the cases of (7.1) and (7.4) we consider a random function θ on $\{-1,1\}^p$, and i.i.d. copies (θ_I) of θ. We define

$$W_I(\boldsymbol{\sigma}) = \theta_I(\sigma_{i(I,1)}, \ldots, \sigma_{i(I,p)}) \tag{7.5}$$

and

$$-H_N(\boldsymbol{\sigma}) = \sum_I \gamma_I \, W_I(\boldsymbol{\sigma}) . \tag{7.6}$$

Here we assume that the r.v. γ_I are as in (7.4), and are independent of the r.v. θ_I. There is no longer a coefficient β since this coefficient can be thought of as part of θ. For example, the case (7.1) is obtained with $p = 2$,

$$\theta(\sigma_1, \sigma_2) = \beta \, g \, \sigma_1 \sigma_2$$

where g is standard Gaussian.

The purpose of this chapter is to describe the behavior of the system governed by the Hamiltonian (7.6) under a "high temperature condition" of the type

$$(\gamma + \gamma^2)\,\mathsf{E}\, S \exp 4S \le \frac{1}{K(p)} \tag{7.7}$$

where

$$S = \sup |\theta(\sigma_1, \dots, \sigma_p)| \,. \tag{7.8}$$

Condition (7.7) allows in particular to reach large values of γ.

The model governed by (7.6) differs in an essential way from the models considered in the previous sections. This is because, since one spin interacts only with finitely many others, the central limit theorem does not apply, and the Gaussian behavior that was ubiquitous is now absent. The methods we will develop to attack this situation will be rather different from those of the previous chapters. Still, even though it might be hard to express this clearly, there is no doubt that many features are shared by this situation and those of the previous chapters.

7.2 Pure State

The purpose of this section is to show that under (7.7) "the system is in a pure state", that is, the spin correlations vanish. In fact we will prove that

$$\mathsf{E}\,|\langle \sigma_1 \sigma_2 \rangle - \langle \sigma_1 \rangle \langle \sigma_2 \rangle| \le \frac{K}{N} \,. \tag{7.9}$$

The proof will consist in proving a certain statement by induction over N, and is similar in spirit to the argument occurring at the end of Section 2.3. In order to make the induction work, it is necessary to carry a suitable induction hypothesis, that will prove a stronger statement than (7.9). This stronger statement will be useful later in its own right.

We say that two functions f, f' on Σ_N^n *depend on k coordinates* if we can find indices $1 \le i_1 < \dots < i_k \le N$ and functions $\overline{f}, \overline{f}'$ from $\{-1,1\}^{kn}$ to $\mathbb{R}$ such that

$$f(\boldsymbol{\sigma}^1, \dots, \boldsymbol{\sigma}^n) = \overline{f}(\sigma^1_{i_1}, \dots, \sigma^1_{i_k}, \sigma^2_{i_1}, \dots, \sigma^2_{i_k}, \dots, \sigma^n_{i_1}, \dots, \sigma^n_{i_k})$$

and similarly for f'. The reason we define this for 2 functions is to stress that both functions depend on the same set of k coordinates.

For $j \le N$, consider the transformation T_j of Σ_N^n that, for a point $(\boldsymbol{\sigma}^1, \dots, \boldsymbol{\sigma}^n)$ of Σ_N^n, exchanges the j-th coordinates of $\boldsymbol{\sigma}^1$ and $\boldsymbol{\sigma}^2$.

Definition 7.2.1. *Given numbers γ_0, B, we say that Property $C(N, \gamma_0, B)$ holds if the following is true. Consider two functions f, f' on Σ_N^n, and assume that they depend on k coordinates. Assume that $f \ge 0$, that*

$$f' \circ T_N = -f' \tag{7.10}$$

and that for a certain number A we have $|f'| \le Af$ at each point of Σ_N^n. Then if $\gamma \le \gamma_0$ we have

$$\mathsf{E}\left|\frac{\langle f' \rangle}{\langle f \rangle}\right| \le \frac{kAB}{N}\,. \tag{7.11}$$

Theorem 7.2.2. *There exists a number $K_1 = K_1(p)$ and a number $K_0 = K_0(p, \gamma_0)$ such that if $\gamma \le \gamma_0$ and*

$$K_1(p)(\gamma_0 + \gamma_0^2)\mathsf{E}\, S \exp 4S(1 + \mathsf{E}\, S \exp 4S) \le 1\,, \tag{7.12}$$

then Property $C(N, \gamma_0, K_0)$ holds for each N.

Research problem 7.2.3. (Level 2) In condition (7.12), is it possible to replace the term $\gamma_0 + \gamma_0^2$ by γ_0?

To rephrase a special case of this problem, if we assume that the function S is constant, and $\gamma \ge 1$, do we need that $K_1 S\gamma^2 \le 1$ in order for $C(N, \gamma_0, K_0)$ to hold for a certain K_0, or is it sufficient to assume that $K_1 S\gamma \le 1$? The same question arises for Theorem 7.4.1 below.

To provide motivation for Theorem 7.2.2, we relate Theorem 7.2.2 and (7.9). We take $n = 2$, $f = 1$,

$$f'(\boldsymbol{\sigma}^1, \boldsymbol{\sigma}^2) = \sigma_1^1(\sigma_N^1 - \sigma_N^2)\,,$$

so that (7.10) holds and $|f'| \le 2f$. Thus under (7.12) we get by (7.11)

$$\mathsf{E}\,|\langle \sigma_1^1(\sigma_N^1 - \sigma_N^2)\rangle| \le \frac{K}{N}$$

i.e.

$$\mathsf{E}\,|\langle \sigma_1 \sigma_N \rangle - \langle \sigma_1 \rangle \langle \sigma_N \rangle| \le \frac{K}{N}\,,$$

which is (7.9). Another important consequence of Theorem 7.2.2 is Proposition 7.2.6 below.

To prove Theorem 7.2.2, we will proceed by induction over N. We will show that if K_0 and K_1 are suitably chosen, then

$$C(N-1, \gamma_0, K_0) \Rightarrow C(N, \gamma_0, K_0)\,. \tag{7.13}$$

Since $C(1, \gamma_0, 1)$ is true, this will prove Theorem 7.2.2.

Before we can prove (7.13), we need to set up the cavity method. For $\boldsymbol{\sigma} = (\sigma_1, \ldots, \sigma_N) \in \Sigma_N$, we write as usual $\boldsymbol{\rho} = (\sigma_1, \ldots, \sigma_{N-1}) \in \Sigma_{N-1}$. We consider the Hamiltonian

$$-H_{N-1}(\boldsymbol{\rho}) = \sum \gamma_I\, W_I(\boldsymbol{\sigma})\,, \tag{7.14}$$

where the summation is restricted to those $I \subset \{1, \ldots N-1\}$, i.e. $I \in [N-1]^p$. Thus (7.14) is the Hamiltonian of a $(N-1)$-spin system, except that we have changed γ into γ_- such that

$$\frac{\gamma_-}{(N-1)^{p-1}} = \frac{\gamma}{N^{p-1}}$$

i.e. $\gamma_- = \gamma(1-1/N)^{p-1} \leq \gamma$. In particular

$$\gamma \leq \gamma_0 \Rightarrow \gamma_- \leq \gamma_0 \,, \tag{7.15}$$

a fact that will greatly help the induction.

From (7.14) we have

$$-H_N(\boldsymbol{\sigma}) = -H_{N-1}(\boldsymbol{\rho}) + \sum_{N \in I} \gamma_I \, W_I(\boldsymbol{\sigma}) \,. \tag{7.16}$$

Here and below, the notation $\sum_{N \in I}$ means that the summation is over all sets $I \in [N]^p$ for which $N \in I$. Since $N \in I$ means that $i(I,p) = N$, we have

$$\sum_{N \in I} \gamma_I \, W_I(\boldsymbol{\sigma}) = \sum_{N \in I} \gamma_I \, \theta_I(\sigma_{i(I,1)}, \ldots, \sigma_{i(I,p-1)}, \sigma_N) \,. \tag{7.17}$$

Given a function f on Σ_N^n, we have the algebraic identity

$$\langle f \rangle = \frac{\langle \mathrm{Av} \, f \, \mathcal{E} \rangle_-}{\langle \mathrm{Av} \, \mathcal{E} \rangle_-} \,. \tag{7.18}$$

Here

$$\mathcal{E} = \mathcal{E}(\boldsymbol{\sigma}^1, \ldots, \boldsymbol{\sigma}^n) = \exp \sum_{\ell \leq n} \sum_{N \in I} \gamma_I \, W_I(\boldsymbol{\sigma}^\ell) \,;$$

as usual Av means average over $\sigma_N^1, \ldots, \sigma_N^n = \pm 1$. Thus $\mathrm{Av} \, f \, \mathcal{E}$ is a function of $(\boldsymbol{\rho}^1, \ldots, \boldsymbol{\rho}^n)$ only, and $\langle \mathrm{Av} \, f \, \mathcal{E} \rangle_-$ means that it is then averaged for Gibbs' measure relative to the Hamiltonian (7.14).

In the right-hand side of (7.18), we have two distinct sources of randomness: the randomness in $\langle \cdot \rangle_-$ and the randomness in $\mathcal{E}$. It will be essential that these randomness be probabilistically independent. In the previous chapters we were taking expectations given $\langle \cdot \rangle_-$. We find it more convenient to now take expectations given $\mathcal{E}$. This expectation is denoted by E_-. According to (7.18) we have

$$\mathsf{E} \left| \frac{\langle f' \rangle}{\langle f \rangle} \right| = \mathsf{E} \left| \frac{\langle \mathrm{Av} \, f' \, \mathcal{E} \rangle_-}{\langle \mathrm{Av} \, f \, \mathcal{E} \rangle_-} \right| = \mathsf{E} \mathsf{E}_- \left| \frac{\langle \mathrm{Av} \, f' \, \mathcal{E} \rangle_-}{\langle \mathrm{Av} \, f \, \mathcal{E} \rangle_-} \right| \,. \tag{7.19}$$

We study this when $|f'| \leq A f$, $f' \circ T_N = -f'$ and f, f' depend on k coordinates. Without loss of generality, we assume that these coordinates are $1, \ldots, k-1, N$. First, we observe that, since we assume $|f'| \leq A f$, we have $|f' \, \mathcal{E}| \leq A f \, \mathcal{E}$, so that $|\mathrm{Av} \, f' \, \mathcal{E}| \leq \mathrm{Av} \, |f' \, \mathcal{E}| \leq A \, \mathrm{Av} \, f \, \mathcal{E}$, and thus

$$\mathsf{E}_-\left|\frac{\langle \mathrm{Av}\, f'\,\mathcal{E}\rangle_-}{\langle \mathrm{Av}\, f\,\mathcal{E}\rangle_-}\right| \le A\,. \tag{7.20}$$

The precise study of the left-hand side of (7.20) is made difficult when certain rare events occur, in which case we will use the crude bound (7.20) rather than the induction hypothesis. We describe them first. Consider the following two events, that depend only on the randomness of $\mathcal{E}$:

$$\Omega_1 : \exists I \in [N]^p\,,\ N \in I\,,\ \gamma_I = 1\,,\ I \cap \{1,\ldots,k-1\} \ne \emptyset\,,$$

$$\Omega_2 : \exists I_1, I_2 \in [N]^p\,,\ N \in I_1\,,\ N \in I_2\,,\ \gamma_{I_1} = \gamma_{I_2} = 1\,,$$
$$I_1 \cap I_2 \cap \{1,\ldots,N-1\} \ne \emptyset\,.$$

These are rare events, since, very crudely, there are at most $(k-1)\,N^{p-2}$ sets $I \in [N]^p$ for which $N \in I$ and $I \cap \{1,\ldots,k-1\} \ne \emptyset$, so that

$$\mathsf{P}(\Omega_1) \le \frac{\gamma(k-1)\,N^{p-2}}{N^{p-1}} \le \frac{\gamma k}{N}\,.$$

Similarly, there are at most N^{2p-3} pairs $I_1, I_2 \in [N]^p$ with $N \in I_1$, $N \in I_2$ and $I_1 \cap I_2 \cap \{1,\ldots,N-1\} \ne \emptyset$, so that

$$\mathsf{P}(\Omega_2) \le \left(\frac{\gamma}{N^{p-1}}\right)^2 N^{2p-3} = \frac{\gamma^2}{N}\,.$$

Thus, if we set $\Omega = \Omega_1 \cup \Omega_2$, we have

$$\mathsf{P}(\Omega) \le \frac{\gamma k + \gamma^2}{N}\,. \tag{7.21}$$

Using (7.19), (7.20) and (7.21), we have

$$\begin{aligned}\mathsf{E}\left|\frac{\langle f'\rangle}{\langle f\rangle}\right| &= \mathsf{E}\left(1_\Omega\, \mathsf{E}_-\left|\frac{\langle \mathrm{Av}\, f'\,\mathcal{E}\rangle_-}{\langle \mathrm{Av}\, f\,\mathcal{E}\rangle_-}\right|\right) + \mathsf{E}\left(1_{\Omega^c}\, \mathsf{E}_-\left|\frac{\langle \mathrm{Av}\, f'\,\mathcal{E}\rangle_-}{\langle \mathrm{Av}\, f\,\mathcal{E}\rangle_-}\right|\right) \\ &\le \frac{\gamma k + \gamma^2}{N} A + \mathsf{E}\left(1_{\Omega^c}\, \mathsf{E}_-\left|\frac{\langle \mathrm{Av}\, f'\,\mathcal{E}\rangle_-}{\langle \mathrm{Av}\, f\,\mathcal{E}\rangle_-}\right|\right)\,. \end{aligned} \tag{7.22}$$

The next task is to use the induction hypothesis to study the last term above. When Ω does not occur, the sets I for which $N \in I$ and $\gamma_I = 1$ do not intersect $\{1,\ldots,k-1\}$, and the intersection of any two of them is $\{N\}$. Let $r = \sum_{N\in I} \gamma_I$ be the number of these sets. Let

$$J = \bigcup \{I;\ N \in I\,,\ \gamma_I = 1\} \backslash \{N\}\,,$$

so that $\operatorname{card} J = r(p-1)$. Moreover

$$J \cap \{1,\ldots,k-1\} = \emptyset\,. \tag{7.23}$$

For $i \le N-1$ let us denote by U_i the transformation of Σ_{N-1}^n that exchanges the coordinates σ_i^1 and σ_i^2 of a point $(\boldsymbol{\rho}^1,\ldots,\boldsymbol{\rho}^n)$ of Σ_{N-1}^n. That is, U_i is to $N-1$ what T_i is to N.

Lemma 7.2.4. *When Ω does not occur we have*

$$(\mathrm{Av}\, f'\,\mathcal{E}) \circ \prod_{i\in J} U_i = -\mathrm{Av}\, f'\,\mathcal{E}\,. \tag{7.24}$$

This (crucial...) "algebraic" lemma means that something of the special symmetry of f' (as in (7.10)) is preserved when one replaces f' by $\mathrm{Av}\, f'\,\mathcal{E}$.

Proof. Let us write $T = \prod_{i\in J} T_i$. We observe that

$$f' \circ T = f'$$

because f' depends only on the coordinates in $\{1,\ldots,k-1,N\}$, a set disjoint from J. Thus

$$f' \circ T \circ T_N = f' \circ T_N = -f' \tag{7.25}$$

by (7.10). On the other hand, $\mathcal{E}$ depends only on the coordinates in $J \cup \{N\}$, and, since $\mathcal{E}(\boldsymbol{\sigma}^1,\boldsymbol{\sigma}^2,\ldots,\boldsymbol{\sigma}^n) = \mathcal{E}(\boldsymbol{\sigma}^2,\boldsymbol{\sigma}^1,\ldots,\boldsymbol{\sigma}^n)$, we have

$$\mathcal{E} \circ T \circ T_N = \mathcal{E}$$

and, combining with (7.25)

$$(f'\,\mathcal{E}) \circ T \circ T_N = -f'\,\mathcal{E}$$

so that, since T_N^2 is the identity,

$$(f'\,\mathcal{E}) \circ T = -(f'\,\mathcal{E}) \circ T_N\,. \tag{7.26}$$

Now,

$$\mathrm{Av}((f'\,\mathcal{E}) \circ T_N) = \mathrm{Av}\, f'\,\mathcal{E}$$

$$\mathrm{Av}((f'\,\mathcal{E}) \circ T) = (\mathrm{Av}\, f'\,\mathcal{E}) \circ \prod_{i\in J} U_i$$

so that applying Av to (7.26) proves (7.24). □

Let us set $k' = r(p-1) = \mathrm{card}\, J$, and let us enumerate as $i_1,\ldots,i_{k'}$ the points of J. From (7.24) we have

$$\mathrm{Av}\, f'\,\mathcal{E} = \frac{1}{2}\left(\mathrm{Av}\, f'\,\mathcal{E} - (\mathrm{Av}\, f'\,\mathcal{E}) \circ \prod_{s\le k'} U_{i_s}\right) = \frac{1}{2}\sum_{1\le s\le k'} f_s \tag{7.27}$$

where

$$f_s = (\mathrm{Av}\, f'\,\mathcal{E}) \circ \prod_{u\le s-1} U_{i_u} - (\mathrm{Av}\, f'\,\mathcal{E}) \circ \prod_{u\le s} U_{i_u}\,.$$

Since U_i^2 is the identity, we have

$$f_s \circ U_{i_s} = -f_s\,. \tag{7.28}$$

In words, (7.27) decomposes $\mathrm{Av}\, f' \mathcal{E}$ in a sum of $k' = r(p-1)$ pieces that possess the symmetry property required to use the induction hypothesis. It remains to establish the property that will play the role of the inequality $|f'| \le A f$. We denote by $I_1, \ldots, I_r$ the sets I for which $N \in I$ and $\gamma_I = 1$. If $v \le r$, we set

$$S_v = \sup |\theta_{I_v}(\sigma_1, \ldots, \sigma_p)| = \sup_{\boldsymbol{\sigma}} |W_{I_v}(\boldsymbol{\sigma})| \,,$$

where we recall that the random function θ_I is such that $W_I(\boldsymbol{\sigma}) = \theta_I(\sigma_{i(I,1)}, \ldots, \sigma_{i(I,p)})$.

Lemma 7.2.5. *Assume that Ω does not occur and that $i_s \in I_v$. Then*

$$|f_s| \le 4\, A\, S_v \exp\left(4 \sum_{u \le r} S_u\right) \mathrm{Av}\, f\, \mathcal{E} \,. \tag{7.29}$$

Proof. Let us write

$$\mathcal{E}' = \exp \sum_{3 \le \ell \le n} \sum_{N \in I} \gamma_I\, W_I(\boldsymbol{\sigma}^\ell) \,.$$

Thus

$$\mathcal{E} \ge \mathcal{E}' \exp\left(-2 \sum_{u \le r} S_u\right)$$

and hence

$$\mathrm{Av}\, f\, \mathcal{E} \ge (\mathrm{Av}\, f\, \mathcal{E}') \exp\left(-2 \sum_{u \le r} S_u\right) \,. \tag{7.30}$$

Next,

$$\begin{aligned} f_s &= (\mathrm{Av}\, f' \mathcal{E}) \circ \prod_{u \le s-1} U_{i_u} - (\mathrm{Av}\, f' \mathcal{E}) \circ \prod_{u \le s} U_{i_u} \\ &= \mathrm{Av}\left((f' \mathcal{E}) \circ \prod_{u \le s-1} T_{i_u} - (f' \mathcal{E}) \circ \prod_{u \le s} T_{i_u}\right) \\ &= \mathrm{Av}\left(f' \left(\mathcal{E} \circ \prod_{u \le s-1} T_{i_u} - \mathcal{E} \circ \prod_{u \le s} T_{i_u}\right)\right) \end{aligned} \tag{7.31}$$

because $f' \circ T_{i_u} = f'$ for each u, since f depends only on the coordinates $1, \ldots, k-1, N$. Let us write

$$\mathcal{E}'' = \exp \sum_{\ell \le 2} \sum_{N \in I} \gamma_I\, W_I(\boldsymbol{\sigma}^\ell)$$

so that $\mathcal{E} = \mathcal{E}'' \mathcal{E}'$. For each i, we have $\mathcal{E}' \circ T_i = \mathcal{E}'$, so that

$$\mathcal{E} \circ \prod_{u \le s-1} T_{i_u} - \mathcal{E} \circ \prod_{u \le s} T_{i_u} = \mathcal{E}' \left(\mathcal{E}'' \circ \prod_{u \le s-1} T_{i_u} - \mathcal{E}'' \circ \prod_{u \le s} T_{i_u} \right) ,$$

and, if we set

$$\Delta = \sup \left| \mathcal{E}'' \circ \prod_{u \le s-1} T_{i_u} - \mathcal{E}'' \circ \prod_{u \le s} T_{i_u} \right| = \sup |\mathcal{E}'' - \mathcal{E}'' \circ T_{i_s}| ,$$

we get from (7.31) that

$$|f_s| \le \Delta \, \mathrm{Av} \, (|f'| \, \mathcal{E}') \le A \, \Delta \, \mathrm{Av} \, f \, \mathcal{E}' \tag{7.32}$$

because $|f'| \le A \, f$.

To bound Δ, we write $\mathcal{E}'' = \prod_{u \le r} \mathcal{E}_u$, where

$$\mathcal{E}_u = \exp \sum_{\ell \le 2} W_{I_u}(\boldsymbol{\sigma}^\ell) .$$

We note that $\mathcal{E}_u \circ T_{i_s} = \mathcal{E}_u$ if $u \ne v$, because then $\mathcal{E}_u$ depends only on the coordinates in I_u, and $i_s \notin I_u$ if $u \ne v$. Thus

$$\mathcal{E}'' - \mathcal{E}'' \circ T_{i_s} = (\mathcal{E}_v - \mathcal{E}_v \circ T_{i_s}) \prod_{u \ne v} \mathcal{E}_u .$$

Now, using the inequality $|e^x - e^y| \le |x - y| \, e^a$ for $|x|, |y| \le a$, we get

$$|\mathcal{E}_v - \mathcal{E}_v \circ T_{i_s}| \le 4 \, S_v \exp 2 \, S_v .$$

Since $\mathcal{E}_u \le \exp 2 \, S_u$, we get $\Delta \le 4 \, S_v \exp 2 \sum_{u \le r} S_u$.

Combining with (7.30) and (7.32) completes the proof. □

Proposition 7.2.6. *Assume that $N \ge 2$ and that condition $C(N-1, \gamma_0, B)$ holds. Consider f', f as in Definition 7.2.1, and assume that $\gamma \le \gamma_0$. Then*

$$\mathsf{E} \left| \frac{\langle f' \rangle}{\langle f \rangle} \right| \le \frac{kA}{N} (\gamma + \gamma^2 + 4 \, p^2 \, BD \exp 4 \, D) \tag{7.33}$$

where

$$D = \left(\frac{\gamma}{(p-1)!} + \left(\frac{\gamma}{(p-1)!} \right)^2 \right) \mathsf{E} \, S \exp 4 \, S \, (1 + 4 \, \mathsf{E} \, S \exp 4 \, S) .$$

Proof. We keep the notation of Lemmas 7.2.4 and 7.2.5. Since $\gamma_- \leq \gamma$, we can use $C(N-1,\gamma_0,B)$ to conclude from (7.28) and (7.29) that, since f_r and $\mathrm{Av}\,\mathcal{E}\,f$ depend on $k-1+r(p-1) \leq k(1+r(p-1))$ coordinates, and since $1/(N-1) \leq 2/N$ because $N \geq 2$, outside Ω we have

$$\mathsf{E}_-\left|\frac{\langle f_s\rangle_-}{\langle \mathrm{Av}\, f\,\mathcal{E}\rangle_-}\right| \leq \frac{8kAB}{N}(1+r(p-1))\,S_v \exp\left(4\sum_{u\leq r} S_u\right).$$

There are two sources of randomness in the right-hand side. First there is the randomness in the γ_I $(N \in I)$ and then there is the randomness in S_I. These are independent. If we denote by E_S expectation in the r.v. S_I, we get

$$\mathsf{E}_S\,\mathsf{E}_-\left|\frac{\langle f_r\rangle_-}{\langle \mathrm{Av}\, f\,\mathcal{E}\rangle_-}\right| \leq \frac{8kAB}{N}(1+r(p-1))\,U\,V^{r-1}$$

where

$$U = \mathsf{E}\,S\exp 4\,S\,; \quad V = \mathsf{E}\exp 4\,S\,.$$

Combining with (7.27), since there are $r(p-1)$ terms we get

$$\mathsf{E}_S\,\mathsf{E}_-\left|\frac{\langle \mathrm{Av}\, f'\,\mathcal{E}\rangle_-}{\langle \mathrm{Av}\, f\,\mathcal{E}\rangle_-}\right| \leq \frac{4kAB(p-1)}{N}\,r(1+r(p-1))\,UV^{r-1} \tag{7.34}$$

where $r = \sum_{N\in I}\gamma_I$. This bound assumes that Ω does not occur; but combining (7.34) and (7.22) we obtain the bound

$$\mathsf{E}\left|\frac{\langle f'\rangle}{\langle f\rangle}\right| \leq \frac{kA}{N}(\gamma+\gamma^2+4B(p-1)\,U\mathsf{E}(r(1+r(p-1))V^{r-1}))\,.$$

Now, r is the sum of the $M = \binom{N-1}{p-1}$ independent r.v. γ_I with $P(\gamma_I = 1) = \gamma/N^{p-1}$ and $P(\gamma_I = 0) = 1-\gamma/N^{p-1}$. Thus

$$\begin{aligned}
\mathsf{E}\,r\,V^{r-1} &= \frac{1}{V}\mathsf{E}\,r\,V^r = \frac{1}{V}\sum_I \mathsf{E}\,\gamma_I\,V^{\sum_J \gamma_J}\\
&= \frac{1}{V}\sum_I\left(\mathsf{E}\,(\gamma_I\,V^{\gamma_I})\prod_{J\neq I}\mathsf{E}\,V^{\gamma_J}\right)\\
&= \frac{M}{V}\left(\frac{\gamma}{N^{p-1}}V\right)\left(1-\frac{\gamma}{N^{p-1}}+\frac{\gamma}{N^{p-1}}V\right)^{M-1}\\
&= \frac{\gamma M}{N^{p-1}}\left(1+\frac{\gamma}{N^{p-1}}(V-1)\right)^{M-1}\\
&\leq \frac{\gamma M}{N^{p-1}}\exp\frac{\gamma M}{N^{p-1}}(V-1)\,.
\end{aligned}$$

Here, I and J range over the M elements of $[N]^p$ that contain N. Now $\gamma M/N^{p-1} \leq \gamma/(p-1)!$, and, since $e^x \leq 1+xe^x$ for $x>0$ (as is obvious on series expansion) we have $V \leq 1+4U$, and thus

$$\mathsf{E}\, r\, V^{r-1} \le \frac{\gamma}{(p-1)!} \exp \frac{4\,\gamma\, U}{(p-1)!}\,.$$

In a similar manner,

$$\begin{aligned}
\mathsf{E}\, r^2\, V^{r-1} &= \frac{1}{V}\,\mathsf{E} \sum_{I_1, I_2} \gamma_{I_1}\, \gamma_{I_2}\, V^{\sum_J \gamma_J} \\
&= \mathsf{E}\, r\, V^{r-1} + \frac{1}{V}\,\mathsf{E} \sum_{I_1 \neq I_2} \gamma_{I_1}\, \gamma_{I_2}\, V^{\sum_J \gamma_J} \\
&= \mathsf{E}\, r\, V^{r-1} + M(M-1)\left(\frac{\gamma}{N^{p-1}}\right)^2 V \left(1 + \frac{\gamma}{N^{p-1}}\,(V-1)\right)^{M-2} \\
&\le \mathsf{E}\, r\, V^{r-1} + \left(\frac{M\gamma}{N^{p-1}}\right)^2 V \exp \frac{M\gamma}{N^{p-1}}\,(V-1) \\
&\le \mathsf{E}\, r\, V^{r-1} + \left(\frac{\gamma}{(p-1)!}\right)^2 (1 + 4U) \exp \frac{4\,\gamma\, U}{(p-1)!}
\end{aligned}$$

so that

$$\mathsf{E}\left((r + r^2(p-1))\, V^{r-1}\right) \le p \left(\frac{\gamma}{(p-1)!} + \left(\frac{\gamma}{(p-1)!}\right)^2 (1 + 4U)\right) \exp \frac{4\,\gamma\, U}{(p-1)!}$$

and the result follows by trivial bounds. □

Proof of Theorem 7.2.2. If

$$D_0 = \left(\frac{\gamma_0}{(p-1)!} + \left(\frac{\gamma_0}{(p-1)!}\right)^2\right) \mathsf{E}\, S \exp 4\, S (1 + 4 \mathsf{E}\, S \exp 4\, S)$$

is small enough that $4\,p^2 D_0 \exp 4\, D_0 \le 1/2$, then (7.33) implies

$$\mathsf{E}\left|\frac{\langle f' \rangle}{\langle f \rangle}\right| \le \frac{kA}{N}\left(\gamma + \gamma^2 + \frac{B}{2}\right). \tag{7.35}$$

Thus, if $B = 2(\gamma + \gamma^2) = K_0$, (7.35) shows that (7.11) holds. □

We explained at length in Section 2.4 that condition (7.9) implies that, as long as one considers a bounded number of spins, Gibbs' measure is nearly a product measure.

Let us denote by $\langle \cdot \rangle_\bullet$ average with respect to the product measure μ on Σ_{N-1} such that

$$\forall i \le N-1\,, \quad \int \sigma_i \,\mathrm{d}\mu(\boldsymbol{\rho}) = \langle \sigma_i \rangle_- \,.$$

Equivalently, for a function f on Σ_{N-1}, we have

$$\langle f \rangle_\bullet = \langle f(\sigma_1^1, \ldots, \sigma_{N-1}^{N-1}) \rangle_-$$

where σ_i^i is the i-th coordinate of the i-th replica $\boldsymbol{\rho}^i$. The following consequence of Theorem 7.2.2 will be used in Section 7.4.

Proposition 7.2.7. *Under (7.12), and if $\gamma \leq \gamma_0$, the following occurs. Consider sets $I_1, \ldots, I_r \in [N]^p$, containing N, and such that $s \neq s' \Rightarrow I_s \cap I_{s'} = \{N\}$. Consider functions W_s on Σ_N $(s \leq r)$ depending only on the coordinates in I_s and let $S_s = \sup |W_s(\boldsymbol{\sigma})|$. Let*

$$\mathcal{E} = \exp \sum_{s \leq r} W_s(\boldsymbol{\sigma}) \,.$$

Then

$$\mathsf{E}_- \left| \frac{\langle \operatorname{Av} \sigma_N \mathcal{E} \rangle_-}{\langle \operatorname{Av} \mathcal{E} \rangle_-} - \frac{\langle \operatorname{Av} \sigma_N \mathcal{E} \rangle_\bullet}{\langle \operatorname{Av} \mathcal{E} \rangle_\bullet} \right| \leq \frac{4r(p-1)K_0}{N-1} \sum_{s \leq r} \exp 2\, S_s \,. \tag{7.36}$$

Proof. Let $f' = \operatorname{Av} \sigma_N \mathcal{E}$ and $f = \operatorname{Av} \mathcal{E}$. For $0 \leq j \leq N-1$, let us define

$$f_j = f_j(\boldsymbol{\rho}^1, \ldots, \boldsymbol{\rho}^{N-1}) = f(\sigma_1^1, \sigma_2^2, \ldots, \sigma_j^j, \sigma_{j+1}^1, \ldots, \sigma_{N-1}^1)$$

and f'_j similarly. Thus

$$\frac{\langle \operatorname{Av} \sigma_N \mathcal{E} \rangle_-}{\langle \operatorname{Av} \mathcal{E} \rangle_-} = \frac{\langle f'_1 \rangle_-}{\langle f_1 \rangle_-} \,; \quad \frac{\langle \operatorname{Av} \sigma_N \mathcal{E} \rangle_\bullet}{\langle \operatorname{Av} \mathcal{E} \rangle_\bullet} = \frac{\langle f'_{N-1} \rangle_-}{\langle f_{N-1} \rangle_-} \,, \tag{7.37}$$

and the left-hand side of (7.36) is bounded by

$$\sum_{1 \leq j \leq N-1} \mathsf{E}_- \left| \frac{\langle f'_{j-1} \rangle_-}{\langle f_{j-1} \rangle_-} - \frac{\langle f'_j \rangle_-}{\langle f_j \rangle_-} \right| \,.$$

The terms in the summation are zero unless j belongs to the union of the sets I_s, $s \leq r$, for otherwise f' and f do not depend on the j-th coordinate. We then try to bound the term of the summation when $j \in I_v$. We write

$$\begin{aligned} \mathsf{E}_- \left| \frac{\langle f'_{j-1} \rangle_-}{\langle f_{j-1} \rangle_-} - \frac{\langle f'_j \rangle_-}{\langle f_j \rangle_-} \right| &\leq \mathsf{E}_- \left| \frac{\langle f'_{j-1} - f'_j \rangle_-}{\langle f_{j-1} \rangle_-} \right| \\ &\quad + \mathsf{E}_- \left| \frac{\langle f'_j \rangle_- \langle f_{j-1} - f_j \rangle_-}{\langle f_{j-1} \rangle_- \langle f_j \rangle_-} \right| \,. \end{aligned} \tag{7.38}$$

We will use $C(N-1, \gamma_0, K_0)$ to bound these terms. First, we observe that the function $f'_{j-1} - f'_j$ changes sign if we exchange σ_j^1 and σ_j^j. Next, if

$$\begin{aligned} \mathcal{E}' &= \exp \sum_{u \neq v} W_u(\sigma_1^1, \ldots, \sigma_j^j, \sigma_{j+1}^1, \ldots, \sigma_N^1) \\ &= \exp \sum_{u \neq v} W_u(\sigma_1^1, \ldots, \sigma_{j-1}^{j-1}, \sigma_j^1, \sigma_{j+1}^1, \ldots, \sigma_N^1) \end{aligned}$$

then

$$f_{j-1} = \operatorname{Av} \mathcal{E}(\sigma_1^1, \ldots, \sigma_{j-1}^{j-1}, \sigma_j^1, \ldots, \sigma_N^1) \geq \exp(-S_v) \operatorname{Av} \mathcal{E}' \,,$$

where Av denotes average over $\sigma_N^1 = \pm 1$. In a similar fashion, we get $|f'_{j-1}| \le \exp S_v \operatorname{Av} \mathcal{E}'$, $|f'_j| \le \exp S_v \operatorname{Av} \mathcal{E}'$, and thus

$$|f'_{j-1} - f'_j| \le (2 \exp 2 S_v)\, f_{j-1} \;,$$

so that from property $C(N-1, \gamma_0, K_0)$ we get

$$\mathsf{E}_- \left| \frac{\langle f'_{j-1} - f'_j \rangle_-}{\langle f_{j-1} \rangle_-} \right| \le \frac{2K_0}{N-1} r(p-1) \exp 2\, S_v \;, \tag{7.39}$$

because these functions depend on $r(p-1)$ coordinates. To handle the last term on the right-hand side of (7.38), we write

$$\langle f'_j \rangle_- \langle f_{j-1} - f_j \rangle_- = \langle f_1 \rangle_-$$

where

$$\begin{aligned} f_1(\boldsymbol{\rho}^1, \ldots, \boldsymbol{\rho}^{2N-2}) = f'_j(\boldsymbol{\rho}^1, \ldots, \boldsymbol{\rho}^{N-1})(f_{j-1}(\boldsymbol{\rho}^N, \ldots, \boldsymbol{\rho}^{2N-2}) \\ - f_j(\boldsymbol{\rho}^N, \ldots, \boldsymbol{\rho}^{2N-2})) \end{aligned}$$

changes signs when one exchanges σ_j^N and σ_j^{N+j-1}. Since $|f'_j| \le f_j$, we proceed as before to get the same bound as in (7.39). This yields the result. □

7.3 The Functional Order Parameter

We expect (say, under (7.12)) that the r.v. $\langle \sigma_1 \rangle$ will converge in law. In the case of the SK model, the limit law was the law of $\operatorname{th}(\beta z \sqrt{q} + h)$ where z is standard Gaussian and thus this law depended only on the simple parameter q. The most striking feature of the present model is that the limiting law is a complicated object, that no longer depends simply on a few parameters. It is therefore reasonable to think of this limiting law μ as being itself a kind of parameter (the correct value of which has to be found). This is what the physicists mean when they say "that the order parameter of the model is a function" because they identify a probability distribution μ on $\mathbb{R}$ with the function $t \to \mu([t, \infty))$.

The purpose of the present section is to find the correct value of this parameter, value that as in the case of the SK model will be given as the solution of a certain equation.

The constructions that we will perform may seem unmotivated at first sight because they are done with the foresight of knowing how the system behaves. Hopefully, the motivations will become apparent in the next section.

In a first step, given a probability measure μ on $[-1, 1]$, and an integer $r \ge 0$, we proceed to the construction of a probability distribution $T_r(\mu)$.

We remind the reader that the quantities $W_I(\boldsymbol{\sigma})$ are given by (7.5). We consider an independent sequence $(\theta_s)_{s\le r}$ of random functions on $\{-1,1\}^p$, with the same distribution as θ. We define the random function $\mathcal{E}_r$ on $\{-1,1\}^{r(p-1)+1}$ by

$$\begin{aligned}\mathcal{E}_r &= \mathcal{E}(\sigma_1,\ldots,\sigma_{r(p-1)},\varepsilon)\\ &= \exp\sum_{1\le s\le r}\theta_s(\sigma_{(s-1)(p-1)+1},\ldots,\sigma_{s(p-1)},\varepsilon)\ .\end{aligned}\tag{7.40}$$

We denote Av the average over $\varepsilon=\pm 1$, so that $\operatorname{Av}\varepsilon\,\mathcal{E}_r$ and $\operatorname{Av}\mathcal{E}_r$ are functions on $\{-1,1\}^{r(p-1)}$.

Given a sequence $\mathbf{x}=(x_1,\ldots,x_{r(p-1)})$ with $|x_i|\le 1$, and a function f on $\{-1,1\}^{r(p-1)}$, we denote by $\langle f\rangle_{\mathbf{x}}$ the average of f with respect to the product measure ν on $\{-1,1\}^{r(p-1)}$ such that $\int\sigma_i\,\mathrm{d}\nu(\boldsymbol{\eta})=x_i$, where $\boldsymbol{\eta}=(\sigma_1,\ldots,\sigma_{r(p-1)})$. Here the dependence on r is kept implicit, and the value of r will be clear from the context.

When $r\ge 1$ we define $T_r(\mu)$ as the law of the r.v.

$$\frac{\langle\operatorname{Av}\varepsilon\,\mathcal{E}_r\rangle_{\mathbf{X}}}{\langle\operatorname{Av}\mathcal{E}_r\rangle_{\mathbf{X}}}\tag{7.41}$$

when the sequence $\mathbf{X}=(X_1,\ldots,X_{r(p-1)})$ is i.i.d. of law μ, independent of the randomness in $\mathcal{E}_r$. When $r=0$, we define $T_r(\mu)$ as the probability concentrated at 0.

Finally, we define

$$T(\mu)=\sum_{r\ge 0}\frac{c^r\,e^{-c}}{r!}\,T_r(\mu)\tag{7.42}$$

where $c=\gamma/(p-1)!$.

Theorem 7.3.1. *Under Condition (7.12) there exists a unique probability distribution μ on $[-1,1]$ such that $\mu=T(\mu)$.*

The proof will consist of showing that for a certain distance on the set of probability measures on $[-1,1]$, T is a contraction. The so-called Monge-Kantorovich transportation-cost distance is ideally adapted to this purpose. Given a compact metric space $(\mathcal{X},d)$, and two probability measures μ_1,μ_2 on $\mathcal{X}$, their transportation-cost distance is defined as

$$d(\mu_1,\mu_2)=\inf\mathsf{E}\,d(X_1,X_2)\ ,\tag{7.43}$$

where the infimum is taken over all pairs (X_1,X_2) of r.v. such that the law of X_j is μ_j $(j=1,2)$. Equivalently,

$$d(\mu_1,\mu_2)=\inf\int d(x_1,x_2)\,\mathrm{d}\xi(x_1,x_2)\ ,$$

where the infimum is over all probability measures ξ on $\mathcal{X}^2$ with marginals μ_1 and μ_2 respectively. It is not immediately clear that the formula (7.43) defines a distance. This is however obvious due to the (fundamental) "duality formula"

$$d(\mu_1, \mu_2) = \sup \left(\int f(x) \, \mathrm{d}\mu_1(x) - \int f(x) \, \mathrm{d}\mu_2(x) \right) \tag{7.44}$$

where the supremum is taken over all functions f from $\mathcal{X}$ to $\mathbb{R}$ with Lipschitz constant 1, i.e. that satisfy $|f(x) - f(y)| \leq d(x, y)$ for all x, y in $\mathcal{X}$. The formula (7.44) is a simple consequence of the Hahn-Banach theorem. We will not use it in this chapter in any essential way, so we refer the reader to Lemma 8.4.3 for the complete proof of a similar result. It is also clear from (7.44) that the distance d defines the weak topology on the set of probability measures on $\mathcal{X}$ (so that this metric space is complete).

For completeness, we prove two elementary facts concerning d.

Lemma 7.3.2. *Consider numbers $(a_r)_{r\geq 1}$, $a_r \geq 0$, $\sum_{r\geq 1} a_r = 1$, and probability measures μ_r, ν_r on $\mathcal{X}$. Then*

$$d\left(\sum_{r\geq 1} a_r \, \mu_r, \sum_{r\geq 1} a_r \, \nu_r \right) \leq \sum_{r\geq 1} a_r \, d(\mu_r, \nu_r) \, .$$

Proof. If for each $r \geq 1$ the probability measure ξ_r on $\mathcal{X}^2$ has marginals μ_r and ν_r, then $\xi = \sum_{r\geq 1} a_r \, \xi_r$ is a probability measure on $\mathcal{X}^2$ with marginals $\sum_{r\geq 1} a_r \, \mu_r$ and $\sum_{r\geq 1} a_r \, \nu_r$ respectively. Thus

$$\begin{aligned} d\left(\sum_{r\geq 1} a_r \, \mu_r, \sum_{r\geq 1} a_r \, \nu_r \right) &\leq \int d(x_1, x_2) \, \mathrm{d}\xi(x_1, x_2) \\ &= \sum_{r\geq 1} a_r \int d(x_1, x_2) \, \mathrm{d}\xi_r(x_1, x_2) \, . \end{aligned}$$

Taking the infimum over ξ_r, $r \geq 1$, yields the result. □

Lemma 7.3.3. *Consider numbers $(a_r)_{r\geq 1}$, $(b_r)_{r\geq 1}$, $a_r, b_r \geq 0$, $\sum_{r\geq 1} a_r = \sum_{r\geq 1} b_r = 1$, and probability measures $(\mu_r)_{r\geq 1}$ on $\mathcal{X}$. Then*

$$d\left(\sum_{r\geq 1} a_r \, \mu_r, \sum_{r\geq 1} b_r \, \mu_r \right) \leq \Delta \sum_{r\geq 1} |a_r - b_r|$$

where Δ is the diameter of $\mathcal{X}$.

Proof. Let

$$c_r = \min(a_r, b_r) \geq a_r - |a_r - b_r| \; ,$$

so that

$$\sum_{r \geq 1} c_r \geq 1 - \sum_{r \geq 1} |a_r - b_r| \; .$$

Consider the positive measure $\xi = \sum_{r \geq 1} c_r \, \mu_r$, so

$$\sum_{r \geq 1} a_r \, \mu_r = \xi + \xi_1 \; , \; \sum_{r \geq 1} b_r \, \mu_r = \xi + \xi_2$$

where ξ_1, ξ_2 are positive measures of mass

$$c = 1 - \sum_{r \geq 1} c_r \leq \sum_{r \geq 1} |a_r - b_r| \; .$$

We then write $\xi + \xi_1 = (1 - c)\, \xi' + c\, \xi_1'$, $\xi + \xi_2 = (1 - c)\, \xi' + c\, \xi_2'$ where ξ', ξ_1', ξ_2' are probabilities, and the result follows from Lemma 7.3.2. □

We now turn to the study of T. Let us recall the notation $\langle \cdot \rangle_{\mathbf{x}}$ introduced just before (7.41).

Lemma 7.3.4. *For a function f on $\{-1, 1\}^{r(p-1)}$, we have*

$$\frac{\partial}{\partial x_i} \langle f \rangle_{\mathbf{x}} = \langle \Delta_i \, f \rangle_{\mathbf{x}} \tag{7.45}$$

where $\Delta_i \, f(\boldsymbol{\eta}) = (f(\boldsymbol{\eta}_i^+) - f(\boldsymbol{\eta}_i^-))/2$, and where $\boldsymbol{\eta}_i^+$ (resp. $\boldsymbol{\eta}_i^-$) is obtained by replacing the i-th coordinate of $\boldsymbol{\eta}$ by 1 (resp. -1).

Proof. The measure ν_x on $\{-1, 1\}$ such that $\int \eta \, \mathrm{d}\nu_x(\eta) = x$ gives mass $(1 + x)/2$ to 1 and mass $(1 - x)/2$ to -1, so that for a function f on $\{-1, 1\}$

$$\langle f \rangle_x = \int f(\eta) \, \mathrm{d}\nu_x(\eta) = \frac{1}{2}\,(f(1) + f(-1)) + \frac{x}{2}\,(f(1) - f(-1)) \; .$$

Thus

$$\frac{\mathrm{d}}{\mathrm{d}x} \langle f \rangle_x = \frac{1}{2}\,(f(1) - f(-1)) = \left\langle \frac{1}{2}\,(f(1) - f(-1)) \right\rangle_x \; . \tag{7.46}$$

Since $\langle \cdot \rangle_{\mathbf{x}}$ is a product measure, integrating first in the coordinates different from i, (7.45) follows from (7.46). □

Lemma 7.3.5. *If $\mathcal{E}_r$ is as in (7.40), if $1 \leq s \leq r$ and if $(s - 1)(p - 1) < i \leq s(p - 1)$, then*

$$\left| \frac{\partial}{\partial x_i} \frac{\langle \mathrm{Av}\, \varepsilon \, \mathcal{E}_r \rangle_{\mathbf{x}}}{\langle \mathrm{Av}\, \mathcal{E}_r \rangle_{\mathbf{x}}} \right| \leq 2 S_s \exp 2 S_s$$

where $S_s = \sup |\theta_s|$.

Proof. By Lemma 7.3.4 we have

$$\frac{\partial}{\partial x_i}\frac{\langle \mathrm{Av}\,\varepsilon\,\mathcal{E}_r\rangle_{\mathbf{x}}}{\langle \mathrm{Av}\,\mathcal{E}_r\rangle_{\mathbf{x}}}=\frac{\langle \Delta_i(\mathrm{Av}\,\varepsilon\,\mathcal{E}_r)\rangle_{\mathbf{x}}}{\langle \mathrm{Av}\,\mathcal{E}_r\rangle_{\mathbf{x}}}-\frac{\langle \mathrm{Av}\,\varepsilon\,\mathcal{E}_r\rangle_{\mathbf{x}}\langle \Delta_i\mathrm{Av}\,\mathcal{E}_r\rangle_{\mathbf{x}}}{\langle \mathrm{Av}\,\mathcal{E}_r\rangle_{\mathbf{x}}^2}\;. \tag{7.47}$$

Now

$$|\Delta_i(\mathrm{Av}\,\varepsilon\,\mathcal{E}_r)|=|\mathrm{Av}\,(\varepsilon\,\Delta_i\,\mathcal{E}_r)|\le \mathrm{Av}\,|\Delta_i\,\mathcal{E}_r|\;.$$

We write $\mathcal{E}_r=\mathcal{E}'\,\mathcal{E}''$, where $\mathcal{E}'=\exp\theta_s(\sigma_{(s-1)(p-1)+1},\dots,\sigma_{s(p-1)},\varepsilon)$, and where $\mathcal{E}''$ does not depend on σ_i. Thus, using that $|e^x-e^y|\le|x-y|\,e^a\le 2ae^a$ for $|x|,|y|\le a$, we get

$$|\Delta_i\,\mathcal{E}_r|=|\mathcal{E}''\,\Delta_i\,\mathcal{E}'|\le(S_s\exp S_s)\,\mathcal{E}''\le(S_s\exp 2S_s)\,\mathcal{E}_r$$

and thus

$$\left|\frac{\langle \Delta_i(\mathrm{Av}\,\varepsilon\,\mathcal{E}_r)\rangle_{\mathbf{x}}}{\langle \mathrm{Av}\,\mathcal{E}_r\rangle_{\mathbf{x}}}\right|\le S_s\exp 2S_s\;.$$

The last term of (7.47) is bounded similarly. □

Proposition 7.3.6. *For any two probability measures μ_1,μ_2 on $[-1,1]$ we have*

$$d(T_r(\mu_1),T_r(\mu_2))\le 2r(p-1)\,d(\mu_1,\mu_2)\,\mathsf{E}\,S\exp 2S\;. \tag{7.48}$$

Proof. If $\mathbf{y}$ is another element of $[-1,1]^{r(p-1)}$, Lemma 7.3.5 implies that

$$\left|\frac{\langle \mathrm{Av}\,\varepsilon\,\mathcal{E}_r\rangle_{\mathbf{x}}}{\langle \mathrm{Av}\,\mathcal{E}_r\rangle_{\mathbf{x}}}-\frac{\langle \mathrm{Av}\,\varepsilon\,\mathcal{E}_r\rangle_{\mathbf{y}}}{\langle \mathrm{Av}\,\mathcal{E}_r\rangle_{\mathbf{y}}}\right|$$
$$\le 2\sum_{s\le r}S_s\exp 2S_s\sum_{(s-1)(p-1)<i\le s(p-1)}|x_i-y_i|\;. \tag{7.49}$$

Consider now a pair (X,Y) of r.v., such that the law of X (resp. Y) is μ_1 (resp. μ_2). Consider independent copies $(X_i,Y_i)_{i\le r(p-1)}$ of this pair, that are independent of the functions θ_s, $s\le r$, and let $\mathbf{X}=(X_i)_{i\le r(p-1)}$, $\mathbf{Y}=(Y_i)_{i\le r(p-1)}$. Then

$$\frac{\langle \mathrm{Av}\,\varepsilon\,\mathcal{E}_r\rangle_{\mathbf{X}}}{\langle \mathrm{Av}\,\mathcal{E}_r\rangle_{\mathbf{X}}}\quad\text{and}\quad\frac{\langle \mathrm{Av}\,\varepsilon\,\mathcal{E}_r\rangle_{\mathbf{Y}}}{\langle \mathrm{Av}\,\mathcal{E}_r\rangle_{\mathbf{Y}}}$$

have respectively laws $T_r(\mu_1)$ and $T_r(\mu_2)$. Using (7.49) for $\mathbf{x}=\mathbf{X}$, $\mathbf{y}=\mathbf{Y}$ and taking expectation we get

$$d(T_r(\mu_1),T_r(\mu_2))\le 2r(p-1)\,\mathsf{E}\,|X-Y|\,\mathsf{E}\,S\exp 2S$$

and (7.48) follows by taking the infimum over the choices of the pair (X,Y). □

Proof of Theorem 7.3.1. Using Proposition 7.3.6 and Lemma 7.3.2, we get

$$d(T(\mu_1),T(\mu_2)) \le \left(2(p-1)\sum_{r\ge 1} r\,\frac{c^r e^{-c}}{r!}\right) \mathsf{E}\, S \exp 2S\, d(\mu_1,\mu_2)$$
$$= (2(p-1)\,c\,\mathsf{E}\, S\exp 2S)\, d(\mu_1,\mu_2)\,.$$

Since $c = \gamma/(p-1)!$, we see that if

$$2\,\frac{\gamma}{(p-2)!}\,\mathsf{E}\, S\exp 2S \le \frac{1}{2} \tag{7.50}$$

then we get

$$d(T(\mu_1),T(\mu_2)) \le \frac{1}{2}\, d(\mu_1,\mu_2)\,, \tag{7.51}$$

and this proves the Theorem. □

The measure μ depends on γ through $c = \gamma/(p-1)!$. Let us indicate this by denoting μ_γ the solution of the equation $\mu = T(\mu)$.

Lemma 7.3.7. *If γ, γ' satisfy (7.50) we have*

$$d(\mu_\gamma,\mu_{\gamma'}) \le 4\frac{|\gamma-\gamma'|}{(p-1)!}\exp\left(\frac{|\gamma-\gamma'|}{(p-1)!}\right)\,.$$

Proof. Since $\mu_\gamma = T_\gamma(\mu_\gamma)$, $\mu_{\gamma'} = T_{\gamma'}(\mu_{\gamma'})$, we have

$$d(\mu_\gamma,\mu_{\gamma'}) \le d(T_\gamma(\mu_\gamma),T_\gamma(\mu_{\gamma'})) + d(T_\gamma(\mu_{\gamma'}),T_{\gamma'}(\mu_{\gamma'}))$$
$$\le \frac{1}{2}\, d(\mu_\gamma,\mu_{\gamma'}) + d(T_\gamma(\mu_{\gamma'}),T_{\gamma'}(\mu_{\gamma'}))\,,$$

using (7.51). Using Lemma 7.3.3, we have

$$d(T_\gamma(\mu_{\gamma'}),T_{\gamma'}(\mu_{\gamma'})) \le 2\sum_{r\ge 0}\left|\frac{c^r\, e^{-c}}{r!} - \frac{c'^r\, e^{-c'}}{r!}\right| \tag{7.52}$$

where $c = \gamma/(p-1)!$ and $c' = \gamma'/(p-1)!$ Assuming without loss of generality that $\gamma' \le \gamma$, for $c' \le x \le c$ we have

$$\frac{\mathrm{d}}{\mathrm{d}x}(x^r\, e^{-x}) \le r\, x^{r-1}\, e^{-x} \le r\, c^{r-1}\, e^{-c'}$$

so that

$$\frac{1}{r!}\,|c^r\, e^{-c} - c'^r\, e^{-c'}| \le |c-c'|\,\frac{c^{r-1}}{(r-1)!}\, e^{-c'}$$

and by summation we get

$$\sum \frac{1}{r!}\,|c^r\, e^{-c} - c'^r\, e^{-c'}| \le |c-c'|\, e^{|c-c'|}\,. \tag{7.53}$$

□

7.4 The Replica-Symmetric Solution

The first goal of this section is to prove the following

Theorem 7.4.1. *There exists a number $K_1(p)$ such that if*

$$K_1(p)(\gamma+\gamma^2)\,\mathsf{E}S\exp 4S(1+\mathsf{E}\exp 4S)\le 1 \tag{7.54}$$

then given any integer k we can find i.i.d. r.v. $z_1,\ldots,z_k$ of law μ_γ such that

$$\mathsf{E}\sum_{i\le k}|\langle\sigma_i\rangle - z_i|\le\frac{K}{N}\,. \tag{7.55}$$

We will then compute $\lim_{N\to\infty} p_N(\gamma)=\lim_{N\to\infty}\frac{1}{N}\mathsf{E}\log Z_N(\gamma)$.

It will be useful to consider the metric space $[-1,1]^k$, provided with the distance d given by

$$d((x_i)_{i\le k},(y_i)_{i\le k})=\sum_{i\le k}|x_i-y_i|\,. \tag{7.56}$$

Its diameter is $2k$. The transportation-cost distance on the space of probability measures on $[-1,1]^k$ that is induced by (7.56) will also be denoted by d. We define

$$D(N,k,\gamma_0)=\sup_{\gamma\le\gamma_0} d(\mathcal{L}(\langle\sigma_1\rangle,\ldots,\langle\sigma_k\rangle),\mu_\gamma^{\otimes k}) \tag{7.57}$$

where $\mathcal{L}(X)$ denotes the law of the r.v. X. We observe that $D(N,k,\gamma_0)\le 2k$.

The content of Theorem 7.4.1 is that under (7.54) we have $D(N,k,\gamma_0)\le K/N$ for each k. This inequality will be proved by obtaining a suitable induction relation on the quantities $D(N,k,\gamma_0)$.

First, we observe that, by symmetry,

$$\mathcal{L}(\langle\sigma_1\rangle,\ldots,\langle\sigma_k\rangle)=\mathcal{L}(\langle\sigma_{N-k+1}\rangle,\ldots,\langle\sigma_N\rangle)\,.$$

The starting point of the proof of Theorem 7.4.1 is a formula similar to (7.18), but where we remove the last k coordinates rather than the last one. Writing now $\boldsymbol{\rho}=(\sigma_1,\ldots,\sigma_{N-k})$, we consider the Hamiltonian

$$-H_{N-k}(\boldsymbol{\rho})=\sum\gamma_I\,W_I(\boldsymbol{\sigma})\,, \tag{7.58}$$

where the summation is restricted to $I\in[N-k]^p$. It is the Hamiltonian of an $(N-k)$-spin system, except that we have replaced γ by $\gamma_-=\gamma(1-k/N)^{p-1}\le\gamma$. In particular $\gamma_-\le\gamma_0$ whenever $\gamma\le\gamma_0$. Let us denote again by $\langle\cdot\rangle_-$ an average for the Gibbs measure with Hamiltonian (7.58). (The value of k will be clear from the context.) Given a function f on Σ_N, we then have

$$\langle f \rangle = \frac{\langle \mathrm{Av}\, f\, \mathcal{E} \rangle_-}{\langle \mathrm{Av}\, \mathcal{E} \rangle_-} \,. \tag{7.59}$$

Here,

$$\mathcal{E} = \exp \sum \gamma_I \, W_I(\boldsymbol{\sigma}) \,, \tag{7.60}$$

where the summation is over the sets I of $[N]^p$ for which $I \cap \{N-k+1, \ldots, N\} \neq \emptyset$, and Av means average over $\sigma_{N-k+1}, \ldots, \sigma_N = \pm 1$. Thus

$$\mathcal{L}(\langle \sigma_{N-k+1} \rangle, \ldots, \langle \sigma_N \rangle) = \mathcal{L}\left(\frac{\langle \mathrm{Av}\, \sigma_{N-k+1}\, \mathcal{E} \rangle_-}{\langle \mathrm{Av}\, \mathcal{E} \rangle_-}, \ldots, \frac{\langle \mathrm{Av}\, \sigma_N\, \mathcal{E} \rangle_-}{\langle \mathrm{Av}\, \mathcal{E} \rangle_-} \right) .$$

Let us observe that there are three sources of randomness in the right-hand side of (7.59): the randomness associated to the $(N-k)$-spin system of Hamiltonian (7.58); the randomness associated to the γ_I occurring in (7.60); and the randomness of the functions $W_I(\boldsymbol{\sigma})$ occurring there. These sources are independent of each other. We denote by $\mathcal{L}_-$ the law of a r.v. given the randomness of the γ_I and the W_I in (7.60); and by $\mathcal{L}_0$ the law of a r.v. given only the randomness of the γ_I.

Let us consider the following (exceptional) events

$$\Omega_1 : \exists I \in [N]^p \,, \ \mathrm{card}\, I \cap \{N-k+1, \ldots, N\} \geq 2 \,, \ \gamma_I = 1 \tag{7.61}$$

$$\begin{aligned} \Omega_2 : \exists I_1, I_2 \in [N]^p \,, \ & I_1 \cap \{N-k+1, \ldots, N\} \neq \emptyset \,, \\ & I_2 \cap \{N-k+1, \ldots, N\} \neq \emptyset \,, \ I_1 \neq I_2 \,, \ I_1 \cap I_2 \cap \{1, \ldots, N-k\} \neq \emptyset \,, \\ & \gamma_{I_1} = \gamma_{I_2} = 1 \,. \end{aligned} \tag{7.62}$$

Since the number of possible sets I in (7.61) is at most $k^2 N^{p-2}$, we have $\mathsf{P}(\Omega_1) \leq \gamma k^2/N$. Since the number of possible pairs I_1, I_2 in (7.62) is at most $k^2 N^{2p-3}$, we have $\mathsf{P}(\Omega_2) \leq \gamma^2 k^2/N$, so that if $\Omega = \Omega_1 \cup \Omega_2$ we have

$$\mathsf{P}(\Omega) \leq \frac{(\gamma + \gamma^2)\, k^2}{N} \,. \tag{7.63}$$

Given a sequence $\mathbf{x} = (x_1, \ldots, x_{N-k})$, $|x_i| \leq 1$ and a function f on Σ_{N-k}, we denote by $\langle f \rangle_{\mathbf{x}}$ the average of f with respect to the product measure ν on Σ_{N-k} such that $\int \sigma_i \, \mathrm{d}\nu(\boldsymbol{\rho}) = x_i$ for $0 \leq i \leq N-k$.

We now start a sequence of lemmas that aim at obtaining the desired induction relations among the quantities $D(N, k, \gamma_0)$.

Lemma 7.4.2. *Consider the sequence*

$$\mathbf{Y} = (\langle \sigma_1 \rangle_-, \ldots, \langle \sigma_{N-k} \rangle_-) \,.$$

Set

$$u_j = \frac{\langle \mathrm{Av}\, \sigma_{N-k+j}\, \mathcal{E} \rangle_-}{\langle \mathrm{Av}\, \mathcal{E} \rangle_-} \,; \ v_j = \frac{\langle \mathrm{Av}\, \sigma_{N-k+j}\, \mathcal{E} \rangle_{\mathbf{Y}}}{\langle \mathrm{Av}\, \mathcal{E} \rangle_{\mathbf{Y}}} \,.$$

Then if Ω does not occur we have

$$d(\mathcal{L}_0(u_1, \ldots, u_k), \mathcal{L}_0(v_1, \ldots, v_k)) \le \frac{kr^2\, K(p, \gamma_0)}{N-k+1} \mathsf{E} \exp 2S \tag{7.64}$$

where

$$r = \sum \gamma_I \mathbf{1}_{\{I \cap \{N-k+1, \ldots, N\} \neq \emptyset\}} \,. \tag{7.65}$$

Proof. If we denote by E_- expectation taken only in the randomness of the $(N-k)$-spin system, we have

$$d(\mathcal{L}_-(u_1, \ldots, u_k), \mathcal{L}_-(v_1, \ldots, v_k)) \le \sum_{j \le k} \mathsf{E}_- |u_j - v_j| \,. \tag{7.66}$$

Let us denote by $I_1, \ldots, I_r$ the sets for which $\gamma_I = 1$, $I \cap \{N-k+1, \ldots, N\} \neq \emptyset$, and for $s \le r$ let us set $S_s = \sup_{\boldsymbol{\sigma}} |W_{I_s}(\boldsymbol{\sigma})|$. Then, if Ω does not occur, the sets $I_s \cap \{1, \ldots, N-k\}$, $s = 1, \ldots, r$, are all disjoint. There is nothing to change to the proof of Proposition 7.2.7 to obtain that

$$\mathsf{E}_- |u_j - v_j| \le \frac{r\, K(p, \gamma_0)}{N-k} \sum_{s \le r} \exp 2S_s \,.$$

Thus, by (7.66),

$$d(\mathcal{L}_-(u_1, \ldots, u_k), \mathcal{L}_-(v_1, \ldots, v_k)) \le \frac{kr\, K(p, \gamma_0)}{N-k} \sum_{s \le r} \exp 2S_s \,.$$

To obtain (7.64), we simply take expectation in the r.v. W_{I_s}, $s \le r$, and we use a "continuous version" of Lemma 7.3.2 (that follows from this lemma by straightforward approximation). □

Lemma 7.4.3. *Consider an independent sequence $\mathbf{X} = (X_1, \ldots, X_{N-k})$ where each X_ℓ is distributed like μ_{γ_-}. Set*

$$w_j = \frac{\langle \mathrm{Av}\, \sigma_{N-k+j}\, \mathcal{E} \rangle_{\mathbf{X}}}{\langle \mathrm{Av}\, \mathcal{E} \rangle_{\mathbf{X}}} \,.$$

Then, if r is as in (7.65), outside Ω we have

$$d(\mathcal{L}_0(v_1, \ldots, v_k), \mathcal{L}_0(w_1, \ldots, w_k)) \le 4\, D(N-k, r(p-1), \gamma_0)\, \mathsf{E}\, S \exp 2S \,. \tag{7.67}$$

Proof. Consider $i \le N-k$ and the unique integer $s \le r$ such that $i \in I_s$. There is nothing to change to the proof of Lemma 7.3.5 to see that (with obvious notation)

$$\left|\frac{\partial}{\partial x_i}\frac{\langle \mathrm{Av}\,\sigma_{N-k+j}\,\mathcal{E}\rangle_{\mathbf{x}}}{\langle \mathrm{Av}\,\mathcal{E}\rangle_{\mathbf{x}}}\right| \le 2S_s \exp 2S_s\,. \tag{7.68}$$

An important observation is that $\langle \mathrm{Av}\,\sigma_{N-k+j}\,\mathcal{E}\rangle_{\mathbf{x}}/\langle \mathrm{Av}\,\mathcal{E}\rangle_{\mathbf{x}}$ can depend on x_i only if $N-k+j \in I_s$. To see this, we set

$$\mathcal{E}_j = \exp \sum W_{I_\ell}(\boldsymbol{\sigma})\;, \tag{7.69}$$

where the summation is over I_ℓ such that $N-k+j \in I_\ell$, $\ell \le r$. Thus $\mathcal{E} = \mathcal{E}_j\,\mathcal{E}'_j$, where $\mathcal{E}'_j$ is defined as in (7.69), but with a summation over I_ℓ such that $N-k+j \notin I_\ell$, $\ell \le r$, and, using that Ω does not occur we see that

$$\mathrm{Av}\,\sigma_{N-k+j}\,\mathcal{E} = (\mathrm{Av}\,\sigma_{N-k+j}\,\mathcal{E}_j)(\mathrm{Av}\,\mathcal{E}'_j) \tag{7.70}$$

$$\mathrm{Av}\,\mathcal{E} = (\mathrm{Av}\,\mathcal{E}_j)(\mathrm{Av}\,\mathcal{E}'_j)\;. \tag{7.71}$$

In both (7.70) and (7.71) the two factors on the right depend on disjoint sets of coordinates. Since $\langle\cdot\rangle_{\mathbf{x}}$ is a product measure, we get

$$\langle \mathrm{Av}\,\sigma_{N-k+j}\,\mathcal{E}\rangle_{\mathbf{x}} = \langle \mathrm{Av}\,\sigma_{N-k+j}\,\mathcal{E}_j\rangle_{\mathbf{x}}\,\langle \mathrm{Av}\,\mathcal{E}'_j\rangle_{\mathbf{x}}$$

and similarly with (7.71), so that

$$\frac{\langle \mathrm{Av}\,\sigma_{N-k+j}\,\mathcal{E}\rangle_{\mathbf{x}}}{\langle \mathrm{Av}\,\mathcal{E}\rangle_{\mathbf{x}}} = \frac{\langle \mathrm{Av}\,\sigma_{N-k+j}\,\mathcal{E}_j\rangle_{\mathbf{x}}}{\langle \mathrm{Av}\,\mathcal{E}_j\rangle_{\mathbf{x}}}$$

can depend on x_i only if $\mathcal{E}_j$ depends on x_i, i.e. only if $N-k+j \in I_s$. Thus, if $\mathbf{X} = (X_i)_{i\le N-k}$, then by (7.68)

$$\left|\frac{\langle \mathrm{Av}\,\sigma_{N-k+j}\,\mathcal{E}\rangle_{\mathbf{X}}}{\langle \mathrm{Av}\,\mathcal{E}\rangle_{\mathbf{X}}} - v_j\right| \le \sum \left(\sum_{i\in I_s} |X_i - \langle\sigma_i\rangle_-|\right) 2S_s \exp 2S_s\;, \tag{7.72}$$

where the first summation is over those $s \le r$ for which $N-k+j \in I_s$. Now, by definition of $D(N-k, r(p-1), \gamma_0)$, we can find the sequence $\mathbf{X}$ such that

$$\mathsf{E}_- \sum |X_i - \langle\sigma_i\rangle_-| \le 2D(N-k, r(p-1), \gamma_0)\;,$$

where the summation is over $i \in \bigcup_{s\le r} I_s$, $i \le N-k$. Using symmetry among coordinates, we can assume that

$$\forall\, i \in \bigcup_{s\le r} I_s\;,\; i \le N-k\;,\; \mathsf{E}_-|X_i - \langle\sigma_i\rangle_-| \le \frac{2}{r(p-1)}\,D(N-k, r(p-1), \gamma_0)\;.$$

Thus, taking expectation in (7.72) we get

$$\mathsf{E}_-\left|\frac{\langle \mathrm{Av}\,\sigma_{N-k+j}\,\mathcal{E}\rangle_{\mathbf{X}}}{\langle \mathrm{Av}\,\mathcal{E}\rangle_{\mathbf{X}}} - v_j\right| \le \frac{2}{r}\,D(N-k, r(p-1), \gamma_0)\sum 2S_s \exp 2S_s\;,$$

where the summation is over those $s \le r$ for which $N - k + j \in I_s$. Thus

$$d(\mathcal{L}_-(v_1, \dots, v_k), \mathcal{L}_-(w_1, \dots, w_k)) \le \frac{4}{r} D(N-k, r(p-1), \gamma_0) \sum_{s \le r} S_s \exp 2S_s \ .$$

To obtain (7.67) we simply average over the randomness of the W_{I_s}, $s \le r$, as in Lemma 7.4.2. □

For $1 \le j \le k$ we set

$$U_j = \{I \in [N]^p \, ; \; I \cap \{N-k+1, \dots, N\} = \{N-k+j\}\}$$

$$r(j) = \operatorname{card} \{I \in U_j \, ; \; \gamma_I = 1\} \ .$$

Lemma 7.4.4. *Outside Ω we have*

$$\mathcal{L}_0(w_1, \dots, w_k) = T_{r(1)}(\mu_{\gamma_-}) \otimes \cdots \otimes T_{r(k)}(\mu_{\gamma_-}) \ .$$

Proof. This follows from the definition of the operator T_r, and the fact that Ω does not occur. □

Given integers $r_1, \dots, r_k$, we set

$$a(r_1, \dots, r_k) = \mathsf{P}\,(\forall j \le k \, , \; r(j) = r_j) \ .$$

(The dependence on N is kept implicit.)

Proposition 7.4.5. *We have*

$$d\left(\mathcal{L}(u_1, \dots, u_k), \sum_{r_1, \dots, r_k \ge 0} a(r_1, \dots, r_k) \, T_{r_1}(\mu_{\gamma_-}) \otimes \cdots \otimes T_{r_k}(\mu_{\gamma_-})\right)$$
$$\le \frac{K(p, \gamma_0) \, k^3}{N-k+1} \mathsf{E} \exp 2S + 4 \, \mathsf{E}\,(S \exp 2S) \, \mathsf{E} \, D(N-k, r(p-1), \gamma_0) \ , \qquad (7.73)$$

where, in the last term, the expectation refers to the r.v. r of (7.65).

Proof. Combining Lemmas 7.4.2 to 7.4.4, we see that outside Ω we have

$$d(\mathcal{L}_0(u_1, \dots, u_k), T_{r(1)}(\mu_{\gamma_-}) \otimes \cdots \otimes T_{r(k)}(\mu_{\gamma_-}))$$
$$\le \frac{k r^2 K(p, \gamma_0)}{N-k} \mathsf{E} \exp 2S + 4 \, \mathsf{E}\,(S \exp 2S) \, D(N-k, r(p-1), \gamma_0) \ , \qquad (7.74)$$

while, in Ω, the left-hand side is bounded by $2k$. Taking expectation, using Lemma 7.3.2, (7.63), and the fact that $\mathsf{E}\, r^2 \le L k^2 (\gamma^2 + \gamma)$ yield the result. □

We set $c_- = \gamma_- / (p-1)!$, $c(r) = c_-^r \, e^{-c_-} / r!$.

Lemma 7.4.6. *We have*

$$\sum_{r_1,\ldots,r_k} |a(r_1,\ldots,r_k) - c(r_1)\ldots c(r_k)| \le \frac{K(p,\gamma_0)\,k}{N} . \tag{7.75}$$

Proof. Let us write

$$a(r) = \mathsf{P}\left(\sum\{\gamma_I, I \in U_j\} = r\right) ,$$

so that since the sets U_j, $j = 1,\ldots,k$ are disjoint, we have

$$a(r_1,\ldots,r_k) = a(r_1)\ldots a(r_k) .$$

We observe that $\sum_{r\ge 0} a(r) = \sum_{r\ge 0} c(r) = 1$, so that

$$\begin{aligned}\sum_{r_1,\ldots,r_k} &|a(r_1)\ldots a(r_\ell)\,c(r_{\ell+1})\ldots c(r_k) \\ &- a(r_1)\ldots a(r_\ell)\,a(r_{\ell+1})\,c(r_{\ell+2})\ldots c(r_k)| \\ &\le \sum_r |c(r) - a(r)|\end{aligned} \tag{7.76}$$

and thus the left-hand side of (7.75) is at most $k\sum_r |c(r) - a(r)|$.

Let

$$M = \operatorname{card} U_j = \binom{N-k}{p-1} = \frac{(N-k)\ldots(N-k-p+2)}{(p-1)!} \tag{7.77}$$

so that

$$a(r) = \left(\frac{\gamma}{N^{p-1}}\right)^r \left(1 - \frac{\gamma}{N^{p-1}}\right)^{M-r} \binom{M}{r} .$$

Now

$$\binom{M}{r} = \frac{M(M-1)\ldots(M-r+1)}{r!} = \frac{M^r}{r!}\left(1 - \frac{1}{M}\right)\cdots\left(1 - \frac{r-1}{M}\right) ,$$

and since for $x, y > 0$ we have $(1-x)(1-y) \ge 1 - (x+y)$, we get

$$\frac{M^r}{r!}\left(1 - \frac{r(r-1)}{2M}\right) \le \binom{M}{r} \le \frac{M^r}{r!} .$$

Since $1 - x \le e^{-x}$, setting $d = \gamma M/N^{p-1}$, we get

$$a(r) \le \frac{d^r}{r!} e^{-d} \left(1 - \frac{\gamma}{N^{p-1}}\right)^{-r} .$$

Also, since $1 + x \le e^x$, we have $e^{-x}(1+x) \le 1$, so that for $x < 1$ we have $e^{-x}(1-x^2) \le 1 - x$. Thus

$$\left(1-\frac{\gamma}{N^{p-1}}\right) \geq \exp\left(-\frac{\gamma}{N^{p-1}}\right)\left(1-\left(\frac{\gamma}{N^{p-1}}\right)^2\right)$$

and

$$\left(1-\frac{\gamma}{N^{p-1}}\right)^M \geq \exp\left(-\frac{M\gamma}{N^{p-1}}\right)\left(1-M\left(\frac{\gamma}{N^{p-1}}\right)^2\right) .$$

It follows that

$$a(r) \geq \frac{d^r}{r!}\, e^{-d}\left(1-\frac{r(r-1)}{2M}-\frac{d\gamma}{N^{p-1}}\right)$$

and thus

$$\left|a(r)-\frac{d^r}{r!}\, e^{-d}\right| \leq \frac{d^r\, e^{-d}}{r!}\left(\left(1-\frac{\gamma}{N^{p-1}}\right)^{-r}-1+\frac{r(r-1)}{2M}+\frac{d\gamma}{N^{p-1}}\right)$$

and, by summation,

$$\sum_{r\geq 0}\left|a(r)-\frac{d^r}{r!}\, e^{-d}\right| \leq \frac{d^2}{2M}+\frac{d\gamma}{N^{p-1}}+\exp\left(d\left(1-\frac{\gamma}{N^{p-1}}\right)^{-1}-d\right)-1 .$$

If $d^2/M \leq 1/4$, then $M(\gamma/N^{p-1})^2 \leq 1/4$, and in particular $\gamma/N^{p-1} \leq 1/2$. Thus

$$d\left(\left(1-\frac{\gamma}{N^{p-1}}\right)^{-1}-1\right) \leq \frac{2d\gamma}{N^{p-1}} = \frac{2d^2}{M} \leq \frac{1}{2}$$

and since $e^x - 1 \leq 2x$ for $x \leq 1/2$ we obtain

$$\sum_r \left|a(r)-\frac{d^r}{r!}\, e^{-d}\right| \leq \frac{8d^2}{M} . \tag{7.78}$$

This remains true if $d^2/M \geq 1/4$ because the left-hand side is at most 2.

Now

$$c_- = \frac{\gamma}{(p-1)!}\left(1-\frac{k}{N}\right)^{p-1} ;$$

$$d = \frac{\gamma}{(p-1)!}\left(1-\frac{k}{N}\right)\left(1-\frac{k+1}{N}\right)\ldots\left(1-\frac{k+p-2}{N}\right)$$

and thus

$$|c_- - d| \leq \frac{L\gamma p^2}{N(p-1)!} .$$

By (7.53) we get, using that $k \leq N$ so that $|c_- - d| \leq K(p,\gamma_0)$,

$$\sum_r \left|c(r)-\frac{d^r}{r!}\, e^{-d}\right| \leq \frac{K(p,\gamma_0)}{N} .$$

Combining with (7.78) yields the result, since $8d^2/M \leq K(p)/N$. □

Proposition 7.4.7. *We have*

$$D(N,k,\gamma_0) \le \frac{K_0(p,\gamma_0)\,k^2}{2(N-k+1)}\,\mathsf{E}\exp 2S + 4\,\mathsf{E}(S\exp 2S)\,\mathsf{E}\,D(N-k, r(p-1),\gamma_0) \tag{7.79}$$

where the last expectation is as in Proposition 7.4.5.

Proof. Since $\mu_{\gamma_-} = \sum_r c(r)\,T_r(\mu_{\gamma_-})$, we get

$$\sum_{r_1,\ldots,r_k} c(r_1)\ldots c(r_k)\,T_{r_1}(\mu_{\gamma_-})\otimes\cdots\otimes T_{r_k}(\mu_{\gamma_-}) = \mu_{\gamma_-}^{\otimes k}\,,$$

so that by Lemmas 7.3.3 and 7.4.6 we have

$$d\left(\sum_{r_1,\ldots,r_k} a(r_1,\ldots,r_k)\,T_{r_1}(\mu_{\gamma_-})\otimes\cdots\otimes T_{r_k}(\mu_{\gamma_-}),\mu_{\gamma_-}^{\otimes k}\right) \le \frac{K(p,\gamma_0)\,k^2}{N-k+1}\,\mathsf{E}\exp 2S\,. \tag{7.80}$$

Now $|\gamma-\gamma_-|\le \gamma k/N$, so that by Lemma 7.3.7,

$$d(\mu_\gamma,\mu_{\gamma_-}) \le k\,\frac{K(p,\gamma_0)}{N}$$

and thus

$$d(\mu_\gamma^{\otimes k},\mu_{\gamma_-}^{\otimes k}) \le k^2\,\frac{K(p,\gamma_0)}{N}\,.$$

Combining with (7.80) and (7.73) yields the result. □

Proof of Theorem 7.4.1. We set $A = \mathsf{E}\exp 2S$. We prove the existence of a number $K_1(p)$ such that under (7.54) we have

$$\forall\, N,k \qquad D(N,k,\gamma_0) \le 2K_0(p,\gamma_0)\,\frac{k^2}{N}\,A\,. \tag{7.81}$$

The proof of (7.81) goes by induction over N. For the induction step from $N-1$ to N, since $D(N,k,\gamma_0)\le 2k$, and since we can assume $K_0(p,\gamma_0)\ge 1$, we can assume $k^2\le N$, and hence $k\le N/2$. By (7.79) we get, since $N-k\ge N/2$,

$$D(N,k,\gamma_0) \le \frac{K_0(p,\gamma_0)\,k^2}{N}\,A + \frac{L}{N}\,\mathsf{E}(S\exp 4S)\,\mathsf{E}(2K_0(p,\gamma_0)(r(p-1))^2)\,.$$

Now, $\mathsf{E}\,r^2 \le K(p)(\gamma+\gamma^2)\,k^2$ so that

$$D(N,k,\gamma_0) \le \frac{K_0(p,\gamma_0)\,k^2}{N}\,(A+K(p)(\gamma_0+\gamma_0^2)\,\mathsf{E}\,(S\exp 4S)) \le 2K_0(p,\gamma_0)\,\frac{k^2}{N}\,A$$

under (7.54). □

We now turn to the computation of $p_N(\gamma)$. We consider i.i.d. copies $(\theta_s)_{s\geq 1}$ of the random function θ, that are independent of θ, and we recall the definition of $\mathcal{E}_r$ in (7.40). We set as usual $c = \gamma/(p-1)!$, and we consider the function

$$F(\gamma) = \log 2 - \frac{c(p-1)}{p} \int \mathsf{E} \log\langle \exp \theta(\sigma_1, \ldots, \sigma_p)\rangle_{\mathbf{x}} \, d\mu_\gamma(x_1) \ldots d\mu_\gamma(x_p)$$
$$+ \sum_{r\geq 1} \frac{c^r e^{-c}}{r!} \int \mathsf{E} \log\langle \mathrm{Av}\, \mathcal{E}_r\rangle_{\mathbf{x}} \, d\mu_\gamma(x_1) \ldots d\mu_\gamma(x_{r(p-1)}) \,. \tag{7.82}$$

Here Av means average over $\varepsilon = \pm 1$. Thus $\mathrm{Av}\, \mathcal{E}_r$ depends on $\sigma_1, \ldots, \sigma_{r(p-1)}$; $\langle \cdot \rangle_{\mathbf{x}}$ means integration with respect to the product measure ν such that $\int \sigma_i \, d\nu = x_i$; and of course μ_γ is as in Section 7.3. In particular $F(\gamma)$ is well defined under (7.54).

Theorem 7.4.8. *Under (7.54), and if $\gamma \leq \gamma_0$ we have*

$$|p_N(\gamma) - F(\gamma)| \leq \frac{K}{N} \tag{7.83}$$

where K does not depend on γ, N.

The proof of Theorem 7.4.8 is based on the following two propositions, that assume (7.54) and $\gamma \leq \gamma_0$.

Proposition 7.4.9. *We have*

$$F'(\gamma) = \frac{1}{p!} \int \mathsf{E} \log\langle \theta(\sigma_1, \ldots, \sigma_p)\rangle_{\mathbf{x}} \, d\mu_\gamma(x_1) \ldots d\mu_\gamma(x_p) \,.$$

Proposition 7.4.10. *We have*

$$\left| p_N'(\gamma) - \frac{1}{p!} \int \mathsf{E} \log\langle \theta(\sigma_1, \ldots, \sigma_p)\rangle_{\mathbf{x}} \, d\mu_\gamma(x_1) \ldots d\mu_\gamma(x_p) \right| \leq \frac{K}{N}$$

where $p_N'(\gamma)$ is the right derivative of $p_N(\gamma)$.

Once these are proved, to obtain Theorem 7.4.8, it suffices to observe that $p_N(0) = F(0) = \log 2$.

The proof of Proposition 7.4.9 will require a significant effort. This proposition is used only to obtain the explicit formula proposed by Theorem 7.4.8; Proposition 7.4.10 is sufficient to testify that we know how to compute $p_N(\gamma)$. We prove it first. It relies on the following.

Lemma 7.4.11. *We have*

$$\left| p_N'(\gamma) - \frac{1}{p!} \mathsf{E} \log\langle \exp \theta(\sigma_1, \ldots, \sigma_p)\rangle \right| \leq \frac{K}{N} \tag{7.84}$$

where $p_N'(\gamma)$ is the right derivative of $p_N(\gamma)$.

Here, $\langle\cdot\rangle$ denotes average for Gibbs' measure with Hamiltonian (7.6), the random function θ is independent of the randomness in $\langle\cdot\rangle$, and K does not depend on γ, N.

Proof. We consider a number $\delta > 0$, and an i.i.d. sequence $(\eta_I)_{I\in[N]^p}$ with $\mathsf{P}(\eta_I = 1) = \delta/N^{p-1}$ and $\mathsf{P}(\eta_I = 0) = 1 - \delta/N^{p-1}$. We assume that this sequence is independent of all random sequences previously considered. We consider two Hamiltonians

$$-H_N^1(\boldsymbol{\sigma}) = \sum_I (\gamma_I + \eta_I)\, W_I(\boldsymbol{\sigma}) \tag{7.85}$$

$$-H_N^2(\boldsymbol{\sigma}) = \sum_I \min(1, \gamma_I + \eta_I)\, W_I(\boldsymbol{\sigma}) \,. \tag{7.86}$$

The r.v. $\gamma'_I = \min(1, \gamma_I + \eta_I)$ are i.i.d., take only the values 0 and 1, and

$$\mathsf{P}(\gamma'_I = 1) = 1 - \left(1 - \frac{\gamma}{N^{p-1}}\right)\left(1 - \frac{\delta}{N^{p-1}}\right) = \frac{\gamma'}{N^{p-1}}$$

where

$$\gamma' = \gamma + \delta - \frac{\gamma\delta}{N^{p-1}} \,. \tag{7.87}$$

Thus the Hamiltonian (7.86) coincides with the Hamiltonian (7.6) if we replace γ by γ' defined above.

For $j = 1, 2$, we define

$$p_N^j(\delta) = \frac{1}{N}\,\mathsf{E}\log\sum_{\boldsymbol{\sigma}} \exp(-H_N^j(\boldsymbol{\sigma})) \,. \tag{7.88}$$

Thus $p_N^2(\delta) = p_N(\gamma')$.

Since $\gamma_I + \eta_I = \min(1, \gamma_I + \eta_I) + \eta_I\,\gamma_I$ we have

$$p_N^1(\delta) = p_N^2(\delta) + \frac{1}{N}\,\mathsf{E}\log\left\langle \exp\sum \eta_I\,\gamma_I\,W_I\right\rangle_2$$

where $\langle\cdot\rangle_2$ denotes average for Gibbs' measure with Hamiltonian (7.86). Hence

$$|p_N^1(\delta) - p_N^2(\delta)| \le \frac{1}{N}\,\mathsf{E}\sum_I \eta_I\,\gamma_I\,S_I \,,$$

where of course $S_I = \sup W_I$. Thus

$$|p_N^1(\delta) - p_N^2(\delta)| \le \frac{1}{N}\binom{N}{p}\frac{\delta\gamma}{N^{2(p-1)}}\,\mathsf{E}\,S \le \frac{\delta\gamma}{N^{p-1}}\,\mathsf{E}\,S \,. \tag{7.89}$$

We have

$$p'_N(\gamma) = \lim_{\gamma'\to\gamma^+}\frac{p_N(\gamma') - p_N(\gamma)}{\gamma' - \gamma} = \lim_{\delta\to 0^+}\frac{p_N^2(\delta) - p_N(\gamma)}{\delta(1 - \gamma/N^{p-1})} \,.$$

Hence by (7.89) to prove (7.84) it suffices to prove that

$$\left|\lim_{\delta\to 0+} \frac{p_N^1(\delta)-p_N(\gamma)}{\delta} - \frac{1}{p!}\,\mathsf{E}\log\langle\exp\theta(\sigma_1,\ldots,\sigma_p)\rangle\right| \le \frac{K}{N}\,. \tag{7.90}$$

We have

$$p_N^1(\delta)-p_N(\gamma) = \frac{1}{N}\,\mathsf{E}\log\left\langle\exp\sum_I \eta_I W_I\right\rangle\,. \tag{7.91}$$

For $r \ge 0$, consider the event

$$\Omega_r = \left\{\sum_I \eta_I = r\right\}\,.$$

We observe that

$$\mathsf{E}\left(1_{\Omega_0}\log\left\langle\exp\sum_I \eta_I W_I\right\rangle\right) = 0\,, \tag{7.92}$$

and that, by symmetry among the sites,

$$\mathsf{E}\left(1_{\Omega_1}\log\left\langle\exp\sum_I \eta_I W_I\right\rangle\right) = \mathsf{P}(\Omega_1)\,\mathsf{E}\log\langle\exp\theta(\sigma_1,\ldots,\sigma_p)\rangle\,. \tag{7.93}$$

For $r \ge 2$, we note the inequality

$$\left|\mathsf{E}\left(1_{\Omega_r}\log\left\langle\exp\sum_I \eta_I W_I\right\rangle\right)\right| \le \mathsf{P}(\Omega_r)\,r\,\mathsf{E}S\,. \tag{7.94}$$

If $M = \binom{N}{p}$, we have

$$\mathsf{P}(\Omega_1) = M\,\frac{\delta}{N^{p-1}}\left(1-\frac{\delta}{N^{p-1}}\right)^{M-1}$$

so that

$$\lim_{\delta\to 0}\frac{\mathsf{P}(\Omega_1)}{\delta} = \frac{M}{N^{p-1}} = \frac{N}{p!}\left(1-\frac{1}{N}\right)\cdots\left(1-\frac{p-1}{N}\right)\,. \tag{7.95}$$

Also,

$$\sum_{r\ge 2} r\,\mathsf{P}(\Omega_r) = \mathsf{E}r - \mathsf{P}(\Omega_1) = \frac{M\delta}{N^{p-1}}\left(1-\left(1-\frac{\delta}{N^{p-1}}\right)^{M-1}\right)\,,$$

so that

$$\lim_{\delta\to 0+} \frac{1}{\delta}\sum_{r\geq 2} r\,\mathsf{P}(\Omega_r) = 0\ ,$$

and hence

$$\lim_{\delta\to 0+} \left|\mathsf{E}\left(1_{\Omega_1^c}\log\left\langle \exp\sum_I \eta_I W_I\right\rangle\right)\right| = 0\ . \tag{7.96}$$

Combining (7.91) with (7.93), (7.95)and (7.96) yields (7.90) and hence (7.84). □

Proof of Proposition 7.4.10. Let us denote by E_0 expectation in the randomness of $\langle\cdot\rangle$. Given θ, it follows from Theorem 7.2.2 (as used in Proposition 7.2.7) that

$$\mathsf{E}_0|\langle\exp\theta(\sigma_1,\ldots,\sigma_p)\rangle - \langle\exp\theta(\sigma_1,\ldots,\sigma_p)\rangle_{\mathbf{X}}| \leq \frac{K(p,\gamma_0)\exp S}{N}\ ,$$

where $S=\sup|\theta|$, $\mathbf{X}=(\langle\sigma_1\rangle,\ldots,\langle\sigma_p\rangle)$. Consider i.i.d. r.v. $z_1,\ldots,z_p$ of law μ_γ such that $\sum_{i\leq p}\mathsf{E}_0|\langle\sigma_i\rangle - z_i|\leq K(p,\gamma_0)/N$, as provided by (7.55). Since

$$\frac{\partial}{\partial x_i}\langle\exp\theta(\sigma_1,\ldots,\sigma_p)\rangle_{\mathbf{x}} \leq \exp S$$

we see as in Section 7.3 that

$$\mathsf{E}_0|\langle\exp\theta(\sigma_1,\ldots,\sigma_p)\rangle_{\mathbf{X}} - \langle\exp\theta(\sigma_1,\ldots,\sigma_p)\rangle_{\mathbf{Y}}| \leq \frac{K(p,\gamma_0)\exp S}{N}$$

where $\mathbf{Y}=(z_1,\ldots,z_N)$. Now,

$$|\log x - \log y| \leq \frac{|x-y|}{\min(|x|,|y|)}$$

so that

$$\mathsf{E}_0|\log\langle\exp\theta(\sigma_1,\ldots,\sigma_p)\rangle - \log\langle\exp\theta(\sigma_1,\ldots,\sigma_p)\rangle_{\mathbf{Y}}| \leq \frac{K(p,\gamma_0)\exp 2S}{N}$$

and hence

$$\left|\mathsf{E}_0\log\langle\exp\theta(\sigma_1,\ldots,\sigma_p)\rangle - \int\log\langle\exp\theta(\sigma_1,\ldots,\sigma_p)\rangle_{\mathbf{x}}\,\mathrm{d}\mu_\gamma(x_1)\ldots\mathrm{d}\mu_\gamma(x_p)\right| \leq \frac{K(p,\gamma_0)\exp 2S}{N}\ .$$

Taking expectation and using Lemma 7.4.11 finishes the proof. □

We now turn to the proof of Proposition 7.4.9. If we view the right-hand side of (7.82) as a function $U(\gamma,\mu)$ of γ and $\mu=\mu_\gamma$, we have $F(\gamma)=U(\gamma,\mu_\gamma)$.

The idea of Proposition 7.4.9 is that it is in some sense true that $\partial U/\partial\mu = 0$, so that

$$F'(\gamma) = \frac{\partial U}{\partial\gamma}(\gamma,\mu_\gamma)\,,$$

but giving a precise meaning to this requires significant work. The special structure of μ will of course be essential. We consider the functions (defined for $-1 \le x_1 \le 1$)

$$\Phi(x_1) = \int \mathsf{E}\log\langle\exp\theta(\sigma_1,\ldots,\sigma_p)\rangle_{\mathbf{x}}\,\mathrm{d}\mu_\gamma(x_2)\ldots\mathrm{d}\mu_\gamma(x_p) \tag{7.97}$$

and, for $r \ge 1$,

$$\Psi_r(x_1) = \int \mathsf{E}\log\langle\mathrm{Av}\mathcal{E}_r\rangle_{\mathbf{x}}\,\mathrm{d}\mu_\gamma(x_2)\ldots\mathrm{d}\mu_\gamma(x_{r(p-1)}) \tag{7.98}$$

where $\mathcal{E}_r$ is given by (7.40). Thus

$$F(\gamma) = \log 2 - \frac{c(p-1)}{p}\int\Phi(x)\,\mathrm{d}\mu_\gamma(x) + \sum_{r\ge1}\frac{c^r\,e^{-c}}{r!}\int\Psi_r(x)\,\mathrm{d}\mu_\gamma(x)\,. \tag{7.99}$$

The special structure of μ is reflected by the following identity.

Proposition 7.4.12. *We have, for $-1 \le y \le 1$*

$$\Phi(y) - \sum_{r\ge1}\frac{c^{r-1}\,e^{-c}}{(r-1)!}\Psi_r(y) = -\sum_{r\ge1}\frac{c^r\,e^{-c}}{r!}\int\Psi_r(x)\,\mathrm{d}\mu_\gamma(x)\,. \tag{7.100}$$

We decompose the proof in several steps.

Lemma 7.4.13. *Consider a function $f : \{-1,1\} \to \mathbb{R}$ and denote by $\langle f\rangle_y$ the integral of f with respect to the probability measure ν_y on $\{-1,1\}$ such that $\int\varepsilon\,\mathrm{d}\nu_y(\varepsilon) = y$. Then if $y = \langle\mathrm{Av}\,\varepsilon\,\mathcal{E}_r\rangle_{\mathbf{x}}/\langle\mathrm{Av}\,\mathcal{E}_r\rangle_{\mathbf{x}}$, we have*

$$\langle f\rangle_y = \frac{\langle\mathrm{Av}\,f(\varepsilon)\,\mathcal{E}_r\rangle_{\mathbf{x}}}{\langle\mathrm{Av}\,\mathcal{E}_r\rangle_{\mathbf{x}}}\,. \tag{7.101}$$

Proof. This formula is obvious if either f is constant or if $f(\varepsilon) = \varepsilon$, so that it holds for all functions by linear combination. □

Lemma 7.4.14. *With the notation of Lemma 7.4.13, assume that $f > 0$. Then we have*

$$\int\log\langle f\rangle_y\,\mathrm{d}\mu_\gamma(y) = e^{-c}\log(\mathrm{Av}f(\varepsilon)) + \sum_{r\ge1}\frac{c^r\,e^{-c}}{r!}\int\mathsf{E}\log\left(\frac{\langle\mathrm{Av}\,f(\varepsilon)\,\mathcal{E}_r\rangle_{\mathbf{x}}}{\langle\mathrm{Av}\,\mathcal{E}_r\rangle_{\mathbf{x}}}\right)\mathrm{d}\mu_\gamma(x_1)\ldots\mathrm{d}\mu_\gamma(x_{r(p-1)})\,. \tag{7.102}$$

Proof. We use the identity

$$\mu_\gamma = \sum_{r\geq 0} \frac{c^r\, e^{-c}}{r!}\, T_r(\mu_\gamma)$$

together with (7.101) and the definition of $T_r(\mu_\gamma)$. (We recall that $T_0(\mu_\gamma)$ is a unit mass at the point 0.) □

Proof of Proposition 7.4.12. We consider a realization of θ, numbers $y_1, \ldots, y_{p-1}$ and the function f on $\{-1,1\}$ given by

$$f(\varepsilon) = \int \exp \theta(\sigma_1, \ldots, \sigma_{p-1}, \varepsilon)\, \mathrm{d}\nu_{y_1}(\sigma_1) \ldots \mathrm{d}\nu_{y_{p-1}}(\sigma_{p-1})$$

where ν_y is defined in Lemma 7.4.13. Writing $\mathbf{x} = (x_1, \ldots, x_{r(p-1)})$, $\mathbf{x}' = (x_1, \ldots, x_{r(p-1)}, y_1, \ldots, y_{p-1})$, we have

$$\langle \mathrm{Av}\, f(\varepsilon)\, \mathcal{E}_r \rangle_{\mathbf{x}} = \langle \mathrm{Av}\, \exp \theta(\sigma_{r(p-1)+1}, \ldots, \sigma_{(r+1)(p-1)}, \varepsilon)\, \mathcal{E}_r \rangle_{\mathbf{x}'} \,. \tag{7.103}$$

This is obvious if one writes the explicit formula for each side. This is also true for $r = 0$, if we interpret $\langle \mathrm{Av} f(\varepsilon)\, \mathcal{E}_r \rangle_{\mathbf{x}}$ as $\mathrm{Av} f(\varepsilon)$. Let us denote by E_0 expectation given θ (that is, averaging only in the randomness occurring in $\mathcal{E}_r$, $r \geq 1$). We apply (7.102) to get

$$\begin{aligned}\int \log \langle f \rangle_y \, \mathrm{d}\mu_\gamma(y) = e^{-c} \log \langle \mathrm{Av} \exp \theta(\sigma_1, \ldots, \sigma_{p-1}, \varepsilon) \rangle_{\mathbf{x}'} + \sum_{r\geq 1} \frac{c^r\, e^{-c}}{r!} \\ \times \Bigg[\int \mathsf{E}_0 \log \langle \mathrm{Av} \exp \theta(\sigma_{r(p-1)+1}, \ldots \sigma_{(r+1)(p-1)}, \varepsilon)\, \mathcal{E}_r \rangle_{\mathbf{x}'} \\ \mathrm{d}\mu_\gamma(x_1) \ldots \mathrm{d}\mu_\gamma(x_{r(p-1)}) \\ - \int \mathsf{E}_0 \log \langle \mathrm{Av}\, \mathcal{E}_r \rangle_{\mathbf{x}}\, \mathrm{d}\mu_\gamma(x_1) \ldots \mathrm{d}\mu_\gamma(x_{r(p-1)}) \Bigg] \,.\end{aligned} \tag{7.104}$$

We observe that

$$\langle f \rangle_y = \langle \exp \theta(\sigma_1, \ldots, \sigma_p) \rangle_{\mathbf{y}} \,,$$

where $\mathbf{y} = (y_1, \ldots, y_{p-1}, y)$. Thus, taking expectation in θ and integrating $y_2, \ldots, y_{p-1}$ independently for μ_γ in (7.104) brings the relation

$$\begin{aligned}\Phi(y_1) &= e^{-c} \Psi_1(y_1) \\ &+ \sum_{r\geq 1} \frac{c^r\, e^{-c}}{r!} \left(\Psi_{r+1}(y_1) - \int \mathsf{E} \log \langle \mathrm{Av}\, \mathcal{E}_r \rangle_{\mathbf{x}}\, \mathrm{d}\mu_\gamma(x_1) \ldots \mathrm{d}\mu_\gamma(x_{r(p-1)}) \right)\end{aligned}$$

and this is (7.100). □

Consider now another number $\gamma' \le \gamma_0$. For simplicity we write $c' = \gamma'/(p-1)!$, $\mu' = \mu_{\gamma'}$ and $\mu = \mu_\gamma$. We write

$$B = \int \Phi(x)\,\mathrm{d}\mu(x)\,; \quad A_r = \int \Psi_r(x)\,\mathrm{d}\mu(x)$$

and we denote by B', A'_r the corresponding quantities when γ is replaced by γ'. Thus

$$F(\gamma) = \log 2 + \sum_{r\ge 1} \frac{c^r\, e^{-r}}{r!} A_r - \frac{c(p-1)}{p} B\,.$$

We write

$$F(\gamma') - F(\gamma) = \mathrm{I} + \mathrm{II} + \mathrm{III} \tag{7.105}$$

where

$$\mathrm{I} = \sum_{r\ge 1} \frac{c'^r\, e^{-c'} - c^r\, e^{-c}}{r!} A_r - (c'-c)\frac{p-1}{p} B \tag{7.106}$$

$$\mathrm{II} = \sum_{r\ge 1} \frac{c^r\, e^{-c}}{r!}(A'_r - A_r) - \frac{c(p-1)}{p}(B' - B) \tag{7.107}$$

$$\mathrm{III} = \sum_{r\ge 1} \frac{c'^r\, e^{-c'} - c^r\, e^{-c}}{r!}(A'_r - A_r) - \frac{(c'-c)(p-1)}{p}(B'-B)\,. \tag{7.108}$$

Lemma 7.4.15. *We have*

$$\lim_{\gamma'\to\gamma} \frac{\mathrm{I}}{\gamma' - \gamma} = \frac{1}{p!} B\,. \tag{7.109}$$

Proof. First, we observe that since $|A_r| \le r\,\mathsf{E}\,S$, there are no problems of convergence of the series $\sum_{r\ge 1}$. Thus,

$$\lim_{\gamma'\to\gamma} \frac{\mathrm{I}}{\gamma'-\gamma} = \frac{1}{(p-1)!}\left(\sum_{r\ge 1}\left(\frac{c^{r-1}\, e^{-c}}{(r-1)!} - \frac{c^r\, e^{-c}}{r!}\right) A_r - \frac{p-1}{p} B\right)\,. \tag{7.110}$$

If we integrate (7.100) with respect to μ, we get the relation

$$\sum_{r\ge 1}\left(\frac{c^{r-1}\, e^{-c}}{(r-1)!} - \frac{c^r\, e^{-c}}{r!}\right) A_r = B\,,$$

and combining with (7.110) this proves the result. □

The following is obvious.

Lemma 7.4.16. *For a function V on $\mathbb{R}^n$ we have*

$$|V(x'_1,\dots,x'_n) - V(x_1,\dots,x_n)| \le \sum_{i\le n} |x'_i - x_i| \sup\left|\frac{\partial V}{\partial x_i}\right|\,. \tag{7.111}$$

Lemma 7.4.17. *For a function V on $\mathbb{R}^n$ we have*

$$\left|\int V(x_1,\ldots,x_n)\,\mathrm{d}\mu'(x_1)\ldots\mathrm{d}\mu'(x_n)-\int V(x_1,\ldots,x_n)\,\mathrm{d}\mu(x_1)\ldots\mathrm{d}\mu(x_n)\right|$$

$$\le d(\mu,\mu')\sum_{i\le n}\sup\left|\frac{\partial V}{\partial x_i}\right| . \tag{7.112}$$

Proof. This is nearly obvious. We consider a pair (X,X') of r.v. with $\mu=\mathcal{L}(X)$, $\mu'=\mathcal{L}(X')$, and i.i.d. copies (X_i,X_i') of this pair. Thus

$$\int V(x_1,\ldots,x_n)\,\mathrm{d}\mu'(x_1)\ldots\mathrm{d}\mu'(x_n)=\mathsf{E}\,V(X_1',\ldots,X_n')$$

$$\int V(x_1,\ldots,x_n)\,\mathrm{d}\mu(x_1)\ldots\mathrm{d}\mu(x_n)=\mathsf{E}\,V(X_1,\ldots,X_n)$$

and the left-hand side of (7.112) is at most

$$\mathsf{E}\,|V(X_1',\ldots,X_n')-V(X_1,\ldots,X_n)|\le\mathsf{E}\,|X-X'|\sum_{i\le n}\sup\left|\frac{\partial V}{\partial x_i}\right|$$

by (7.110). We then take the infimum over the choices of X and X'. □

Lemma 7.4.18. *If*

$$V(x_1,\ldots,x_{r(p-1)})=\log\langle \mathrm{Av}\mathcal{E}_r\rangle_{\mathbf{x}} ,$$

then we have

$$\sum_{i\le r(p-1)}\sup\left|\frac{\partial V}{\partial x_i}\right|\le(p-1)\sum_{s\le r}\exp 2S_s \tag{7.113}$$

$$\sum_{i,j\le r(p-1)}\sup\left|\frac{\partial^2 V}{\partial x_i\,\partial x_j}\right|\le(p-1)^2\left(\sum_{s\le r}\exp 2S_s\right)^2 . \tag{7.114}$$

Proof. It is very similar to that of Lemma 7.3.5 using simply that $|e^x-e^y|\le 2e^a$ if $|x|,|y|\le a$. □

Lemma 7.4.19. *We have*

$$\lim_{\gamma'\to\gamma}\frac{\mathrm{III}}{\gamma'-\gamma}=0 .$$

Proof. We combine (7.112) and (7.113) to obtain

$$|A_r'-A_r|\le(p-1)\,r\,d(\mu,\mu')\,\mathsf{E}\exp 2S$$

$$|B'-B|\le p\,d(\mu,\mu')\,\mathsf{E}\exp 2S .$$

The result then follows from Lemma 7.3.7. □

It is harder to control the term II.

Lemma 7.4.20. *For a function V on $\mathbb{R}^n$, we have*

$$\Bigl|V(x'_1,\ldots,x'_n)-\sum_{\ell\le n}V(x_1,\ldots,x_{\ell-1},x'_\ell,x_{\ell+1},\ldots,x_n)$$
$$+(n-1)\,V(x_1,\ldots,x_n)\Bigr|$$
$$\le\sum_{i<j}|x'_i-x_i||x'_j-x_j|\sup\left|\frac{\partial^2 V}{\partial x_i\,\partial x_j}\right| . \tag{7.115}$$

Proof. We proceed by induction over n. Both sides of (7.115) are 0 for $n=1$. For the induction step from $n-1$ to n, we consider the quantity

$$V(x'_1,\ldots,x'_n)-\sum_{\ell\le n}V(x_1,\ldots,x_{\ell-1},x'_\ell,x_{\ell+1},\ldots,x_n)+(n-1)\,V(x_1,\ldots,x_n)$$

as a function $g(x'_n)$. Then

$$g'(x'_n)=\frac{\partial V}{\partial x_n}(x'_1,\ldots,x'_n)-\frac{\partial V}{\partial x_n}(x_1,\ldots,x_{n-1},x'_n)$$

and thus by (7.111)

$$|g'(x'_n)|\le\sum_{i\le n-1}|x'_i-x_i|\sup\left|\frac{\partial^2 V}{\partial x_i\,\partial x_n}\right| ,$$

and

$$|g(x'_n)-g(x_n)|\le\sum_{i\le n-1}|x'_i-x_i||x'_n-x_n|\sup\left|\frac{\partial^2 V}{\partial x_i\,\partial x_n}\right| .$$

Using the induction hypothesis to control $g(x_n)$ finishes the proof. □

Lemma 7.4.21. *Consider a function V on $\mathbb{R}^n$. Then*

$$\Bigl|\int V(x_1,\ldots,x_n)\,\mathrm{d}\mu'(x_1)\ldots\mathrm{d}\mu'(x_n) \tag{7.116}$$
$$-\sum_{\ell\le n}\int V(x_1,\ldots,x_n)\,\mathrm{d}\mu(x_1)\ldots\mathrm{d}\mu(x_{\ell-1})\,\mathrm{d}\mu'(x_\ell)\,\mathrm{d}\mu(x_{\ell+1})\ldots\mathrm{d}\mu(x_n)$$
$$+(n-1)\int V(x_1,\ldots,x_n)\,\mathrm{d}\mu(x_1)\ldots\mathrm{d}\mu(x_n)\Bigr|\le d(\mu,\mu')^2\sum_{i<j}\sup\left|\frac{\partial^2 V}{\partial x_i\,\partial x_j}\right| .$$

Proof. Using Lemma 7.4.20, this is very similar to the proof of Lemma 7.4.14. The key point is that in the right-hand side of (7.115) only the terms $|x'_i-x_i||x'_j-x_j|$ for $i\ne j$ occur; and that, by independence, with the notation of the proof of Lemma 7.4.17 we have

$$\mathsf{E}\,|X'_i-X_i||X'_j-X_j|=(\mathsf{E}\,|X'_1-X_1|)^2 .$$

□

Proposition 7.4.22. *We have*

$$\left|A'_r - A_r - r(p-1)\left(\int \Psi_r(x)\,\mathrm{d}\mu'(x) - \int \Psi_r(x)\,\mathrm{d}\mu(x)\right)\right|$$
$$\le Ld(\mu,\mu')^2\, r^2\, p^2\, \mathsf{E}\exp 4S\ . \tag{7.117}$$

$$\left|B' - B - p\left(\int \Phi(x)\,\mathrm{d}\mu'(x) - \int \Phi(x)\,\mathrm{d}\mu(x)\right)\right|$$
$$\le Ld(\mu,\mu')^2\, p^2\, \mathsf{E}\exp 4S\ . \tag{7.118}$$

Proof. To prove (7.117) we use (7.116) with $n = r(p-1)$, $V(x_1,\ldots,x_{r(p-1)}) = \log\langle \mathrm{Av}\mathcal{E}_r\rangle_{\mathbf{x}}$. In that case the right-hand side of (7.116) is bounded by

$$d(\mu,\mu')^2\,(p-1)^2\left(\sum_{s\le r}\exp 2S_s\right)^2 ,$$

using (7.114). We then take expectation in the randomness of $\mathcal{E}_r$. We observe that

$$\mathsf{E}\int V(x_1,\ldots,x_n)\,\mathrm{d}\mu(x_1)\ldots\mathrm{d}\mu(x_n) = A_r = \int \Psi_r(x)\,\mathrm{d}\mu(x)$$

and that, by symmetry between coordinates,

$$\mathsf{E}\int V(x_1,\ldots,x_n)\,\mathrm{d}\mu(x_1)\ldots\mathrm{d}\mu(x_{\ell-1})\,\mathrm{d}\mu'(x_\ell)\,\mathrm{d}\mu(x_{\ell+1})\ldots\mathrm{d}\mu(x_n)$$
$$= \int \Psi_r(x)\,\mathrm{d}\mu'(x)\ .$$

This proves (7.117), and (7.118) is similar. □

Proof of Proposition 7.4.9. It remains only to prove that

$$\lim_{\gamma'\to\gamma}\frac{\mathrm{II}}{\gamma'-\gamma} = 0\ . \tag{7.119}$$

Using (7.117) and (7.118), we see that

$$\Bigg|\mathrm{II} - c(p-1)\ \Bigg[\sum_{r\ge 1}\frac{c^{r-1}\,e^{-c}}{(r-1)!}\left(\int \Psi_r(x)\,\mathrm{d}\mu'(x) - \int \Psi_r(x)\,\mathrm{d}\mu(x)\right)$$
$$-\left(\int \Phi(x)\,\mathrm{d}\mu'(x) - \int \Phi(x)\,\mathrm{d}\mu(x)\right)\Bigg]\Bigg| \le Kd(\mu,\mu')^2$$

where K does not depend on γ'. Thus

$$|\text{II}| \le c(p-1)\left|\int \Xi(x)\,\mathrm{d}\mu'(x) - \int \Xi(x)\,\mathrm{d}\mu(x)\right| + Kd(\mu',\mu)^2 \qquad (7.120)$$

where

$$\Xi(x) = \sum_{r\ge 1} \frac{c^{r-1}\,e^{-c}}{(r-1)!}\,\Psi_r(x) - \Phi(x)\;.$$

By Proposition 7.4.12, $\Xi(x)$ is independent of x, so that $\int \Xi(x)\,\mathrm{d}\mu(x) = \int \Xi(x)\,\mathrm{d}\mu'(x)$. Thus (7.119) follows from (7.120) and Lemma 7.3.7. □

8. An Assignment Problem

8.1 Introduction

Given positive numbers $c(i,j)$, $i,j \le N$, the assignment problem is to find

$$\min_{\sigma} \sum_{i \le N} c(i, \sigma(i)) \,, \tag{8.1}$$

where σ ranges over all permutations of $\{1, \ldots, N\}$. In words, if $c(i,j)$ represents the cost of assigning job j to worker i, we want to minimize the total cost when exactly one job is assigned to each worker.

We will be interested in the random version of the problem, where the numbers $c(i,j)$ are independent uniformly distributed over $[0,1]$.

Mézard and Parisi [97], [98] studied (8.1) by introducing a suitable Hamiltonian, and conjectured that

$$\lim_{N \to \infty} \mathsf{E} \min_{\sigma} \sum_{i \le N} c(i, \sigma(i)) = \frac{\pi^2}{6} \,. \tag{8.2}$$

This was proved by D. Aldous [2]. Aldous takes advantage of a feature of the present model, that makes it rather special among the various models we studied: the existence of a "limiting object" (which he discovered [1]).

Related to (8.2) is the following remarkable conjecture, due to G. Parisi.

Conjecture 8.1.1. (Level 2) If the r.v. $c(i,j)$ are independent and satisfy $\mathsf{P}(c(i,j) \ge x) = e^{-x}$ for $x \ge 0$, then we have

$$\mathsf{E} \min_{\sigma} \sum_{i \le N} c(i, \sigma(i)) = 1 + \frac{1}{2^2} + \cdots + \frac{1}{N^2} \,. \tag{8.3}$$

The link with (8.2) is that it can be shown that if the r.v. $c(i,j)$ are i.i.d., and their common distribution has a density f on $\mathbb{R}^+$ with respect to Lebesgue measure, then if f is continuous in a neighborhood of 0, the limit in (8.2) depends only on $f(0)$. (Thus it makes no difference to assume that $c(i,j)$ is uniform over $[0,1]$ or is exponential of mean 1 as in Conjecture 8.1.1.)

Despite Aldous' result, (and even if one could prove Conjecture 8.1.1) the disordered system considered by Mézard and Parisi remains of interest. It is obviously related to the other models we have considered, and in particular to the model of Chapter 7. Yet it is rather different. In fact, it is different enough that we could not prove anything about it. We will consider a slightly different model. In this model, we consider two integers M, N, $M \geq N$. We consider independent r.v. $(c(i,j))_{i\leq N, j\leq M}$ that are uniform over $[0,1]$. The configuration space is the set $\Sigma_{N,M}$ of all one-to-one maps σ from $\{1,\ldots,N\}$ to $\{1,\ldots,M\}$. On this space we consider the Hamiltonian

$$H_{N,M}(\sigma) = \beta N \sum_{i\leq N} c(i,\sigma(i)) \,, \tag{8.4}$$

where β is a parameter. (The reader observes that there is no $-$ sign.) Given a number $\alpha > 0$, we will study the system for $N \to \infty$, $M = \lfloor N(1+\alpha) \rfloor$, and our results will hold for $\beta \leq \beta(\alpha)$.

Research problem 8.1.2. (Level 2) Extend the results of the present chapter to the case $\beta \leq \beta_0$, where β_0 is independent of α.

8.2 The Cavity Method

We consider the partition function

$$Z_{N,M} = \sum_{\sigma} \exp(-H_{N,M}(\sigma)) \,, \tag{8.5}$$

where the summation is over all possible values of σ in $\Sigma_{N,M}$. Through the chapter we write

$$a(i,j) = \exp(-\beta N c(i,j)) \,. \tag{8.6}$$

Thus

$$Z_{N,M} = \sum_{\sigma} \prod_{i\leq N} a(i,\sigma(i)) \,.$$

The cavity method will require removing elements from $\{1,\ldots,N\}$ and $\{1,\ldots,M\}$. Given a set $A \subset \{1,\ldots,N\}$ and a set $B \subset \{1,\ldots,M\}$ such that $N - \operatorname{card} A \leq M - \operatorname{card} B$, we write

$$Z_{N,M}(A;B) = \sum_{\sigma} \prod a(i,\sigma(i)) \,.$$

The product is taken over $i \in \{1,\ldots,N\}\backslash A$ and the sum is taken over the one-to-one maps σ from $\{1,\ldots,N\}\backslash A$ to $\{1,\ldots,M\}\backslash B$. Thus $Z_{N,M} = Z_{N,M}(\emptyset;\emptyset)$.

Lemma 8.2.1. *If $i \notin A$, we have*

$$Z_{N,M}(A;B) = \sum_{\ell \notin B} a(i,\ell)\, Z_{N,M}(A \cup \{i\};\ B \cup \{\ell\}) \,. \tag{8.7}$$

If $j \notin B$, we have

$$Z_{N,M}(A;B) = Z_{N,M}(A;B\cup\{j\}) + \sum_{k \notin A} a(k,j)\, Z_{N,M}(A\cup\{k\};\ B\cup\{j\}) \,. \tag{8.8}$$

Proof. One replaces each occurrence of $Z_{N,M}(\cdot;\cdot)$ by its value and one checks that the same terms occur in the left-hand and right-hand sides. □

The following deserves no proof.

Lemma 8.2.2. *If $M \notin B$, we have*

$$Z_{N,M}(A; B \cup \{M\}) = Z_{N,M-1}(A;B) \,. \tag{8.9}$$

If $N \notin A$, we have

$$Z_{N,M}(A \cup \{N\}; B) = Z_{N-1,M}(A;B) \,. \tag{8.10}$$

In (8.10), and in similar situations below, we make the convention that $Z_{N-1,M}(\cdot;\cdot)$ is considered for a parameter λ' such that $\lambda'(N-1) = \lambda N$.

When $A = \{i_1, i_2, \ldots\}$, $B = \{j_1, j_2, \ldots\}$, we write

$$Z_{N,M}(A;B) = Z_{N,M}(i_1, i_2, \ldots; j_1, j_2, \ldots) \,.$$

The following is also obvious from the definitions.

Lemma 8.2.3. *We have*

$$\sum_{\ell \le M} Z_{N,M}(\emptyset; \ell) = (M - N)\, Z_{N,M} \,. \tag{8.11}$$

From the study of the previous models, we would expect that (for small β) "the spin at site i is nearly independent from the spin at site j". Here this should mean that if $i_1 \neq i_2$, under Gibbs' measure the variables $\sigma \to \sigma(i_1)$ and $\sigma \to \sigma(i_2)$ are nearly independent. It is however not obvious how to quantify this phenomenon in a way suitable for a proof by induction. To overcome this obstacle, we will consider a different property. We will prove that

$$\frac{Z_{N,M}(i;j)}{Z_{N,M}} \simeq \frac{Z_{N,M}(\emptyset;j)}{Z_{N,M}} \frac{Z_{N,M}(i;\emptyset)}{Z_{N,M}} \,. \tag{8.12}$$

It should be obvious that this is a very strong constraint, and that it deals with independence. One can also get convinced that it deals with Gibbs' measure by observing that if one multiplies the left-hand side of (8.12) by $a(i,j)$, one obtains the Gibbs' probability of the event $\sigma(i) = j$.

We consider the quantities

$$u_{N,M}(j) = \frac{Z_{N,M}(\emptyset; j)}{Z_{N,M}} \; ; \; w_{N,M}(i) = \frac{Z_{N,M}(i;\emptyset)}{Z_{N,M}} \,. \tag{8.13}$$

These quantities occur in the right-hand side of (8.12). They are closely related, as the following shows.

Lemma 8.2.4. *We have*

$$u_{N,M}(M) = \frac{Z_{N,M-1}}{Z_{N,M}} = \frac{1}{1 + \sum_{k\le N} a(k,M)\, w_{N,M-1}(k)} \tag{8.14}$$

$$w_{N,M}(N) = \frac{Z_{N-1,M}}{Z_{N,M}} = \frac{1}{\sum_{\ell\le M} a(N,\ell)\, u_{N-1,M}(\ell)} \,. \tag{8.15}$$

Proof. To prove (8.14) we use (8.8) with $A = B = \emptyset$ and $j = M$ and (8.9) with $A = \{k\}$ and $B = \emptyset$ to obtain

$$\begin{aligned} Z_{N,M} &= Z_{N,M-1} + \sum_{k\le N} a(k,M)\, Z_{N,M-1}(k;\emptyset) \\ &= Z_{N,M-1}\left(1 + \sum_{k\le N} a(k,M)\, w_{N,M-1}(k)\right) \,. \end{aligned} \tag{8.16}$$

The proof of (8.15) is similar, using now (8.7) and (8.10). □

To prove (8.12) we will consider the following quantities, where $i \le N$, $j \le M-1$

$$L_{N,M}(i) = \frac{Z_{N,M}\, Z_{N,M-1}(i;\emptyset) - Z_{N,M}(i;\emptyset)\, Z_{N,M-1}}{Z^2_{N,M}} \tag{8.17}$$

$$R_{N,M}(j) = \frac{Z_{N,M}\, Z_{N,M-1}(\emptyset;j) - Z_{N,M}(\emptyset;j)\, Z_{N,M-1}}{Z^2_{N,M}} \,. \tag{8.18}$$

The idea is that

$$L_{N,M}(i) = \frac{Z_{N,M}\, Z_{N,M}(i;M) - Z_{N,M}(i;\emptyset)\, Z_{N,M}(\emptyset;M)}{Z^2_{N,M}}$$

so that (8.12) means that $\mathsf{E} L^2_{N,M}(i)$ is small. (This expectation does not depend on i.) It is good to notice that $|L_{N,M}(i)| \le 2$ and $|R_{N,M}(j)| \le 2$.

Lemma 8.2.5. *We have*

$$L_{N,M}(N) = -\frac{\sum_{\ell \le M-1} a(N,\ell)\, R_{N-1,M}(\ell) - a(N,M)\, u^2_{N-1,M}(M)}{\left(\sum_{\ell \le M} a(N,\ell)\, u_{N-1,M}(\ell)\right)^2} \tag{8.19}$$

$$R_{N,M}(M-1) = -\frac{\sum_{k \le N} a(k,M)\, L_{N,M-1}(k)}{\left(1 + \sum_{k \le N} a(k,M)\, w_{N,M-1}(k)\right)^2} \,. \tag{8.20}$$

Proof. As in (8.16) we obtain that

$$\begin{aligned} Z_{N,M}(\emptyset; M-1) &= Z_{N,M-1}(\emptyset; M-1) \\ &\quad + \sum_{k \le N} a(k,M)\, Z_{N,M-1}(k; M-1) \,. \end{aligned} \tag{8.21}$$

Using this and (8.16) in the numerator of (8.18), and (8.16) in the denominator, and regrouping the terms yields (8.20). The proof of (8.19) is similar. □

We end this section by a technical but essential fact.

Lemma 8.2.6. *We have*

$$\sum_{\ell \le M} u_{N,M}(\ell) = M - N \tag{8.22}$$

$$\sum_{j \le M-1} R_{N,M}(j) = -u_{N,M}(M) + u^2_{N,M}(M) \,. \tag{8.23}$$

Proof. First, (8.22) follows from (8.11). From (8.11) we have

$$\sum_{j \le M-1} Z_{N,M}(\emptyset; j) = (M-N)\, Z_{N,M} - Z_{N,M}(\emptyset; M) = (M-N)\, Z_{N,M} - Z_{N,M-1},$$

and changing M into $M-1$ in (8.11) we get

$$\sum_{j \le M-1} Z_{N,M-1}(\emptyset, j) = (M-1-N)\, Z_{N,M-1} \,.$$

These two relations imply (8.23) in a straightforward manner. □

8.3 Decoupling

In this section, we prove (8.12) and, more precisely, the following.

Theorem 8.3.1. *Given $\alpha > 0$, there exists $\beta(\alpha) > 0$ such that if $\beta \leq \beta(\alpha)$ and $M = \lfloor N(1+\alpha) \rfloor$, then*

$$\mathsf{E}\, L_{N,M}^2(N) \leq \frac{K(\alpha)}{N} \tag{8.24}$$

$$\mathsf{E}\, R_{N,M}^2(M-1) \leq \frac{K(\alpha)}{N} . \tag{8.25}$$

The method of proof consists of using Lemma 8.2.5 to relate $\mathsf{E}\, R_{N,M}^2(M-1)$ with $\mathsf{E}\, L_{N,M-1}^2(N)$ and $\mathsf{E}\, L_{N,M}^2(N)$ with $\mathsf{E}\, R_{N-1,M}^2(M-1)$, and to iterate these relations. In the right-hand sides of (8.19) and (8.20), we will first integrate in the quantities $a(N,\ell)$ and $a(k,M)$, that are probabilistically independent of the other quantities (an essential fact). Our first task is to learn how to do this.

We consider an independent sequence of r.v. $(X_k)_{k\geq 1}$ uniformly distributed on $[0,1]$, and we set $a(k) = \exp(-\beta\, N X_k)$. The following is obvious.

Lemma 8.3.2. *We have*

$$\mathsf{E}\, a(k)^p = \frac{1}{\beta\, p\, N}(1 - \exp(-\beta\, p\, N)) \leq \frac{1}{\beta\, p\, N} . \tag{8.26}$$

Lemma 8.3.3. *Consider numbers $(x_k)_{k\leq N}$. Then we have*

$$\mathsf{E}\left(\sum_{k\leq N} a(k)\, x_k\right)^2 \leq \left(\frac{1}{\beta^2 N} + \frac{1}{2\beta N}\right) \sum_{k\leq N} x_k^2 . \tag{8.27}$$

Proof. We have

$$\begin{aligned}\mathsf{E}\left(\sum_{k\leq N} a(k)\, x_k\right)^2 &= \sum_{k\leq N} x_k^2\, \mathsf{E}\, a^2(k) + \sum_{k\neq\ell} x_k\, x_\ell\, \mathsf{E}\, a(k)\, \mathsf{E}\, a(\ell) \\ &\leq \frac{1}{2\beta N} \sum_{k\leq N} x_k^2 + \left(\frac{1}{\beta N}\right)^2 \sum_{k\neq\ell} |x_k|\,|x_\ell| .\end{aligned}$$

Now, by the Cauchy-Schwarz inequality we have

$$\sum_{k\neq\ell} |x_k|\,|x_\ell| \leq \left(\sum_{k\leq N} |x_k|\right)^2 \leq N \sum_{k\leq N} x_k^2 .$$

□

Corollary 8.3.4. *If $\beta \leq 1$ we have*

$$\mathsf{E}\, R_{N,M}^2(M-1) \leq \frac{2}{\beta^2}\, \mathsf{E}\, L_{N,M-1}^2(N) .$$

Proof. From (8.20) we have

$$\mathsf{E}\, R^2_{N,M}(M-1) \le \mathsf{E}\left(\sum_{k\le N} a(k,M)\, L_{N,M-1}(k)\right)^2 .$$

The sequence $(a(k,M))_{k\le N}$ has the same distribution as the sequence $(a(k))_{k\le N}$, so that integrating first in this sequence and using (8.27) we get

$$\mathsf{E}\, R^2_{N,M}(M-1) \le \frac{2}{\beta^2 N}\sum_{k\le N} \mathsf{E}\, L^2_{N,M}(k) = \frac{2}{\beta^2}\,\mathsf{E}\, L^2_{N,M}(N)$$

by symmetry. □

Of course this is very crude; but trying to do better runs in technical difficulties that we could not overcome. (See Problem 8.1.1.)

For $v \ge 0$, we define the function $\psi(v)$ by

$$\psi(v) = \begin{cases} v/e & \text{if} \quad v \le e \\ \log v & \text{if} \quad v \ge e\,. \end{cases}$$

Note in particular that $\psi(v)/v$ is non-increasing.

Lemma 8.3.5. *If $\beta N \ge 1$, then, for $v \ge 0$ we have*

$$\mathsf{E}\exp(-va(1)) \le \exp\left(-\frac{\psi(v)}{L\beta N}\right) . \tag{8.28}$$

Proof. Since $\beta N \ge 1$, by (8.26) we have $\mathsf{E}\, a(1) \ge 1/2\beta N$, so that, if $v \le e$, since $a(1) \le 1$, and since $e^{-x} \le 1 - x/L$ for $x \le e$, we have

$$\mathsf{E}\exp(-va(1)) \le 1 - \frac{1}{L}\,\mathsf{E}\,(va(1)) \le 1 - \frac{v}{L\beta N} \le \exp\left(-\frac{v}{L\beta N}\right) .$$

If $e \le v \le e^{\beta N}$, we write, since $\exp(-x) \le 1/2$ for $x \ge 1$,

$$\begin{aligned} \mathsf{E}\exp(-va(1)) &\le 1 - \frac{1}{2}\,\mathsf{P}(va(1) \ge 1) \\ &\le \exp\left(-\frac{1}{2}\,\mathsf{P}(\exp(\beta N X_1) \le v)\right) \\ &= \exp\left(-\frac{1}{2\beta N}\log v\right) . \end{aligned}$$

Finally, if $v \ge e^{\beta N}$, since $a(1) \ge e^{-\beta N}$ we write

$$\mathsf{E}\exp(-va(1)) \le \exp(-ve^{-\beta N}) \le \exp\left(-\frac{\log v}{\beta N}\right) .$$

The last inequality uses that the function $\theta(v) = v/\log v$ increases for $v \ge e$, and thus $\theta(v) \ge \theta(e^{\beta N})$. □

Lemma 8.3.6. *Consider numbers $(u_m)_{m\le M}$, $0\le u_m\le 1$, and assume that $\sum_{m\le M} u_m\ge 4$. Then, for $t\ge 0$ we have*

$$\mathsf{E}\exp\left(-\sum_{m\le M, m\ne k,\ell} ta(m)\,u_m\right)\le \exp\left(-\frac{\psi(t)}{L\beta N}\left(\sum_{m\le M} u_m\right)\right). \tag{8.29}$$

Proof. By (8.28), and since $\psi(u_m t)\ge u_m\psi(t)$,

$$\mathsf{E}\exp(-t\,a(m)\,u_m)\le \exp\left(-\frac{\psi(u_m t)}{L\beta N}\right)\le \exp\left(-\frac{u_m}{L\beta N}\psi(t)\right).$$

Moreover we have

$$\sum_{m\ne k,\ell} u_m\ge \sum_{m\le M} u_m-2\ge \sum_{m\le M} u_m/2\,.$$

The result follows by independence. □

Lemma 8.3.7. *If $n\le 5$ and $a\ge 7$ we have*

$$\int_0^\infty t^n\exp(-a\,\psi(t))\,\mathrm{d}t\le \frac{L}{a^{n+1}}\,. \tag{8.30}$$

Proof. We have

$$\int_0^\infty t^n\exp\left(-\frac{at}{e}\right)\mathrm{d}t\le\frac{L}{a^{n+1}}\,,$$

$$\int_e^\infty t^n\exp(-a\,\psi(t))\,\mathrm{d}t=\int_e^\infty t^{n-a}\,\mathrm{d}t=\frac{1}{a-n-1}\,e^{n+1-a}\le L\,e^{-a}\le\frac{L}{a^{n+1}}\,.$$

□

Proposition 8.3.8. *Consider numbers $0\le u_\ell,\ u'_\ell\le 1$, $\ell\le M$. Assume that $\sum_{\ell\le M} u_\ell\ge 4$, $\sum_{\ell\le M} u'_\ell\ge 4$ and consider $b\le N^{-1}\sum_{\ell\le M} u_\ell$, $b\le N^{-1}\sum_{\ell\le M} u'_\ell$. Consider numbers $(y_\ell)_{\ell\le M}$. Then, if $\beta N\ge 1$, $\beta\le b/L$, we have*

$$\mathsf{E}\frac{\left(\sum_{\ell\le M} a(\ell)\,y_\ell\right)^2}{\left(\sum_{\ell\le M} a(\ell)\,u_\ell\right)^2\left(\sum_{\ell\le M} a(\ell)\,u'_\ell\right)^2}\le\frac{L\beta^2}{b^4}\left(\frac{1}{N}\sum_{\ell\le M} y_\ell\right)^2 + \frac{L\beta^3(M+N)}{b^5N^2}\sum_{\ell\le M} y_\ell^2\,. \tag{8.31}$$

Moreover we have

$$\mathsf{E}\left(\sum_{\ell\le M} a(\ell)\,u_\ell\right)^{-4}\le\frac{L\beta^4}{b^4}\,. \tag{8.32}$$

Proof. We write

$$\frac{1}{\left(\sum_{\ell\leq M} a(\ell)\, u_\ell\right)^2 \left(\sum_{\ell\leq M} a(\ell)\, u'_\ell\right)^2} \leq \frac{2}{\left(\sum_{\ell\leq M} a(\ell)\, u_\ell\right)^4} + \frac{2}{\left(\sum_{\ell\leq M} a(\ell)\, u'_\ell\right)^4}$$

to reduce (8.31) to the case where $u'_\ell = u_\ell$.

Let us write $\dot{a}(\ell) = a(\ell) - \mathsf{E}\, a(\ell)$, so that, since $\mathsf{E}\, a(\ell) = \mathsf{E}\, a(1)$,

$$\sum_{\ell\leq M} a(\ell)\, y_\ell = (\mathsf{E}\, a(1)) \sum_{\ell\leq M} y_\ell + \sum_{\ell\leq M} \dot{a}(\ell)\, y_\ell \tag{8.33}$$

and, since $\mathsf{E}\, a(1) \leq 1/(\beta N)$,

$$\left(\sum_{\ell\leq M} a(\ell)\, y_\ell\right)^2 \leq \frac{2}{\beta^2}\left(\frac{1}{N}\sum_{\ell\leq M} y_\ell\right)^2 + 2\left(\sum_{\ell\leq M} \dot{a}(\ell)\, y_\ell\right)^2 .$$

Thus to prove (8.31) it suffices to prove it in the case where $\sum_{\ell\leq M} y_\ell = 0$ and to prove (8.32). We first prove (8.31) when $\sum_{\ell\leq M} y_\ell = 0$. We use the magic formula

$$\frac{1}{A^4} = \int_0^\infty \frac{t^3}{3!} \exp(-At)\, \mathrm{d}t \tag{8.34}$$

and we get from (8.33) that the left-hand side of (8.31) is $\int_0^\infty (t^3/3!)\, \Phi(t) \mathrm{d}t$ where

$$\begin{aligned}\Phi(t) &= \mathsf{E}\left(\left(\sum_{\ell\leq M} \dot{a}(\ell)\, y_\ell\right)^2 \exp\left(-t\sum_{m\leq M} a(m)\, u_m\right)\right)\\ &= \sum_{k,\ell\leq M} y_k\, y_\ell\, U_t(k,\ell)\end{aligned}$$

where

$$U_t(k,\ell) = \mathsf{E}\left(\dot{a}(k)\, \dot{a}(\ell) \exp\left(-t\sum_{m\leq M} a(m)\, u_m\right)\right) .$$

We bound these quantities. If $k = \ell$, since $\mathsf{E}\, \dot{a}(k)^2 \leq \mathsf{E}\, a^2(k) \leq 1/(2\beta N)$, we have by Lemma 8.3.6 that

$$\begin{aligned}U_t(k,k) &= \mathsf{E}\left(\dot{a}(k)^2 \exp(-ta(k)u_k)\right) \mathsf{E}\exp\left(-t\sum_{m\leq M, m\neq k} a(m)u_m\right)\\ &\leq \frac{1}{2\beta N} \exp\left(-\frac{b}{L\beta}\,\psi(t)\right) .\end{aligned}$$

If $k \neq \ell$, we write $U_t(k,\ell) = B_k\, B_\ell\, C_{k\ell}$, where

$$B_k = \mathsf{E}\,(\dot{a}(k)\exp(-t\,a(k)\,u_k))\,,\ C_{k\ell} = \mathsf{E}\exp\left(-t\sum_{m\neq k,\ell} a(m)\,u_m\right)\,.$$

Since $\mathsf{E}\,\dot{a}(k) = 0$ and since $|e^{-x} - 1| \le x$ for $x \ge 0$, we have

$$\begin{aligned} |B_k| &= |\mathsf{E}\,(\dot{a}(k)(\exp(-t\,a(k)\,u_k) - 1)| \\ &\le t\,\mathsf{E}\,|\dot{a}(k)\,a(k)| \le t(\mathsf{E}\,\dot{a}(k)^2)^{1/2}(\mathsf{E}\,a^2(k))^{1/2} \\ &\le t\,\mathsf{E}\,a^2(k) \le \frac{t}{2\beta N}\,. \end{aligned}$$

Using Lemma 8.3.6 to control $C_{k\ell}$, we then get

$$|U_t(k,\ell)| \le \frac{t^2}{4\beta^2 N^2}\exp\left(-\frac{b}{L\beta}\,\psi(t)\right)$$

and thus

$$\begin{aligned} \int_0^\infty \frac{t^3}{3!}\,\Phi(t)\,\mathrm{d}t \le \frac{1}{2\beta N}\sum_{k\le M} y_k^2 \int_0^\infty t^3\exp\left(-\frac{b}{L\beta}\,\psi(t)\right)\mathrm{d}t \\ + \frac{1}{4\beta^2 N^2}\sum_{k\neq\ell} |y_k||y_\ell| \int_0^\infty t^5\exp\left(-\frac{b}{L\beta}\,\psi(t)\right)\mathrm{d}t\,. \end{aligned}$$

We then use that $\sum_{k\neq\ell}|y_k||y_\ell| \le M\sum_{k\le M} y_k^2$, that $\beta \le Lb$, that $b \le 1$ and (8.30) to conclude the proof of (8.31). The proof of (8.32) is much easier and follows the same line. It is left to the reader.

Corollary 8.3.9. *If $\beta \le \alpha/L$, $\beta N \ge 1$, $M \ge \lfloor N(1+\alpha)\rfloor$, $M \le 3N$, we have*

$$\mathsf{E}\,L_{N,M}^2(N) \le \frac{L\beta^3}{\alpha^5}\,\mathsf{E}\,(R_{N-1,M}^2(M-1)) + \frac{K(\alpha)}{N}\,. \tag{8.35}$$

Proof. We will use (8.31) and (8.19). For $\ell \le M$, let $u_\ell = u_{N,M}(\ell)$, so that $0 \le u_\ell \le 1$ and by (8.22), we have $\sum_{\ell\le M} u_\ell = M - N := Nb$. We have $M - N = \lfloor N\alpha\rfloor$ so that if $\beta \le \alpha/4$ and $\beta N \ge 1$, we have $\alpha N \ge 4$ and hence $M - N \ge 4$. (This is needed in Lemma 8.3.6.) We also have $\lfloor N\alpha\rfloor \ge N\alpha/2$ so that $b \ge \alpha/2$. We write $y_\ell = R_{N-1,M}(\ell)$ for $\ell \le M-1$, $y(M) = -u_{N,M}^2(M)$, so that $\sum_{\ell\le M} y_\ell = -u_{N,M}(M)$ by (8.23). By (8.19) we have

$$L_{N,M}^2(N) = \frac{\left(\sum_{\ell\le M} a(N,\ell)\,y_\ell\right)^2}{\left(\sum_{\ell\le M} a(N,\ell)\,u_\ell\right)^4}\,.$$

We then take expectation in $a(N,\ell)$, $\ell \le M$ only, we use (8.31), and we observe that $M/Nb = (b+1)/b \le L/\alpha$ since $b \le 2$. □

Proof of Theorem 8.3.1. To avoid trivial complications, we assume $\alpha \le 1$. Assume $M = \lfloor N(1+\alpha) \rfloor$, and, for $k \le N/2$ let

$$V(k) = \mathsf{E}\, R^2_{N-k,M-k}(M-k-1)\,.$$

Since $M - k \ge \lfloor (N-k)(1+\alpha) \rfloor$ and $M - k \le 3(N-k)$, combining Corollaries 8.3.9 and 8.3.4, we see that if $\beta(N-k) \ge 1$,

$$V(k) \le \frac{L\beta}{\alpha^5} V(k+1) + \frac{K(\alpha)}{N-k}\,.$$

Thus if $L\beta/\alpha^5 \le 1/2$, $k \le N/2$ and $2\beta N \ge 1$, combining these relations yields

$$V(0) \le 2^{-k} V(k) + \frac{K(\alpha)}{N} \le 2^{-k+2} + \frac{K(\alpha)}{N}\,,$$

since $V(k) \le 4$. Taking $k \simeq \log N$ proves (8.25), and (8.24) follows by (8.35). □

Theorem 8.3.10. *Under the conditions of Theorem 8.3.1, for $j \le M-1$, $i \le N-1$ we have*

$$\mathsf{E}\,((u_{N,M}(j) - u_{N,M-1}(j))^2) \le \frac{K(\alpha)}{N} \tag{8.36}$$

$$\mathsf{E}\,((u_{N,M}(j) - u_{N-1,M}(j))^2) \le \frac{K(\alpha)}{N} \tag{8.37}$$

$$\mathsf{E}\,((w_{N,M}(i) - w_{N,M-1}(i))^2) \le \frac{K(\alpha)}{N} \tag{8.38}$$

$$\mathsf{E}\,((w_{N,M}(i) - w_{N-1,M}(i))^2) \le \frac{K(\alpha)}{N}\,. \tag{8.39}$$

Proof. The proofs are similar, so we prove only (8.36). We can assume $j = M-1$. Using (8.16), (8.21) we get

$$\begin{aligned} u_{N,M}(M-1) &= \frac{Z_{N,M}(\emptyset; M-1)}{Z_{N,M}} \\ &= \frac{Z_{N,M-2}}{Z_{N,M-1}} \left(\frac{1 + \sum_{k \le N} a(k,M)\, w_{N,M-2}(k)}{1 + \sum_{k \le N} a(k,M)\, w_{N,M-1}(k)} \right) \end{aligned}$$

so that

$$\begin{aligned} &u_{N,M}(M-1) - u_{N,M-1}(M-1) \\ &= \frac{Z_{N,M-2}}{Z_{N,M-1}} \left(\frac{1 + \sum_{k \le N} a(k,N)\, w_{N,M-2}(k)}{1 + \sum_{k \le N} a(k,N)\, w_{N,M-1}(k)} - 1 \right) \\ &= \frac{\sum_{k \le N} a(k,N)\, L_{N,M-1}(k)}{1 + \sum_{k \le N} a(k,N)\, w_{N,M-1}(k)}\,, \end{aligned}$$

using the identity

$$L_{N,M}(i) = \frac{Z_{N,M-1}}{Z_{N,M}} \left(w_{N,M-1}(i) - w_{N,M}(i) \right) ,$$

which is obvious from (8.17). Thus (8.37) follows from (8.24). □

8.4 Empirical Measures

One of our goals is to show that as $N \to \infty$ (and $M = \lfloor N(1+\alpha) \rfloor$) the random measure

$$\mu_{N,M} = \frac{1}{M} \sum_{j \le M} \delta_{u_{N,M}(j)} \tag{8.40}$$

becomes close to a non-random measure μ_α. The probability μ_α will occur as the fixed point of a certain operator just as in Section 7.3. Unfortunately, we do not know how to prove directly the existence of this fixed point, and this forces us to make some detours. We will first prove that $\mu_{N,M}$ is essentially non-random. We consider an independent copy $u'_{N,M}(j)$ of the variables $u_{N,M}(j)$. That is, these numbers are defined as the numbers $u_{N,M}(j)$, excepts that we replace the costs $c(i,j)$ by an independent family $c'(i,j)$. We define

$$\nu_{N,M} = \frac{1}{N} \sum_{i \le N} \delta_{w_{N,M}(i)} \,. \tag{8.41}$$

Given two probability measures μ, ν on $\mathbb{R}$, we define the quantity $\Delta(\mu, \nu)$ as

$$\Delta(\mu,\nu) = \inf \mathsf{E}\,(X-Y)^2 \,, \tag{8.42}$$

where the infimum is over all pairs (X, Y) of r.v. such that X has law μ and Y has law ν. (The quantity $\Delta^{1/2}(\mu,\nu)$ is called the Wasserstein distance between μ and ν.) It is good to note, since $\int x \, \mathrm{d}\mu(x) - \int y \, \mathrm{d}\nu(y) = \mathsf{E}(X-Y)$, that by the Cauchy-Schwarz inequality we have

$$\left(\int x \, \mathrm{d}\mu(x) - \int y \, \mathrm{d}\nu(y) \right)^2 \le \Delta(\mu,\nu) \,. \tag{8.43}$$

Theorem 8.4.1. *Under the conditions of Theorem 8.3.1 we have*

$$\lim_{N \to \infty} \mathsf{E}\,\Delta(\mu_{N,M}, \mu'_{N,M}) = 0$$

$$\lim_{N \to \infty} \mathsf{E}\,\Delta(\nu_{N,M}, \nu'_{N,M}) = 0 \,.$$

Here of course $\mu'_{N,M}$ is an independent copy of $\mu_{N,M}$, that is $\mu'_{N,M} = M^{-1} \sum_{j \le M} \delta_{u'_{N,M}(j)}$, and similarly for $\nu'_{N,M}$.

Lemma 8.4.2. *We have*

$$\Delta\left(\frac{1}{N}\sum_{i\le N}\delta_{x_i}\,,\ \frac{1}{N}\sum_{i\le N}\delta_{y_i}\right)=\inf_{\sigma}\frac{1}{N}\sum_{i\le N}(x_i-y_{\sigma(i)})^2\,, \tag{8.44}$$

where the infimum is over all permutations σ of $\{1,\ldots,N\}$.

We will use this lemma when $x_i = w_{N,M}(i)$, and almost surely any two of these points are distinct. For this reason, we will give the proof only in the (easier) case where any two of the points x_i (resp. y_i) are distinct.

Proof. The inequality $\le$ should be obvious. To prove the converse inequality, we observe that if X has law $N^{-1}\sum_{i\le N}\delta_{x_i}$ and Y has law $N^{-1}\sum_{i\le N}\delta_{y_i}$, then

$$\mathsf{E}\,(X-Y)^2=\sum_{i,j\le N}\mathsf{P}\,(X=x_i\,,\ Y=y_i)\,(x_i-y_j)^2\,.$$

We observe that the bistochastic matrices are exactly the matrices $a_{ij} = N\,\mathsf{P}\,(X=x_i\,,\ Y=y_j)$. Thus the left-hand side of (8.44) is

$$\frac{1}{N}\inf\sum_{i,j\le N}a_{ij}(x_i-y_j)^2\,,$$

where the infimum is over all bistochastic matrices (a_{ij}). The infimum is attained at an extreme point, and this extreme point is a permutation matrix. □

Lemma 8.4.3. *If μ and ν have compact support, then*

$$\Delta(\mu,\nu)=\sup\left\{\int f\,\mathrm{d}\mu+\int g\,\mathrm{d}\nu\,;\ f,g\ \textit{continuous}\right.$$
$$\left.\forall x,y,f(x)+g(y)\le(x-y)^2\right\}. \tag{8.45}$$

Proof. Let us denote by a the right-hand side of (8.45), and I a compact interval with $\mu(I)=\nu(I)=1$. We consider the subset $\mathcal{S}$ of the set $\mathcal{C}(I\times I)$ of continuous functions on $I\times I$ that consists of the functions $w(x,y)$ such that there exists continuous functions f and g on I for which

$$\int f\,\mathrm{d}\mu+\int g\,\mathrm{d}\nu=a \tag{8.46}$$

and

$$\forall x,y\in I\ ,\ w(x,y)>f(x)+g(y)-(x-y)^2\,. \tag{8.47}$$

It follows from the definition of a that for each function w in $\mathcal{S}$ there exist x and y with $w(x,y)>0$. Since $\mathcal{S}$ is convex and open, the Hahn-Banach theorem asserts that we can find a linear funcitional Φ on $\mathcal{C}(I\times I)$ such that

$\Phi(w) > 0$ for each w in $\mathcal{S}$. It is obvious from the definition of $\mathcal{S}$ that Φ is positive, and, since it is a matter of normalization, we can assume that it is a probability, which we denote by θ. If f and g are as in (8.46), then for each $\varepsilon > 0$ we see by (8.47) that the function $f(x) + g(y) - (x-y)^2 + \varepsilon$ belongs to $\mathcal{S}$ and thus

$$\int (x-y)^2 \, \mathrm{d}\theta(x,y) \le \int (f(x) + g(y)) \, \mathrm{d}\theta(x,y) \,. \tag{8.48}$$

Now this holds true if we replace f by $f + f'$ where $\int f' \, \mathrm{d}\mu = 0$. Thus this latter condition must imply that $\int f'(x,y) \, \mathrm{d}\theta(x,y) = 0$, and this means that μ is the first marginal of θ. Similarly, ν is the second marginal of θ and (8.48) then implies that $\int (x-y)^2 \, \mathrm{d}\theta(x,y) \le a$. □

Throughout the rest of this section we assume the conditions of Theorem 8.3.1, i.e. $M = \lfloor N(1+\alpha) \rfloor$ and $\beta \le \beta(\alpha)$.

Lemma 8.4.4. *Consider continuous functions f and g on $\mathbb{R}$ such that*

$$\forall x, y, \; f(x) + g(y) \le (x-y)^2 \,. \tag{8.49}$$

Then, given any $\varepsilon > 0$, the probability of the event

$$\int f \, \mathrm{d}\mu_{N,M} + \int g \, \mathrm{d}\mu'_{N,M} \le \varepsilon + \frac{6}{\beta^2} \Delta(\nu_{N,M}, \nu'_{N,M}) \tag{8.50}$$

goes to 1 as $N \to \infty$.

Consider an integer r. For a subset J of $\{1, \ldots, N\}$, with $\operatorname{card} J = r$, we set

$$S(f,J) = \frac{1}{r} \sum_{j \in J} f(u_{N,M}(j)) \,; \; S'(g,J) = \frac{1}{r} \sum_{j \in J} g(u'_{N,M}(j)) \,.$$

Lemma 8.4.5. *Consider functions f and g as in (8.49). Then there exists a sequence $\varepsilon(r)$ with $\lim_{r \to \infty} \varepsilon(r) = 0$ and a sequence $\varepsilon(r,N)$ such that $\lim_{N \to \infty} \varepsilon(r,N) = \varepsilon(r)$ for which*

$$S(f,J) + S'(g,J) \le \frac{6}{\beta^2} \Delta(\nu_{N,M}, \nu'_{N,M}) + W(N,J) \tag{8.51}$$

where $W(N,J) \ge 0$ and $\mathsf{E}\, W(N,J) \le \varepsilon(r,N)$.

Proof of Lemma 8.4.4. We average (8.51) over J to get

$$\int f \, \mathrm{d}\mu_{N,M} + \int g \, \mathrm{d}\mu'_{N,M} \le \frac{6}{\beta^2} \Delta(\nu_{N,M}, \nu'_{N,M}) + W(r,N)$$

where $\mathsf{E}\, W(r,N) \le \varepsilon(r,N)$, so that $\mathsf{P}\left(W(r,N) \ge \varepsilon\right) \le \varepsilon(r,N)/\varepsilon$. Thus if Ω_N denotes the event (8.50), we have $\limsup_{N \to \infty} \mathsf{P}\left(\Omega_N^c\right) \le \varepsilon(r)/\varepsilon$. Since this holds for all r, and since $\lim_{r \to \infty} \varepsilon(r) = 0$, the result follows. □

For a subset J of $\{1, \ldots, M\}$, with $\operatorname{card} J \le M - N$, we set $w_{N,M\setminus J}(k) = Z_{N,M}(k; J)/Z_{N,M}(\emptyset; J)$. By (8.14) and symmetry, we have

$$u_{N,M}(j) = \frac{1}{1 + \sum_{k \le N} a(k,j)\, w_{N,M\setminus\{j\}}(k)} .$$

We consider a subset J of $\{1, \ldots, M\}$ with $\operatorname{card} J = r$, $j \in J$ and we set

$$\overline{u}_{N,M}(j) = \frac{1}{1 + \sum_{k \le N} a(k,j)\, w_{N,M\setminus J}(k)} . \tag{8.52}$$

Lemma 8.4.6. *We have*

$$\mathsf{E}\,(u_{N,M}(j) - \overline{u}_{N,M}(j))^2 \le \frac{K(r,\alpha,\beta)}{N} .$$

Proof. Using (8.27) we have

$$\begin{aligned} \mathsf{E}\,(u_{N,M}(j) - \overline{u}_{N,M}(j))^2 &\le \frac{L}{\beta^2}\,\mathsf{E}\,(w_{N,M\setminus\{j\}}(N) - w_{N,M\setminus J}(N))^2 \\ &= \frac{L}{\beta^2}\,\mathsf{E}\,(w_{N,M-1}(N) - w_{N,M-r}(N))^2 , \end{aligned}$$

and the result follows from (8.38). □

Proof of Lemma 8.4.5. As a consequence of Lemma 8.4.6, to prove (8.51) we can replace $u_{N,M}(j)$ by $\overline{u}_{N,M}(j)$ in the definition of $S(f,J)$, and similarly for $S'(g,J)$.

We denote by E_a expectation in the r.v. $a(k,j)$, $k \le N$, $j \in J$ only. These r.v. are independent of the r.v. $w_{N,M\setminus J}(k)$. Given the r.v. $w_{N,M\setminus J}(k)$, the variables $\overline{u}_{N,M}(j)$, $j \in J$ are independent (as shown by (8.52)) and thus, if A is such that $|f(x)| \le A$ for $x \in [0,1]$, we have

$$\mathsf{E}_a \left(\frac{1}{r} \sum_{j \in J} f(\overline{u}_{N,M}(j)) - \frac{1}{r} \sum_{j \in J} \mathsf{E}_a\, f(\overline{u}_{N,M}(j)) \right)^2 \le \frac{A^2}{r} ,$$

so that

$$\frac{1}{r} \sum_{j \in J} f(\overline{u}_{N,M}(j)) \le \frac{1}{r} \sum_{j \in J} \mathsf{E}_a\, f(\overline{u}_{N,M}(j)) + W_1(N,J) , \tag{8.53}$$

where $\mathsf{E}\, W_1(N,J) \le A/\sqrt{r}$. A similar result holds for g and $\overline{u}'_{N,M}(j)$. Next, we prove that, for $j \in J$, we have

$$\mathsf{E}_a\, f(\overline{u}_{N,M}(j)) + \mathsf{E}_a\, g(\overline{u}'_{N,M}(j)) \le \frac{2}{\beta^2}\,\Delta \tag{8.54}$$

where

$$\Delta = \Delta\left(\frac{1}{N}\sum_{k\le N}\delta_{w(k)}, \frac{1}{N}\sum_{k\le N}\delta_{w'(k)}\right),$$

for $w(k) = w_{N,M\setminus J}(k)$, $w'(k) = w'_{N,M\setminus J}(k)$. Indeed by Lemma 8.4.2 we can find a permutation σ of $\{1,\ldots,N\}$ such that $\Delta = N^{-1}\sum_{k\le N}(w(k) - w'(\sigma(k)))^2$. We have

$$\mathsf{E}_a\, f(\overline{u}_{N,M}(j)) = \mathsf{E}_a\, f\left(\frac{1}{1+\sum_{k\le N} a(k,j)\, w(k)}\right)$$

$$\mathsf{E}_a\, g(\overline{u}'_{N,M}(j)) = \mathsf{E}_a\, g\left(\frac{1}{1+\sum_{k\le N} a(k,j)\, w'(\sigma(k))}\right).$$

Using (8.49) and (8.27) we get

$$\mathsf{E}_a\, f(\overline{u}_{N,M}(j)) + \mathsf{E}_a\, g(\overline{u}'_{N,M}(j)) \le \frac{2}{\beta^2 N}\sum_{k\le N}(w(k) - w(\sigma(k)))^2,$$

and this proves (8.54).

Finally, since Δ is the square of a distance, and since $\left(\sum_{i\le 3} x_i\right)^2 \le 3\sum_{i\le 3} x_i^2$, we have

$$\Delta \le 3\Delta\left(\frac{1}{N}\sum_{k\le N}\delta_{w_{N,M}(k)}, \frac{1}{N}\sum_{k\le N}\delta_{w'_{N,M}(k)}\right) + 3W_2(N,J)$$

where

$$\begin{aligned} W_2(N,J) &= \frac{1}{N}\sum_{k\le N}(w_{N,M}(k) - w_{N,M\setminus J}(k))^2 \\ &\quad + \frac{1}{N}\sum_{k\le N}(w'_{N,M}(k) - w'_{N,M\setminus J}(k))^2, \end{aligned}$$

so that, by (8.38), $\mathsf{E}\, W_2(N,J) = 2\,\mathsf{E}\,(w_{N,M}(N) - w_{N,M-r}(N))^2$. This quantity depends on J only through r and tends to 0 as $N \to \infty$. Combining with (8.53) and (8.54) finishes the proof. □

Proposition 8.4.7. *We have*

$$\limsup_{N\to\infty} \mathsf{E}\Delta(\mu_{N,M}, \mu'_{N,M}) \le \frac{L}{\beta^2}\limsup_{N\to\infty}\mathsf{E}\Delta(\nu_{N,M}, \nu'_{N,M}).$$

Proof. From Lemma 8.4.3 and a compactness argument, we see that given $\varepsilon > 0$, there exists a finite family $\mathcal{F}$ of couples (f,g) of continuous functions that satisfy (8.49) and such that, given two probability measures μ, μ' on $[0,1]$ we have

$$\Delta(\mu,\mu') \le \varepsilon + \sup_{(f,g)\in\mathcal{F}} \left(\int f\,\mathrm{d}\mu + \int g\,\mathrm{d}\mu'\right).$$

Using this for $\mu = \mu_{N,M}$, $\mu' = \mu'_{N,M}$ the conclusion follows from Lemma 8.4.4 in a straightforward manner. □

To complement Proposition 8.4.7, we will prove the following.

Proposition 8.4.8. *Under the conditions of Theorem 8.3.1 we have*

$$\limsup_{N\to\infty} \mathsf{E}\Delta(\nu_{N,M},\nu'_{N,M}) \le \frac{L\beta^3}{\alpha^5} \limsup_{N\to\infty} \mathsf{E}\Delta(\mu_{N,M},\mu'_{N,M}). \tag{8.55}$$

Together with Proposition 8.4.7, this proves Theorem 8.4.1, since $\Delta(\mu_{N,M}, \mu'_{N,M}) \le 1$. The proof of Proposition 8.4.8 resembles that of Proposition 8.4.7, but an extra difficulty occurs from the fact that the numbers $w_{N,M}(k)$ need not be bounded. We will overcome this difficulty through a truncation argument. We observe that, from (8.15), (8.22) and (8.32) we have $\mathsf{E}\,w^4_{N,M}(i) \le K(\alpha)$ for $i \le N$. For a number $b > 0$, let us write $w_{N,M,b}(i) = \min(b, w_{N,M}(i))$, so that

$$\mathsf{E}\left((w_{N,M,b}(i) - w_{N,M}(i))^2\right) \le \mathsf{E}\left(w^2_{N,M}(i)\,\mathbf{1}_{\{w_{N,M}(i)\ge b\}}\right) \le \frac{K(\alpha)}{b^2},$$

and if $\nu_{N,M,b} = N^{-1}\sum_{i\le N}\delta_{w_{N,M,b}(i)}$, we have

$$\mathsf{E}\,\Delta(\nu_{M,N},\nu_{N,M,b}) \le \mathsf{E}\left((w_{N,M,b}(N) - w_{N,M}(N))^2\right) \le \frac{K(\alpha)}{b^2}.$$

So, to prove (8.55), it suffices to prove that for each b we have

$$\limsup_{N\to\infty} \mathsf{E}\Delta(\nu_{N,M,b},\nu'_{N,M,b}) \le \frac{L\beta^3}{\alpha^5} \limsup_{N\to\infty} \mathsf{E}\Delta(\mu_{N,M},\mu'_{N,M}).$$

This is very similar to the proof of Proposition 8.4.7, except that instead of using (8.27) we use (8.31). □

8.5 Structure Theorems

Let us denote by $\mathcal{L}(X)$ the law of a r.v. X. Since $\nu_{N,M-1}$ is essentially nonrandom (by Theorem 8.4.1) we can expect from (8.14) that $\mathcal{L}(u_{N,M}) := \mathcal{L}(u_{N,M}(M))$ is determined by $\mathcal{L}(w_{N,M-1}(M-1))$, and hence by $\mathcal{L}(w_{N,M}) := \mathcal{L}(w_{N,M}(M))$. (And the other way around with (8.15).) However making this idea precise requires some technicalities. We will take care of these first. We consider an independent sequence $(X_\ell)_{\ell\ge1}$ uniform over $[0,1]$, and we set $a_N(\ell) = \exp(-\beta N X_\ell)$. This is what was denoted by $a(\ell)$ in the previous section, but we change notation because we need to use different values of N.

Proposition 8.5.1. *Consider integers* N, M, N', M', $M \le 2N$, $M' \le 2N$, *and numbers* $(u_\ell)_{\ell \le M}$, $(u'_\ell)_{\ell \le M'}$, $0 \le u_\ell,\ u'_\ell \le 1$. *Assume that* $\sum_{\ell \le M} u_\ell \ge M\,\alpha/2$ *and* $\sum_{\ell \le M'} u'_\ell \ge M'\,\alpha/2$. *Then*

$$\begin{aligned}
&\Delta\left(\mathcal{L}\left(\frac{1}{\sum_{\ell\le M} a_N(\ell)\,u_\ell}\right), \mathcal{L}\left(\frac{1}{\sum_{\ell\le M'} a_{N'}(\ell)\,u'_\ell}\right)\right)\\
&\le K(\alpha)\left(\frac{1}{N} + \frac{1}{N'} + \left|\frac{M}{N} - \frac{M'}{N'}\right|\right)\\
&+ \frac{L\beta^3}{\alpha^5}\,\Delta\left(\frac{1}{M}\sum_{\ell\le M}\delta_{u_\ell}, \frac{1}{M'}\sum_{\ell\le M'}\delta_{u'_\ell}\right)\\
&+ \frac{L\beta^2}{\alpha^4}\left(\frac{1}{M}\sum_{\ell\le M} u_\ell - \frac{1}{M'}\sum_{\ell\le M'} u'_\ell\right)^2 .
\end{aligned} \tag{8.56}$$

As a consequence, if a sequence μ_r of probability measures of the type $M_r^{-1}\sum_{\ell\le M_r}\delta_{u(r,\ell)}$ with $0 \le u(r,\ell) \le 1$ converges for Wasserstein's distance to a probability measure μ on $[0,1]$ with $\int x\,\mathrm{d}\mu(x) > \alpha/2$, and if $M_r \to \infty$, $M_r/N_r \to 1+\alpha$, the sequence

$$\mathcal{L}\left(\frac{1}{\sum_{\ell\le M_r} a_{N_r}(\ell)\,u(r,\ell)}\right)$$

is a Cauchy sequence for Wasserstein's distance; its limit $A(\mu)$ is independent of the sequence μ_r. (The proof that Wasserstein's distance defines a complete metric space requires a few lines of work.)

Proposition 8.5.2. *Given a number* $\alpha > 0$, *there exists a number* $\beta(\alpha) > 0$ *such that if* $\beta \le \beta(\alpha)$ *to each probability measure* μ *on* $[0,1]$ *with* $\int x\,\mathrm{d}\mu(x) \ge \alpha/2$, *we can associate a probability measure* $A(\mu)$ *on* $\mathbb{R}^+$ *such that, for numbers* $0 \le u_\ell \le 1$, *if* $\sum_{\ell\le M} u_\ell \ge \alpha\,M/2$, *then, given an integer* N,

$$\begin{aligned}
&\Delta\left(A(\mu), \mathcal{L}\left(\frac{1}{\sum_{\ell\le M} a_N(\ell)\,u_\ell}\right)\right) \le K(\alpha)\left(\frac{1}{N} + \left|\frac{M}{N} - (1+\alpha)\right|\right)\\
&+ \frac{L\beta^2}{\alpha^4}\left(\int x\,\mathrm{d}\mu(x) - \frac{1}{M}\sum_{\ell\le M} u_\ell\right)^2 + \frac{L\beta^3}{\alpha^5}\,\Delta\left(\mu, \frac{1}{M}\sum_{\ell\le M}\delta_{u_\ell}\right) .
\end{aligned} \tag{8.57}$$

Moreover,

$$\Delta(A(\mu), A(\mu')) \le \frac{L\beta^2}{\alpha^4}\left(\int x\,\mathrm{d}\mu(x) - \int x\,\mathrm{d}\mu'(x)\right)^2 + \frac{L\beta^3}{\alpha^5}\,\Delta(\mu,\mu') . \tag{8.58}$$

The operator $A(\mu)$ was constructed above, and (8.57), (8.58) follow from (8.56) by taking limits. The operator $A(\mu)$ can be given a nice probabilistic interpretation as follows. Consider a Poisson point process on $\mathbb{R}^+$ of intensity measure Lebesgue's measure; and number the points it creates in non-decreasing order as $(\xi_i)_{i\geq 1}$. Consider an independent sequence $(Y_i)_{i\geq 1}$ distributed like μ, and independent of the sequence (ξ_i). Then $A(\mu)$ is the law of $\left(\sum_{i\geq 1} \exp(-\beta\xi_i/(1+\alpha))\, Y_i\right)^{-1}$. This, however, does not seem so helpful so we do not give the details. The main idea is that if the variables $(X_\ell)_{\ell\leq M}$ are independent uniform over $[0,1]$, the variables $(NX_\ell)_{\ell\leq M}$, when reordered in non-decreasing order resemble the beginning of the sequence $(\xi_i/(1+\alpha))$.

Proposition 8.5.3. *Under the conditions of Theorem 8.3.1 we have*

$$\lim_{N\to\infty} \Delta(\mathcal{L}(w_{N,M}), A(\mathcal{L}(u_{N,M}))) = 0\,.$$

This makes precise the idea that the law of $u_{N,M}$ determines the law of $w_{N,M}$.

Proof. Let us denote by $\mathcal{L}_a$ the law given the sequence $(u_{N-1,M}(\ell))$. Then, using (8.15) in the first line and (8.45) in the second line, (and since integrating outside a supremum yields a larger result than integrating inside it) we get

$$\begin{aligned}
&\Delta(\mathcal{L}(w_{N,M}), A(\mathcal{L}(u_{N,M})))\\
&= \Delta\left(\mathsf{E}\,\mathcal{L}_a\left(\frac{1}{\sum_{\ell\leq M} a_N(\ell)\, u_{N-1,M}(\ell)}\right), A(\mathcal{L}(u_{N,M}))\right)\\
&\leq \mathsf{E}\,\Delta\left(\mathcal{L}_a\left(\frac{1}{\sum_{\ell\leq M} a_N(\ell)\, u_{N-1,M}(\ell)}\right), A(\mathcal{L}(u_{N,M}))\right)\,.
\end{aligned}$$

Using (8.57) it suffices to prove that

$$\lim_{N\to\infty} \mathsf{E}\,\Delta\left(\frac{1}{M}\sum_{\ell\leq M} \delta_{u_{N-1,M}(\ell)}, \mathcal{L}(u_{N,M})\right) = 0$$

since by (8.43) this implies that

$$\lim_{N\to\infty} \mathsf{E}\left|\frac{1}{M}\sum_{\ell\leq M} u_{N-1,M}(\ell) - \mathsf{E}u_{N,M}\right| = 0\,.$$

So, by (8.37), it suffices to prove that

$$\lim_{N\to\infty} \mathsf{E}\,\Delta(\mu_{N,M}, \mathcal{L}(u_{N,M})) = 0\,.$$

But, by (8.40), we have $\mathcal{L}(u_{N,M}) = \mathcal{L}(u'_{N,M}) = \mathsf{E}\,\mu'_{N,M}$ and hence

$$\mathsf{E}\,\Delta(\mu_{N,M}, \mathcal{L}(u_{N,M})) \le \mathsf{E}\,\Delta(\mu_{N,M}, \mu'_{N,M}) .$$

The result follows by Theorem 8.4.1. □

We turn to the proof of Proposition 8.5.1. The main difficulty is that one can have $M \ne M'$. This will be addressed with the following

Lemma 8.5.4. *Consider independent r.v. X_ℓ, X, uniform over $[0,1]$. Consider an integer $R \ge 1$, a number $\gamma \ge 2$ and the r.v.*

$$a = \exp(-\gamma X)\,;\ a' = \sum_{\ell \le R} \exp(-\gamma R\, X_\ell) .$$

Then we can find a pair of r.v. (Y, Y') such that Y has the same law as a and Y' has the same law as a' with

$$\mathsf{E}\,|Y - Y'| \le \frac{L}{\gamma^2}\,,\ \mathsf{E}\,(Y - Y')^2 \le \frac{L}{\gamma^2} . \tag{8.59}$$

Proof. Consider the function Y on $[0,1]$ given by

$$Y(x) = \inf\{t;\ \mathsf{P}(a \ge t) \le x\}$$

and define Y' similarly. The law of Y (resp. Y') under Lebesgue's measure is the law of a (resp. a') and $\int |Y(x) - Y'(x)|\,dx$ is the area between the graphs of Y and Y'. This is also the area between the graphs of the functions $t \mapsto \mathsf{P}(a \ge t)$ and $t \mapsto \mathsf{P}(a' \ge t)$, i.e.

$$\mathsf{E}\,|Y - Y'| = \int_0^\infty |\mathsf{P}(a \ge t) - \mathsf{P}(a' \ge t)|\,dt . \tag{8.60}$$

We proceed to bound the right-hand side. For $t \le 1$ we have

$$\mathsf{P}(a \ge t) = \min\left(1, \frac{1}{\gamma}\log\frac{1}{t}\right)$$

$$\mathsf{P}(\exp(-\gamma\, R\, X_\ell) \ge t) = \min\left(1, \frac{1}{\gamma\, R}\log\frac{1}{t}\right) .$$

Since $a' \ge t$ as soon as one of the summands exceeds t, we have, by independence,

$$\mathsf{P}(a' \ge t) \ge 1 - \left(1 - \min\left(1, \frac{1}{\gamma\, R}\log\frac{1}{t}\right)\right)^R := \psi(t) .$$

Since $(1-x)^R \ge 1 - Rx$ for $x \ge 0$, we have

$$\psi(t) \le \min\left(R, \frac{1}{\gamma}\log\frac{1}{t}\right) ,$$

and since $\psi(t) \le 1$, we have in fact

$$\psi(t) \le \min\left(1, \frac{1}{\gamma}\log\frac{1}{t}\right) = \mathsf{P}(a \ge t) .$$

Since

$$x \ge 0 \Rightarrow (1-x))^R \le e^{-Rx} \le 1 - Rx + \frac{R^2x^2}{2} ,$$

we have

$$0 \le \mathsf{P}(a \ge t) - \psi(t) \le \frac{1}{2}\left(\frac{1}{\gamma}\log\frac{1}{t}\right)^2 . \tag{8.61}$$

Thus,

$$\begin{aligned}
&\int_0^1 |\mathsf{P}(a \ge t) - \mathsf{P}(a' \ge t)|\,\mathrm{d}t \\
&\le \int_0^1 |\mathsf{P}(a \ge t) - \psi(t)|\,\mathrm{d}t + \int_0^1 |\mathsf{P}(a' \ge t) - \psi(t)|\,\mathrm{d}t \\
&\le \int_0^1 |\mathsf{P}(a \ge t) - \psi(t)|\,\mathrm{d}t + \int_0^1 \mathsf{P}(a' \ge t) - \psi(t)\,\mathrm{d}t \\
&\le 2\int_0^1 |\mathsf{P}(a \ge t) - \psi(t)|\,\mathrm{d}t + \int_0^1 (\mathsf{P}(a' \ge t) - \mathsf{P}(a \ge t))\,\mathrm{d}t .
\end{aligned}$$

Hence, using (8.61), and since $a \le 1$,

$$\begin{aligned}
&\int_0^\infty |\mathsf{P}(a \ge t) - \mathsf{P}(a' \ge t)|\,\mathrm{d}t \\
&= \int_0^1 |\mathsf{P}(a \ge t) - \mathsf{P}(a' \ge t)|\,\mathrm{d}t + \int_1^\infty \mathsf{P}(a' \ge t)\,\mathrm{d}t \\
&\le 2\int_0^1 |\mathsf{P}(a \ge t) - \psi(t)|\,\mathrm{d}t + \int_0^\infty \mathsf{P}(a' \ge t)\,\mathrm{d}t - \int_0^1 \mathsf{P}(a \ge t)\,\mathrm{d}t \\
&\le \frac{L}{\gamma^2} + \mathsf{E}\,a' - \mathsf{E}\,a \le \frac{L}{\gamma^2} .
\end{aligned} \tag{8.62}$$

Thus we have shown that $\mathsf{E}\,|Y - Y'| \le L/\gamma^2$. Since $Y \le 1$, we have

$$\begin{aligned}
\mathsf{E}\,(Y - Y')^2 &\le 2\,\mathsf{E}\,((Y - \min(Y',2))^2 + 2\,\mathsf{E}\,((\min(Y',2) - Y')^2) \\
&\le 4\,\mathsf{E}\,|Y - Y'| + 4\int_0^\infty t\,\mathsf{P}(a' \ge t+2)\,\mathrm{d}t \\
&\le \frac{L}{\gamma^2} + 4\int_0^\infty t\,\mathsf{P}(a' \ge t+2)\,\mathrm{d}t .
\end{aligned} \tag{8.63}$$

Now, for $\lambda > 0$

$$\begin{aligned}
\mathsf{P}(a' \ge t) &\le \exp(-\lambda t)\,\mathsf{E}\exp\lambda\,a' \\
&= \exp(-\lambda t)(\mathsf{E}\exp(\lambda\exp(-\gamma\,R\,X)))^R
\end{aligned}$$

and

$$\begin{aligned}\mathsf{E}\exp(\lambda\exp(-\gamma\, R\, X)) &= \sum_{p\geq 0}\frac{\lambda^p}{p!}\,\mathsf{E}\exp(-\gamma\, R\, p\, X)\\ &\leq 1+\sum_{p\geq 1}\frac{\lambda^p}{p!\, p\,\gamma\, R}\leq 1+\frac{e^\lambda}{\gamma\, R}\\ &\leq \exp\left(\frac{e^\lambda}{\gamma\, R}\right)\end{aligned}$$

so that

$$\mathsf{P}(a'\geq t)\leq \exp\left(\frac{e^\lambda}{\gamma}-\lambda\, t\right)\,.$$

Taking $\lambda=\log\gamma$, we get

$$\mathsf{P}(a'\geq t)\leq L\,\gamma^{-t}$$

so that since $\gamma\geq 2$ we have

$$\int_0^\infty t\,\mathsf{P}(a'\geq t+2)\,\mathrm{d}t\leq \frac{L}{\gamma^2}\,.$$

□

Lemma 8.5.5. *If the r.v. X is uniform over $[0,1]$, then for $p\geq 1$ we have*

$$\mathsf{E}\,|\exp(-\gamma\, X)-\exp(-\gamma'\, X)|^p\leq L\left|\frac{1}{\gamma}-\frac{1}{\gamma'}\right|\,. \tag{8.64}$$

Proof. Assuming e.g. $\gamma\leq\gamma'$, we have $0\leq\exp(-\gamma\, X)-\exp(-\gamma\, X')\leq 1$, so that

$$\begin{aligned}\mathsf{E}\,|\exp(-\gamma\, X)-\exp(-\gamma'\, X)|^p &\leq \mathsf{E}\,(\exp(-\gamma\, X)-\exp(-\gamma'\, X))\\ &=\frac{1}{\gamma}\,(1-\exp(-\gamma))-\frac{1}{\gamma'}\,(1-\exp(-\gamma'))\,.\end{aligned}$$

□

Proof of Proposition 8.5.1. The idea is to reduce to the case $M=M'$ by using Lemma 8.5.4. Applying this lemma for $\gamma=\beta N$, we can find an independent sequence $a'(\ell)$, $\ell\leq M$, with

$$\mathsf{E}\,|a_N(\ell)-a'(\ell)|\leq\frac{L}{\beta^2\, N^2}\,,\quad \mathsf{E}\,((a_N(\ell)-a'(\ell))^2)\leq\frac{L}{\beta^2\, N^2} \tag{8.65}$$

where $a'(\ell)=\sum_{k\leq M'}\exp(-\beta\, N\, M' X_{\ell,k})$, and where the r.v. $X_{\ell,k}$ are independent uniform over $[0,1]$. We have

$$\Delta\left(\mathcal{L}\left(\frac{1}{\sum_{\ell\le M} a_N(\ell)\, u_\ell}\right), \mathcal{L}\left(\frac{1}{\sum_{\ell\le M} a'(\ell)\, u_\ell}\right)\right)$$

$$\le \mathsf{E}\left(\frac{\sum_{\ell\le M}(a_N(\ell)-a'(\ell))\, u_\ell}{\left(\sum_{\ell\le M} a_N(\ell)\, u_\ell\right)\left(\sum_{\ell\le M} a'(\ell)\, u_\ell\right)}\right)^2 \le \mathrm{I}+\mathrm{II} \tag{8.66}$$

where

$$\mathrm{I} = 2\,\mathsf{E}\frac{\left(\sum_{\ell\le M} |a_N(\ell)-a'(\ell)|\, u_\ell\right)^2}{\left(\sum_{\ell\le M} a_N(\ell)\, u_\ell\right)^4}\ ;\ \mathrm{II} = 2\,\mathsf{E}\frac{\left(\sum_{\ell\le M} |a_N(\ell)-a'(\ell)|\, u'_\ell\right)^2}{\left(\sum_{\ell\le M} a(\ell)\, u_\ell\right)^4}\,.$$

Using (8.65), a small variation of the proof of (8.31) shows that I and II are at most $K(\alpha,\beta)/N$. The sequence $(a'(\ell))_{\ell\ge1}$ has the same distribution as the sequence $(a_{NM'}(\ell))_{\ell\ge1}$. Thus (repeating each value u_ℓ exactly M' times) we have succeeded in replacing N by NM' and M by MM'. In a similar manner we replace N' by $N'M$ and M' by MM', and thus we can assume $M = M'$. It then suffices to bound

$$\mathrm{III} - \Delta\left(\mathcal{L}\left(\frac{1}{\sum_{\ell\le M} a_N(\ell)\, u_\ell}\right)\ ,\ \mathcal{L}\left(\frac{1}{\sum_{\ell\le M} a_{N'}(\ell)\, u'_\ell}\right)\right) \le \mathrm{IV}+\mathrm{V}\ ,$$

where

$$\mathrm{IV} = 2\Delta\left(\mathcal{L}\left(\frac{1}{\sum_{\ell\le M} a_N(\ell)\, u_\ell}\right)\ ,\ \mathcal{L}\left(\frac{1}{\sum_{\ell\le M} a_N(\ell)\, u'_\ell}\right)\right)$$

$$\mathrm{V} = 2\Delta\left(\mathcal{L}\left(\frac{1}{\sum_{\ell\le M} a_N(\ell)\, u'_\ell}\right)\ ,\ \mathcal{L}\left(\frac{1}{\sum_{\ell\le M} a_{N'}(\ell)\, u'_\ell}\right)\right)\,.$$

It follows from Lemma 8.4.2 and (8.31) that

$$\mathrm{IV} = \frac{L\beta^2}{\alpha^4}\left(\frac{1}{M}\sum_{\ell\le M} u_\ell - \frac{1}{M}\sum_{\ell\le M} u'_\ell\right)^2 + \frac{L\beta^3}{\alpha^5}\Delta\left(\frac{1}{M}\sum_{\ell\le M}\delta_{u_\ell}, \frac{1}{M}\sum_{\ell\le M}\delta_{u'_\ell}\right)\,.$$

Now

$$\mathrm{V} \le 2\,\mathsf{E}\frac{\left(\sum_{\ell\le M}(a_N(\ell)-a_{N'}(\ell))\, u'_\ell\right)^2}{\left(\sum_{\ell\le M} a_N(\ell)\, u'_\ell\right)^2\left(\sum_{\ell\le M} a_{N'}(\ell)\, u'_\ell\right)^2}\,.$$

Using (8.64) and a small variation of the proof of (8.31) we see that $\mathrm{V} \le K(\alpha)\, M\, |1/N - 1/N'|$. (It is here that the condition $M \le 2N$, $M' \le 2N'$ is used. When $M = M'$, it implies that $M\,|1/N - 1/N'| \le 2$, and $(M\,|1/N - 1/N'|)^2 \le 2M\,|1/N - 1/N'|$.) □

Proposition 8.5.6. *To each probability measure ν on $\mathbb{R}^+$ we can associate a probability measure $B(\nu)$ on $[0,1]$ with the following property. Given numbers $w(k) \geq 0$, $k \leq N$, we have*

$$\Delta\left(B(\nu), \mathcal{L}\left(\frac{1}{1+\sum_{k\leq N} a(k)\, w(k)}\right)\right) \leq \frac{K(\alpha)}{N} + \frac{L}{\beta^2}\Delta\left(\nu, \frac{1}{N}\sum_{k\leq N}\delta_{w(k)}\right) .$$

Moreover, for two probability measures ν, ν' on $\mathbb{R}$ we have

$$\Delta(B(\nu), B(\nu')) \leq \frac{L}{\beta^2}\,\Delta(\nu, \nu') . \tag{8.67}$$

Proof. Similar to Proposition 8.5.2, but easier. □

Proposition 8.5.7. *We have*

$$\lim_{N\to\infty} \Delta(\mathcal{L}(u_{N,M}), B(\mathcal{L}(w_{N,M}))) = 0 .$$

Proof. Similar to Proposition 8.5.3. □

Theorem 8.5.8. *Under the conditions of Theorem 8.3.1, the limits*

$$\mu = \lim_{N\to\infty} \mathcal{L}(u_{N,M})\,; \ \ \nu = \lim_{N\to\infty} \mathcal{L}(w_{N,M}) \tag{8.68}$$

exist. Moreover

$$\int x\, \mathrm{d}\mu\,(x) = \alpha\,, \ \ \mu = B(\nu)\,, \ \ \nu = A(\mu)\,. \tag{8.69}$$

Proof. It follows from (8.58) and (8.67) that there can be at most one couple (μ, ν) that satisfies (8.69). Propositions 8.5.3 and 8.5.7 show that any cluster point (μ, ν) of the sequence $(\mathcal{L}(u_{N,M}), \mathcal{L}(w_{N,M}))$ satisfies these relations; so the cluster point is unique and the limits exist. □

Comment. We do not know how to prove directly the existence of a solution to the relations (8.69).

Research problem 8.5.9. (Level 2) Is it true that given an integer n, there exists a constant $K(\alpha, n)$, and independent r.v. $U_1, \ldots, U_n$ of law μ with

$$\mathsf{E}\sum_{i\leq r}(u_{N,M}(i) - U_i)^2 \leq \frac{K(\alpha, n)}{N} \ ? \tag{8.70}$$

The problem is to get a strong rate of convergence. We have already essentially proved that we can find the r.v. U_i such that the left-hand side of (8.70) goes to 0 as $N \to \infty$.

Theorem 8.5.10. *Under the conditions of Theorem 8.3.1, we have*

$$\frac{1}{N}\lim_{N\to\infty} \mathsf{E}\log Z_{N,M} = -\int \log x \, d\mu(x) - (1+\alpha)\int \log x \, d\nu(x) \, .$$

Proof. We will stay somewhat informal in this proof. We write $A_{N,M} = \mathsf{E}\log Z_{N,M}$, so that

$$A_{N,M} - A_{N,M-1} = \mathsf{E}\log\frac{Z_{N,M}}{Z_{N,M-1}} = -\mathsf{E}\log u_{N,M}(M-1)$$

$$A_{N,M} - A_{N-1,M} = \mathsf{E}\log\frac{Z_{N,M}}{Z_{N-1,M}} = -\mathsf{E}\log w_{N,M}(N) \, .$$

By Theorem 8.5.8, these quantities have limits $-\int \log x \, d\mu(x)$ and $-\int \log x \, d\nu(x)$ respectively. (To obtain the required tightness, we observe that from (8.14), (8.15) and Markov's inequality we have $\mathsf{P}(u_{N,M}(M-1) < t) \le Kt$ and $\mathsf{P}(w_{N,M}(N) < t) < Kt$.) Setting $M(R) = \lfloor R(1+\alpha) \rfloor$, we write $A_{N,M} - A_{1,1}$ as the sum of the $N-1$ quantities $A_{R,M(R)} - A_{R-1,M(R)}$ for $2 \le R \le M$ and of about $(1+\alpha)N$ quantities $A_{R-1,M(R)-\ell} - A_{R-1,M(R)-\ell-1}$ for $\ell = 0$ or $\ell = 1$. (Observe that $M(R) - 2 \le M(R-1) \le M(R) - 1$.) □

It is probably easy at this stage to study in detail Gibbs' measure G, but the utter state of exhaustion that the author has now reached does not allow us to say more than a few informal words. If G denotes Gibbs' measure, the relation

$$G(\{\sigma(i) = j\}) = a(i,j)\,\frac{Z_{N,M}(i;j)}{Z_{N,M}} \simeq a(i,j)\, w_{N,M}(i)\, u_{N,M}(j) \tag{8.71}$$

should allow a precise description of the law of $\sigma \mapsto \sigma(i)$ under G. If $i_1 \ne i_2$, $j_1 \ne j_2$, we have

$$G(\{\sigma(i_1) = j_1 \,;\ \sigma(i_2) = j_2\}) = a(i_1,j_1)\, a(i_2,j_2)\, \frac{Z_{N,M}(i_1,i_2;j_1,j_2)}{Z_{N,M}} \, . \tag{8.72}$$

One can generalize (8.12) to show that

$$\frac{Z_{N,M}(i_1,i_2;j_1,j_2)}{Z_{N,M}} \simeq w_{N,M}(i_1)\, w_{N,M}(i_2)\, u_{N,M}(j_1)\, u_{N,M}(j_2)$$

and thus (8.71) and (8.72) show that, as expected

$$G(\{\sigma(i_1) = j_1, \sigma(i_2) = j_2\}) \simeq G(\{\sigma(i_1) = j_1\})\, G(\{\sigma(i_2) = j_2\}) \, .$$

A. Appendix. Elements of Probability Theory

A.1 Tail Estimates

Given any r.v. X, we have

$$\mathsf{P}(X \geq t)\, e^t \leq \mathsf{E} \exp X \,.$$

Thus, if $\lambda > 0$, since $\mathsf{P}(X \geq t) = \mathsf{P}(\lambda X \geq \lambda t)$,

$$\mathsf{P}(X \geq t) \leq e^{-\lambda t}\, \mathsf{E} \exp \lambda X$$

and, changing X into $-X$ and t into $-t$,

$$\mathsf{P}(X \leq t) \leq e^{\lambda t}\, \mathsf{E} \exp(-\lambda X) \,.$$

If $X = \sum_{i \leq N} X_i$ where $(X_i)_{i \leq N}$ are independent,

$$\mathsf{P}(X \geq t) \leq e^{-\lambda t} \prod_{i \leq N} \mathsf{E} \exp \lambda X_i = \exp\left(-\lambda t + \sum_{i \leq N} \log \mathsf{E} \exp \lambda X_i \right) . \quad \text{(A.1)}$$

If $(\eta_i)_{i \leq N}$ are independent Bernoulli r.v., i.e. $\mathsf{P}(\eta_i = \pm 1) = 1/2$, then $\mathsf{E} \exp \lambda\, a_i\, \eta_i = \operatorname{ch} \lambda\, a_i$, and thus

$$\mathsf{P}\left(\sum_{i \leq N} a_i\, \eta_i \geq t \right) \leq \exp\left(-\lambda t + \sum_{i \leq N} \log \operatorname{ch} \lambda\, a_i \right) . \quad \text{(A.2)}$$

It is obvious on power series expansion that $\operatorname{ch} t \leq \exp(t^2/2)$, so that

$$\mathsf{P}\left(\sum_{i \leq N} a_i\, \eta_i \geq t \right) \leq \exp\left(-\lambda t + \frac{\lambda^2}{2} \sum_{i \leq N} a_i^2 \right) ,$$

and by optimization over λ, for all $t \geq 0$,

$$\mathsf{P}\left(\sum_{i \leq N} a_i\, \eta_i \geq t \right) \leq \exp\left(-\frac{t^2}{2 \sum_{i \leq N} a_i^2} \right) . \quad \text{(A.3)}$$

This inequality is often called the subgaussian inequality. By symmetry, $\mathsf{P}\left(\sum_{i\geq N} a_i\,\eta_i \leq -t\right)$ is bounded by the same expression, so that

$$\mathsf{P}\left(\left|\sum_{i\leq N} a_i\,\eta_i\right| \geq t\right) \leq 2\exp\left(-\frac{t^2}{2\sum_{i\leq N} a_i^2}\right) . \tag{A.4}$$

Related to (A.3) is the fact that

$$\mathsf{E}\exp\frac{1}{2}\left(\sum_{i\leq N} a_i\,\eta_i\right)^2 \leq \frac{1}{\sqrt{1-\sum_{i\leq N} a_i^2}} . \tag{A.5}$$

Equivalently,

$$\sum \exp\frac{1}{2}\left(\sum_{i\leq N} a_i\,\sigma_i\right)^2 \leq \frac{2^N}{\sqrt{1-\sum_{i\leq N} a_i^2}} ,$$

where the summation is over all sequences $(\sigma_i)_{i\leq N}$ with $\sigma_i = \pm 1$. To prove (A.5) we consider a standard Gaussian r.v. g independent of the r.v. η_i and, using (A.36), we have, denoting by E_g expectation in g only,

$$\begin{aligned}
\mathsf{E}\exp\frac{1}{2}\left(\sum_{i\leq N} a_i\,\eta_i\right)^2 &= \mathsf{E}\,\mathsf{E}_g \exp\sum_{i\leq N} g\,a_i\,\eta_i \\
&= \mathsf{E}_g \exp\sum_{i\leq N} \log\operatorname{ch} g\,a_i \\
&\leq \mathsf{E}_g \exp\frac{g^2}{2}\sum_{i\leq N} a_i^2 \\
&= \frac{1}{\sqrt{1-\sum_{i\leq N} a_i^2}} ,
\end{aligned}$$

where we have used again that $\log\operatorname{ch} t \leq t^2/2$.

Going back to (A.2), if $a_i = 1$ for each $i \leq N$, changing t into Nt, we get

$$\mathsf{P}\left(\sum_{i\leq N} \eta_i \geq Nt\right) \leq \exp N(-\lambda t + \log\operatorname{ch}\lambda) .$$

If $0 \leq t < 1$, the exponent is minimized for $\operatorname{th}\lambda = t$, i.e.

$$\frac{e^{\lambda}-e^{-\lambda}}{e^{\lambda}+e^{-\lambda}} = \frac{e^{2\lambda}-1}{e^{2\lambda}+1} = t ,$$

so that $e^{2\lambda} = (1+t)/(1-t)$ and

$$\lambda = \frac{1}{2}(\log(1+t) - \log(1-t)) .$$

Also, $\mathrm{ch}^{-2}\lambda = 1 - \mathrm{th}^2\lambda = 1 - t^2$, so that

$$\log \mathrm{ch}\,\lambda = -\frac{1}{2}\log(1-t^2) ,$$

and

$$\begin{aligned}\min_{\lambda}\,(-\lambda t + \log \mathrm{ch}\,\lambda) &= -\frac{1}{2}(t\log(1+t) - t\log(1-t)) \\ &\quad -\frac{1}{2}\log(1-t) - \frac{1}{2}\log(1+t) \\ &= -\mathcal{I}(t) \end{aligned} \tag{A.6}$$

where

$$\mathcal{I}(t) = \frac{1}{2}((1+t)\log(1+t) + (1-t)\log(1-t)) . \tag{A.7}$$

The function $\mathcal{I}(t)$ is probably better understood by noting that

$$\mathcal{I}(0) = \mathcal{I}'(0) = 0\,,\ \mathcal{I}''(t) = \frac{1}{1-t^2} . \tag{A.8}$$

It follows from (A.6) that

$$\mathsf{P}\left(\sum_{i\le N}\eta_i \ge Nt\right) \le \exp(-N\mathcal{I}(t)) ,$$

or, equivalently, that

$$\mathrm{card}\left\{\boldsymbol{\sigma}\in\Sigma_N\,;\ \sum_{i\le N}\sigma_i \ge tN\right\} \le 2^N\exp(-N\mathcal{I}(t)) . \tag{A.9}$$

If k is an integer, then $\sum_{i\le N}\sigma_i = k$ provided the sequence $(\sigma_i)_{i\le N}$ contains $(N+k)/2$ times 1 and $(N-k)/2$ times -1. Using Stirling's formula $n! \sim n^n\,e^{-n}\sqrt{2\pi n}$, we get

$$\begin{aligned}\mathrm{card}\left\{\boldsymbol{\sigma}\in\Sigma_N\,;\ \sum_{i\le N}\sigma_i = k\right\} &= \binom{N}{\frac{N+k}{2}} = \frac{N!}{\left(\frac{N+k}{2}\right)!\left(\frac{N-k}{2}\right)!} \\ &\ge \frac{1}{L}\frac{\sqrt{N}}{\sqrt{(N-k)(N+k)}}\frac{N^N}{\left(\frac{N+k}{2}\right)^{(N+k)/2}\left(\frac{N-k}{2}\right)^{(N-k)/2}} \\ &\ge \frac{2^N}{L\sqrt{N}}\frac{1}{\left(1+\frac{k}{N}\right)^{(N+k)/2}\left(1-\frac{k}{N}\right)^{(N-k)/2}} \\ &= \frac{2^N}{L\sqrt{N}}\exp\left(-N\mathcal{I}\left(\frac{k}{N}\right)\right) . \end{aligned} \tag{A.10}$$

Since by Lemma 5.3.6 the function $t \to \log \mathrm{ch}\sqrt{t}$ is concave, we have from (A.2) that

$$\mathsf{P}\left(\sum_{i\le N} a_i\,\eta_i \ge t\,\sqrt{N}\right) \le \exp N\left(-\lambda t + \log \mathrm{ch}\,\lambda\sqrt{\sum_{i\le N} a_i^2}\right)$$

and, using (A.6)

$$\mathsf{P}\left(\sum_{i\le N} a_i\,\eta_i \ge t\,\sqrt{N}\right) \le \exp\left(-N\mathcal{I}\left(\frac{t}{\sqrt{\sum_{i\le N} a_i^2}}\right)\right) . \tag{A.11}$$

A.2 Bernstein's Inequality

Theorem A.2.1. *Consider a r.v. X with $\mathsf{E}X = 0$ and an independent sequence $(X_i)_{i\le N}$ distributed like X. Assume that, for a certain number A, we have*

$$\mathsf{E}\exp\frac{|X|}{A} \le 2 . \tag{A.12}$$

Then, for all $t > 0$ we have

$$\mathsf{P}\left(\sum_{i\le N} X_i \ge t\right) \le \exp\left(-\min\left(\frac{t^2}{4NA^2}, \frac{t}{2A}\right)\right) \tag{A.13}$$

$$\mathsf{P}\left(\sum_{i\le N} X_i \ge t\right) \le \exp\left(-\frac{t^2}{2N\mathsf{E}X^2}\left(1 - \frac{4A^3 t}{N(\mathsf{E}X^2)^2}\right)\right) . \tag{A.14}$$

Proof. From (A.1) we have

$$\mathsf{P}\left(\sum_{i\le N} X_i \ge t\right) \le \exp(-\lambda t + N\log \mathsf{E}\exp\lambda X) . \tag{A.15}$$

We have

$$\mathsf{E}\exp\lambda X = 1 + \mathsf{E}\,\varphi(\lambda X) \tag{A.16}$$

where $\varphi(x) = e^x - x - 1$. Now $\varphi(x) \le \varphi(|x|)$ and for $x > 0$, the function $\lambda \to \varphi(\lambda x)/\lambda^2$ increases. This is obvious on power series expansions. Thus, for $\lambda \le 1/A$, we have

$$\mathsf{E}\,\varphi(\lambda X) \le \lambda^2 A^2 \mathsf{E}\,\varphi(|X|/A) \le \lambda^2 A^2 (2-1) \le \lambda^2 A^2 .$$

Since $\log(1+x) \le x$, we have $\log \mathsf{E}\exp\lambda X \le \lambda^2 A^2$, and by (A.15),

$$\mathsf{P}\left(\sum_{i\le N} X_i \ge t\right) \le \exp(-\lambda t + N\lambda^2 A^2)\,.$$

We take $\lambda = t/2NA^2$ if $t \le 2NA$. When $t \ge 2NA$, we take $\lambda = 1/A$, and then

$$-\lambda t + N\lambda^2 A^2 = -\frac{t}{A} + N \le -\frac{t}{2A}\,.$$

This proves (A.13). To prove (A.14) we replace (A.16) by

$$\mathsf{E}\exp\lambda X = 1 + \frac{\lambda^2 \mathsf{E}X^2}{2} + \mathsf{E}\varphi_1(\lambda X)$$

where $\varphi_1(x) = e^x - x^2/2 - x - 1$. Then $\varphi_1(x) \le \varphi_1(|x|)$, and for $x > 0$, the function $\lambda \to \varphi_1(\lambda x)/\lambda^3$ increases. Thus, if $\lambda \le 1/A$, we get

$$\mathsf{E}\varphi_1(\lambda X) \le \lambda^3 A^3 \mathsf{E}\varphi_1(|X|/A) \le \lambda^3 A^3$$

so that $\log \mathsf{E}\exp\lambda X \le \lambda^2 \mathsf{E}X^2/2 + \lambda^3 A^3$ and we take $\lambda = t/N\mathsf{E}X^2$ to obtain (A.14). The restriction $\lambda \le 1/A$ is irrelevant. Indeed, if $\lambda \ge 1/A$, then $t \ge N\mathsf{E}X^2/A$, and

$$\frac{2A^3 t}{N(\mathsf{E}X^2)^2} \ge \frac{2A^2}{\mathsf{E}X^2} \ge 1$$

because $\mathsf{E}X^2/2A^2 \le \mathsf{E}\exp|X|/A \le 2$. Thus (A.14) is automatically satisfied. □

Another important version of Bernstein's inequality assumes

$$|X| \le A\,. \tag{A.17}$$

In that case we have

$$\mathsf{E}\varphi(\lambda X) = \sum_{p\ge 2}\frac{\lambda^p}{p!}\,\mathsf{E}X^p \le \lambda^2\,\mathsf{E}X^2 \sum_{p\ge 2}\frac{(\lambda A)^{p-2}}{p!} \le \lambda^2\,\mathsf{E}X^2$$

if $|\lambda|\,A \le 1$. Proceeding as before, and taking now $\lambda = \min(t/\mathsf{E}X^2, 1/A)$, we get

$$\mathsf{P}\left(\sum_{i\le N} X_i \ge t\right) \le \exp\left(-\min\left(\frac{t^2}{4N\mathsf{E}X^2}, \frac{t}{2A}\right)\right)\,. \tag{A.18}$$

We will also need a version of (A.13) for martingale difference sequences. Assume that we are given an increasing sequence $(\Xi_i)_{0\le i\le N}$ of σ-algebras. A sequence $(X_i)_{1\le i\le N}$ is called a martingale difference sequence if X_i is Ξ_i-measurable and $E_{i-1}(X_i) = 0$, where E_{i-1} denotes conditional expectation given Ξ_{i-1}. Let us assume that for a certain number A we have

$$\forall i \le N\,,\ E_{i-1} \exp \frac{|X_i|}{A} \le 2\,. \tag{A.19}$$

Exactly as before, this implies that for $|\lambda|\, A \le 1$ we have $E_{i-1} \exp \lambda X_i \le \exp \lambda^2 A^2$. Thus

$$\begin{aligned} E_{k-1} \exp \lambda \sum_{i \le k} X_i &= \exp\left(\lambda \sum_{i \le k-1} X_i\right) E_k \exp \lambda\, X_k \\ &\le \exp\left(\lambda \sum_{i \le k-1} X_i + \lambda^2 A^2\right)\,. \end{aligned}$$

By decreasing induction over k, this shows that

$$E_{k-1} \exp \lambda \sum_{i \le N} X_i \le \exp\left(\lambda \sum_{i \le k-1} X_i + (N-k+1)\,\lambda^2 A^2\right)\,.$$

Taking $k=1$, taking expectation yields $\mathsf{E} \exp \lambda \sum_{i \le N} X_i \le \exp N \lambda^2 A^2$, and use of Chebyshev inequality as before gives

$$\mathsf{P}\left(\sum_{i \le N} X_i \ge t\right) \le \exp\left(-\min\left(\frac{t^2}{4NA}, \frac{1}{2A}\right)\right)\,. \tag{A.20}$$

A.3 ε-Nets

A *ball* of $\mathbb{R}^M$ is a convex balanced set with non-empty interior. The convex hull of a set A is denoted by $\operatorname{conv} A$.

Proposition A.3.1. *Given a ball B of $\mathbb{R}^M$, we can find a subset A of B such that*

$$\operatorname{card} A \le \left(1 + \frac{1}{\varepsilon}\right)^M \tag{A.21}$$

$$\forall x \in B\,,\ A \cap (x + 2\,\varepsilon B) \ne \emptyset \tag{A.22}$$

$$\operatorname{conv} A \supset (1 - 2\,\varepsilon) B\,. \tag{A.23}$$

Given a linear functional φ on $\mathbb{R}^M$,

$$\sup_{x \in A} \varphi(x) \ge (1 - 2\,\varepsilon) \sup_{x \in B} \varphi(x)\,. \tag{A.24}$$

As a corollary, we can find a subset A of $(1-2\varepsilon)^{-1}B$ such that $\operatorname{card} A \le (1+\varepsilon^{-1})^M$ and $B \subset \operatorname{conv} A$. The case $\varepsilon = 1/4$ is of interest: $\operatorname{card} A \le 5^M$, $\sup_{x\in A}\varphi(x) \ge (1/2)\sup_{x\in B}\varphi(x)$.

Proof. We simply take for A a maximal subset of B such that the sets $x+\varepsilon B$ are disjoint for $x \in A$. These sets are of volume $\varepsilon^M \operatorname{Vol} B$, and are entirely contained in the set $(1+\varepsilon)B$, which is of volume $(1+\varepsilon)^M \operatorname{Vol} B$, and this proves (A.21).

Given x in B, we can find y in A with $(x+\varepsilon B)\cap(y+\varepsilon B) \ne \emptyset$, for otherwise this would contradict the maximality of A. Thus $y \in (x+2\varepsilon B)\cap A$, and this proves (A.22).

Given x in B, we can find y_0 in A with $x - y_0 \in 2\varepsilon B$. Applying this to $(x-y_0)/2\varepsilon$, we find y_1 in A with $x - y_0 - 2\varepsilon y_1 \in (2\varepsilon)^2 B$, and in this manner we find a sequence (y_i) in A with

$$y = \sum_{i\ge 0}(2\varepsilon)^i y_i \in (1-2\varepsilon)^{-1}\operatorname{conv} A\ ,$$

since A is finite. This proves (A.23), of which (A.24) is an immediate consequence. □

A.4 Random Matrices

In this section we get some control of the norm of certain random matrices. Much more detailed (and difficult) results are known.

Lemma A.4.1. *Consider independent standard Gaussian r.v.* $(g_{ij})_{1\le i<j\le N}$. *Then, with overwhelming probability (that is, with probability at least* $1 - L\exp(-N/L)$*) we have*

$$\forall\,(x_i)_{i\le N}\,,\ \forall\,(y_i)_{i\le N}\,,\ \left|\sum_{i<j} g_{ij}\,x_i\,y_j\right| \le L\sqrt{N}\left(\sum_{i\le N} x_i^2 \sum_{i\le N} y_i^2\right)^{1/2}. \tag{A.25}$$

Proof. Let us denote by B the Euclidean ball of $\mathbb{R}^N$, and by A a subset of $2B$ with $\operatorname{card} A \le 5^N$ and $\operatorname{conv} A \supset B$, as provided by Proposition A.3.1. If $(x_i)_{i\le N}$ and $(y_i)_{i\le N}$ belong to A, then $\mathsf{E}\left(\sum_{i<j} g_{ij}\,x_i\,y_j\right)^2 \le \sum_{i\le N} x_i^2 \sum_{j\le N} y_j^2 \le 16$ and by (A.37) we have

$$\mathsf{P}\left(\left|\sum_{i<j} g_{ij}\,x_i\,y_j\right| \ge t\right) \le 2\exp\left(-\frac{t^2}{32}\right)\,,$$

so that with probability at most $2(25)^N\exp(-8N)$ we have

$$\forall (x_i)_{i\le N}\,,\ \forall (y_i)_{i\le N} \in A\,,\ \left|\sum_{i<j} g_{ij}\, x_i\, y_j\right| \le 16\sqrt{N}$$

and this implies (A.25). □

We consider independent Bernoulli r.v. $(\eta_{i,k})_{i\le N, k\le M}$, that is, $\mathsf{P}(\eta_{i,k} = \pm 1) = 1/2$.

Lemma A.4.2. *Consider numbers* $(\alpha_{k,k'})_{k,k'\le M}$ *with* $\sum \alpha_{k,k'}^2 \le 1$. *Then, for* $t > 0$ *we have*

$$\mathsf{P}\left(\sum_{k\ne k'}\sum_{i\le N} \alpha_{k,k'}\,\eta_{i,k}\,\eta_{i,k'} \ge t\right) \le \exp\left(-\min\left(\frac{t^2}{NL}, \frac{t}{L}\right)\right) \tag{A.26}$$

$$\mathsf{P}\left(\sum_{k\ne k'}\sum_{i\le N} \alpha_{k,k'}\,\eta_{i,k}\,\eta_{i,k'} \ge t\right) \le \exp\left(-\frac{t^2}{2N}\left(1-\frac{Lt}{N}\right)\right)\,. \tag{A.27}$$

Proof. The r.v. $X_i = \sum_{k\ne k'} \alpha_{k,k'}\,\eta_{i,k}\,\eta_{i,k'}$ are i.i.d., and obviously $\mathsf{E}X_i = 0$, $\mathsf{E}X_i^2 = \sum \alpha_{k,k'}^2 \le 1$. By an important result of C. Borell [13], we have $\mathsf{E}\exp(|X_i|/L) \le 2$ so that (A.26) is a consequence of (A.13) and (A.27) is a consequence of (A.14). □

Proposition A.4.3. *Consider a number* $0 < a \le 1$. *If* $n\log(LM/n) \le Na^2$, *the following event occurs with probability at least* $1 - L\exp(-a^2N/L)$. *Given any subset* I *of* $\{1,\ldots,M\}$ *with* $\operatorname{card} I = n$, *and any sequences* $(x_k)_{k\le M}$, $(y_k)_{k\le M}$, *we have*

$$\begin{aligned}&\sum_{i\le N}\left(\sum_{k\in I} x_k\,\eta_{i,k}\right)\left(\sum_{k\in I} y_k\,\eta_{i,k}\right)\\ &\le N\sum_{k\in I} x_k\,y_k + NLa\left(\sum_{k\in I} x_k^2\right)^{1/2}\left(\sum_{k\in I} y_k^2\right)^{1/2}\,.\end{aligned} \tag{A.28}$$

Corollary A.4.4. *If* $a \le 1$ *and* $M \le Na^2/L$, *then with probability at least* $1 - L\exp(-a^2N/L)$ *we have*

$$\begin{aligned}&\sum_{i\le N}\left(\sum_{k\le M} x_k\,\eta_{i,k}\right)\left(\sum_{k\le M} y_k\,\eta_{i,k}\right)\\ &\le N\sum_{k\le M} x_k\,y_k + NLa\left(\sum_{k\le M} x_k^2\right)^{1/2}\left(\sum_{k\le M} y_k^2\right)^{1/2}\,,\end{aligned} \tag{A.29}$$

and

$$\sum_{i\le N}\left(\sum_{k\le M} x_k\,\eta_{i,k}\right)^2 \le N(1+La)\left(\sum_{k\le M} x_k^2\right)\,. \tag{A.30}$$

Proof of Proposition A.4.3. We rewrite (A.28) as

$$\sum_{i\le N}\ \sum_{k\ne k',k,k'\in I} x_k\,y_{k'}\,\eta_{i,k}\,\eta_{i,k'} \le LNa\left(\sum_{k\in I} x_k^2\right)^{1/2}\left(\sum_{k\in I} y_k^2\right)^{1/2}\,. \tag{A.31}$$

Consider a subset A of $\mathbb{R}^n$, with $\operatorname{card} A \le 5^n$, $A\subset 2B$ and $\operatorname{conv} A\supset B$, where B is the Euclidean ball $\sum_{k\le n} x_k^2 = 1$. To ensure (A.31) it suffices that

$$\sum_{i\le N}\ \sum_{k\ne k',k,k'\in I} x_k\,y_{k'}\,\eta_{i,k}\,\eta_{i,k'} \le La \tag{A.32}$$

whenever $(x_k)_{k\in I}\in A$, $(y_k)_{k\in I}\in A$. Now, by (A.26), we have

$$\begin{aligned}\mathsf{P}&\left(\sum_{i\le N}\ \sum_{k\ne k',k,k'\in I} x_k\,y_{k'}\,\eta_{i,k}\,\eta_{i,k'} \ge Nu\right)\\ &\le \exp\left(-\frac{N}{L}\min(u^2,u)\right)\,. \end{aligned}\tag{A.33}$$

The number of possible choices for I and the sequences $(x_k)_{k\in I}$, $(y_k)_{k\in I}$ is at most

$$\binom{M}{n}(\operatorname{card} A)^2 \le \left(\frac{eM}{n}\right)^n 25^n = \exp\left(n\log\left(\frac{LM}{n}\right)\right) \le \exp Na^2$$

so that taking $u = L'a$ where L' large enough, all the events (A.33) simultaneously occur with a probability at least $1-\exp(-Na^2)$. □

Here is another nice consequence of Lemma A.4.2.

Lemma A.4.5. *If* $\varepsilon > 0$ *we have*

$$\begin{aligned}\mathsf{P}&\left(\sum_{1\le k<k'\le M} NR_{k,k'}^2 \ge (1-2\varepsilon)^{-2}\,u\right)\\ &\qquad\le \left(1+\frac{1}{\varepsilon}\right)^{M^2}\exp\left(-\frac{u}{2}\left(1-L\sqrt{\frac{u}{N}}\right)\right)\end{aligned}$$

where $R_{k,k'} = N^{-1}\sum_{i\le N}\eta_{i,k}\,\eta_{i,k'}$.

Proof. We have

$$\Big(\sum_{k<k'} R_{k,k'}^2\Big)^{1/2} = \sup \sum_{k<k'} \alpha_{k,k'}\, R_{k,k'}$$

where the supremum is taken over the subset B of $\mathbb{R}^{M(M-1)/2}$ of sequences $\alpha_{k,k'}$ with $\sum_{k<k'} \alpha_{k,k'}^2 \le 1$. We use Proposition A.3.1 to find a subset A of B with $\operatorname{card} A \le (1+\varepsilon^{-1})^{M^2}$ such that

$$\sup_A \sum \alpha_{k,k'}\, R_{k,k'} \ge (1-2\,\varepsilon)\Big(\sum_{k<k'} R_{k,k'}^2\Big)^{1/2} .$$

Thus

$$\begin{aligned}
&\mathsf{P}\Big(\sum_{k<k'} N R_{k,k'}^2 \ge (1-2\,\varepsilon)^{-2}\, u\Big)\\
&= \mathsf{P}\bigg(\Big(\sum_{k<k'} R_{k,k'}^2\Big)^{1/2} \ge (1-2\,\varepsilon)^{-1}\sqrt{\frac{u}{N}}\bigg)\\
&\le \mathsf{P}\Big(\sup_A \sum_{k<k'} \alpha_{k,k'}\, R_{k,k'} \ge \sqrt{\frac{u}{N}}\Big)\\
&\le \Big(1+\frac{1}{\varepsilon}\Big)^{M^2} \exp\Big(-\frac{u}{2}\Big(1-L\sqrt{\frac{u}{N}}\Big)\Big) ,
\end{aligned}$$

we use (A.27) in the last line. □

Corollary A.4.6. *We have*

$$\begin{aligned}
&2^{-Nn}\operatorname{card}\Big\{(\boldsymbol{\sigma}^1,\ldots,\boldsymbol{\sigma}^n)\,;\ \sum_{\ell<\ell'\le n} N R_{\ell,\ell'}^2 \ge (1-2\,\varepsilon)^{-2}\, u\Big\}\\
&\le \Big(1+\frac{1}{\varepsilon}\Big)^{n^2} \exp\Big(-\frac{u}{2}\Big(1-L\sqrt{\frac{u}{N}}\Big)\Big)
\end{aligned}$$

where $R_{\ell,\ell'} = N^{-1}\sum_{i\le N} \sigma_i^\ell\, \sigma_i^{\ell'}$.

Proof. This is another way to formulate Lemma A.4.5 when $M=n$. □

A.5 Poisson Point Processes

A Poisson random variable X of expectation a is an integer-valued r.v. such that, for $k=0\,,1\,,\ldots$

$$\mathsf{P}(X = k) = \frac{a^k}{k!}\, e^{-a}$$

so that

$$\mathsf{E} \exp \lambda X = \sum_{k \geq 0} \frac{a^k}{k!}\, e^{\lambda k - a} = \exp a(e^{\lambda} - 1)\,. \tag{A.34}$$

If X_1, X_2 are independent Poisson r.v., $X_1 + X_2$ is a Poisson r.v.

Given a positive measure μ of finite total mass $|\mu|$ (say on $\mathbb{R}^3$) a Poisson point process of intensity measure μ is a random finite subset $\Pi = \Pi_\mu$ with the following properties:

1. $\operatorname{card} \Pi$ is a Poisson r.v. of expectation $|\mu|$.
2. Given that $\operatorname{card} \Pi = k$, Π is distributed like the set $\{X_1, \ldots, X_k\}$ where $X_1, \ldots, X_k$ are i.i.d. r.v. of law $\mu/|\mu|$.

We list without proof some of the main properties of Poisson point processes. (The proofs are all very easy.)

Given two disjoint Borel sets, A, B, $\Pi \cap A$ and $\Pi \cap B$ are independent Poisson point processes. As a consequence of (A.34), if a Borel function f takes only finitely many values

$$\mathsf{E} \exp\left(\sum_{x \in \Pi} f(x)\right) = \exp\left(\int (\exp f(x) - 1)\, \mathrm{d}\mu(x)\right)\,. \tag{A.35}$$

Given two finite measures μ_1, μ_2, if Π_{μ_1} and Π_{μ_2} are independent, $\Pi_{\mu_1} \cup \Pi_{\mu_2}$ is a Poisson point process of intensity measure $\mu_1 + \mu_2$.

Given a (continuous) map φ, $\varphi(\Pi)$ is a Poisson point process of intensity measure $\varphi(\mu)$, the image measure of μ by φ.

If ν is a probability (say on $\mathbb{R}^3$), and $(U_\alpha)_{\alpha \geq 1}$ are i.i.d. r.v. of law ν, we can construct a Poisson point process of intensity measure $\mu \otimes \nu$ as follows. We number in a random order the points of Π as $x_1, \ldots, x_k$, and we consider the couples $(x_1, U_1), \ldots, (x_k, U_k)$.

Consider now a positive measure μ on $\mathbb{R}^+$. We do not assume that μ is finite, but we assume that $\mu([a, \infty))$ is finite for each $a \geq 0$. We denote by μ_0 the restriction of μ to $[1, \infty)$, by μ_k its restriction to $[2^{-k}, 2^{-k+1}[$, $k \geq 1$. Consider for $k \geq 0$ a Poisson point process Π_k of intensity measure μ_k, and assume that these are independent. We can define a Poisson point process of intensity measure μ as $\Pi = \cup_{k \geq 0} \Pi_k$. Then for each a, $\Pi \cap [a, \infty)$ is a Poisson point process, the intensity measure of which is the restriction of μ to $[a, \infty)$.

A.6 Gaussian r.v.

A (centered) Gaussian r.v. g has a density of the type

$$\frac{1}{\sqrt{2\pi}\, \tau} \exp\left(-\frac{t^2}{2\tau^2}\right)$$

so that $\mathsf{E}\, g^2 = \tau^2$. When $\tau = 1$, g is called standard Gaussian. We hardly ever use non centered Gaussian r.v., so that the expression "Gaussian r.v." means for us "centered Gaussian r.v.". A fundamental fact is that

$$\mathsf{E} \exp a\, g = \exp \frac{a^2 \tau^2}{2} \,. \tag{A.36}$$

Indeed,

$$\begin{aligned} \mathsf{E} \exp a\, g &= \frac{1}{\sqrt{2\pi}\,\tau} \int_{-\infty}^{\infty} \exp\left(at - \frac{t^2}{2\,\tau^2}\right) \mathrm{d}t \\ &= \left(\exp \frac{a^2\tau^2}{2}\right) \frac{1}{\sqrt{2\pi}\,\tau} \int_{-\infty}^{\infty} \exp\left(-\frac{(t - a\tau^2)^2}{2\,\tau^2}\right) \mathrm{d}t \\ &= \exp \frac{a^2\tau^2}{2} \,. \end{aligned}$$

Using (A.1) we get, for $t \geq 0$

$$\mathsf{P}(g \geq t) \leq \exp\left(-\frac{t^2}{2\,\tau^2}\right) \,. \tag{A.37}$$

Elementary estimates show that for $t > 0$ we have

$$\mathsf{P}(g \geq t) \geq \frac{1}{L(1 + t/\tau)} \exp\left(-\frac{t^2}{2\tau^2}\right) \,. \tag{A.38}$$

There is of course a more precise understanding of the tails of g than (A.37), (A.38); but these will suffice here.

Given independent standard Gaussian r.v. $z_1, \ldots, z_M$, their joint law has density $(2\pi)^{-M/2} \exp(-\|\mathbf{x}\|^2/2)$, where $\|\mathbf{x}\|^2 = \sum_{i \leq M} x_i^2$. This density is invariant by rotation, and, as a consequence, the law of every linear combination $z = \sum_{i \leq M} \alpha_i\, z_i$ is Gaussian. The set $\mathcal{G}$ of these linear combinations is a vector space, each element of which is a Gaussian r.v. Such a space is often called a Gaussian space. It has a natural dot product, given, with obvious notation by $\mathsf{E}\, z\, z' = \sum_{k \leq M} \alpha_k\, \alpha_{k'}$. Given two linear subspaces F_1, F_2 of F, if these spaces are orthogonal, i.e. $\mathsf{E}\, z_1\, z_2 = 0$ whenever for $z_1 \in F_1$, $z_2 \in F_2$, these spaces are probabilistically independent. This is obvious from rotational invariance: since after a suitable rotation that these spaces are spanned by two disjoint subsets of $z_1, \ldots, z_M$.

Considering g, $g_1, \ldots, g_n$ in $\mathcal{G}$, the r.v.

$$g'_\ell = g_\ell - g\, \frac{\mathsf{E}(g_\ell\, g)}{\mathsf{E}(g^2)} \tag{A.39}$$

is such that $\mathsf{E}\, g'_\ell\, g = 0$: thus g is independent of the family $(g'_1, \ldots, g'_n)$.

Given a function F on $\mathbb{R}$ (of moderate growth at infinity) we have the integration by parts formula

$$\mathsf{E}\, g F(g) = \mathsf{E}\, g^2\, \mathsf{E}\, F'(g)\ . \tag{A.40}$$

To see this, if $\mathsf{E}\, g^2 = \tau^2$, we have

$$\begin{aligned}\mathsf{E}\, g F(g) &= \frac{1}{\sqrt{2\pi}\tau}\int_{\mathbb{R}} t \exp\left(-\frac{t^2}{2\,\tau^2}\right) F(t)\, \mathrm{d}t \\ &= \frac{\tau^2}{\sqrt{2\pi}\tau}\int_{\mathbb{R}} \exp\left(-\frac{t^2}{2\,\tau^2}\right) F'(t)\, \mathrm{d}t \\ &= \mathsf{E}\, g^2\, \mathsf{E}\, F'(g)\end{aligned}$$

provided $\lim_{|t|\to\infty} F(t)\exp(-t^2/2\,\tau^2) = 0$. This formula generalizes as follows. Given g, $g_1, \ldots, g_n$ in $\mathcal{G}$, and a function F of n variables, we have

$$\mathsf{E}\, g F(g_1, \ldots, g_n) = \sum_{\ell \le n} \mathsf{E}(g\, g_\ell)\, \mathsf{E}\frac{\partial F}{\partial x_\ell}(g_1, \ldots, g_n)\ . \tag{A.41}$$

To see this, we use the variables g'_ℓ of (A.39) and we apply (A.40) at $(g'_\ell)_{\ell \le n}$ given. Since $g_\ell = g'_\ell + g\, \mathsf{E}(g\, g_\ell)/\mathsf{E}(g^2)$, (A.41) follows.

Another interesting fact is the following. If $(q_{u,v})_{u,v\le n}$ is a symmetric positive definite matrix, there exists jointly Gaussian r.v. $(Y_u)_{u\le n}$ such that $\mathsf{E}\, Y_u\, Y_v = q_{u,v}$. This is obvious when the matrix $(q_{u,v})$ is diagonal; the general case follows from the fact that a symmetric matrix diagonalizes in an orthogonal basis.

A.7 Integration by Parts

If Y is a r.v., $Y \ge 0$, then

$$\mathsf{E}(Y) = \int_0^\infty \mathsf{P}(Y \ge t)\, \mathrm{d}t\ . \tag{A.42}$$

If F is a non decreasing function on $\mathbb{R}$, $F \ge 0$, then

$$\mathsf{E}\, F(X) = \int_0^\infty \mathsf{P}(F(X) \ge u)\, \mathrm{d}u$$

and, by change of variable, setting $u = F(t)$ and assumming $F(-\infty) = 0$,

$$\mathsf{E}\, F(X) = \int_{-\infty}^\infty F'(t)\, \mathsf{P}(X \ge t)\, \mathrm{d}t\ . \tag{A.43}$$

If now $0 \le a < b$ we have

$$\begin{aligned}\mathsf{E}\left(Y\, \mathbf{1}_{\{a\le Y\le b\}}\right) &= a\, \mathsf{P}(Y \ge a) - b\, \mathsf{P}(Y > b) + \int_a^b \mathsf{P}(Y \ge t)\, \mathrm{d}t \\ &\le a\, \mathsf{P}(Y \ge a) + \int_a^b \mathsf{P}(Y \ge t)\, \mathrm{d}t\end{aligned}$$

and, as in (A.43), if $a < b$,

$$\mathsf{E}\, F(X)\, \mathbf{1}_{\{a \le X \le b\}} \le F(a)\, \mathsf{P}(X \ge a) + \int_a^b F'(t)\, \mathsf{P}(X \ge t)\, \mathrm{d}t\,. \tag{A.44}$$

References

1. Aldous D. (1992) Asymptotics in the random assignment problem. *Probab. Theory Related Fields* **93**, no. 4, pp. 507-534.
2. Aldous D. (2001) The $\zeta(2)$ limit in the random assignment problem. *Random Structures and Algorithms* **18**, no. 4, pp. 381-418.
3. Aizenman M., Contucci P. (1998) On the stability of the quenched state in mean-field spin-glass models. *J. Statist. Phys.* **92**, no. 5-6, pp. 765-783.
4. Aizenman M., Lebowitz J.L., Ruelle D. (1987) Some rigorous results on the Sherrington-Kirkpatrick model. *Comm. Math. Phys.* **112**, pp. 3-20.
5. Albeverio S., Tirozzi B., Zegarlinski B. (1992) Rigorous results for the free energy in the Hopfield model. *Comm. Math. Phys.* **150**, no. 2, pp.337-373 .
6. Almeida J.R.L., Thouless D.J. (1978) Stability of the Sherrington-Kirkpatrick solution of a spin glass model. *J. Phys. A: Math. Gen.* **II**, pp. 983-990.
7. Amit D.J., Gutfreund H., Sompolinsky H. (1987) Statistical mechanics of neural networks near saturation. *Annals of Physics* **173**, pp. 30-67.
8. Barbina S., Márquez-Carreras X., Rovira C., Tindel S. (2002) Higher order expansions for the overlap of the SK model, Manuscript.
9. Baffioni F., Rosati F. (2000) On the Ultrametric Overlap Distribution for Mean Field Spin Glass Models (I), *The European Physical Journal B* **17**, pp. 439-447.
10. Bobkov S. G., Ledoux M. (2000) From Brunn-Minkowski to Brascamp-Lieb and to logarithmic Sobolev inequalities. *Geom. Funct. Anal.* **10**, no. 5, pp. 1028-1052.
11. Bollobás B. (2001) Random graphs. Second edition. *Cambridge Studies in Advanced Mathematics* **73**. Cambridge University Press, Cambridge.
12. Bolthausen E., Sznitman A.-S. (1998) On Ruelle's probability cascades and an abstract cavity method, *Comm. Math. Phys.* **197**, no. 2, pp. 247–276.
13. Borell C. (1984) On polynomial chaos and integrability, *Probab. Math. Statist.* **3**, no. 2, pp. 191-203.
14. Bouchaud J.-P., Mézard M. (1997) Universality classes for extreme values statistics. *J. Phys. A. Math. Gen* **30**, pp. 7997-8015.
15. Bouten M. (1988) Replica symmetry instability in perceptron models. Comment on: "Optimal storage properties of neural network models" , *J. Phys. A* **21**, no. 1, pp. 271–284, by E. Gardner and B. Derrida. With a reply by Derrida. *J. Phys. A* **27** (1994), no. 17, pp. 6021-6025.
16. Bovier A. (1994) Self-averaging in a class of generalized Hopfield models. *J. Phys. A* **27**, no. 21, pp. 7069-7077.
17. Bovier A. (1997) Comment on "capacity of Hopfield model". *J. Phys. A* **30**, pp. 7993-7996.
18. Bovier A. (1999) Sharp upper bounds on perfect retrieval in the Hopfield model. *J. Appl. Probab.* **36**, no. 3, pp. 941-950.

19. Bovier A. Statistical mechanics of disordered systems, *MaPhySto Lecture Notes*, 192 pp.
20. Bovier A., van Enter A. C.D., Niederhauser B. (1999) Stochastic symmetry-breaking in a Gaussian Hopfield model. *J. Statist. Phys.* **95**, no. 1-2, pp. 181-213.
21. Bovier A., Gayrard V. (1992) Rigorous bounds on the storage capacity of the dilute Hopfield model. *J. Statist. Phys.* **69**, no. 3-4, pp. 597-627.
22. Bovier A., Gayrard V. (1993) Rigorous results on the thermodynamics of the dilute Hopfield model. *J. Statist. Phys.* **72**, no. 1-2, pp. 79-112.
23. Bovier A., Gayrard V. (1993) Lower bounds on the memory capacity of the dilute Hopfield model. Cellular automata and cooperative systems (Les Houches, 1992), pp. 55-66. *NATO Adv. Sci. Inst. Ser. C Math. Phys. Sci.* **396**, Kluwer Acad. Publ., Dordrecht.
24. Bovier A., Gayrard V. (1993) The Hopfield model on random graphs: lower bounds on the storage capacity. Dynamics of complex and irregular systems (Bielefeld, 1991), pp. 277-288. *Bielefeld Encount. Math. Phys.* VIII, World Sci. Publishing, River Edge, NJ.
25. Bovier A., Gayrard V., Picco P. (1994) Gibbs states of the Hopfield model in the regime of perfect memory. *Probab. Theory Related Fields* **100**, no. 3, pp. 329-363.
26. Bovier A., Gayrard V., Picco P. (1995) Gibbs states of the Hopfield model with extensively many patterns. *J. Statist. Phys.* **79**, no. 1-2, pp. 395-414.
27. Bovier A., Gayrard V., Picco P. (1995) Large deviation principles for the Hopfield model and the Kac-Hopfield model. *Probab. Theory Related Fields* **101**, no. 4, pp. 511-546.
28. Bovier A., Gayrard V. (1996) An almost sure large deviation principle for the Hopfield model. *Ann. Probab.* **24**, no. 3, pp. 1444-1475.
29. Bovier A., Gayrard V. (1997) The retrieval phase of the Hopfield model: a rigorous analysis of the overlap distribution. *Probab. Theory Related Fields* **107**, no. 1, pp. 61-98.
30. Bovier A., Gayrard V. (1997) An almost sure central limit theorem for the Hopfield model. *Markov Process. Related Fields* **3**, no. 2, pp. 151-173.
31. Bovier A., Gayrard V. (1998) Hopfield models as generalized random mean field models. Mathematical aspects of spin glasses and neural networks. *Progr. Probab.* **41**, pp. 3-89. Birkhäuser Boston, Inc., Boston, MA.
32. Bovier A., Gayrard V. (1998) Metastates in the Hopfield model in the replica symmetric regime. *Math. Phys. Anal. Geom.* **1**, no. 2, pp. 107-144.
33. Bovier A., Gayrard V., Picco P. (1994) Gibbs states for the Hopfield model in the regime of perfect memory. *Prob. Theory Relat. Fields* **100**, pp. 1329-363.
34. Bovier A., Gayrard V., Picco P. (1995) Gibbs states for the Hopfield model with extensively many patterns. *J. Stat. Phys.* **79**, pp. 395-414.
35. Bovier A., Kurkova L., Loewe M. (2002) Fluctuations of the free energy in the REM and the p-spin SK model, *Ann. Probab.* **30**, pp. 605-651.
36. Bovier A., Mason D. M. (2001) Extreme value behavior in the Hopfield model. *Ann. Appl. Probab.* **11**, no. 1, pp. 91–120.
37. Bovier A., Niederhauser B. The spin-glass phase-transition in the Hopfield model with p-spin interactions. *Adv. Theor. Math. Phys.*, to appear.
38. Bovier A., Picco P. (editors) (1997) Mathematical aspects of spin glasses and Neural networks. *Progress in Probability* **41**, Birkhäuser, Boston.
39. Brascamp H., Lieb E. (1976) On extension of the Brunn-Minkowski and the Prékopa-Leindler Theorems, including inequalities for Log-Concave functions, and with an application to the Diffusion equation, *J. Funct. Anal.* **22**, pp. 366-389.

40. Canisius J., van Enter A. C. D., van Hemmen J. L. (1983) On a classical spin glass model. *Z. Phys. B* **50**, no. 4, pp. 311-336.
41. Carmona P. (2002) High temperature Sherrington-Kirkpatrick models for general spins, to appear.
42. Catoni O. (1996) The Legendre transform of two replicas of the Sherrington-Kirkpatrick spin glass model. A free energy inequality. *Probab. Theory Related Fields* **105**, no. 3, pp. 369-392.
43. Comets F. (1996) A spherical bound for the Sherrington-Kirkpatrick model. Hommage à P.A. Meyer et J. Neveu. *Astérisque* **236**, pp. 103-108.
44. Comets F. (1998) The martingale method for mean-field disordered systems at high temperature. Mathematical aspects of spin glasses and neural networks. *Progr. Probab.* **41**, pp. 91-113. Birkhäuser Boston, Inc., Boston, MA.
45. Comets F., Giacomin G., Lebowitz J.L. (1999) The Sherrington-Kirkpatrick model with short range ferromagnetic interactions. *C.R. Acad. Sci. Paris Sér. I Math.* **328**, no. 1, pp. 57-62.
46. Comets F., Neveu J. (1995) The Sherrington-Kirkpatrick model of spin glasses and stochastic calculus: the high temperature case. *Comm. Math. Phys.* **166**, no. 3, pp. 549-564.
47. Contucci P. (2002) Towards a classification of stochastically stable quenched measures. Manuscript.
48. Derrida B. (1981) Random energy model: An exactly solvable model of disordered systems. *Phys. Rev. B* **24**, no. 5, pp. 2613-2626.
49. Derrida B., Gardner E. (1988) Optimal storage properties of neural network models. *J. Phys. A* **21**, pp. 271-284.
50. Derrida B., Gardner E. (1989) Three unfinished works on the optimal storage capacity of networks. Special issue in memory of Elizabeth Gardner, pp. 1957-1988. *J. Phys. A* **22**, no. 12, pp. 1983-1994.
51. Ellis R.S. (1985) Entropy, large Deviations, and Statistical Mechanics, Springer-Verlag.
52. van Enter A. C. D., van Hemmen J. L., Pospiech C.(1988) Mean-field theory of random-site q-state Potts models. *J. Phys. A* **21**, no. 3, pp. 791-801.
53. van Enter A.C.D., Schaap H.G. (2002) Infinitely many states and stochastic symmetry in a Gaussian Potts-Hopfield model. *J.Phys. A* **35** pp. 2581-2592.
54. Feng J., Shcherbina M., Tirozzi B. (2001) On the critical capacity of the Hopfield model. *Comm. Math. Phys.* **216**, no. 1, pp. 139-177.
55. Feng J., Tirozzi B. (1995) The SLLN for the free-energy of a class of neural networks. *Helv. Phys. Acta* **68**, no. 4, pp. 365-379.
56. Feng J., Tirozzi B. (1997) Capacity of the Hopfield model. *J. Phys. A* **30**, no. 10, pp. 3383-3391.
57. Fröhlich J., Zegarlinski B. (1987) Some comments on the Sherrington-Kirkpatrick model of spin glasses. *Comm. Math. Phys.* **112**, pp. 553-566.
58. Gardner E. (1985) Spin glasses with p-spin interactions. *Nuclear Phys. B* **257**, no. 6, pp. 747-765.
59. Gardner E. (1988) The space of interactions in neural network models. *J. Phys. A* **21**, pp. 257-270.
60. Gentz B. (1996) An almost sure central limit theorem for the overlap parameters in the Hopfield model. *Stochastic Process. Appl.* **62**, no. 2, pp. 243-262.
61. Gentz B. (1996) A central limit theorem for the overlap in the Hopfield model. *Ann. Probab.* **62**, no. 4, pp. 1809-1841.
62. Gentz B. (1998) On the central limit theorem for the overlap in the Hopfield model. Mathematical aspects of spin glasses and neural networks *Progr. Probab.* **41**, pp. 115-149, Birkhäuser Boston, Boston, MA.

63. Gentz B., Löwe M. (1999) The fluctuations of the overlap in the Hopfield model with finitely many patterns at the critical temperature. *Probab. Theory Related Fields* **115**, no. 3, pp. 357-381.
64. Gentz B., Löwe M. (1999) Fluctuations in the Hopfield model at the critical temperature. *Markov Process. Related Fields* **5**, no. 4, pp. 423-449.
65. Ghirlanda S., Guerra F. (1998) General properties of overlap probability distributions in disordered spin systems. Towards Parisi ultrametricity. *J. Phys. A* **31**, no. 46, pp. 9149-9155.
66. Gross D., Mézard M. (1984) The simplest spin glass. *Nuclear Phys. B* **240**, pp. 431-452.
67. Guerra F. (1995) *Fluctuations and thermodynamic variables in mean field spin glass models* in Stochastic Processes, Physics and Geometry, S. Albeverio et. al. editors, World Scientific, Singapore.
68. Guerra F. (1996) About the overlap distribution in mean field spin glass models. *Int. J. Modern Physics B*, pp. 1675-1684.
69. Guerra F. (1995) *Functional Order Parameters for the Quenched Free Energy in Mean Field Spin Glass Models* in *Field Theory and Collective Phenomena*, S. De Lillo, P. Sodano, F.C. Khanna, G.W. Semenov, eds, World Scientific, Singapore.
70. Guerra F. (1996) About the overlap distribution in mean field spin glass models. *International Journal of Modern Physics B* **10**, pp. 1675-1684.
71. Guerra F. (2001) Sum rules for the free energy in the mean field spin glass model. *Fields Institute Communications* **30**, 161.
72. Guerra F. (2003) Replica Broken Bounds in the Mean Field Spin Glass Model. *Communications Math. Phys.***233**, pp. 1-12.
73. Guerra F., Toninelli F.L. (2002) Quadratic replica coupling for the Sherrington-Kirkpatrick mean field spin glass model, J. Math. Phys. **43**,no. 7, pp. 3704-3716.
74. Guerra F., Toninelli F.L. (2002) Central limit theorem for fluctuations in the high temperature region of the Sherrington-Kirkpatrick spin glass model, *J. Math. Phys.***43**, no.12, pp. 6224-6237.
75. Guerra F., Toninelli F.L. (2002) The Thermodynamic Limit in Mean Field Spin Glass Models, Commun. Math. Phys. **230**, pp. 71-79.
76. Guerra F., Toninelli F.L. The Infinite Volume Limit in Generalized Mean Field Disordered Models, to appear in Markov Processes and Related Fields.
77. van Hemmen J. L., Palmer R. G. (1979) The replica method and a solvable spin glass system. *J. Phys. A* **12**, no. 4, pp. 563-580.
78. van Hemmen J. L., Palmer R. G. (1982) The thermodynamic limit and the replica method for short-range random systems. *J. Phys. A* **15**, no. 12, pp. 3881-3890.
79. Hertz J., Krogh A., Palmer R.C. (1991) *Introduction to the theory of neural computation* Addison Wesley.
80. Hoeffding W. (1963) Probability inequalities for sums of bounded random variables, *J. Amer. Statist. Assoc.* **58**, pp. 13-30.
81. Hopfield J.J. (1982) Neural networks and physical systems with emergent collective computational abilities. *Proc. Natl. Acad. Sci. USA* **79**, pp. 1554-2558.
82. Hopfield J.J. (1982) Neural networks and physical systems with emergent collective computational abilities. *Proc. Natl. Acad. Sci. USA* **81**, pp. 3088.
83. Ibragimov A., Sudakov V.N., Tsirelson B.S. (1976) Norms of Gaussian sample functions. Proceedings of the Third Japan USSR Symposium on Probability theory. *Lecture Notes in Math.* **550**, Springer Verlag, pp. 20-41.

84. Kahane J.-P. (1986) Une inégalité du type de Slepian et Gordon sur les processus gaussiens, *Israel J. Math.* **55**, no. 1, pp. 109-110.
85. Kim J.H., Roche J.R. (1998) Covering cubes by random half cubes, with applications to binary neural networks: rigorous combinatorial approaches. Eight annual Workshop on Computatioanl Learning Theory, Santa Cruz, 1995 *J. Comput. System Sci.* **56**, no. 2, pp. 223-252.
86. Kingman J. F. C. (1970) The representation of partition structures, *J. London Math Soc.* **18**, pp. 374-380.
87. Kingman J. F. C. (1982) Exchangeability and the evolution of large partitions. In G. Koch and F. Spizzichino, editors, Exchangeability in Probability and statistics, (Rome, 1981) pp. 97-112, North Holland, Amsterdam.
88. Krauth W., Mézard M. (1989) Storage capacity of memory network with binary couplings. *J. Phys.* **50**, pp. 3057-3066.
89. R. Latala (2002) Exponential inequalities for the SK model of spin glasses, extending Guerra's method. Manuscript.
90. Ledoux M. (2001) The concentration of measure phenomenon. *Mathematical Surveys and Monographs* **89**, American Mathematical Society, Providence, RI, x+181 pp. ISBN: 0-8218-2864-9.
91. Ledoux M. (2000) On the distribution of overlaps in the Sherrington-Kirkpatrick spin glass model, *J. Statist. Phys.* **100**, no. 5-6, pp. 871-892.
92. Ledoux M., Talagrand M. (1991) *Probability in Banach Spaces* Springer-Verlag.
93. Loukianova D. (1997) Lower bounds on the restitution error of the Hopfield model. *Probab. Th. Relat. Fields* **107**, pp. 161-176.
94. Löwe M. (1998) On the storage capacity of Hopfield models with correlated patterns. *Ann. Appl. Probab.* **8**, no. 4, pp. 1216-1250.
95. Löwe M. (1999) The storage capacity of generalized Hopfield models with semantically correlated patterns. *Markov Process. Related Fields* **5**, no. 1, pp. 1-19.
96. Mézard M. (1988) The space of interactions in neural networks: Gardner's computation with the cavity method. *J. Phys. A* **22**, pp. 2181-2190.
97. Mézard M., Parisi G. (1985) Replicas and optimization. *J. Physique. Lett.* **46**, L771.
98. Mézard M., Parisi G. (1986) Mean field equations for the matching and the traveling salesman problem. *Europhys. lett.* **916**.
99. Mézard M., Parisi G., Virasoro M. (1987) *Spin glass theory and beyond*, World Scientific, Singapore.
100. Milman V. (1988) The heritage of P. Lévy in functional analysis. Colloque P. Lévy sur les processus stochastiques, *Astérisque* **157-158**, pp. 273-301.
101. Milman V., Schechtman G. (1986) Asymptotic theory of finite dimensional normed spaces. *Lecture Notes in Math.* **1200**, Springer Verlag.
102. Monasson R., Zecchina R. (1997) Statistical mechanics of the random K-sat model. *Phys. Rev. E* pp. 1357-1370.
103. Neveu J. (1989) A continuous state branching process in relation with the GREM model of spin glass theory. Rapport interne 267, Ecole Polytechnique, Centre de Mathématiques Appliquèes.
104. Newman C. (1988) Memory capacity in neural network models: rigorous lower bounds. *Neural Networks* **I**, pp. 223-238.
105. Newman C., Stein D. (1993) Chaotic size dependence in spin glasses. Cellular automata and cooperative systems, (Les Houches, 1992). *NATO Adv. Sci. Inst. Ser. C Math. Phys. Sci* **396**, Kluwer Acad. Publ. Dordrecht.
106. Newman C., Stein D. (1997) Non-mean-field behavior of realistic spin glasses. *Phys. Rev. Letters* **76**, pp. 515-518.

107. Newman C., Stein D. (1997) Spatial inhomogeneity and thermodynamic chaos. *Phys. Rev. Letters* **76**, pp. 4821-4824.
108. Newman C., Stein D. (1998) Thermodynamic chaos and the structure of short-range spin glasses, mathematical aspects of spin glasses and neural networks. A. Bovier and P. Picco editors, *Progress in Probability* **41**, pp. 243-287, Birkhäuser, Boston.
109. Parisi G. (1992) Field theory, disorder, simulation. *World Scientific Lecture Notes in Physics* **45**, World Scientific, Singapore.
110. Pastur L.A. (1992) Statistical physics and spectral theory of disordered systems: some recent developments. *Mathematical Physics* **X**, (Leipzig, 1991), pp. 70-86, Springer, Berlin.
111. Pastur L.A., Figotin A. (1977) Exactly soluble model of a spin glass. *Sov. J. Low Temp. Phys.* **3**, pp. 378-383.
112. Pastur L.A., Figotin A. (1978) On the theory of disordered spin systems. *Theor. Math. Phys.* **35**, pp. 403-414.
113. Pastur L.A., Shcherbina M.V. (1991) Absence of self-averaging of the order parameter in the Sherrington-Kirkpatrick model. *J. Statist. Phys.* **62**, no. 1-2, pp. 1-19.
114. Pastur L.A., Shcherbina M.V., Tirozzi B. (1994) The replica-symmetric solution without replica trick for the Hopfield model. *J. Statist. Phys.* **74**, no. 5-6, pp. 1161-1183.
115. Pastur L.A., Shcherbina M.V., Tirozzi B. (1999) On the replica-symmetric equations for the Hopfield model. *J. Math. Phys.* **40**, no. 8, pp. 3930-3947.
116. Petritris D. (1994) Thermodynamic formalism of neural computing. Dynamics of Complex interacting systems (Santiago). *Nonlinear Phenom. Complex Systems* **2**, pp. 81-146, Kluwer Acad. Publ.
117. Petritris D. (1996) Equilibrium statistical mechanics of frustrated spin glasses: a survey of mathematical results. *Ann. Inst. H. Poincaré Phys. Théor.* **64**, pp. 255-288.
118. Pisier G. (1986) Probabilistic methods in the geometry of Banach spaces. Probability and analysis (Varenna, 1985), pp. 167-241, Lecture Notes in Math., 1206, Springer, Berlin.
119. Pitman J. Combinatorial Stochastic Processes, Lecture Notes for the St. Flour Summer School, July 2002, to appear in *Springer Lecture Notes in Math.*
120. Pitman J., Yor M. (1997) The two-parameter Poisson-Dirichlet distribution derived from a stable subordinator. *Ann. Probab.* **25**, pp. 855-900.
121. Ruelle D. (1987) A mathematical reformulation of Derrida's REM and GREM. *Comm. Math. Phys.* **108**, pp. 225-239.
122. Shcherbina M.V. (1991) More about absence of self averaging of the order parameter in the Sherrington-Kirkpatrick model. *CARR Reports in Mathematical Physics* no 3/91, Department of Mathematics, University of Rome "la Sapienza".
123. Shcherbina M.V. (1997) On the replica-symmetric solution for the Sherrington-Kirkpatrick model. *Helv. Phys. Acta* **70**, pp. 838-853.
124. Shcherbina M.V. (1999) Some estimates for the critical temperature of the Sherrington-Kirkpatrick model with magnetic field. In: *Mathematical Results in Statistical Mechanics*, World Scientific, Singapore, p. 455-474.
125. Shcherbina M.V., Tirozzi B. (1994) The free energy of a class of Hopfield models. *J. Statist. Phys.* **72**, no. 1-2, pp. 113-125.
126. Shcherbina M.V., Tirozzi B. (1995) A perturbative expansion for the Hopfield model. *Helv. Phys. Acta* **68**, no. 5, pp. 470-491.
127. Shcherbina M.V., Tirozzi B. (2001) On the critical capacity of the Hopfield model. *Comm. Math. Phys.* **216**, no. 1, pp. 139-177.

128. Shcherbina M.V., Tirozzi B. On the rigorous solution of Gardner's problem *Comm. Math. Phys*, to appear.
129. Sherrington D., Kirkpatrick S. (1972) Solvable model of a spin glass. *Phys. Rev. Lett.* **35**, pp. 1792-1796.
130. Talagrand M. (1987) Regularity of Gaussian processes. *Acta. Math.* **159**, pp. 99-149.
131. Talagrand M. (1993) Regularity of infinitely divisible processes, *Ann. Probab.* **21**, no. 1, pp. 362-432.
132. Talagrand M. (1995) Concentration of measure and isoperimetric inequalities in product spaces. *Publ. Math. I.H.E.S.* **81**, pp. 73-205.
133. Talagrand M. (1996) A new look at independence. *Ann. Probab.* **24**, pp. 1-34.
134. Talagrand M. (1998) The Sherrington-Kirkpatrick model: a challenge to mathematicians. *Prob. Th. Relat. Fields* **110**, pp. 109-176.
135. Talagrand M. (1998) Rigorous results for the Hopfield model with many patterns. *Prob. Th. Relat. Fields* **110**, pp. 177-276.
136. Talagrand M. (1998) Huge random structures and mean field models for spin glasses, Proceedings of the Berlin International Congress of mathematicians. *Documenta Math.*, Extra Vol. I, pp. 507-536.
137. Talagrand M. (1999) Self-averaging and the space of interactions in neural networks. *Random Structures and Algorithms* **14**, pp. 199-213.
138. Talagrand M. (1999) Intersecting random half cubes. *Random Structures and Algorithms* **15**, pp. 436-449.
139. Talagrand M. (2000) Verres de spin et optimisation combinatoire, Séminaire Bourbaki, Vol.1998-99. *Astérisque* **15**, pp. 436-449.
140. Talagrand M. (2000) Rigorous low temperature results for the mean field p-spin interaction model. *Probab. Theor. Relat. Fields* **117**, pp. 303-360.
141. Talagrand M. (2000) Multiple levels of symmetry breaking. *Probab. Theor. Relat. Fields* **117**, pp. 449-466.
142. Talagrand M. (2000) Intersecting random half-spaces: towards the Derrida-Gardner formula. *Ann. Probab.* **28**, pp. 725-758.
143. Talagrand M. (2000) Exponential inequalities and replica-symmetry breaking for the Sherrington-Kirkpatrick model. *Ann. Probab.* **28**, pp. 1018-1062.
144. Talagrand M. (2000) Exponential inequalities and convergence of moments in the replica-symmetric phase of the Hopfield model. *Ann. Probab.* **28**, pp. 1393-1469.
145. Talagrand M. (2000) Large deviation principles and generalized Sherrington-Kirkpatrick models. *Ann. Fac. Sci. Toulouse Math.* **9**, pp. 203-244.
146. Talagrand M. (2001) The high temperature phase of the random K-sat problem. *Probab. Theor. Relat. Fields* **119**, pp. 187-212.
147. Talagrand M. (2000) Spin glasses: a new direction for probability theory? Mathematics towards the third millennium (Rome, 1999).*Atti Accad. Naz. Lincei Cl. Sci. Fis. Mat. Natur. Rend. Lincei* **9**, Mat. Appl. Special Issue, pp. 127–146.
148. Talagrand M. (2001) The Hopfield model at the critical temperature. *Probab. Theor. Relat. Fields* **121**, pp. 237-268.
149. Talagrand M. (2002) On the high temperature phase of the Sherrington-Kirkpatrick model. *Ann. Probab.* **30**, pp. 364-381.
150. Talagrand, M. (2002) On the Gaussian perceptron at high temperature.*Math. Phys. Anal. Geom.***5**, no. 1, pp. 77-99.
151. Talagrand M. (2002) Gaussian averages, Bernoulli averages and Gibbs' measures *Random Structures and Algorithms***21**, pp. 197-204.
152. Talagrand M. (2003) Self organization in the low temperature phase of a spin glass model*Review Math. Phys.***14**, pp. 1-78.

153. Talagrand M. (2000) Rigorous results for mean field spin glasses models: a first course, Saint Flour Summer school in Probability, August 2000, *Springer Lectures Notes in Math.*, to appear.
154. Talagrand M (2003) The generalized Parisi formula *Comptes Rendus Acad. Sci. Paris*, to appear.
155. Tindel S. (2003) Quenched large deviation principle for the overlap of a p-spins system *J. Stat. Phys.* **110**, pp. 51-72.
156. Toninelli F. (2002) Private communication.
157. Toninelli F. (2002) About the Almeida-Thouless transition line in the Sherrington-Kirkpatrick mean field spin glass model. *Europhysics letters* **60**, no. 5, pp. 764-767.
158. Toubol A. (1998) High temperature regime for a multidimensional Sherrington-Kirkpatrick model of spin glass. *Probab. Theory Related Fields* **110**, no. 4, pp. 497-534.
159. Toubol A. (1999) Small random perturbation of a classical mean field model. *Stochastic Process. Appl.* **81**, no. 1, pp. 1-24.
160. Toulouse G. (1983) Frustration and disorder, new problems in statistical mechanics, spin glasses in a historical perspective. Heidelberg Colloquium on Spin Glasses, J.L. van Hemmen and I. Morgenstern eds. *Lecture Notes in Physics* **192**, Springer Verlag.

Index

GPSR Compliance

The European Union's (EU) General Product Safety Regulation (GPSR) is a set of rules that requires consumer products to be safe and our obligations to ensure this.

If you have any concerns about our products, you can contact us on ProductSafety@springernature.com

In case Publisher is established outside the EU, the EU authorized representative is:

Springer Nature Customer Service Center GmbH
Europaplatz 3
69115 Heidelberg, Germany

Batch number: 10755248

Printed by Printforce, the Netherlands